P9-DIY-145

CHARLES E. MERRILL PUBLISHING COMPANY
A Bell & Howell Company
Columbus Toronto London Sydney

Thomas L. Floyd
Mayland Technical College

2ND EDITION

DIGITAL FUNDAMENTALS

Published by Charles E. Merrill Publishing Company
A Bell & Howell Company
Columbus, Ohio 43216

This book was set in Times Roman and Memphis.
Production Coordination: Cherlyn B. Paul
Text Designer: Ann Mirels
Cover Photograph: Courtesy of Digital Equipment Corporation
The first edition was published under the title **Digital Logic Fundamentals.**

Library of Congress Catalog Card Number: 81–82715
International Standard Book Number: 0–675–09876–9
Printed in the United States of America
6 7 8 9 10—86 85

To my mother, Mary Floyd, and
to my daughters, Debbie and Cindy.

Daniel D. Ramos

MERRILL'S INTERNATIONAL SERIES IN ELECTRICAL AND ELECTRONICS TECHNOLOGY

BATESON
Introduction to Control System Technology, 2nd Edition, 8255-2

BOYLESTAD
Introductory Circuit Analysis, 4th Edition, 9938-2
Student Guide to Accompany Introductory Circuit Analysis, 4th Edition, 9856-4

BOYLESTAD/KOUSOUROU
Experiments in Circuit Analysis, 4th Edition, 9858-0

NASHELSKY/BOYLESTAD
BASIC Applied to Circuit Analysis, 20161-6

BREY
Microprocessor/hardware Interfacing and Applications, 20158-6

FLOYD
Digital Fundamentals, 2nd Edition, 9876-9
Electronic Devices, 20157-8
Essentials of Electronic Devices, 20062-8
Principles of Electric Circuits, 8081-9
Electric Circuits: Electron Flow Version, 20037-7

STANLEY, B.H.
Experiments in Electric Circuits, 9805-X

GAONKAR
Microprocessor Architecture, Programming and Applications With the 8085/8080A, 20159-4

ROSENBLATT/FRIEDMAN
Direct and Alternating Current Machinery, 2nd Edition, 20160-8

SCHWARTZ
Survey of Electronics, 2nd Edition, 8554-3

SEIDMAN/WAINTRAUB
Electronics: Devices, Discrete and Inegrated Circuits, 8494-6

STANLEY, W.D.
Operational Amplifiers With Linear Integrated Circuits, 20090-3

TOCCI
Fundamentals of Electronic Devices, 3rd Edition, 9887-4
Electronic Devices, 3rd Edition, Conventional Flow Version, 20063-6
Fundamentals of Pulse and Digital Circuits, 3rd Edition, 20033-4
Introduction to Electric Circuit Analysis, 2nd Edition, 20002-4

WARD
Applied Digital Electronics, 9925-0

PREFACE

Because of the importance of digital techniques in a wide range of applications, it is imperative that the electronics technician and technologist be well-versed in digital fundamentals.

This text provides a thorough introduction to digital concepts and microprocessor-related topics for the electronics technologist and the electrical engineering technology student. While retaining the essential qualities of the first edition, this second edition has been enhanced and updated.

As in the first edition, basic concepts and functional and logical operation are emphasized. Coverage of "inside-the-chip" circuitry has been removed from the main body of the text and given chapter-like coverage in the appendices. This treatment makes circuit coverage optional; those who wish to include it may insert it at any point in the chapter sequence. For those who do not cover circuit technology or want only to reference it, the material does not break the flow of topics, and a student background in device and circuit theory is not required.

Specific integrated circuit devices are introduced as examples of digital functions only after the basic concepts have been covered. This approach retains the general nature of the first edition and adds some specific coverage of the 54/74 TTL series of digital integrated circuits, as well as other technologies, for the purpose of illustration.

Coverage of some troubleshooting techniques for digital systems has been added to this edition, as well as additional examples of applications.

Chapter 9, "Memories," is completely revised to emphasize semiconductor memories, including the concept of *stacks* as associated with microprocessors. Magnetic storage devices, including the floppy disk and magnetic bubble memories, are also discussed. Chapter 10 on interfacing covers buses and bus standards, three-state devices, and other data transfer topics, including greatly expanded coverage of D/A and A/D conversion methods. Chapter 11 has been extensively revised yet retains some features of the discussion of arithmetic processes in the first edition. Chapter 12 is a completely new treatment of microprocessor fundamentals; the coverage is oriented around the 6800 and provides an excellent foundation from which to launch into a detailed microprocessor course.

ACKNOWLEDGMENTS

As always, many people have helped make this second edition a reality.

First, let me express my gratitude to Chris Conty, Cher Paul, Jan Hall, Cathy White, Tony Samuolis, and all the other people at Merrill for their cooperation and their efforts to produce a quality product.

Also, my appreciation to Sam Oppenheimer, Broward Community College, John Canavan, DeVry Institute of Technology/Dallas, and David M. Hata, Portland Community College, who reviewed the manuscript and provided many valuable suggestions.

My thanks also to Bill Oltman for his very thorough work in checking the manuscript, and to the Hewlett-Packard Company, Intel Corporation, Motorola Incorporated, and Texas Instruments Incorporated for providing the data sheets and photographs for use in this book.

Finally, I am grateful to these people, who volunteered their time and advice: John L. Morgan DeVry Institute of Technology/Dallas; Mike Ewald, J. M. Perry Institute; Bob Kutz, Institute of Electronic Science/Texas A & M University; Johnny Wortham, State Technical Institute at Memphis; Ruth Seeley, Technical/Vocational Institute/New Mexico; Salvatore Angelastro, Suffolk Community College; and Richard Maxey, Chabot College.

Thomas L. Floyd

CONTENTS

1
INTRODUCTION

NUMBER SYSTEMS AND CODES

3
LOGIC GATES

4
BOOLEAN ALGEBRA

5
COMBINATIONAL LOGIC

6
FUNCTIONS OF COMBINATIONAL LOGIC

7 FLIP-FLOPS AND OTHER MULTIVIBRATORS

8 COUNTERS AND REGISTERS

9 MEMORIES

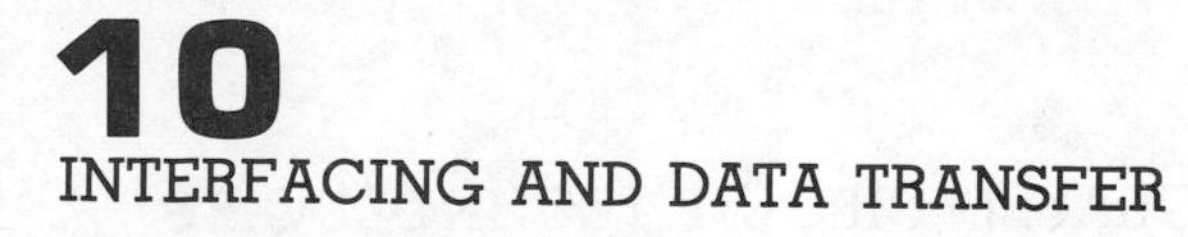

INTERFACING AND DATA TRANSFER

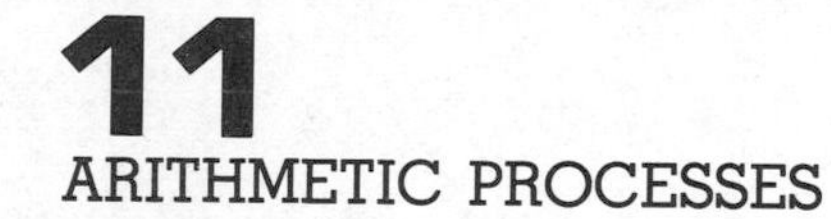

ARITHMETIC PROCESSES

12

THE MICROPROCESSOR

A

INTEGRATED CIRCUIT TECHNOLOGIES

DATA SHEETS

ERROR DETECTION AND CORRECTION

DIGITAL
FUNDAMENTALS

Digital Electronics began in 1946 with the first *electronic* digital computer implemented with vacuum-tube circuitry, the Electronic Numerical Integrator and Computer (ENIAC). The concept can be traced to Charles Babbage, who developed a mechanical digital computer in the 1830's but never completed a working machine. The first digital computer was built around 1944 by a Harvard University professor; it was essentially electromechanical, not electronic.

The term *digital* refers to the way these computers perform their operations by counting numbers (digits).

For years applications of digital electronics were confined to computer systems. Recently, however, digital techniques have been applied in such areas as telephone systems, telemetry and communication data processing, radar systems, missile guidance, navigation systems, television, and many military and consumer products. The technology has progressed from vacuum-tube circuitry through discrete semiconductor circuitry to increasingly complex integrated circuits.

Digital electronics typically involves circuits and *systems*—a combination of circuits connected in some way to perform a specified function—in which only two states are utilized. These two states are normally represented within the circuitry by two different voltage levels; other circuit conditions—current values, open or closed switch, and on or off lamp—can also represent the two states. In digital systems, combinations of the two states are used to represent symbols, characters, numbers, and other types of information. The binary number system can readily be accommodated to digital electronics because it is composed of only two digits, 0 and 1, called *bits,* a contraction of the term *binary digit.* These two digits can be represented by the two voltage levels of a digital circuit. The combination of states produced by several digital circuits can represent combinations of binary digits that form numbers or represent other information.

1 INTRODUCTION

1–1 LOGIC LEVELS AND PULSE WAVEFORMS

In digital systems two voltage levels represent the two binary digits, 1 and 0. If the higher of the two voltages represents a 1 and the lower voltage represents a 0, this is called *positive logic*. On the other hand, if the lower of the two voltages represents a 1 and the higher voltage represents a 0, we have a *negative logic* system. To illustrate, suppose that we have +5 V and 0 V as our logic-level voltages. We will designate the +5 V as the HIGH level and the 0 V as the LOW level, so positive and negative logic can be defined as

Positive logic	*Negative logic*
HIGH = 1	HIGH = 0
LOW = 0	LOW = 1

Both positive and negative logic are used in digital systems, but positive logic is the more common. For this reason we will use only positive logic in this text.

Pulses are very important in the operation of digital circuits and systems because the voltage levels are normally changing back and forth between the HIGH and LOW states. Figure 1–1 (a) shows that a single *positive* pulse is generated when the voltage (or current) goes from its *normally* LOW level to its HIGH level and then back to its LOW level. The *negative* pulse in Figure 1–1 (b) is generated when the voltage goes from its *normally* HIGH level to its LOW level and back to its HIGH level.

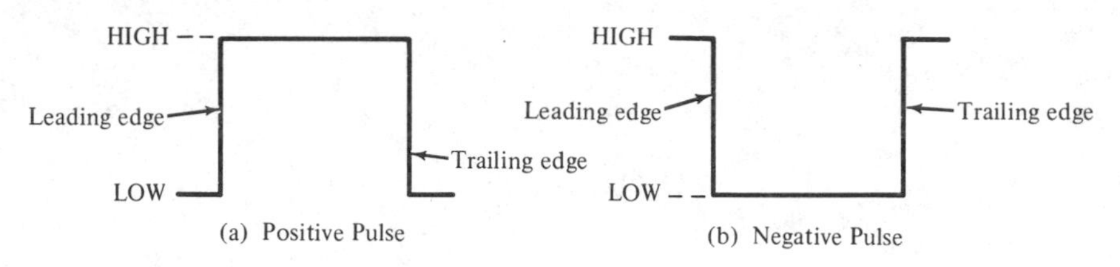

FIGURE 1–1 *Ideal pulses.*

As indicated in Figure 1–1, the pulse is composed of two edges: a *leading* edge and a *trailing* edge. For a positive pulse, the leading edge is a *positive-going* transition (*rising* edge) and the trailing edge is a *negative-going* transition (*falling* edge). The pulses in Figure 1-1 are ideal because the rising and falling edges change in zero time (instantaneously). Actually, these transitions take time and never occur instantaneously, although for most digital logic work we can assume ideal pulses. Figure 1–2 shows a nonideal pulse with finite rising and falling edges. The time required for the pulse to go from its LOW level to its HIGH level is called the *rise time* (t_r), and the time required for the transition from the HIGH level to the LOW level is called the *fall time* (t_f). In actual practice it is common to measure rise time from 10 percent of the pulse amplitude to 90 percent of the pulse amplitude, and to measure the fall time from 90 to 10 percent of the pulse amplitude. This is due to nonlinearities that commonly occur near the bottom and the top of the pulse, as indicated in Figure 1–2.

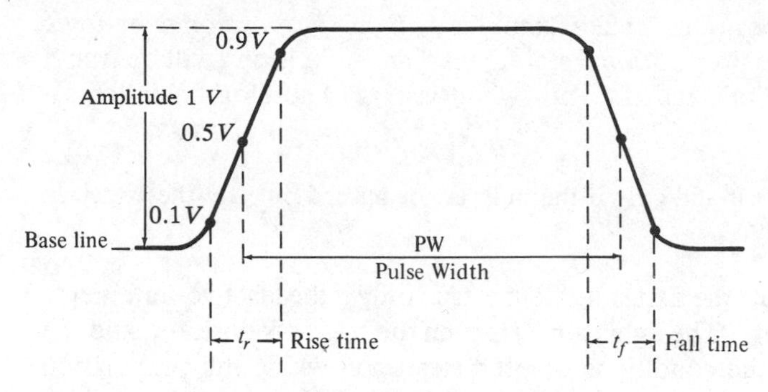

FIGURE 1–2 *Nonideal pulse characteristics.*

The pulse width (t_w) is a measure of the duration of the pulse and is typically defined as the time between the 50 percent points on the rising and falling edges, as indicated in Figure 1–2.

Most waveforms encountered in digital systems are composed of series of pulses and can be classified as *periodic* or *nonperiodic*. A periodic pulse waveform is one that repeats itself at a fixed interval called the *period*. The frequency is the rate at which it repeats itself and is measured in pulses per second (pps) or Hertz (Hz). A nonperiodic pulse waveform, of course, does not repeat itself at fixed intervals and may be composed of pulses of differing pulse widths and/or differing time intervals between the pulses. An example of each type is shown in Figure 1–3.

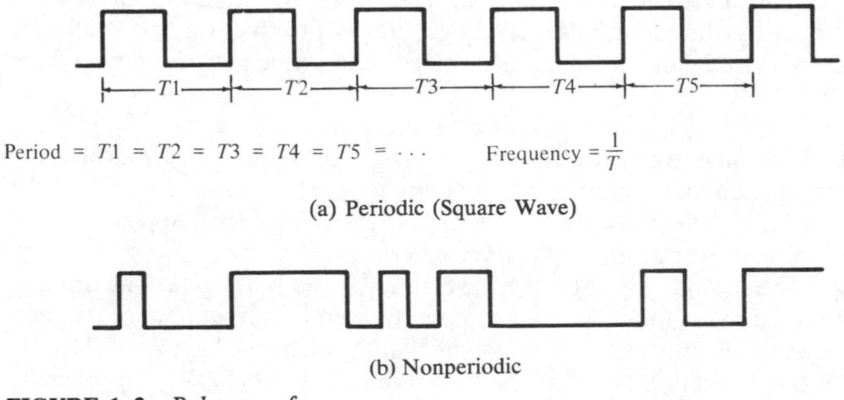

FIGURE 1–3 *Pulse waveforms.*

ELEMENTS OF DIGITAL LOGIC 1–2

In its basic form, logic is that realm of human reasoning that tells us a certain proposition is true if certain conditions or premises are true. "The light is on" is an example of a proposition (declarative statement) that can be true or false. "The bulb is not burned out" and "the switch is on" are other examples of propositions that can be classified as true or false.

Several propositions, when combined, form *propositional* or *logic* functions. For example, the propositional statement "The light is on" will be true if "The bulb is not burned out" and if "The switch is on." Therefore, this logical statement can be made:

> The light is on if and only if the bulb is not burned out and the switch is on.

In this example the first statement is true only if the last two statements are true. The first statement ("The light is on") is then the basic proposition, and the other two statements are the conditions or premises upon which the proposition depends.

Many situations, problems, and processes that we encounter in our daily lives can be expressed in the form of propositional or logic functions and can be readily automated. Since these functions are true/false or yes/no statements, digital circuits with their two-state characteristics are extremely applicable.

A mathematical system for formulating logical statements with symbols, so problems can be written and solved in a manner similar to ordinary algebra, was developed by the Irish logician and mathematician George Boole in the 1850s. *Boolean algebra,* as it is known today, finds application in the design and analysis of digital systems and will be covered in Chapter 4.

The term *logic* is applied to digital circuits used to implement logical functions. Several basic digital circuits are the basic *elements* forming the building blocks for complex digital systems such as the computer. We will now look at these elements and discuss their functions in a very general way to give you an overall view of the primary ingredients of digital electronics. Later chapters will cover these circuits in full detail.

The first basic type of logic element is the NOT circuit. The primary function of this circuit is to produce one logic level from the opposite logic level. If the higher level is applied to the input of the NOT circuit, the lower level appears on the output. If the lower level is applied to the input, the higher level appears on the output. This circuit is commonly called an *inverter.*

The second type of basic logic element is the AND gate. The primary function of this circuit is to produce a true condition on its output if *all* of its input conditions are true. For instance, let us assume that the higher voltage level (HIGH) represents a true condition and that the lower voltage level (LOW) represents a not-true (false) condition. Let us also use the example statement "The light is on" as our logic function. The AND gate can be used to tell us if this statement is true or false by applying the conditional functions "The switch is on" and "The bulb is not burned out" to the gate inputs. If both of these conditional statements are true, then both inputs to the AND gate are HIGH and, as a result, the output is also HIGH. If either condition is false as represented by a LOW, then the output is LOW. In other words, the AND gate "tells" us that *the light is on if and only if the switch is on AND the bulb is not burned out.*

The third basic type of logic element is the OR gate. The primary function of this circuit is to produce a true indication on its output when *one or more* of its input conditions are true. Again, let us assume that the higher voltage level represents a true condition and that the lower voltage level represents a false condition. Let us take as an example a door that will open automatically and that can be controlled from a remote transmitter or from a wall switch. The higher voltage level is required to energize the motor in order to open the door. Our propositional statement for this example is "The door opens" and the conditional statements are "The transmitter is on" and "The wall switch is on." The OR gate can be used to "tell" us when the propositional statement is true or when it is false by applying the conditional functions to its inputs. If either one or both of these conditional statements are true, then the output of the OR gate will be HIGH. If neither conditional statement is true, the output of the gate is LOW. For this example, the OR gate "tells" us that *the door opens if the transmitter is on, OR the wall switch is on, OR both are on.* The gate output can be used to open the door by properly energizing the activating motor.

The fourth basic type of logic element, which belongs to a class of circuits called multivibrators, is the bistable multivibrator, or *flip-flop*. The primary function of this circuit is to store or "memorize" a binary digit for an indefinite period of time. This device is distinguished from those previously discussed by its ability to retain either logic level after input conditions have been removed. Actually, the flip-flop can be constructed from combinations of basic gates, but it is treated as a distinct logic element because of its importance in digital systems.

The basic elements—the gates, inverter, and flip-flop—are used to construct more complex logic circuits such as counters, registers, decoders, and memories. These more complex logic functions are then combined to form complete digital systems designed to perform specified tasks.

DIGITAL LOGIC FUNCTIONS 1–3

The inverter, the basic gates, and the flip-flop combine to form more complex logic circuits that perform many operations. Some of the more common logic functions are comparison, arithmetic operations, decoding, encoding, code conversion, counting, memorizing, and multiplexing. We will see what these functions mean in terms of their "block" operations; many are already quite familiar to you because they are common operations.

Comparison

The *comparison* function is performed by a logic circuit called a *comparator*. Its function is to compare two quantities and to indicate if they are equal or not equal. For example, suppose we have two numbers and we wish to know if they are equal or not equal, and, if unequal, which is greater. Figure 1–4 is a block diagram of a comparator. One number is applied to input *A* and the other to input *B*. The outputs indicate the relation of the two numbers by producing a HIGH level on the proper output line. Suppose a binary representation of the number 2 is applied to input A, and a binary representation of the number 5 is applied to input B. (We will discuss the representation of numbers and symbols in detail in later). A HIGH level will appear on the "$A < B$" output, indicating the relationship between the two numbers.

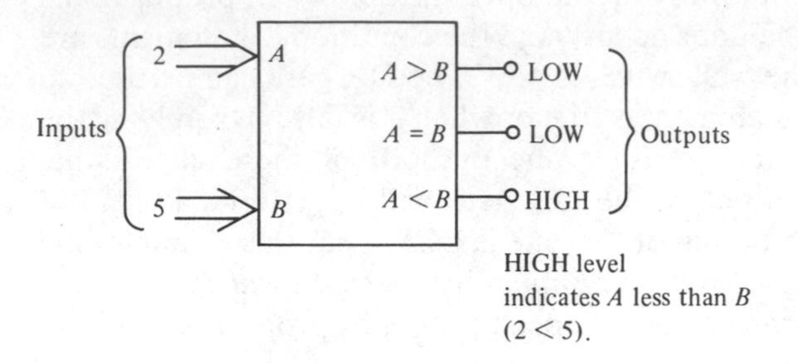

FIGURE 1–4 *The comparator.*

Arithmetic operations

The *addition* function is performed by a logic circuit called an *adder*. Its function is to add two numbers with a carry and generate a *sum* and an *output carry*. The block diagram in Figure 1–5 indicates the addition of the digit 3 and the digit 9. We all know that the sum is 12, and the adder indicates this result by producing the digit 2 on the *sum* output and the digit 1 on the *carry* output. Note that we assume the input carry in this example to be 0. In later chapters we will learn how numbers can be represented by the logic levels of a digital circuit.

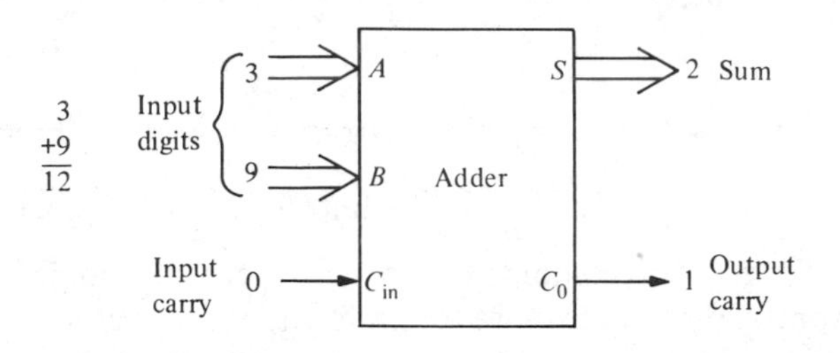

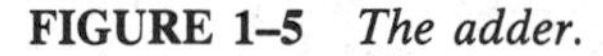

FIGURE 1–5 *The adder.*

Subtraction is the second arithmetic function that can be performed by digital logic circuits. A *subtractor* requires three inputs: the two numbers to be subtracted and an *input borrow*. The two outputs are the *difference* and the *output borrow*. When, for instance, 5 is subtracted from 8 with no input borrow, the difference is 3 with no output borrow. We will see later how subtraction can actually be performed by an adder, because subtraction is simply a special case of addition.

The third arithmetic operation that can be performed by logic circuits is *multiplication*. Since numbers are always multiplied two at a time, two inputs are required. The output of the multiplier is the *product*. Since multiplication is simply a series of additions with shifts in the positions of the partial products, it can be performed using an adder.

Division, the fourth type of arithmetic operation, turns out to be a series of subtractions, comparisons, and shifts and so can also be performed using an adder. Two inputs to the *divider* are required, and the outputs generated are called the *quotient* and the *remainder.*

Decoding

The *decoder* converts coded information, such as binary, into a recognizable form, such as decimal. For example, a particular type of decoder converts a four-bit binary code into the appropriate decimal digit, as illustrated in Figure 1–6.

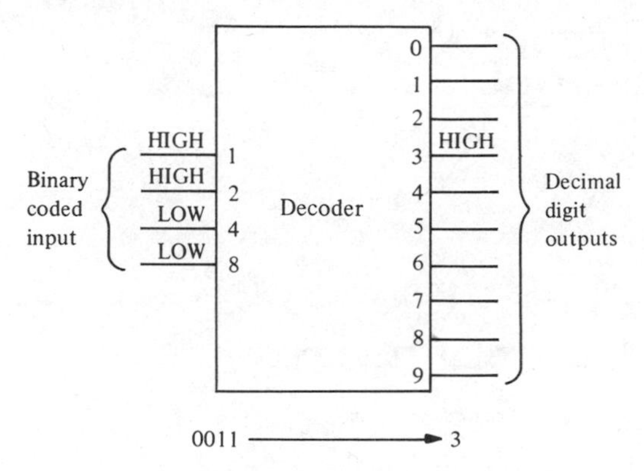

FIGURE 1–6 *Example of a decoder.*

Encoding

The *encoder* converts information, such as a decimal number or an alphabetic character, into some coded form. For example, a certain type of encoder converts each of the decimal digits, 0 through 9, to a binary code as shown in Figure 1–7. A HIGH level on a given input corresponding to a specific decimal digit produces the proper four-bit code on the output lines.

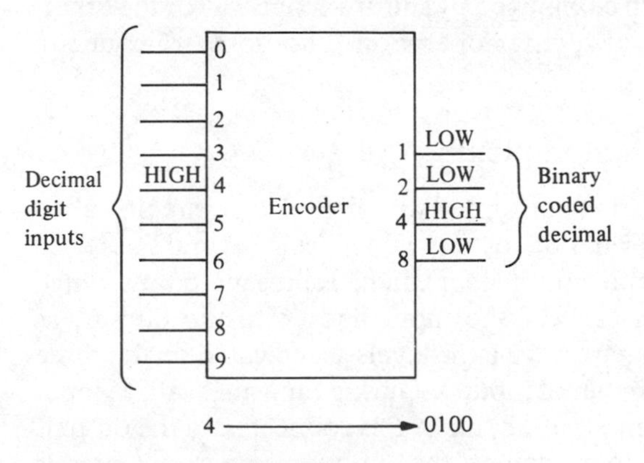

FIGURE 1–7 *Example of an encoder.*

Counting

The *counting* operation is very important in digital systems. There are many types of digital counters, but their basic function is to count events represented by changing levels or pulses or to generate a particular sequence of numbers. In order to count, the counter must "remember" the present number so that it can go to the next proper number in sequence. Therefore, storage or memory capability is an important characteristic of all counters, and flip-flops are generally used to implement these devices. Another important counter application is frequency division. Figure 1–8 illustrates two simple counter operations.

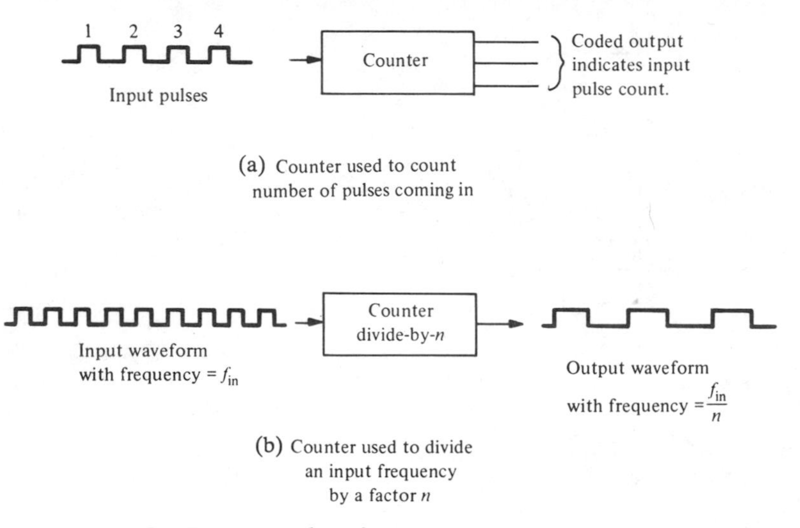

FIGURE 1–8 *Two examples of counter operations.*

Registers

Registers are digital circuits used for the temporary storage and shifting of information. For instance, a number in binary form can be stored in a register, and then its position within the register can be changed by shifting it one way or the other. Figure 1–9 illustrates a simple form of register operation. Flip-flops are common storage elements used in registers.

Multiplexing and demultiplexing

Multiplexing is an operation performed with digital logic circuits called *multiplexers*. Multiplexing allows information to be switched from several lines onto a single line in some sequence. A simple multiplexer can be represented by a switch operation that sequentially contacts each of the input lines with the output, as illustrated in Figure 1–10. Assume that we have logic levels as indicated on the three inputs (a multiplexer can have any number of inputs). During time interval $T1$, input A is connected to the output; during interval $T2$, input B is connected to the output; and during interval $T3$, input C is connected to the output. As a result of this *multiplexing action,* we have the three logic levels (information) on the inputs appearing in sequence on the output line. The switching action, of course, is accomplished with logic circuits, as we will learn later. It might be well to mention at this point that the inverse of the multiplexing function is called *demultiplexing*. Here, logic data from a single input line are sequentially switched onto several output lines, as shown in Figure 1–11.

Digital circuits can be used to perform a large variety of tasks limited only by the imagination. Here, in order to give you a general picture of some of the basic aspects of digital logic, we have touched on only a few of the more basic and common functions that can be used as building blocks for increasingly complex systems.

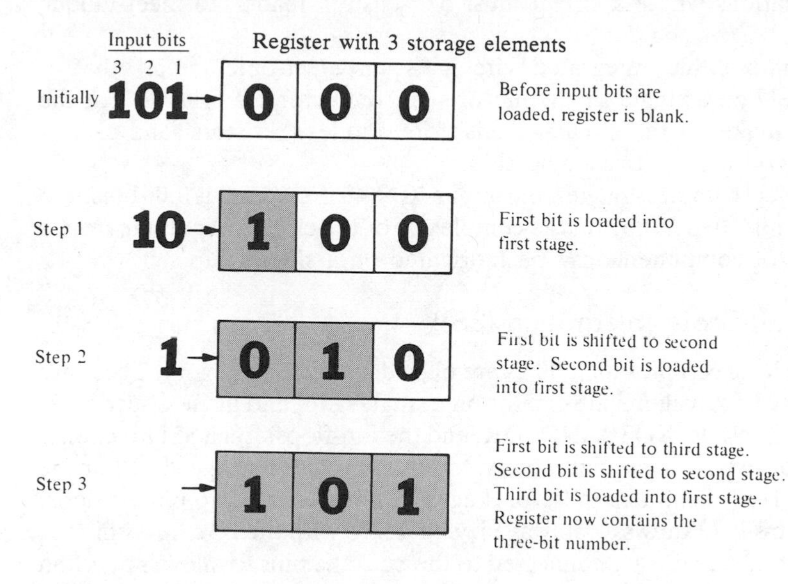

FIGURE 1–9 *Example of a simple shift register operation.*

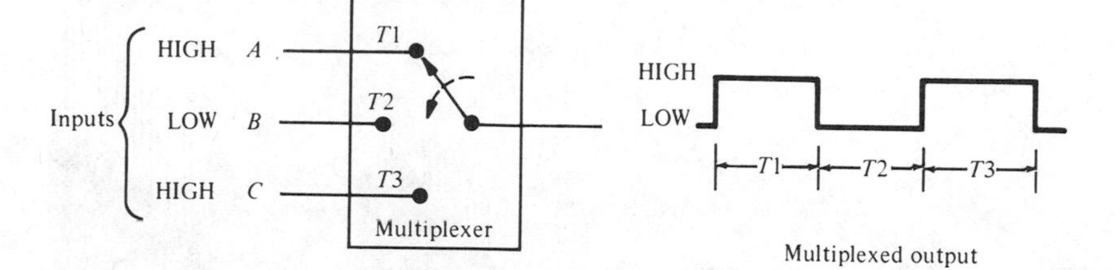

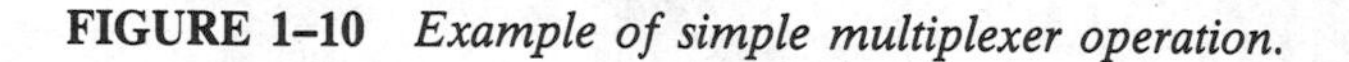

FIGURE 1–10 *Example of simple multiplexer operation.*

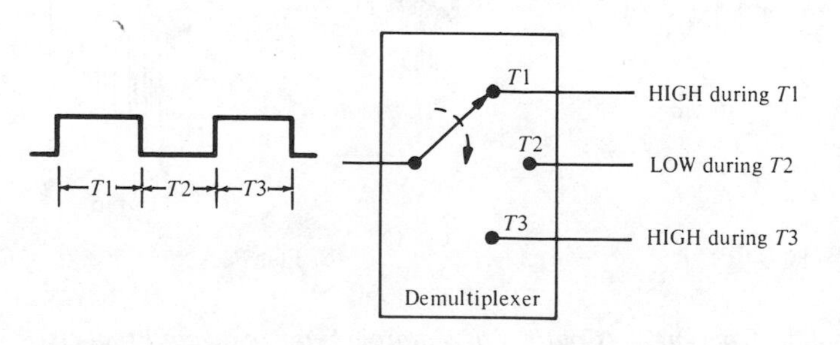

FIGURE 1–11 *Example of simple demultiplexer operation.*

1–4 DIGITAL INTEGRATED CIRCUITS

All of the logic functions that we have discussed and many more are available in some type of integrated circuit (IC) form and in varying levels of complexity. Modern digital systems utilize ICs to a large extent in their designs. In most cases, ICs have size, power, reliability, and cost advantages over discrete circuitry (except in very specialized applications where a circuit must be "custom made" to meet unique requirements).

A monolithic integrated circuit is an electronic circuit that is constructed entirely on a single small chip of semiconductor material. All of the components that make up the circuit—transistors, diodes, resistors, and capacitors—are an integral part of this single chip.

Typical chip sizes range from about 40 × 40 mils (a mil is 0.001 inch) to about 250 × 250 mils, depending on the complexity of the circuit. Anything from a few to thousands of components can be fabricated on a single chip.

Small-Scale Integration (SSI)

The least complex digital ICs are placed in the SSI category. These are circuits with up to 12 equivalent gate circuits on a single chip, and include such basic functions as NAND, NOR, NOT, AND, OR, and the flip-flops. Each SSI function is packaged in one of two main configurations, the dual-in-line package (DIP) or the flat pack. Figure 1–12 illustrates these packages in both 14- and 16-pin versions.

Figure 1–13 shows a cutaway view of a DIP with the IC chip within the package. Leads from the chip are connected to the package pins to allow inputs and outputs to the circuit.

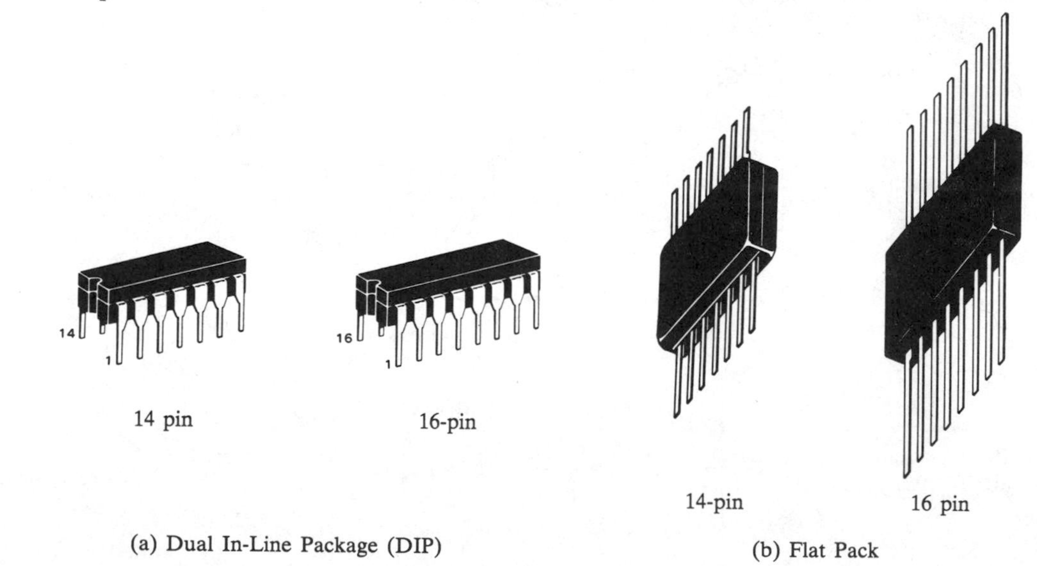

FIGURE 1–12 *Some common IC packages. (Courtesy of Motorola Semiconductor Products)*

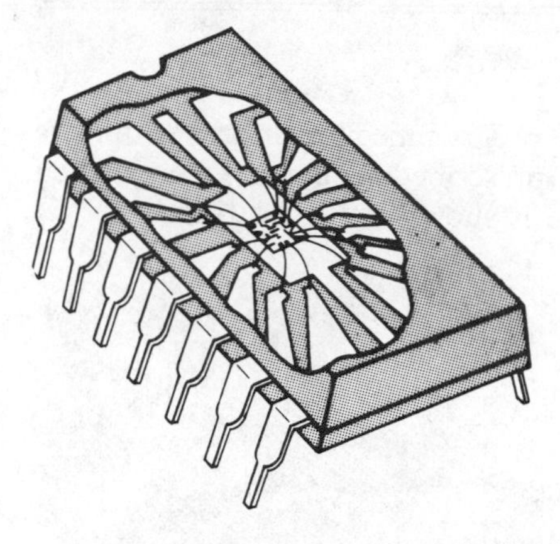

FIGURE 1–13 *Cutaway view of a dual-in-line 14-pin package showing the IC chip mounted inside with leads to input and output pins.*

Medium-Scale Integration (MSI)

The next classification according to digital circuit complexity is called medium-scale integration (MSI). These are circuits with complexities ranging from 12 to 100 equivalent gates on a chip. MSI circuits include the more complex logic functions such as encoders, decoders, counters, registers, multiplexers, arithmetic circuits, small memories, and others. MSI functions are available in 16- or 24-pin DIPS and some flat packs. Figure 1–14 illustrates 24-pin packages.

Large-Scale Integration (LSI)

Circuits with complexities of greater than 100 equivalent gates per chip, including large memories and microprocessors, generally fall into the LSI category.

(a) 24-pin Dual-in-Line

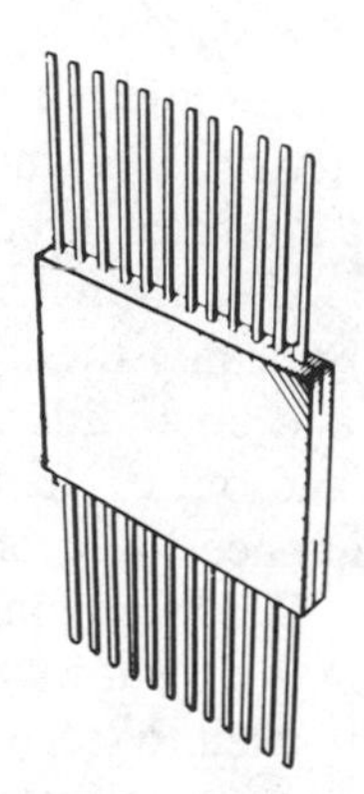

(b) 24-pin Flat Pack

FIGURE 1–14 *Two types of 24-pin ICs. (Courtesy of Motorola Semiconductor Products)*

1–5 MICROPROCESSORS

The microprocessor is an LSI device that can be programmed to perform arithmetic and logic operations and other functions in a prescribed sequence for any given application. A typical microprocessor package is shown in Figure 1–15.

FIGURE 1–15 *Microprocessor in a 40-pin dual-in-line package. (Courtesy of Motorola)*

The microprocessor is used as the central processing unit (CPU) in microcomputer systems where it is connected with other ICs such as memories and input/output interface circuits.

The arrangement of circuits within the microprocessor (called its *architecture*) permits the system to respond correctly to each of many different *instructions* and, in addition to arithmetic and logic operations, controls the flow of signals into and out of the computer, routing each to its proper destination in the required sequence to accomplish a specified task.

The interconnections or paths along which signals flow are are called *buses*. Figure 1–16 shows a simplified diagram of a microcomputer system. The memories are used to store binary information before and after processing, and the required sequence of instructions is called the *program*. The input/output (I/0) interface block properly connects external devices—keyboards, video terminals, and printers—to the microcomputer.

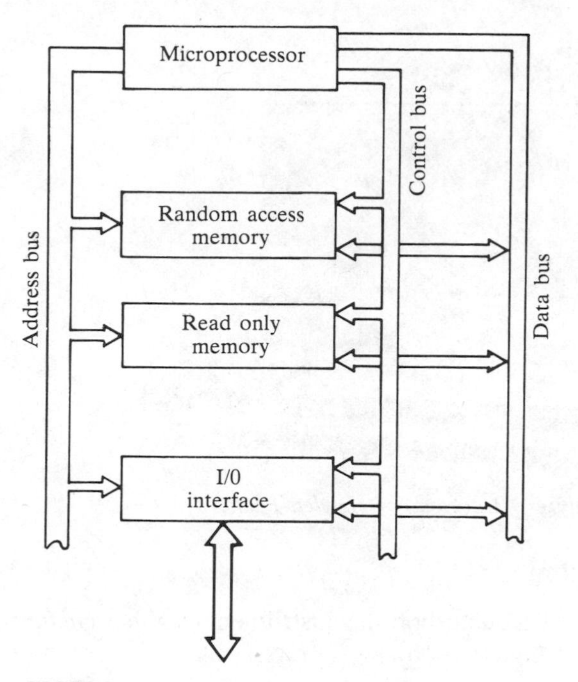

FIGURE 1–16 *Microcomputer block diagram*

DIGITAL TROUBLESHOOTING INSTRUMENTS 1–6

Troubleshooting is the technique of systematically isolating and identifying a fault in a circuit or system. A variety of special instruments is available for use in digital troubleshooting. Some typical equipment is presented in this section.

Logic Analyzers

Figure 1–17 shows a typical *logic analyzer*. This is basically a multichannel oscilloscope type of instrument with the ability to detect and display logic levels in several forms.

Most logic analyzers can display data in several formats. The *timing diagram* format displays several *pulse waveforms* with the proper time relationships. The *bit* format displays a bit pattern of 1s and 0s on the screen. Bit patterns from a functioning unit can be displayed and compared to the patterns of a faulty unit to detect errors.

A third format is a display using hexadecimal code. The *hexadecimal code,* as you will learn later, is a convenient way to represent binary information. Some analyzers also provide for octal displays. Octal is another code representation for binary and is discussed in the next chapter.

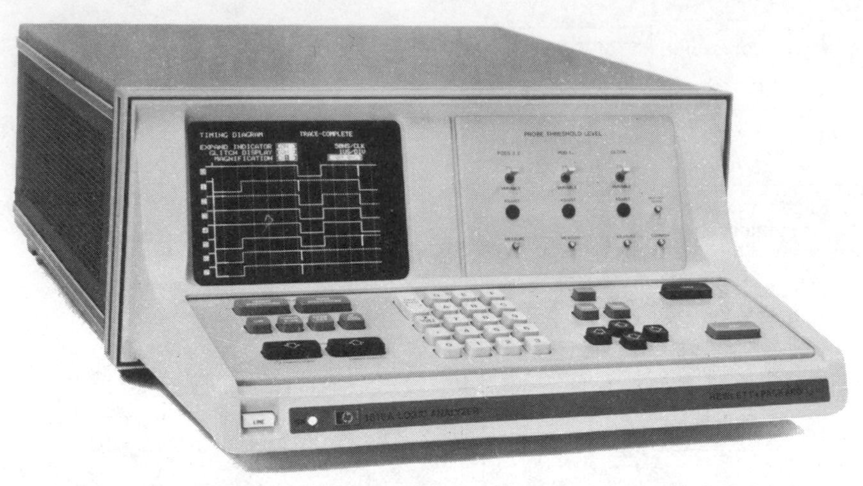

FIGURE 1–17 *The 1615A logic analyzer. (Courtesy of Hewlett-Packard)*

Signature Analyzers

Another useful digital troubleshooting instrument is the *signature analyzer* such as the one shown in Figure 1–18.

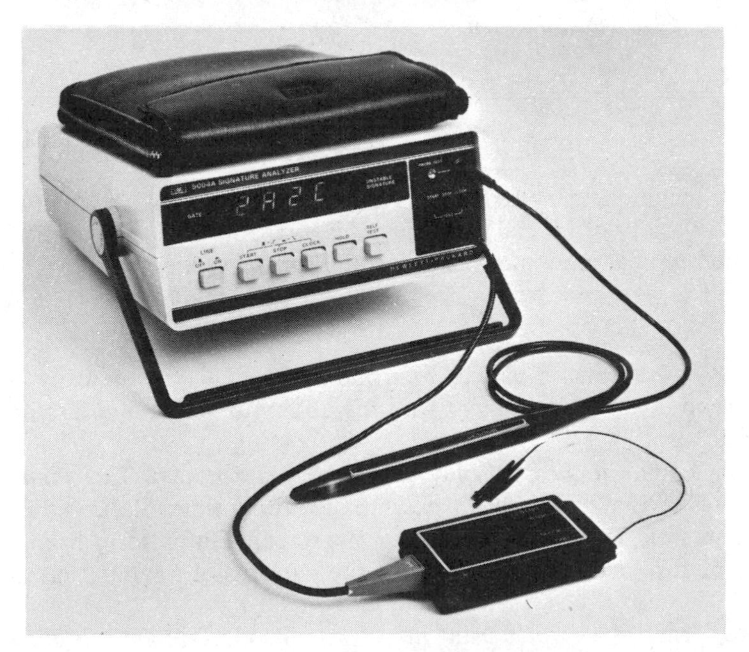

FIGURE 1–18 *The 5004A signature analyzer. (Courtesy of Hewlett-Packard)*

Signature analysis is a troubleshooting technique that locates faults to the component level in microprocessor-based systems.

A signature analyzer converts a pattern of 1s and 0s at a given test point in the circuit under test and displays the pattern as a hexadecimal code called the *signature* for that particular point. By comparing the indicated signature with a known correct signature, a technician can identify a faulty device.

Logic probes

The *logic probe* provides a simple means for digital troubleshooting by detecting voltage levels or pulses at a given point. A typical logic probe is shown in Figure 1–19.

FIGURE 1–19 *The 545A logic probe. (Courtesy of Hewlett-Packard)*

The logic probe uses a single lamp to indicate the various states possible on a digital signal path, such as HIGHs, LOWs, single pulses, pulse trains, and open circuits. A typical indication format is

Lamp on:	HIGH
Lamp off:	LOW
Lamp dim:	open or bad level
One flash:	single pulse
Repetitive flashes:	Pulse train

Although the logic probe is a very useful instrument, it alone cannot solve all troubleshooting problems. Often is must be used in conjunction with logic analyzers, signature analyzers, and other types of probes.

Current Tracer and Logic Pulser

These instruments are shown in Figure 1–20. *Current tracing* is a very effective troubleshooting technique in many cases. It is particularly difficult to isolate a bad element when a given circuit node (point of connection) is *stuck* in one logic state, and several elements are common to the node. This is where current tracing is very useful.

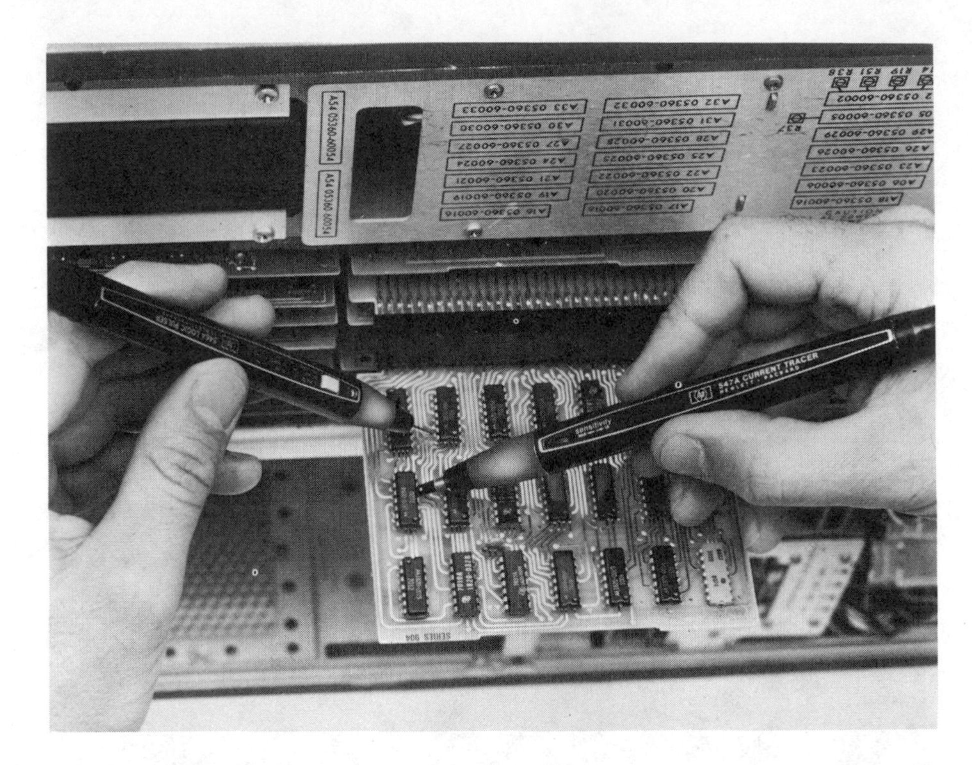

FIGURE 1–20 *The 546A logic pulser and the 547A current tracer. (Courtesy of Hewlett-Packard)*

The hand-held *current tracer* has a one-lamp indicator that glows when its tip is held over a *pulsing* current path. This instrument can detect whether current is flowing at all and most importantly, *where* the current is flowing. For instance, if a node is stuck in the LOW state due to a shorted input in one of the devices connected to the node, a very strong current exists between the circuit driving the node and the faulty component. For purpuses of detection, the current has to be *pulsing;* if it isn't, a *logic pulser* can be used to force pulses on the node and the current can then be traced to the bad component.

Logic Clip

A typical *logic clip* is shown in Figure 1–21. It is "clipped" to an integrated circuit where it makes contact with each pin on the IC. The lamps on the clip then indicate the logic conditions of each pin.

Lamp on:	HIGH
Lamp off:	LOW
Lamp dim:	pulses

FIGURE 1–21 *The 548A logic clip. (Courtesy of Hewlett-Packard)*

Problems

Section 1–1

1–1 Name the binary digits.

1–2 Define the term *bit*.

1–3 In a positive logic system, the following sequence of levels represents what sequence of binary digits:
HIGH, HIGH, LOW, HIGH, LOW, LOW, LOW, HIGH.

1–4 For the pulse shown in Figure 1–22, determine the following: (a) rise time (b) fall time (c) pulse width (d) amplitude.

1–5 Determine the period of the waveform from the graph in Figure 1–23.

1–6 What is the frequency for the pulse waveform in Figure 1–23?

1–7 Is the pulse waveform in Figure 1–23 periodic or nonperiodic?

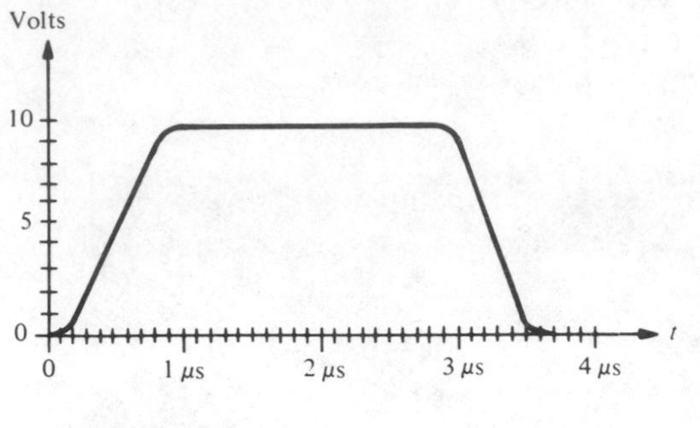

FIGURE 1–22

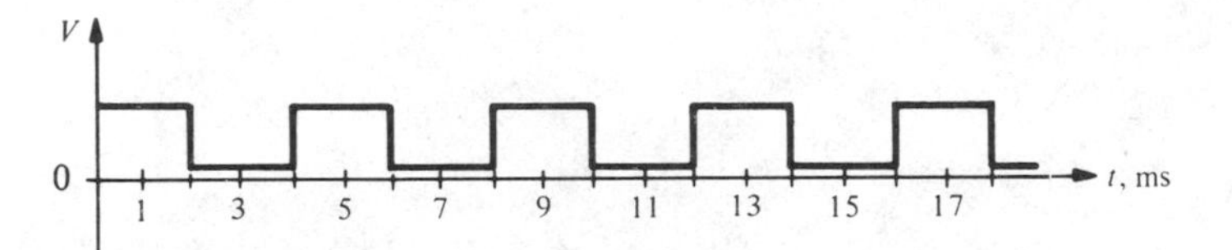

FIGURE 1–23

Section 1–2

1–8 Explain the difference between an AND gate and an OR gate.

1–9 Define one characteristic that distinguishes a flip-flop from a gate.

1–10 A basic logic element requires HIGHs on all of the inputs to make the output HIGH. What type of logic circuit is this?

1–11 A basic 2-input logic element has a HIGH on one input, a LOW on the other input, and the output is LOW. Identify the element.

1–12 A basic 2-input logic element has a HIGH on one input, a LOW on the other input, and the output is HIGH. What is the logic element?

Section 1–3

1–13 List and briefly describe seven logic functions.

1–14 What is a common storage element used in registers?

1–15 Explain the essential difference between multiplexing and demultiplexing.

Section 1–4

1–16 Define SSI, MSI, and LSI and identify each in terms of equivalent gate complexities.

1–17 What does DIP stand for?

Section 1–5

1–18 List three block functions found in a microcomputer system.

1–19 What is a *bus*?

Section 1–6

1–20 Define the term *troubleshooting*.

1–21 List at least five types of digital troubleshooting instruments.

The binary number system is fundamental to the study of digital electronics. In this chapter we will examine binary numbers and their relationship to other number systems such as decimal, octal, and hexadecimal. Arithmetic operations with binary numbers are covered to provide a basis for understanding how computers and other digital systems operate.

Several important digital codes are also introduced in this chapter. These are BCD, excess -3, and two alphanumeric codes (ASCII and EBCDIC). Further descriptions of digital codes are found at appropriate points throughout the text.

NUMBER SYSTEMS

2–1 DECIMAL NUMBERS

We will begin our study of number systems with the one with which we are all familiar, the decimal system. Because human anatomy is characterized by four fingers and a thumb on each hand, it was only natural that our method of counting involve the use of ten digits, that is, a system with a *base of ten.* Each of the ten decimal digits, 0 through 9, represents a certain quantity. The ten symbols (digits) do not limit us to expressing only ten different quantities because we use the various digits in appropriate positions within a number to indicate the magnitude of the quantity. We can express quantities up through nine before we run out of digits; if we wish to express a quantity greater than nine we use two or more digits, and the position of each digit within the number tells us the magnitude it represents. If, for instance, we wish to express the quantity twenty-three, we use (by their respective positions in the number) the digit 2 to represent the quantity twenty and the digit 3 to represent the quantity three. Therefore, the position of each of the digits in the decimal number indicates the magnitude of the quantity represented and can be assigned a "weight."

The value of a decimal number is the sum of the digits times their respective weights. The following examples will illustrate.

Example 2–1

$$23 = 2 \times 10 + 3 \times 1 = 20 + 3$$

The digit 2 has a weight of 10, as indicated by its position, and the digit 3 has a weight of 1, as indicated by its position.

Example 2–2

$$568 = 5 \times 100 + 6 \times 10 + 8 \times 1 = 500 + 60 + 8$$

The digit 5 has a weight of 100, the digit 6 has a weight of 10, and the digit 8 has a weight of 1.

2–2 9'S AND 10'S COMPLEMENTS

9's Complement

The 9's and 10's complements of decimal numbers will be defined in this section. The 9's complement of a decimal number is found by subtracting each digit in the number from 9. The 9's complement of each of the decimal digits is as follows:

Digit	*9's complement*
0	9
1	8
2	7
3	6
4	5
5	4
6	3
7	2
8	1
9	0

The following examples will illustrate how to determine the 9's complement of a decimal number.

Example 2–3

Find the 9's complement of each of the following decimal numbers: 12, 28, 56, 115, 562, and 3497.

Solutions:

To get the 9's complement of a decimal number, we subtract *each* digit in the number from 9.

(a) $\begin{array}{r} 99 \\ -\ 12 \\ \hline 87 \end{array}$ 9's complement of 12

(b) $\begin{array}{r} 99 \\ -\ 28 \\ \hline 71 \end{array}$ 9's complement of 28

(c) $\begin{array}{r} 99 \\ -56 \\ \hline 43 \end{array}$ 9's complement of 56

(d) $\begin{array}{r} 999 \\ -115 \\ \hline 884 \end{array}$ 9's complement of 115

(e) $\begin{array}{r} 999 \\ -562 \\ \hline 437 \end{array}$ 9's complement of 562

(f) $\begin{array}{r} 9999 \\ -3497 \\ \hline 6502 \end{array}$ 9's complement of 3497

9's Complement Subtraction

The usefulness of the 9's complement stems from the fact that subtraction of a smaller decimal number from a larger one can be accomplished by *adding* the 9's complement of the subtrahend (in this case the smaller number) to the minuend and then adding the carry to the result (end-around carry). When subtracting a larger number from a smaller one, there is no carry, and the result is in 9's complement form and negative. This procedure has a distinct advantage in certain types of arithmetic logic. A few examples will demonstrate decimal subtraction using the 9's complement method.

Example 2–4

Perform the following subtractions using the 9's complement method:

	Regular Subtraction	9's Complement Subtraction	
(a)	8	8	
	−3	+6	9's complement of 3
	5	(1) 4	
		→ +1	add carry to result
		5	
(b)	13	13	
	−7	+92	9's complement of 07
	6	(1) 05	
		→ +1	add carry to result
		6	
(c)	54	54	
	−21	+78	9's complement of 21
	33	(1) 32	
		→ +1	add carry to result
		33	
(d)	15	15	
	−28	+71	9's complement of 28
	−13	86	9's complement of result
		↓	(no carry indicates that
		−13	the answer is negative
			and in complement form)

10's Complement

The 10's complement of a decimal number is equal to the *9's complement plus 1.* This is illustrated in the following example.

Example 2–5

Convert the following decimal numbers to their 10's complement form: 8, 17, 52, and 428.

Solutions:

First we find the 9's complement, and then we add 1.

(a)	9		(b)	99	
	−8			−17	
	1	9's complement of 8; add 1		82	9's complement of 17; add 1
	+1			+1	
	2	10's complement of 8		83	10's complement of 17
(c)	99		(d)	999	
	−52			−428	
	47	9's complement of 52; add 1		571	9's complement of 428; add 1
	+1			+1	
	48	10's complement of 52		572	10's complement of 428

10's Complement Subtraction

The 10's complement, like the 9's complement, can be used to perform subtraction by adding the minuend to the 10's complement of the subtrahend and *dropping* the carry. This is illustrated in the following example:

Example 2–6

Perform the following subtractions using the 10's complement method:

	Regular Subtraction	10's Complement Subtraction	
(a)	8	8	
	−3	+7	10's complement of 3
	5	~~1~~5	drop carry
(b)	13	13	
	−7	+93	10's complement of 07
	6	~~1~~06	drop carry
(c)	54	54	
	−21	+79	10's complement of 21
	33	~~1~~33	drop carry
(d)	196	196	
	−155	+845	10's complement of 155
	41	~~1~~041	drop carry

Note that the 10's complement has an advantage over the 9's complement in that we do not have to add the carry to the result.

Example 2–6, continued
However, the 10's complement is not as easy to generate as the 9's complement.

2–3 BINARY NUMBERS

The binary number system is simply another way to count. It is less complicated than the decimal system because it is composed of only *two* digits. It may seem more difficult at first because it is unfamiliar to you.

Just as the decimal system with its ten digits is a base-ten system, the binary system with its two digits is a *base-two* system. The two binary digits (bits) are 1 and 0. The position of the 1 or 0 in a binary number indicates its "weight" or value within the number, just as the position of a decimal digit determines the magnitude of that digit. The weight of each successively higher position (to the left) in a binary number is an increasing power of two.

Counting in Binary

To learn to count in binary, let us first look at how we count in decimal. We start at 0 and count up to 9 before we run out of digits. We then start another digit position (to the left) and continue counting 10 through 99. At this point we have exhausted all two-digit combinations, so a third digit is needed in order to count from 100 through 999.

A comparable situation occurs when counting in binary, except that we have only two digits. We begin counting—0, 1; at this point we have used both digits, so we include another digit position and continue—10, 11. We have now exhausted all combinations of two digits, so a third is required. With three digits we can continue to count—100, 101, 110, and 111. Now we need a fourth digit to continue, and so on. A binary count of 0 through 31 is shown in Table 2–1.

An easy way to remember how to write a binary sequence such as in Table 2–1 for a 5-bit example is as follows:

1. The right-most column in the binary number begins with a 0 and alternates each bit.
2. The next column begins with two 0s and alternates every two bits.
3. The next column begins with four 0s and alternates every four bits.
4. The next column begins with eight 0s and alternates every eight bits.
5. The next column begins with sixteen 0s and alternates every sixteen bits.

TABLE 2–1

Count	Decimal Number	Binary Number
zero	0	00000
one	1	00001
two	2	00010
three	3	00011
four	4	00100
five	5	00101
six	6	00110
seven	7	00111
eight	8	01000
nine	9	01001
ten	10	01010
eleven	11	01011
twelve	12	01100
thirteen	13	01101
fourteen	14	01110
fifteen	15	01111
sixteen	16	10000
seventeen	17	10001
eighteen	18	10010
nineteen	19	10011
twenty	20	10100
twenty-one	21	10101
twenty-two	22	10110
twenty-three	23	10111
twenty-four	24	11000
twenty-five	25	11001
twenty-six	26	11010
twenty-seven	27	11011
twenty-eight	28	11100
twenty-nine	29	11101
thirty	30	11110
thirty-one	31	11111

As you have seen, it takes at least five binary digits (bits) to count from 0 to 31. The following formula tells us how high we can count in decimal, beginning with zero, with n bits:

$$\text{highest decimal number} = 2^n - 1 \qquad \textbf{(2–1)}$$

For instance, with two bits we can count from 0 through 3.

$$2^2 - 1 = 4 - 1 = 3$$

With four bits, we can count from 0 to 15.

$$2^4 - 1 = 16 - 1 = 15$$

A table of powers of two (2^n) is shown in Appendix D.

Evaluating a Binary Number by Conversion to Decimal

A binary number is a weighted number, as mentioned previously. The value of a given binary number in terms of its decimal equivalent can be determined by adding the product of each bit and its weight. The right-most bit is the *least significant bit (LSB)* in a binary number and has a weight of $2^0 = 1$. The weights increase by a power of two for each bit from right to left. The method of evaluating a binary number is illustrated by the following example:

Example 2–7

Evaluate the binary number 1101101.

Solution:

Binary weight:	2^6	2^5	2^4	2^3	2^2	2^1	2^0
Weight value:	64	32	16	8	4	2	1
Binary number:	1	1	0	1	1	0	1

$$\begin{aligned} & 1 \times 64 + 1 \times 32 + 0 \times 16 + 1 \times 8 + 1 \times 4 + 0 \times 2 + 1 \times 1 \\ = & \quad 64 + 32 + 0 + 8 + 4 + 0 + 1 \\ = & \quad 109_{10} \end{aligned}$$

This is the equivalent decimal value of the binary number. The subscript *10* identifies this as a decimal number to avoid confusion with other number systems.

The binary numbers we have seen so far have been whole numbers. Fractional numbers can also be represented in binary by placing bits to the right of the *binary point* just as fractional decimal digits are placed to the right of the decimal point.

The general form of a binary number can be expressed as

$$2^n \ldots 2^3 2^2 2^1 2^0 . 2^{-1} 2^{-2} \ldots 2^{-n}$$

↑binary point

This indicates that all the bits to the left of the binary point have weights that are positive powers of two, as we have previously discussed. All bits to the right of the binary point have weights that are negative powers of two, or fractional weights as illustrated in the following example:

Example 2–8

Determine the value of the binary fractional number 0.1011.

Solution:

First, we determine the weight of each bit and then sum the weights times the bit.

Binary weight:	2^{-1}	2^{-2}	2^{-3}	2^{-4}
Weight value:	0.5	0.25	0.125	0.0625
Binary number:	0.1	0	1	1

$$1 \times 0.5 + 0 \times 0.25 + 1 \times 0.125 + 1 \times 0.0625$$
$$= 0.5 + 0 + 0.125 + 0.0625$$
$$= 0.6875_{10}$$

It should be pointed out here that to determine the decimal value of a binary number, fractional or whole, we simply *add the weights of each 1 and ignore each 0* because the product of a 0 and its weight is 0.

Another method of evaluating a binary fraction is to determine the *whole-number value* of the bits and divide by the *total possible combinations* of the number of bits appearing in the fraction. For instance, for the binary fraction in Example 2–8.

If we neglect the binary point, the value of 1011_2 is 11_{10} in decimal. With four bits, there are 16 possible combinations. Dividing, we obtain $11/16 = 0.6875_{10}$.

The following examples will illustrate the evaluation of binary numbers in terms of their equivalent decimal values. The subscript *2* identifies binary numbers.

Example 2–9

Determine the decimal value of 11101.011_2.

Solution:

$$(16 + 8 + 4 + 1).(0.25 + 0.125) = 29.375_{10}$$

Using the alternate method to evaluate the fraction, we have $3/8 = 0.375_{10}$ (same result).

Example 2–10

Determine the decimal value of 110101.11_2.

Solution:

$$(32 + 16 + 4 + 1).(0.5 + 0.25) = 53.75_{10}$$

Using the alternate method to evaluate the fraction gives us $3/4 = 0.75_{10}$ (same result).

2–4 BINARY ADDITION

There are four basic rules for adding binary digits. They are:

$0 + 0 = 0$	0 plus 0 equals 0
$0 + 1 = 1$	0 plus 1 equals 1
$1 + 0 = 1$	1 plus 0 equals 1
$1 + 1 = 10$	1 plus 1 equals 0 with a carry of 1 (binary two)

Notice that three of the addition rules result in a single bit, and that the addition of two 1s yields a binary two (10_2). When adding binary numbers, the latter condition creates a sum of 0 in a given column and a carry of 1 over to the next higher column, as illustrated below.

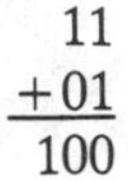

$$\begin{array}{r} 11 \\ +01 \\ \hline 100 \end{array}$$

In the right column, $1 + 1 = 0$ with a carry of 1 to the next column to the left. In the next column, $1 + 1 + 0 = 0$ with a carry of 1 to the next column to the left. In the left column, $1 + 0 + 0 = 1$.

When there is a carry, we have a situation where three bits are being added (a bit in each of the two numbers and a carry bit). The rules for this are as follows:

Carry bits	$1 + 0 + 0 = 1$	1 with no carry
	$1 + 1 + 0 = 10$	0 with a carry of 1
	$1 + 0 + 1 = 10$	0 with a carry of 1
	$1 + 1 + 1 = 11$	1 with a carry of 1

The following example will illustrate binary addition with the equivalent decimal addition shown.

Example 2–11

(a) $\begin{array}{r} 3 \\ +3 \\ \hline 6_{10} \end{array}$ $\begin{array}{r} 11 \\ +11 \\ \hline 110_2 \end{array}$ (b) $\begin{array}{r} 4 \\ +2 \\ \hline 6_{10} \end{array}$ $\begin{array}{r} 100 \\ +10 \\ \hline 110_2 \end{array}$

(c) $\begin{array}{r} 7 \\ +3 \\ \hline 10_{10} \end{array}$ $\begin{array}{r} 111 \\ +11 \\ \hline 1010_2 \end{array}$ (d) $\begin{array}{r} 6 \\ +4 \\ \hline 10_{10} \end{array}$ $\begin{array}{r} 110 \\ +100 \\ \hline 1010_2 \end{array}$

(e) $\begin{array}{r} 15 \\ +12 \\ \hline 27_{10} \end{array}$ $\begin{array}{r} 1111 \\ +1100 \\ \hline 11011_2 \end{array}$ (f) $\begin{array}{r} 28 \\ +19 \\ \hline 47_{10} \end{array}$ $\begin{array}{r} 11100 \\ +10011 \\ \hline 101111_2 \end{array}$

BINARY SUBTRACTION 2–5

There are four basic rules for subtracting binary digits, which are as follows:

$0 - 0 = 0$	0 minus 0 equals 0
$1 - 1 = 0$	1 minus 1 equals 0
$1 - 0 = 1$	1 minus 0 equals 1
$10_2 - 1 = 1$	10_2 minus 1 equals 1

When subtracting numbers, we sometimes have to borrow from the next higher column. The only time a borrow is required in binary is when we try to subtract a 1 from a 0. In this case, when a 1 is borrowed from the next higher column, a 10_2 is created in the column being subtracted, and the last of the four basic rules listed above must be applied. The following examples illustrate binary subtraction:

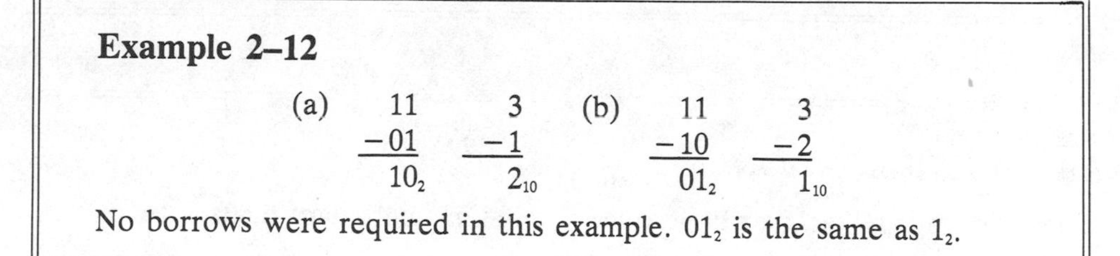

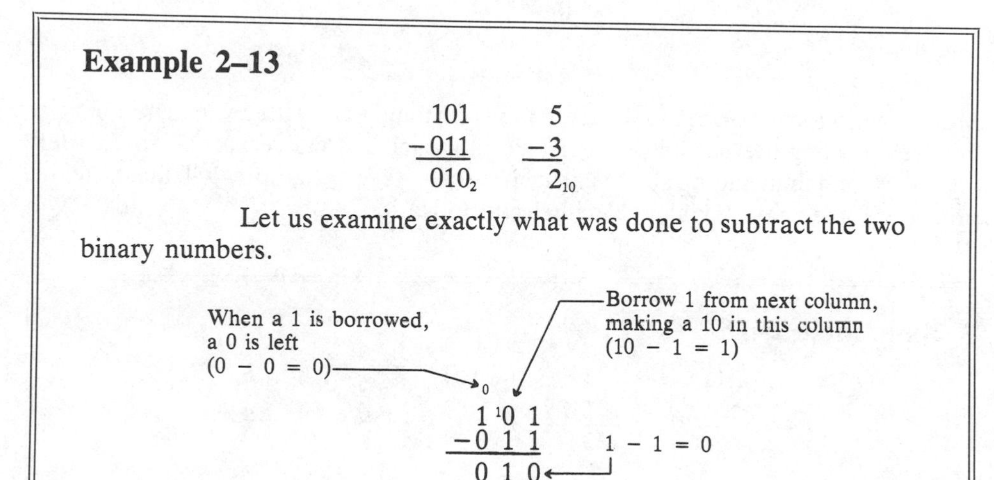

Example 2–14

$$
\begin{array}{rr}
1001 & 9 \\
-0110 & -6 \\
\hline
0011 & 3
\end{array}
$$

Example 2–14, continued

Let us look at the subtraction process for these two numbers.

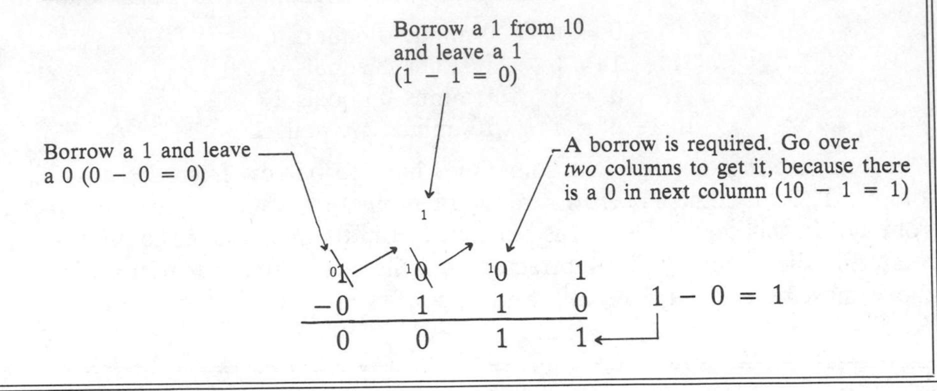

2–6 BINARY MULTIPLICATION

The following are the four basic rules for multiplying binary digits:

$$0 \times 0 = 0$$
$$0 \times 1 = 0$$
$$1 \times 0 = 0$$
$$1 \times 1 = 1$$

Multiplication is performed in binary in the same manner as with decimal numbers. It involves forming the partial products, shifting each successive partial product left one place, and then adding all the partial products. A few examples will illustrate the procedure and the equivalent decimal multiplication.

Example 2–15

(a)
```
  11        3
 ×1        ×1
 ---       --
  11₂       3₁₀
```

(b)
```
   11       3
  ×11      ×3
  ---      --
   11       9₁₀
  11
 ----
 1001₂
```

(c)
```
    111      7
   ×101     ×5
   ----     --
    111     35₁₀
   000
  111
 ------
 100011₂
```

(d)
```
     1011     11
    ×1001     ×9
    -----     --
     1011     99₁₀
    0000
   0000
  1011
 -------
 1100011₂
```

BINARY DIVISION 2–7

Division in binary follows the same procedure as division in decimal.

Example 2–16

```
            10      2                    11        3
(a)  11)110     3)6        (b)  10)110         2)6
        11        6                 10           6
        000       0                 10           0
                                    10
                                    00

             11      3                  10.1        2.5
(c)  100)1100   4)12       (d)  110)1111.0     6)15.0
         100       12               110          12
         100        0                11 0         3 0
         100                         11 0         3 0
         000                         00 0           0
```

1'S COMPLEMENT 2–8

Obtaining the 1's complement of a binary number

The 1's complement of a binary number is found by simply changing all 1s to 0s and all 0s to 1s, as illustrated by a few examples.

Binary number	*1's complement*
10101	01010
10111	01000
111100	000011
11011011	00100100

1's Complement Subtraction

Subtraction of binary numbers can be accomplished by the direct method described in Section 2–5 or by using the 1's complement method, which *allows us to subtract using only addition.*

When subtracting a smaller number from a larger number, the 1's complement method is as follows:

1. Determine the 1's complement of the smaller number.
2. Add the 1's complement to the larger number.
3. Remove the carry and add it to the result. This is called *end-around carry.*

Example 2–17

Subtract 10011_2 from 11001_2 using the 1's complement method. Show direct subtraction for comparison.

Solution:

Direct Subtraction	*1's Complement Method*	
11001	11001	
−10011	+01100	1's complement of 10011
00110	(1)00101	
	└→ +1	add end-around carry
	00110	final answer

When subtracting a larger number from a smaller, the 1's complement method is as follows:

1. Determine the 1's complement of the larger number.
2. Add the 1's complement to the smaller number.
3. The answer is negative and is the 1's complement of the result. There is no carry.

Example 2–18

Subtract 1101_2 from 1001_2 using the 1's complement method. Show direct subtraction for comparison.

Solution:

Direct Subtraction	*1's Complement Method*	
1001	1001	
−1101	+0010	1's complement of 1101
−0100	1011	answer in 1's complement form and negative
	└→	−0100 final answer

The 1's complement method is particularly useful in arithmetic logic circuits because subtraction can be accomplished with an adder. Also, the 1's complement of a number is easily obtained by inverting each bit in the number.

2–9 2'S COMPLEMENT

Obtaining the 2's Complement of a Binary Number

The 2's complement of a binary number is found by adding 1 to the 1's complement. An example will show how this is done.

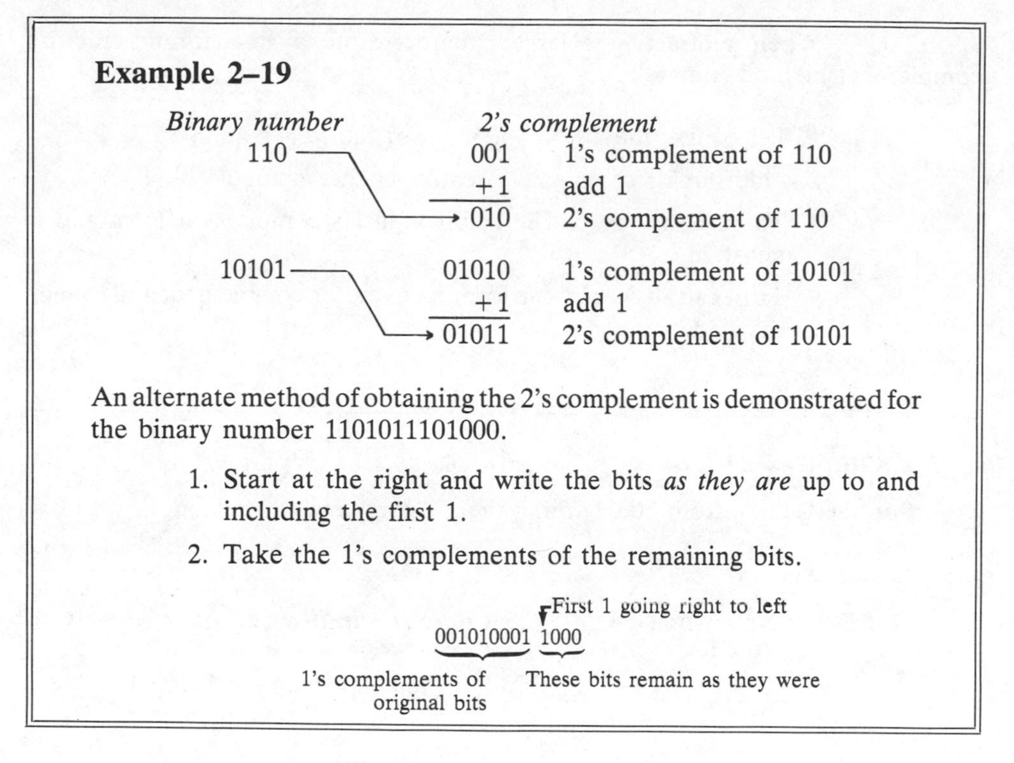

Example 2–19

Binary number	*2's complement*	
110	001	1's complement of 110
	+1	add 1
	→ 010	2's complement of 110
10101	01010	1's complement of 10101
	+1	add 1
	→ 01011	2's complement of 10101

An alternate method of obtaining the 2's complement is demonstrated for the binary number 1101011101000.

1. Start at the right and write the bits *as they are* up to and including the first 1.
2. Take the 1's complements of the remaining bits.

2's Complement Subtraction

Subtraction can be done using the 2's complement method as indicated below.

When subtracting a smaller number from a larger, the 2's complement method is applied as follows:

1. Determine the 2's complement of the smaller number.
2. Add the 2's complement to the larger number.
3. Discard the carry (there is always a carry in this case).

Example 2–20

Subtract 1011_2 from 1100_2 using the 2's complement method. Show direct subtraction for comparison.

Solution:

Direct Subtraction	*2's Complement Method*	
1100	1100	
−1011	+0101	2's complement of 1011
0001	~~1~~0001	discard carry
	→	0001 final answer

When subtracting a larger number from a smaller one, the 2's complement method is as follows:

1. Determine the 2's complement of the larger number.
2. Add the 2's complement to the smaller number.
3. There is no carry. The result is in 2's complement form and is negative.
4. To get an answer in true form, take the 2's complement and change the sign.

Example 2-21

Subtract 11100_2 from 10011_2 using the 2's complement method.

Solution:

Direct Subtraction	*2's Complement Method*	
10011	10011	
−11100	+00100	2's complement of 11100
−01001	10111	no carry; 2's complement of answer
	→	−01001 final answer

At this point, both the 1's and 2's complement methods of subtraction may seem excessively complex compared with direct subtraction. However, as mentioned before, they both have distinct advantages when implemented with logic circuits because they allow subtraction to be done using only addition. Both the 1's and the 2's complements of a binary number are relatively easy to accomplish with logic circuits; and the 2's complement has an advantage over the 1's complement in that an end-around carry operation does not have to be performed. We will see later how arithmetic operations are implemented with logic circuits using both methods.

2–10 DECIMAL-TO-BINARY CONVERSION

In Section 2–3 we discussed how to determine the equivalent decimal value of a binary number. Now, you will learn two ways of converting from a decimal to a binary number.

Sum-of-Weights Method

One way to find the binary number equivalent to a given decimal number is to determine the set of binary weight values whose sum is equal to the decimal number. For instance, the decimal number 9 can be expressed as the sum of binary weights as follows:

$$9 = 8 + 1 = 2^3 + 2^0$$

By placing a 1 in the appropriate weight positions, 2^3 and 2^0, and a 0 in the other positions, we have the binary number for decimal 9.

2^3	2^2	2^1	2^0	
1	0	0	1	binary nine

Example 2–22

Convert the decimal numbers 12, 25, 58, and 82 to binary.

Solutions:

$12_{10} = 8 + 4 = 2^3 + 2^2 \longrightarrow 1100_2$

$25_{10} = 16 + 8 + 1 = 2^4 + 2^3 + 2^0 \longrightarrow 11001_2$

$58_{10} = 32 + 16 + 8 + 2 = 2^5 + 2^4 + 2^3 + 2^1 \longrightarrow 111010_2$

$82_{10} = 64 + 16 + 2 = 2^6 + 2^4 + 2^1 \longrightarrow 1010010_2$

Repeated Division-by-2 Method

A more systematic method of converting from decimal to binary is the repeated *division by two* process. For example, to convert the decimal number 12 to binary we begin by dividing 12 by 2, and then we divide each resulting quotient by 2 until we have a 0 quotient. The remainder generated by each division forms the binary number. The first remainder to be produced is the least significant bit (LSB) in the binary number.

Example 2–23

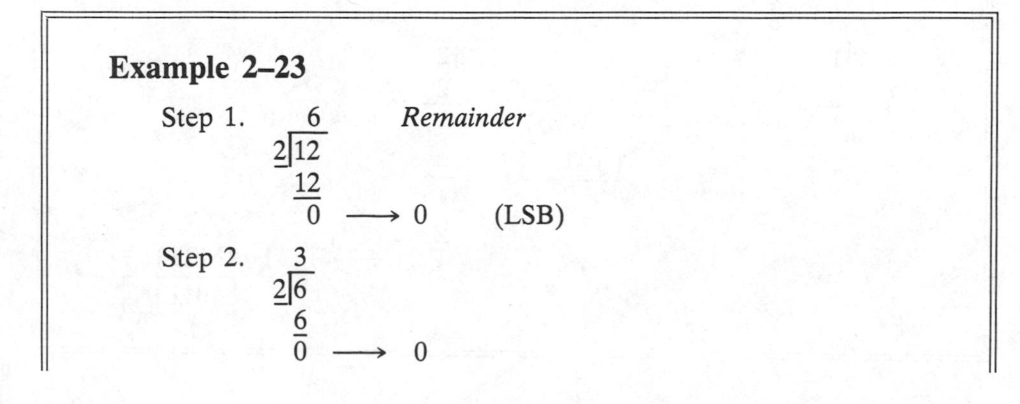

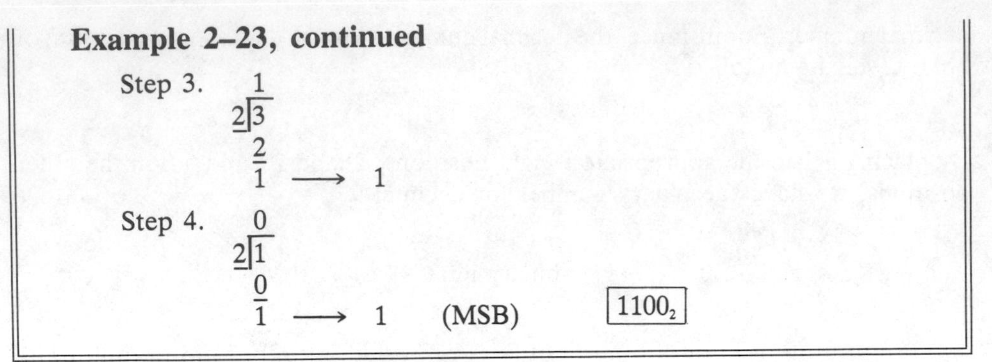

Example 2–24

Convert the decimal numbers 19 and 45 to binary.

Solutions:

(a)

Division		*Remainder*	
19 ÷ 2 = 9; 19 − 18 = 1	⟶	1	(LSB)
9 ÷ 2 = 4; 9 − 8 = 1	⟶	1	
4 ÷ 2 = 2; 4 − 4 = 0	⟶	0	
2 ÷ 2 = 1; 2 − 2 = 0	⟶	0	
1 ÷ 2 = 0; 1 − 0 = 1	⟶	1	(MSB)

10011_2

(b)

Division		*Remainder*	
45 ÷ 2 = 22; 45 − 44 = 1	⟶	1	(LSB)
22 ÷ 2 = 11; 22 − 22 = 0	⟶	0	
11 ÷ 2 = 5; 11 − 10 = 1	⟶	1	
5 ÷ 2 = 2; 5 − 4 = 1	⟶	1	
2 ÷ 2 = 1; 2 − 2 = 0	⟶	0	
1 ÷ 2 = 0; 1 − 0 = 1	⟶	1	(MSB)

101101_2

OCTAL NUMBERS 2–11

The octal number system is composed of eight digits, which are

0, 1, 2, 3, 4, 5, 6, 7

To count above 7, we begin another column and start over.

10, 11, 12, 13, 14, 15, 16, 17, 20, 21, etc.

Counting in octal is the same as counting decimal, except any number with an 8 or 9 is omitted. Table 2–2 shows the equivalent octal numbers for decimal numbers through 31.

TABLE 2-2

Decimal	Octal	Decimal	Octal
0	0	16	20
1	1	17	21
2	2	18	22
3	3	19	23
4	4	20	24
5	5	21	25
6	6	22	26
7	7	23	27
8	10	24	30
9	11	25	31
10	12	26	32
11	13	27	33
12	14	28	34
13	15	29	35
14	16	30	36
15	17	31	37

To distinguish octal numbers from decimal numbers, we will use the subscript 8 to indicate an octal number. For instance, 15_8 is equivalent to 13_{10}.

Evaluating an Octal Number by Conversion to Decimal

Since the octal number system has a base of eight, each successive digit position is an increasing power of eight, beginning in the right-most column with 8^0. The evaluation of an octal number in terms of its *decimal eqivalent* is accomplished by multiplying each digit by its weight and summing the products, as illustrated below for octal 2374.

Weight:	8^3	8^2	8^1	8^0
Decimal value:	512	64	8	1
Octal number:	2	3	7	4

$$\begin{aligned} 2374_8 &= 2 \times 8^3 + 3 \times 8^2 + 7 \times 8^1 + 4 \times 8^0 \\ &= 2 \times 512 + 3 \times 64 + 7 \times 8 + 4 \times 1 \\ &= 1024 + 192 + 56 + 4 = 1276_{10} \end{aligned}$$

Decimal-to-Octal Conversion

A method of converting a decimal number into an octal number is the repeated *division by eight* method, which is similar to the method used in conversion of decimal to binary. To show how it works, we will convert the decimal number 359 to octal. Each successive division by 8 yields a remainder which is a digit in the equivalent octal number. The first remainder generated is the least significant digit. (LSD).

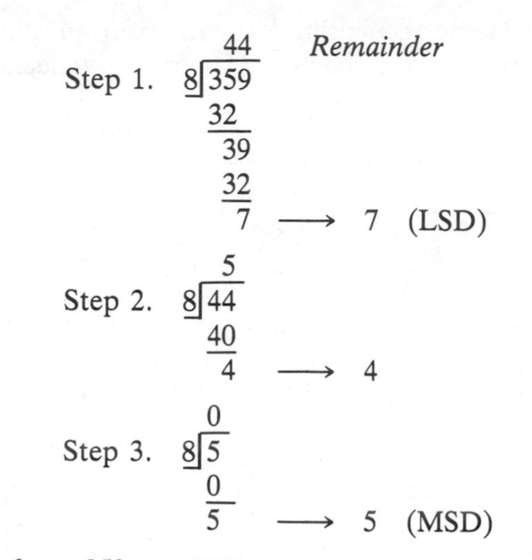

Therefore, $359_{10} = 547_8$

The octal numbers that we have seen to this point have been whole numbers. Fractional octal numbers are represented by digits to the right of the octal point.

The general form of an octal number can be expressed as

$$8^n \ldots 8^3 8^2 8^1 8^0 . 8^{-1} 8^{-2} 8^{-3} \ldots 8^{-n}$$

This shows that all digits to the left of the octal point have weights that are positive powers of eight, as we have seen previously. All digits to the right of the octal point have fractional weights, or negative powers of eight as illustrated in the following example.

Example 2–25

Determine the decimal value of the octal fraction 0.325_8.

Solution:

First, we determine the weights of each digit and then sum the weight times the digit.

Octal weight:	8^{-1}	8^{-2}	8^{-3}
Decimal value:	0.125	0.015625	0.001953
Octal number:	0.3	2	5

$$0.325_8 = 3(0.125) + 2(0.015625) + 5(0.001953)$$
$$= 0.375 + 0.03125 + 0.009765 = 0.416015_{10}$$

Octal-to-Binary Conversion

Because all three-bit binary numbers are required to represent the eight octal digits, it is very easy to convert from octal to binary and from binary to octal. For this reason, the octal number system is used in some digital systems especially for input/output applications.

Each octal digit can be represented by three bits as indicated.

Octal Digit	*Binary*
0	000
1	001
2	010
3	011
4	100
5	101
6	110
7	111

To convert an octal number to a binary number, simply replace each octal digit by the appropriate three bits. This is illustrated in the following example:

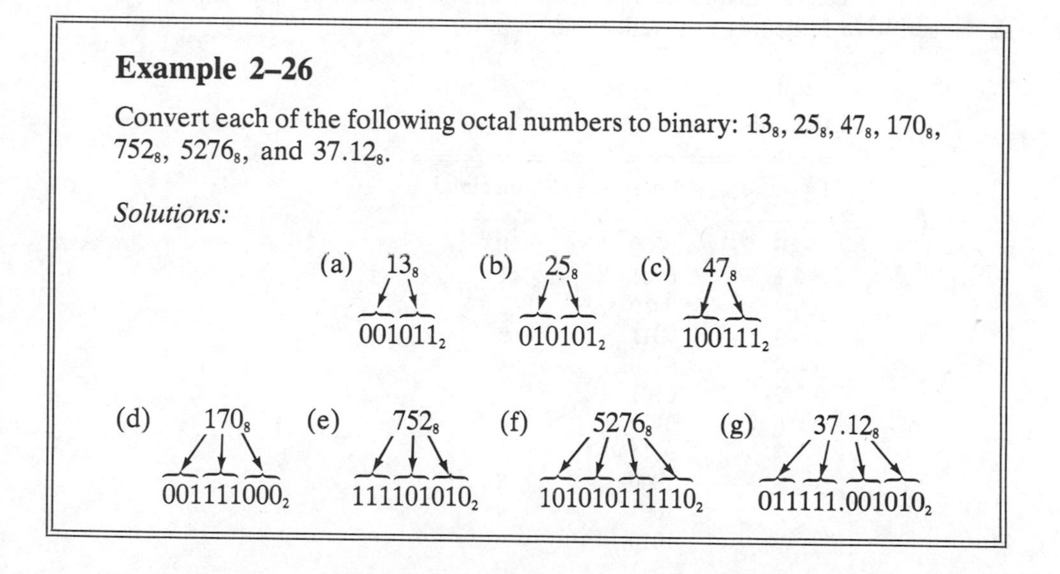

Binary-to-Octal Conversion

Conversion of a binary number to an octal number is also a straightforward process. Beginning at the right-most bit, simply break the binary number into groups of three bits and convert each group into the appropriate octal digit.

Example 2–27

Convert each of the following binary numbers into its octal equivalent: 110101_2, 101111001_2, 1011100110_2, and 1001101.1011_2.

Solutions:

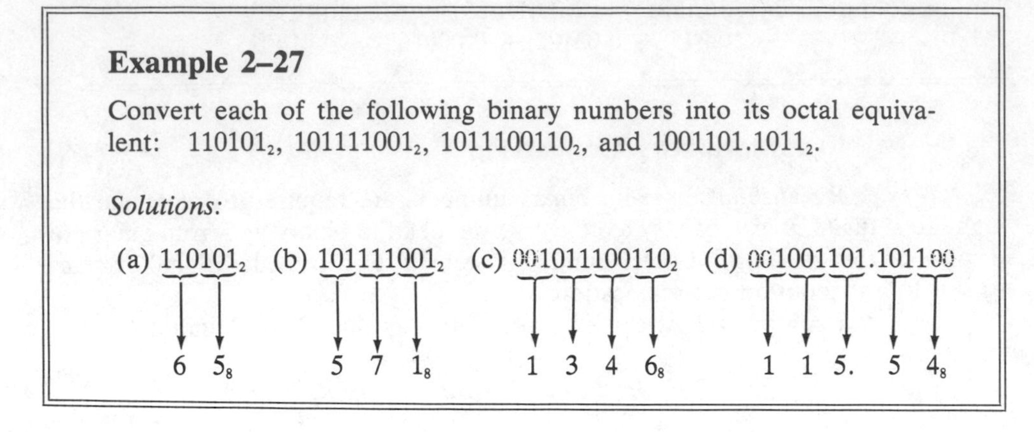

2–12 HEXADECIMAL NUMBERS

The hexadecimal system has a base sixteen; that is, it is composed of 16 digits and characters. Many digital systems process binary data in groups that are multiples of four bits, making the hexadecimal number very convenient because each hexadecimal digit represents a four-bit binary number (as listed in Table 2–3).

Ten digits and six alphabetic characters make up this number system. A subscript *16* indicates a hexadecimal number.

TABLE 2–3

Decimal	Binary	Hexadecimal
0	0000	0
1	0001	1
2	0010	2
3	0011	3
4	0100	4
5	0101	5
6	0110	6
7	0111	7
8	1000	8
9	1001	9
10	1010	A
11	1011	B
12	1100	C
13	1101	D
14	1110	E
15	1111	F

Binary-to-Hexadecimal Conversion

Converting a binary number to hexadecimal is a straightforward procedure. Simply break the binary number into four-bit groups starting at the right and replace each group by the equivalent hexadecimal symbol.

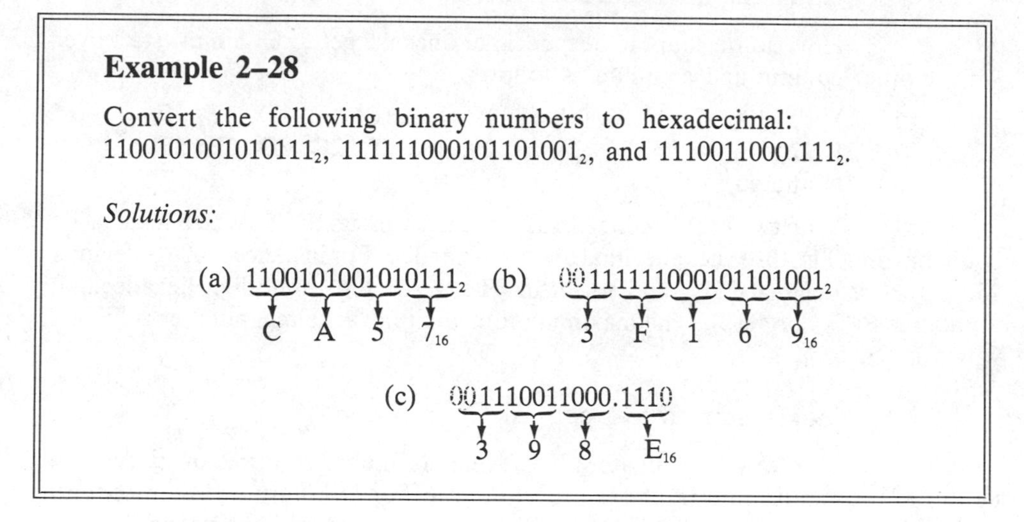

Example 2–28

Convert the following binary numbers to hexadecimal: 1100101001010111_2, 111111000101101001_2, and 1110011000.111_2.

Solutions:

(a) $1100101001010111_2 \rightarrow \text{CA57}_{16}$ (b) $00111111000101101001_2 \rightarrow \text{3F169}_{16}$

(c) $001110011000.1110 \rightarrow \text{398.E}_{16}$

Hexadecimal-to-Binary Conversion

To go from a hexadecimal number to a binary number, we reverse the process and replace each hexadecimal symbol with the appropriate four bits.

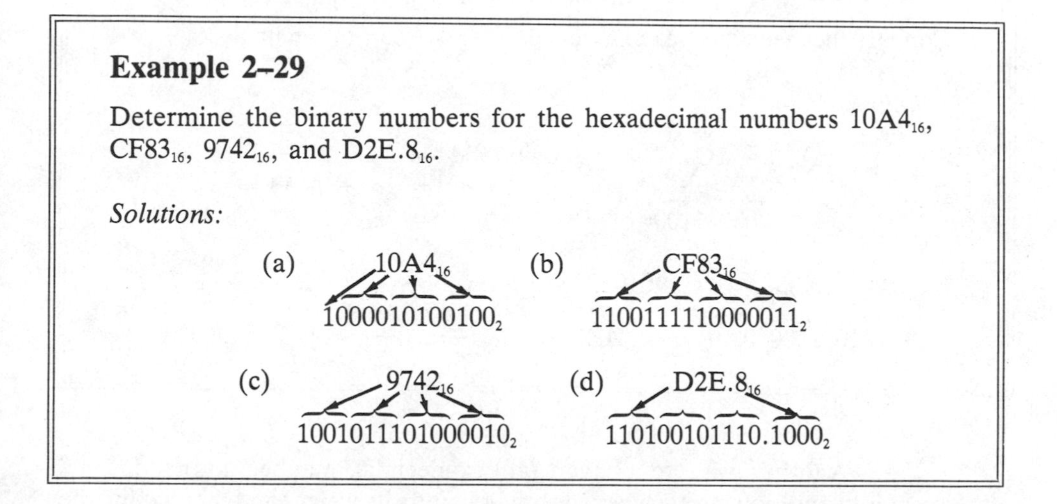

Example 2–29

Determine the binary numbers for the hexadecimal numbers $10A4_{16}$, $CF83_{16}$, 9742_{16}, and $D2E.8_{16}$.

Solutions:

(a) $10A4_{16} \rightarrow 1000010100100_2$ (b) $CF83_{16} \rightarrow 1100111110000011_2$

(c) $9742_{16} \rightarrow 1001011101000010_2$ (d) $D2E.8_{16} \rightarrow 110100101110.1000_2$

The use of letters to represent quantities may seem strange at first, but keep in mind that *any* number system is only a set of symbols. If we understand what these symbols mean in terms of quantities represented, then the form of the symbols themselves is unimportant once we get accustomed to using them.

It should be clear that it is much easier to write the hexadecimal number than the equivalent binary number, and since conversion is so easy, the hexadecimal system is a "natural" for communicating with digital systems utilizing four-bit groupings of binary numbers.

Counting in Hexadecimal

How do we count in hexadecimal once we get to F? Simply start over with another column and continue as follows:

10, 11, 12, 13, 14, 15, 16, 17, 18, 19, 1A, 1B, 1C, 1D, 1E, 1F,
20, 21, 22, 23, 24, 25, 26, 27, 28, 29, 2A, 2B, 2C, 2D, 2E, 2F,
30, 31, and so forth

With two hexadecimal digits, we can count up to FF_{16} which is 255_{10}. To count beyond this, three hexadecimal digits are needed. For instance, 100_{16} is decimal 256_{10}, 101_{16} is decimal 257_{10}, and so forth. The maximum three digit hexadecimal number is FFF_{16}, or 4095_{10}. The maximum four-digit hexadecimal number is $FFFF_{16}$ which is $65,535_{10}$.

Hexadecimal-to-Decimal Conversion

One way to evaluate a hexadecimal number in terms of its decimal equivalent is to first convert the hexadecimal number to binary and then convert from binary to decimal. The following example illustrates this procedure.

Example 2–30

Convert the hexadecimal numbers $1C_{16}$ and $A85_{16}$ to decimal.

Solution:

(a) $1C_{16} \rightarrow 0001\ 1100_2$

$$00011100_2 = 2^4 + 2^3 + 2^2 = 16 + 8 + 4 = 28_{10}$$

(b) $A85_{16} \rightarrow 1010\ 1000\ 0101_2$

$$101010000101_2 = 2^{11} + 2^9 + 2^7 + 2^2 + 2^0 = 2048 + 512 + 128 + 4 + 1 = 2693_{10}$$

Another way to convert a hexadecimal number to its decimal equivalent is by multiplying each hexadecimal digit by its weight and then taking the sum of these products. The weights of a hexadecimal number are increasing powers of 16 (from right to left). For a 4-digit hexadecimal number the weights are:

16^3	16^2	16^1	16^0
4096	256	16	1

The following example will show this conversion method.

Example 2–31

Convert $E5_{16}$ and $B2F8_{16}$ to decimal.

Solution:

(a) $E5_{16} = E \times 16 + 5 \times 1 = 14 \times 16 + 5 \times 1 = 224 + 5 = 229_{10}$

(b)
$$\begin{aligned} B2F8_{16} &= B \times 4096 + 2 \times 256 + F \times 16 + 8 \times 1 \\ &= 11 \times 4096 + 2 \times 256 + 15 \times 16 + 8 \times 1 \\ &= 45{,}056 + 512 + 240 + 8 \\ &= 45{,}816_{10} \end{aligned}$$

Decimal-to-Hexadecimal Conversion

Repeated division of a decimal number by 16 will produce the equivalent hexadecimal number formed by the remainders of each division. This is similar to the repeated division by 2 for decimal to binary conversion and repeated division by 8 for decimal to octal conversion.

The following example will illustrate this procedure.

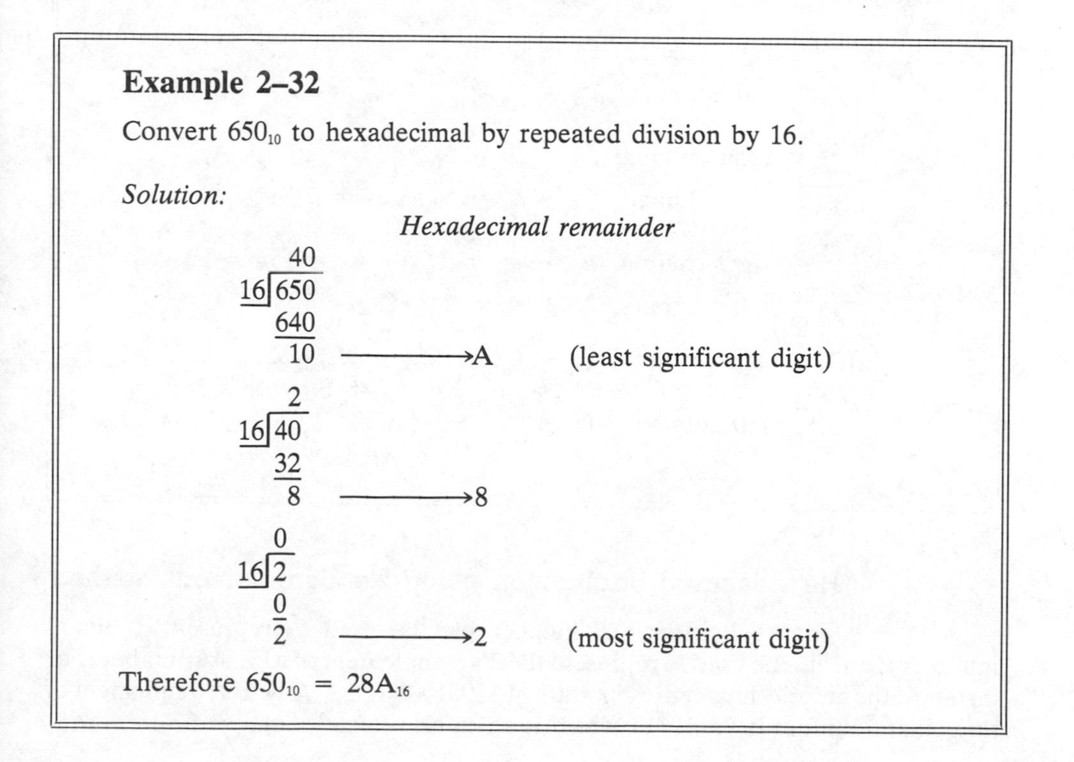

Example 2–32

Convert 650_{10} to hexadecimal by repeated division by 16.

Solution:

Therefore $650_{10} = 28A_{16}$

Hexadecimal Addition

Addition can be done directly with hexadecimal numbers by remembering that the hexadecimal digits 0 through 9 are equivalent to decimal digits 0 through 9 and that hexadecimal digits A through F are equivalent to decimal numbers 10 through 15.

When adding hexadecimal numbers, the following rules should be followed.

1. In any given column of an addition problem, think of the two hexadecimal digits in terms of their decimal value. For instance $5_{16} = 5_{10}$ and $C_{16} = 12_{10}$.
2. If the sum of these two digits is 15_{10} or less, bring down the corresponding hexadecimal digit.
3. If the sum of these two digits is greater than 15_{10}, bring down the amount of the sum that exceeds 16_{10} and carry a 1 to the next column.

Example 2–33

Add the following hexadecimal numbers:
(a) $23_{16} + 16_{16}$ (b) $58_{16} + 22_{16}$ (c) $2B_{16} + 84_{16}$ (d) $DF_{16} + AC_{16}$

Solution:

(a) $\begin{array}{r} 23_{16} \\ +\ 16_{16} \\ \hline 39_{16} \end{array}$
right column: $3_{16} + 6_{16} = 3_{10} + 6_{10} = 9_{10} = 9_{16}$
left column: $2_{16} + 1_{16} = 2_{10} + 1_{10} = 3_{10} = 3_{16}$

(b) $\begin{array}{r} 58_{16} \\ +\ 22_{16} \\ \hline 7A_{16} \end{array}$
right column: $8_{16} + 2_{16} = 8_{10} + 2_{10} = 10_{10} = A_{16}$
left column: $5_{16} + 2_{16} = 5_{10} + 2_{10} = 7_{10} = 7_{16}$

(c) $\begin{array}{r} 2B_{16} \\ +\ 84_{16} \\ \hline AF_{16} \end{array}$
right column: $B_{16} + 4_{16} = 11_{10} + 4_{10} = 15_{10} = F_{16}$
left column: $2_{16} + 8_{16} = 2_{10} + 8_{10} = 10_{10} = A_{16}$

(d) $\begin{array}{r} DF_{16} \\ +\ AC_{16} \\ \hline 18B_{16} \end{array}$
right column: $F_{16} + C_{16} = 15_{10} + 12_{10} = 27_{10}$
$27_{10} - 16_{10} = 11_{10} = B_{16}$ with a 1 carry
left column: $D_{16} + A_{16} + 1_{16} = 13_{10} + 10_{10} + 1_{10} = 24_{10}$
$24_{10} - 16_{10} = 8_{10} = 8_{16}$ with a 1 carry

Hexadecimal Subtraction using 2's Complement Method

Since a hexadecimal number can be used to represent a binary number, it can also be used to represent the 2's complement of a binary number. For instance, the hexadecimal representation of 11001001_2 is $C9_{16}$. The 2's complement of this binary number is 00110111 which is written in hexadecimal as 37_{16}.

As you have learned, the 2's complement allows us to subtract by *adding* binary numbers. We can also use this method for hexadecimal subtraction as the following example will show.

Example 2–34

Subtract the following hexadecimal numbers:

(a) $84_{16} - 2A_{16}$ (b) $C3_{16} - 0B_{16}$

Solution:

(a)
$$\begin{array}{r} 84_{16} \\ -\ 2A_{16} \\ \hline 5A_{16} \end{array}$$
$2A_{16} = 00101010_2$

2's complement of $2A_{16} = 11010110 = D6_{16}$

$$\begin{array}{r} 84_{16} \\ +\ D6_{16} \\ \hline \not{1}5A_{16} \end{array}$$
Add by rules of previous section.

$5A_{16}$ Drop carry as in 2's complement addition.

(b)
$$\begin{array}{r} C3_{16} \\ -\ 0B_{16} \\ \hline B8_{16} \end{array}$$
$0B_{16} = 00001011_2$

2's complement of $0B_{16} = 11110101 = F5_{16}$

$$\begin{array}{r} C3_{16} \\ +\ F5_{16} \\ \hline \not{1}B8_{16} \end{array}$$
Add by rules of previous section.

$B8_{16}$ Drop carry.

BINARY CODED DECIMAL (BCD) 2–13

As you learned in the last chapter, decimal, octal, and hexadecimal numbers can be represented by binary digits. Not only numbers, but letters and other symbols, can be represented by 1s and 0s. In fact, *any* entity expressible as numbers, letters, or other symbols can be represented by binary digits, and therefore can be processed by digital logic circuits.

Combinations of binary digits that represent numbers, letters, or symbols are *digital codes.* In many applications, special codes are used for such auxiliary functions as error detection.

The 8421 Code

The 8421 code is a type of *binary coded decimal* (BCD) code and is composed of four bits representing the decimal digits 0 through 9. The designation "8421" indicates the binary weights of the four bits (2^3, 2^2, 2^1, 2^0). The ease of conversion between 8421 code numbers and the familiar decimal numbers is the main advantage of this code. The 8421 code is the predominant BCD code, and when we refer to BCD we always mean the 8421 code unless otherwise stated.

"Binary coded decimal" means that *each* decimal digit is represented by a binary code of four bits. All you have to remember are the ten binary combinations that represent the ten decimal digits as shown in Table 2–4.

You should realize that with four bits, sixteen numbers (2^4) can be represented, and that in the 8421 code only ten of these are used. The six code combinations that are not used—1010, 1011, 1100, 1101, 1110, and 1111—are *invalid* in the 8421 BCD code.

To express any decimal number in BCD, simply replace each decimal digit by the appropriate four-bit code, as shown by the following example:

TABLE 2–4 *The 8421 BCD code.*

8421 (BCD)	Decimal
0000	0
0001	1
0010	2
0011	3
0100	4
0101	5
0110	6
0111	7
1000	8
1001	9

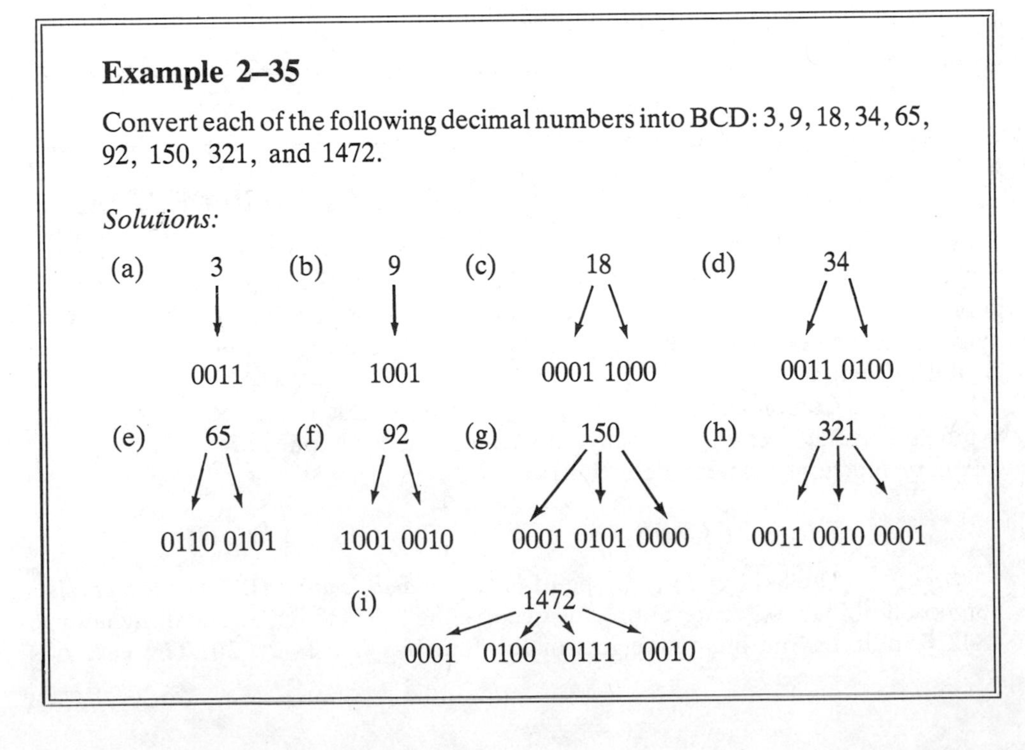

Example 2–35

Convert each of the following decimal numbers into BCD: 3, 9, 18, 34, 65, 92, 150, 321, and 1472.

Solutions:

(a) 3 → 0011

(b) 9 → 1001

(c) 18 → 0001 1000

(d) 34 → 0011 0100

(e) 65 → 0110 0101

(f) 92 → 1001 0010

(g) 150 → 0001 0101 0000

(h) 321 → 0011 0010 0001

(i) 1472 → 0001 0100 0111 0010

It is equally easy to determine a decimal number from a BCD number. Start at the right of the integer number and break the code into groups of four bits, and then write the decimal digit represented by each four-bit group. An example will illustrate.

Example 2–36

Find the decimal numbers represented by the following BCD codes: 10000110, 00110001, 1010011, 100101110100, and 1100001100000.

Solutions:

(a) 1000 0110 → 8 6

(b) 0011 0001 → 3 1

(c) 0101 0011 → 5 3

(d) 1001 0111 0100 → 9 7 4

(e) 1 1000 0110 0000 → 1 8 6 0

Note: If there are *not* four bits in the left-most group, zeros are implied.

BCD Addition

BCD is a numerical code, and many applications require that arithmetic operations be performed. Addition is the most important operation because the other three operations (subtraction, multiplication, and division) can be accomplished using addition. Here is how to add two BCD numbers:

1. Add the two numbers, using the rules for binary addition in Section 2–4.
2. If a four-bit sum is equal to or less than 9, it is a *valid* BCD number.
3. If a four-bit sum is greater than 9, or if a carry is generated, it is an *invalid* result. Add 6 (0110) to the four-bit sum in order to skip the six invalid states and return the code to 8421. If a carry results when 6 is added, simply add it to the next four-bit group.

Several examples will illustrate BCD addition for the case where the sum of any four-bit column does not exceed 9.

Example 2–37

Add the following BCD numbers (the decimal addition is shown for comparison):

(a)
```
  0011      3
 +0100     +4
  0111      7₁₀
```

(b)
```
  0110      6
 +0010     +2
  1000      8₁₀
```

Example 2–37, continued

(c)

	BCD	Decimal
	0010 0011	23
+	0001 0101	+15
	0011 1000	38_{10}

(d)

	BCD	Decimal
	1000 0110	86
+	0001 0011	+13
	1001 1001	99_{10}

(e)

	BCD	Decimal
	0100 0101 0000	450
+	0100 0001 0111	+417
	1000 0110 0111	867_{10}

(f)

	BCD	Decimal
	1000 0111 0011	873
+	0001 0001 0010	+112
	1001 1000 0101	985_{10}

Note that in each case the sum in any four-bit column does not exceed 9, and the results are valid BCD numbers.

Next, we will deal with the case of an invalid sum (greater than 9 or a carry) by illustrating the procedure with several examples.

Example 2–38

Add the following BCD numbers.

(a)

BCD		Decimal
1001		9
+0100		+ 4
1101	invalid BCD number	13_{10}
+0110	add 6	
0001 0011	valid BCD number	
↓ ↓		
1 3_{10}		

(b)

BCD		Decimal
1001		9
+1001		+ 9
1 0010	invalid because of carry	18_{10}
+0110	add 6	
0001 1000	valid BCD number	
↓ ↓		
1 8_{10}		

(c)

BCD		Decimal
0001 0110		16
+0001 0101		+15
0010 1011	right group is invalid, left group valid, add 6	31_{10}
+0110	to invalid code (add carry to next group)	
0011 0001	valid BCD number	
↓ ↓		
3 1_{10}		

(d)

```
          0110    0111                                  67
         +0101    0011                                 +53
          1011    1010    both groups are invalid      120₁₀
         +0110   +0110    add 6 to both groups
   0001   0010    0000    valid BCD number
    ↓      ↓       ↓
    1      2       0₁₀
```

THE EXCESS–3 CODE 2–14

The Excess–3 is an important digital code that is derived by adding 3 to *each* decimal digit and then converting the result to four-bit binary. Since no definite weights can be assigned to the four digit positions, Excess–3 is an unweighted code. For instance, the Excess–3 code for the decimal 2 is

$$\begin{array}{r} 2 \\ +3 \\ \hline 5 \end{array} \rightarrow 0101$$

The Excess–3 code for the decimal 9 is

$$\begin{array}{r} 9 \\ +3 \\ \hline 12 \end{array} \rightarrow 1100$$

The Excess–3 code for each decimal digit is found by the same procedure, and the entire code is shown in Table 2–5.

Notice that ten of a possible 16 code combinations are used in the Excess–3 code. The six *invalid* combinations are 0000, 0001, 0010, 1101, 1110, and 1111.

TABLE 2–5 *Excess–3 code.*

Decimal	Binary	Excess–3
0	0000	0011
1	0001	0100
2	0010	0101
3	0011	0110
4	0100	0111
5	0101	1000
6	0110	1001
7	0111	1010
8	1000	1011
9	1001	1100

Self-Complementing Property

The key feature of the Excess–3 code is that it is *self-complementing.* This means that the *1's complement of an Excess–3 number is the Excess–3 code for the 9's complement of the corresponding decimal number.* For example, the Excess–3 code for decimal 4 is 0111. The 1's complement of this is 1000, which is the Excess–3 code for the decimal 5 (and 5 is the 9's complement of 4).

It should be noted that the 1's complement is easily produced with digital logic circuits by simply inverting each bit. The self-complementing property makes the Excess–3 code useful in some arithmetic operations, because subtraction can be performed using the 9's complement method.

Table 2–6 shows that the 1's complement of each Excess–3 number is the Excess–3 code for the 9's complement of the corresponding decimal digit.

Table 2–6

Excess–3	*1's Complement*	*Decimal Digit*	*9's Complement*
0011	1100	0	9
0100	1011	1	8
0101	1010	2	7
0110	1001	3	6
0111	1000	4	5
1000	0111	5	4
1001	0110	6	3
1010	0101	7	2
1011	0100	8	1
1100	0011	9	0

Example 2–39

Convert each of the following decimal numbers to Excess–3 code: 13, 35, 87, 159, and 430.

Solutions:

First, add 3 to *each* digit in the decimal number, and then convert each sum to its equivalent binary code.

(a)
```
 1    3
+3   +3
──   ──
 4    6
 ↓    ↓
0100 0110   Excess–3 for 13₁₀
```

(b)
```
 1    5    9
+3   +3   +3
──   ──   ──
 4    8   12
 ↓    ↓    ↓
0100 1000 1100   Excess–3 for 159₁₀
```

(c)
```
 3    5
+3   +3
──   ──
 6    8
 ↓    ↓
0110 1000   Excess–3 for 35₁₀
```

(d)
```
 4    3    0
+3   +3   +3
──   ──   ──
 7    6    3
 ↓    ↓    ↓
0111 0110 0011   Excess–3 for 430₁₀
```

(e)
```
   8     7
  +3    +3
  11    10
   ↓     ↓
1011  1010  Excess–3 for 87₁₀
```

Excess–3 Addition

In this section we will cover addition using Excess–3 numbers. The three rules for Excess–3 addition are

1. Add the Excess–3 numbers using the rules for binary addition (Section 2–4).
2. If there is *no* carry from a four-bit group, *subtract* 3 (0011) from that group to get the Excess–3 code for the digit.
3. If there is a carry from a four-bit group, *add* 3 (0011) to that group to get the Excess–3 code for the digit, and add 3 to any *new* column (digit) generated by the last carry.

No carry from a particular four-bit group indicates that the sum is Excess–6 because we have added *two* Excess–3 numbers. Therefore, we have to subtract 3 to get back to Excess–3. A carry indicates an invalid result that requires the addition of 3 in order to skip the invalid states and return the digit to Excess–3 form.

Rule 2 is illustrated in Example 2–40, and rule 3 in Example 2–41.

Example 2–40

Convert each of the decimal numbers to Excess–3 and add as indicated.

(a)
```
  8          1011   Excess–3 for 8
 +1         +0100   Excess–3 for 1
  9          1111   no carry
              -11   subtract 3 (11₂)
             1100   Excess–3 for 9
```

(b)
```
  15      0100  1000   Excess–3 for 15
 +12     +0100  0101   Excess–3 for 12
  27      1000  1101   no carries
           -11   -11   subtract 3 (11₂) from each group
          0101  1010   Excess–3 for 27
```

(c)
```
  35      0110  1000   Excess–3 for 35
 +24     +0101  0111   Excess–3 for 24
  59      1011  1111   no carries
           -11   -11   subtract 3 (11₂) from each group
          1000  1100   Excess–3 for 59
```

Example 2–40, continued

(d)	273	0101	1010	0110	Excess–3 for 273
	+126	+0100	0101	1001	Excess–3 for 126
	399	1001	1111	1111	no carries
		−11	−11	−11	subtract 3 (11_2) from each group
		0110	1100	1100	Excess–3 for 399

Example 2–41

Convert each decimal number to Excess–3 and add as indicated.

(a)	7			1010	Excess–3 for 7
	+6			+1001	Excess–3 for 6
	13		1	0011	there is a carry
			+11	+11	add 3 to both groups
			0100	0110	Excess–3 for 13
(b)	15		0100	1000	Excess–3 for 15
	+15		+0100	1000	
	30		1001	0000	carry out of right-most group only
			−11	+11	subtract 3 from left; add 3 to right
			0110	0011	Excess–3 for 30
(c)	98		1100	1011	Excess–3 for 98
	+86		+1011	1001	Excess–3 for 86
	184	1	1000	0100	carry out of both groups
		+11	+11	+11	add 3 to each group
		0100	1011	0111	Excess–3 for 184

2–15 ALPHANUMERIC CODES

In order to communicate, we need not only numbers, but also letters and other symbols. In the strictest sense, codes that represent numbers and alphabetic characters (letters) are called *alphanumeric* codes. Most of these codes, however, also represent symbols and various instructions necessary for conveying intelligible information.

At a minimum, an alphanumeric code must represent ten decimal digits and 26 letters of the alphabet, for a total of 36 items. This requires six bits in each code combination because five bits are insufficient ($2^5 = 32$). There are 64 total combinations of six bits, so we have 28 unused code combinations. Obviously, in many applications symbols other than just numbers and letters are necessary to communicate completely. We need spaces to separate words, periods to mark the end of sentences or for decimal points, instructions to tell the receiving system what to do with the information, and more. So, with codes that are six bits long, we can handle decimal numbers, the alphabet, and 28 other symbols. This should give you an idea of the requirements for a basic alphanumeric code.

ASCII

One standardized alphanumeric code, called the *American Standard Code for Information Interchange* (ASCII), is perhaps the most widely used type. This is a seven-bit code where the decimal digits are represented by the 8421 BCD code preceded by 011. The letters of the alphabet and other symbols and instructions are represented by other code combinations as shown in Table 2–7. For instance, the letter *A* is represented by 1000001, the letter *B* by 1000010, the comma by 0101100, and ETX (end of text) by 0000011.

TABLE 2–7 *American Standard Code for Information Interchange.*

	000	001	010	011	100	101	110	111
0000	NUL	DLE	SP	0	@	P	`	p
0001	SOH	DC_1	!	1	A	Q	a	q
0010	STX	DC_2	"	2	B	R	b	r
0011	ETX	DC_3	#	3	C	S	c	s
0100	EOT	DC_4	$	4	D	T	d	t
0101	ENQ	NAK	%	5	E	U	e	u
0110	ACK	SYN	&	6	F	V	t	v
0111	BEL	ETB	'	7	G	W	g	w
1000	BS	CAN	(	8	H	X	h	x
1001	HT	EM	)	9	I	Y	i	y
1010	LF	SUB	*	:	J	Z	j	z
1011	VI	ESC	+	;	K	[	k	{
1100	FF	FS	,	<	L	\	l	|
1101	CR	GS	–	=	M	]	m	}
1110	SO	RS	.	>	N	↑	n	~
1111	SI	US	/	?	O	——	o	DEL

Definitions of control abbreviations:

ACK	Acknowledge	DLE	Data link escape
BEL	Bell	EM	End of Medium
BS	Backspace	ENQ	Enquiry
CAN	Cancel	EOT	End of transmission
CR	Carriage return	ESC	Escape
DC_1–DC_4	Direct control	ETB	End of transmission block
DEL	Delete idle	ETX	End text

FF	Form feed	SI	Shift in
FS	Form separator	SO	Shift out
GS	Group separator	SOH	Start of heading
HT	Horizontal tab	STX	Start text
LF	Line feed	SUB	Substitute
NAK	Negative acknowledge	SYN	Synchronous idle
NUL	Null	US	Unit separator
RS	Record separator	VT	Vertical tab

Example of code format:

B_7 B_1
1000100 is the code for D
three-bit group four-bit group

EBCDIC

Another alphanumeric code also frequently encountered is called the *Extended Binary Coded Decimal Interchange Code* (EBCDIC). This is an eight-bit code in which the decimal digits are represented by the 8421 BCD code preceded by 1111. Both lowercase and uppercase letters are represented in addition to numerous other symbols and commands, as shown in Table 2–8. For example, uppercase *A* is 11000001, and lowercase *a* is 10000001.

TABLE 2–8 *Partial EBCDIC table.*

Bit positions 0,1 →	00				01				10				11			
Bit positions 2,3 →	00	01	10	11	00	01	10	11	00	01	10	11	00	01	10	11
Bit positions 4, 5, 6, 7 → 0000	NUL		DS		SP	&	-									0
0001			SOS				/		a	j			A	J		1
0010			FS						b	k	s		B	K	S	2
0011		TM							c	l	t		C	L	T	3
0100	PF	RES	BYP	PN					d	m	u		D	M	U	4
0101	HT	NL	LF	RS					e	n	v		E	N	V	5
0110	LC	BS	EOB	UC					f	o	w		F	O	W	6
0111	DL	IL	PRE	EOT					g	p	x		G	P	X	7
1000									h	q	y		H	Q	Y	8
1001									i	r	z		I	R	Z	9
1010		CC	SM		¢	!		:								
1011						$	,	#								
1100					<	*	%	@								
1101					(	)	–	'								
1110					+	;	>	=								
1111	CU1	CU2	CU3	¬	|		?	"								

Definitions of control abbreviations:

BS	Backspace	LC	Lowercase	
BYP	Bypass	LF	Line feed	
CC	Cursor control	NL	New line	
CU1	Customer use	PF	Punch off	
CU2	Customer use	PN	Punch on	
CU3	Customer use	PRE	Prefix	
DL	Delete	RES	Restore	
DS	Digit select	RS	Reader stop	
EOB	End of block	SM	Set mode	
EOT	End of transmission	SP	Space	
FS	Field separator	TM	Tape mark	
HT	Horizontal tab	UC	Uppercase	
IL	Idle			

Meanings of unfamiliar symbols

|	Vertical bar: logical OR
¬	Logical NOT
-	Hyphen or minus sign
—	Underscore (01101101)

Example of code format:

01234567 Bit positions
11000110 is F

Problems

Section 2–2

2–1 Determine the 9's complement of each decimal number.

(a) 3 (b) 5 (c) 8 (d) 12
(e) 17 (f) 25 (g) 49 (h) 86
(i) 127 (j) 381 (k) 690 (l) 1354

2–2 Perform the following subtractions using the 9's complement method:

(a) 6 − 2 (b) 15 − 7 (c) 23 − 14 (d) 48 − 33
(e) 69 − 68 (f) 91 − 70 (g) 98 − 59 (h) 100 − 82

2–3 Perform the following subtractions using the 9's complement method:

(a) 115 − 92 (b) 159 − 125 (c) 298 − 200
(d) 561 − 443 (e) 846 − 709 (f) 1024 − 837

2–4 Determine the 10's complement of each decimal number.

(a) 7 (b) 19 (c) 36
(d) 52 (e) 84 (f) 90

Section 2–3

2–5 Convert the following binary numbers to decimal:

(a) 11 (b) 100 (c) 111 (d) 1000
(e) 1001 (f) 1100 (g) 1011 (h) 1111

2–6 Convert the following binary numbers to decimal:

(a) 1110 (b) 1010 (c) 11100 (d) 10000
(e) 10101 (f) 11101 (g) 10111 (h) 11111

2–7 Convert each binary number to decimal.
(a) 110011.11 (b) 101010.01 (c) 1000001.111
(d) 1111000.101 (e) 1011100.10101 (f) 1110001.0001
(g) 1011010.1010 (h) 1111111.11111

2–8 What is the highest decimal number that can be represented by the following number of binary digits (bits)?
(a) 2 (b) 3 (c) 4 (d) 5
(e) 6 (f) 7 (g) 8 (h) 9
(i) 10 (j) 11

2–9 How many bits are required to represent the following decimal numbers?
(a) 17 (b) 35 (c) 49 (d) 68
(e) 81 (f) 114 (g) 132 (h) 205
(i) 271

2–10 Generate the following binary count sequences:
(a) 0 through 7 (b) 8 through 15 (c) 16 through 31
(d) 32 through 63 (e) 64 through 75

Section 2–4

2–11 Add the binary numbers.
(a) 11 + 1 (b) 10 + 10 (c) 101 + 11
(d) 111 + 110 (e) 1001 + 101 (f) 1101 + 1011

Section 2–5

2–12 Use direct subtraction on the following binary numbers:
(a) 11 − 1 (b) 101 − 100 (c) 110 − 101
(d) 1110 − 11 (e) 1100 − 1001 (f) 11010 − 10111

Section 2–6

2–13 Perform the following binary multiplications:
(a) 11 × 11 (b) 100 × 10 (c) 111 × 101
(d) 1001 × 110 (e) 1101 × 1101 (f) 1110 × 1101

Section 2–7

2–14 Divide the binary numbers as indicated.
(a) 100 ÷ 10 (b) 1001 ÷ 11 (c) 1100 ÷ 100

Section 2–8

2–15 Determine the 1's complement of each binary number.
(a) 101 (b) 110 (c) 1010
(d) 11010111 (e) 1110101 (f) 00001

2–16 Perform the following subtractions using the 1's complement method:
(a) 11 − 10 (b) 100 − 11 (c) 1010 − 111
(d) 1101 − 1010 (e) 11100 − 1101 (f) 100001 − 1010
(g) 1001 − 1110 (h) 10111 − 11111

Section 2–9

2–17 Determine the 2's complement of each binary number.
(a) 10 (b) 111 (c) 1001
(d) 1101 (e) 11100 (f) 10011

2–18 Perform the following subtractions using the 2's complement method:
(a) 10 − 01 (b) 111 − 110 (c) 1101 − 1001
(d) 1111 − 1101 (e) 10111 − 10011 (f) 10001 − 11100
(g) 10101 − 10111 (h) 1111000 − 1111111

Section 2–10

2–19 Convert each decimal number to binary using the sum of weights method.
(a) 10 (b) 17 (c) 24 (d) 48
(e) 61 (f) 93 (g) 125 (h) 186
(i) 298

2–20 Convert each decimal number to binary by repeated division by two.
(a) 15 (b) 21 (c) 28 (d) 34
(e) 40 (f) 59 (g) 65 (h) 73
(i) 99

Section 2–11

2–21 Convert each octal number to decimal.
(a) 12_8 (b) 27_8 (c) 56_8
(d) 64_8 (e) 103_8 (f) 557_8
(g) 163_8 (h) 1024_8 (i) 7765_8

2–22 Convert each decimal number to octal by repeated division by eight.
(a) 15 (b) 27 (c) 46 (d) 70
(e) 100 (f) 142 (g) 219 (h) 435
(i) 791

2–23 Convert each octal number to binary.
(a) 13_8 (b) 57_8 (c) 101
(d) 321_8 (e) 540_8 (f) 4653_8
(g) 13271_8 (h) 45600_8 (i) 100213_8
(j) 103.45_8

2–24 Convert each binary number to octal.
(a) 111 (b) 10 (c) 110111
(d) 101010 (e) 1100 (f) 1011110
(g) 101100011001 (h) 10110000011 (i) 111111101111000
(j) 10011.011

Section 2–12

2–25 Convert each hexadecimal number to binary.
(a) 38_{16} (b) 59_{16} (c) $A14_{16}$ (d) $5C8_{16}$
(e) 4100_{16} (f) $FB17_{16}$ (g) $8A.9_{16}$

2–26 Convert each binary number to hexadecimal.
(a) 1110 (b) 10 (c) 10111
(d) 10100110 (e) 1111110000 (f) 100110000010

2–27 Convert each hexadecimal number to decimal.
(a) 23_{16} (b) 92_{16} (c) $1A_{16}$ (d) $8D_{16}$
(e) $F3_{16}$ (f) EB_{16} (g) $5C2_{16}$ (h) 700_{16}

2–28 Convert each decimal number to hexadecimal.
(a) 8_{10} (b) 14_{10} (c) 33_{10} (d) 52_{10}
(e) 284_{10} (f) 2890_{10} (g) 4019_{10} (h) 6500_{10}

2–29 Perform the following additions.
(a) $37_{16} + 29_{16}$ (b) $A0_{16} + 6B_{16}$ (c) $FF_{16} + BB_{16}$

2–30 Perform the following subtractions.
(a) $51_{16} - 40_{16}$ (b) $C8_{16} - 3A_{16}$ (c) $FD_{16} - 88_{16}$

Section 2–13

2–31 Convert each of the following decimal numbers to 8421 BCD:
(a) 10 (b) 13 (c) 18
(d) 21 (e) 25 (f) 36
(g) 44 (h) 57 (i) 69
(j) 98 (k) 125 (l) 156

2–32 Convert each of the decimal numbers in Problem 2-31 to straight binary and compare the number of bits required with that required in BCD.

2–33 Convert the following decimal numbers to BCD:
(a) 104 (b) 128 (c) 132
(d) 150 (e) 186 (f) 210
(g) 359 (h) 547 (i) 1051
(j) 2563

2–34 Convert each of the BCD code numbers to decimal.
(a) 0001 (b) 0110 (c) 1001
(d) 00011000 (e) 11001 (f) 00110010
(g) 1000101 (h) 10011000 (i) 100001110000
(j) 011000011001

2–35 Convert each of the BCD code numbers to decimal.
(a) 10000000 (b) 1000110111
(c) 1101000110 (d) 10000100001
(e) 11101010100 (f) 100000000000
(g) 100101111000 (h) 1011010000011
(i) 1001000000011000 (j) 0110011001100111

2–36 Add the following BCD numbers:
(a) 0010 + 0001 (b) 0101 + 0011
(c) 0111 + 0010 (d) 1000 + 0001
(e) 00011000 + 00010001 (f) 01100100 + 00110011
(g) 01000000 + 01000111 (h) 10000101 + 00010011

2–37 Add the following BCD numbers:

(a) 1000 + 0110 **(b)** 0111 + 0101
(c) 1001 + 1000 **(d)** 1001 + 0111
(e) 00100101 + 00100111 **(f)** 01010001 + 01011000
(g) 10011000 + 10010111 **(h)** 010101100001 + 0011100001000

2–38 Convert each pair of decimal numbers to BCD and add as indicated.

(a) 4 + 3 **(b)** 5 + 2 **(c)** 6 + 4
(d) 17 + 12 **(e)** 28 + 23 **(f)** 65 + 58
(g) 113 + 101 **(h)** 295 + 157

Section 2–14

2–39 Convert each of the following decimal numbers to Excess–3 code.

(a) 1 **(b)** 3 **(c)** 6
(d) 10 **(e)** 18 **(f)** 29
(g) 56 **(h)** 75 **(i)** 107
(j) 149 **(k)** 231 **(l)** 500
(m) 1251 **(n)** 2379 **(o)** 6841

2–40 Convert each Excess–3 code number to decimal.

(a) 0011 **(b)** 1001
(c) 0111 **(d)** 01000110
(e) 01111100 **(f)** 10000101
(g) 10010101011 **(h)** 110000110110
(i) 101001011000 **(j)** 101101000111

2–41 Perform each of the following Excess–3 additions:

(a) 0011 + 0011 **(b)** 0101 + 0100
(c) 0111 + 0101 **(d)** 1001 + 0110
(e) 1011 + 1000 **(f)** 01010101 + 00110110
(g) 01011011 + 01001100 **(h)** 01110011 + 01100101
(i) 10011010 + 10011100 **(j)** 11001011 + 10110011

2–42 Decode the following ASCII coded message. Bit 7 of the first character is the left-most bit in the first row.

1001001101010000110001001000100000011010011001100
1000010100010110001011001110001100010100111000001
1001001100010000110001010100100100010000011010100
0011000100011110001011001110100100110101011010011
0011000100100110100110011000011100101110010011000
0100101001100010100001000101101001010100111010000
1001001101001010000011010100100100110011111001110
0011000100000110011101000100001100001100010011000
0100101001100010010011001110101001110100001001001
1010010100000110101001001001100111110011100000011

2–43 Convert your name and address to ASCII code.

2–44 Convert your name and address to EBCDIC.

In this chapter we will study the various types of logic gates that make up a typical digital system. The emphasis is on the logical operation of the circuits and the limitations and considerations involved in their operation. It is very important to know what the output of a gate is for various combinations of inputs and to understand how the electrical characteristics affect its operation so that we can predict and analyze how a circuit will perform in a given system.

Because of the wide use of integrated circuits, detailed knowledge of circuit design is of less importance for those involved in the development, application, or maintenance of digital equipment. However, basic circuit implementations for each type of gate are discussed in Appendix A to give you a better feel for what a gate is made of and to introduce you to important integrated circuit (IC) technologies.

The American National Standard Institute's (ANSI) graphic symbols for logic diagrams are used in this chapter and throughout the book.

3 LOGIC GATES

3–1 THE INVERTER

The inverter (NOT circuit) performs a basic logic function called *inversion* or *complementation.* The purpose of the inverter is to convert one logic level to the other logic level. In terms of the binary digits, it changes a 1 to a 0 or a 0 to a 1. The standard logic symbol for an inverter is shown in Figure 3–1. The small circle or "bubble" on the symbols is called a *negation indicator* by ANSI, and it indicates an *active 0.* The absence of the bubble indicates an *active 1.*

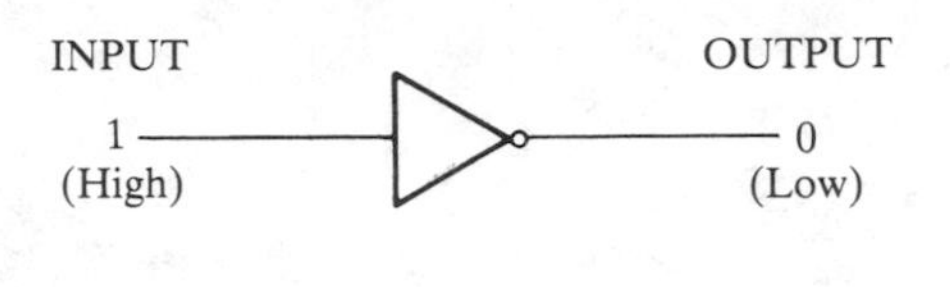

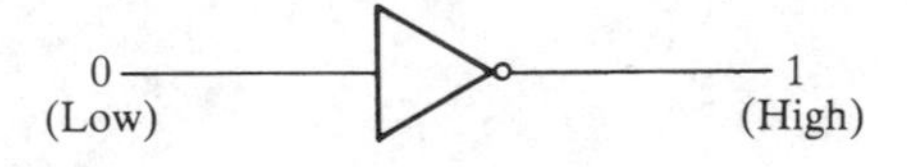

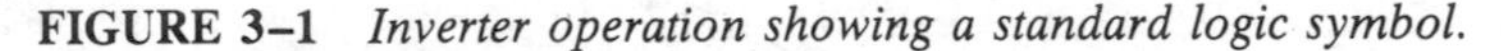

FIGURE 3–1 *Inverter operation showing a standard logic symbol.*

Because in *positive logic* a HIGH level corresponds to a 1 and a LOW level to a 0, we will also use the bubble to indicate an active LOW and the absence of a bubble to indicate an active HIGH.

Another symbol used to indicate levels as specified by ANSI is called the *polarity indicator.* This is a small triangle used to indicate an active LOW level. An inverter using this symbol is shown in Figure 3–2.

In this book, the negation indicator or bubble symbol will be consistently used because the industry seems to prefer this approach at the present time.

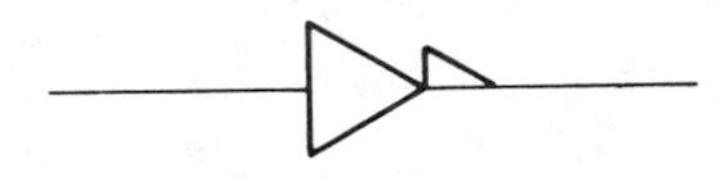

FIGURE 3–2 *Inverter using the polarity indicator.*

Inverter Operation

When a HIGH level is applied to an inverter input, a LOW level will appear on its output. When a LOW level is applied to its input, a HIGH will appear on its output. This operation is summarized in Table 3–1, which shows the output for each possible input in terms of levels and bits. These are called *truth tables.*

INPUT	OUTPUT
Low	High
High	Low

(a)

INPUT	OUTPUT
0	1
1	0

(b)

Table 3–1 *Inverter truth tables.*

Pulsed Operation

Figure 3–3 shows the output of an inverter for a pulse input where t_1 and t_2 indicate the corresponding points on the input and output pulse waveforms. Note that when the input is LOW, the output is HIGH and when the input is HIGH, the output is LOW, thereby producing an inverted output pulse.

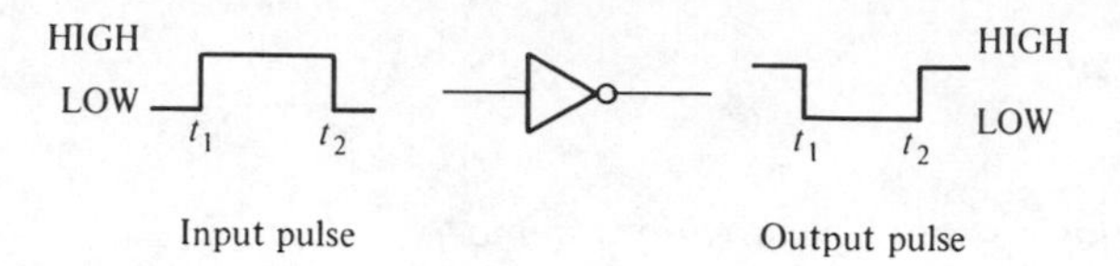

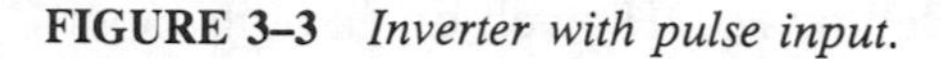

FIGURE 3–3 *Inverter with pulse input.*

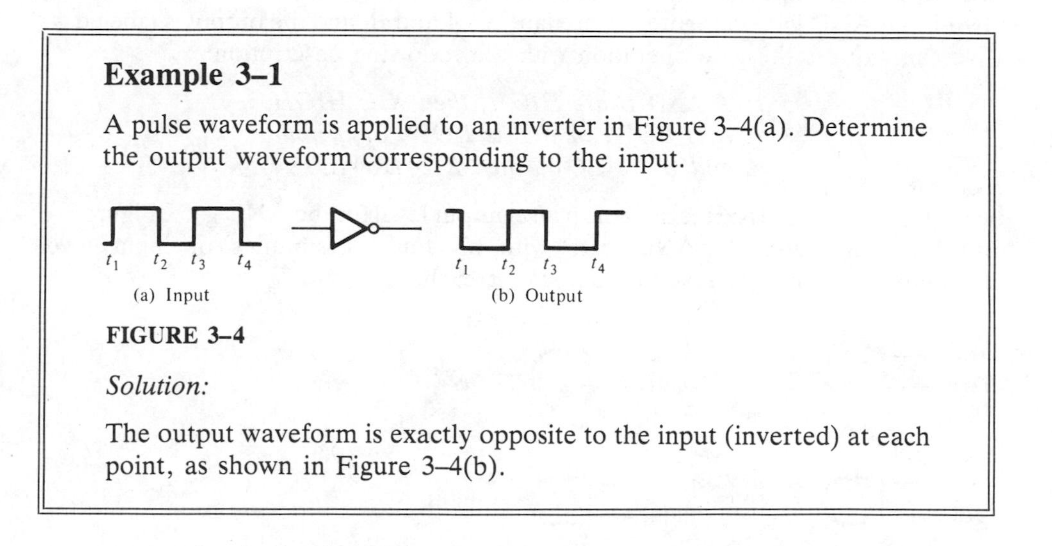

Example 3–1

A pulse waveform is applied to an inverter in Figure 3–4(a). Determine the output waveform corresponding to the input.

(a) Input (b) Output

FIGURE 3–4

Solution:

The output waveform is exactly opposite to the input (inverted) at each point, as shown in Figure 3–4(b).

THE AND GATE 3–2

The AND gate performs the basic operation of logic multiplication more commonly known as the AND function. The mathematical aspects of this are discussed in Chapter 4.

The AND gate is composed of two or more inputs and a single output, as indicated by the standard logic symbols shown in Figure 3–5. Inputs are on the left and the output is on the right in each symbol. Gates with two, three, four, and eight inputs are shown; however, an AND gate can have any number of inputs greater than one.

Logical Operation of the AND Gate

The operation of the AND gate is such that the output is HIGH only when *all* of the inputs are HIGH. When *any* of the inputs are LOW, the output is LOW. Therefore, the basic purpose of an AND gate is to determine when certain

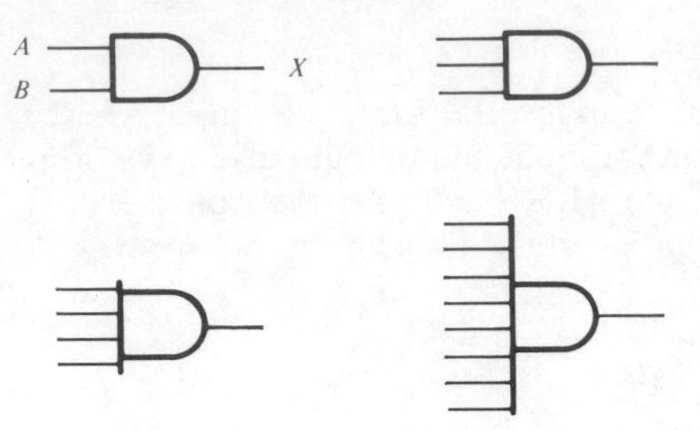

FIGURE 3–5 *Standard AND gate logic symbols with various numbers of inputs.*

conditions are simultaneously true as indicated by HIGH levels on all of its inputs and to produce a HIGH on its output indicating this condition. The inputs of the two-input AND gate in Figure 3–5 are labeled *A* and *B*, and the output is labeled *X*. We can express the gate operation with the following description:

If A **AND** *B are HIGH, then X is HIGH.*
If A is LOW, or if B is LOW, or if both
A and B are LOW, then X is LOW.

The HIGH level is the *active* output level for the AND gate. Figure 3–6 illustrates a two-input AND gate with all four possibilities of input level combinations, and the resulting output for each.

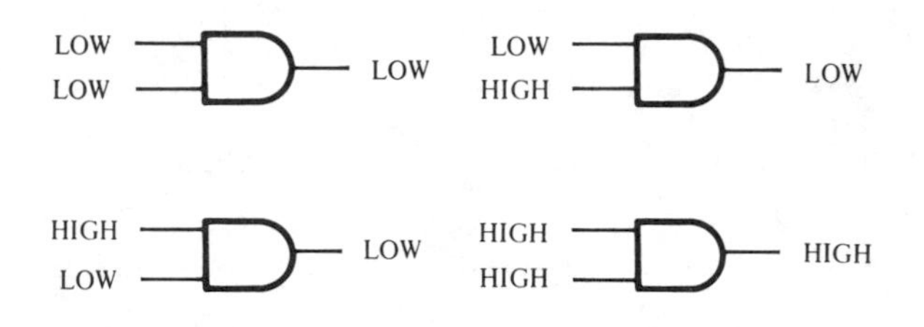

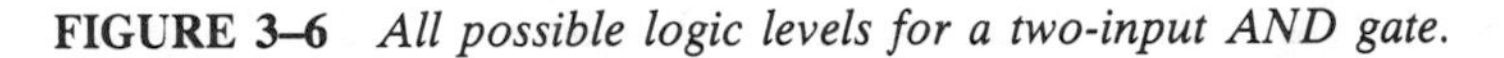

FIGURE 3–6 *All possible logic levels for a two-input AND gate.*

We generally express the logical operation of a gate with a table that lists all input combinations and the corresponding outputs. This table of combinations is also called a *truth table,* and is illustrated in Table 3–2 for a two-input AND gate. Tables 3–3 and 3–4 are the truth tables for a three-input AND gate and a

TABLE 3–2 *Truth table for a two-input AND gate.*

Inputs		Output
A	*B*	*X*
LOW	LOW	LOW
LOW	HIGH	LOW
HIGH	LOW	LOW
HIGH	HIGH	HIGH

four-input AND gate, respectively. Notice that for each gate, regardless of the number of inputs, the output is HIGH *only* when *all* inputs are HIGH.

The total number of possible combinations of binary inputs is determined by the following formula:

$$N = 2^n \qquad (3\text{–}1)$$

where N is the total possible combinations and n is the number of input variables. To illustrate, the following calculations are made using Equation (3–1):

For two input variables: $2^2 = 4$
For three input variable: $2^3 = 8$
For four input variables: $2^4 = 16$

This is how we determine the number of combinations for each of the truth tables.

TABLE 3–3 *Truth table for a three-input AND gate.*

Inputs			Output
A	*B*	*C*	*X*
LOW	LOW	LOW	LOW
LOW	LOW	HIGH	LOW
LOW	HIGH	LOW	LOW
LOW	HIGH	HIGH	LOW
HIGH	LOW	LOW	LOW
HIGH	LOW	HIGH	LOW
HIGH	HIGH	LOW	LOW
HIGH	HIGH	HIGH	HIGH

TABLE 3–4 *Truth table for a four-input AND gate.*

Inputs				Output
A	*B*	*C*	*D*	*X*
LOW	LOW	LOW	LOW	LOW
LOW	LOW	LOW	HIGH	LOW
LOW	LOW	HIGH	LOW	LOW
LOW	LOW	HIGH	HIGH	LOW
LOW	HIGH	LOW	LOW	LOW
LOW	HIGH	LOW	HIGH	LOW
LOW	HIGH	HIGH	LOW	LOW
LOW	HIGH	HIGH	HIGH	LOW
HIGH	LOW	LOW	LOW	LOW
HIGH	LOW	LOW	HIGH	LOW
HIGH	LOW	HIGH	LOW	LOW
HIGH	LOW	HIGH	HIGH	LOW
HIGH	HIGH	LOW	LOW	LOW
HIGH	HIGH	LOW	HIGH	LOW
HIGH	HIGH	HIGH	LOW	LOW
HIGH	HIGH	HIGH	HIGH	HIGH

Pulsed Operation

In a majority of applications, the inputs to a gate are not stationary levels, but are voltages that change frequently between two logic levels and that can be classified as pulse waveforms. We will now look at the operation of AND gates with pulsed input waveforms. Keep in mind that an AND gate obeys the truth table operation regardless of whether its inputs are constant levels or pulsed levels.

In examining the pulsed operation of the AND gate, we will look at the input levels with respect to each other in order to determine the output level at any given time. For example, in Figure 3–7 the inputs are both HIGH during the interval *T*1, making the output HIGH during this interval. During interval *T*2, input *A* is LOW and input *B* is HIGH, so the output is LOW. During interval *T*3, both inputs are HIGH again, and therefore the output is HIGH. During interval *T* 4, input *A* is HIGH and input *B* is LOW, resulting in a LOW output. Finally, during interval *T*5, input *A* is LOW, input *B* is LOW, and the output is therefore LOW.

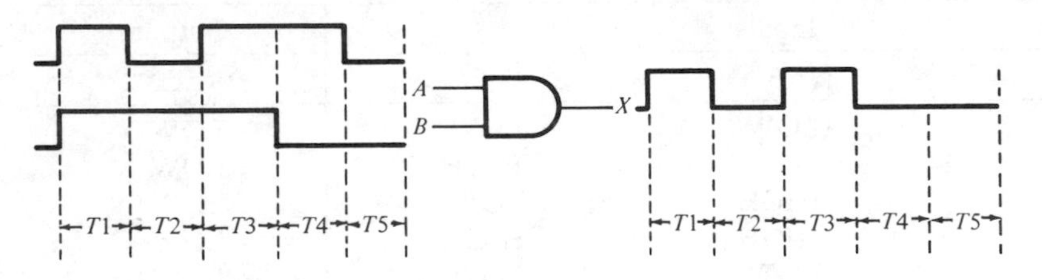

FIGURE 3–7 *Example of pulsed AND gate operation.*

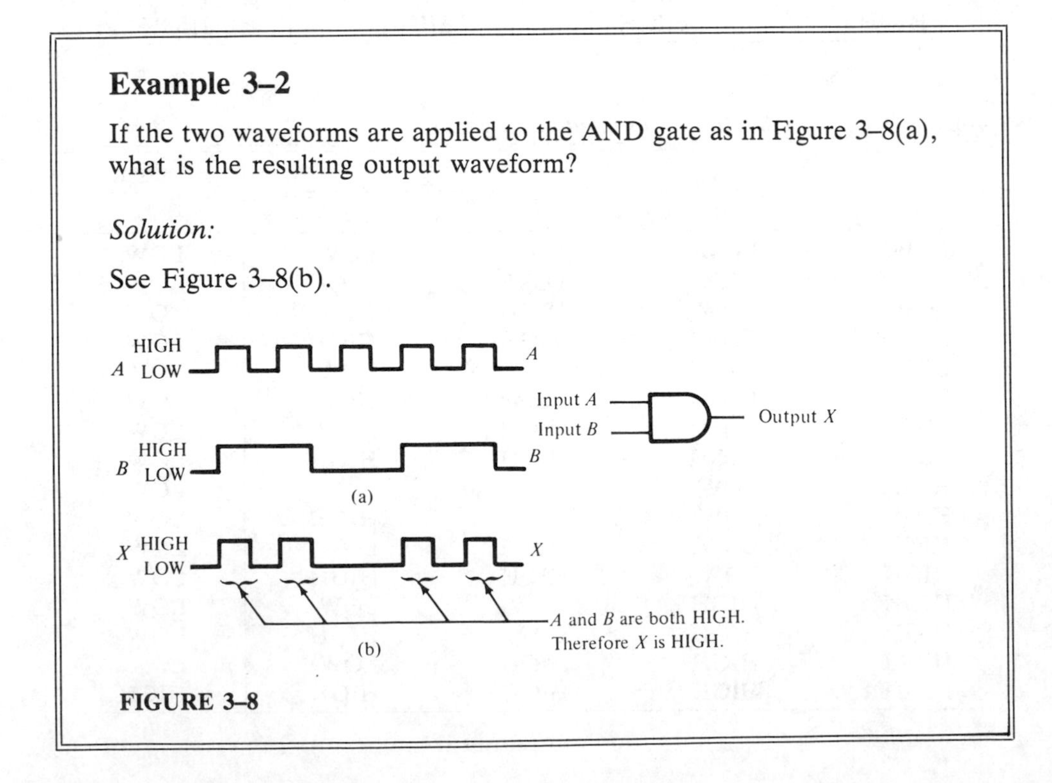

Example 3–2

If the two waveforms are applied to the AND gate as in Figure 3–8(a), what is the resulting output waveform?

Solution:

See Figure 3–8(b).

FIGURE 3–8

It is very important, when analyzing the pulsed operation of logic gates, to pay very careful attention to the time relationships of all the inputs with respect to each other and with respect to the output.

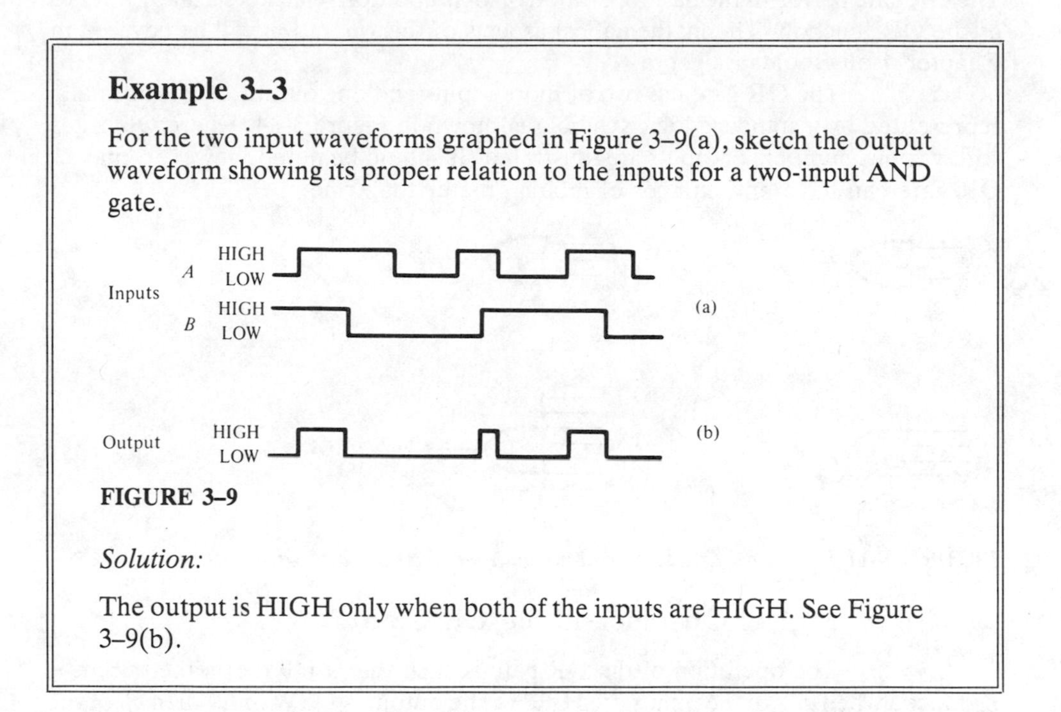

Example 3–3

For the two input waveforms graphed in Figure 3–9(a), sketch the output waveform showing its proper relation to the inputs for a two-input AND gate.

FIGURE 3–9

Solution:

The output is HIGH only when both of the inputs are HIGH. See Figure 3–9(b).

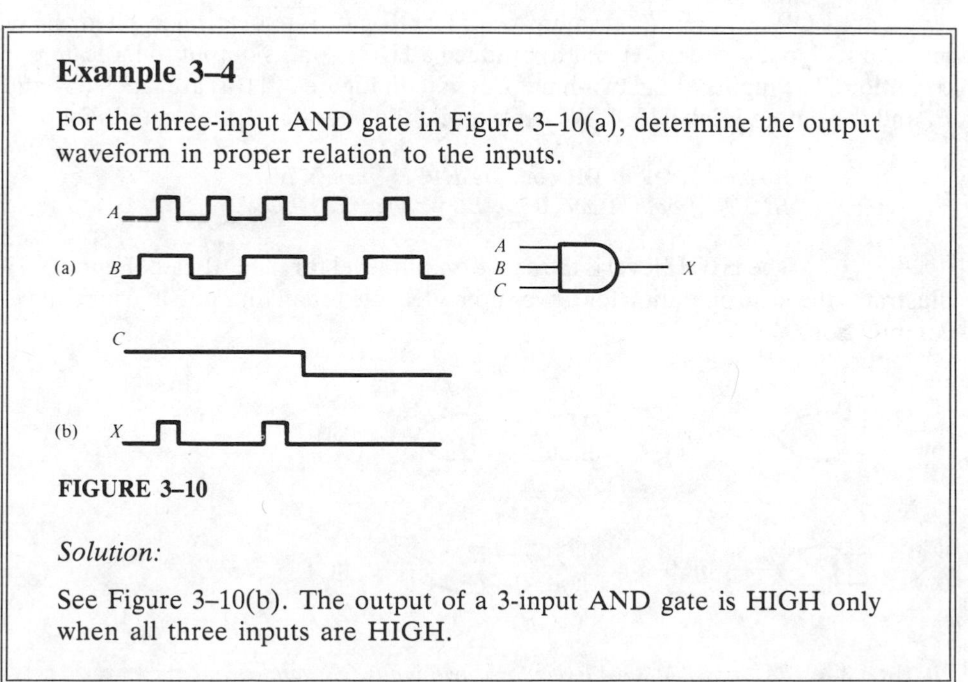

Example 3–4

For the three-input AND gate in Figure 3–10(a), determine the output waveform in proper relation to the inputs.

FIGURE 3–10

Solution:

See Figure 3–10(b). The output of a 3-input AND gate is HIGH only when all three inputs are HIGH.

3–3 THE OR GATE

The OR gate performs the basic operation of logic addition, which we usually refer to as the OR function. The mathematical aspects of this operation will be covered in Chapter 4 on Boolean algebra.

The OR gate has two or more inputs and one output, and is normally represented by a standard logic symbol, as shown in Figure 3–11, where OR gates with various numbers of inputs are illustrated. It should be noted, however, that an OR gate can have any number of inputs greater than one.

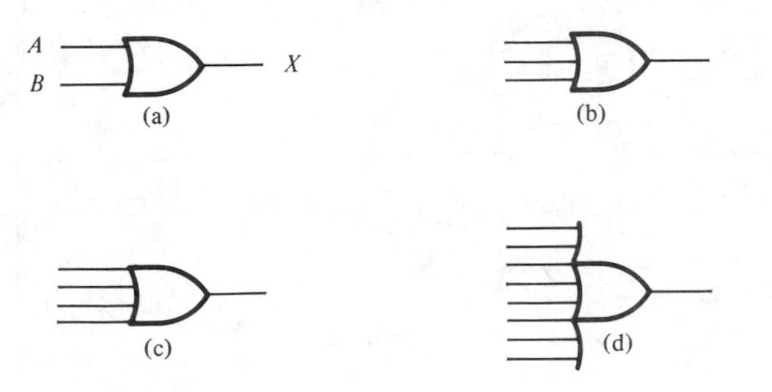

FIGURE 3–11 *Standard OR gate symbols with various numbers of inputs.*

Logical Operation of the OR Gate

The operation of the OR gate is such that a HIGH on the output is produced when *any* of the inputs are HIGH. The output is LOW only when *all* of the inputs are LOW. Therefore, the purpose of an OR gate is to determine when one or more of its inputs are HIGH and to produce a HIGH on its output to indicate this condition. The inputs of the two-input OR gate in Figure 3–11(a) are labeled *A* and *B*, and the output is labeled *X*. We can express the operation of the gate as follows:

If either A **OR** *B* **OR** *both are HIGH, then X is HIGH. If both A and B are LOW, then X is LOW.*

The HIGH level is the *active* output level for the OR gate. Figure 3–12 illustrates the logic operation for a two-input OR gate for all four possible input level combinations.

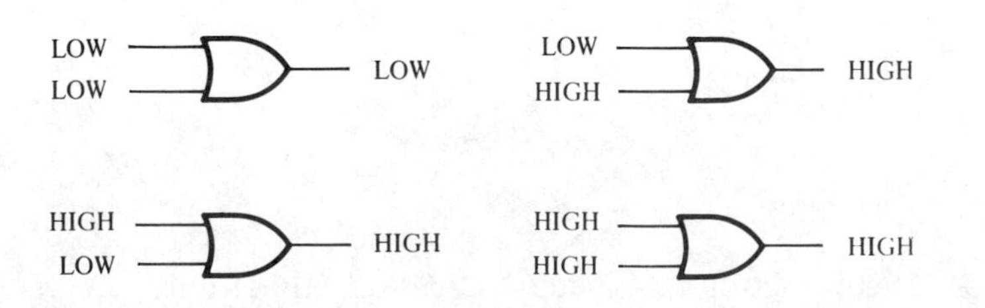

FIGURE 3–12 *All possible logic levels for a two-input OR gate.*

The logical operation of the two-input OR gate can be described in the truth table form shown in Table 3–5. Truth tables for three-input OR gates and four-input OR gates are given in Tables 3–6 and 3–7, respectively. Notice that for each gate, regardless of the number of inputs, the output is HIGH when *any* of the inputs are HIGH.

TABLE 3–5 *Truth table for a two-input OR gate.*

Inputs		Output
A	*B*	*X*
LOW	LOW	LOW
LOW	HIGH	HIGH
HIGH	LOW	HIGH
HIGH	HIGH	HIGH

TABLE 3–6 *Truth table for a three-input OR gate.*

Inputs			Output
A	*B*	*C*	*X*
LOW	LOW	LOW	LOW
LOW	LOW	HIGH	HIGH
LOW	HIGH	LOW	HIGH
LOW	HIGH	HIGH	HIGH
HIGH	LOW	LOW	HIGH
HIGH	LOW	HIGH	HIGH
HIGH	HIGH	LOW	HIGH
HIGH	HIGH	HIGH	HIGH

TABLE 3–7 *Truth table for a four-input OR gate.*

Inputs				Output
A	*B*	*C*	*D*	*X*
LOW	LOW	LOW	LOW	LOW
LOW	LOW	LOW	HIGH	HIGH
LOW	LOW	HIGH	LOW	HIGH
LOW	LOW	HIGH	HIGH	HIGH
LOW	HIGH	LOW	LOW	HIGH
LOW	HIGH	LOW	HIGH	HIGH
LOW	HIGH	HIGH	LOW	HIGH
LOW	HIGH	HIGH	HIGH	HIGH
HIGH	LOW	LOW	LOW	HIGH
HIGH	LOW	LOW	HIGH	HIGH
HIGH	LOW	HIGH	LOW	HIGH
HIGH	LOW	HIGH	HIGH	HIGH
HIGH	HIGH	LOW	LOW	HIGH
HIGH	HIGH	LOW	HIGH	HIGH
HIGH	HIGH	HIGH	LOW	HIGH
HIGH	HIGH	HIGH	HIGH	HIGH

Pulsed Operation

Let us now turn our attention to the operation of an OR gate with pulsed inputs, keeping in mind what we have learned about its logical operation.

Again, the important thing in analysis of gate operation with pulsed waveforms is the relationship of all the waveforms involved. For example, in Figure 3–13, the inputs A and B are both HIGH during interval $T1$, making the output

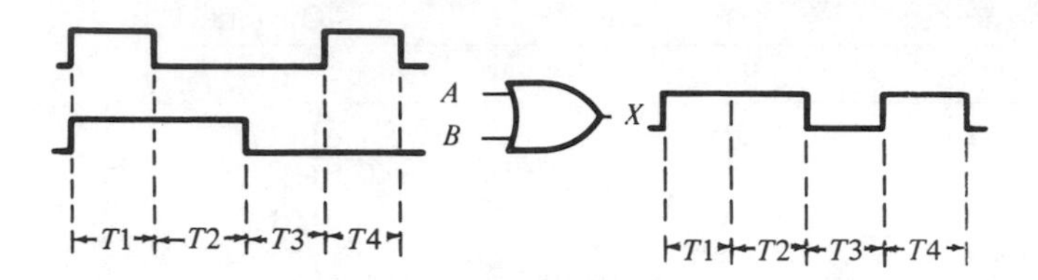

FIGURE 3–13 *An example of pulsed OR gate operation.*

HIGH. During interval $T2$, input A is LOW, but because input B is HIGH, the output is HIGH. Both inputs are LOW during interval $T3$, and we have a LOW output during this time. During $T4$, the output is HIGH because input A is HIGH.

In this illustration, we have simply applied the truth table operation of the OR gate to each of the intervals during which the levels are nonchanging. A few examples will further illustrate OR gate operation with pulse waveforms on the inputs.

Example 3–5

If the two waveforms are applied to the OR gate as in Figure 3–14(a), what is the resulting output waveform?

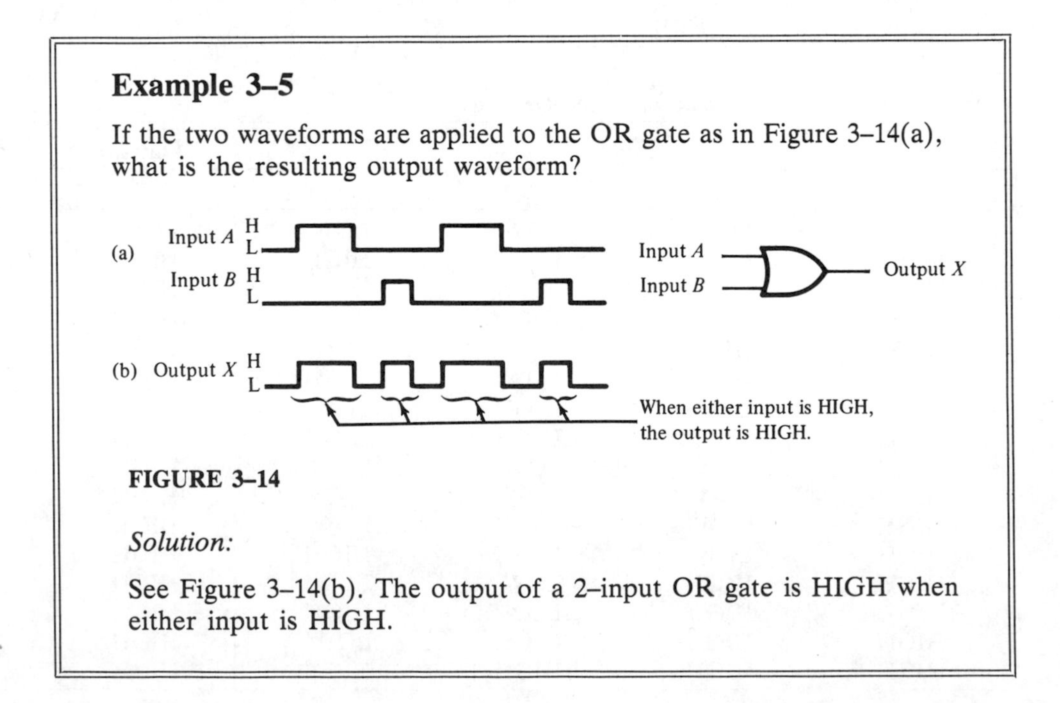

FIGURE 3–14

Solution:

See Figure 3–14(b). The output of a 2-input OR gate is HIGH when either input is HIGH.

Example 3–6

For the two input waveforms in Figure 3–15(a), sketch the output waveform showing its proper relation to the inputs for a two-input OR gate.

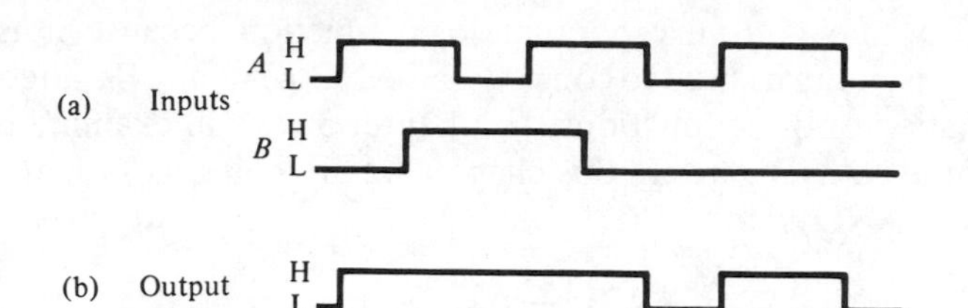

FIGURE 3-15

Solution:

When either input or both inputs are HIGH, the output is HIGH. See Figure 3–15(b).

Example 3–7

For the three-input OR gate in Figure 3–16(a), determine the output waveform in proper relation to the inputs.

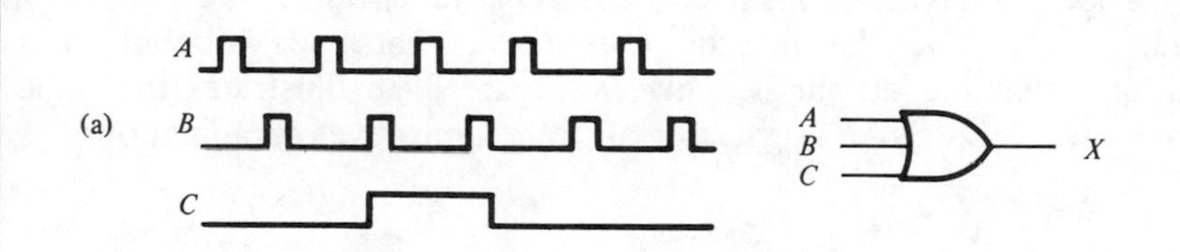

FIGURE 3–16

Solution:

See Figure 3–16(b). The output is HIGH when any of the inputs are HIGH.

3–4 THE NAND GATE

The term NAND is a contraction of NOT–AND, and implies an AND function with a complemented (inverted) output. A standard logic symbol for a two-input NAND gate and its equivalency to an AND gate followed by an inverter are shown in Figure 3–17.

The NAND gate is a very popular logic function because it is a "universal" function; that is, it can be used to construct an AND gate, an OR gate, an inverter, or any combination of these functions. In Chapter 5 we will examine this "universal" property of the NAND gate. In this chapter we are going to look at the logical operation of the NAND gate.

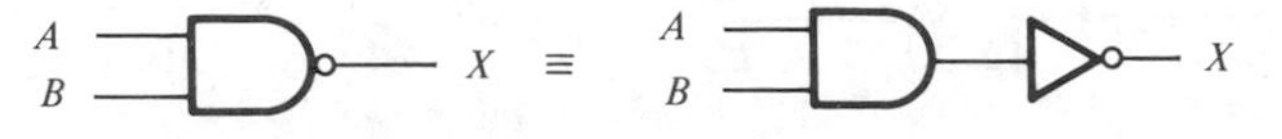

FIGURE 3–17 *Standard NAND gate logic symbol and its NOT/AND equivalent.*

Logical Operation of the NAND Gate

The logical operation of the NAND gate is such that a LOW output occurs only if *all* inputs are HIGH. If *any* of the inputs are LOW, the output will be HIGH. For the specific case of a two-input NAND gate as shown in Figure 3–17, with the inputs labeled A and B and the output labeled X, we can state the operation as follows:

> *If A* **AND** *B are HIGH, then X is LOW. If A is LOW, or B is LOW, or if both A and B are LOW, then X is HIGH.*

Note that this operation is the opposite of the AND as far as output is concerned. In a NAND gate, the LOW level is the *active* output level. The circle or bubble on the output indicates that the output is *active 0.* Figure 3–18 illustrates the logical operation of a two-input NAND gate for all four input level combinations.

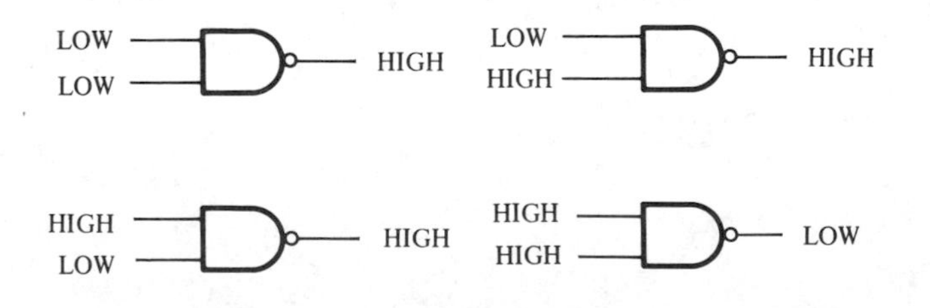

FIGURE 3–18 *Logical operation of a two-input NAND gate.*

The truth table summarizing the logical operation of the two-input NAND gate is shown in Table 3–8. The extensions of the truth table to three- and four-input NAND gates appear as problems at the end of this chapter.

TABLE 3–8 *Truth table for a two-input NAND gate.*

Inputs		Output
A	*B*	*X*
LOW	LOW	HIGH
LOW	HIGH	HIGH
HIGH	LOW	HIGH
HIGH	HIGH	LOW

The NAND gate as an Active LOW-Input OR

Inherent in the NAND gate's operation are the conditions where one or more LOWs produce a HIGH output. If you will look at Table 3–8, you will see that the output is HIGH when any of the inputs are LOW. These conditions can be stated as follows:

> *If A is LOW* **OR** *B is LOW* **OR** *if both A and B are LOW, then X is HIGH.*

Here we have an OR operation that requires LOW inputs to produce a HIGH output and that is referred to as *negative-OR*. This function of the NAND gate can be considered as a secondary aspect of its operation, and is represented by the standard logic symbol in Figure 3–19(b). The two symbols in Figure 3–19 represent the same gate, but they also serve to define its role in a particular application, as illustrated by the following examples.

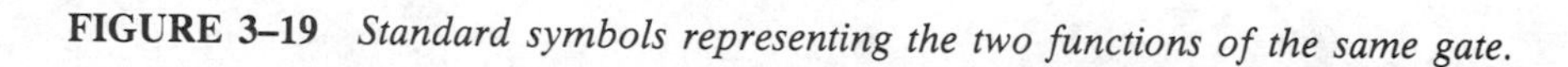

(a) NAND (b) Negative-OR

FIGURE 3–19 *Standard symbols representing the two functions of the same gate.*

Example 3–8

The simultaneous occurrence of two HIGH level voltages must be detected and indicated by a LOW level output that is used to illuminate an indicator lamp. Sketch the operation.

Solution:

This application requires a NAND function, since the output has to be *active LOW* in order to produce current through the bulb when two *HIGHs* occur on its intputs. The *NAND* symbol is therefore used to show the operation. See Figure 3–20.

Example 3–8, continued

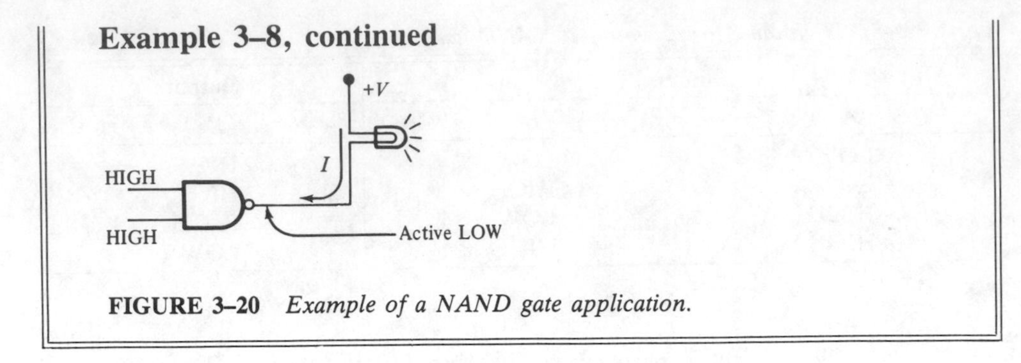

FIGURE 3–20 *Example of a NAND gate application.*

Example 3–9

A gate is required to monitor two lines, to detect the occurrence of LOW level voltages on either or both lines, and to generate a HIGH level output used to illuminate an indicator lamp. Sketch the operation.

Solution:

This application requires an *active LOW* input OR function because the output has to be *active HIGH* in order to produce an indication of the occurrence of one or more LOW levels on its inputs. In this case, the gate functions in the *negative-OR* mode, and is represented by the appropriate symbol shown in Figure 3–21. A LOW on either input or both inputs causes an active-HIGH output to activate the lamp.

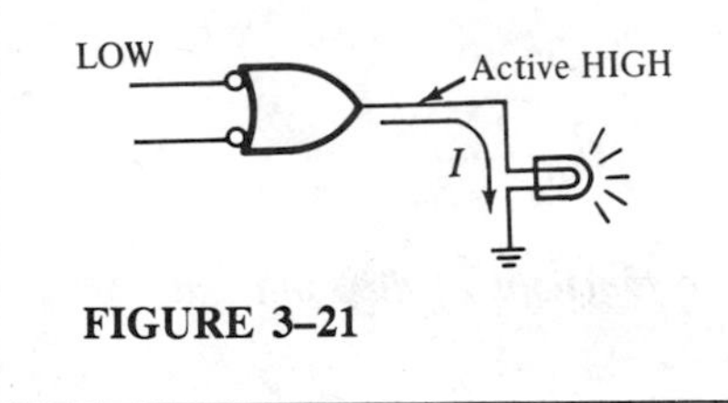

FIGURE 3–21

Pulsed Operation

We will now look at the pulsed operation of the NAND gate. Remember from the truth table that any time *all* of the inputs are HIGH, the output will be LOW, and this is the *only* time a LOW output occurs. A few examples will serve to illustrate pulsed operation.

Example 3–10

If the two waveforms shown in Figure 3–22(a) are applied to the NAND gate, determine the resulting output waveform.

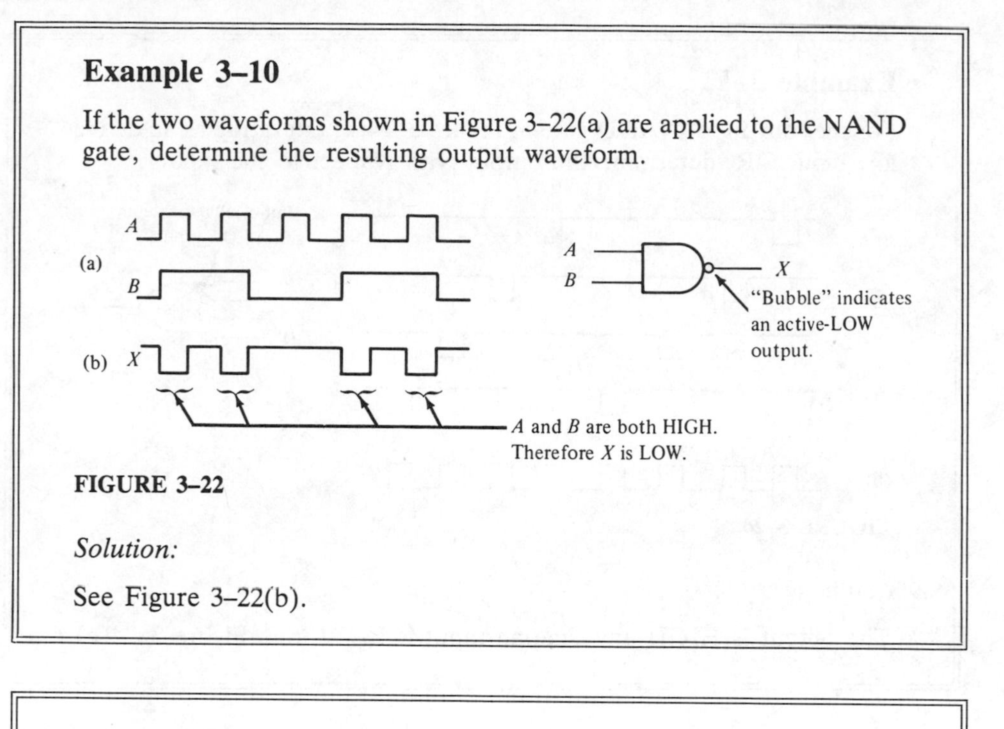

FIGURE 3–22

Solution:

See Figure 3–22(b).

Example 3–11

Sketch the output waveform for the three-input NAND gate in Figure 3–23(a), showing its proper relationship to the inputs.

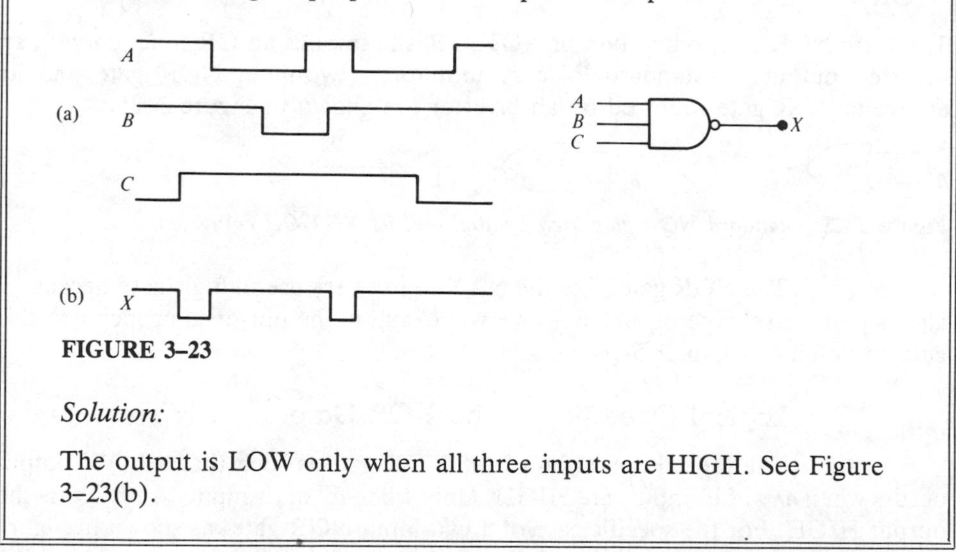

FIGURE 3–23

Solution:

The output is LOW only when all three inputs are HIGH. See Figure 3–23(b).

Example 3–12

For the four-input NAND gate in Figure 3–24(a) operating as an active low-input OR, determine the output with respect to the inputs.

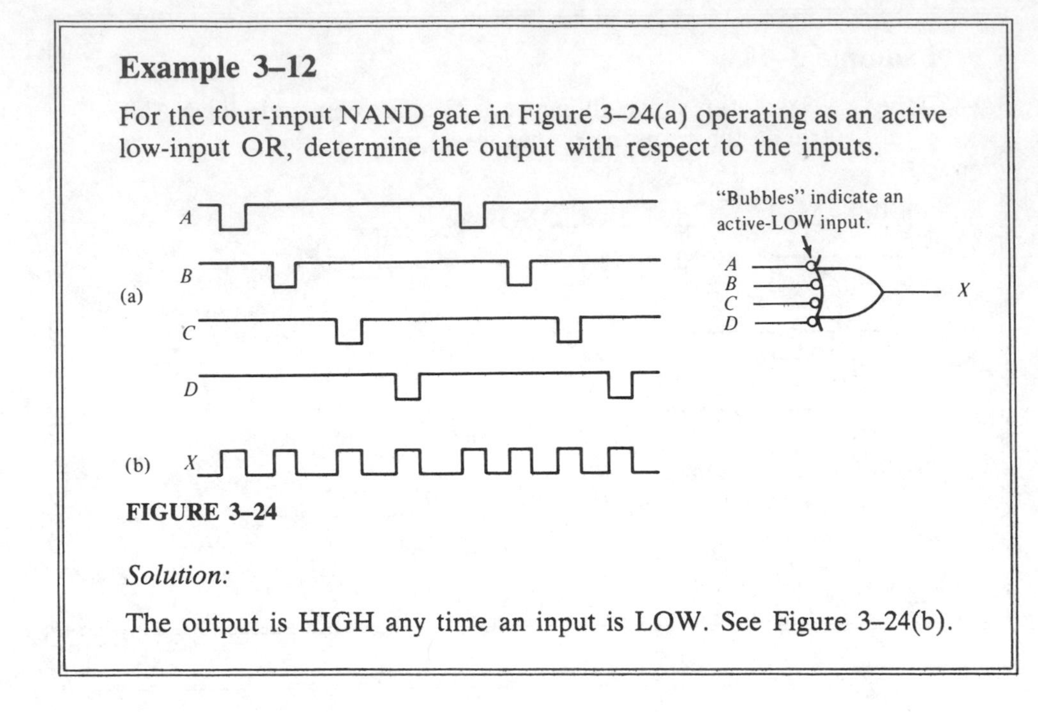

FIGURE 3–24

Solution:

The output is HIGH any time an input is LOW. See Figure 3–24(b).

3–5 THE NOR GATE

The term NOR is a contraction of NOT–OR and implies an OR function with an inverted output. A standard logic symbol for a two-input NOR gate and its equivalent OR gate followed by an inverter are shown in Figure 3–25.

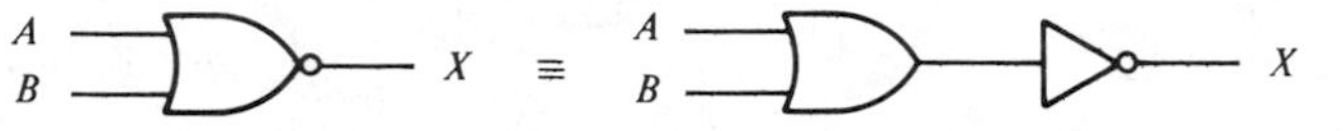

Figure 3–25 *Standard NOR gate logic symbol and its NOT/OR equivalent.*

The NOR gate, like the NAND, is a very useful logic gate because it also is a universal type of function. We will examine the universal property of this gate in detail in Chapter 5.

Logical Operation of the NOR Gate

The logical operation of the NOR gate is such that a LOW output occurs when *any* of its inputs are HIGH. Only when *all* of its inputs are LOW is the output HIGH. For the specific case of a two-input NOR gate, as shown in Figure 3–25 with the inputs labeled A and B and the output labeled X, we can state the operation as follows:

If A **OR** *B* **OR** *both are HIGH, then X is LOW.*
If both A and B are LOW, then X is HIGH.

Note that this operation results in an output opposite that of the OR gate. In a NOR gate, the LOW output is the *active* output level. As was pointed out for the NAND gate, the bubble on the output indicates that the function is *active 0*. Figure 3–26 illustrates the logical operation of a two-input NOR gate for all four possible input combinations.

The truth table for the two-input NOR gate is given in Table 3–9. The extensions of the table to three- and four-input NOR gates appear as problems at the end of this chapter.

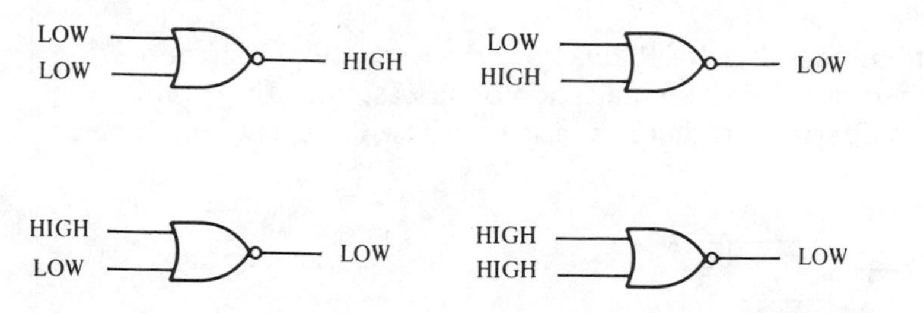

FIGURE 3–26 *Logical operation of a two-input NOR gate.*

TABLE 3–9 *Truth table for a two-input NOR gate.*

Inputs		Output
A	*B*	*X*
LOW	LOW	HIGH
LOW	HIGH	LOW
HIGH	LOW	LOW
HIGH	HIGH	LOW

The NOR Gate as an Active LOW-Input AND

The NOR gate, like the NAND, also displays another mode of operation that is inherent in the way it logically functions. Table 3–9 shows that a HIGH is produced on the gate output only if *all* of the inputs are LOW. In reference to Figure 3–27(a), this aspect of NOR operation is stated as follows:

If both A **AND** *B are LOW, then X is HIGH.*

We have essentially an AND operation that requires all *LOW* inputs to produce a HIGH output. This is called *negative-AND,* and can be considered a secondary aspect of NOR gate operation. The standard symbol for the negative-AND function of the NOR gate is shown in Figure 3–27(b). It is important to remember that the two symbols in this figure represent the same gate and serve only to distinguish between the two facets of logical operation. The following two examples will illustrate this.

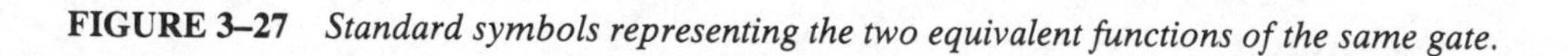

FIGURE 3–27 *Standard symbols representing the two equivalent functions of the same gate.*

Example 3–13

A certain application requires that two lines be monitored for the occurrence of a HIGH level voltage on either or both. Upon detection of a HIGH level, the circuit must provide a LOW voltage to energize a particular indicating device. Sketch the operation.

Solution:

This application requires a NOR function, since the output has to be *active LOW* in order to give an indication of at least one HIGH on its inputs. The NOR symbol is therefore used to show the operation. See Figure 3–28.

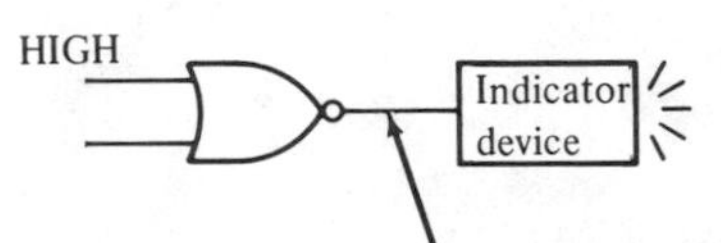

Active LOW energizes the device.

FIGURE 3–28 *Example of a NOR gate application.*

Example 3–14

A device is needed to indicate when two LOW levels occur simultaneously on its inputs and to produce a HIGH output as an indication. Sketch the operation.

Solution:

Here, an active low-input AND function is required, as shown in Figure 3–29.

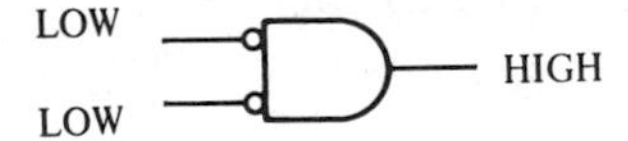

FIGURE 3–29

Pulsed Operation

The next three examples will illustrate the logical operation of the NOR gate with pulsed inputs. Again, as with the other types of gates, we will simply follow the truth table operation in order to determine the output waveforms.

Example 3–15

If the two waveforms shown in Figure 3–30(a) are applied to the NOR gate, what is the resulting output waveform?

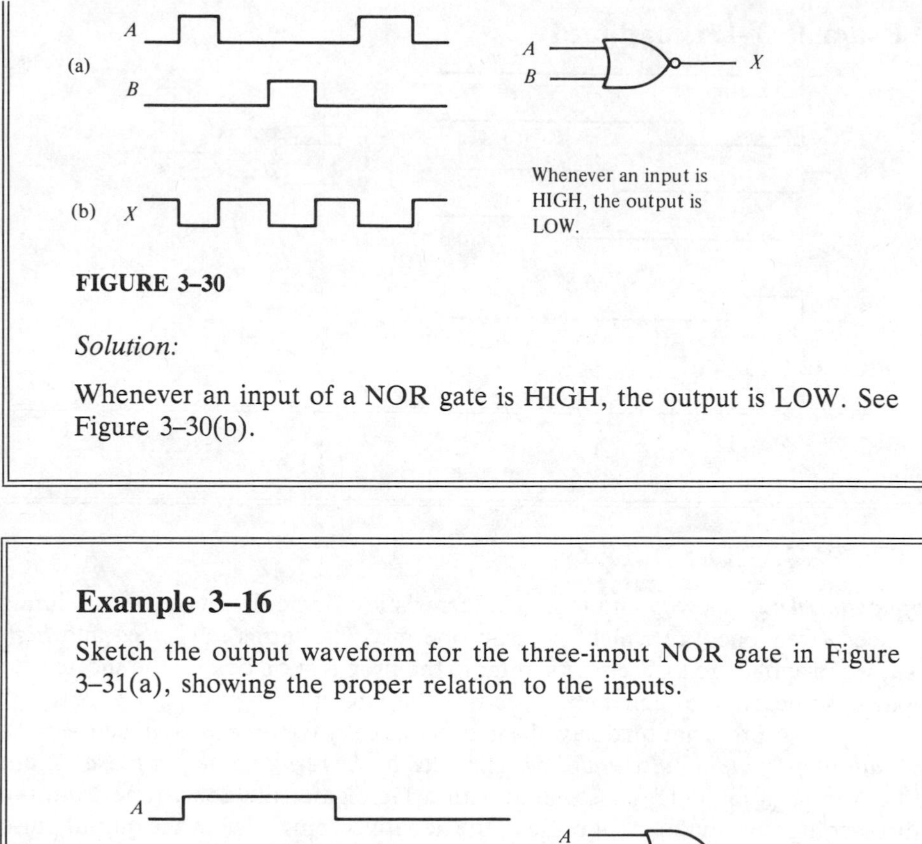

FIGURE 3–30

Solution:

Whenever an input of a NOR gate is HIGH, the output is LOW. See Figure 3–30(b).

Example 3–16

Sketch the output waveform for the three-input NOR gate in Figure 3–31(a), showing the proper relation to the inputs.

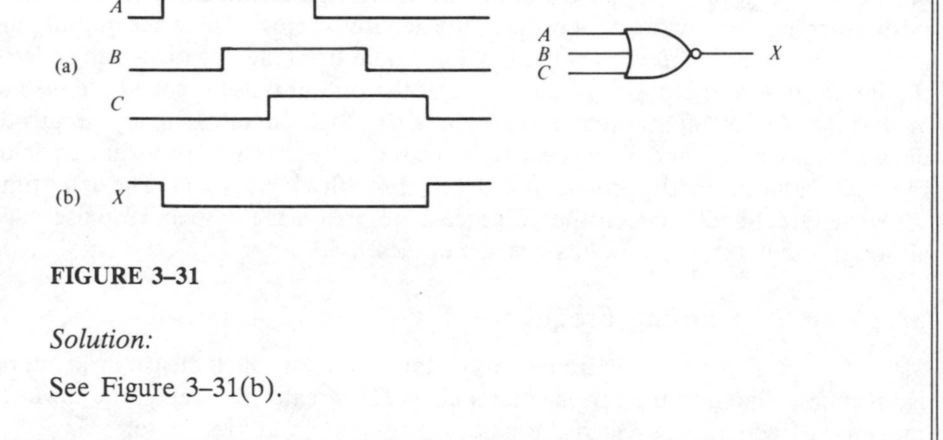

FIGURE 3–31

Solution:

See Figure 3–31(b).

Example 3–17

For the four-input NOR gate operating as an active low-input AND in Figure 3–32(a), determine the output relative to the inputs.

Solution:

Any time *all* of the inputs are LOW, the output is HIGH. See Figure 3–32(b).

Example 3–17, continued

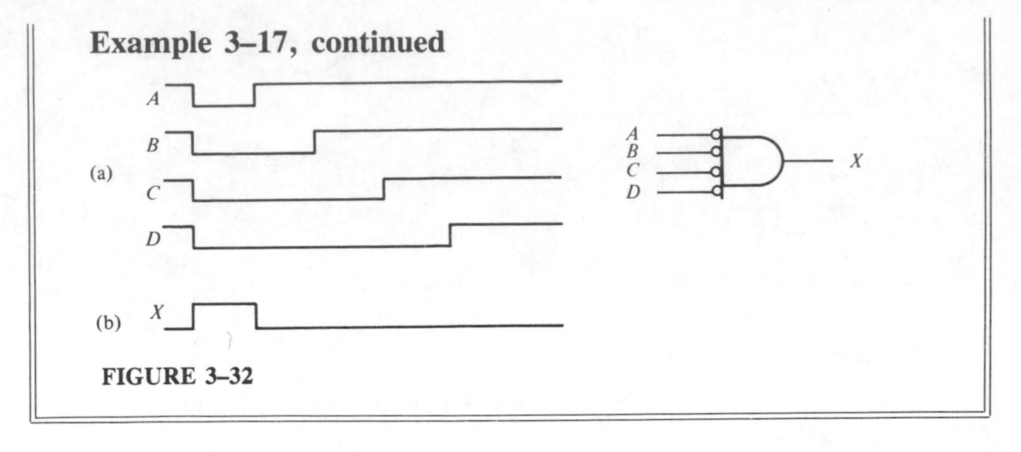

FIGURE 3–32

3–6 GATE PROPAGATION DELAY

Propagation delay is a very important characteristic of logic circuits because it limits the speed (frequency) at which they can operate. The terms *low speed* and *high speed,* when applied to logic circuits, refer to the propagation delays; the shorter the propagation delay, the higher the speed of the circuit.

A propagation delay of a gate is basically the *time interval between the application of an input pulse and the occurrence of the resulting output pulse.* There are two propagation delays associated with a logic gate: the delay time from the positive-going edge of the input pulse to the negative-going edge of the output pulse is called the *turn-on delay*(t_{pHL}), and the delay time from the negative-going edge of the input pulse to the positive-going edge of the output pulse is called the *turn-off delay*(t_{pLH}). These definitions apply to an *inverting* logic circuit. In many circuits the delays are not equal, and the larger of the two is the "worst case" propagation delay. Figure 3–33 illustrates the propagation delays through a logic circuit. The delay times are usually defined between the 50 percent points on the respective pulse edges although other reference points are sometimes used.

Operating Frequency

The maximum frequency of input pulses at which the gate will operate is inversely related to the propagation delay. The greater the delay, the lower the maximum frequency at which the gate will function and visa versa.

Typically, the propagation delay of a gate is considerably less than the period of the maximum frequency of operation.

TTL *vs* CMOS

A comparison of two important integrated circuit (IC) technologies, TTL (transistor–transistor logic) and CMOS (complementary metal-oxide semiconductor) shows that CMOS is much slower than TTL—typically three to six times slower. For example, a family of standard TTL gates has a typical propagation delay

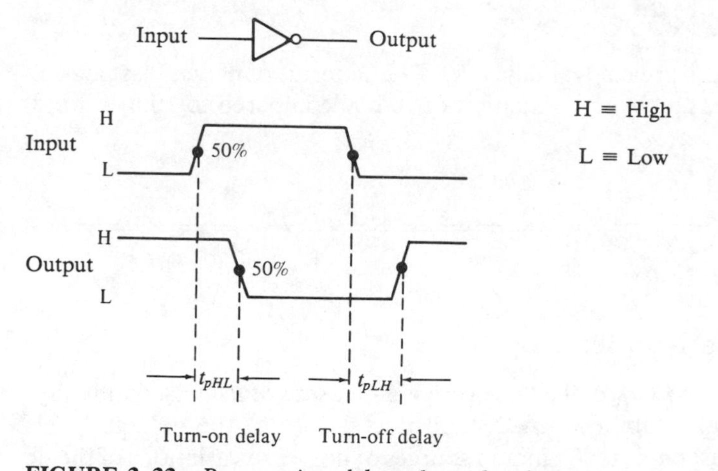

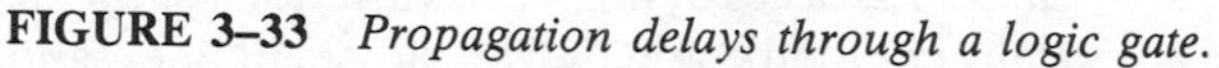

FIGURE 3–33 *Propagation delays through a logic gate.*

of 10 ns, and a comparable family of CMOS has a propagation delay of 40 ns.

This, of course, means that CMOS cannot be operated at frequencies as high as standard TTL. Refer to Appendix A for a detailed coverage of IC technologies.

3–7 POWER DISSIPATION

The power dissipation of a logic gate equals the dc supply voltage V_{CC} times the *average* supply current I_{CC}. Normally, the value of I_{CC} for a LOW gate output is higher than for a HIGH output. The manufacturer's data sheet usually specifies both of these values as I_{CCL} and I_{CCH}. The average I_{CC} is then determined based on a 50 percent duty cycle operation of the gate (LOW half of the time and HIGH half of the time).

Example 3–18

A certain gate draws 2 mA when its output is HIGH and 3.5 mA when its output is LOW. What is its average power dissipation if V_{CC} is 5V and it is operated on a 50 percent duty cycle?

Solution:

The average I_{CC} is

$$I_{CC} = \frac{I_{CCH} + I_{CCL}}{2} = \frac{2\text{mA} + 3.5\text{mA}}{2} = 2.75 \text{ mA}$$

The average power is

$$P_{\text{AVG}} = V_{CC}\, I_{CC} = (5\text{V})(2.75 \text{ mA}) = 13.75 \text{ mW}$$

TTL *vs* CMOS

CMOS has a great advantage over TTL in terms of power dissipation. A typical CMOS gate has a power dissipation of 10 nW compared to 10 mW for a typical standard TTL gate.

3–8 NOISE IMMUNITY

dc Noise Margins

The *dc noise margin* of a logic gate is a measure of its noise immunity. Noise immunity is a gate's ability to withstand fluctuations of the voltage levels (noise) with which it must operate. Common sources of noise are variations of the dc supply voltage, ground noise, magnetically coupled voltages from adjacent lines, and radiated signals. The term *dc* noise margin applies to noise voltages of relatively long duration compared to a gate's response time.

To learn about dc noise margins, we will define several voltages that are typically specified on a manufacturer's data sheet for an IC logic gate:

V_{IL} = LOW level input voltage
V_{IH} = HIGH level input voltage
V_{OL} = LOW level output voltage
V_{OH} = HIGH level output voltage

Every logic circuit has certain limits on the values of these voltages within which it will operate properly. To illustrate this, let us discuss a logic gate that operates with 0 V as its ideal LOW and +5 V as its ideal HIGH. These voltages will vary because of circuit parameters, and the gate must be able to tolerate variations within certain specified limits. If, for example, a LOW level voltage range of 0 V to 0.5 V is acceptable to a gate, then the *maximum* LOW level input voltage [$V_{IL(max)}$] for that gate is +0.5 V. Any voltage above this value appears as a possible HIGH to the gate. The effect of a fluctuation due to noise of the LOW level input voltage on the output of the gate is illustrated in Figure 3–34.

Now, if the type of logic gate we are using has a maxmium LOW level *output* voltage [$V_{OL(max)}$] of +0.2 V, then there is a "safety" margin of 0.3 V between

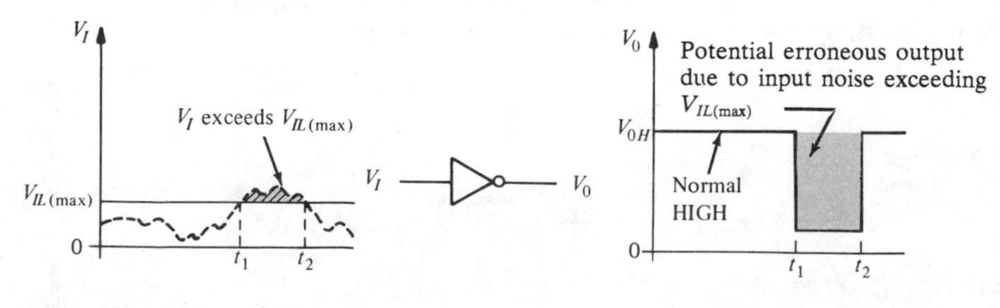

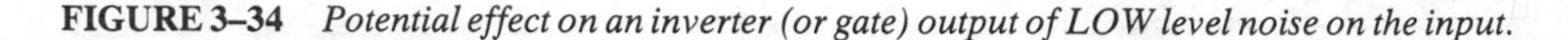

FIGURE 3–34 *Potential effect on an inverter (or gate) output of LOW level noise on the input.*

the maximum LOW level that a gate puts out and the maximum LOW level input that a gate being driven can tolerate. This is called the LOW level *noise margin* and is expressed as

$$V_{NL} = V_{IL(\text{max})} - V_{OL(\text{max})} \tag{3–2}$$

Figure 3–35 illustrates the LOW level dc noise margin. In essence, there can be fluctuations due to noise on the line between the two gates without affecting the output of the second gate, as long as the peak value of these fluctuations does not exceed the LOW level noise margin (V_{NL}).

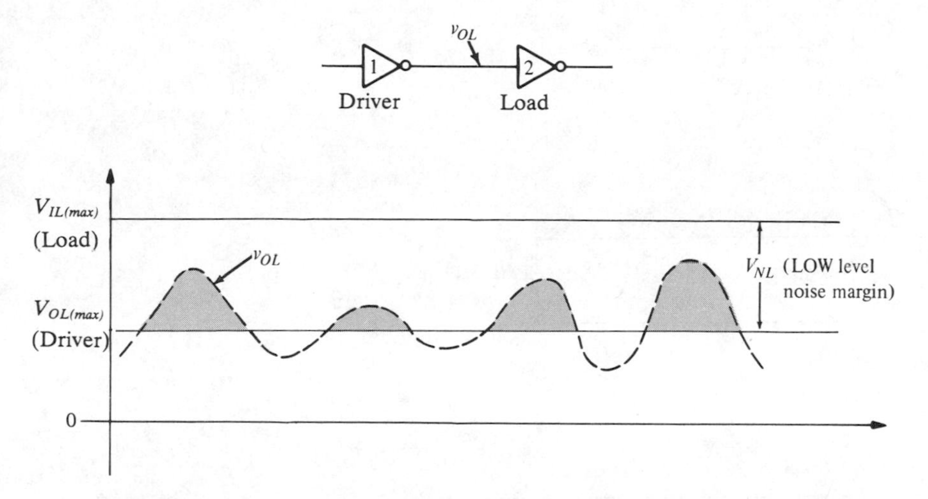

FIGURE 3–35 *LOW level noise margin.*

Now let us consider the HIGH level dc noise margin of a gate. If, for example, a HIGH level input voltage between +5 V and +4 V is acceptable as a HIGH level to the gate, then the *minimum* HIGH level input voltage $[V_{IH(\text{min})}]$ is +4.0 V. Any voltage above this value is acceptable as a HIGH to the gate and any voltage below this value would appear as a possible LOW, as illustrated in Figure 3–36.

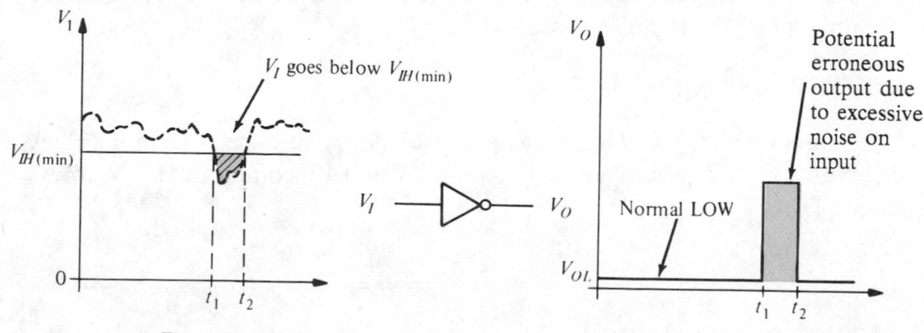

FIGURE 3–36 *Potential effect on an inverter (or gate) output of HIGH level noise on the input.*

Now, if the driving gate has a minimum HIGH level output voltage [$V_{OH(\min)}$] of +4.5 V, then there is a HIGH level dc noise margin of 0.5 V, expressed as

$$V_{NH} = V_{OH(\min)} - V_{IH(\min)} \tag{3-3}$$

Figure 3–37 illustrates the HIGH level noise margin, and shows that there can be fluctuations due to noise on the line between the two gates without affecting the output of the second gate, as long as the peak value of the noise does not exceed V_{NH}, the HIGH level noise margin.

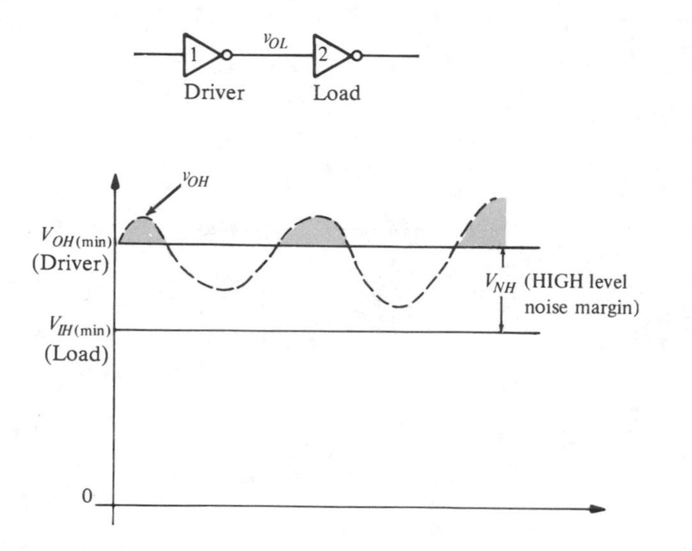

FIGURE 3–37 *HIGH level noise margin.*

ac Noise Margin

The term *ac noise margin* applies to the noise immunity of a gate to noise of very short duration. A typical gate is much more immune to this type of noise because it cannot respond fast enough to be affected. Therefore ac noise margins are considerably higher than dc noise margins.

TTL *vs* CMOS

Generally, CMOS logic has a higher dc noise margin than TTL. A typical CMOS gate has a dc noise margin between 2 V and 4 V compared to 1 V or less for standard TTL.

3–9 LOADING CONSIDERATIONS

In a digital system, you will typically find many types of digital ICs interconnected to perform various functions. In these situations, the output of a logic gate may be connected to the inputs of several other similar gates so the *load* on the driving gate becomes an important factor.

Fan out

The *fan out* of a gate is the maximum number of inputs of the same IC family that the gate can drive while maintaining its output levels within specified limits. That is, the fan out specifies the maximum loading that a given gate is capable of handling.

We will use TTL to illustrate the concept of fan out. A TTL gate is classified as *current-sinking* because when its output state is LOW, it *accepts* current from the input of the gate which it is driving.

This is illustrated in Figure 3–38 where one TTL gate is driving another TTL gate. Only the output and input portions of typical TTL gate circuits are shown for simplicity. For further discussion of TTL circuits see Appendix A.

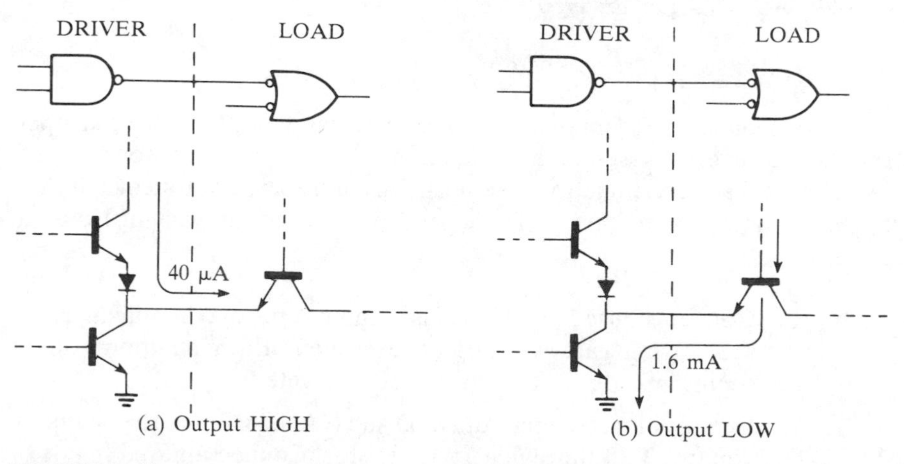

FIGURE 3–38

When the driving gate is in its HIGH output state, the input to the load gate is like a reverse biased diode. This means that there is practically no current required from the driving gate. Actually, for a typical TTL gate, about 40 μA of reverse leakage current flows from the driving gate into the load gate input as shown in Figure 3–38(a).

When the driving gate is in its LOW output state, the input to the load gate is like a forward biased diode. This means that there is a certain amount of current flowing *out* of the load gate input into the driving gate output as shown in Figure 3–38(b). In this case, the driving gate is *sinking* current from the load. For a typical TTL gate, this current is about 1.6 mA.

There is a limitation on the total amount of current that a TTL gate can sink, and this sets the limit on the number of other TTL gate inputs that it can drive. One TTL input is termed a *unit load* (UL) and represents 1.6 mA that the driving gate must sink in the LOW state. The maximum number of unit loads that a typical TTL gate can handle is called its *fan out.*

For a standard TTL gate, the fan out is 10 UL (except for *buffer* gates, which have a fan out of 30). This means that it can drive no more than 10 inputs to other TTL gates and still operate reliably. If the fan out is exceeded, specified operation is not guaranteed. Figure 3–39 shows a gate driving 10 other gates.

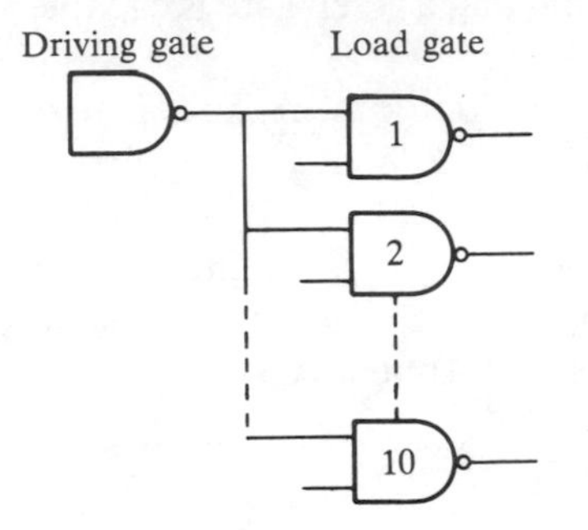

FIGURE 3–39 *A NAND gate driving 10 UL.*

Unused Inputs

An unconnected input on a TTL gate acts as a HIGH because an open input results in a reverse biased emitter junction on the input transistor just as a HIGH level does. However, due to noise sensitivity, it is best not to leave unused TTL inputs unconnected (or open). There are several possible ways of handling unused inputs:

1. Connect unused inputs to a used input if the maximum fan out of the driving gate will not be exceeded. Each additional input represents a unit load to the driving gate.
2. Connect unused inputs of AND and NAND gates to the dc supply voltage (V_{CC}) through a 1 kΩ resistor. Connect unused inputs of OR and NOR gates to ground.
3. Connect unused inputs to the output of a gate that is not being used. The unused gate output must be a constant HIGH for unused inputs of AND and NAND gates.

Unused inputs of CMOS gates must be either connected to the dc supply voltage or ground depending on the type of gate.

TTL *vs* CMOS

CMOS has a much higher fan out than TTL. The only limitation on CMOS loading is the switching speed. The switching time tends to increase as the fan out increases because of the increased input capacitance of the load gates. As mentioned, a standard TTL gate has a maximum fan out of 10 UL. A typical CMOS gate has a maximum fan out exceeding 50 UL.

54/74 FAMILY OF TTL GATES 3–10

In this section, some specific TTL gates are presented. One of the most popular and widely used families of TTL integrated circuits is known as the 54/74 series.

The gate functions are classed as SSI (small-scale integration) and are available in a wide variety of types and combinations. We will examine the packaging arrangements and electrical characteristics of several as specified on the manufacturer's data sheets. The 54/74 family is not the exclusive product of one company, but is a type of circuit produced by several semiconductor firms.

The 54 or 74 is the prefix number of the devices within the family: 54 specifies that the device is capable of operating over the temperature range −55°C to 125°C, and 74 specifies a temperature range of 0°C to 70°C. In this section, the 74 prefix is used for purposes of discussion. Within the 54/74 family there are several variations of the circuit technology such as 74L (low power), 74S (Schottky), 74LS (low power Schottky), and 74H (high speed). These are discussed in Appendix A; in this section we will concentrate on the *standard* 74 gates.

Hex Inverters

The 7404 IC consists of six inverters in a 14-pin single package. The pin connections are shown in Figure 3–40.

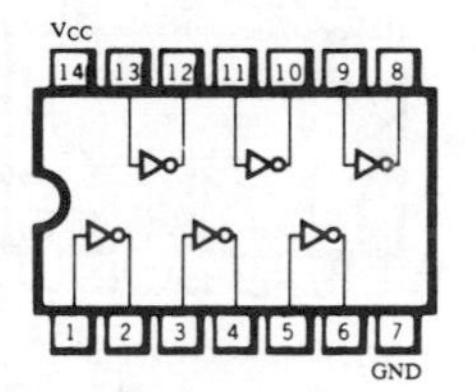

FIGURE 3–40 *7404/7405 hex inverters*

Notice that pin 1 is to the left of the notched end of the dual in-line package (DIP) looking from the top. Ground and V_{CC} are pins 7 and 14 respectively. There are other arrangements, but this is the most common.

The 7405 is a hex inverter IC with identical pin arrangement but with open collector outputs on the inverters. Open collector gates are covered in Appendix A.

AND Gates

There are several types of IC AND gate packages in this family. The 7408 has four 2-input AND gates, the 7409 has four 2-input AND gates with open collector outputs, the 7411 has three 3-input AND gates, the 7415 has three 3-input AND gates with open collector outputs, and the 7421 has two 4-input AND gates. The connection diagrams are shown in Figure 3–41.

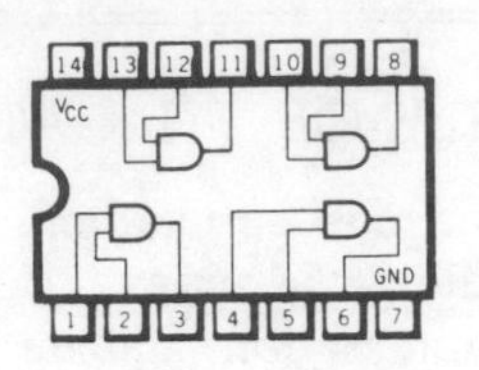

(a) 7408/7409
Quad 2-input AND

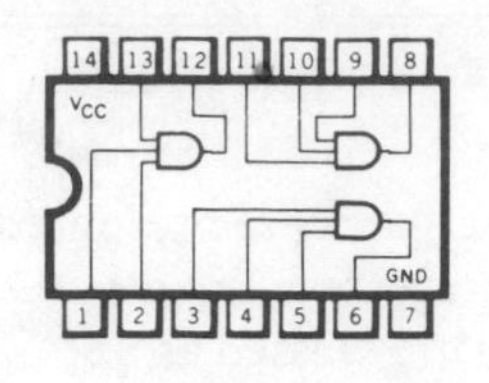

(b) 7411/7415
Triple 3-input AND

(c) 7421
Dual 4-input AND

FIGURE 3–41

NAND Gates

A variety of TTL NAND gates are available. These include the 7400 with four 2-input gates, the 7401 with four 2-input open collector gates, the 7410 with three 3-input gates, the 7412 with three 3-input open collector gates, the 7420 with two 4-input gates, the 7422 with two 4-input open collector gates, the 7430 with one 8-input gate, and the 74133 with one 13-input gate. Figure 3–42 shows the connection diagrams for these devices. Notice that the 74133 is in a 16-pin DIP where pin 8 is ground and pin 16 is V_{CC}.

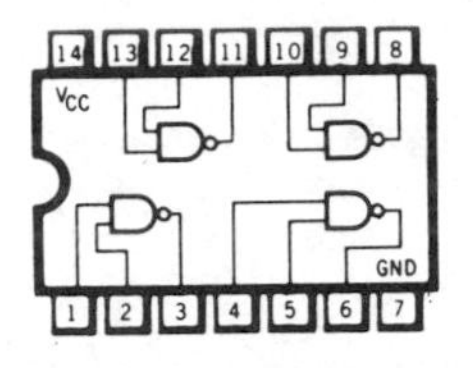

(a) 7400/7401
Quad 2-input NAND

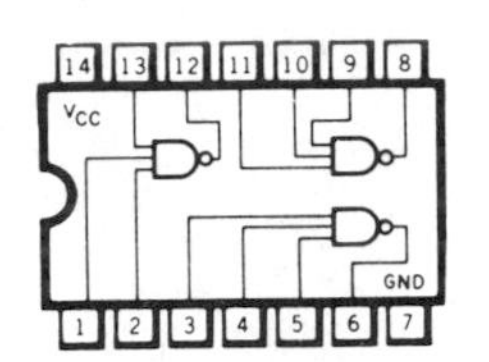

(b) 7410/7412
Triple 3-input NAND

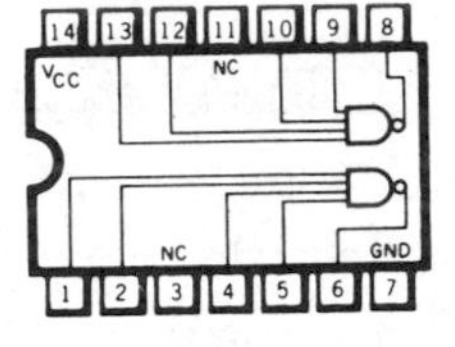

(c) 7420/7422
Dual 4-input NAND

(d) 7430
Single 8-input NAND

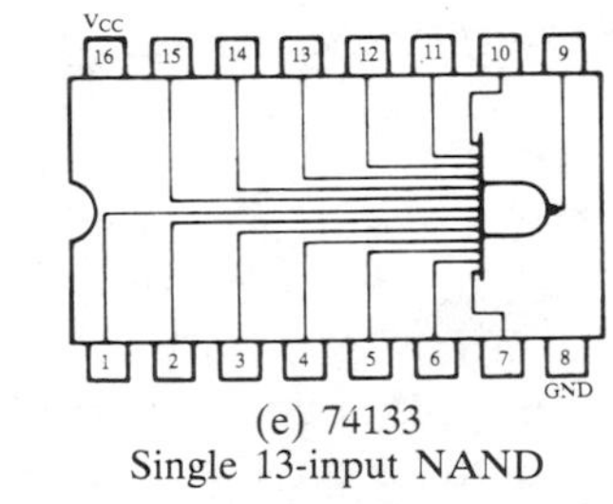

(e) 74133
Single 13-input NAND

FIGURE 3–42 *Some standard TTL NAND gates.*

OR Gates

The 7432 is a package with four 2-input OR gates as shown in Figure 3-43.

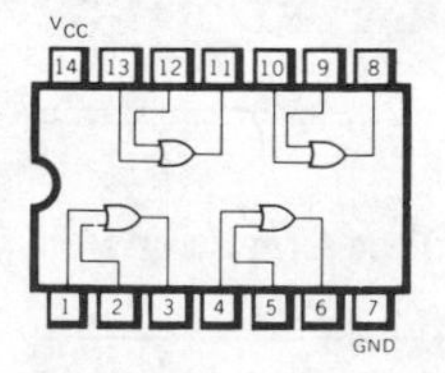

FIGURE 3–43 *A TTL 7432 quad two-input OR gate 14-pin DIP.*

NOR Gates

The 7402 has four 2-input gates, the 7427 has three 3-input gates, and the 74260 has two 5-input gates as shown in Figure 3–44.

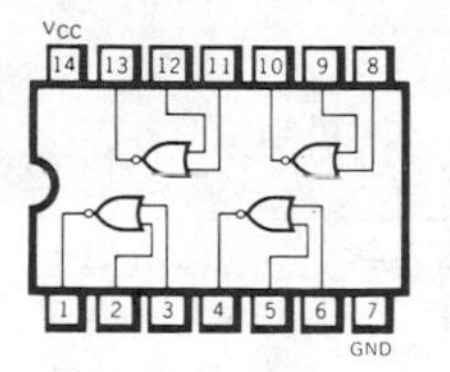

(a) 7402
Quad 2-input NOR

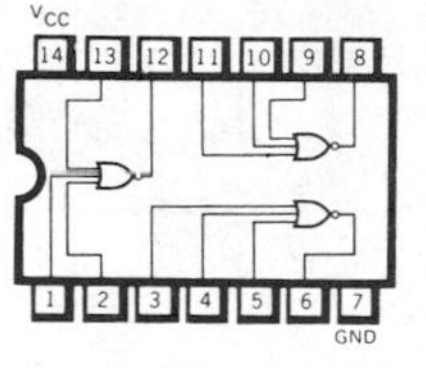

(b) 7427
Triple 3-input NOR

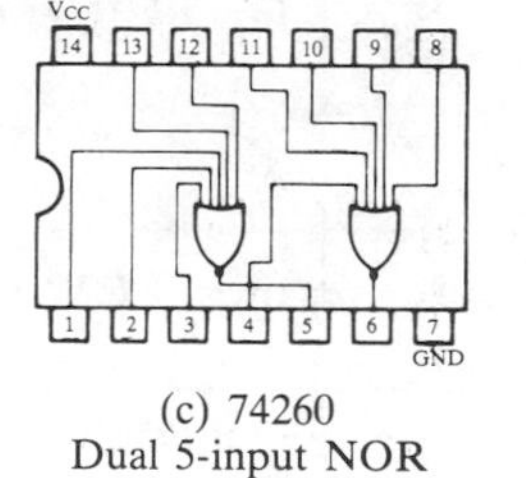

(c) 74260
Dual 5-input NOR

FIGURE 3–44 *Some standard TTL NOR gates.*

Data Sheets

Specific information about operating characteristics of a particular IC family can be determined from *data sheets* published by the manufacturer.

A typical data sheet is divided into three main sections: *recommended operating conditions, electrical characteristics,* and *switching characteristics.* As an example, Figure 3–45 shows the arrangement of a typical data sheet for the 5400/7400 quad 2-input NAND gate. For additional data sheets, see Appendix B.

RECOMMENDED OPERATING CONDITIONS

Parameter	5400			7400			Units
	Minimum	Typical	Maximum	Minimum	Typical	Maximum	
Supply voltage (V_{CC})	4.5	5.0	5.5	4.75	5.0	5.25	Volts
Operating free-air temperature range	−55	25	125	0	25	70	°C
Normalized fan-out from each output			10			10	UL

FIGURE 3–45

ELECTRICAL CHARACTERISTICS OVER OPERATING TEMPERATURE RANGE (unless otherwise noted)

Symbol	Parameter	Limits: Minimum	Limits: Typical (note 2)	Limits: Maximum	Units	Test Conditions (note 1)
V_{IH}	Input HIGH voltage	2.0			Volts	Guaranteed input HIGH voltage
V_{IL}	Input LOW voltage			0.8	Volts	Guaranteed input LOW voltage
V_{OH}	Output HIGH voltage	2.4	3.3		Volts	V_{CC} = Min., I_{OH} = 0.4 mA, V_{IN} = 0.8 V
V_{OL}	Output LOW voltage		0.22	0.4	Volts	V_{CC} = Min., I_{OL} = 16 mA, V_{IN} = 2.0 V
I_{IH}	Input HIGH current			40	μA	V_{CC} = Max., V_{IN} = 2.4. V
				1.0	mA	V_{CC} = max., V_{IN} = 5.5 V
I_{IL}	Input LOW current			−1.6	mA	V_{CC} = Max., V_{IN} = 0.4 V
I_{OS}	Output short circuit current	−20		−55	mA	5400 V_{CC} = Max.
	(note 3)	−18		−55	mA	7400 V_{CC} = Max.
I_{CCH}	Supply current HIGH		4.0	8.0	mA	V_{CC} = Max., V_{IN} = OV
I_{CCL}	Supply current LOW		12	22	mA	V_{CC} = Max., V_{IN} = 5.0 V

SWITCHING CHARACTERISTICS (T_A = 25°C)

Symbol	Parameter	Limits: Minimum	Limits: Typical	Limits: Maximum	Units	Test Conditions
t_{PLH}	Turn-off delay input to output		11	22	ns	V_{CC} = 5.0 V; C_L = 15 pF
t_{PHL}	Turn-on delay input to output		7.0	15	ns	R_L = 400 Ω

Notes:
1. For conditions shown as Min. or Max., use the appropriate value specified under recommended operating conditions for the applicable device type.
2. Typical limits are at V_{CC} = 5.0 V, 25° C.
3. Not more than one output should be shorted at a time.

FIGURE 3–45, continued

Example 3–19

Determine the LOW level and HIGH level noise margins for the 7400 gate using information from the data sheet.

Solution:

LOW level noise margin:

$$V_{NL} = V_{IL(\max)} - V_{OL(\max)}$$

From the data sheet:

$$V_{IL(\max)} = 0.8 \text{ V and } V_{OL(\max)} = 0.4 \text{ V}$$
$$V_{NL} = 0.8 \text{ V} - 0.4 \text{ V} = 0.4 \text{ V}$$

HIGH level noise margin:

$$V_{NH} = V_{OH(\min)} - V_{IH(\min)}$$

From the data sheet:

$$V_{OH(\min)} = 2.4 \text{ V and } V_{IH(\min)} = 2.0 \text{ V}$$
$$V_{NH} = 2.4 \text{ V} - 2.0 \text{ V} = 0.4 \text{ V}$$

3–11 LOGIC GATE APPLICATIONS

In this section we will look at two simple examples to illustrate how logic gates might be applied to practical situations.

AND Gate Application

In a simple application, an AND gate can be used to detect the existence of a specified number of conditions and, in response, to activate an appropriate action.

For example, an automobile's safety system may require that an audible signal be produced to warn the driver that the seat belt is not engaged. The conditions for this are that the ignition switch be on, the seat belt be unbuckled, and that the warning signal last for a specified time and then turn off automatically. The first two conditions can be represented by switch positions and the third by a timer circuit.

Figure 3–46 shows an AND gate whose HIGH output activates a buzzer when these three conditions are met on its inputs. When the ignition switch represented by S_1 in the figure is on, a HIGH is connected to the gate input *A*. When the belt is not properly buckled, switch S_2 is off and a HIGH is connected to the gate input *B*. At the instant the ignition switch is turned on, the timer is activated and produces a HIGH on gate input *C*. The resulting HIGH gate output activates the alarm. After a specified time the timer circuit's output goes LOW, disabling the AND gate and turning off the alarm.

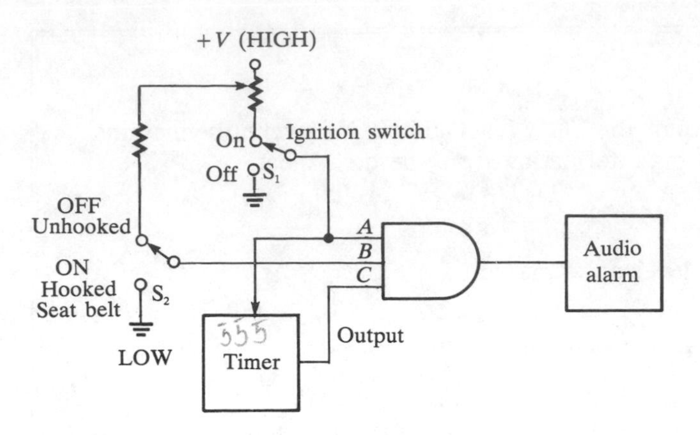

FIGURE 3–46 *Example of an AND gate application.*

OR Gate Application

As an example application of an OR gate, let us assume that in a room with three doors an indicator lamp must be turned on when any of the doors is not completely closed.

The sensors are switches that are open when a door is ajar or open. This open switch creates the HIGH level for the OR gate input as shown in Figure 3-47. If any or all of the doors are open, the gate output is HIGH. This HIGH level is then used to illuminate the indicator lamp.

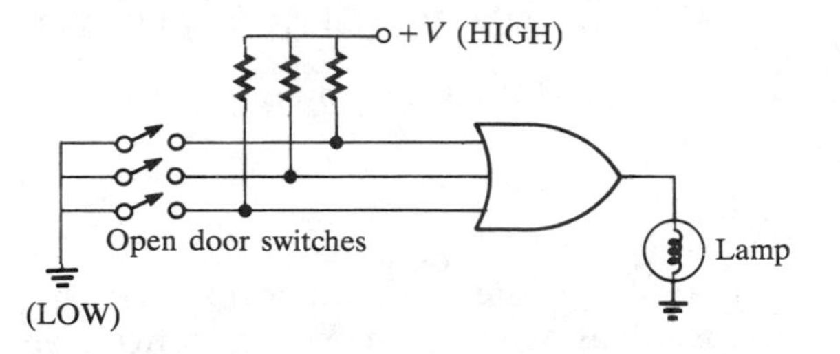

FIGURE 3–47 *Example of an OR gate application.*

3–12 TROUBLESHOOTING LOGIC GATES

Open gate inputs and outputs represent a large percentage of all digital IC failures. Various types of *shorts* account for the remaining failure modes. These include shorts to ground or supply voltage, shorts between traces on printed circuit boards, and shorted inputs and output due to internal gate failures.

Figure 3–48 illustrates an *open* failure of a gate input and how to check for it. Part (a) assumes one of the inputs of a 2-input NAND gate is open. Troubleshooting this type of failure can be accomplished with a *logic pulser* and *logic probe.* Start by pulsing one of the inputs and observing the output activity with the logic probe as in part (b). If activity is observed then that particular input and output are okay. Next, pulse the other input. No activity on the output indicates the input is open as shown in part (c).

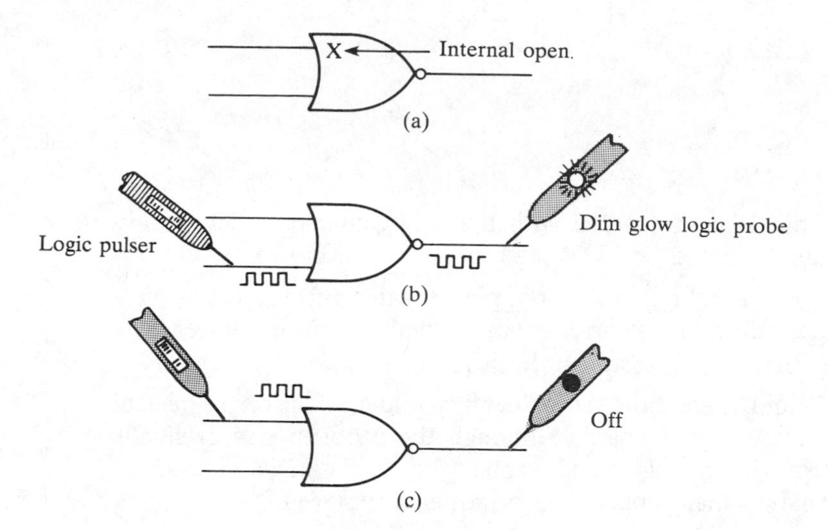

FIGURE 3–48 *Troubleshooting an open input.*

Figure 3–49 illustrates troubleshooting an open output. In part (a) one of the inputs is pulsed and no activity is observed on the output. In part (b), the other input is pulsed and, again, no activity is observed on the output. This test indicates that the gate output is open.

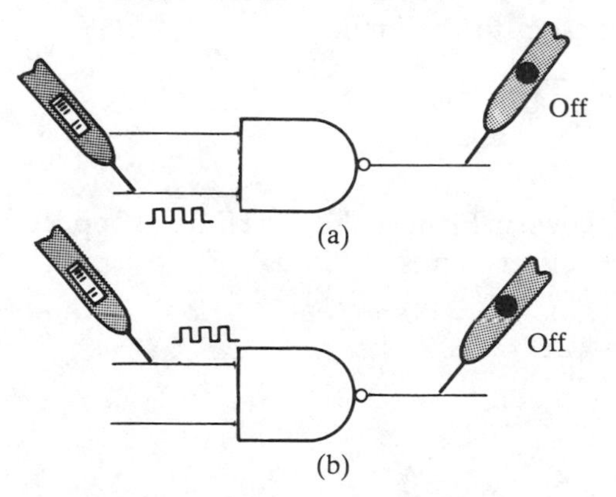

FIGURE 3–49 *Troubleshooting an open output.*

Example 3–20

The gate in Figure 3–50 is found to be defective by using a logic probe and pulser. Find out the nature of the fault (internal to gate or on pc board) before removing what appears to be an internally defective IC. Pin 11 is HIGH.

Example 3–20, continued

FIGURE 3–50

Solution:

(a) Pulse pin 12 and observe with the probe that pin 13 has pulse activity. It should be LOW due to HIGH on pin 11.

(b) Pulse pin 12 and read current at pin 13 with a current tracer. Now pulse pin 13 and read current at pin 12 with the current tracer. The same current is observed in both readings.

(c) Pins 12 and 13 are shorted together by a solder bridge on the back of the printed circuit board. Although the problem was originally located with a pulser and probe, the tracer added important information that kept the IC from being removed.

Problems

Section 3–1

3–1 The input waveform shown in Figure 3–51 is applied to an inverter. Sketch the output waveform in proper relationship to the input.

FIGURE 3–51

3–2 A network of cascaded inverters is shown in Figure 3–52. If a HIGH is applied to point *A*, determine the logic levels at points *B* through *F*.

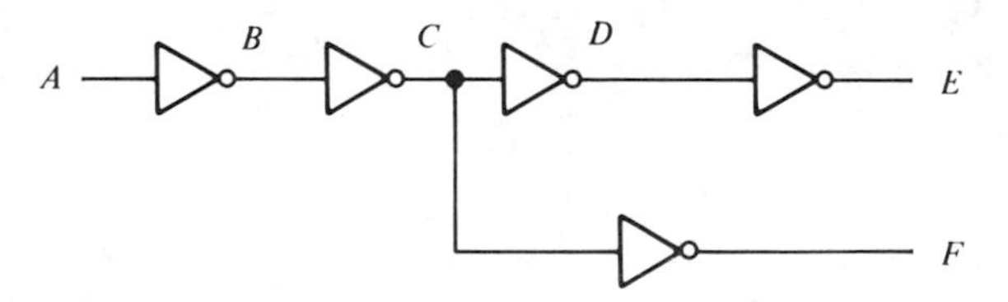

FIGURE 3–52

Section 3–2

3–3 Determine the output, *X*, for a two-input AND gate with the input waveforms shown in Figure 3–53.

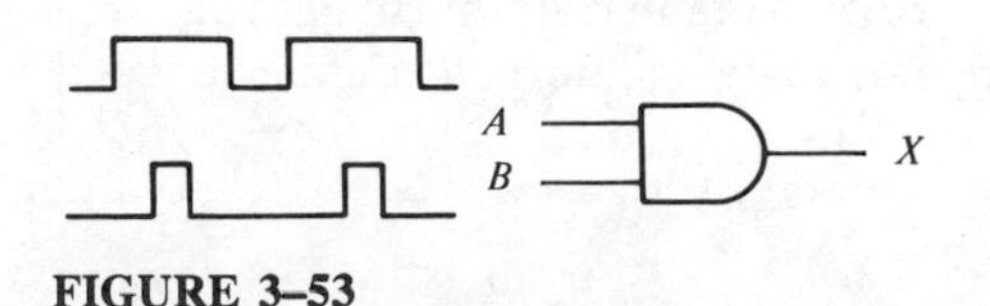

FIGURE 3–53

3–4 Repeat Problem 3–3 for the waveforms in Figure 3–54.

FIGURE 3–54

3–5 The input waveforms are applied to a three-input AND gate as indicated in Figure 3–55. Determine the output waveform in proper relationship to the inputs.

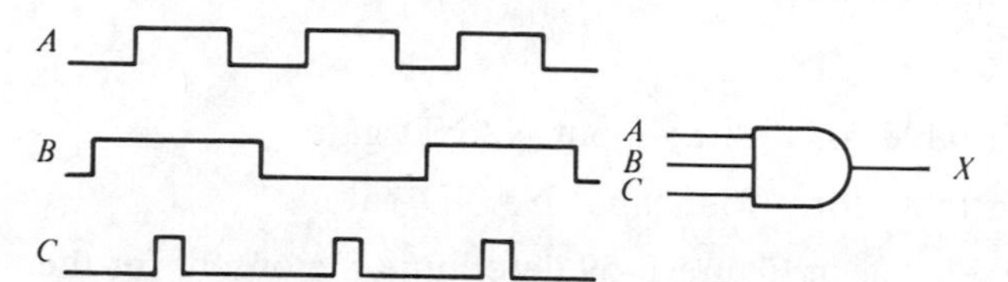

FIGURE 3–55

3–6 The input waveforms are applied to a four-input gate as indicated in Figure 3–56. Determine the output waveform in proper relation to the inputs.

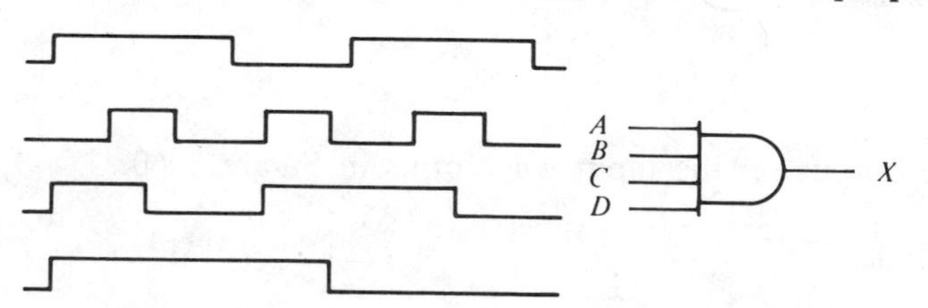

FIGURE 3–56

3–7 Determine a second input waveform that can be used with input A in Figure 3-57 to produce the specified output from a two-input AND gate. (Choose an input unlike the output.)

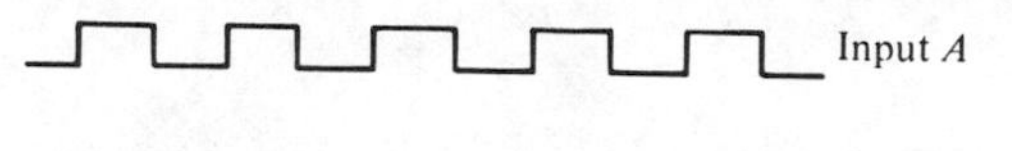

FIGURE 3–57

Section 3–3

3–8 Repeat Problem 3–7 for a two-input OR gate.

3–9 Determine the output for a two-input OR gate using the input waveforms in Problem 3–4.

3–10 Repeat Problem 3-5 for a three-input OR gate.

3–11 Repeat Problem 3-6 for a four-input OR gate.

3–12 For the five input waveforms in Figure 3–58, determine the output for an AND gate, and the output for an OR gate.

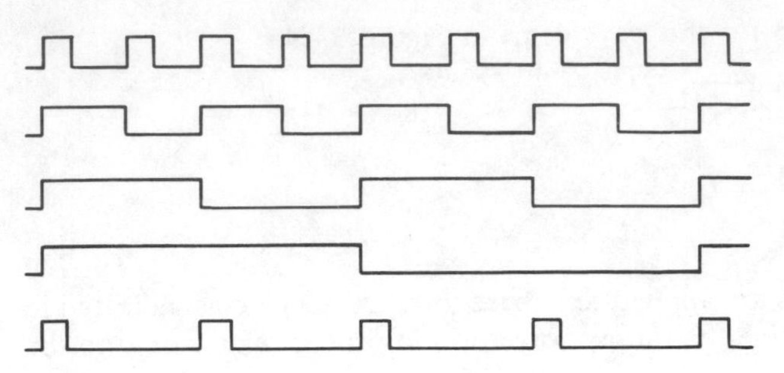

FIGURE 3–58

Section 3–4

3–13 Make a complete truth table for a three-input NAND gate.

3–14 Make a complete truth table for a four-input NAND gate.

3–15 For the set of input waveforms in Figure 3–59 determine the output for the gate shown.

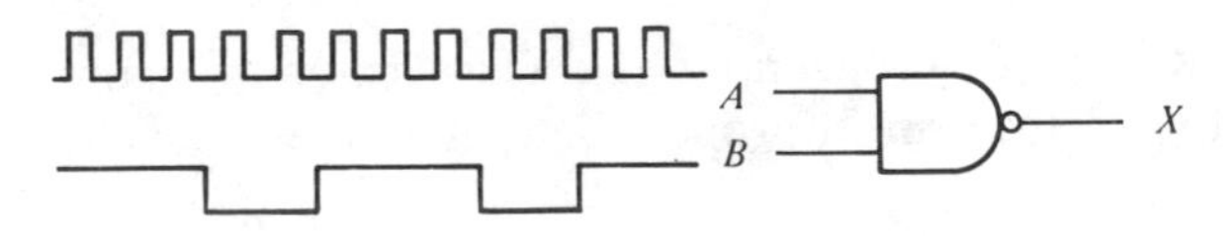

FIGURE 3–59

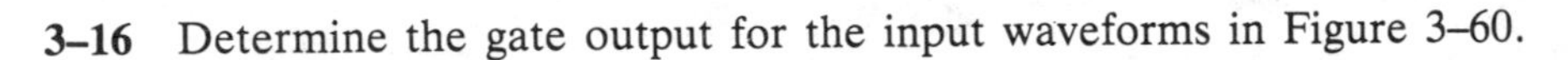

3–16 Determine the gate output for the input waveforms in Figure 3–60.

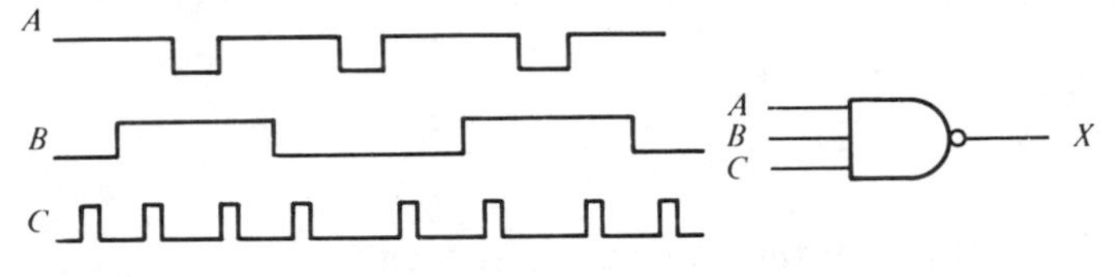

FIGURE 3–60

3–17 Determine the output waveform in Figure 3–61.

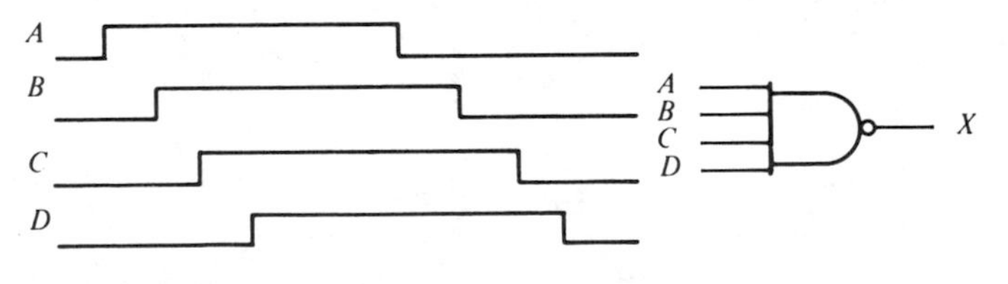

FIGURE 3–61

3–18 As we learned, the two logic symbols shown in Figure 3–62 represent the same circuit. The difference between the two is strictly a matter of how we look at them from a functional viewpoint. For the NAND symbol we are looking for two HIGHs on the inputs to give us a LOW output. For the negative-OR, we are looking for at least one LOW on the inputs to give us a HIGH on the output. Using these two functional viewpoints, show that each gate will produce the same output for the given inputs.

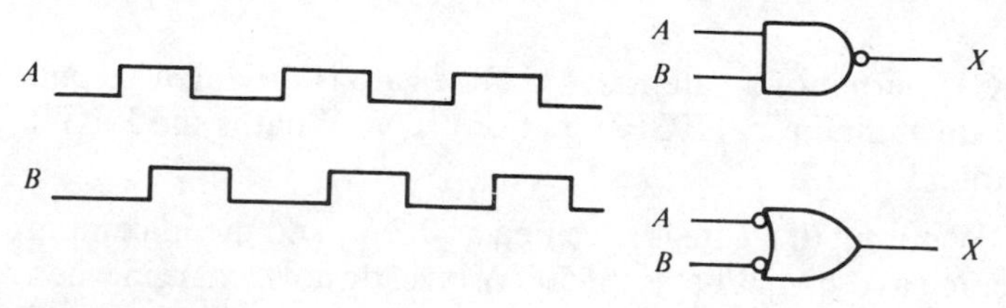

FIGURE 3–62

Section 3–5

3–19 Make a complete truth table for a three-input NOR gate.

3–20 Make a complete truth table for a four-input NOR gate.

3–21 Repeat Problem 3–15 for a two-input NOR gate.

3–22 Determine the output waveform in Figure 3–63.

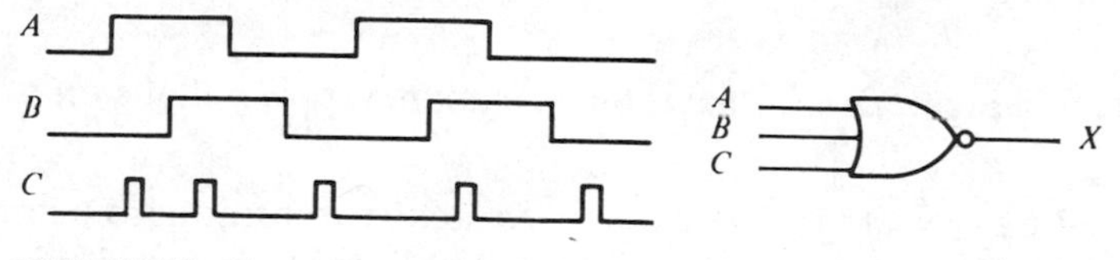

FIGURE 3–63

3–23 Repeat Problem 3–17 for a four-input NOR gate.

3–24 For the NOR symbol, we are looking for at least one HIGH on the inputs to give us a LOW on the output. For the negative-AND we are looking for two LOWs on the inputs to give us a HIGH output. Using these two functional points of view, show that both gates in Figure 3–64 will produce the same output for the given inputs.

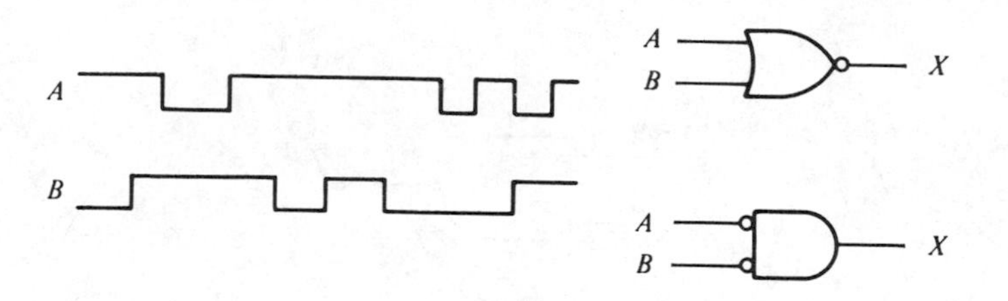

FIGURE 3–64

Section 3–7

3–25 If a logic gate operates on a dc supply voltage of +5 V and draws an average current of 4 mA, what is its power dissipation?

3–26 I_{CCH} is the dc supply current from V_{CC} when all outputs of an IC are HIGH. I_{CCL} is the dc supply current when all outputs are LOW. For a 7400 IC determine the *typical* power dissipation when all four gate outputs are HIGH.

3–27 Determine the typical 7400 power dissipation when two outputs are HIGH and two are LOW.

Section 3–8

3–28 The minimum HIGH output of a gate is 2.4 V. The gate is driving a second gate that can tolerate a minimum HIGH input of 2.1 V. What is the HIGH level dc noise margin?

3–29 The maximum LOW output of a gate is specified as 0.3 V, and the maximum LOW input is specified as 0.6 V. What is the LOW level dc noise margin when these gates work together?

Section 3–9

3–30 If a 7400 NAND gate is driving six unit loads, determine the amount of current it sinks in the LOW output state.

3–31 A 7400 NAND gate is driving ten unit loads. How much current must it source in the HIGH state?

Section 3–10

3–32 From the data sheet in Figure 3–45, what is the maximum supply voltage for a 5400 gate?

3–33 How long does it take a typical 54/74 gate output to make a transition from its LOW state to its HIGH state in response to an input? Refer to Figure 3–45.

3–34 Determine the maximum LOW output voltage from the data sheet in Figure 3-45.

Section 3–12

3–35 Using a logic probe and pulser, a technician makes the observations indicated in Figure 3–65. For each observation determine the most likely gate failure.

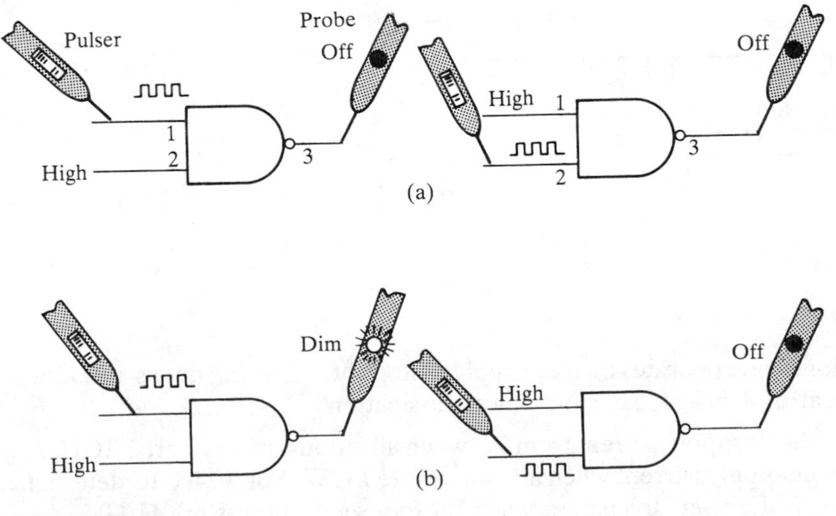

FIGURE 3–65

Boolean algebra is a set of rules, laws, and theorems by which logical operations can be expressed symbolically in equation form and manipulated mathematically. We are interested in its applications as a convenient and systematic way of expressing and analyzing the operation of digital circuits and systems. In 1854, George Boole published a classic book entitled *An Investigation of the Laws of Thought on Which are Founded the Mathematical Theories of Logic and Probabilities.* It was in this publication that a "logical algebra" that is commonly known today as Boolean algebra was developed.

The application of Boolean algebra to the analysis and design of digital logic circuits was first explored by Claude Shannon at MIT in a 1938 thesis entitled *A Symbolic Analysis of Relay and Switching Circuits.* Essentially, this paper described a method by which any circuit consisting of combinations of switches and relays could be represented by mathematical expressions. Today, of course, semiconductor circuits have for the most part replaced mechanical switches and relays. However, the same logical analysis is still valid, and a basic knowledge in this area is essential to the study of digital logic.

4

BOOLEAN ALGEBRA

4–1 BOOLEAN OPERATIONS

Symbology

In the applications of Boolean algebra in this book, we will use *capital letters* to represent variables and functions of variables. Any single variable or a function of several variables can have either a 1 or a 0 value. In Boolean algebra, the binary digits are utilized to represent the two levels that occur within digital logic circuits. A binary 1 will represent a HIGH level, and a binary 0 will represent a LOW level in Boolean equations. This is in keeping with our use in this text of positive logic as explained in Chapter 1.

The complement of a variable is represented by a "bar" over the letter. For instance, for a variable represented by A, the complement of A is $\overline{A}$. So if $A = 1$, then $\overline{A} = 0$; or if $A = 0$, then $\overline{A} = 1$. The complement of a variable A is usually read "A bar" or "Not A". Sometimes a prime symbol rather than the bar symbol is used to denote the complement. For example the complement of A can be written as A'.

The logical AND function of two variables is represented either by a "dot" between the two variables, such as $A \cdot B$, or by simply writing the adjacent letters without the "dot", such as AB. We will normally use the latter notation because it is easier to write. The logical OR function of two variables is represented by a "+" between the two variables, such as $A + B$.

Boolean Addition and Multiplication

Addition in Boolean algebra involves variables having values of either a binary 1 or a binary 0. The basic rules for Boolean addition are listed below.

$$0 + 0 = 0$$
$$0 + 1 = 1$$
$$1 + 0 = 1$$
$$1 + 1 = 1$$

In the application of Boolean algebra to logic circuits, *Boolean addition is the same as the OR*. Notice that it differs from binary addition in the case where two 1s are added.

Multiplication in Boolean algebra follows the same basic rules governing binary multiplication which were discussed in Chapter 2 and which are listed below:

$$0 \cdot 0 = 0$$
$$0 \cdot 1 = 0$$
$$1 \cdot 0 = 0$$
$$1 \cdot 1 = 1$$

Boolean multiplication is the same as the AND.

LOGIC EXPRESSIONS 4–2

NOT

The operation of an inverter (NOT circuit) can be expressed with symbols as follows: if the input variable is called A and the output variable is called X, then $X = \overline{A}$. This expression "says" that the output is the complement of the input, so that if $A = 0$, then $X = 1$, and if $A = 1$, then $X = 0$. Figure 4–1 illustrates this.

A — $X = \overline{A}$

FIGURE 4–1

AND

The operation of a two-input AND gate can be expressed in equation form as follows: if one input variable is A, the other input variable is B, and the output variable is X, then the Boolean expression for this basic gate function is $X = AB$. Figure 4–2(a) shows the gate with the input and output variables indicated.

To extend the AND expression to more than two input variables, we simply use a new letter for each input variable. The function of a three-input AND gate, for example, can be expressed as $X = ABC$, where A, B, and C are the input variables. The expression for a four-input AND gate can be $X = ABCD$, and so on. Figures 4–2(b) and 4–2(c) show AND gates with three and four input variables, respectively.

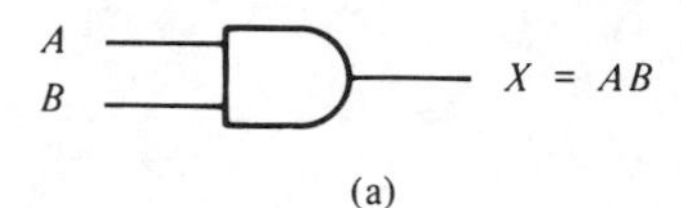

(a)

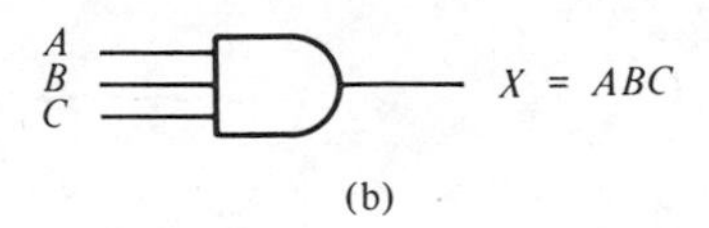

(b)

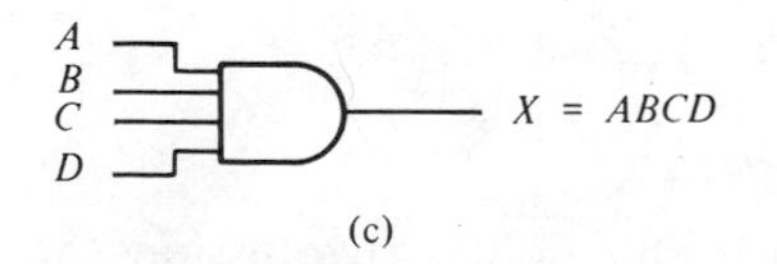

(c)

FIGURE 4–2

An evaluation of AND gate operation can be made using the Boolean expressions for the output. For example, each variable on the inputs can be either a 1 or a 0, so for the two-input AND gate we can make the following substitutions in the equation for the output X:

$$A = 0, B = 0: X = AB = 0 \cdot 0 = 0$$
$$A = 0, B = 1: X = AB = 0 \cdot 1 = 0$$
$$A = 1, B = 0: X = AB = 1 \cdot 0 = 0$$
$$A = 1, B = 1: X = AB = 1 \cdot 1 = 1$$

The evaluation of this equation simply tells us that the output X of an AND gate is a 1 (HIGH) only when both inputs are 1s (HIGHs). A similar analysis can be made for any number of input variables.

OR

The operation of a two-input OR gate can be expressed in equation form as follows: if one input is A, the other input B, and the output is X, then the Boolean expression is $X = A + B$. Figure 4–3(a) shows the gate logic symbol, with input and output variables labeled.

To extend the OR expression to more than two input variables, a new letter is used for each additional variable. For instance, the function of a three-input OR gate can be expressed as $X = A + B + C$. The expression for a four-input OR gate can be written as $X = A + B + C + D$, and so on. Figures 4–3(b) and 4–3(c) show OR gate logic symbols with three and four input variables, respectively.

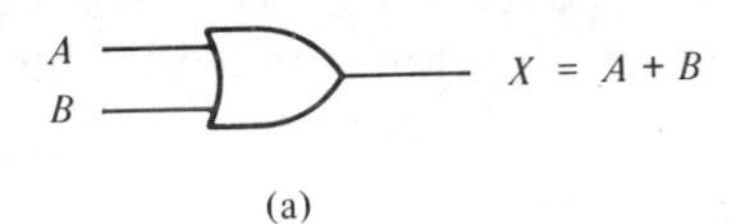

(a)

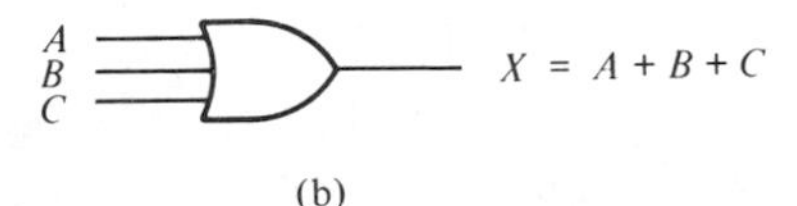

(b)

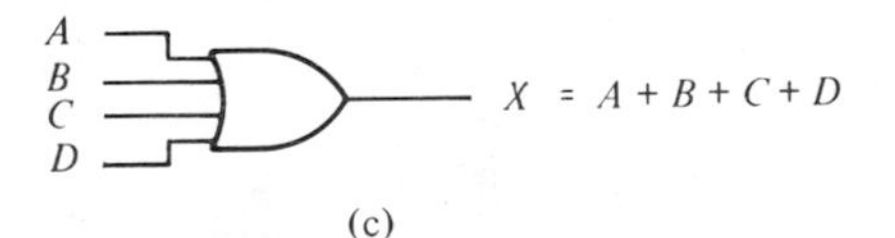

(c)

FIGURE 4–3

OR gate operation can be evaluated using the Boolean expressions for the output X by substituting all possible combinations of 1 and 0 values for the input variables, as shown below for a two-input OR gate:

$$A = 0, B = 0: X = A + B = 0 + 0 = 0$$
$$A = 0, B = 1: X = A + B = 0 + 1 = 1$$
$$A = 1, B = 0: X = A + B = 1 + 0 = 1$$
$$A = 1, B = 1: X = A + B = 1 + 1 = 1$$

This evaluation shows that the output of an OR gate is a 1 (HIGH) when any one or more of the inputs are 1 (HIGH). A similar analysis can be extended to OR gates with any number of input variables.

NAND

The Boolean expression for a two-input NAND gate is $X = \overline{AB}$. This expression says that the two input variables, A and B, are first ANDed and then complemented, as indicated by the "bar" over the AND expression. This is a logical description in equation form of the operation of a NAND gate with two inputs. If we evaluate this expression for all possible values of the two input variables, the results are as follows:

$$A = 0, \quad B = 0: \quad X = \overline{A \cdot B} = \overline{0 \cdot 0} = \overline{0} = 1$$
$$A = 0, \quad B = 1: \quad X = \overline{A \cdot B} = \overline{0 \cdot 1} = \overline{0} = 1$$
$$A = 1, \quad B = 0: \quad X = \overline{A \cdot B} = \overline{1 \cdot 0} = \overline{0} = 1$$
$$A = 1, \quad B = 1: \quad X = \overline{A \cdot B} = \overline{1 \cdot 1} = \overline{1} = 0$$

So, you see that once a Boolean expression is determined for a given logic function, then that function can be evaluated for all possible values of the variables. The evaluation tells us exactly what the output of the logic circuit is for each of the input conditions, and therefore gives us a complete description of the circuit's logical operation. The NAND expression can be extended to more than two input variables by including additional letters to represent all of the variables.

NOR

Finally, the expression for a two-input NOR gate can be written as $X = \overline{A + B}$. This equation says that the two input variables are first ORed and then complemented, as indicated by the "bar" over the OR expression. Evaluating this expression, we get the following results:

$$A = 0, \quad B = 0: \quad X = \overline{A + B} = \overline{0 + 0} = \overline{0} = 1$$
$$A = 0, \quad B = 1: \quad X = \overline{A + B} = \overline{0 + 1} = \overline{1} = 0$$
$$A = 1, \quad B = 0: \quad X = \overline{A + B} = \overline{1 + 0} = \overline{1} = 0$$
$$A = 1, \quad B = 1: \quad X = \overline{A + B} = \overline{1 + 1} = \overline{1} = 0$$

RULES AND LAWS OF BOOLEAN ALGEBRA 4–3

As in other areas of mathematics, there are certain well-developed rules and laws that must be followed in order to properly apply Boolean algebra. The most important of these are presented in this section.

Three of the basic laws of Boolean algebra are the same as in ordinary algebra: the *commutative laws,* the *associative laws,* and the *distributive laws.*

Commutative Laws

The commutative law of addition for two variables is written algebraically as

$$A + B = B + A \qquad \textbf{(4–1)}$$

This states that the order in which the variables are ORed makes no difference. Remember, in Boolean algebra terminology as applied to logic circuits, addition and the OR function are the same. Figure 4–4 illustrates the commutative law as applied to the OR gate.

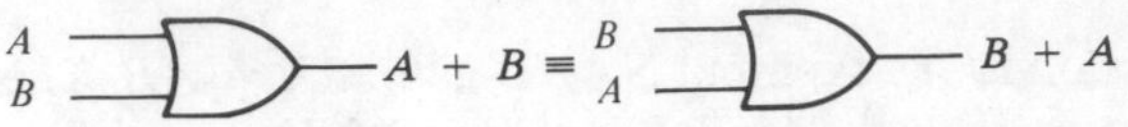

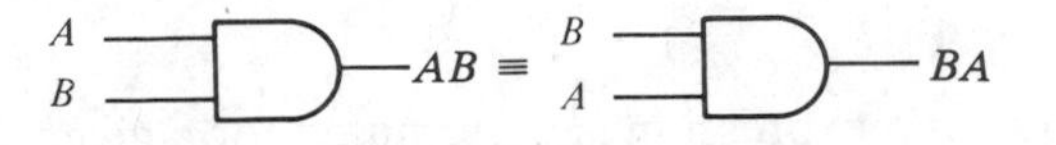

FIGURE 4–4 *Application of commutative law of addition.*

The commutative law of multiplication of two variables is

$$AB = BA \tag{4–2}$$

This states that the order in which the variables are ANDed makes no difference. Figure 4–5 illustrates this law as applied to the AND gate.

A
B
AB ≡
B
A
BA

FIGURE 4–5 *Application of commutative law of multiplication.*

Associative Laws

The associative law of addition is stated as follows for three variables:

$$A + (B + C) = (A + B) + C \tag{4–3}$$

This law states that in the ORing of several variables, the result is the same regardless of the grouping of the variables. Figure 4–6 illustrates this law as applied to OR gates.

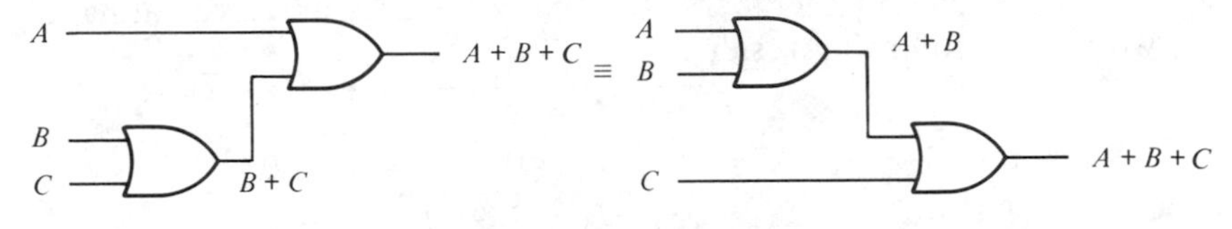

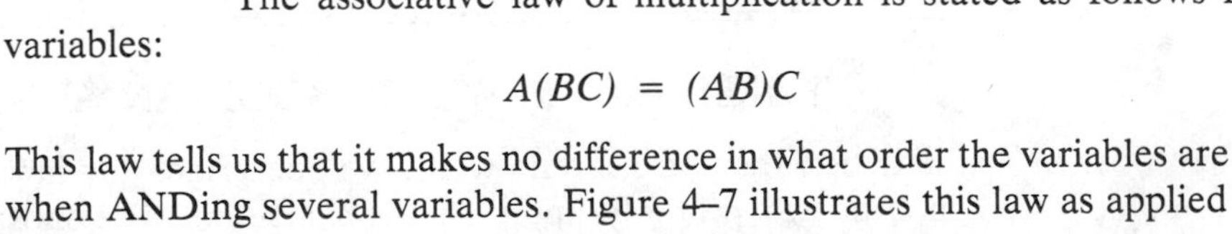

FIGURE 4–6 *Application of associative law of addition.*

The associative law of multiplication is stated as follows for three variables:

$$A(BC) = (AB)C \tag{4–4}$$

This law tells us that it makes no difference in what order the variables are grouped when ANDing several variables. Figure 4–7 illustrates this law as applied to AND gates.

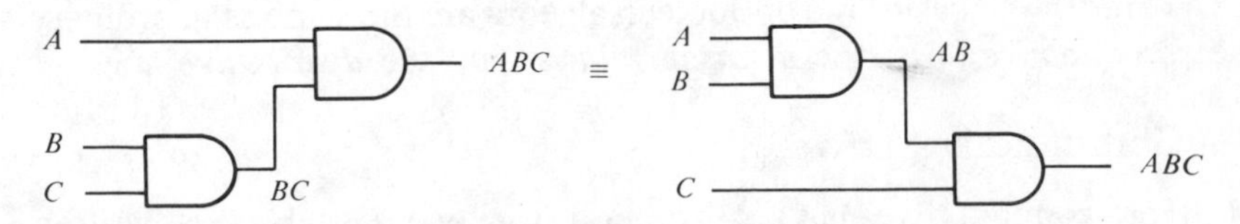

FIGURE 4–7 *Application of associative law of multiplication.*

Distributive Law

The distributive law is written for three variables as follows:

$$A(B + C) = AB + AC \tag{4–5}$$

This law states that ORing several variables and ANDing the result with a single variable is equivalent to ANDing the single variable with each of the several variables and then ORing the products. This law and the ones previously discussed should be familiar because they are the same as in ordinary algebra. Keep in mind that each of these laws can be extended to include any number of variables. Figure 4–8 illustrates this law in terms of gate implementation.

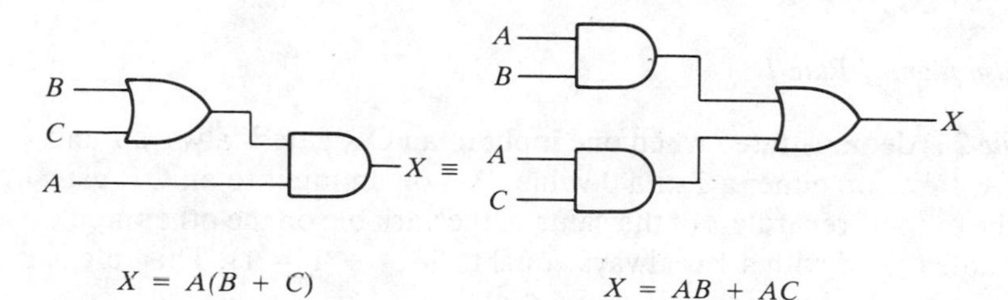

FIGURE 4–8 *Application of distributive law.*

Rules for Boolean Algebra

Table 4–1 lists several basic rules that are useful in manipulating and simplifying Boolean algebra expressions.

TABLE 4–1 *Basic rules of Boolean algebra.*

Rule
1. $A + 0 = A$
2. $A + 1 = 1$
3. $A \cdot 0 = 0$
4. $A \cdot 1 = A$
5. $A + A = A$
6. $A + \overline{A} = 1$
7. $A \cdot A = A$
8. $A \cdot \overline{A} = 0$
9. $\overline{\overline{A}} = A$
10. $A + AB = A$
11. $A + \overline{A}B = A + B$
12. $(A + B)(A + C) = A + BC$

NOTE: A can represent a single variable or combination of variables.

We will now look at rules 1 through 9 of Table 4–1 in terms of their application to logic gates. Rules 10 through 12 will be derived in terms of the simpler rules and the laws previously discussed.

Rule 1 can be understood by observing what happens when one input to an OR gate is always 0 and the other input, A, can take on either a 1 or a 0 value. If

A is a 1, it is obvious that the output is a 1, which is equal to A. If A is a 0, the output is a 0, which is also equal to A. Therefore, it follows that a variable ORed with a 0 is equal to the value of the variable $(A + 0 = A)$. This rule is further demonstrated in Figure 4–9 where input B is fixed at 0.

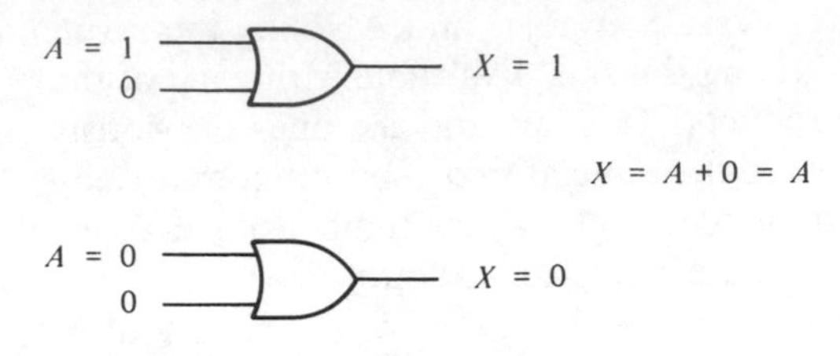

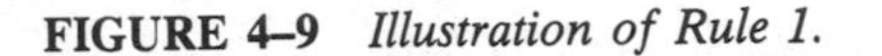
FIGURE 4–9 *Illustration of Rule 1.*

Rule 2 is demonstrated when one input to an OR gate is always 1 and the other input, A, takes on either a 1 or a 0 value. A 1 on an input to an OR gate produces a 1 on the output, regardless of the value of the variable on the other input. Therefore, a variable ORed with a 1 is always equal to 1 $(A + 1 = 1)$. This rule is illustrated in Figure 4–10 where input B is fixed at 1.

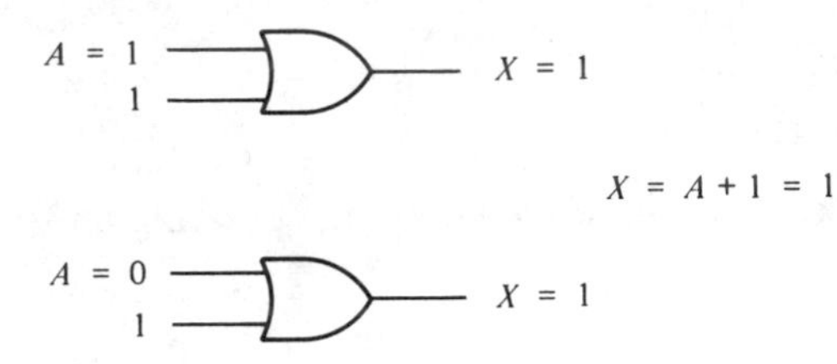

FIGURE 4–10 *Illustration of Rule 2.*

Rule 3 is demonstrated when a 0 is ANDed with a variable. Of course, any time one input to an AND gate is 0, the output is 0, regardless of the value of the variable on the other input. A variable ANDed with a 0 always produces a 0 $(A \cdot 0 = 0)$. This rule is illustrated in Figure 4–11 where input B is fixed at 0.

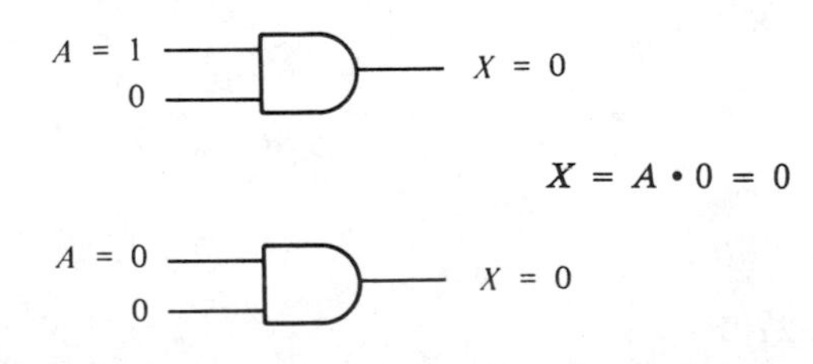

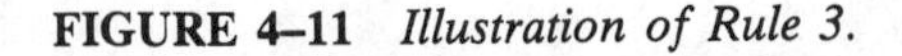
FIGURE 4–11 *Illustration of Rule 3.*

Rule 4 can be verified by ANDing a variable with a 1. If the variable A is a 0, the output of the AND gate is a 0. If the variable A is a 1, the output of the AND gate is a 1 because both inputs are now 1s. Therefore, the AND function of a variable and a 1 is equal to the value of the variable $(A \cdot 1 = A)$. This is shown in Figure 4–12 where input B is fixed at 1.

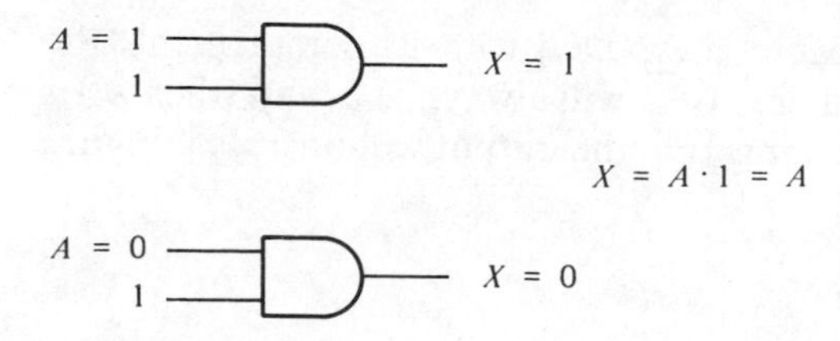

FIGURE 4–12 *Illustration of Rule 4.*

Rule 5 states that if a variable is ORed with itself, the output is equal to the variable. For instance, if A is a 0, then $0 + 0 = 0$, and if A is a 1 then $1 + 1 = 1$. This is shown in relation to OR gate application in Figure 4–13 where both inputs are the same variable.

A = 1, A = 1 → X = 1

$$X = A + A = A$$

A = 0, A = 0 → X = 0

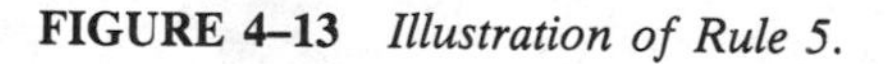

FIGURE 4–13 *Illustration of Rule 5.*

Rule 6 can be explained as follows: if a variable and its complement are ORed, the result is always a 1. If A is a 0, then $0 + \overline{0} = 0 + 1 = 1$. If A is a 1, then $1 + \overline{1} = 1 + 0 = 1$. See Figure 4–14 for further illustration of this rule where one input is the complement of the other.

A = 1, $\overline{A}$ = 0 → X = 1

$$X = A + \overline{A} = 1$$

A = 0, $\overline{A}$ = 1 → X = 1

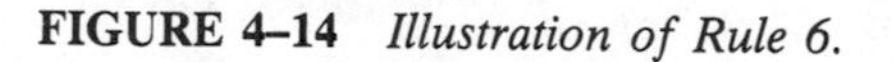

FIGURE 4–14 *Illustration of Rule 6.*

Rule 7 states that if a variable is ANDed with itself, the result is equal to the variable. For example, if $A = 0$, then $0 \cdot 0 = 0$, and if $A = 1$, then $1 \cdot 1 = 1$. For either case, the output of an AND gate is equal to the value of the input variable A. Figure 4–15 illustrates this rule.

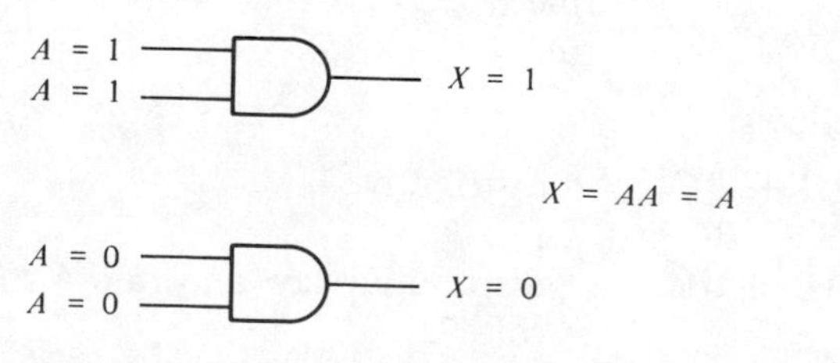

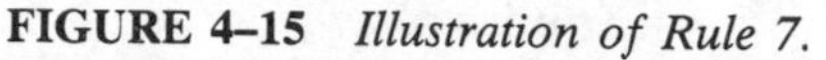

FIGURE 4–15 *Illustration of Rule 7.*

Rule 8 states that if a variable is ANDed with its complement, the result is 0. This is readily seen because either A or $\overline{A}$ will always be 0, and when a 0 is applied to the input of an AND gate, it insures that the output will be 0 also. Figure 4–16 will help illustrate this rule.

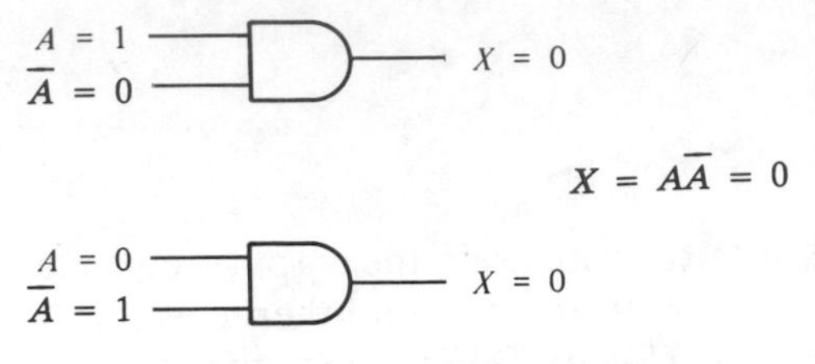

FIGURE 4–16 *Illustration of Rule 8.*

Rule 9 simply says that if a variable is complemented twice, the result is the variable itself. If we start with the variable A and complement (invert) it once, we get $\overline{A}$. If we then take $\overline{A}$ and complement (invert) it, we get A, which is the original variable. This is shown in Figure 4–17, using inverters.

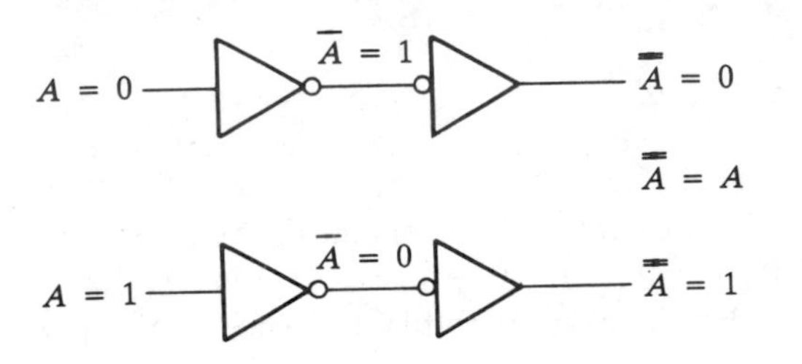

FIGURE 4–17 *Illustration of Rule 9.*

Rule 10 is proved by using the distributive law, rule 2, and rule 4 as follows:

$$\begin{aligned} A + AB &= A(1 + B) && \text{distributive law} \\ &= A \cdot 1 && \text{rule 2} \\ &= A && \text{rule 4} \end{aligned}$$

Rule 11 is proved as follows:

$$\begin{aligned} A + \overline{A}B &= (A + AB) + \overline{A}B && \text{rule 10} \\ &= (AA + AB) + \overline{A}B && \text{rule 7} \\ &= AA + AB + A\overline{A} + \overline{A}B && \text{rule 8 (adding } A\overline{A} = 0) \\ &= (A + \overline{A})(A + B) && \text{by factoring} \\ &= 1 \cdot (A + B) && \text{rule 6} \\ &= A + B && \text{rule 4} \end{aligned}$$

Rule 12 is proved as follows:

$$\begin{aligned} (A + B)(A + C) &= AA + AC + AB + BC && \text{distributive law} \\ &= A + AC + AB + BC && \text{rule 7} \\ &= A(1 + C) + AB + BC && \text{distributive law and rule 4} \\ &= A \cdot 1 + AB + BC && \text{rule 2} \\ &= A(1 + B) + BC && \text{distributive law} \\ &= A \cdot 1 + BC && \text{rule 2} \\ &= A + BC && \text{rule 4} \end{aligned}$$

4–4 DEMORGAN'S THEOREMS

DeMorgan, a logician and mathematician who was acquainted with Boole, proposed two theorems that are an important part of Boolean algebra. They are stated as follows in equation form:

$$\overline{AB} = \overline{A} + \overline{B} \tag{4–6}$$

$$\overline{A + B} = \overline{A}\,\overline{B} \tag{4–7}$$

The theorem expressed in Equation (4–6) can be stated as *The complement of a product is equal to the sum of the complements.* This really says that the complement of two or more variables ANDed is the same as the OR of the complements of each individual variable.

The theorem expressed in Equation (4–7) can be stated as *The complement of a sum is equal to the product of the complements.* This says that the complement of two or more variables ORed is the same as the AND of the complements of each individual variable.

These theorems are illustrated by the gate equivalencies in Figure 4–18.

A	B	$\overline{AB}$	$\overline{A} + \overline{B}$
0	0	1	1
0	1	1	1
1	0	1	1
1	1	0	0

A	B	$\overline{A + B}$	$\overline{A}\,\overline{B}$
0	0	1	1
0	1	0	0
1	0	0	0
1	1	0	0

FIGURE 4–18 *Gate equivalencies illustrating DeMorgan's theorems.*

Example 4–1

Express the following complement-of-product terms as sum-of-complements using DeMorgan's theorem:

$$\overline{ABC}, \quad \overline{ABCD}, \quad \overline{ABCDEF}$$

Solutions:

Use DeMorgan's theorem as expressed in Equation (4–6).

$$\overline{ABC} = \overline{A} + \overline{B} + \overline{C}$$

$$\overline{ABCD} = \overline{A} + \overline{B} + \overline{C} + \overline{D}$$

$$\overline{ABCDEF} = \overline{A} + \overline{B} + \overline{C} + \overline{D} + \overline{E} + \overline{F}$$

Example 4–2

Express the following complement-of-sum terms as product-of-complements using DeMorgan's theorem:

$$\overline{A + B + C}$$
$$\overline{A + B + C + D}$$
$$\overline{A + B + C + D + E + F}$$

Solutions:

Use DeMorgan's theorem as expressed in Equation (4–7).

$$\overline{A + B + C} = \overline{A}\,\overline{B}\,\overline{C}$$
$$\overline{A + B + C + D} = \overline{A}\,\overline{B}\,\overline{C}\,\overline{D}$$
$$\overline{A + B + C + D + E + F} = \overline{A}\,\overline{B}\,\overline{C}\,\overline{D}\,\overline{E}\,\overline{F}$$

4–5 BOOLEAN EXPRESSIONS FOR GATE NETWORKS

The form of a given Boolean (logic) expression indicates the type of gate network it describes. For example, let us take the expression $A(B + CD)$ and determine what sort of logic circuit it represents. First, there are four variables: A, B, C, and D. C is ANDed with D, giving CD; then CD is ORed with B, giving $(B + CD)$. Then this is ANDed with A to produce the final function. Figure 4-19 illustrates the gate network represented by this particular Boolean expression $A(B + CD)$.

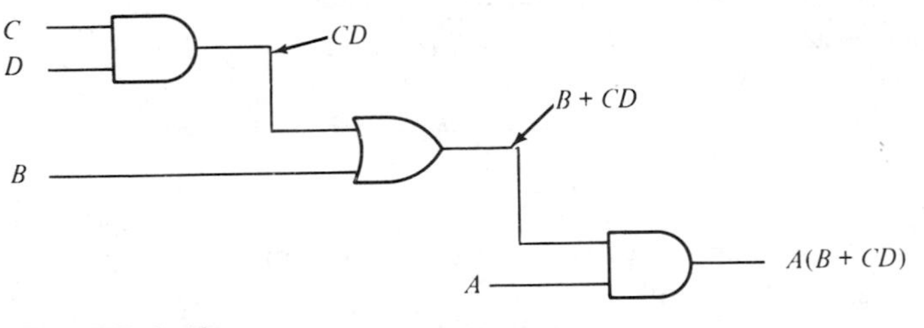

FIGURE 4–19

As you have seen, the form of the Boolean expression does determine how many logic gates are used, what type of gates are needed, and how they are connected together; this will be explored further in Chapter 5. The more complex an expression, the more complex the gate network will be. It is therefore an advantage to simplify an expression as much as possible in order to have the simplest gate network. We will cover simplification methods in a later section of this chapter. There are also certain forms of Boolean expressions that are more commonly used than others; the two most important of these are the *sum-of-products* and the *product-of-sums* forms.

Sum-of-Products Form

What does the *sum-of-products* form mean? First, let us review products in Boolean algebra. A product of two or more variables or their complements is simply the AND function of those variables. The product of two variables can be expressed as AB, the product of three variables as ABC, the product of four variables as $ABCD$, etc. Recall that a sum in Boolean algebra is the same as the OR function, so *a sum-of-products expression is two or more AND functions ORed together.* For instance, $AB + CD$ is a sum-of-products expression. Several other examples of expressions in sum-of-product form are as follows:

$$AB + BCD$$
$$ABC + DEF$$
$$A\overline{B}C + D\overline{E}FG + AEG$$
$$AB\overline{C} + \overline{D}EF + FGH + A\overline{F}G$$

A sum-of-products form can also contain a term with a single variable, such as $A + BCD + EFG$.

One reason the sum-of-products is a useful form of Boolean expression is the straightforward manner in which it can be implemented with logic gates. We have AND functions that are ORed, as Example 4–3 illustrates.

Example 4–3

Implement the expression $AB + BCD + EFGH$ with logic gates.

Solution:

See Figure 4–20.

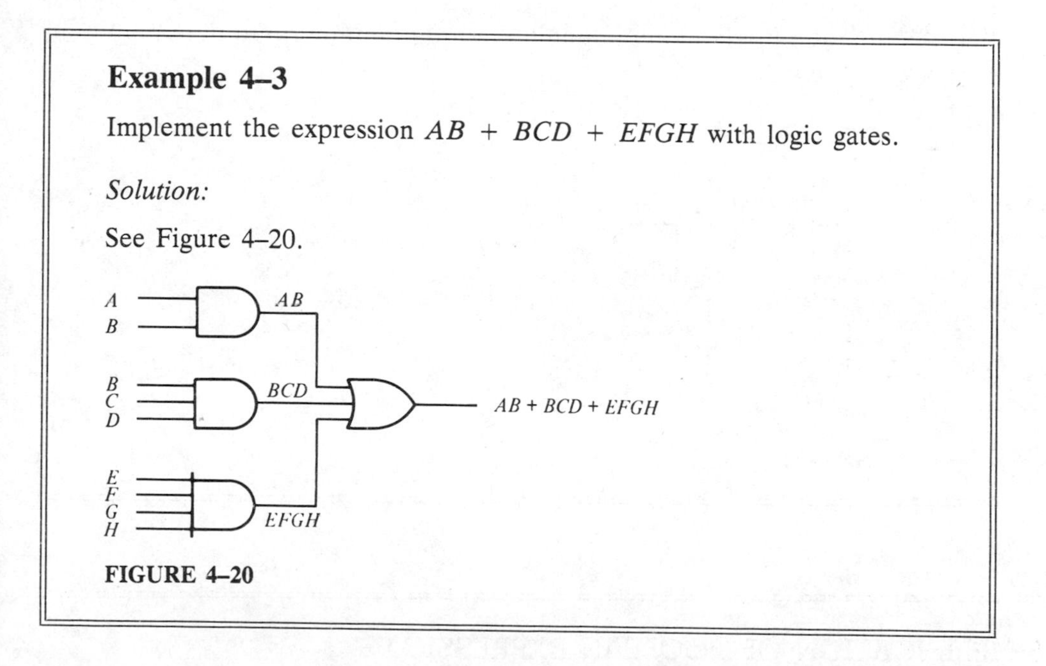

FIGURE 4–20

An important characteristic of the sum-of-products form is that the corresponding implementation is always a *two-level* gate network; that is, the maximum number of gates through which a signal must pass in going from an input to the output is *two*, excluding inversions.

Product-of-Sums Form

The product-of-sums form can be thought of as the dual of the sum-of-products. It is, in terms of logic functions, the AND of two or more OR functions. For instance, $(A + B)(B + C)$ is a product-of-sums expression. Several other examples are:

$$(A + B)(B + C + D)$$
$$(A + B + C)(D + E + F)$$
$$(A + B + \overline{C})(\overline{D} + E + F + G)(A + E + G)$$
$$(A + B + \overline{C})(\overline{D} + E + F)(F + G + H)(A + \overline{F} + G)$$

A product-of-sums expression can also contain a single variable term such as $A(B + C + D)(E + F + G)$.

This form also lends itself to straightforward implementation with logic gates because it simply involves ANDing two or more OR terms. A two-level gate network will always result, as the following example will show:

Example 4–4

Construct the following function with logic gates:

$$(A + B)(C + D + E)(F + G + H + I)$$

Solution:

See Figure 4–21.

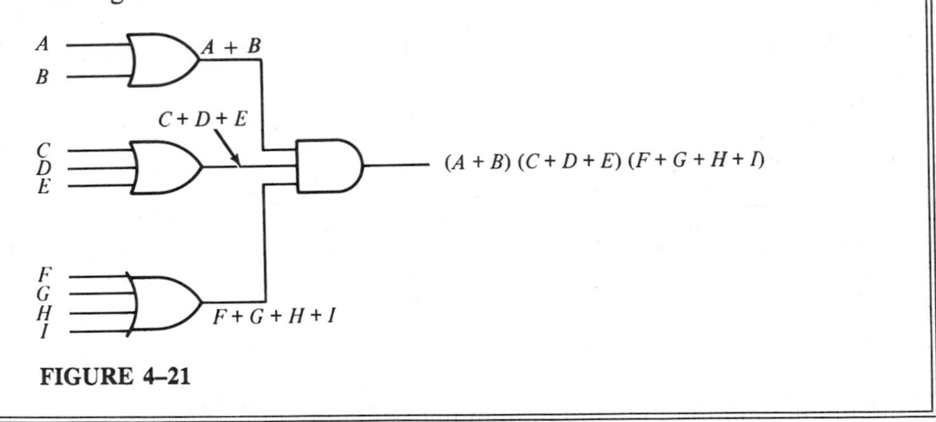

FIGURE 4–21

4–6 SIMPLIFICATION OF BOOLEAN EXPRESSIONS

Many times in the application of Boolean algebra, we have to reduce a particular expression to its simplest form or change its form to a more convenient one in order to implement the expression most efficiently. The approach taken in this section is to use the basic laws, rules, and theorems of Boolean algebra to manipulate and simplify an expression. This method depends on a thorough knowledge of Boolean algebra, and considerable practice in its application. Examples will serve to illustrate the technique.

Example 4–5

Simplify the expression $AB + A(B + C) + B(B + C)$ using Boolean algebra techniques.

Solution:

(This is not necessarily the only approach.)

Step 1. Apply the distributive law to the *second* and *third* terms in the expression, as follows:

$$AB + AB + AC + BB + BC$$

Step 2. Apply rule 7 $(BB = B)$:

$$AB + AB + AC + B + BC$$

Step 3. Apply rule 5 $(AB + AB = AB)$:

$$AB + AC + B + BC$$

Step 4. Factor B out of the *last two* terms, as follows:

$$AB + AC + B(1 + C)$$

Step 5. Apply the commutative law and rule 2 $(1 + C = 1)$:

$$AB + AC + B \cdot 1$$

Step 6. Apply rule 4 $(B \cdot 1 = B)$:

$$AB + AC + B$$

Step 7. Factor B out of the *first* and *third* terms, as follows:

$$B(A + 1) + AC$$

Step 8. Apply rule 2 $(A + 1 = 1)$:

$$B \cdot 1 + AC$$

Step 9. Apply rule 4 $(B \cdot 1 = B)$:

$$B + AC$$

At this point we have simplified the expression as much as possible. It should be noted that once you gain experience in applying Boolean algebra, many individual steps can be combined. See if you can find an alternate approach.

Example 4–6

Simplify the expression $[A\overline{B}(C + BD) + \overline{A}\overline{B}]C$ as much as possible. Note that brackets and parentheses mean the same thing: simply that the terms are multiplied (ANDed) with each other.

Solution:

Step 1. Apply the distributive law (multiply out) to the terms within the brackets:

$$(A\overline{B}C + A\overline{B}BD + \overline{A}\overline{B})C$$

Step 2. Apply rule 8 to the second term in the parentheses:

$$(A\overline{B}C + A \cdot 0 \cdot D + \overline{A}\overline{B})C$$

Step 3. Apply rule 3 to the second term:

$$(A\overline{B}C + 0 + \overline{A}\overline{B})C$$

Step 4. Apply rule 1 within the parentheses:

$$(A\overline{B}C + \overline{A}\overline{B})C$$

Step 5. Apply the distributive law:

$$A\overline{B}CC + \overline{A}\overline{B}C$$

Step 6. Apply rule 7 to the first term:

$$A\overline{B}C + \overline{A}\overline{B}C$$

Step 7. Factor out $\overline{B}C$:

$$\overline{B}C(A + \overline{A})$$

Step 8. Apply rule 6:

$$\overline{B}C \cdot 1$$

Step 9. Apply rule 4:

$$\overline{B}C$$

4–7 KARNAUGH MAP SIMPLIFICATION

Another approach for reducing a Boolean expression to its simplest or minimum form is known as the *Karnaugh map* method, named after its originator. The method is systematic and easily applied. When properly used, it will *always* result in the minimum expression possible.

The effectiveness of algebraic simplification as discussed in Section 4–6 is dependent on your familiarity with the rules, laws, and theorems of Boolean algebra and on your ability and ingenuity in applying them. The Karnaugh map approach, therefore, has a distinct advantage, especially for more complex expressions where algebraic simplification is not immediately obvious or is extremely involved.

A Karnaugh map is composed of a number of adjacent "cells." Each cell represents one particular combination of variable values. Since the total number of possible combinations of n variables is 2^n, there must be 2^n cells in the Karnaugh map. For instance, for a function of two variables, the Karnaugh map has $2^2 = 4$ cells, and is constructed as shown in Figure 4–22.

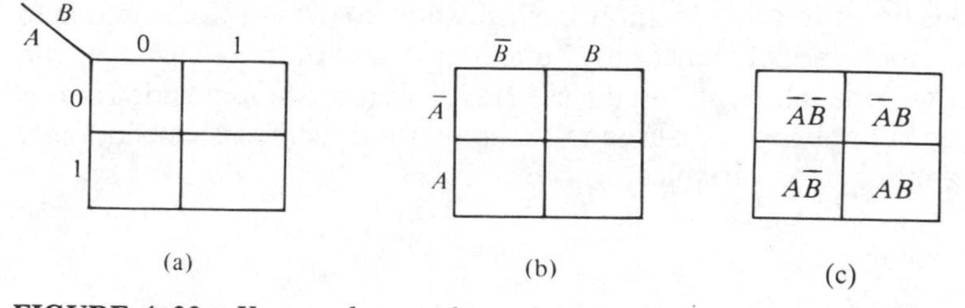

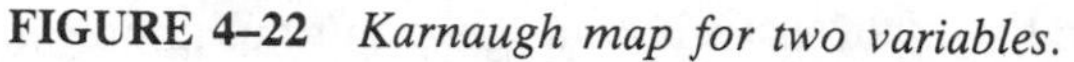

FIGURE 4–22 *Karnaugh map for two variables.*

The variables are A and B, and the labeling of the map is as follows for Figure 4–22(a): the upper left cell is for the case where $A = 0$ and $B = 0$; the upper right cell is for $A = 0$ and $B = 1$; the lower left cell is for $A = 1$ and $B = 0$; and the lower right cell is for $A = 1$ and $B = 1$. An alternate way of drawing the map is shown in Figure 4–22(b), and is essentially the same as that in (a) except that the cells are labeled with variable terms rather than 1 and 0 combinations. The upper left cell corresponds to the combination of variables $\bar{A}\bar{B}$, the upper right cells corresponds to $\bar{A}B$, and so on as shown in Figure 4–22(c).

Now consider a Karnaugh map for a three-variable function. There must be $2^3 = 8$ cells, as shown in Figure 4–23. The variables are A, B, and C. In Figure 4–23(a), the upper left cell is for $A = 0$, $B = 0$, and $C = 0$; the lower left cell is for $A = 1$, $B = 0$, and $C = 0$, etc. In Figure 4–23(b), the upper left cell is for $\bar{A}\bar{B}\bar{C}$; the lower left cell is for $A\bar{B}\bar{C}$, etc. Both notations mean the same thing.

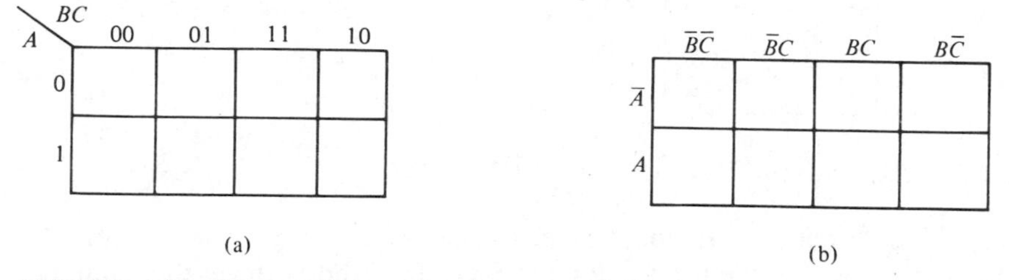

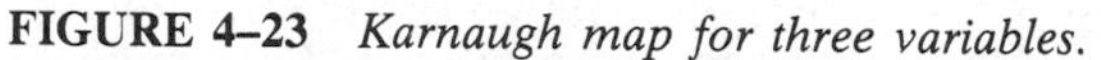

FIGURE 4–23 *Karnaugh map for three variables.*

A map for a four-variable function requires $2^4 = 16$ cells, and is constructed as shown in Figure 4–24.

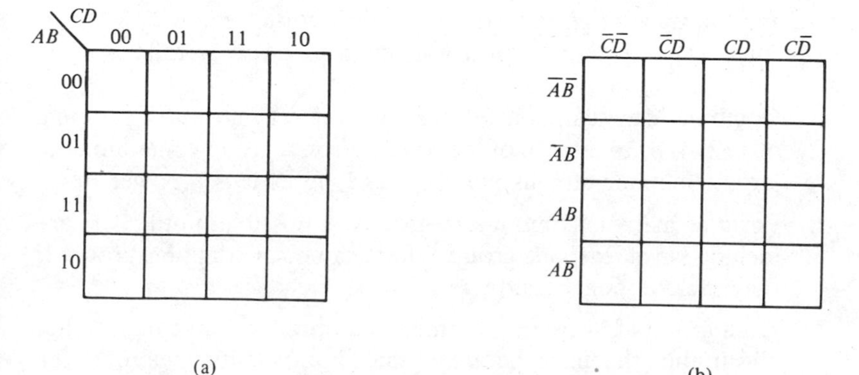

FIGURE 4–24 *Karnaugh map for four variables.*

At this point you should know basically how to draw a Karnaugh map for two-, three-, and four-variable functions. Functions of five or more variables can also be handled with Karnaugh maps, but the coverage of these is not included in this book. The difficulty of applying the Karnaugh map method increases considerably with an increasing number of variables.

Minimization

The next step is to learn how to use the Karnaugh map to minimize a Boolean expression. First, we will discuss how to "plot" a function on the map, and then cover the method of "factoring" the plotted function in order to get the minimum expression.

To plot a Boolean function on a Karnaugh map, 1s are placed in the cells corresponding to the terms that make the entire function a 1. To do this, it is very helpful to get the expression in a sum-of-products form. For example, let us take the expression $X = \overline{A}BC + AB(\overline{C} + C)$. Converting this to a sum-of-products form gives $X = \overline{A}BC + AB\overline{C} + ABC$.

Figure 4–25 shows the three-variable map with 1s plotted in the cells corresponding to each of the terms in the expression.

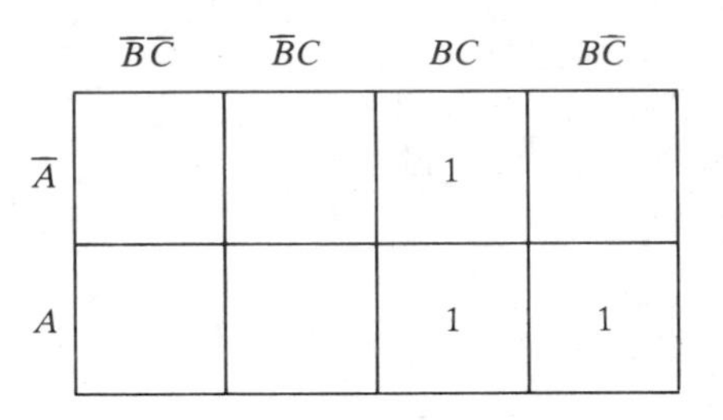

FIGURE 4–25 *1s plotted for* $X = \overline{A}BC + AB\overline{C} + ABC$.

Now that the 1s for this particular function are plotted on the Karnaugh map, we will turn our attention to "factoring" the 1s to get the minimum expression. But first the concept of *adjacent cells* needs to be defined. Adjacent cells on a Karnaugh map are those that differ by only the value of a *single* variable. For instance, in the map of Figure 4–25, two cells that are not adjacent are in the upper left and the lower right corners because two variables differ between these two cells. The same is true for the upper right and the lower left corner cells.

The rules for factoring 1s on a Karnaugh map are as follows:

1. Combine 1s appearing in adjacent cells into a *maximum* grouping of 1, 2, 4, 8, 16, etc. In other words, form a group containing the largest number of cells possible, as long as it is a power of 2.
2. Form as many maximum-size groups as possible, until all 1s are included in at least one group. There can be overlapping groups if they include noncommon 1s.
3. Each group of 1s creates a minimized product term composed of all variables that have the *same* value (1 or 0) within the group. If a

true variable appears in all cells in the group, the *true* form of the variable appears in the product term; if a complemented variable appears in all cells within the group, the *complement* form of the variable appears in the product term; and if both true and complement form of the variable appear within the group, it is omitted from the product term.

4. The product terms represented by each group on the map are now summed (ORed), producing a minimum sum-of-products expression.

The function plotted in Figure 4–25 is factored in Figure 4–26 to produce the minimized expression. Notice that two groupings are required and the term for each group is found by omitting the variable that appears in both true and complement form. The resulting minimum expression is $BC + AB$. See if you can arrive at the same result using the algebraic methods.

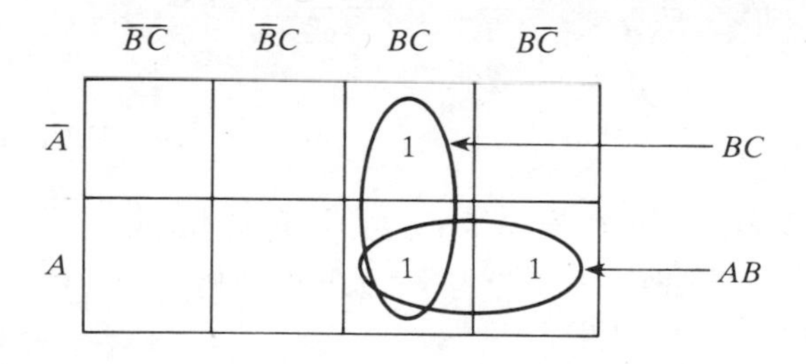

FIGURE 4–26

We will now work through several examples to illustrate the application of Karnaugh maps.

Example 4–7

Use the Karnaugh map method to simplify the expression $X = A(A\overline{B} + AB) + \overline{A}B$ and produce a minimum sum-of-products form.

Solution:

This is a two-variable function, so a map with four cells will be required. In sum-of-products form the expression is $X = A\overline{B} + AB + \overline{A}B$.

The next step is to construct the Karnaugh map and insert the 1s in the proper cells, as shown in Figure 4–27. The 1s are combined in two groups as indicated, each representing the terms as shown in Figure 4–27. The resulting minimum expression is $X = A + B$.

Example 4–7, continued

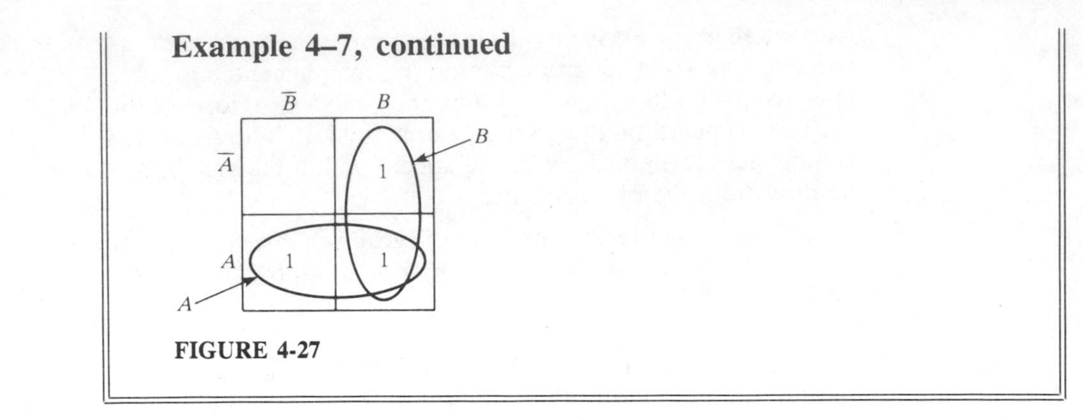

FIGURE 4-27

Example 4–8

Minimize the expression $X = A\overline{B}C + \overline{A}BC + \overline{A}\,\overline{B}C + \overline{A}\,\overline{B}\,\overline{C} + A\overline{B}\,\overline{C}$ in sum-of-products form.

Solution:

Notice that this expression is already in a sum-of-products form from which the 1s can be plotted very easily, as shown in Figure 4–28.

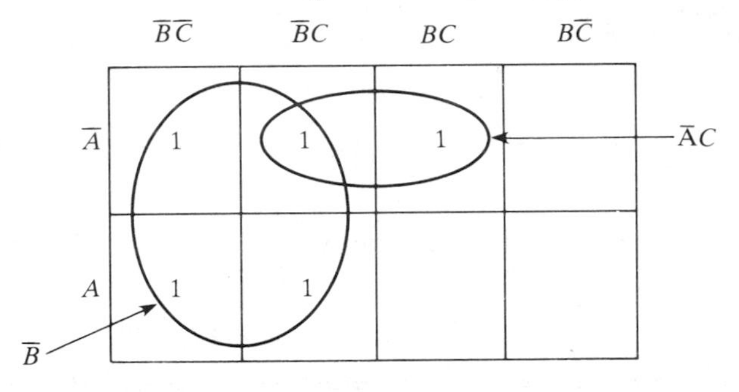

FIGURE 4–28

Four of the 1s appearing in adjacent cells can be grouped. The remaining 1 is absorbed in an overlapping group. The group of four 1s produces a single variable term, $\overline{B}$. This is determined by observing that, within the group, $\overline{B}$ is the only variable that does not change from cell to cell. The group of two 1s produces a two-variable term, $\overline{A}C$. This is determined by observing that, within the group, the variables $\overline{A}$ and C do not change from one cell to the next. To get the minimized function, the two terms that are produced are summed (ORed) as $X = \overline{B} + \overline{A}C$.

Before proceeding to the next example, some additional discussion of adjacent cells is necessary. The Karnaugh map can be visualized as being "rolled" to form a cylinder in first the horizontal and then the vertical direction. If this is done, you can see that each cell along the outer perimeter of the map is adjacent to the corresponding cell on the opposite side of the map. Note that there is a change of only one variable between such cells. For instance, in Figure 4–29 the $\overline{A}\overline{B}\overline{C}\overline{D}$ cell is adjacent to the $\overline{A}\overline{B}C\overline{D}$, the $\overline{A}\overline{B}CD$ cell is adjacent to the $A\overline{B}CD$ cell, and so on. The diagrams in Figure 4–29 illustrate the cell adjacency concept for a three- and a four-variable Karnaugh map.

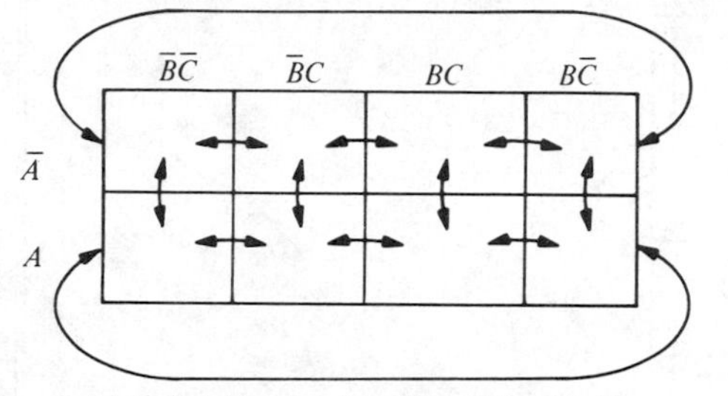

(a) Cell-Adjacency for a Three-Variable Map

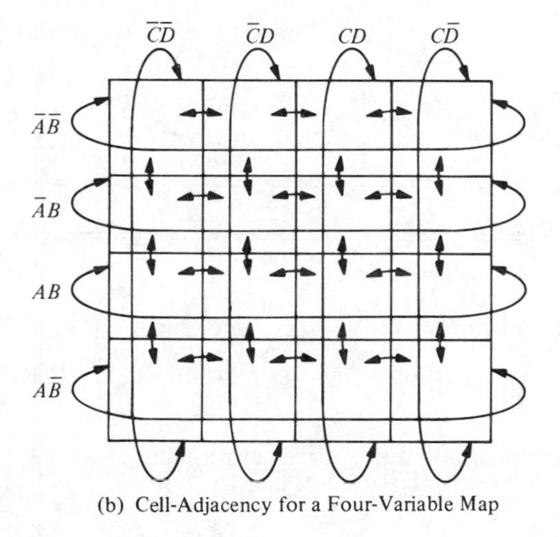

(b) Cell-Adjacency for a Four-Variable Map

FIGURE 4–29

Example 4–9

Reduce the following four-variable function to its minimum sum-of-products form:

$$X = \overline{A}\overline{B}\overline{C}\overline{D} + \overline{A}B\overline{C}\overline{D} + AB\overline{C}\overline{D} + A\overline{B}\overline{C}\overline{D} + \overline{A}\overline{B}CD$$
$$+ A\overline{B}CD + \overline{A}\overline{B}C\overline{D} + \overline{A}BC\overline{D} + ABC\overline{D} + A\overline{B}C\overline{D}$$

If all variables and their components are available, this function would take ten four-input AND gates and one ten-input OR gate.

Example 4–9, continued

Solution:

A group of eight 1s can be factored as shown in Figure 4–30 because the 1s in the outer columns are adjacent. A group of four is formed by the "wrap-around" adjacency of the cells to pick up the remaining two 1s. The minimum form of the original equation is $X = \overline{D} + \overline{B}C$.

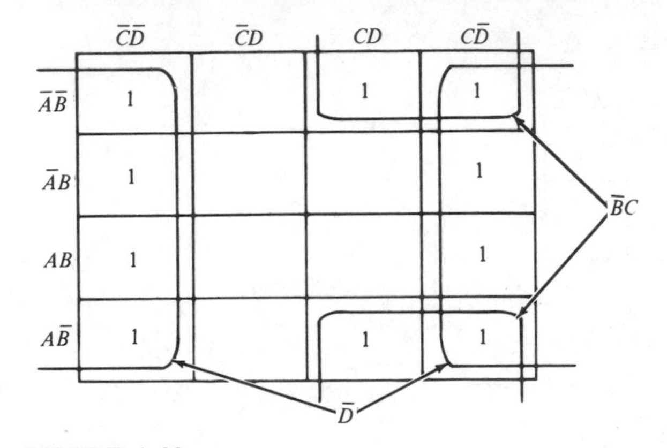

FIGURE 4–30

$X = \overline{D} + \overline{B}C$ takes one two-input AND gate and one two-input OR gate. Compare this to the implementation of the original function.

Example 4–10

Reduce the following function to its minimum sum-of-products form:

$$X = \overline{A}\,\overline{B}\,\overline{C}\,\overline{D} + \overline{A}\,\overline{B}C\overline{D} + A\overline{B}\,\overline{C}\,\overline{D} + \overline{A}CD + A\overline{B}C\overline{D}$$

Solution:

The function is plotted on the four-variable map and factored as indicated in Figure 4–31.

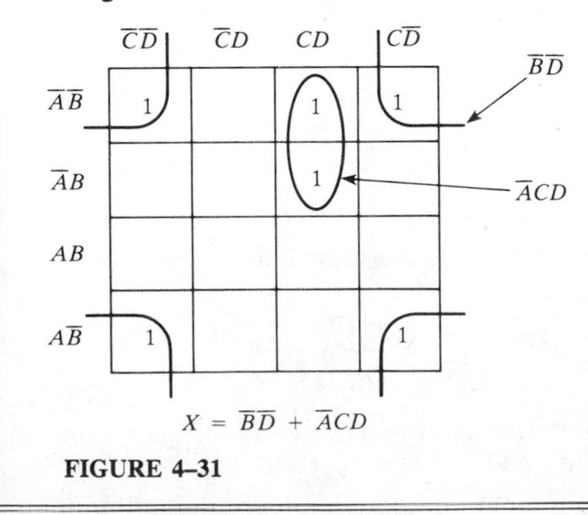

FIGURE 4–31

Problems

Section 4–2

4–1 Evaluate the following expressions for all possible values of the variables:

(a) $X = AB$ (b) $X = ABC$ (c) $X = A + B$
(d) $X = A + B + C$ (e) $X = AB + C$ (f) $X = \overline{A} + B$
(g) $X = A\overline{B}\,\overline{C}$ (h) $X = AB + \overline{A}C$ (i) $X = A(B + C)$
(j) $X = \overline{A}(\overline{B} + \overline{C})$

4–2 Evaluate the following expressions for all possible values of the variables:

(a) $X = (A + B)C + B$ (b) $X = (\overline{A + B})C$
(c) $X = A\overline{B}C + AB$ (d) $X = (A + B)(\overline{A} + B)$
(e) $X = (A + BC)(\overline{B} + \overline{C})$

4–3 Write the Boolean expression for each of the logic circuits in Figure 4-32.

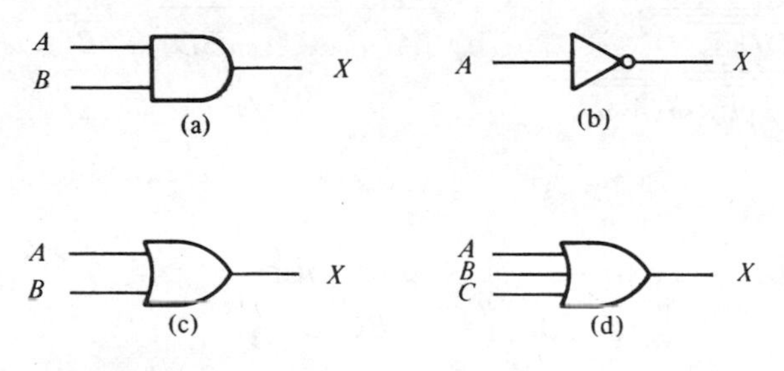

FIGURE 4–32

4–4 Repeat Problem 4–3 for the circuits in Figure 4–33.

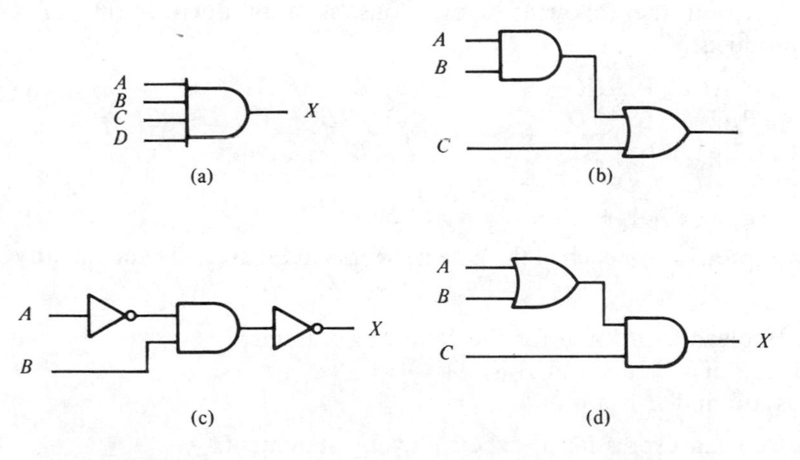

FIGURE 4–33

4–5 Construct a truth table for each of the following Boolean expressions:

(a) $A + B$ (b) AB (c) $AB + BC$
(d) $(A + B)C$ (e) $(A + B)(\overline{B} + C)$

Section 4–4

4–6 Apply DeMorgan's theorems to each expression.

(a) $\overline{A + \overline{B}}$ (b) $\overline{\overline{A}B}$ (c) $\overline{A + B + C}$
(d) $\overline{ABC}$ (e) $\overline{A(B + C)}$ (f) $\overline{AB} + \overline{CD}$
(g) $\overline{AB + CD}$ (h) $\overline{(A + \overline{B})(\overline{C} + D)}$

4–7 Apply DeMorgan's theorems to each expression.

(a) $\overline{A\overline{B}(C + \overline{D})}$ (b) $\overline{AB(CD + EF)}$
(c) $\overline{(A + \overline{B} + C + \overline{D})} + \overline{ABC\overline{D}}$ (d) $\overline{\overline{(\overline{A} + B + C + D)}\,\overline{(A\overline{B}\,\overline{C}D)}}$
(e) $\overline{\overline{AB}(CD + \overline{E}F)(\overline{AB} + \overline{CD})}$

4–8 Apply DeMorgan's theorems to the following:

(a) $\overline{\overline{\overline{(ABC)}\,\overline{(EFG)}} + \overline{\overline{(HIJ)}\,\overline{(KLM)}}}$ (b) $\overline{(A + \overline{B\overline{C}} + CD)} + \overline{\overline{BC}}$
(c) $\overline{\overline{\overline{(A + B)}\,\overline{(C + D)}\,\overline{(E + F)}\,\overline{(G + H)}}}$

Section 4–5

4–9 Convert the following expressions to sum-of-product forms:

(a) $(A + B)(C + D)$ (b) $(A + \overline{B}C)D$
(c) $(A + C)(ABC + ACD)$

4–10 Convert the following expressions to sum-of-product forms:

(a) $AB + CD(A\overline{B} + CD)$ (b) $AB(\overline{B}\,\overline{C} + BC)$
(c) $A + B[AC + (B + \overline{C})D]$

4–11 Identify each of the following expressions as a product-of-sums or a sum-of-products:

(a) $ABC + \overline{A}BC + ABC$ (b) $A + B\overline{C}$
(c) $(A + B)(\overline{B} + \overline{C} + D)$ (d) $\overline{A}(B + C)(A + C)$
(e) $ABCDE + AD + \overline{A}E$ (f) $B + C$
(g) $\overline{A} + B + C + D + E + BC$
(h) $(A + B + C)(\overline{A} + B + \overline{C})(A + B + D)$

4–12 Write an expression for each of the gate networks in Figure 4–34 and identify the form.

4–13 Write a Boolean expression for the following statement:
X is a 1 only if A is a 1 and B is a 1
or if A is a 0 and B is a 0.

4–14 Write a Boolean expression for the following statement:
X is a 1 only if A, B, and C are all 1s
or if only one of the variables is a 0.

4–15 Write a Boolean expression for the following conditions:
X is a 0 if any two of the three variables A, B, and C are 1s. X is a 1 for all other conditions.

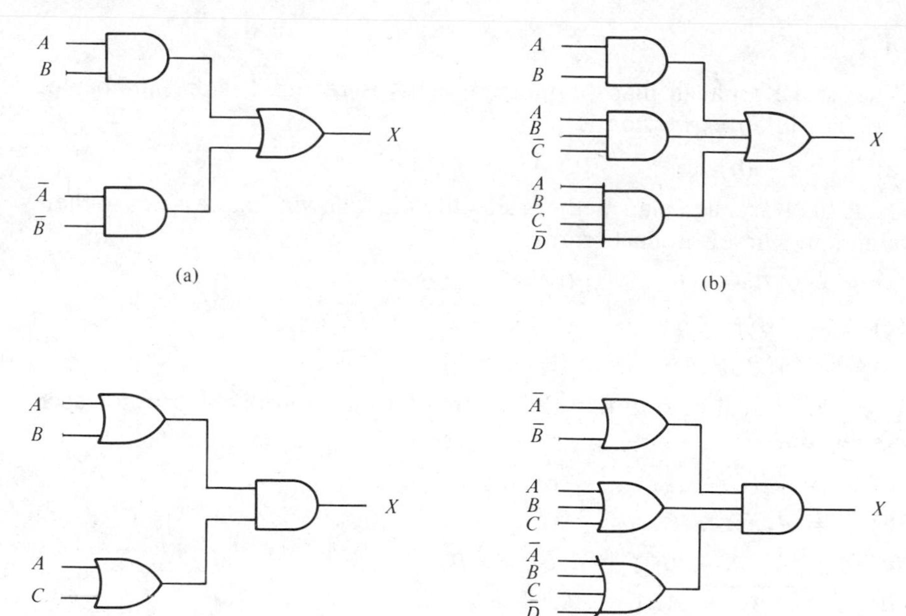

FIGURE 4–34

4–16 Draw the logic circuit represented by each of the following expressions:

(a) $A + B + C$ **(b)** ABC **(c)** $AB + C$
(d) $AB + CD$ **(e)** $\overline{A}B(C + \overline{D})$

4–17 Draw the logic circuit represented by each expression.

(a) $A\overline{B} + \overline{A}B$ **(b)** $AB + \overline{A}\,\overline{B} + \overline{A}BC$ **(c)** $A + B[C + D(B + \overline{C})]$

Section 4–6

4–18 Simplify as much as possible the following expressions, using Boolean algebra techniques:

(a) $A(A + B)$ **(b)** $A(\overline{A} + AB)$ **(c)** $BC + \overline{B}C$
(d) $A(A + \overline{A}B)$ **(e)** $A\overline{B}C + \overline{A}BC + \overline{A}\,\overline{B}C$

4–19 Simplify the following expressions, using Boolean algebra:

(a) $(A + \overline{B})(A + C)$ **(b)** $\overline{A}B + \overline{A}B\overline{C} + \overline{A}BCD + \overline{A}B\overline{C}\,\overline{D}E$
(c) $AB + \overline{AB}C + A$ **(d)** $(A + \overline{A})(AB + AB\overline{C})$
(e) $AB + (\overline{A} + \overline{B})C + AB$

4–20 Simplify each expression, using Boolean algebra:

(a) $BD + B(D + E) + \overline{D}(D + F)$
(b) $\overline{A}\,\overline{B}C + \overline{(A + B + \overline{C})} + \overline{A}\,\overline{B}\,\overline{C}D$
(c) $(B + BC)(B + \overline{B}C)(B + D)$
(d) $ABCD + AB(\overline{CD}) + (\overline{AB})CD$
(e) $ABC[AB + \overline{C}(BC + AC)]$

Section 4–7

4–21 What size Karnaugh map (number of cells) is required if the number of variables in an expression is

(a) 2 (b) 3 (c) 4 (d) 5

4–22 Using the Karnaugh map method, simplify the following expressions to their minimum sum-of-products form:

(a) $X = \overline{A}\,\overline{B} + A\overline{B}$ (b) $X = \overline{A}\,\overline{B} + \overline{A}B$

(c) $X = A\overline{B} + AB$ (d) $X + \overline{A}\,\overline{B} + A\overline{B} + AB$

(e) $X = A(\overline{B} + AB)$ (f) $X = \overline{A}\,\overline{B} + AB$

4–23 Use a Karnaugh map to find the minimum sum-of-products form for each expression.

(a) $X = \overline{A}\,\overline{B}\,\overline{C} + \overline{A}\,\overline{B}C + A\overline{B}C$

(b) $X = AC(\overline{B} + C)$

(c) $X = \overline{A}(BC + B\overline{C}) + A(BC + B\overline{C})$

(d) $X = \overline{A}\,\overline{B}\,\overline{C} + A\overline{B}\,\overline{C} + \overline{A}B\overline{C} + AB\overline{C}$

(e) $X = A + \overline{B}C$

4–24 Use a Karnaugh map to simplify each function to a minimum sum-of-products form.

(a) $X = \overline{A}\,\overline{B}\,\overline{C} + A\overline{B}C + \overline{A}BC + AB\overline{C}$

(b) $X = AC[\overline{B} + A(B + \overline{C})]$

(c) $X = DE\overline{F} + \overline{D}E\overline{F} + \overline{D}\,\overline{E}\,\overline{F}$

4–25 Use a Karnaugh map to reduce each expression to a minimum sum-of-products.

(a) $X = A + B\overline{C} + CD$

(b) $X = \overline{A}\,\overline{B}\,\overline{C}\,\overline{D} + \overline{A}\,\overline{B}\,\overline{C}D + ABCD + ABC\overline{D}$

(c) $X = \overline{A}B(\overline{C}\,\overline{D} + \overline{C}D) + AB(\overline{C}\,\overline{D} + \overline{C}D) + A\overline{B}\,\overline{C}D$

(d) $X = (\overline{A}\,\overline{B} + A\overline{B})(CD + C\overline{D})$

(e) $X = \overline{A}\,\overline{B} + A\overline{B} + \overline{C}\,\overline{D} + C\overline{D}$

4–26 Repeat Problem 4–25 for the following expressions:

(a) $X = \overline{A}(\overline{B}\,\overline{C}\,\overline{D} + \overline{B}C\overline{D}) + A(\overline{B}\,\overline{C}\,\overline{D} + \overline{B}C\overline{D})$

(b) $X = B[\overline{A}\,\overline{C}D + \overline{A}CD) + (A\overline{C}D + ACD)]$

(c) $X = \overline{A}B + A\overline{B} + \overline{A}B\overline{C}D + \overline{A}BCD$

In Chapter 3, logic gates were studied on an individual basis. In this chapter we will combine various types of gates in order to produce specified logic functions. When logic gates are connected to generate a specified output for certain combinations of input variables with no storage involved, the resulting network is called *combinational* logic because it combines the input variables in such a way that the output level is at all times dependent on the combination of input levels.

In analyzing a combinational logic function, we are concerned with determining in what way the output is dependent on the inputs: i.e., for what combinations of inputs is the output a HIGH or a LOW? The analysis is approached in two ways: first, the output of the combinational logic is determined for a given set of input values, based on the logical operation of the individual gates in the network; second, a logic equation for the output function is derived and, using this equation, the output value is determined for various combinations of input values, using Boolean algebra.

5
COMBINATIONAL LOGIC

5–1 ANALYSIS OF COMBINATIONAL NETWORKS

AND–OR Logic

Figure 5–1 shows a combinational logic circuit consisting of two AND gates and one OR gate. Each of the three gates has two input variables as indicated.

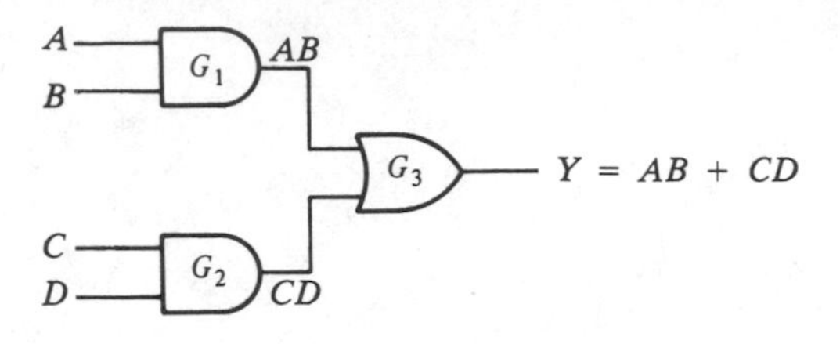

FIGURE 5–1

Each of the input variables can be either a HIGH (1) or a LOW (0). Because there are four input variables, there are sixteen possible combinations of the input variables ($2^4 = 16$). To illustrate an analysis procedure, we will assign one of the sixteen possible input combinations and see what the corresponding output value is.

First, we make each input variable a LOW, and examine the output of each gate in the network in order to arrive at the final output, Y. If the inputs to gate G_1 are both LOW, the output of gate G_1 is LOW. Also, the output of gate G_2 is LOW because its inputs are LOW. As a result of the LOWs on the outputs of gates G_1 and G_2, both inputs to gate G_3 are LOW and, therefore, its output is LOW. We have determined that the output function of the logic circuit of Figure 5–1 is LOW when all of its inputs are LOW. This condition is illustrated in the first row of Table 5–1 along with the remaining fifteen input combinations. You should verify each of these conditions on the logic diagram.

Now, as a second method of analyzing the logical operation of the circuit of Figure 5–1, we can develop a logic equation for the output function and, using Boolean algebra, evaluate the equation for each of the sixteen combinations of input variables.

Since gate G_1 is an AND gate and its two inputs are A and B, its output can be expressed as AB. Gate G_2 is an AND gate and its two inputs are C and D, so its output can be expressed as CD. Gate G_3 is an OR gate and its two inputs are the outputs of gates G_1 and G_2, so its output can be expressed as $AB + CD$. The output of gate G_3 is the output function of the logic network, so $Y = AB + CD$. Figure 5–2 shows the logic functions at each point in the circuit.

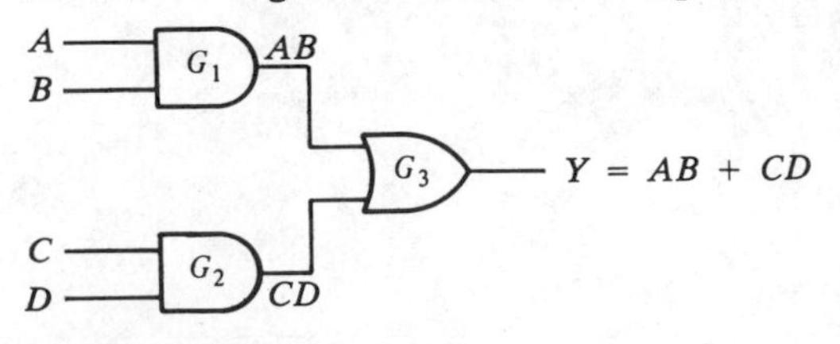

FIGURE 5–2

TABLE 5–1 *Truth table for Figure 5–1 (1 = HIGH, 0 = LOW).*

Inputs				G_1 output	G_2 output	G_3 output
A	B	C	D	(AB)	(CD)	Y
0	0	0	0	0	0	0
0	0	0	1	0	0	0
0	0	1	0	0	0	0
0	0	1	1	0	1	1
0	1	0	0	0	0	0
0	1	0	1	0	0	0
0	1	1	0	0	0	0
0	1	1	1	0	1	1
1	0	0	0	0	0	0
1	0	0	1	0	0	0
1	0	1	0	0	0	0
1	0	1	1	0	1	1
1	1	0	0	1	0	1
1	1	0	1	1	0	1
1	1	1	0	1	0	1
1	1	1	1	1	1	1

We can now evaluate the output equation by substituting into it the various combinations of input variable values. For example, when $A = 1, B = 1, C = 1$, and $D = 0$, the output expression is evaluated as follows.

$$Y = AB + CD = 1 \cdot 1 + 1 \cdot 0 = 1 + 0 = 1$$

The same procedure can be used for any of the other input variable combinations.

Example 5–1

A 74H52 TTL AND–OR logic and pin diagram is shown in Figure 5–3. Write the output expression and determine the logic level at pin 8 for the following input logic levels: pins 3, 4, 13, and 11 are HIGH, and pins 1, 2, 5, 9, 10, and 12 are LOW.

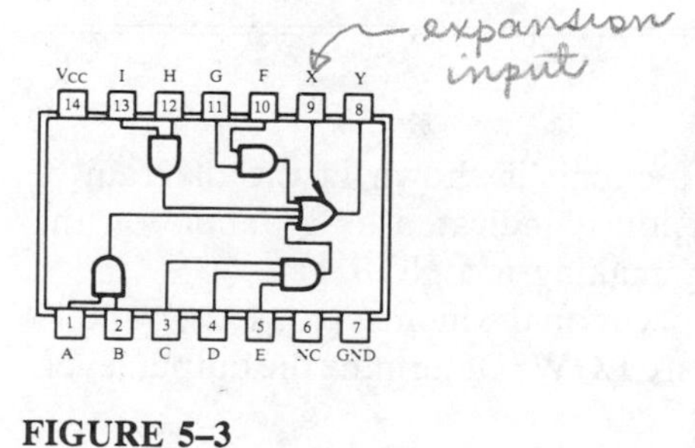

FIGURE 5–3

Example 5–1, continued

Solution:

The output expression is $Y = AB + CDE + FG + HI + X$. The X input is for the connection of an *expander* AND gate when extra inputs must be handled; X is the output of the expander gate.

The output expression is evaluated for the given input levels as follows:

$$Y = AB + CDE + FG + HI + X = 0 \cdot 0 + 1 \cdot 1 \cdot 0 + 0 \cdot 1 + 0 \cdot 1 + 0$$
$$= 0 + 0 + 0 + 0 + 0 = 0$$

AND–OR–INVERT Logic

Figure 5–4(a) shows a combinational logic circuit consisting of two AND gates, one OR gate, and an inverter. As you can see, the operation is the same as for the AND–OR circuit in Figure 5–1 except that the output is inverted. The output expression is $Y = \overline{AB + CD}$. An evaluation of this for the inputs $A = 1, B = 1, C = 1, D = 0$ is as follows:

$$Y = \overline{AB + CD} = \overline{1 \cdot 1 + 1 \cdot 0} = \overline{1 + 0} = \overline{1} = 0$$

Notice that DeMorgan's theorem can be applied to this output expression as follows.

$$Y = \overline{AB + CD} = (\overline{A} + \overline{B})(\overline{C} + \overline{D})$$

The resulting equivalent circuit is shown in Figure 5–4(b).

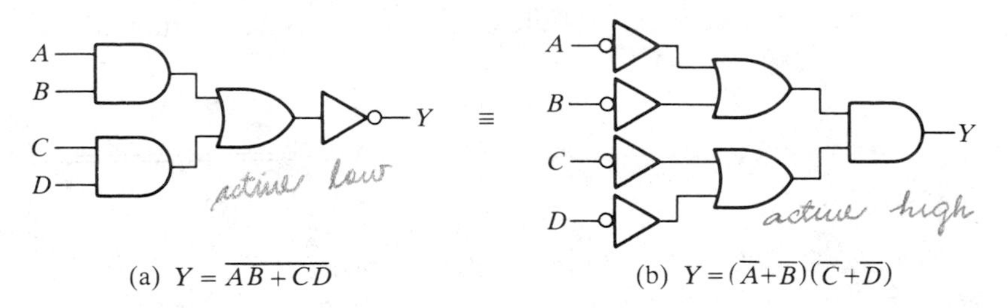

(a) $Y = \overline{AB + CD}$ (b) $Y = (\overline{A} + \overline{B})(\overline{C} + \overline{D})$

FIGURE 5–4 *AND–OR–INVERT circuit and one of its equivalent circuits.*

Example 5–2

A 7451 dual AND–OR–INVERT circuit is shown in the diagram of Figure 5–5. Notice that the inversion is indicated by a bubble on the output of the OR gate effectively making it a NOR gate.

The following levels occur on the inputs: pins 1, 2, 4, and 13 are HIGH, and pins 3, 5, 9, and 10 are LOW. Determine the output levels and indicate the pin designations.

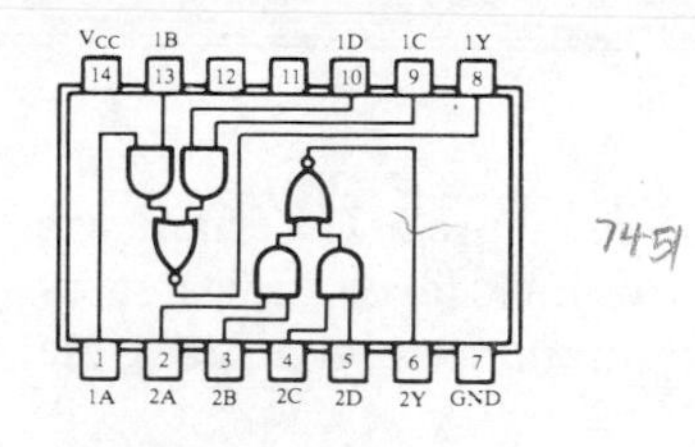

FIGURE 5–5

Solution:

The 1Y output at pin 8 is LOW because both pins 1 and 13 are HIGH. The output 2Y at pin 6 is HIGH because neither AND gate in this circuit has both of its inputs HIGH.

Exclusive-OR Logic

Figure 5–6(a) shows a combinational logic circuit known as an *exclusive-OR*. It is a widely used function because of special arithmetic properties which will be discussed in a later chapter. Because of its wide applications it has a special symbol shown in Figure 5–6(b).

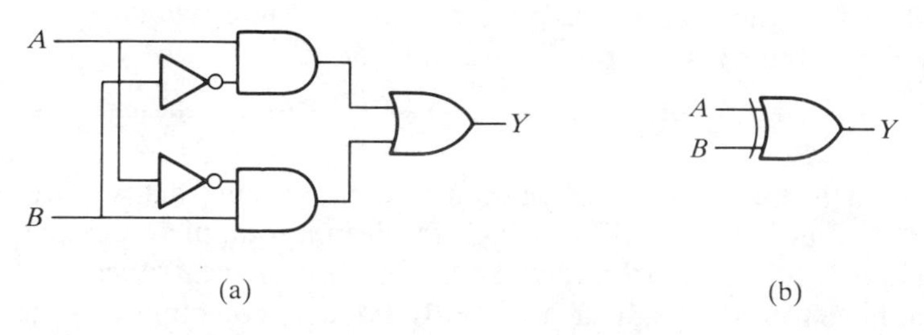

FIGURE 5–6 *Exclusive-OR logic and symbol.*

The output expression for this circuit is $Y = A\overline{B} + \overline{A}B$. Evaluation of this expression results in the truth table in Table 5–2. *Notice that the output is HIGH only when the two inputs are at opposite levels.* A special exclusive-OR operation symbol, $\oplus$, is often used so that the expression $Y = A\overline{B} + \overline{A}B$ can also be written as $Y = A \oplus B$, stated as "$Y = A$ exclusive-OR B."

TABLE 5–2 *Truth table for an exclusive-OR.*

A	B	Y
0	0	0
0	1	1
1	0	1
1	1	0

Example 5–3

The 7486 quad exclusive-OR gate IC is shown in Figure 5-7. Certain pins are connected externally as shown with positive logic levels indicated by 1s and 0s on the inputs. Determine the output at pin 6 for these conditions.

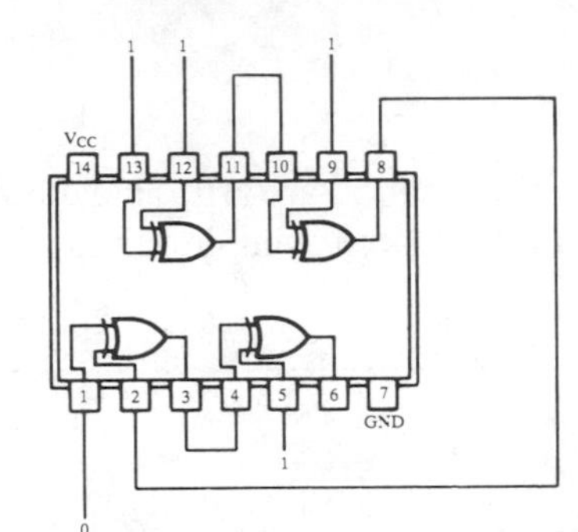

FIGURE 5–7

Solution:

The final output on pin 6 is LOW. This is determined by tracing through the logic with the use of the exclusive-OR truth table. The intermediate levels are as follows: pin 11 = 0, pin 8 = 1, and pin 3 = 1.

Figure 5–8 is the fourth combinational logic circuit that we will analyze. This circuit has four input variables, which result in a total of 16 possible input combinations ($2^4 = 16$). In order to determine the complete logical function of this circuit, we would have to go through each of the 16 possible combinations and determine the resulting output for each. As you might think, this is a quite lengthy process, so we will assign only two out of the 16 possible combinations and analyze

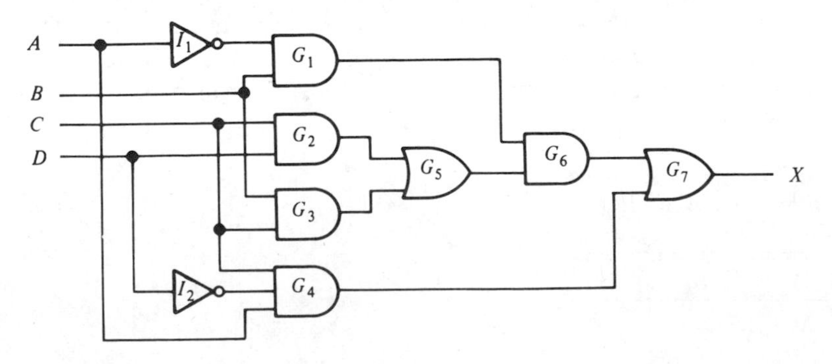

FIGURE 5–8

the circuit operation for each. The remaining combinations will be left for you as a problem at the end of the chapter.

For the first input combination let us use the following: $A = 0, B = 1, C = 0, D = 1$. Since A is a LOW, the output of inverter I_1 is a HIGH. This makes both inputs to gate G_1 HIGH and causes its output to be HIGH. The C input to gate G_2 is LOW, making the output of G_2 LOW. The C input to gate G_3 is LOW, causing its output to be LOW. Because the output of gates G_2 and G_3 are both LOW, the output of gate G_5 is also LOW. The output of gate G_1 and the output of gate G_5 become the inputs to gate G_6, making its output LOW. The outputs of gates G_6 and G_4 are inputs to gate G_7 and, since both are LOW, the output of G_7 is LOW. These logic levels are indicated on the logic diagram of Figure 5–9.

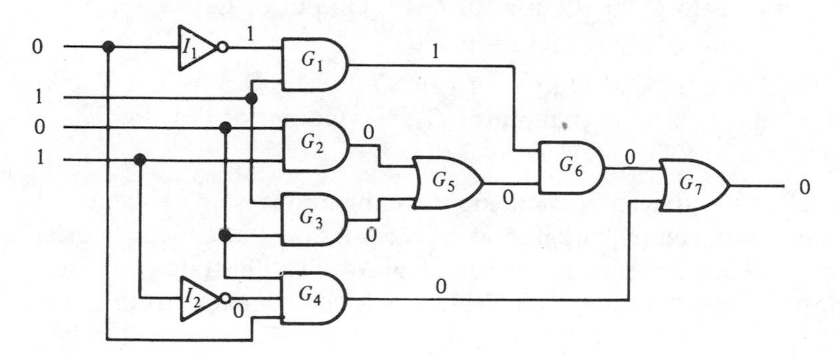

FIGURE 5–9

Next, we set the inputs to the following levels: $A = 1, B = 0, C = 1, D = 0$. If A is HIGH, the output of inverter I_1 is LOW, making the output of gate G_1 LOW. Since D is LOW, the output of gate G_2 is LOW, and since B is LOW, the output of gate G_3 is LOW. The output of inverter I_2 is HIGH because D is LOW; this means that all of the inputs to gate G_4 are HIGH, making its output HIGH. The output of gate G_4 is fed into gate G_7, causing the output to be HIGH. These logic levels are indicated in Figure 5–10. If we continued this analysis for the 14 remaining input combinations, the result would be a complete definition of the logical operation of this circuit.

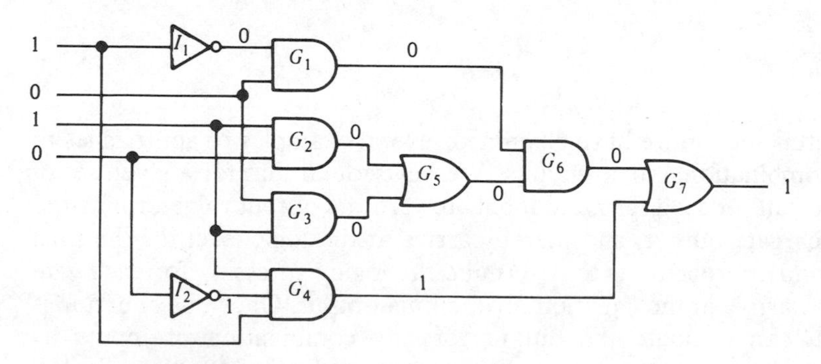

FIGURE 5–10

Having gone through the logic level analysis for two possible input combinations, we now develop the logic equation.

$$
\begin{aligned}
\text{Output of inverter } I_1 &= \overline{A} \\
\text{Output of gate } G_1 &= \overline{A}B \\
\text{Output of gate } G_2 &= CD \\
\text{Output of gate } G_3 &= BC \\
\text{Output of inverter } I_2 &= \overline{D} \\
\text{Output of gate } G_4 &= AC\overline{D} \\
\text{Output of gate } G_5 &= (\text{Output of } G_2) + (\text{Output of } G_3) \\
&= CD + BC \\
\text{Output of gate } G_6 &= (\text{Output of } G_1) \cdot (\text{Output of } G_5) \\
&= \overline{A}B(CD + BC) \\
\text{Final output } X &= \text{Output of gate } G_7 \\
&= (\text{Output of } G_4) + (\text{Output of } G_5) \\
&= AC\overline{D} + \overline{A}B(CD + BC)
\end{aligned}
$$

If each of the sixteen possible input combinations is now substituted into the equation for this circuit (equation for the final output X), a complete logical description or truth table operation is derived. Figure 5–11 illustrates the logic functions at each point in the circuit, and Table 5–3 is the complete truth table.

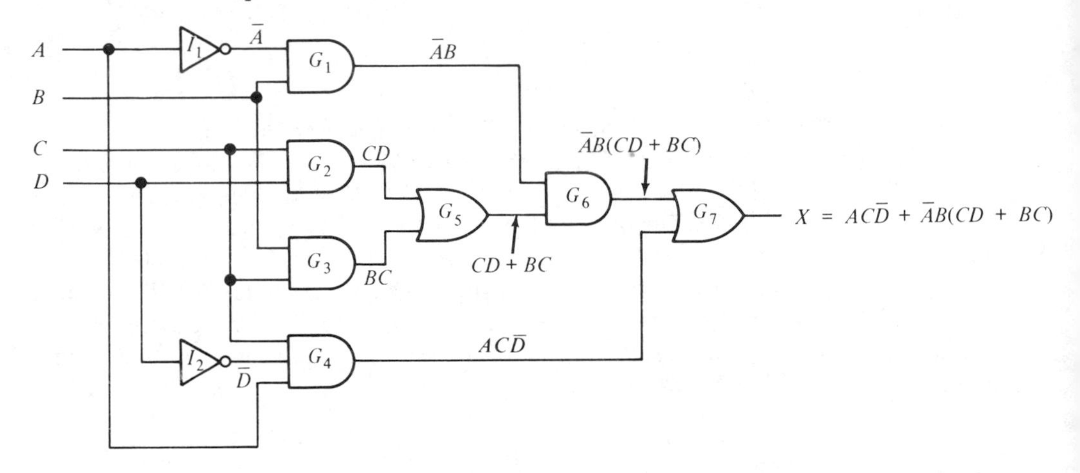

FIGURE 5–11

In this section we have examined several examples of approaches to the analysis of combinational logic circuits. We have found that for a given set of input values, you can proceed from the inputs toward the output to determine the logic level at each gate output, and thereby arrive at the logic level for the final output. The second approach was to determine the logic expression for each gate output in order to arrive at the equation for the final output. Once this equation is determined, you can evaluate the output for any combination of inputs by substituting the appropriate values into the equation and using Boolean algebra rules. Both ways produce the same results.

TABLE 5–3 *Truth table for the circuit of Figure 5-8.*

Inputs				Output
A	B	C	D	X
0	0	0	0	0
0	0	0	1	0
0	0	1	0	0
0	0	1	1	0
0	1	0	0	0
0	1	0	1	0
0	1	1	0	1
0	1	1	1	1
1	0	0	0	0
1	0	0	1	0
1	0	1	0	1
1	0	1	1	0
1	1	0	0	0
1	1	0	1	0
1	1	1	0	1
1	1	1	1	0

IMPLEMENTING COMBINATIONAL NETWORKS 5–2

In this section we will start with an equation that describes a logic function and from it determine the circuit required to implement the function. As before, several examples will be used to illustrate a general procedure.

First, let us review two basic conventions regarding logic equations:

1. Whenever a + appears between two or more terms, such as $X + Y + Z$, this means these terms are ORed.
2. Whenever two or more terms appear as $X \cdot Y \cdot Z$ or simply XYZ, this means that they are ANDed.

Equation (5–1) is the first in the series of examples that will be considered.

$$X = AB + CDE \tag{5–1}$$

A brief inspection reveals that this function is composed of two terms, AB and CDE, with a total of five variables. The first term is formed by ANDing A with B, and the second term is formed by ANDing C, D, and E. These two terms are then ORed to form the function X. These operations are indicated in the structure of the equation as follows:

$$X = AB + CDE$$

AND (second-level operation)

OR (first-level operation)

It should be noted that, in this particular equation, the AND operations forming the two individual terms, AB and CDE, must be performed *before* the terms can be ORed. The OR operation is the last to be performed before the final output function is produced; therefore, it is called a *first-level* operation, meaning that it is performed by the first-level gate, starting at the output and working back toward the inputs. The AND operations are performed by the second-level gates from the output, and are therefore *second-level* operations.

To implement the logic function, a two-input AND gate is required to form the term AB and a three-input AND gate is needed to form the term CDE. A two-input OR gate is then required to combine the two AND terms. The resulting logic circuit is shown in Figure 5–12.

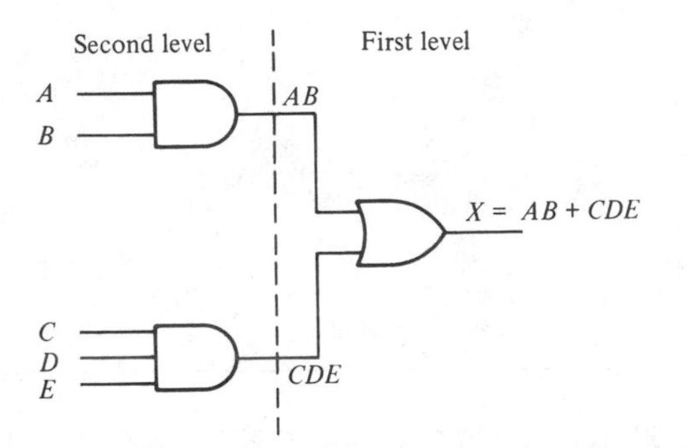

FIGURE 5–12 *Logic circuit for Equation (5–1).*

Now, let us look at a second example expressed in Equation (5–2):

$$X = \overline{A}BC + A\overline{B}\,\overline{C} \qquad \textbf{(5–2)}$$

This equation requires three levels of logic gates and inverters. How do we know this? First, observe that the equation is composed of two terms in sum-of-products form (the two AND terms, $\overline{A}BC$ and $A\overline{B}\,\overline{C}$, are ORed). The OR operation requires one level of logic (one OR gate). The first term in the equation is formed by ANDing the three variables $\overline{A}$, B, and C. The second term is formed by ANDing the three variables, A, $\overline{B}$, and $\overline{C}$. These two operations are performed at the second level of logic (the two AND gates whose outputs are connected to the first level OR gate inputs). The formation of the complement of each variable requires a third level of logic consisting of inverters; these inverters are at the third level because the complements must be formed before each term is formed. The logic gates required to implement this function are as follows: three inverters to form the $\overline{A}$, $\overline{B}$, and $\overline{C}$ variables, two three-input AND gates to form the terms $\overline{A}BC$ and $A\overline{B}\,\overline{C}$ and one two-input OR gate to form the final output function, $\overline{A}BC + A\overline{B}\,\overline{C}$. Figure 5–13 illustrates the implementation of this logic function. Do not confuse levels of logic we have discussed here with input and output levels of a gate (HIGH or LOW).

In this case where there is a three-level logic function, notice that between input A and the output X there are three logic gates; the same is true for input B and input C. In general, *the number of logic levels in a given network is the greatest number of gates and inverters through which a signal has to pass in going from an input to the output.*

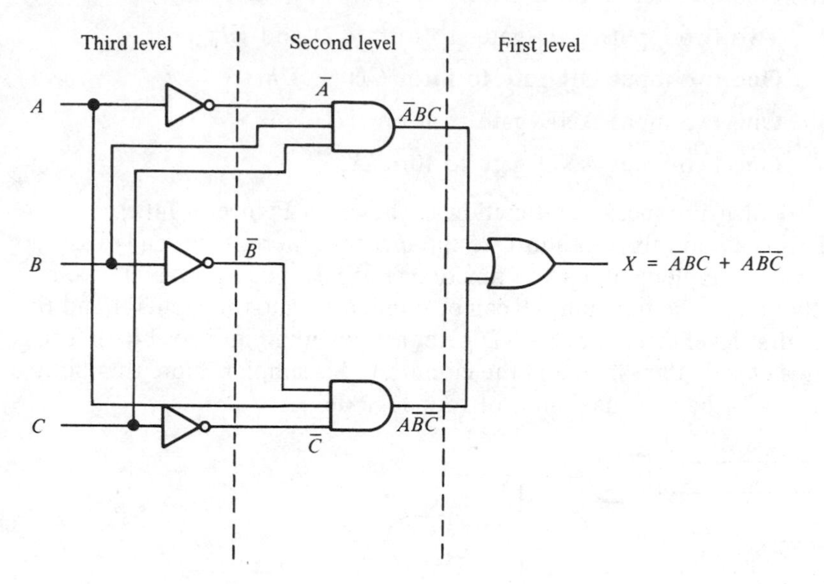

FIGURE 5–13 *Logic circuit for Equation (5–2).*

As a third and final example, we will implement Equation (5–3):

$$X = AB(C\bar{D} + EF) \qquad \textbf{(5–3)}$$

A breakdown of this equation shows that the term AB and the term $C\bar{D} + EF$ are ANDed. The term AB is formed by ANDing the variables A and B. The term $C\bar{D} + EF$ is formed by first ANDing C and $\bar{D}$, ANDing E and F, and then ORing these two terms. This structure is indicated in relation to the equation as follows:

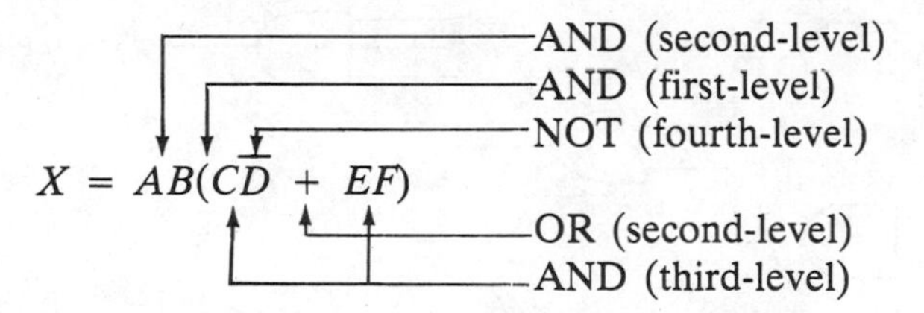

Before the output function X can be formed, we must have the term AB, which is formed by ANDing input A with input B. Also, we must have the term $C\bar{D} + EF$; but before we can get this term, we must have the terms $C\bar{D}$ and EF; but before we can get the term $C\bar{D}$, we must have $\bar{D}$. So, as you can see, there is a "chain" of logic operations that must be done in the proper order before the output function itself is realized. In other words, the output function is composed of several subfunctions.

The logic gates required to implement Equation (5–3) are as follows:

1. One inverter to form $\overline{D}$.
2. Two two-input AND gates to form $C\overline{D}$ and EF.
3. One two-input OR gate to form $C\overline{D} + EF$.
4. One two-input AND gate to form AB.
5. One two-input AND gate to form X.

The logic circuit which produces this function is shown in Figure 5–14(a).

This is a good time to illustrate the fact that there is, in many cases, more than one way to implement a given function. For instance, the AND gate in Figure 5-14(a) that forms the function AB can be eliminated and the inputs A and B brought into the first-level three-input AND gate, as shown in Figure 5–14(b). The resulting output is exactly the same, and the circuit used is simpler. Note that both circuits in Figure 5–14 have a maximum of four logic levels.

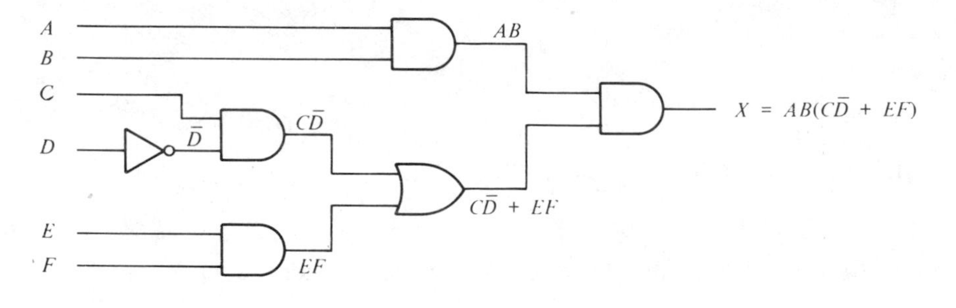

(a)

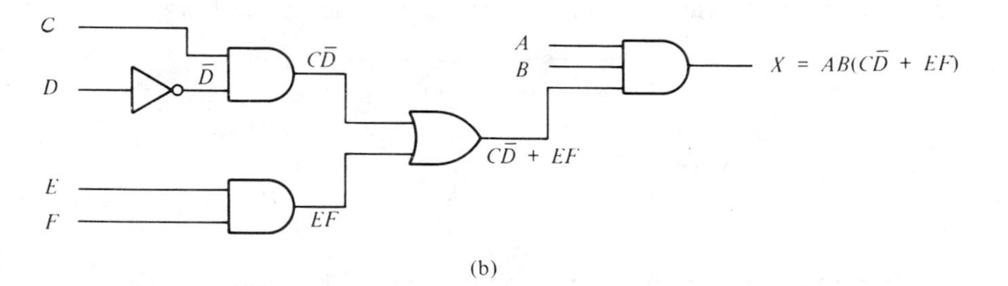

(b)

FIGURE 5–14 *Logic circuit for Equation (5–3).*

Example 5–4

Implement the logic circuit in Figure 5–14(a) using standard 54/74 TTL.

Solution:

The following ICs are selected to implement this function: 7405 hex inverter, 7408 quad two-input AND gates, and 7432 quad two-input OR gates. The interconnections and pin diagrams are shown in Figure 5–15. As you can see, there are spare gates in both packages in this case.

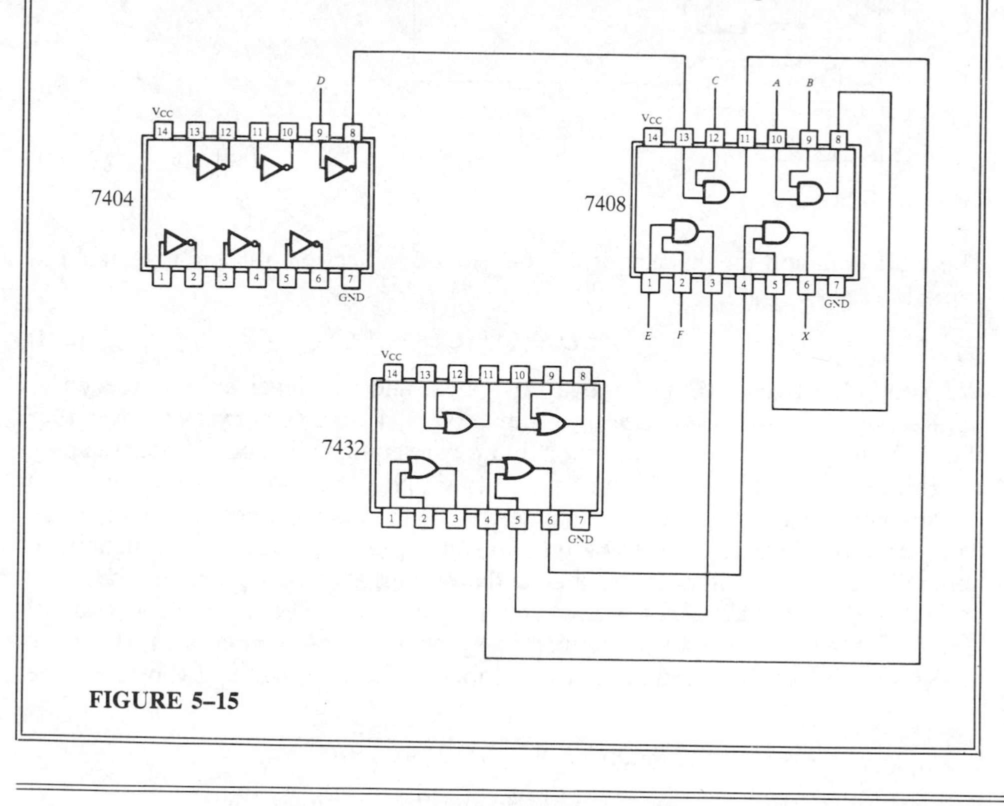

FIGURE 5–15

GATE MINIMIZATION 5–3

In this section we will be concerned with reducing the number of logic gates required to produce a given function. In many applications it is desirable to use the minimum number of gates in the simplest configuration possible to implement a given logic function. This simplification may be desirable for several reasons, such as economy or cost, limitations of available power, minimization of delay times by reduction of logic levels, or maximum utilization of chip area in the case of IC designs.

Here we will examine two basic methods of gate minimization. First, the rules and laws of Boolean algebra will be applied in order to simplify the logic equation for the circuit in question, and thereby reduce the number of gates required to implement the function. Second, the Karnaugh map method will be used to minimize the logic function. Both of these methods were studied in the last chapter.

To begin, Boolean algebra will be applied to the equation for the circuit of Figure 5–8, which is redrawn for convenience in Figure 5–16.

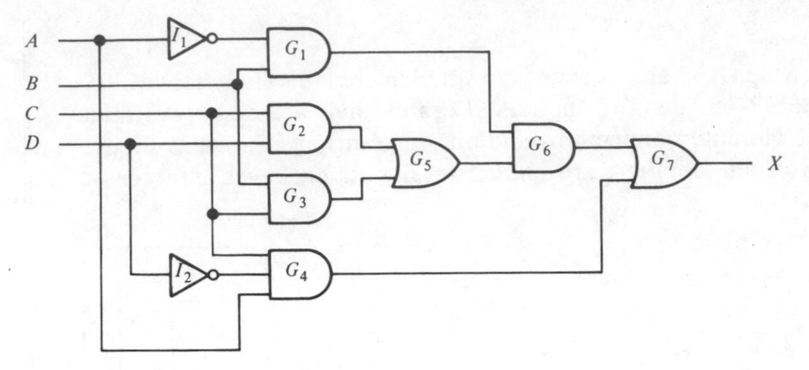

FIGURE 5–16

The logic equation for this circuit was developed in Section 5–1 and is restated in Equation (5–4):

$$X = AC\overline{D} + \overline{A}B(CD + BC) \tag{5–4}$$

We know that five AND gates, two OR gates, and two inverters are needed to implement this function *as it appears* in Equation (5–4). We do not yet know whether this is the simplest form of the equation, but we are going to find out. When we apply the rules of Boolean algebra in an attempt to simplify the equation, our cleverness and imagination are major factors in determining the outcome because an equation can normally be written in many ways and still express the same logic function. In general, we must use all the rules, laws, and theorems at our command in such a way as to arrive at what appears to be the simplest form of the equation. Inspection of Equation (5–4) shows that a possible first step in simplification is to apply the distributive law to the second term by multiplying the term $CD + BC$ by $\overline{A}B$. The result is

$$X = AC\overline{D} + \overline{A}BCD + \overline{A}BBC$$

The rule that $BB = B$ can be applied to the third term, yielding

$$X = AC\overline{D} + \overline{A}BCD + \overline{A}BC$$

Notice that C is common to each term, so it can be factored out using the distributive law:

$$X = C\,(A\overline{D} + \overline{A}BD + \overline{A}B)$$

Now notice that $\overline{A}B$ appears in the last two terms within the parentheses and can be factored out of those two terms:

$$X = C\,[A\overline{D} + \overline{A}B(D + 1)]$$

Since $D + 1 = 1$,

$$X = C\,(A\overline{D} + \overline{A}B)$$

It appears that this equation cannot be reduced any further, although it can be written in a slightly different way (sum-of-products form) by application of the distributive law:

$$X = AC\overline{D} + \overline{A}BC \tag{5–5}$$

This equation can be implemented with two three-input AND gates, two inverters, and one two-input OR gate, as shown in Figure 5–17.

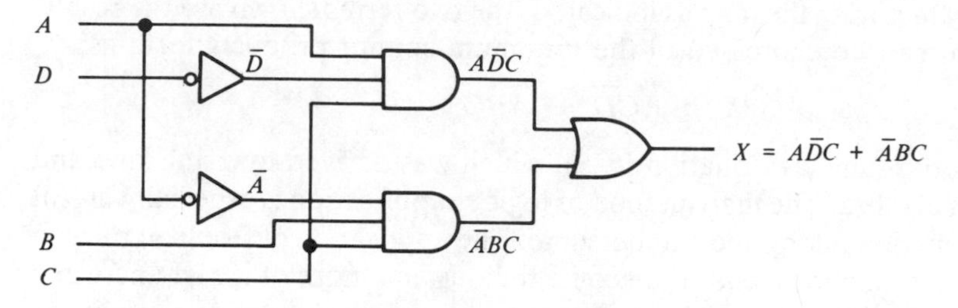

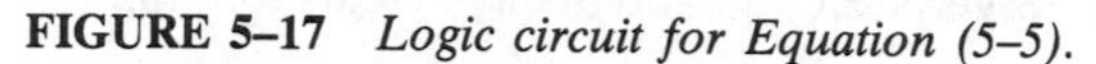
FIGURE 5–17 *Logic circuit for Equation (5–5).*

Compare this circuit with the circuit in Figure 5–16 that has nine logic gates including inverters. The minimized circuit is equivalent, but has only five gates including two inverters. At this point you should verify that the logic circuit of Figure 5–17 is indeed equivalent to the logic circuit of Figure 5–16 in terms of the logical operation. Also, the number of gate *levels* has been reduced from a maximum of five in Figure 5–16 to three in Figure 5–18. Fewer levels mean a shorter propagation delay through the circuit.

Karnaugh Map Approach

The second approach to minimization of Equation (5–4) is the Karnaugh map method. This method is a more systematic means of arriving at a minimum expression and, if factoring is done properly, we can be assured that the result is a minimum expression. The function of Equation (5–4) is rewritten here for convenience:

$$X = AC\bar{D} + \bar{A}B(CD + BC)$$

To minimize this expression, it is converted to a sum-of-products form and plotted on the Karnaugh in Figure 5–18.

$$X = AC\bar{D} + \bar{A}BCD + \bar{A}BC$$

Notice that both the first and third terms include two cells on the map: $AC\bar{D}$ includes

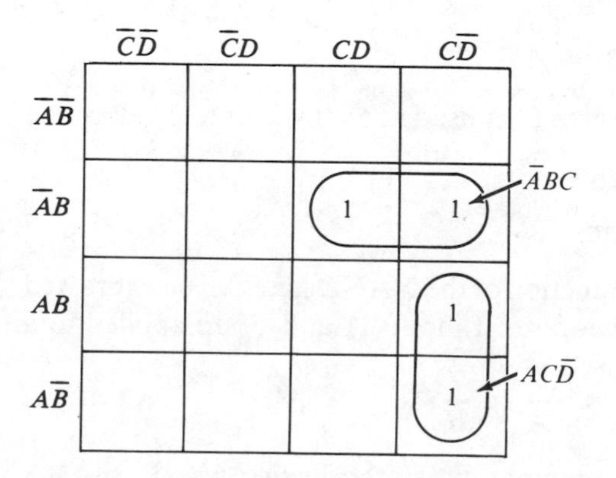

FIGURE 5–18 *Karnaugh map for the function of Equation (5–4).*

cells corresponding to $AB C\overline{D}$ and $A\overline{B}C\overline{D}$. $\overline{A}BC$ includes cells corresponding to $\overline{A}BCD$ and $\overline{A}BC\overline{D}$.

By factoring the map as indicated, the two terms shown are the result, and the function can be expressed in the minimum sum-of-products form as

$$X = AC\overline{D} + \overline{A}BC$$

This equation is the same as Equation (5–5), which was derived by using laws and rules of Boolean algebra. The map method of logic simplification is especially useful when the logic function is long and cumbersome. The application of Boolean rules to extremely complex functions tends to become tedious and more of a trial-and-error process, whereas the map method is very systematic and always yields a minimum result.

5–4 THE UNIVERSAL PROPERTY OF THE NAND GATE

Up to this point only combinational circuits using AND and OR gates and inverters have been considered. To implement many functions, both AND and OR gates in combination are required. In this section we will see that NAND gates can be used to produce *any* logic function. For this reason, they are referred to as *universal gates.*

The NAND gate can be used to generate the NOT function, the AND function, the OR function, and the NOR function. An inverter can be made from a NAND gate by connecting all of the inputs and creating, in effect, a single common input, as shown in Figure 5–19(a) for a two-input gate. An AND function can be generated using only NAND gates, as shown in Figure 5–19(b). Also, an OR function can be produced with NAND gates, as illustrated in Figure 5–19(c). Finally, a NOR function is produced as shown in Figure 5–19(d).

In Figure 5–19(b), a NAND gate is used to invert (complement) a NAND output to form the AND function, as indicated in the following equation:

$$X = \overline{\overline{AB}} = AB$$

In Figure 5–19(c), NAND Gates G_1 and G_2 are used to invert the two input variables before they are applied to NAND gate G_3. The final output is derived as follows by application of DeMorgan's theorem:

$$X = \overline{\overline{A}\,\overline{B}} = A + B$$

In Figure 5–19(d), NAND gate G_4 is used as an inverter to produce the NOR function $\overline{A + B}$.

Multilevel NAND Logic

As you already know, the function of the NAND gate can be expressed in two ways by applying DeMorgan's theorem. If inputs A and B are applied to a NAND gate, the output can be expressed as

$$X = \overline{AB} = \overline{A} + \overline{B}$$

Therefore, the function of the NAND gate can be stated in two ways:

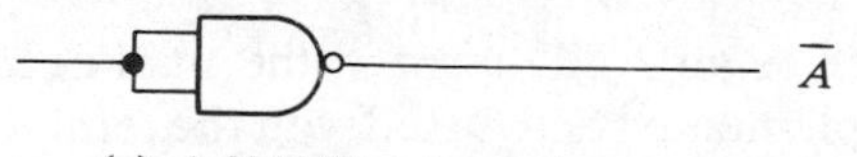

(a) A NAND gate used as an inverter

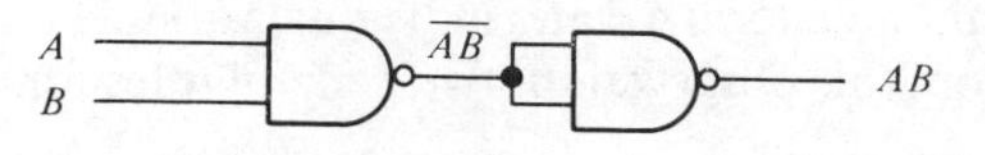

(b) Two NAND gates used as an AND gate

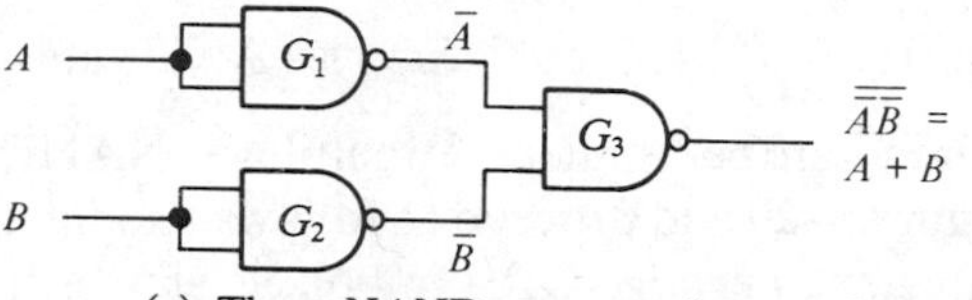

(c) Three NAND gates used as an OR gate

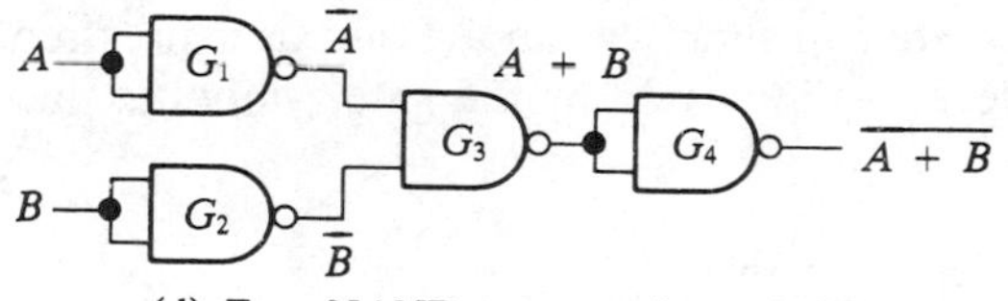

(d) Four NAND gates used as a NOR gate

FIGURE 5–19 *Universal application of NAND gates.*

1. The output of a NAND gate is equal to the complement of the AND of the input variables.
2. The output of a NAND gate is equal to the OR of the complements of the input variables corresponding to a negative-OR.

Rules for multilevel NAND logic can be developed, using these two statements of NAND operation as the basis. To develop a set of rules, we will begin with a two-level NAND circuit, as shown in Figure 5–20.

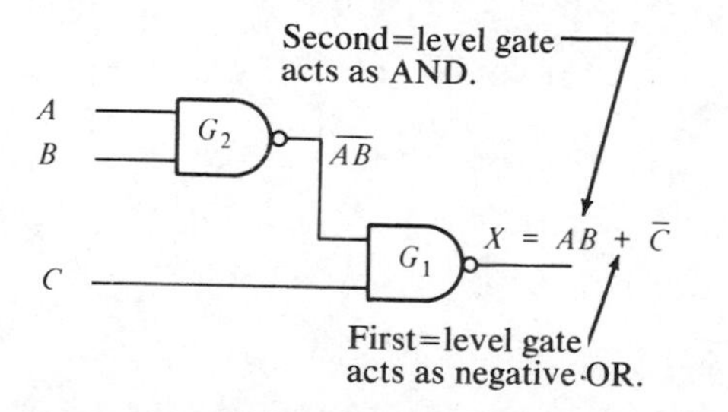

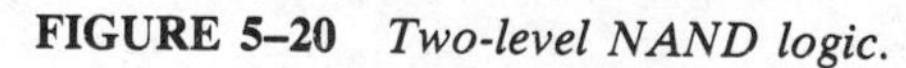

FIGURE 5–20 *Two-level NAND logic.*

For this circuit, the logic equation is developed as follows by repeated application of DeMorgan's theorem:

$$X = \overline{(\overline{AB})C} = \overline{(\overline{A} + \overline{B})C} = AB + \overline{C}$$

Notice that input variables A and B appear as an AND term in the final output equation, and input variable C appears complemented and ORed with the term AB. Gate G_2 is a second-level gate (second gate back from the output) and gate G_1 is a first-level gate (first gate from output). Notice also that individual variables A and B are inputs to the second-level gate G_2, and that the term $\overline{AB}$ and individual variable C are the inputs to the first-level gate G_1. From this observation, two general rules for multilevel NAND logic can be stated:

1. All odd-numbered logic levels (1, 3, 5, etc.) "act" as OR gates with single input variables complemented.
2. All even-numbered logic levels (2, 4, 6, etc.) "act" as AND gates.

To illustrate how this method can be applied to multilevel NAND logic, let us take the network shown in Figure 5–20 and proceed as follows. Replace each *odd*-level NAND gate with its equivalent *negative*-OR symbol, as shown in Figure 5–21. Remember, nothing has been changed operationally in the circuit; we are simply replacing the NAND gate symbol with an *equivalent* one. Now, rather than having to remember the two rules, we can simply write the output expression based on the logical function at each level, as indicated by the gate symbol at that level.

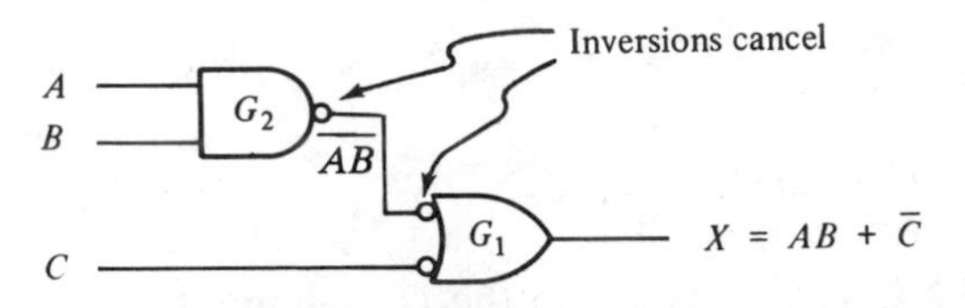

FIGURE 5–21 *NAND/Negative-OR equivalent of two-level NAND logic.*

Notice that the bubble output of G_2 is connected to the bubble input of G_1. This, of course, indicates a double inversion that effectively cancels the inversion at each of the levels. In this approach, all connections between levels are "bubble-to-bubble"

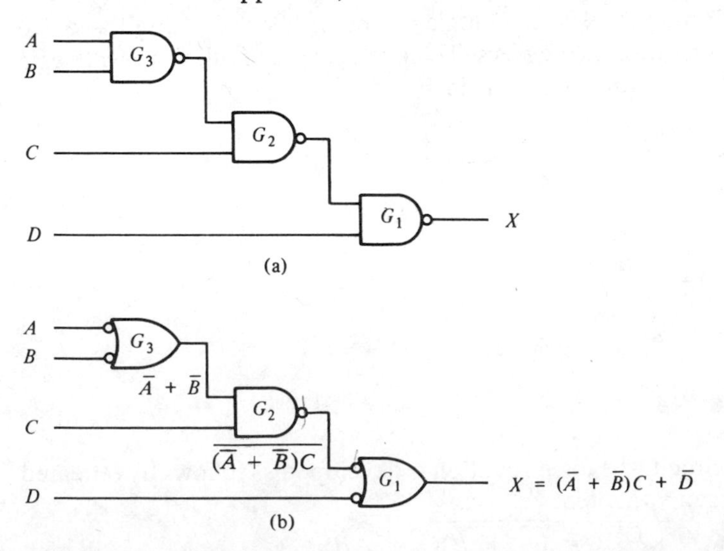

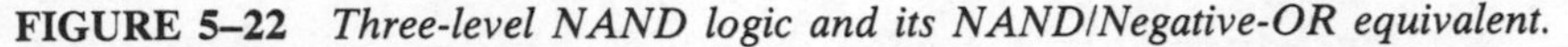
FIGURE 5–22 *Three-level NAND logic and its NAND/Negative-OR equivalent.*

going from even levels to odd levels, and straight connections going from odd levels to even levels. Thus, all inversions can be discarded when writing the output expression, except for single-input variables at odd levels.

Next, let us take the three-level network of Figure 5–22(a) to further illustrate this method. By replacing gates G_1 and G_3 (odd-level gates) with their negative-OR equivalents, we have the equivalent network of Figure 5–22(b). From this we can easily write the output expression as follows:

$$X = (\overline{A} + \overline{B})C + \overline{D}$$

Example 5–5

Write the output expression for the circuit in Figure 5–23 using the procedure just discussed.

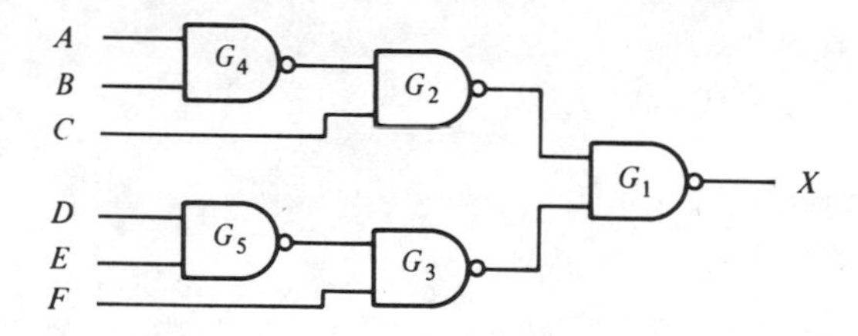

FIGURE 5–23

Solution:

Redrawing the network in Figure 5–23 with equivalent negative-OR symbols at the odd levels, we get Figure 5–24. Writing the expression for X directly from the indicated logic operation at each level gives us

$$X = (\overline{A} + \overline{B})C + (\overline{D} + \overline{E})F$$

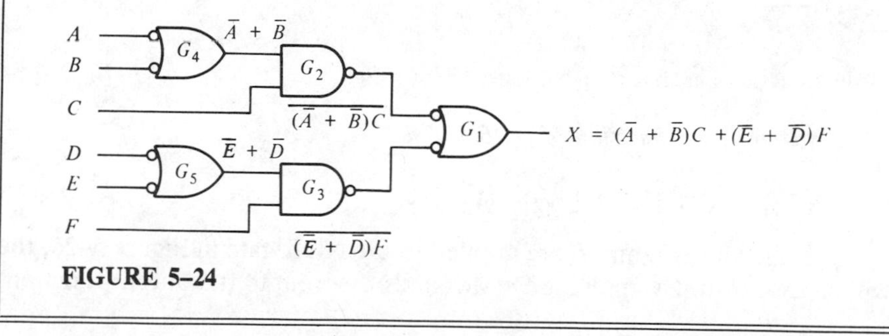

FIGURE 5–24

5–5 THE UNIVERSAL PROPERTY OF THE NOR GATE

As with the NAND gate, the NOR gate can be used to generate the NOT, AND, OR, and NAND functions. A NOT circuit or inverter can be made from a NOR gate by connecting all of its inputs to effectively create a single input, as shown in Figure 5–25(a) with a two-input example. Also, an OR gate can be produced from NOR gates, as illustrated in Figure 5–25(b).

An AND gate can be constructed using NOR gates, as shown in Figure 5–25(c). In this case the NOR gates G_1 and G_2 are used as inverters, and the final output is derived using DeMorgan's theorem as follows:

$$X = \overline{\overline{A} + \overline{B}} = AB$$

Figure 5–25(d) shows NOR gates used to form a NAND function.

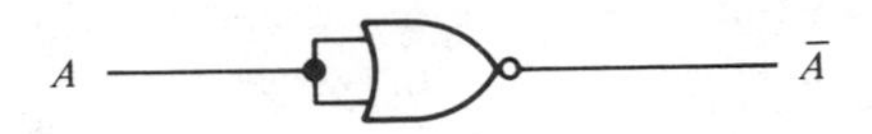

(a) A NOR gate used as an inverter

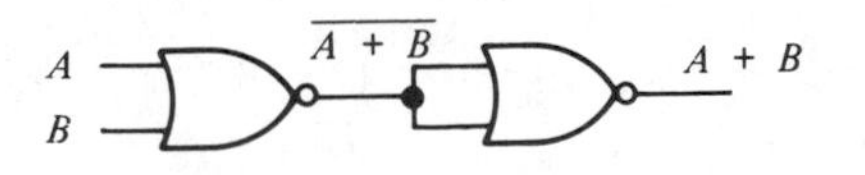

(b) Two NOR gates used as an OR gate

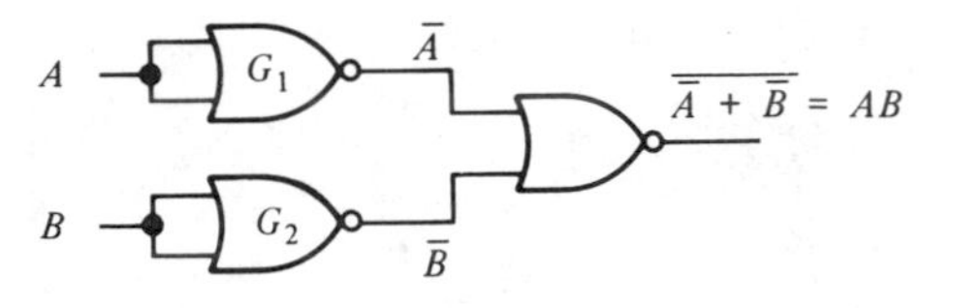

(c) Three NOR gates used as an AND gate

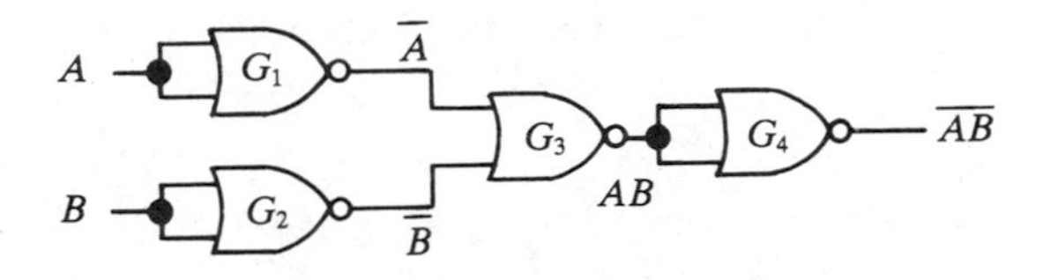

(d) Four NOR gates used as a NAND gate

FIGURE 5–25 *Universal application of NOR gates.*

Multilevel NOR Logic

If the inputs *A* and *B* are applied to the NOR gate in Figure 5–26, the output will be as shown. By applying DeMorgan's theorem to the output function, we get:

$$X = \overline{A + B} = \overline{A}\overline{B}$$

Notice that the output can be expressed in two ways:

1. The output of a NOR gate is equal to the complement of the OR of the input variables.

2. The output of the NOR gate is equal to the AND of the complements of input variables corresponding to a negative AND.

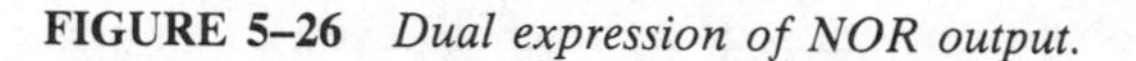

FIGURE 5–26 *Dual expression of NOR output.*

Rules for multilevel NOR logic can be developed, using these two statements of NOR operation as the basis. To develop a set of rules, we will begin with a two-level NOR circuit, as shown in Figure 5–27.

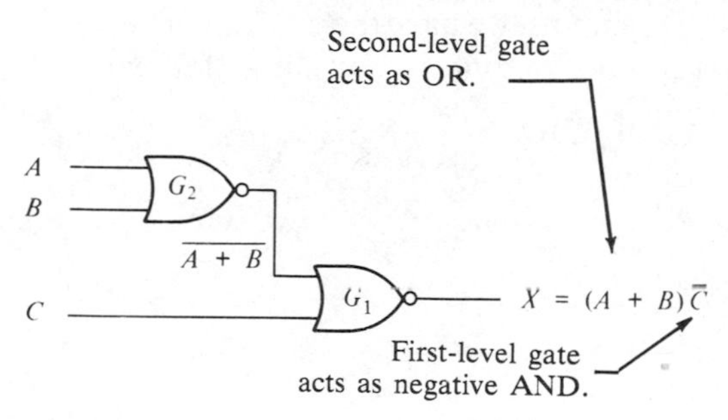

FIGURE 5–27 *Two-level NOR logic.*

For this circuit, the logic equation is developed as follows by application of DeMorgan's theorem:

$$X = \overline{\overline{(A + B)} + C} = \overline{\overline{(A + B)}}\;\overline{C} = (A + B)\overline{C}$$

Notice that the input variables A and B appear as an OR term in the final output equation, and the input variable C appears complemented and ANDed with the term $A + B$. Gate G_2 is a second-level gate (second back from output), and gate G_1 is a first-level gate (first gate from output). Notice also that the individual variables A and B are inputs to the second-level gate, and that the term $\overline{A + B}$ and the individual variable C are the inputs to the first-level gate. From this observation, two rules for multilevel NOR logic can be stated:

1. All odd-numbered logic levels (1, 3, 5, etc.) act as AND gates with single input variables complemented.
2. All even-numbered levels (2, 4, 6, etc.) act as OR gates.

To illustrate how this method can be applied to multilevel NOR logic, let us take the network shown in Figure 5–27 and proceed as follows. Replace each

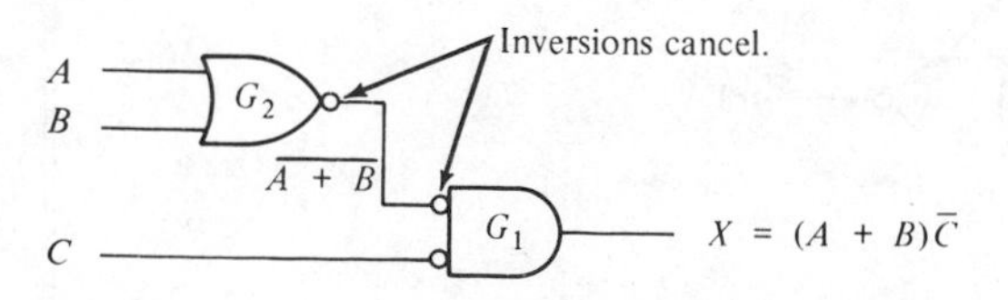

FIGURE 5–28 *NOR/Negative–AND equivalent of two-level NOR logic.*

odd-level NOR gate with its equivalent *negative*-AND symbol, as shown in Figure 5–28. Keep in mind that nothing has been changed operationally in the circuit; we are simply replacing the NOR gate symbol with an *equivalent* one. Now, we can write the expression for the output based on the logical function at each level, as indicated by the gate symbol at that level.

Notice that the bubble output of gate G_2 is connected to the bubble input of gate G_1. The bubble-to-bubble connection going from an even to an odd level effectively cancels the inversions. So, in this case, gate G_2 effectively looks like an OR gate. The only inversions occur for single-input variables at odd levels.

Next, let us take the three-level network of Figure 5–29(a) in order to further illustrate this method of analysis. By replacing gates G_1 and G_3 odd-level gates) with their equivalent negative-AND symbols, we have the equivalent network of Figure 5–29(b). From this we can write the output expression as follows:

$$X = (\overline{A}\overline{B} + C)\overline{D}$$

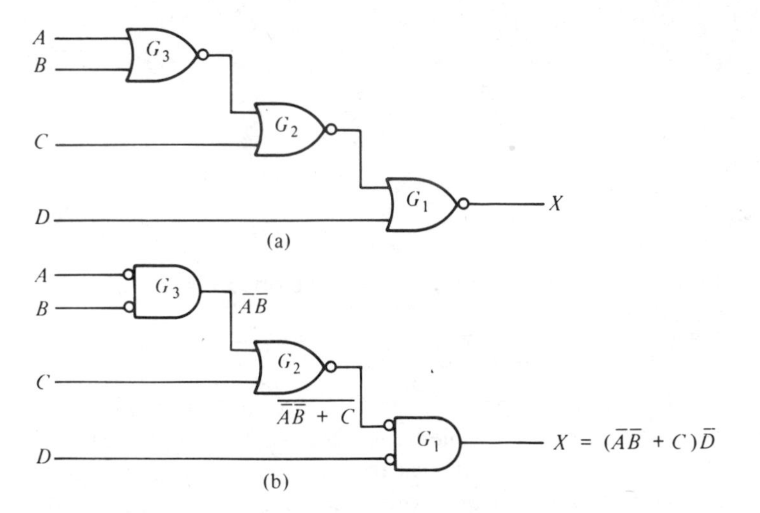

FIGURE 5–29 *Three-level NOR logic and its NOR/Negative-AND equivalent.*

Example 5–6

Write the output expression for the circuit in Figure 5–30, using the method just discussed.

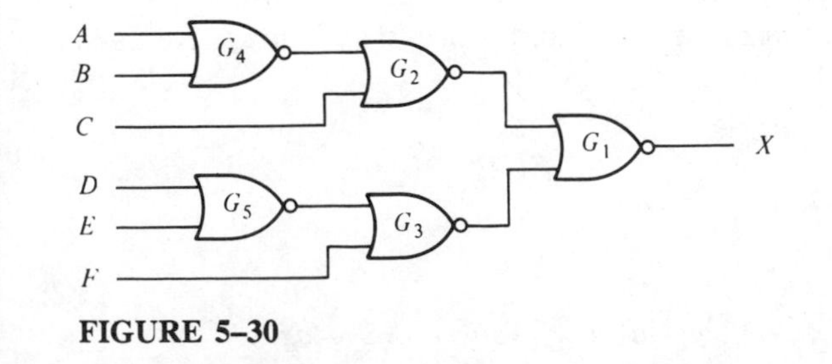

FIGURE 5–30

Solution:

Redraw the network with the equivalent negative-AND symbols at the odd levels, as shown in Figure 5–31. Writing the expression for X directly from the indicated logic operation at each level, we get

$$X = (\overline{A}\overline{B} + C)(\overline{D}\overline{E} + F)$$

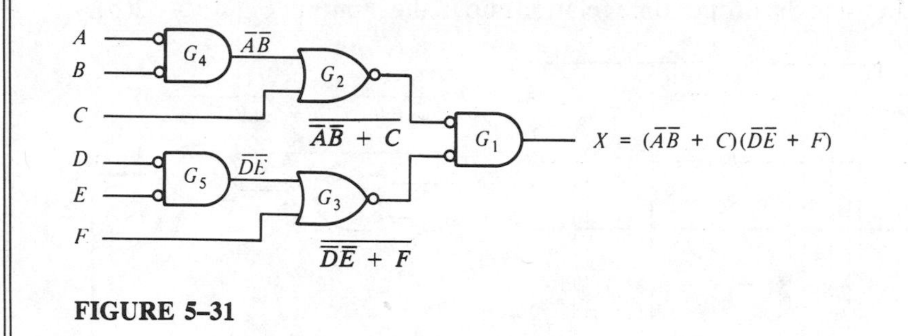

FIGURE 5–31

PULSED OPERATION 5–6

The operation of combinational logic circuits with pulsed inputs follows the same logical operation we have discussed in the previous sections. The output of the logic network is dependent on the states of the inputs at any given time; if the inputs are changing levels according to a time-varying pattern (waveform), the output depends on what states the inputs happen to be at any instant. The following is a summary of the operation of individual gates for use in analyzing combinational circuits with time-varying inputs.

1. The only time that the output of an AND gate is HIGH is when *all* inputs are HIGH *at the same time.*
2. The output of an OR gate is HIGH any time at least *one* of its inputs is HIGH. The output is LOW only when *all* inputs are LOW *at the same time.*
3. The only time the output of a NAND gate is LOW is when *all* inputs are HIGH *at the same time.*
4. The output of a NOR gate is LOW any time at least *one* of its inputs is HIGH. The output is HIGH only when *all* inputs are LOW *at the same time.*

The logical operation of any gate is the same regardless of whether its inputs are pulsed or constant levels. The nature of the inputs (pulsed or constant levels) does not alter the truth table operation of a gate. A few examples will serve to illustrate the analysis of combinational logic circuits with pulsed inputs.

Example 5–7

Determine the output waveform for the circuit in Figure 5–32(a), with the inputs as shown.

Solution:

X is shown in the proper time relationship to the inputs in Figure 5–32(b).

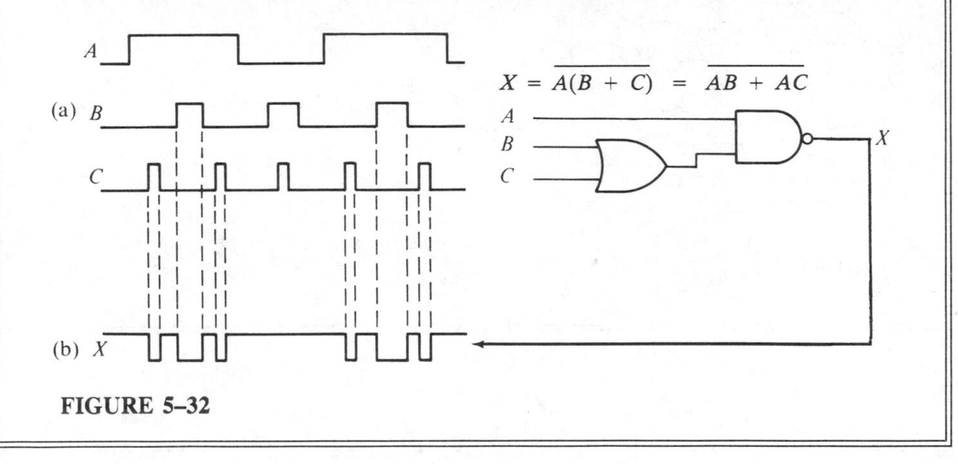

FIGURE 5–32

Example 5–8

Determine the output waveform for the circuit in Figure 5–33(a), if the input waveforms are as indicated.

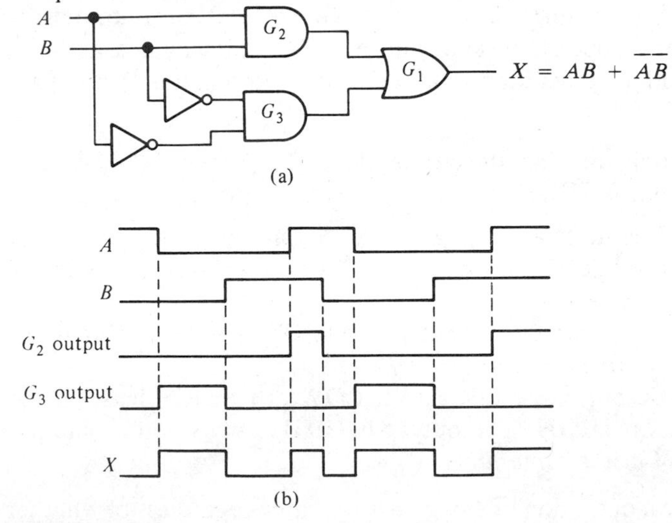

FIGURE 5–33

Solution:

When both inputs are HIGH or when both inputs are LOW, the output is HIGH. This is sometimes called a coincidence circuit or exclusive-NOR.

The intermediate outputs of gates G_2 and G_3 are used to develop the final output. See Figure 5–33(b).

Example 5–9

If the pulse waveforms shown in Figure 5–34(a) are applied to the 7451 AND–OR–INVERT circuit in (b), sketch the output waveform at pin 6.

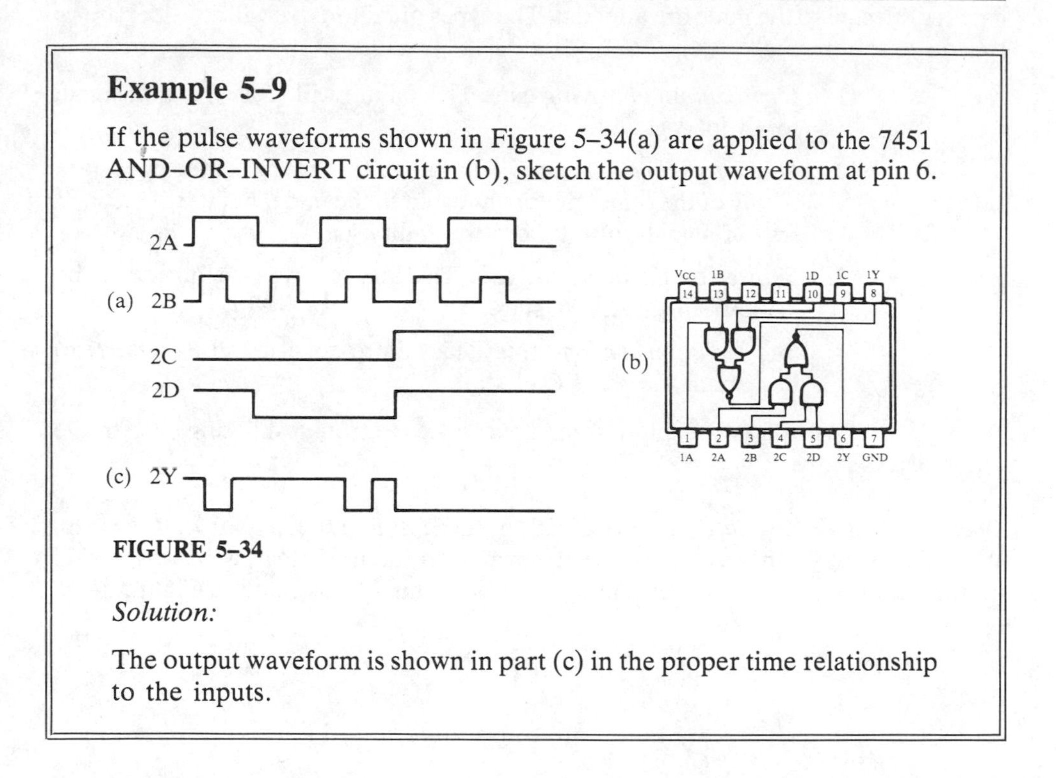

FIGURE 5–34

Solution:

The output waveform is shown in part (c) in the proper time relationship to the inputs.

TROUBLESHOOTING GATE NETWORKS 5–7

In a typical combinational logic network, the output of one gate is connected to one or more gate inputs as shown in Figure 5–35. The interconnecting paths represent a common electrical point known as a *node.*

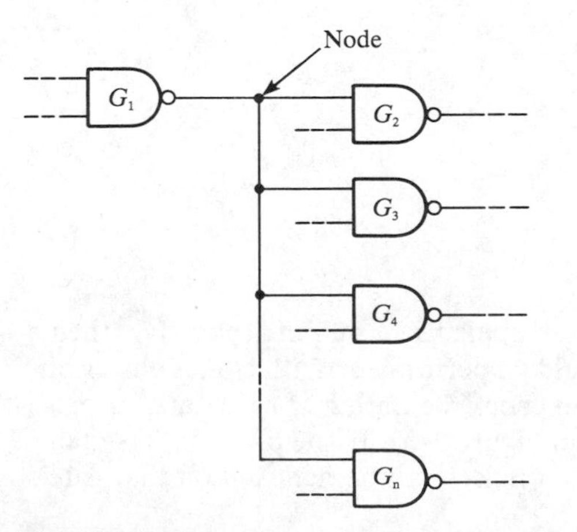

FIGURE 5–35

Gate G_1 in the figure is *driving* the node and the other gates represent *loads* connected to the node. Several types of failures are possible in this situation. Some of these failure modes are difficult to isolate down to a bad gate because all the gates connected to the node are affected. The types of failures we will consider are as follows:

1. *Open output in driving gate.* This failure will cause a loss of signal to all load gates.
2. *Open input in a load gate.* This failure will not affect the operation of any of the other gates connected to the node; but it will result in loss of signal output from the faulty gate.
3. *Shorted output in driving gate.* This failure can cause the node to be "stuck" in the LOW state.
4. *Shorted input in a load gate.* This failure can also cause the node to be "stuck" in the LOW state.

We will now explore some approaches to troubleshooting each of the faults listed above.

Open output in driving gate In this situation, there is no pulse activity on the node. With circuit power on, an open node will normally result in a "floating" level and will be indicated by a *dim lamp* on the *logic probe.* This is illustrated in Figure 5–36.

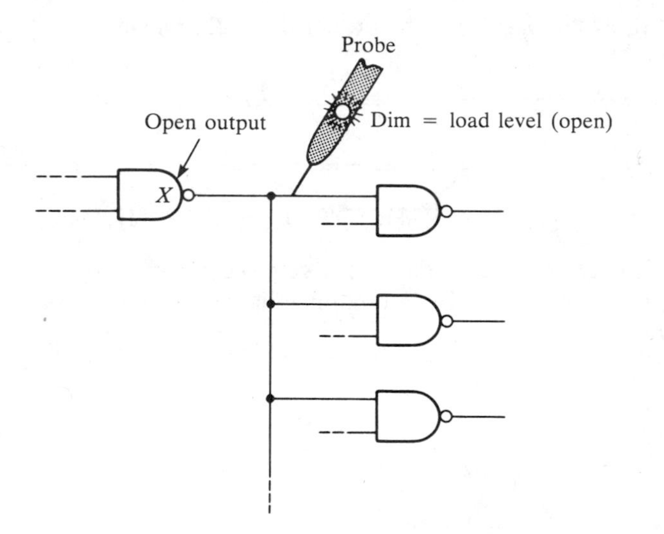

FIGURE 5–36

Open input in a load gate If the check for an open driver output is negative, then a check for an open input in a load gate should be performed. With circuit power off, apply the *logic pulser* tip to the node. Then check the output of each gate for pulse activity with the *logic probe* as illustrated in Figure 5–37. If one of the inputs that is normally connected to the node is open, no pulses will be detected on that gate's output.

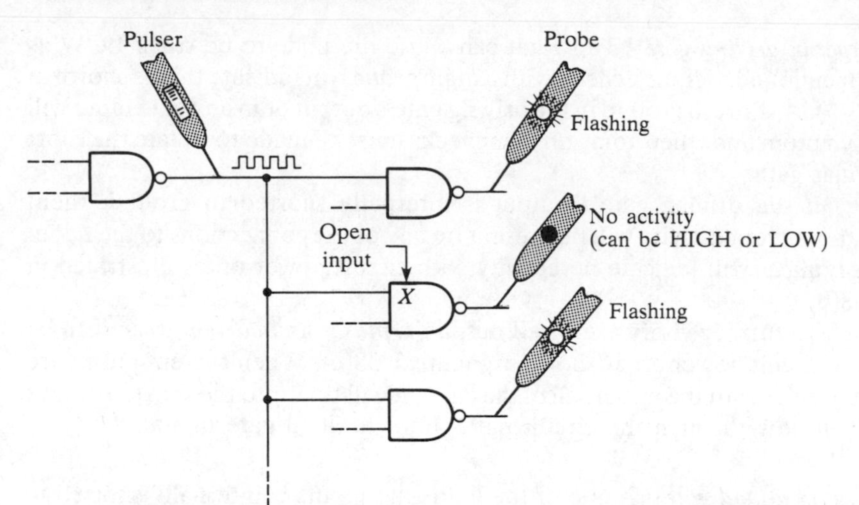

FIGURE 5–37

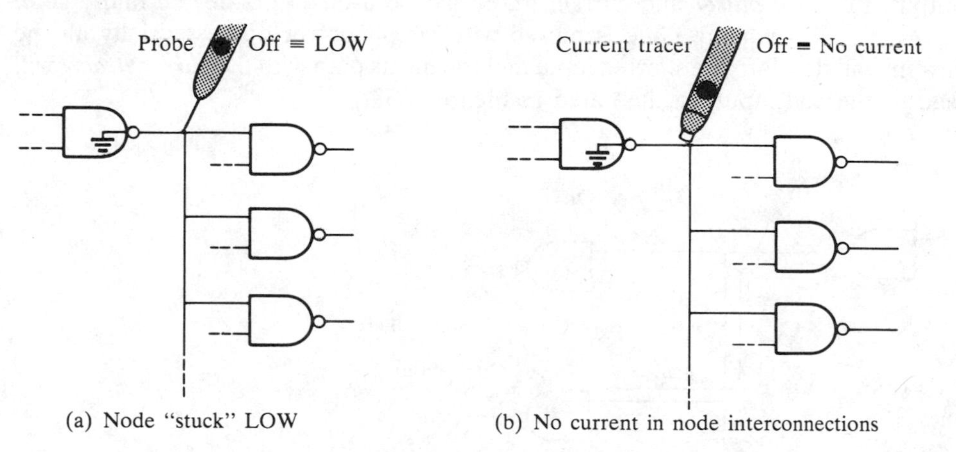

(a) Node "stuck" LOW

(b) No current in node interconnections

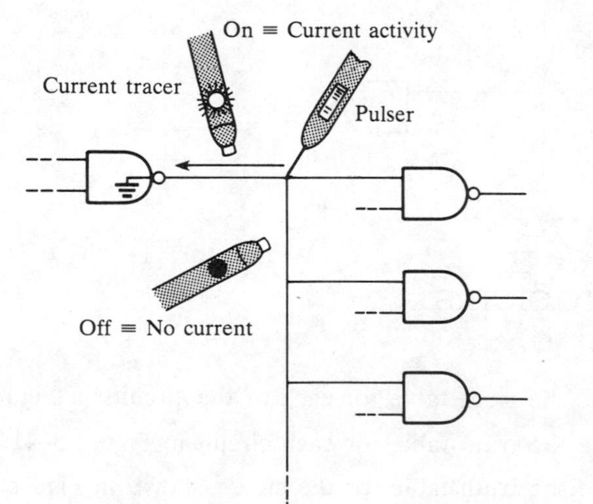

(c) Tracing current along path into short

FIGURE 5–38

Shorted output in driving gate This fault can cause the node to be stuck LOW as previously mentioned. A quick check with a *logic probe* will indicate this as shown in Figure 5–38(a). A short to ground in the driving gate's output or in any gate input will cause this symptom and, therefore, further checks must be made to isolate the short to a particular gate.

If the driving gate's output is internally shorted to ground, then, essentially, no current activity will be present on any of the connections to the node. So, a *current tracer* will indicate no activity with circuit power on as illustrated in Figure 5–38(b).

To further verify a shorted output, a *pulser* and *current tracer* can be used with the circuit power off as shown in Figure 5–38(c). When current pulses are applied to the node with the *pulser,* all of the current will flow into the shorted output and none will flow through the circuit paths into the load gate inputs.

Shorted input in a load gate If one of the load gate inputs is internally shorted to ground, the node will be stuck in the LOW state. Again, as in the case of a shorted output, the *logic pulser* and *current tracer* can be used to isolate the faulty gate.

When the node is pulsed with circuit power off, essentially all the current will flow into the shorted input and tracing its path with the *current tracer* will lead to the bad input as illustrated in Figure 5–39.

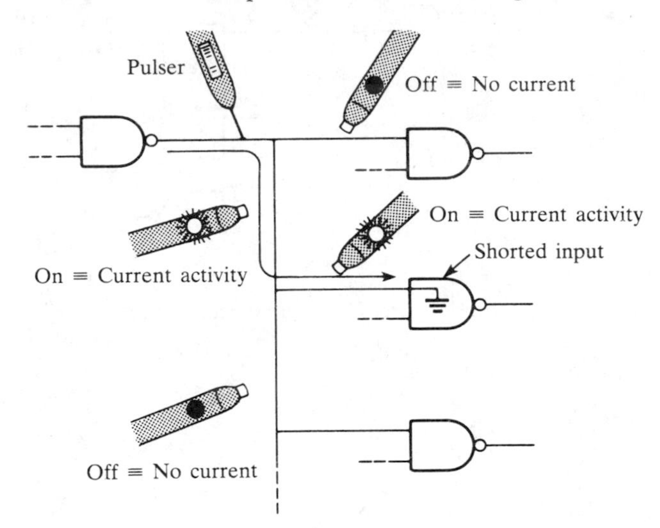

FIGURE 5–39

Problems

Section 5–1

5–1 Determine the truth table for each of the circuits in Figure 5–40.

5–2 Determine the truth table for each circuit in Figure 5–41.

5–3 Determine the truth table for the circuit shown in Figure 5–42.

5–4 Write the logic expression for the output of each of the circuits in Figure 5–43.

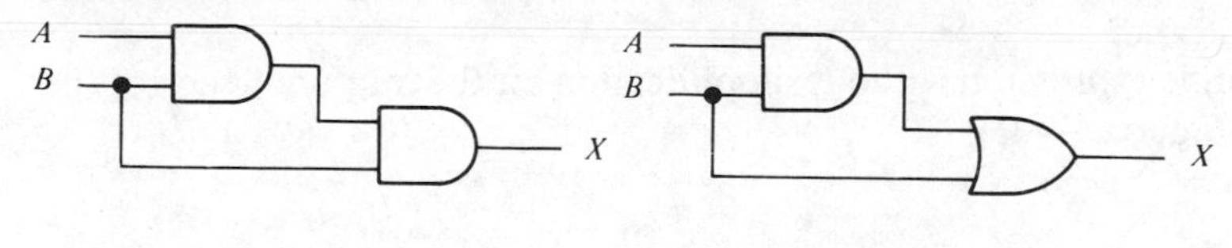

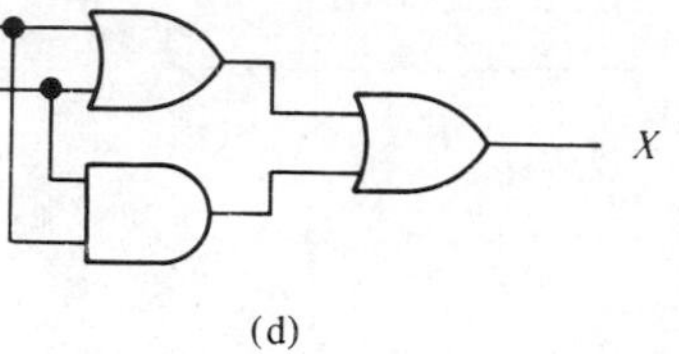

(c) (d)

FIGURE 5–40

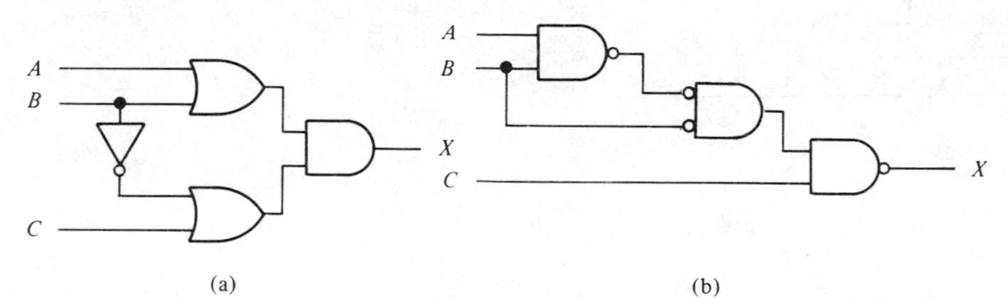

FIGURE 5–41

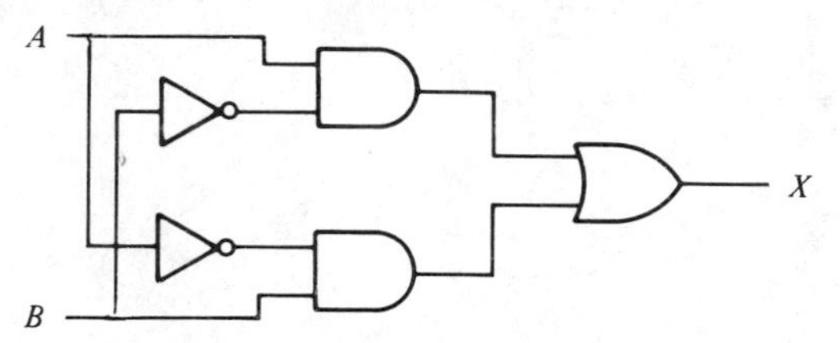

FIGURE 5–42

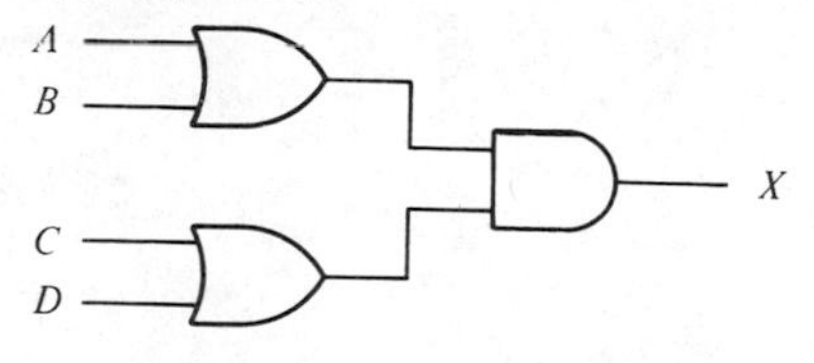

(a)

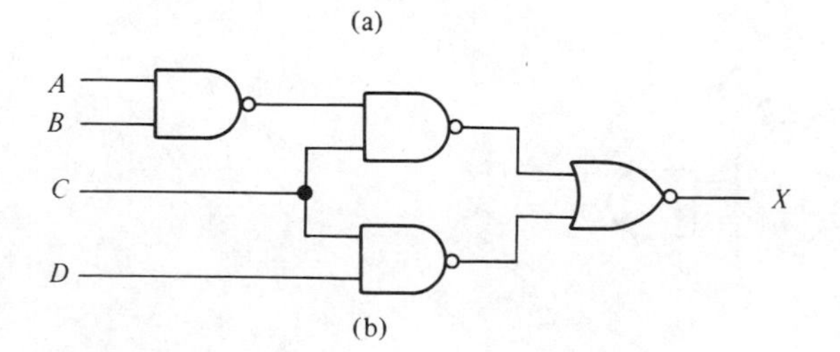

(b)

FIGURE 5–43

5–5 Write the logic expression without simplification for the output of each of the circuits in Figure 5–44.

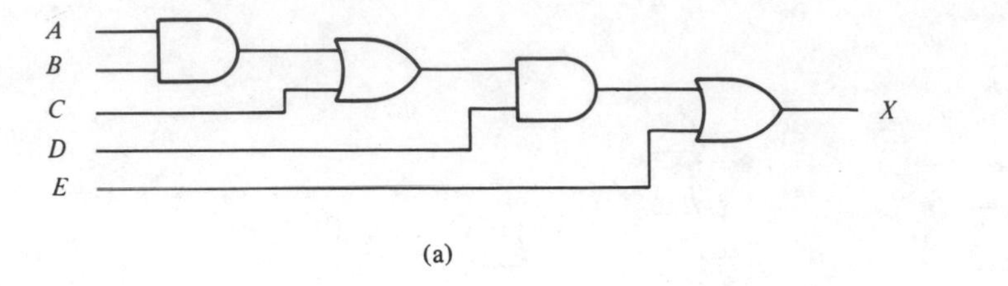

(a)

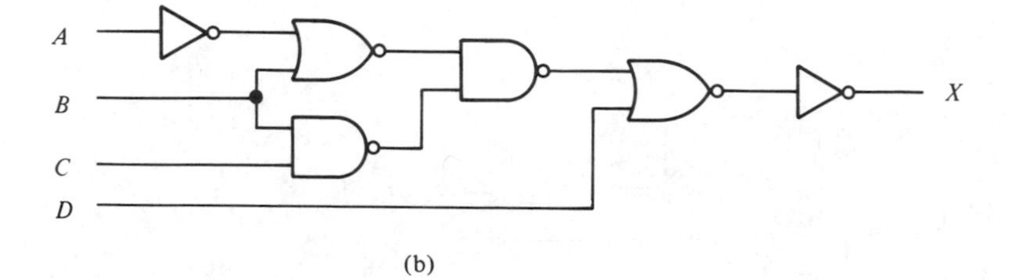

(b)

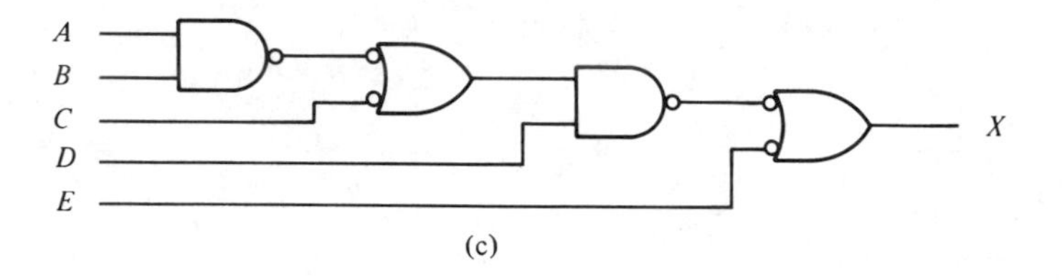

(c)

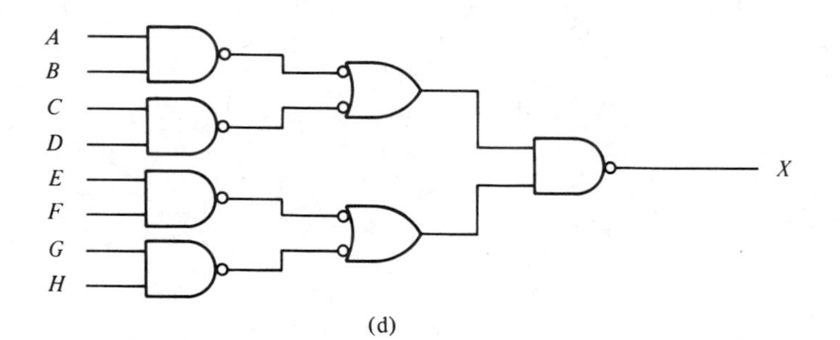

(d)

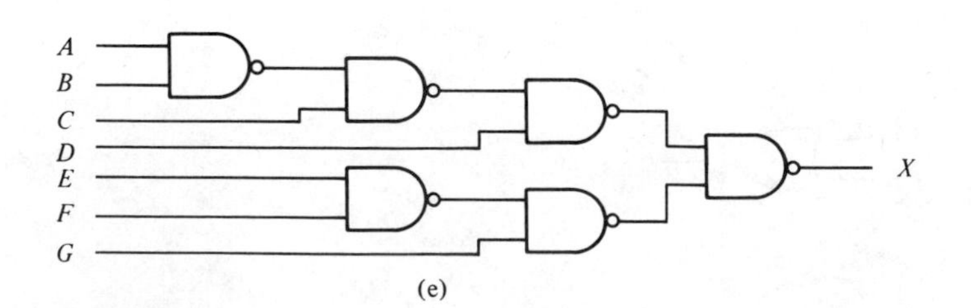

(e)

FIGURE 5–44

5–6 Develop the truth table for the circuit in Figure 5–45 by writing and evaluating the output expression for each input variable combination.

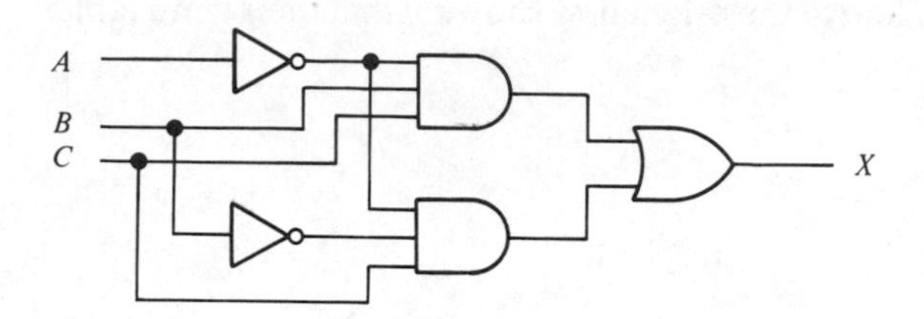

FIGURE 5–45

5–7 Develop the truth table for the circuit of Figure 5–16.

5–8 Repeat Problem 5-6 if each AND gate is replaced by a NAND gate and each OR gate is replaced by a NOR gate.

Section 5–2

5–9 Use AND gates, OR gates, or combinations of both to implement the following logic expressions:

(a) $X = AB$ (b) $X = A + B$
(c) $X = AB + C$ (d) $X = ABC + D$
(e) $X = A + B + C$ (f) $X = ABCD$
(g) $X = A(CD + B)$ (h) $X = AB(C + DEF) + CE(A + B + F)$

5–10 Use AND gates, OR gates, and inverters as needed to implement the following logic expressions:

(a) $X = AB + \overline{B}C$
(b) $X = A(B + \overline{C})$
(c) $X = A\overline{B} + AB$
(d) $X = \overline{ABC} + B(EF + \overline{G})$
(e) $X = A[BC(A + B + C + D)]$
(f) $X = B(C\overline{D}E + \overline{E}FG)(\overline{AB} + C)$

5–11 Use NAND gates, NOR gates, or combinations of both to implement the following logic expressions:

(a) $X = \overline{A}B + CD + (\overline{A + B})(ACD + \overline{BE})$
(b) $X = AB\overline{C}\overline{D} + D\overline{E}F + \overline{AF}$
(c) $X = \overline{A}[B + \overline{C}(D + E)]$

5–12 Show how the following expressions can be implemented using only NAND gates.

(a) $X = ABC$ (b) $X = \overline{ABC}$
(c) $X = A + B$ (d) $X = A + B + \overline{C}$
(e) $X = \overline{AB} + \overline{CD}$ (f) $X = (A + B) \cdot (C + D)$
(g) $X = AB[C(\overline{DE} + \overline{AB}) + \overline{BCE}]$

5–13 Repeat Problem 5–12 using only NOR gates.

Section 5–3

5–14 Simplify the circuit in Figure 5–46 as much as possible, and verify that the simplified circuit is equivalent to the original by showing that their truth tables are identical.

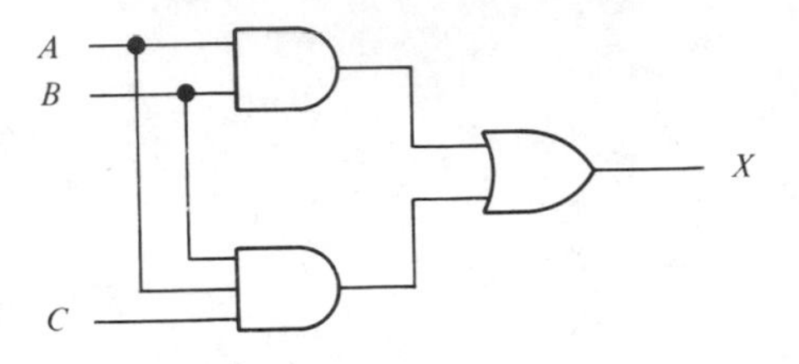

FIGURE 5–46

5–15 Repeat Problem 5–14 for the circuit in Figure 5–47.

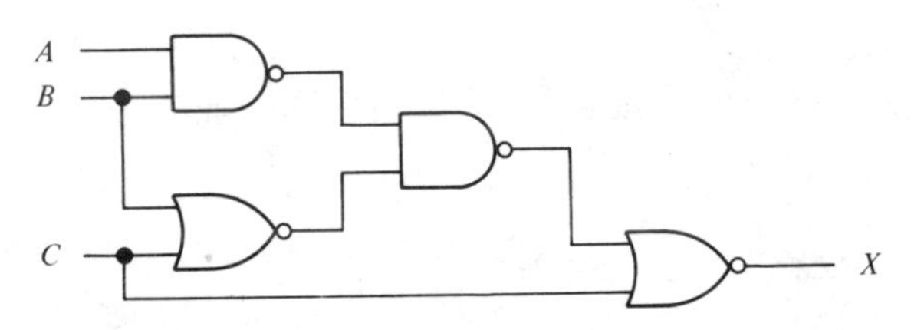

FIGURE 5–47

5–16 Minimize the gates required to implement the functions in each part of Problem 5–10.

5–17 Minimize the gates required to implement the functions in each part of Problem 5–11.

5–18 Minimize the gates required to implement the functions in each part of Problem 5–12.

5–19 Use NAND gates to implement the function plotted on the Karnaugh map in Figure 5–48.

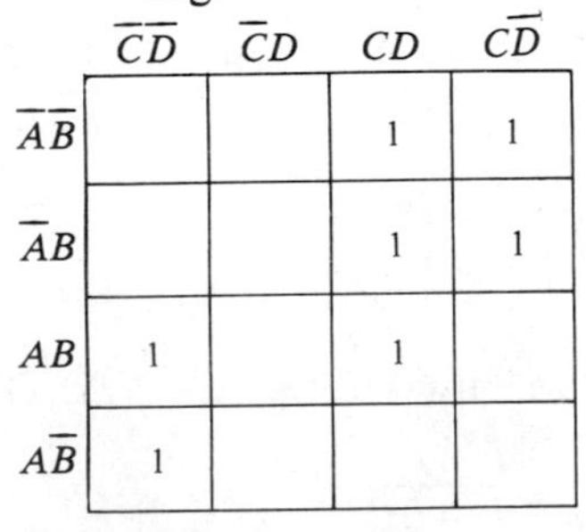

	$\overline{C}\overline{D}$	$\overline{C}D$	CD	$C\overline{D}$
$\overline{A}\overline{B}$			1	1
$\overline{A}B$			1	1
AB	1		1	
$A\overline{B}$	1			

FIGURE 5–48

Section 5–6

5–20 Given the logic circuit and the input waveforms in Figure 5–49 sketch the output waveform.

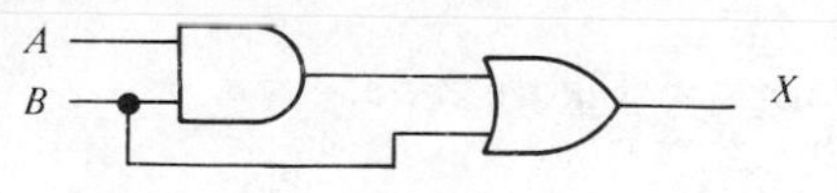

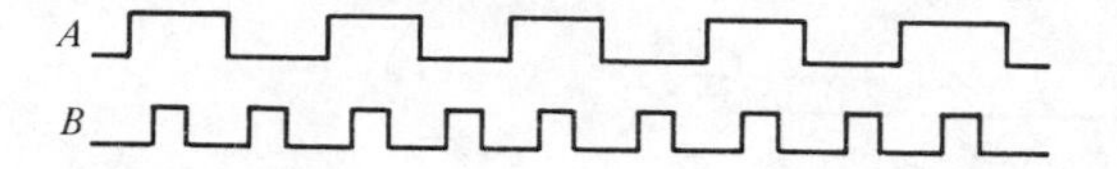

FIGURE 5–49

5–21 For the logic circuit in Figure 5–50, sketch the output waveform in proper relation to the inputs.

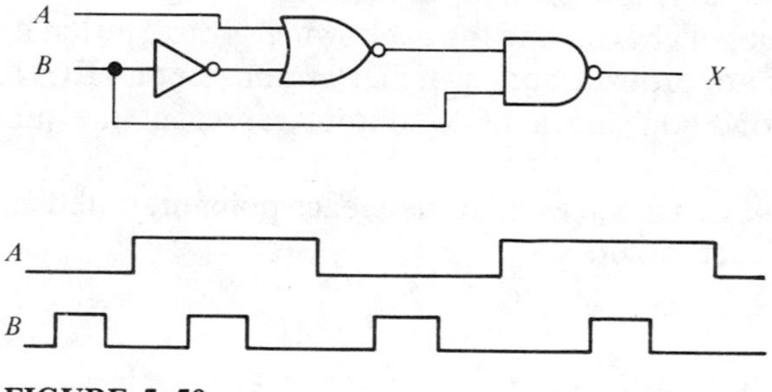

FIGURE 5–50

5–22 Assume that you require the output waveform in Figure 5–51, and one input is given as indicated. Determine the other input needed to generate the desired output for the logic circuit shown.

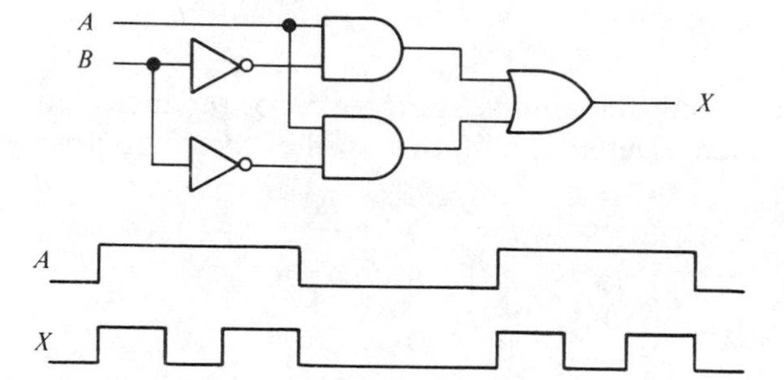

FIGURE 5–51

5–23 For the input waveforms in Figure 5–52, what logic circuit would generate the output waveform shown?

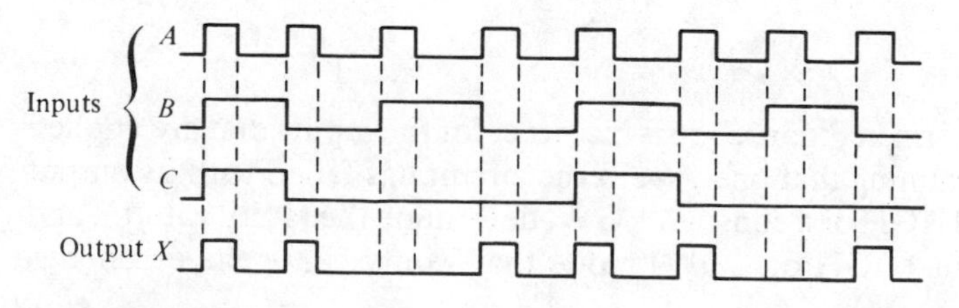

FIGURE 5–52

5–24 Repeat Problem 5–23 for the waveforms in Figure 5-53.

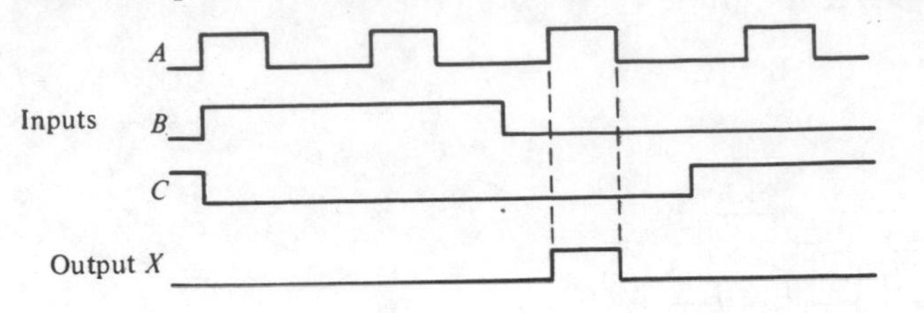

FIGURE 5–53

5–25 Develop the logic circuit required to meet the following requirements: A lamp in a room is to be operated from two switches, one at the back door and one at the front door. The lamp is to be on if the front switch is on and the back switch off, or if the front switch is off and the back switch is on. The lamp is to be off if both switches are off or if both switches are on. Let a HIGH output represent the on condition and a LOW output represent the off condition.

5–26 For the circuit in Figure 5–54, sketch the waveforms at each point indicated in the proper relationship to each other.

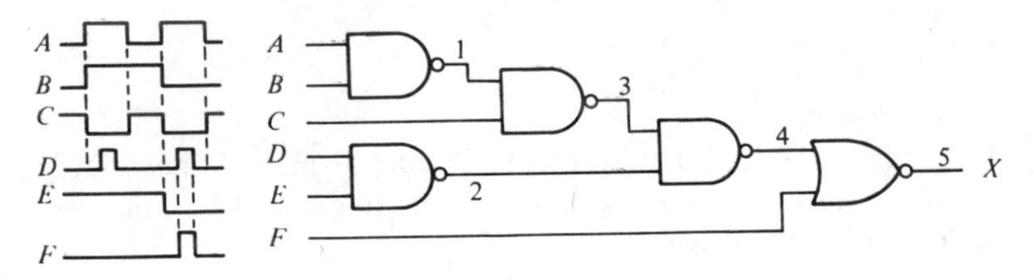

FIGURE 5–54

5–27 For the logic circuit and the input waveforms in Figure 5–55, the indicated output waveform is observed. Determine if this is the correct output waveform.

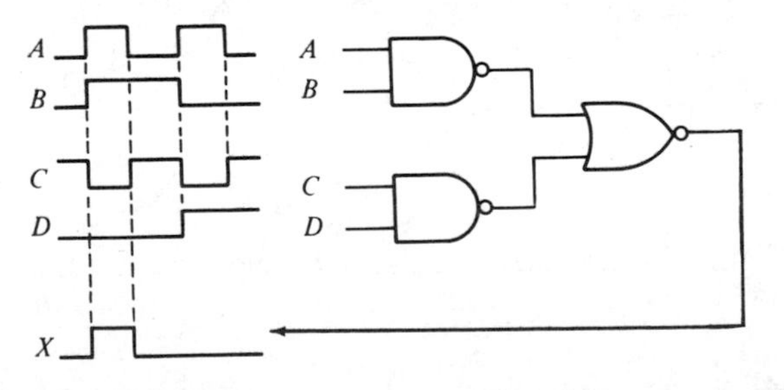

FIGURE 5–55

5–28 The output waveform in Figure 5–56 is incorrect for the inputs that are applied to the circuit. Assuming that one gate in the circuit has failed with its output either a constant HIGH or a constant LOW, determine the faulty gate(s), and the failure (output HIGH or LOW) mode that would cause the erroneous output.

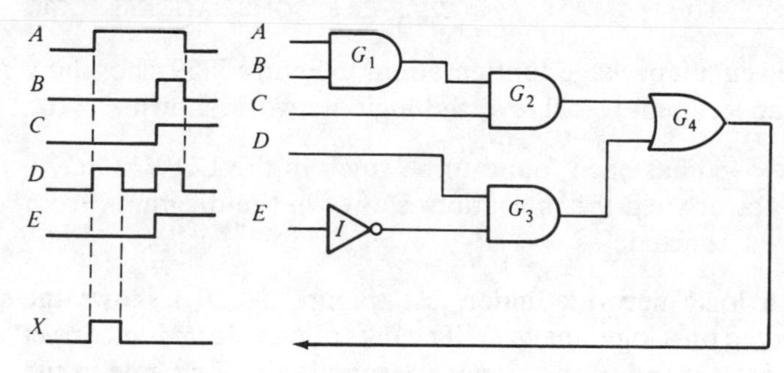

FIGURE 5–56

5–29 Repeat Problem 5–28 for the circuit in Figure 5–57, with input and output waveforms as shown.

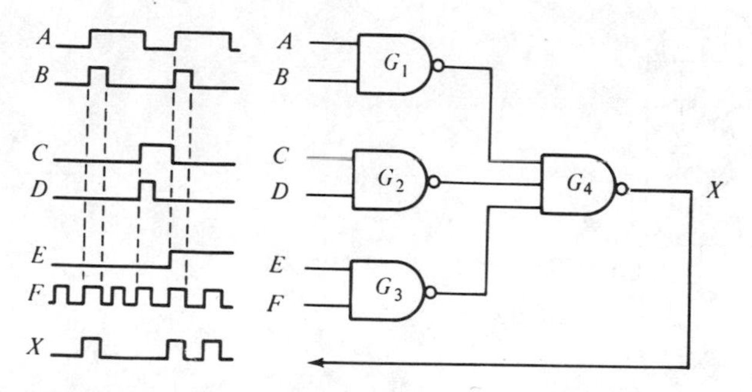

FIGURE 5–57

5–30 Assuming a propagation delay through each gate of 10 nanoseconds (ns), determine if the *desired* output waveform in Figure 5–58 (a pulse with a minimum t_w = 25 ns positioned as shown) will be generated properly with the given inputs.

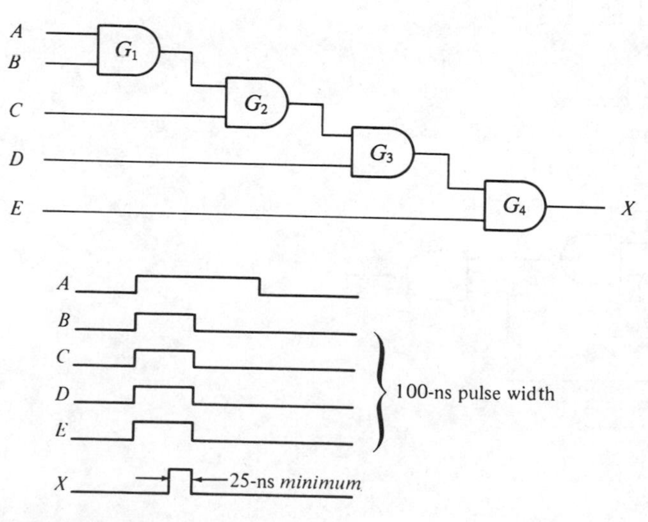

FIGURE 5–58

5–31 Given the integrated circuit package configurations in Figure 5–59, show how you would interconnect them to achieve the logic network shown.

5–32 The node in Figure 5–60 has been found to be stuck in the LOW state. A pulser and current tracer yield the indications shown in the diagram. From this, what would you conclude?

5–33 Figure 5–61(a) is a logic network under test. Figure 5–61(b) shows the waveforms as observed on a logic analyzer. The output waveform is incorrect for the inputs that are applied to the circuit. Assuming that one gate in the network has failed with its output either a constant HIGH or a constant LOW, determine the faulty gate and the failure mode that would cause the erroneous output.

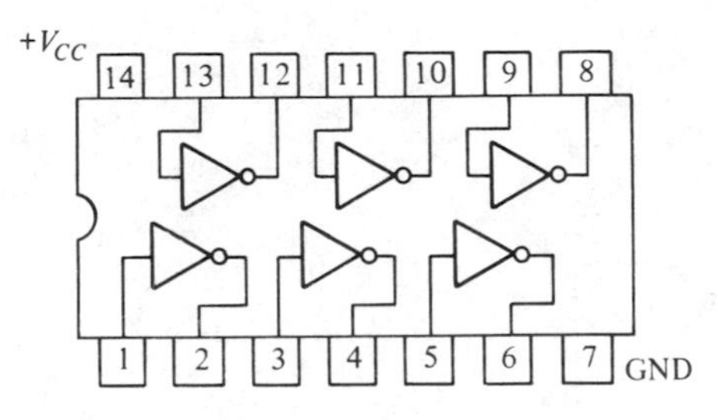

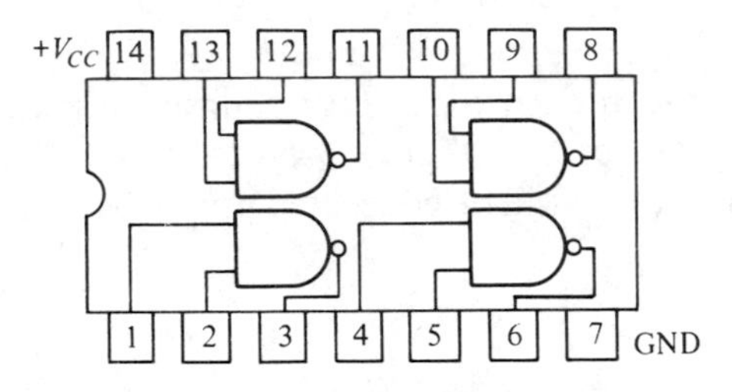

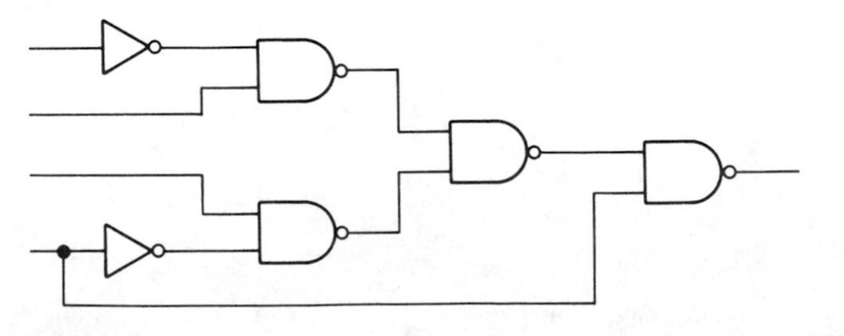

FIGURE 5–59

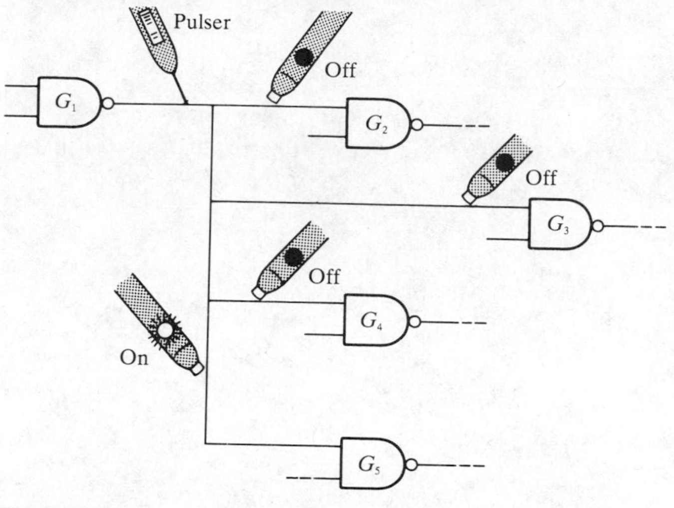

FIGURE 5–60

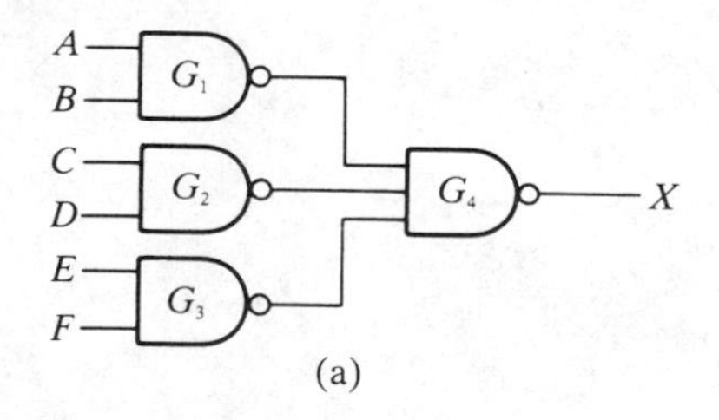

(a)

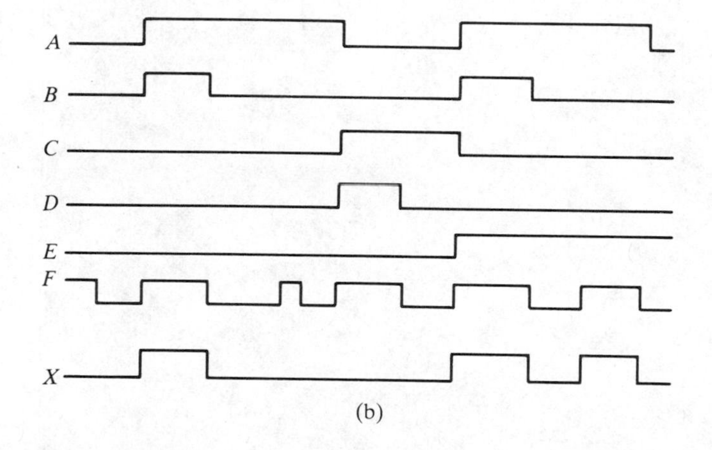

(b)

FIGURE 5–61

In this chapter you will study several important categories of combinational logic. These include adders, comparators, decoders, encoders, code converters, and parity logic. These particular functions are stressed because they are applied in a variety of digital systems and are available in integrated circuit form, including the 54/74 family.

6

FUNCTIONS OF COMBINATIONAL LOGIC

6–1 ADDERS

Recall the basic rules for binary addition as stated in Chapter 2:

$$0 + 0 = 0$$
$$0 + 1 = 1$$
$$1 + 0 = 1$$
$$1 + 1 = 10_2$$

These operations are performed by a logic circuit called a *half-adder*. The half-adder accepts two binary digits on its inputs and produces two binary digits on its outputs, a *sum* bit and a *carry* bit, as shown by the block diagram in Figure 6–1.

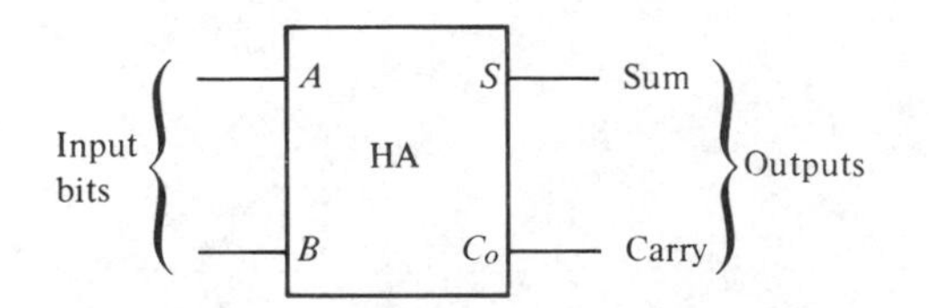

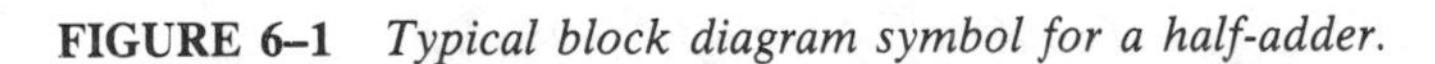

FIGURE 6–1 *Typical block diagram symbol for a half-adder.*

TABLE 6-1
Half-adder truth table.

A	B	C_o	S
0	0	0	0
0	1	0	1
1	0	0	1
1	1	1	0

S = sum
C_o = carry
A and B = input variables

From the logical operation of the half-adder as expressed in Table 6–1, expressions can be derived for the sum and carry outputs as functions of the inputs. Notice that the carry output C_o is a 1 only when both A and B are 1s; therefore, C_o can be expressed as the AND of the input variables.

$$C_o = AB \qquad \textbf{(6–1)}$$

Now observe that the sum output S is a 1 only if the input variables are not equal. The sum can therefore be expressed as the exclusive-OR of the input variables.

$$S = A \oplus B \qquad \textbf{(6–2)}$$

From these two expressions, the implementation required for the half-adder function is apparent. The carry output is produced with an AND gate with A and B on the inputs, and the sum output is generated with an exclusive-OR gate as shown in Figure 6–2.

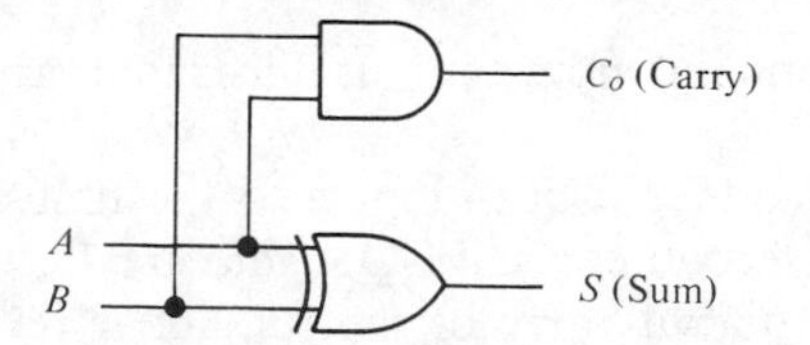

FIGURE 6–2 *Half-adder logic diagram.*

Full Adder

The second basic category of adder is the *full-adder.* The full-adder accepts *three* inputs, and generates a sum output and a carry output. So, the basic difference in a full-adder and a half-adder is that the full-adder accepts an additional input that allows for handling input carries. A block diagram illustrates the full-adder in Figure 6–3, and the truth table in Table 6–2 shows its operation.

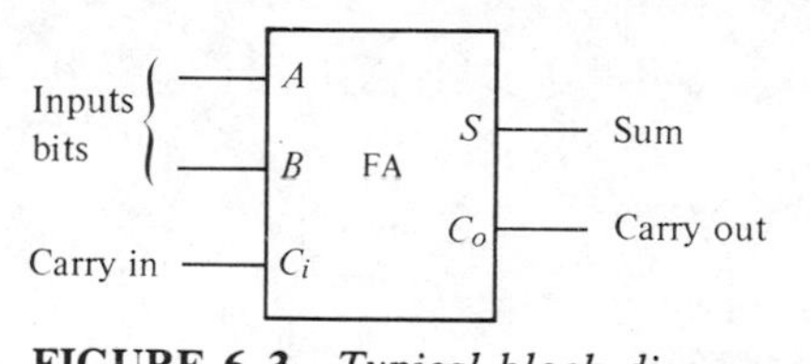

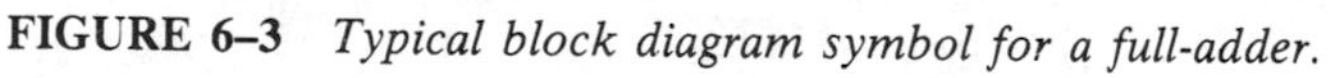

FIGURE 6–3 *Typical block diagram symbol for a full-adder.*

TABLE 6–2 *Truth table for a full adder.*

A	B	C_i	C_o	S
0	0	0	0	0
0	0	1	0	1
0	1	0	0	1
0	1	1	1	0
1	0	0	0	1
1	0	1	1	0
1	1	0	1	0
1	1	1	1	1

C_i = input carry
C_o = output carry
S = sum
A and B = input variables

The full-adder must sum the two input bits and the input carry bit. From the half-adder we know that the sum of the input bits A and B is the exclusive-OR of those two variables, $A \oplus B$. In order to add the input carry C_i to the input bits, it must be exclusive-ORed with $A \oplus B$, yielding the equation for the *sum* output of the full-adder.

$$S = (A \oplus B) \oplus C_i \tag{6–3}$$

This means that to implement the full-adder sum function, two exclusive-OR gates can be used. The first must generate the term $A \oplus B$, and the second has as its input the output of the first gate and the input carry as illustrated in Figure 6–4(a).

The output *carry* is a 1 for the full-adder if both inputs to the first exclusive-OR gate are 1s or if both inputs to the second exclusive-OR gate are 1s. You can verify this by studying Table 6–2. The output carry of the full-adder is therefore A ANDed with B, $A \oplus B$ ANDed with C_i, and these two terms ORed as expressed in Equation (6–4). This function is implemented and combined with the sum logic to form a complete full-adder circuit, as shown in Figure 6–4(b).

$$C_o = AB + (A \oplus B) \cdot C_i \tag{6–4}$$

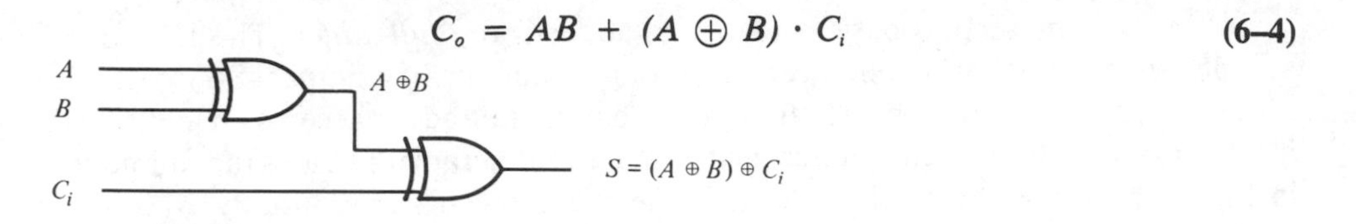

(a) Logic Required to Form the Sum of Three Bits

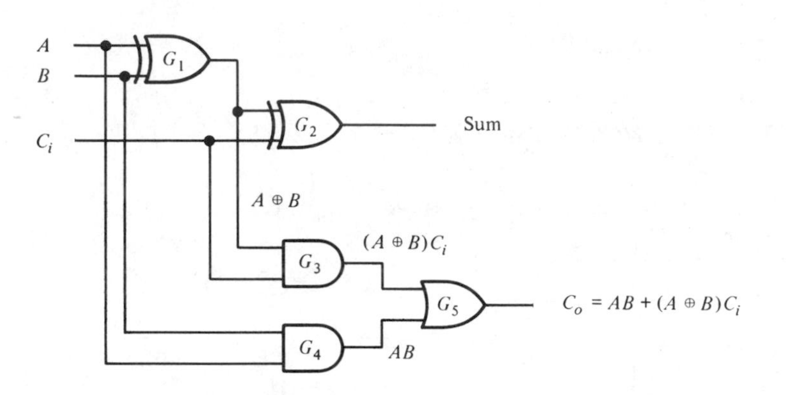

(b) Complete Logic Circuitry for a Full-Adder

FIGURE 6–4 *Full-adder logic.*

Note that in Figure 6–4(b) there are two half-adders, connected as shown in the block diagram of Figure 6–5(a), with their output carries ORed. A simplified block diagram as shown in Figure 6–5(b) will normally be used to represent the full-adder (FA).

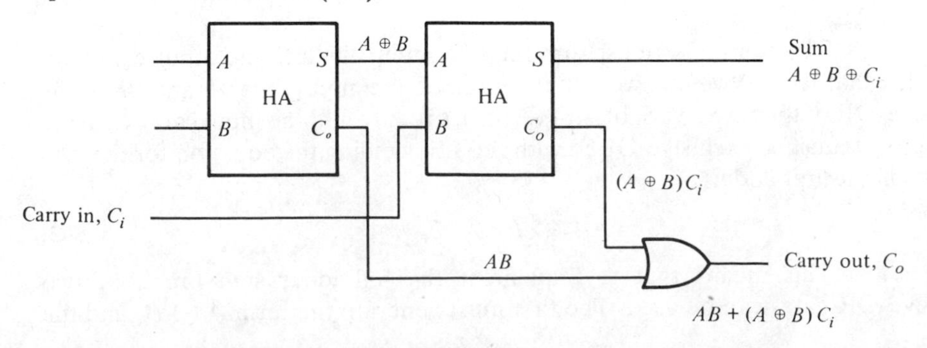

(a) Arrangement of Two Half-Adders to Form a Full-Adder

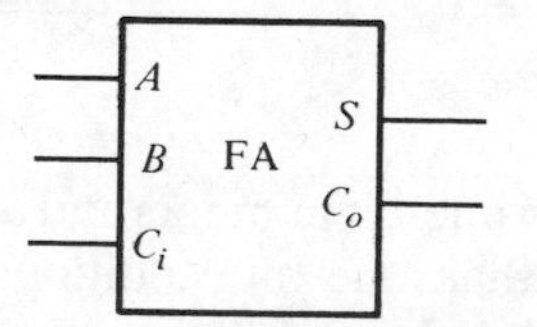

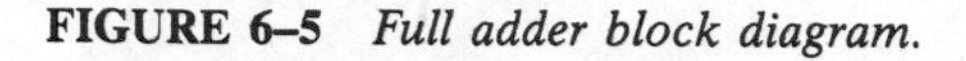

(b) Full-Adder Block Diagram

FIGURE 6–5 *Full adder block diagram.*

Example 6–1

Determine an alternate method for implementing the full-adder.

Solution:

Going back to Table 6–2, we can write sum-of-product expressions for both S and C_o by observing the input conditions that make them 1s. The expressions are as follows:

$$S = \overline{A}\,\overline{B}C_i + \overline{A}B\overline{C}_i + A\overline{B}\,\overline{C}_i + ABC_i$$
$$C_o = \overline{A}BC_i + A\overline{B}C_i + AB\overline{C}_i + ABC_i$$

These two functions are implemented with AND/OR logic as shown in Figure 6–6 without simplification. The C_o function can be simplified; see if you can do it.

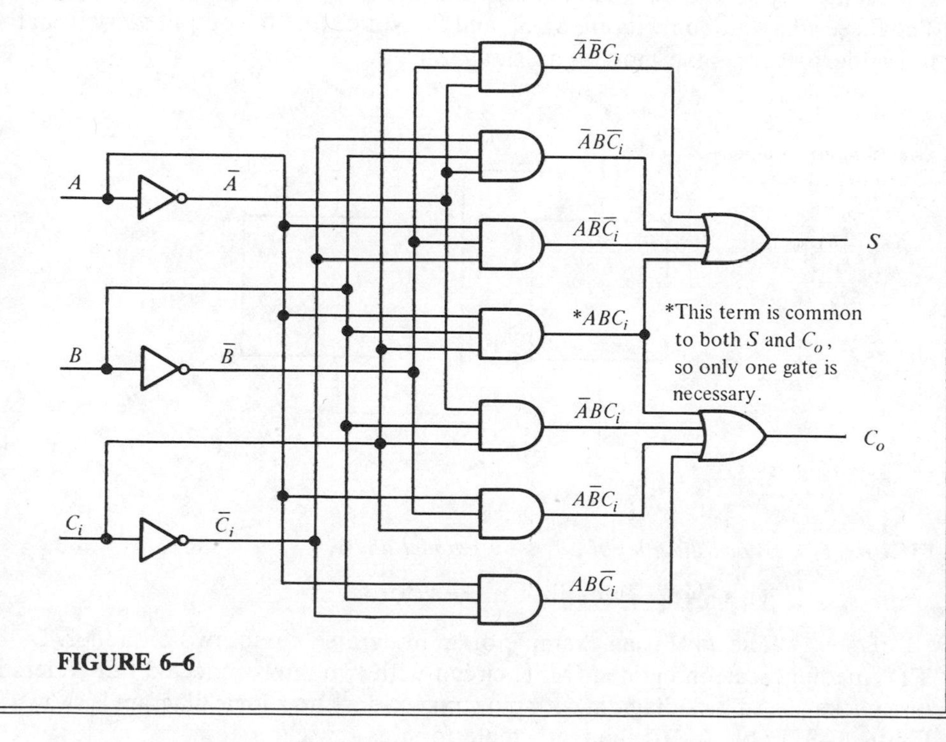

FIGURE 6–6

6–2 PARALLEL BINARY ADDERS

As we have seen, a single full-adder is capable of adding two one-bit numbers and an input carry. In order to add binary numbers with more than one bit, additional full-adders must be employed. When one binary number is added to another, each column generates a sum and a 1 or 0 carry to the next higher order column, as illustrated with two-bit numbers:

```
                       ┌Carry from right column
                       ↓
                       1
                       1 1
                     + 0 1
                     -----
Carry from           1 0 0
second column        ↑
becomes a sum bit.───┘
```

To implement the addition of binary numbers with logic circuits, a full-adder is required for each column. So, for two-bit numbers, two adders are needed; for three-bit numbers, three adders are used, and so on. The carry output of each adder is connected to the carry input of the next higher order adder, as shown in Figure 6–7 for a two-bit adder. It should be pointed out that either a half-adder can be used for the least signficant position or the carry input of a full-adder is made 0 because there is no carry into the least significant bit position.

In Figure 6–7, the least significant bits of the two numbers are represented by A_0 and B_0. The next higher order bits are represented by A_1 and B_1. The three complete sum bits are S_0, S_1, and S_2. Notice that the output carry from the full-adder becomes the most significant sum bit.

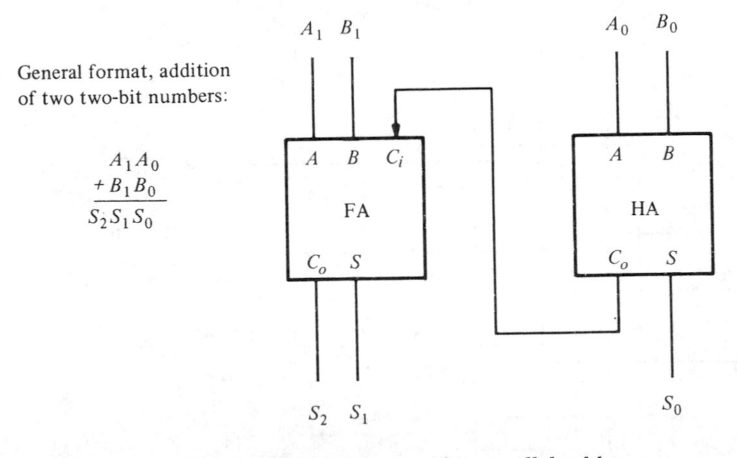

FIGURE 6–7 *Block diagram of a two-bit parallel adder.*

The 7482 Two-Bit Full Adder

The 7482 is an example of an integrated circuit two-bit adder. It is a TTL medium scale integrated (MSI) circuit with two interconnected full-adders in one package. A block diagram with pin connections and a logic diagram is shown in Figure 6–8. Table 6–3 is the truth table for this device.

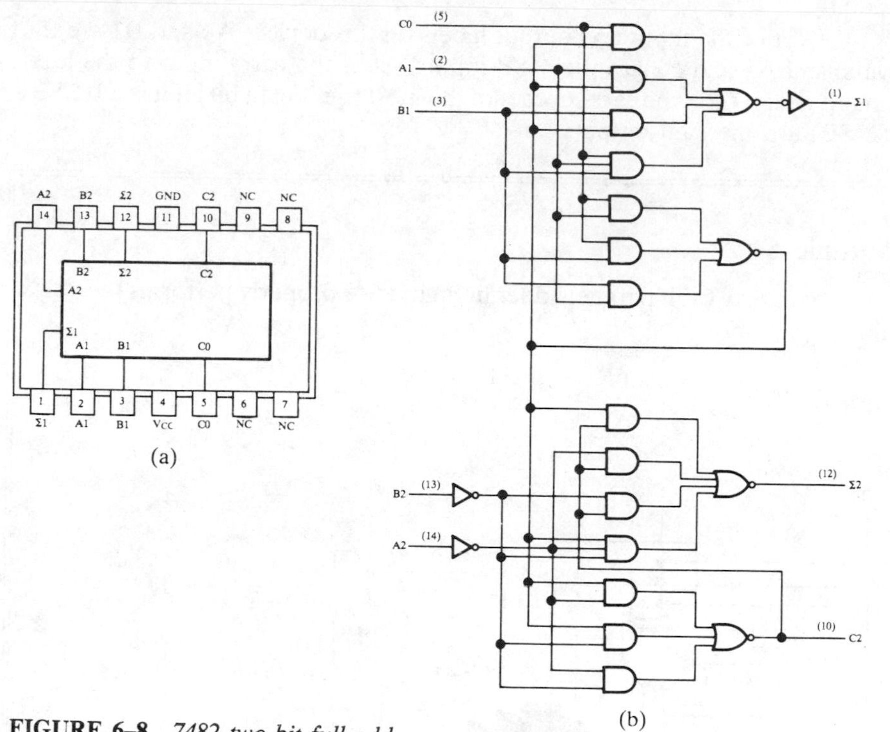

FIGURE 6–8 *7482 two-bit full adder.*

TABLE 6–3 *Truth table for a 7482 two-bit full adder.*

Inputs				Outputs					
				When C0=L			When C0=H		
A_1	B_1	A_2	B_2	Σ1	Σ2	C_2	Σ1	Σ2	C_2
L	L	L	L	L	L	L	H	L	L
H	L	L	L	H	L	L	L	H	L
L	H	L	L	H	L	L	L	H	L
H	H	L	L	L	H	L	H	H	L
L	L	H	L	L	H	L	H	H	L
H	L	H	L	H	H	L	L	L	H
L	H	H	L	H	H	L	L	L	H
H	H	H	L	L	L	H	H	L	H
L	L	L	H	L	H	L	H	H	L
H	L	L	H	H	H	L	L	L	H
L	H	L	H	H	H	L	L	L	H
H	H	L	H	L	L	H	H	L	H
L	L	H	H	L	L	H	H	L	H
H	L	H	H	H	L	H	L	H	H
L	H	H	H	H	L	H	L	H	H
H	H	H	H	L	H	H	H	H	H

H=high level, L=low level

Notice the input and output labels on this device: A1 and B1 are the LSB inputs, and A2 and B2 are the MSB inputs. C0 is the carry input to the least significant bit adder; C2 is the carry output of the most significant bit adder; Σ1(LSB) and Σ2(MSB) are the sum outputs.

Example 6–2

Verify that the two-bit parallel adder in Figure 6–9 properly performs the following addition:

$$\begin{array}{r} 11 \\ +10 \\ \hline 101 \end{array}$$

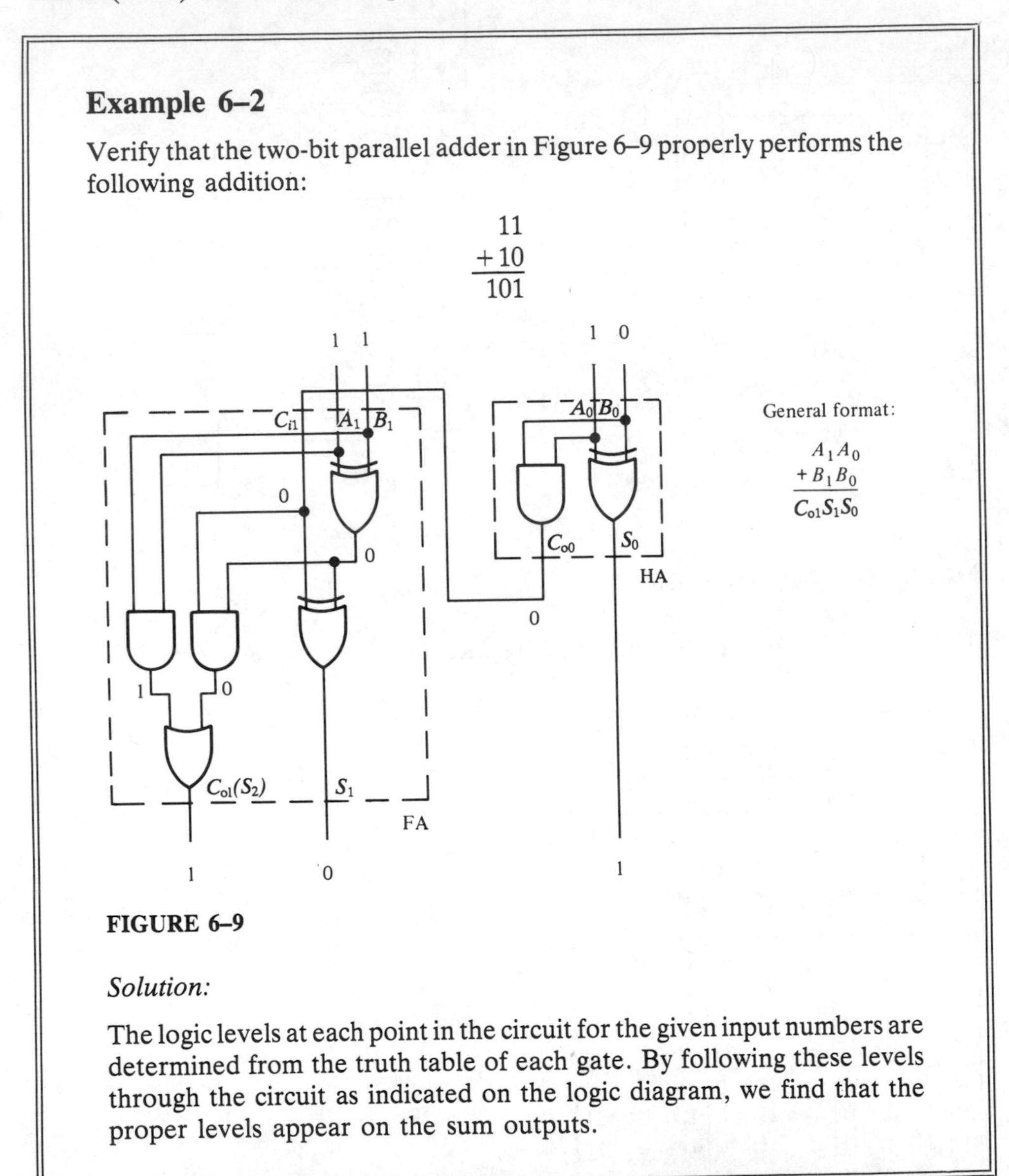

FIGURE 6–9

Solution:

The logic levels at each point in the circuit for the given input numbers are determined from the truth table of each gate. By following these levels through the circuit as indicated on the logic diagram, we find that the proper levels appear on the sum outputs.

Four-Bit and Eight-Bit Parallel Adders

A four-bit parallel adder is shown in Figure 6–10, and an eight-bit parallel adder in Figure 6–11. Again, the least significant bits in each number being added go into the right-most adder; the higher order bits are applied as shown to the successively higher order adders, with the most significant bits in each number being applied to the left-most full-adder. The carry output of each adder is connected to the carry input of the next higher order adder.

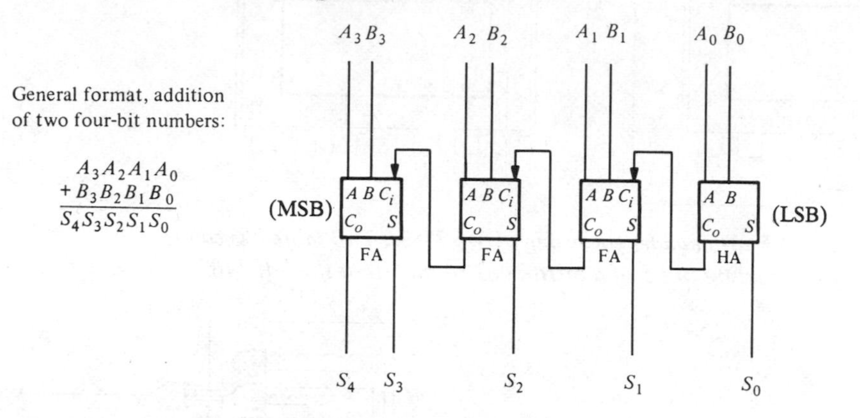

FIGURE 6–10 *Block diagram of a four-bit parallel adder.*

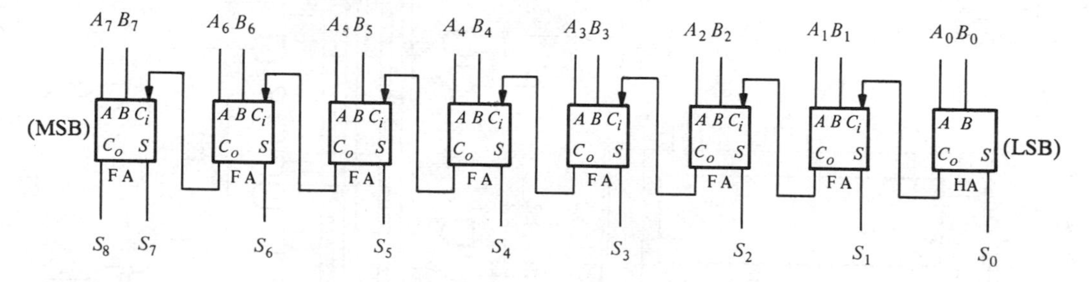

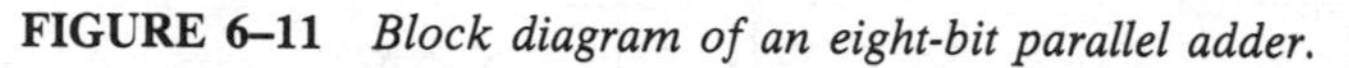

FIGURE 6–11 *Block diagram of an eight-bit parallel adder.*

Example 6–3

Show how to connect two 7482 adders to form a four-bit adder.

Solution:

Figure 6–12 is a connection diagram showing two 7482s used as a four-bit adder.

The 7483A Four-Bit Full Adder

The 7483A is an example of an integrated circuit four-bit full adder. It is an MSI TTL circuit. A block diagram with pin connections is shown in Figure 6–13(a) and the logic diagram in Figure 6–13(b).

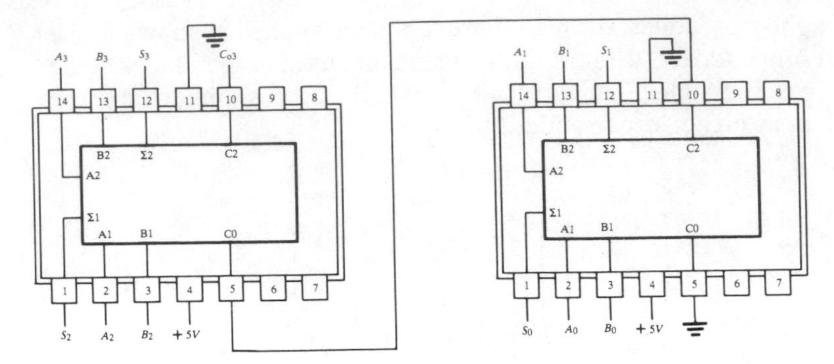

FIGURE 6–12 *A four-bit adder consisting of two 7482s. Two more bits can be added by connecting pin 5 of a third 7482 to pin 10 of the left 7482.*

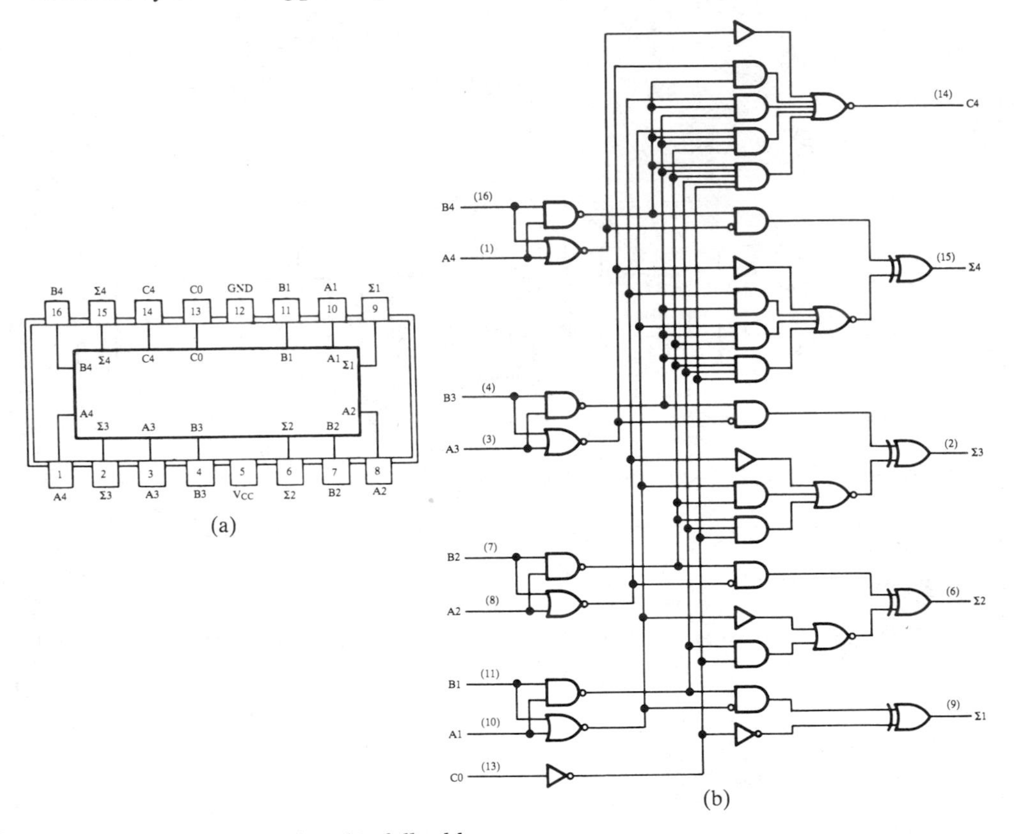

FIGURE 6–13 *7483A four-bit full adder.*

Example 6–4

Show how two 7483A adders can be connected to form an eight-bit parallel adder.

Solution:

The only connection between the two 7483As is the carry out (C4) of the lower order adder to the carry in (C0) of the higher order adder as shown in Figure 6–14. Pin 13 of the right-most 7483A is grounded so that the LSB circuit functions as a half adder (no input carry).

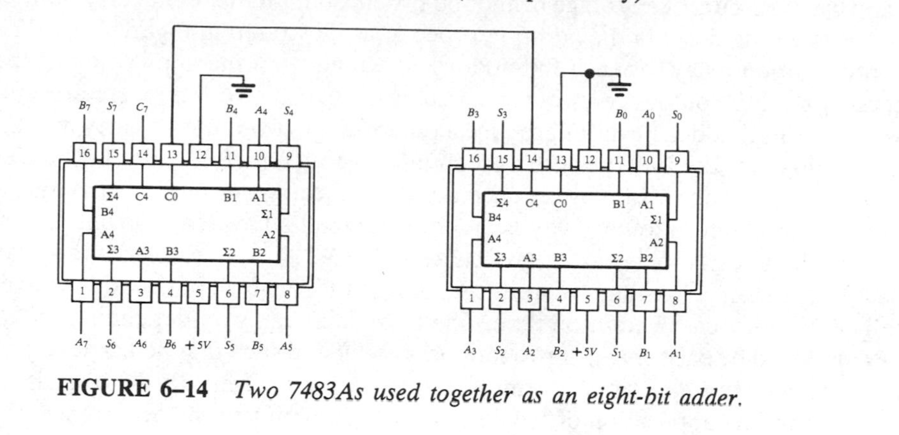

FIGURE 6–14 *Two 7483As used together as an eight-bit adder.*

Example 6–5

For the four-bit parallel adder shown in Figure 6–15, in block diagram form, determine the sum and carry outputs of each adder for the inputs indicated.

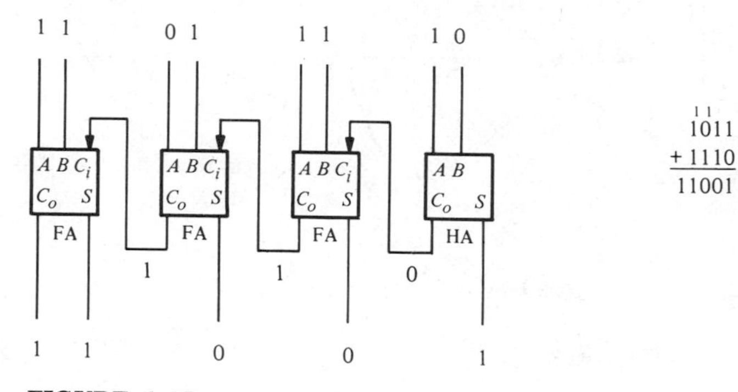

FIGURE 6–15

Solution:

According to the truth table operation of the adders, the sum and carry outputs are as shown. This total sum output agrees with the sum arrived at by long hand addition.

6–3 METHODS OF SPEEDING ADDITION

The parallel adders covered so far are *ripple carry* types where the carry out of each full-adder stage is connected to the carry input of the next higher order stage. The sum and carry outputs of any stage cannot be produced until the input carry occurs; this leads to a time delay in the addition process, as illustrated in Figure 6–16. The carry propagation delay for each full-adder is the time from the application of the input carry until the output carry occurs, assuming the A and B inputs are present.

Full-adder 1 cannot produce a potential carry out until a carry input is applied. Full-adder 2 cannot produce a potential carry out until full-adder 1 produces a carry out. Full-adder 3 cannot produce a potential carry out until a carry out is produced by full-adder 1 followed by a carry out from full-adder 2, and so on. As you can see, the input carry to the least significant stage (FA1) has to "ripple" through all of the adders before a final sum is produced. A delay through all of the adder stages actually is a "worst case" addition time. The total delay can vary depending on the carries produced by each stage. If two numbers are added such that no carries occur between stages, the add time is simply the propagation time through a single full-adder from the application of the data bits on the inputs to the occurrence of a sum output.

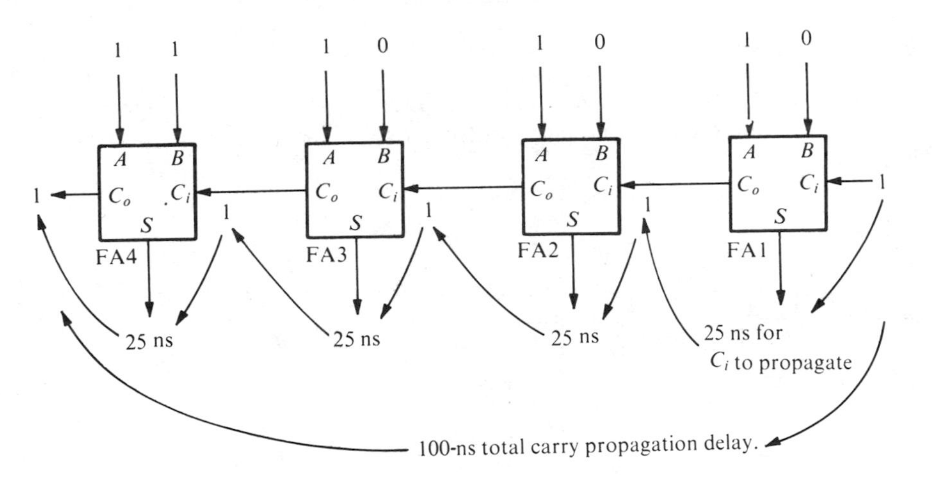

FIGURE 6–16 *A four-bit parallel binary ripple-carry adder showing a typical carry propagation delay.*

Carry-Look-Ahead Adder (CLA)

As you have seen in the discussion of the parallel adder, the speed with which an addition can be performed is limited by the time required for the carries to propagate or ripple through all of the stages of the adder. One method of speeding up this process by eliminating this carry ripple delay is called *carry-look-ahead* addition; this method is based on two functions of the full-adder called the *carry generate* and the *carry propagate* functions.

The *carry generate* function indicates when an output carry is produced (generated) by the full-adder. A carry is *generated* only when *both input bits are 1s*. This condition is expressed as

$$G = AB$$

A carry input is *propagated* by the full-adder when either or both of the input bits are 1s. This condition is expressed as

$$P = A + B$$

The *carry generate* and *carry propagate* conditions are illustrated in Figure 6–17.

How can the carry output of a full-adder be expressed in terms of the carry generate (G) and the carry propagate (P)? The output carry (C_o) is a 1 if the carry generate is a 1 OR if the carry propagate is a 1 AND the input carry (C_i) is a 1. In other words, we get an output carry of 1 if it is *generated* by the full-adder $(A = 1$ AND $B = 1)$ or if the adder can *propagate* the input carry $(A = 1$ OR $B = 1)$ AND $C_i = 1$. This relationship is expressed as

$$C_o = G + PC_i$$

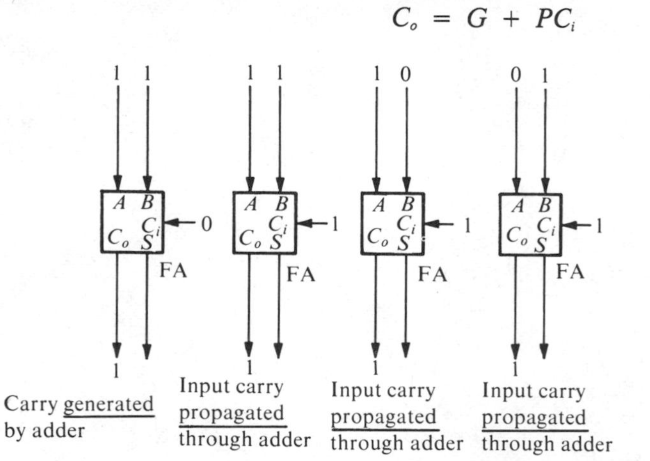

FIGURE 6–17 *Illustration of the carry generate and carry propagate conditions.*

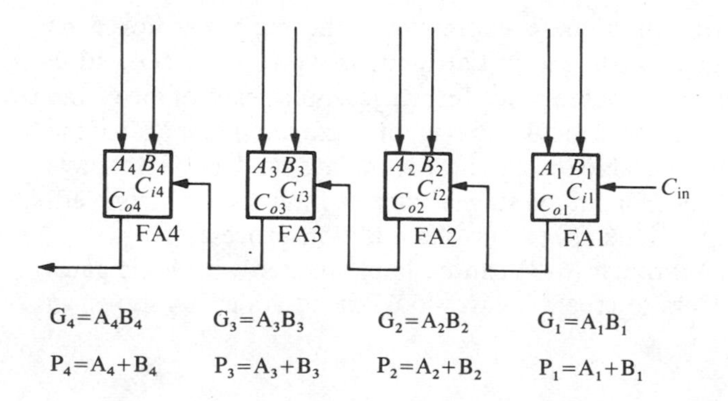

FIGURE 6–18 *Carry generate and carry propagate functions in terms of the input bits to a four-bit adder.*

Now we will see how this concept can be applied to a parallel adder whose individual stages are shown in Figure 6–18 for a four-bit example. For each full-adder, the output carry is dependent on its carry generate *(G)*, its carry propagate *(P)*, and its carry input (C_i). The *G* and *P* functions for each stage are immediately available as soon as the input bits *A* and *B* and the input carry to the LSB adder are applied because they are dependent only on these two bits. The carry input to each stage is the carry output of the previous stage. Based on this, we will now develop expressions for the carry out, C_o, of each full-adder stage for the four-bit example.

Carry out of FA1

$$C_{o1} = G_1 + P_1C_{i1} \tag{6–5}$$
$$C_{i2} = C_{o1}$$

Carry out of FA2

$$\begin{aligned} C_{o2} &= G_2 + P_2C_{i2} \\ &= G_2 + P_2C_{o1} \\ &= G_2 + P_2(G_1 + P_1C_{i1}) \\ &= G_2 + P_2G_1 + P_2P_1C_{i1} \end{aligned} \tag{6–6}$$
$$C_{i3} = C_{o2}$$

Carry out of FA3

$$\begin{aligned} C_{o3} &= G_3 + P_3C_{i3} \\ &= G_3 + P_3C_{o2} \\ &= G_3 + P_3(G_2 + P_2G_1 + P_2P_1C_{i1}) \\ &= G_3 + P_3G_2 + P_3P_2G_1 + P_3P_2P_1C_{i1} \end{aligned} \tag{6–7}$$
$$C_{i4} = C_{o3}$$

Carry out of FA4

$$\begin{aligned} C_{o4} &= G_4 + P_4C_{i4} \\ &= G_4 + P_4C_{o3} \\ &= G_4 + P_4(G_3 + P_3G_2 + P_3P_2G_1 + P_3P_2P_1C_{i1}) \\ &= G_4 + P_4G_3 + P_4P_3G_2 + P_4P_3P_2G_1 + P_4P_3P_2P_1C_{i1} \end{aligned} \tag{6–8}$$

Notice that in each of these expressions, the carry out for each full-adder stage is dependent only on the initial input carry (C_{i1}), its *G* and *P* functions, and the *G* and *P* functions of the preceding stages. Since each of the *G* and *P* functions can be expressed in terms of the *A* and *B* input to the full-adders, all of the output carries are immediately available (except for gate delays) and we do not have to wait for a carry to ripple through all of the stages before a final result is achieved. Thus the carry-look-ahead technique speeds up the addition process.

Equations (6–5) through (6–8) can be implemented with logic gates and connected to the full-adders to create a carry-look-ahead adder, as shown in Figure 6–19.

Carry-Save Adder (CSA)

Up to this point the methods discussed have involved the addition of two numbers at a time. As we have seen, a string of numbers can be added two at a

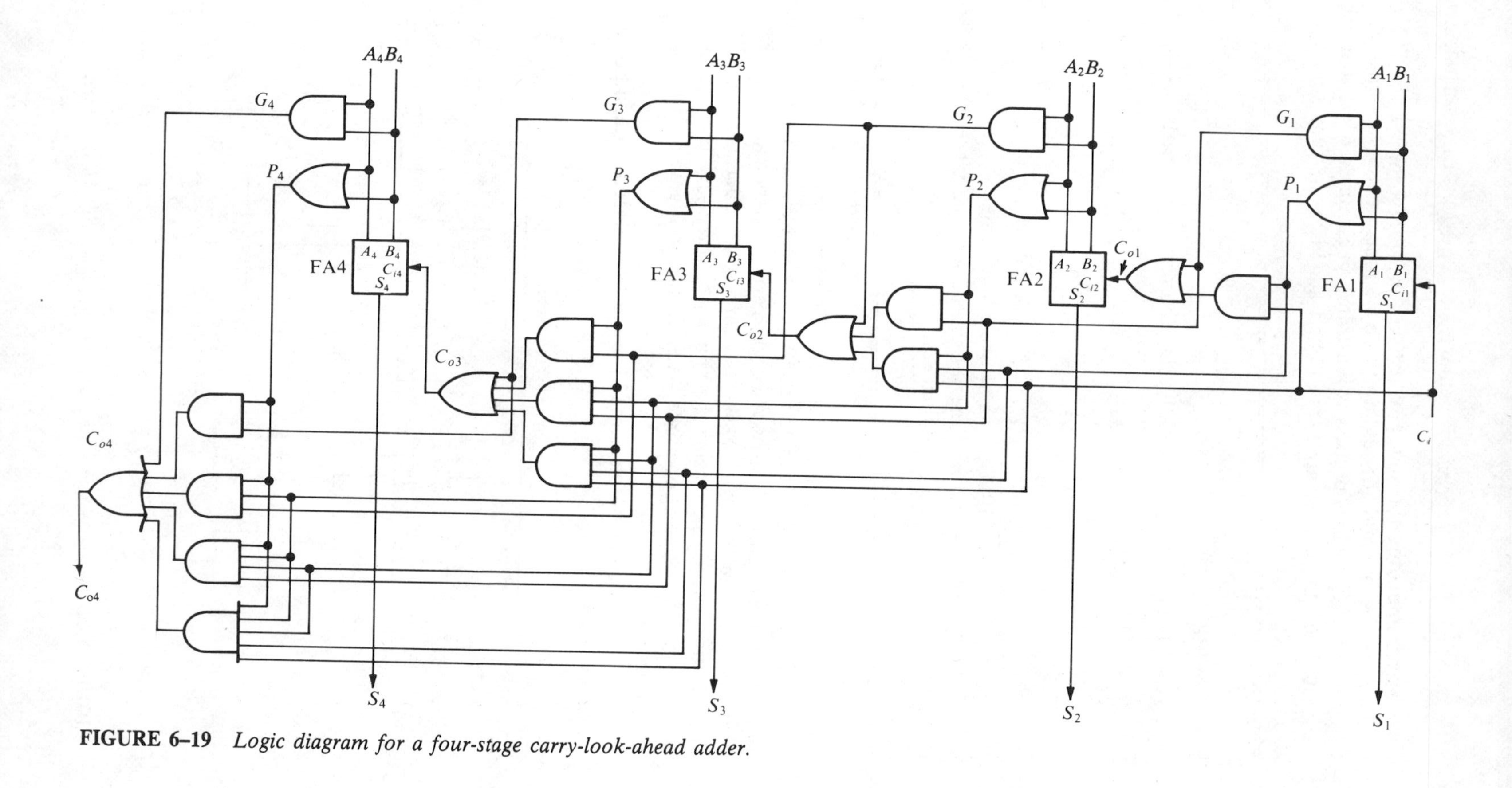

FIGURE 6–19 *Logic diagram for a four-stage carry-look-ahead adder.*

time by accumulating the sum after each addition and adding the next number.

A method for adding three or more numbers at a time is called *carry-save* addition. This process is illustrated in Example 6–6, where the sum is generated and the carries are "saved" and added to the final sum.

Example 6–6

Add 00011, 00001, and 01001 using the carry-save method.

Solution:

00011	*A*
00001	*B*
+01001	*C*
01011	sum, excluding carries
+0001	carries shifted left one place
01101	final sum

A full-adder is used to add a bit from each of the three numbers, using the carry input as one of the bit inputs. A sum from the addition of each of the three bits is formed along with a carry. All of the carries are then added to the sum bits to form the final sum. Figure 6–20 shows a carry-save adder for the addition of three five-bit numbers. By increasing the number of carry-save levels, more than three numbers can be added at one time, as illustrated in Figure 6–21 for the addition of four numbers.

The carry-save method is a way to speed up addition at a cost of increased complexity.

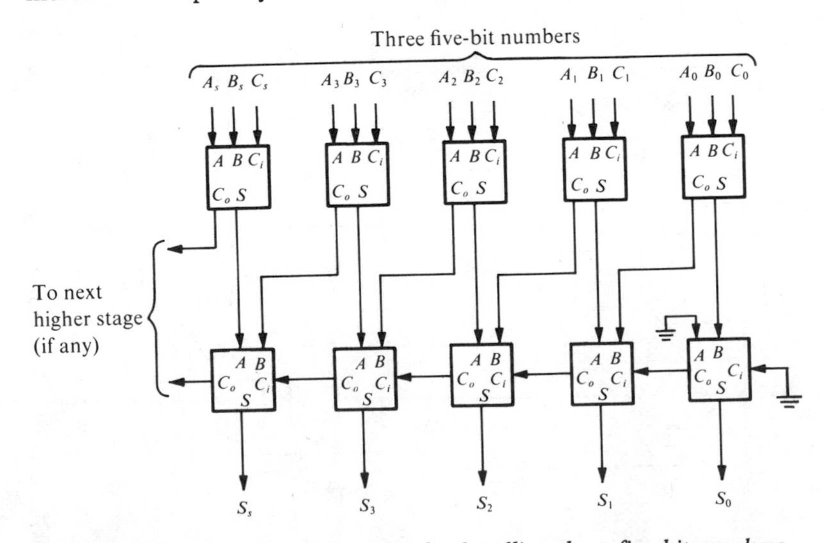

FIGURE 6–20 *A carry-save adder for handling three five-bit numbers.*

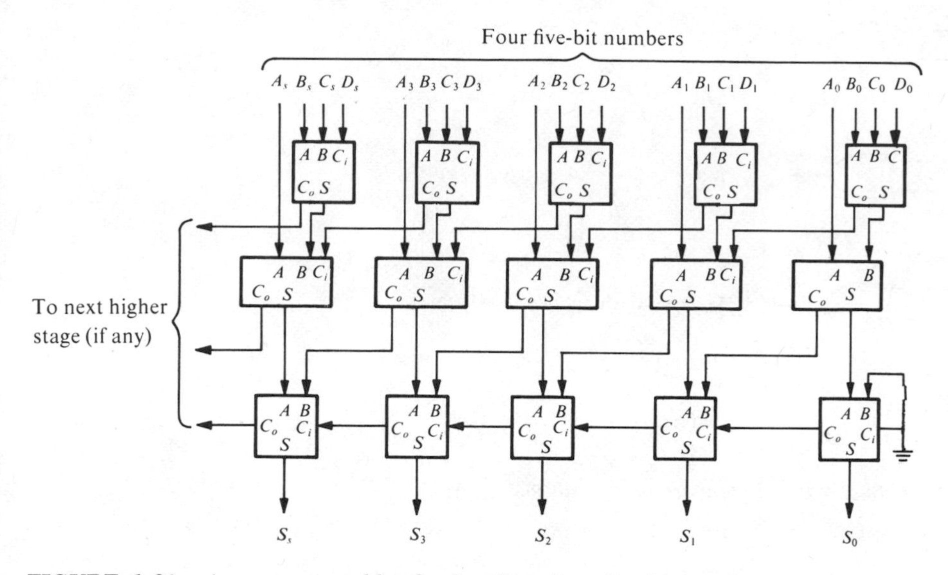

FIGURE 6–21 *A carry-save adder for handling four five-bit numbers.*

COMPARATORS 6–4

The basic function of a *comparator* is to *compare* the magnitudes of two quantities in order to determine the relationship of those quantities. In its simplest form, a comparator circuit determines if two numbers are equal.

The exclusive-OR gate is a basic comparator because its output is a 1 if its two input bits are not equal and a 0 if the inputs are equal. Figure 6–22 shows the exclusive-OR as a two-bit comparator.

In order to compare numbers containing two bits each, an additional exclusive-OR gate is necessary. The two least significant bits (LSBs) of the two numbers are compared by gate G_1, and the two most significant bits (MSBs) are compared by gate G_2, as shown in Figure 6–23. If the two numbers are equal, their corresponding bits are the same and the output of each exclusive-OR gate is a 0. If either of the corresponding sets of bits are not equal, a 1 occurs on that exclusive-OR gate output. In order to produce a *single* output indicating an equality or inequality of two numbers, two inverters and an AND gate can be employed as shown in Figure 6–23. The output of each exclusive-OR gate is inverted and applied to the AND gate input. When the two input bits for each exclusive-OR are equal, the corresponding bits of each number are equal, producing a 1 on both inputs to the AND gate and thus a 1 on the output. When the two numbers are not equal, one or both sets of corresponding bits are unequal, and a 0 appears on at least one input to the AND gate to produce a 0 on its output. Thus the output of the AND gate indicates equality (1) or nonequality (0) of the two numbers. Example 6–7 illustrates this operation for two specific cases. The exclusive-OR gate and inverter are replaced by an exclusive-NOR symbol.

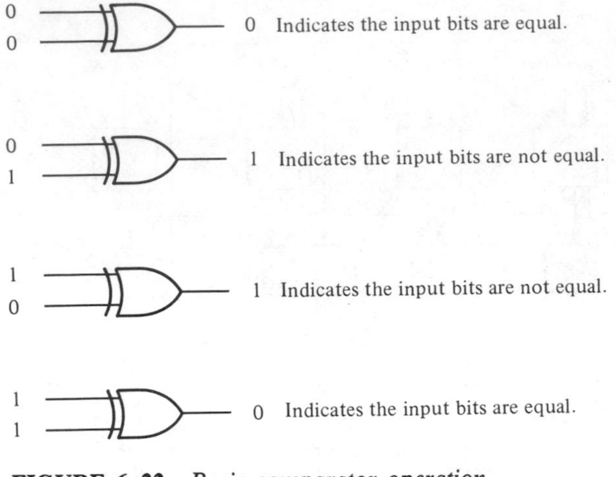

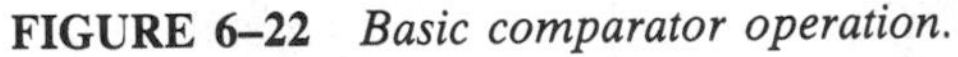

FIGURE 6–22 *Basic comparator operation.*

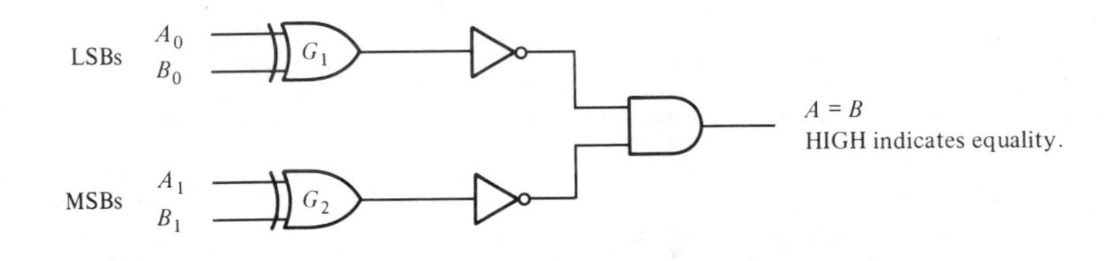

General format: Number $A \rightarrow A_1A_0$
Number $B \rightarrow B_1B_0$

FIGURE 6–23 *Logic diagram for comparison of two two-bit numbers.*

Example 6–7

Apply each of the following sets of numbers to comparator inputs, and determine the output by following the logic levels through the circuit: 10 and 10; 11 and 10.

Solutions:

The output is 1 for inputs 10 and 10, as shown in Figure 6–24(a). The output is 0 for inputs 11 and 10, as shown in Figure 6–24(b).

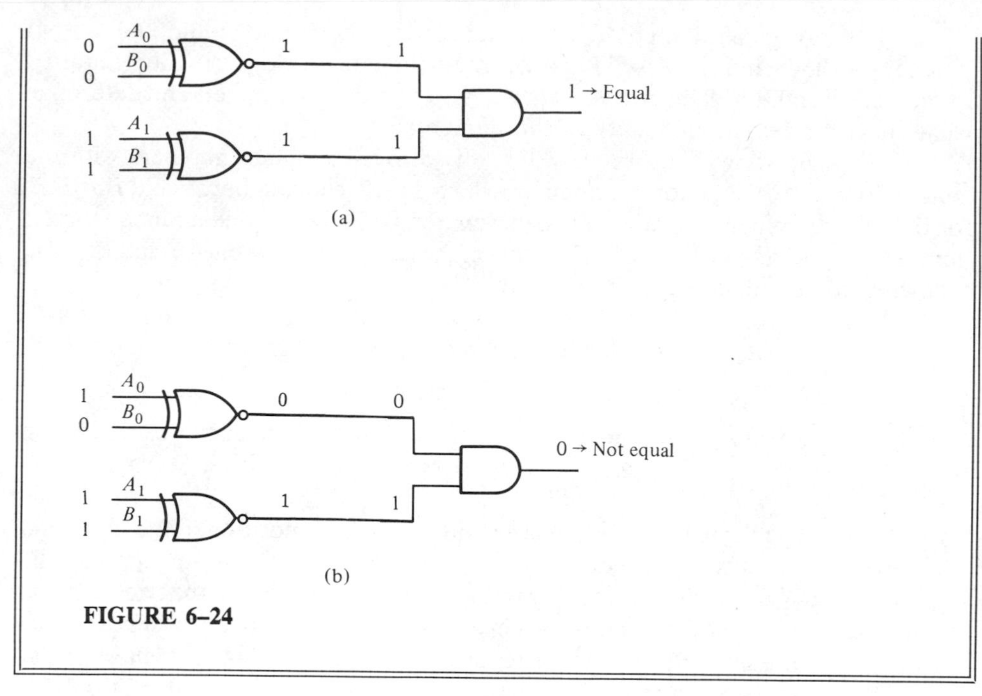

FIGURE 6–24

The basic comparator circuit can be expanded to any number of bits, as illustrated in Figure 6–25 for two four-bit numbers. The AND gate sets the condition that all corresponding bits of the two numbers must be equal if the two numbers themselves are equal.

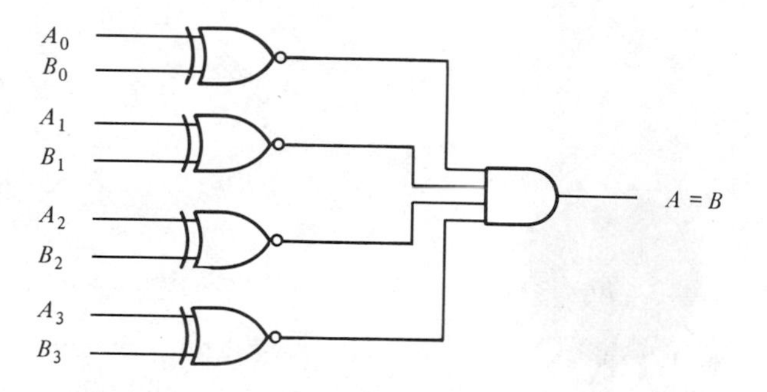

FIGURE 6–25 *Logic diagram for the comparison of two four-bit numbers, $A_3A_2A_1A_0$ and $B_3B_2B_1B_0$.*

Most integrated circuit comparators provide additional outputs that indicate which of the two numbers being compared is the larger. That is, there is an output that indicates when number A is greater than number B ($A > B$) and an output that indicates when number A is less than number B ($A < B$), as shown in the block diagram of Figure 6–26 for a four-bit comparator.

A general method of implementing these two additional output functions is shown in Figure 6–27. In order to understand the logic circuitry required for the $A > B$ and $A < B$ functions, let us examine two binary numbers and determine what characterizes an inequality of the numbers.

For our purposes we will use two four-bit binary numbers with the general format $A_3A_2A_1A_0$ for one number, which we will call number A, and $B_3B_2B_1B_0$ for the other number, which we will call number B. To determine an inequality of numbers A and B, we first examine the highest order bit in each number. The following conditions are possible:

1. $A_3 = 1$ and $B_3 = 0$ indicates number A is greater than number B.
2. $A_3 = 0$ and $B_3 = 1$ indicates number A is less than number B.
3. If $A_3 = B_3$, then we must examine the next lower bit position for an inequality.

The three observations are valid for each bit position in the numbers. The general procedure is to check for an inequality in a bit position, starting with the highest order. When such an inequality is found, the relationship of the two numbers is established and any other inequalities in lower order bit positions *must be ignored* because it is possible for an opposite indication to occur—the highest order indication *must take precedence.* To illustrate, let us assume that number A is 0111, and number B is 1000. Comparison of bits A_3 and B_3 indicates that $A < B$ because $A_3 = 0$ and $B_3 = 1$. However, comparison of bits A_2 and B_2 indicates $A > B$ because $A_2 = 1$ and $B_2 = 0$. The same is true for the remaining lower order bits. In this case priority must be given to A_3 and B_3 because they determine the proper inequality condition.

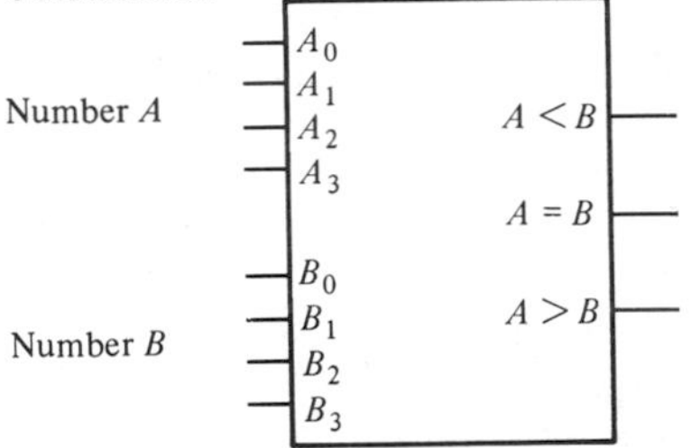

FIGURE 6–26 *Typical block diagram for a four-bit comparator with inequality indication.*

Figure 6–27 shows a method of comparing two four-bit numbers and generating an $A > B$ or an $A < B$ or an $A = B$ output. The $A > B$ condition is determined by gates G_6 through G_{10}. Gate G_6 checks for $A_3 = 1$ and $B_3 = 0$, and its function is expressed as $A_3\overline{B}_3$; gate G_7 checks for $A_2 = 1$ and $B_2 = 0$ ($A_2\overline{B}_2$); gate G_8 checks for $A_1 = 1$ and $B_1 = 0$ ($A_1\overline{B}_1$); gate G_9 checks or $A_0 = 1$ and $B_0 = 0$ ($A_0\overline{B}_0$). These conditions all indicate that number A is greater than number B. The output of each of these gates is ORed by gate G_{10} to produce the $A > B$ output.

Notice that the output of gate G_1 is connected to inputs of gates G_7, G_8, and G_9. This provides a priority inhibit so that if the proper inequality occurs in bits A_3 and B_3, the lower order bit checks will be inhibited. A priority inhibit is also provided by gate G_2 to gates G_8 and G_9, and by gate G_3 to gate G_9.

Gates G_{11} through G_{15} check for an $A < B$ condition. Each AND gate checks a given bit position for the occurrence of a 0 in the number A and a 1 in

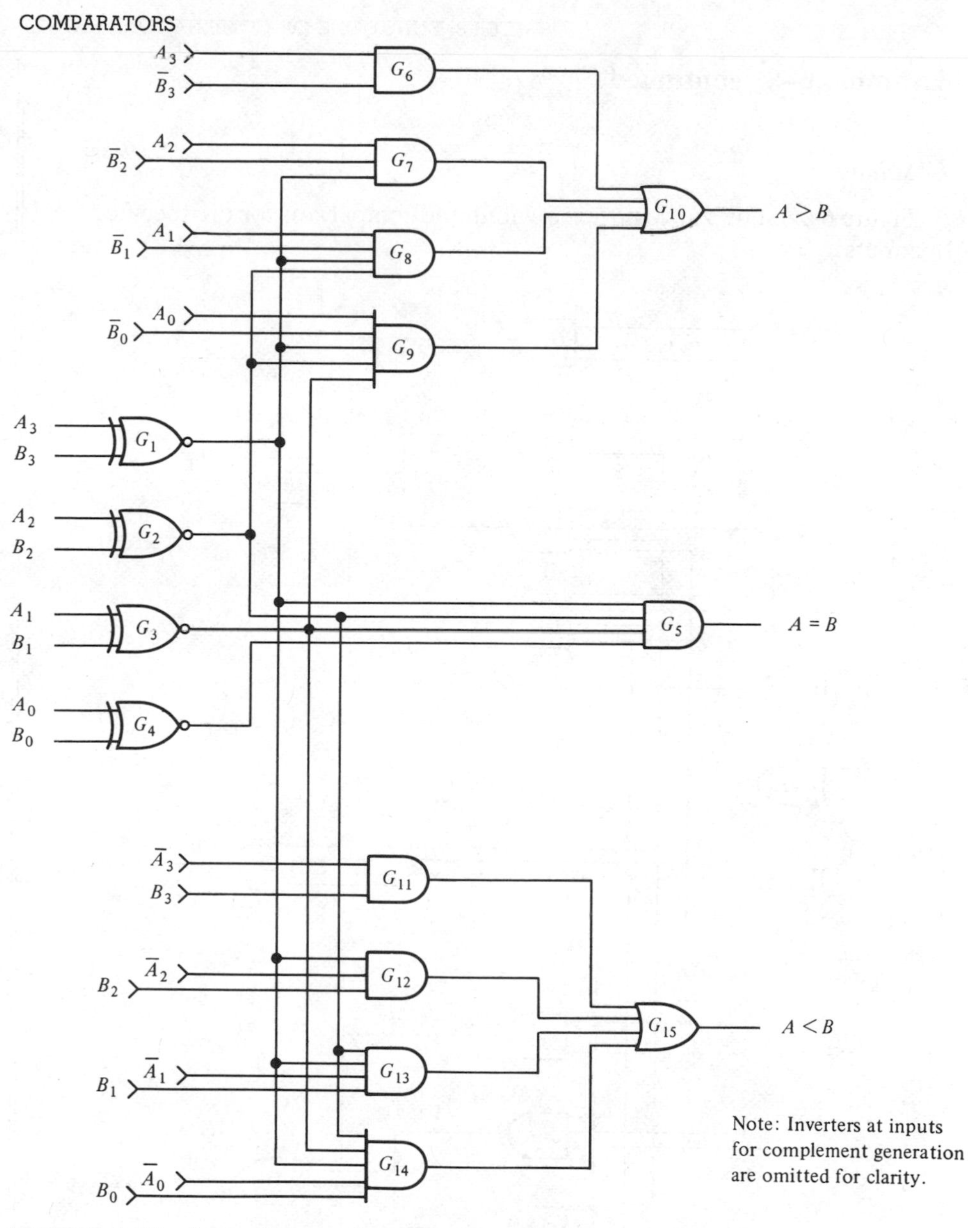

FIGURE 6–27 *Logic diagram for a four-bit comparator.*

number B. Each AND gate output is ORed by gate G_{15} to provide the $A < B$ output. Priority inhibiting is provided as previously discussed.

Example 6–8 shows the comparison of specific numbers and indicates the logic levels throughout the circuitry. You should also go through the analysis with numbers of your own choosing to verify the operation.

Example 6–8

Analyze the comparator operation for the numbers $A = 1010$ and $B = 1001$.

Example 6–8, continued

Solutions:

Figure 6–28 shows all logic levels within the comparator for the specified inputs.

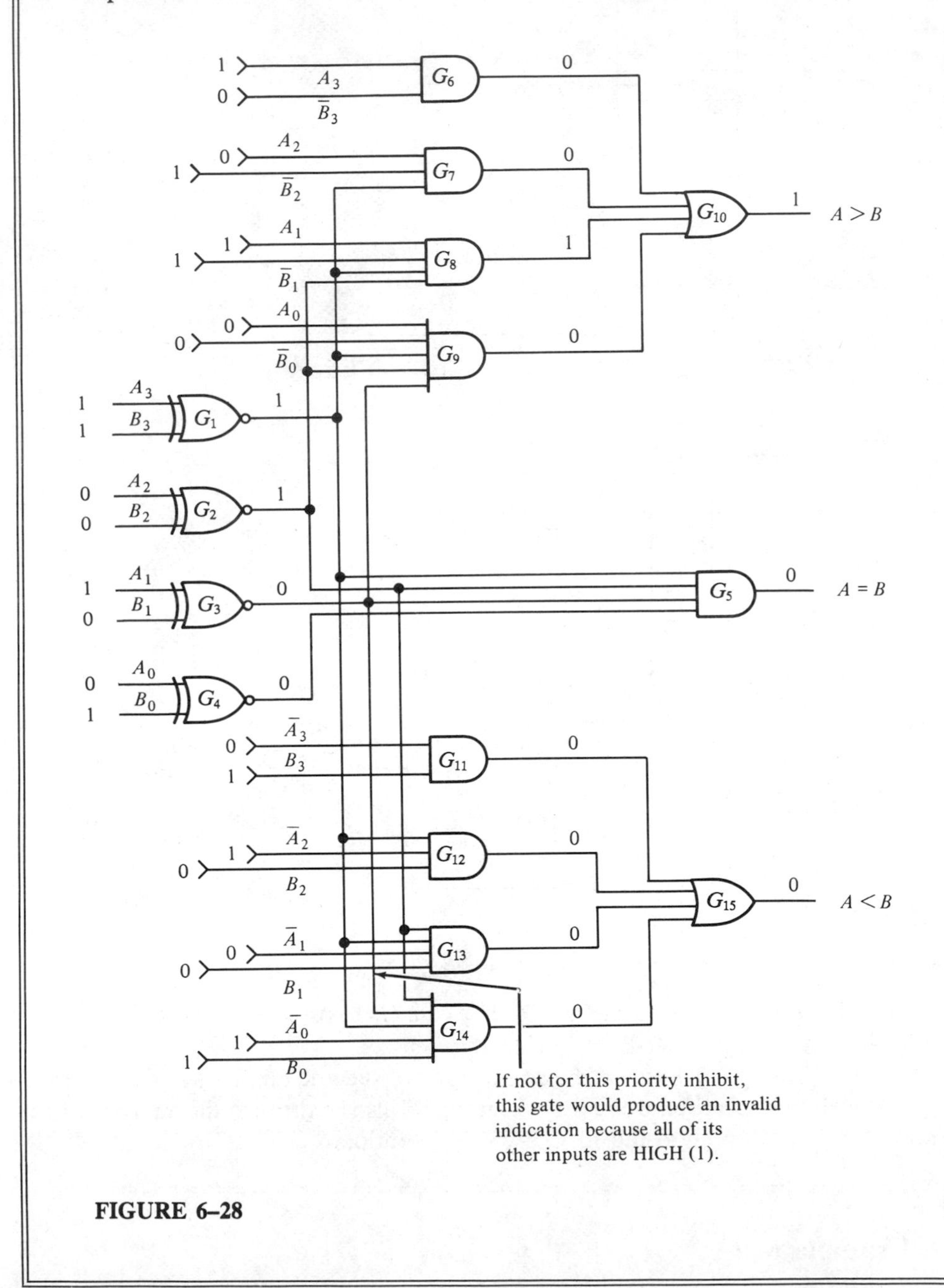

FIGURE 6–28

The 7485 Four-Bit Magnitude Comparator

The 7485 is a representative integrated circuit comparator in the 54/74 family. The block and pin diagram is shown in Figure 6–29.

Notice that this device has all of the inputs and outputs of the generalized comparator just discussed and in addition it has three *cascading inputs*. These inputs allow several comparators to be cascaded for comparison of any number of bits. To expand the comparator, the $A < B, A = B, A > B$ outputs of the less significant comparator are connected to the corresponding cascading inputs of the next higher comparator.

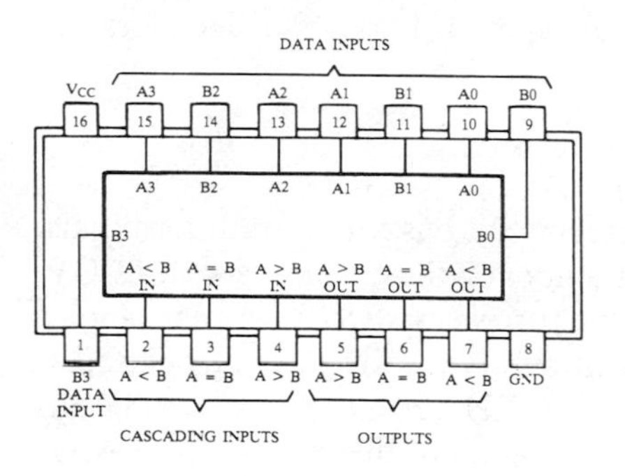

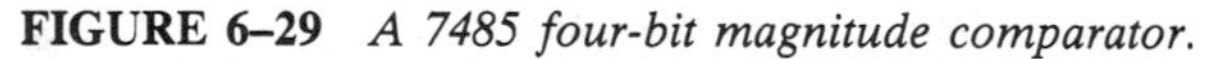

FIGURE 6–29 *A 7485 four-bit magnitude comparator.*

Example 6–9

Use 7485 comparators to compare the magnitudes of eight-bit numbers. Show the comparators with proper interconnections.

Solution:

Two 7485s are required for eight bits. They are connected as shown in Figure 6–30 in a cascaded arrangement.

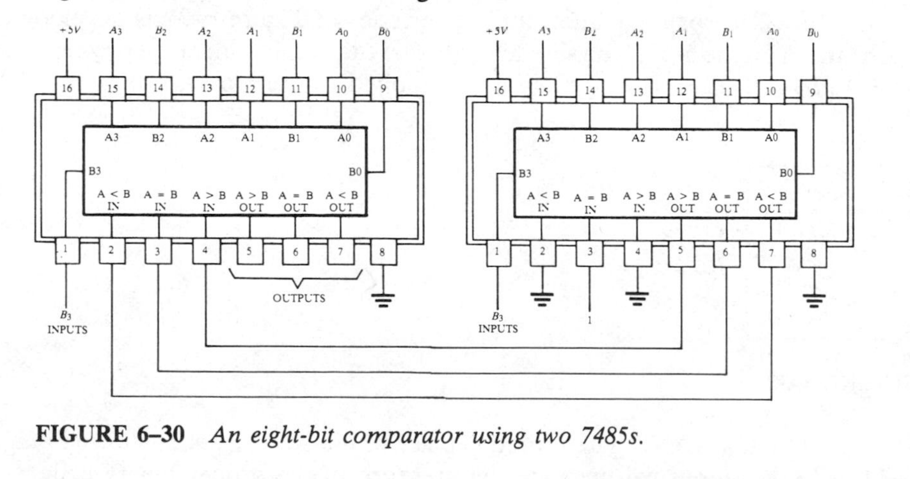

FIGURE 6–30 *An eight-bit comparator using two 7485s.*

6–5 DECODERS

The process of taking some type of code and determining what it represents in terms of a recognizable number or character is called *decoding*. A decoder is a combinational logic circuit that performs the decoding function, and produces an output that indicates the "meaning" of the input code.

A common example of the decoding function occurs when we are given a binary number and determine the decimal number it represents. We have, in essence, "decoded" the binary number because we have converted it into a more familiar form (decimal). In this section, several types of important decoders are examined.

The Basic Binary Decoder

Let us imagine that a certain digital circuit has four output terminals, and is capable of producing any four-bit binary number on these outputs with a LOW representing a 0 and a HIGH representing a 1. Suppose we wish to determine when a binary 1001 occurs on the outputs of this digital circuit. An AND gate can be used as the basic decoding element because it produces a HIGH output only when all of its inputs are HIGH. Therefore, we must make sure that all of the inputs to the AND gate are HIGH when the binary number 1001 occurs; this can be done by inverting the two middle bits (the 0s), as shown in Figure 6–31.

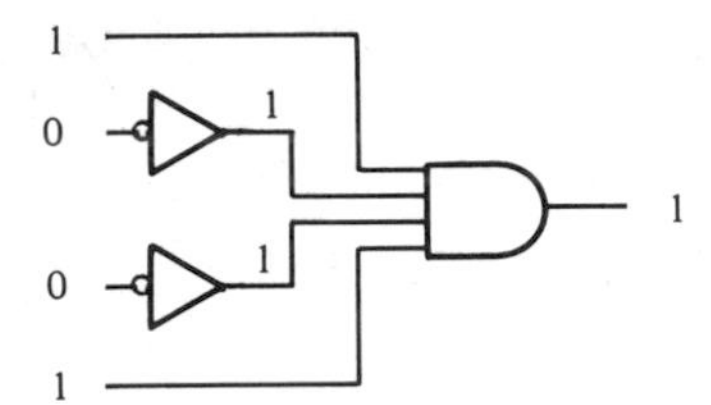

FIGURE 6–31 *Decoding logic for 1001 with an active HIGH output.*

The logic equation for the decoder of Figure 6–31 is developed as illustrated in Figure 6–32. You should verify that the output function is 0 except when $A = 1$, $B = 0$, $C = 0$, and $D = 1$ are applied to the inputs.

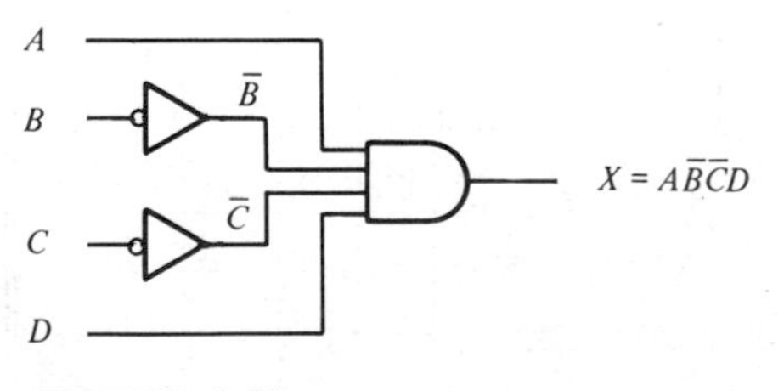

FIGURE 6–32

If a NAND gate is used in place of the AND gate, as shown in Figure 6–33, a LOW output will indicate the presence of the proper binary code.

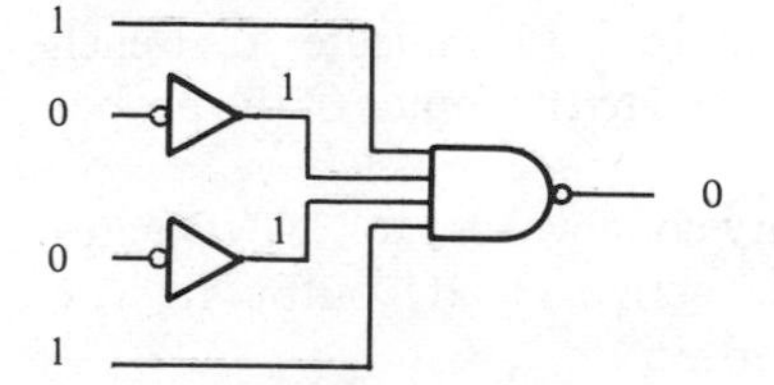

FIGURE 6–33 *Decoding logic for 1001 with an active LOW output.*

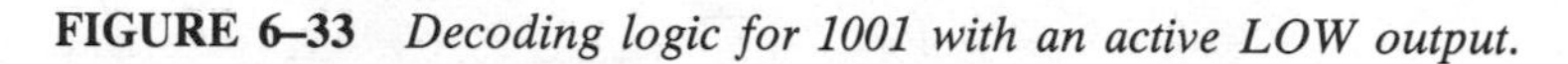

Example 6–10

Determine the logic required to decode the binary number 1011 (decimal 11) by producing a HIGH indication.

Solution:

The decoding function can be formed by complementing only the variables that appear as 0 in the binary number as follows:

$$X = A\overline{B}CD$$

This function can be implemented by connecting the true (uncomplemented) variables *A, C,* and *D* directly to the inputs of an AND gate, and inverting the variable *B* before applying it to the AND gate input. The decoding logic is shown in Figure 6–34.

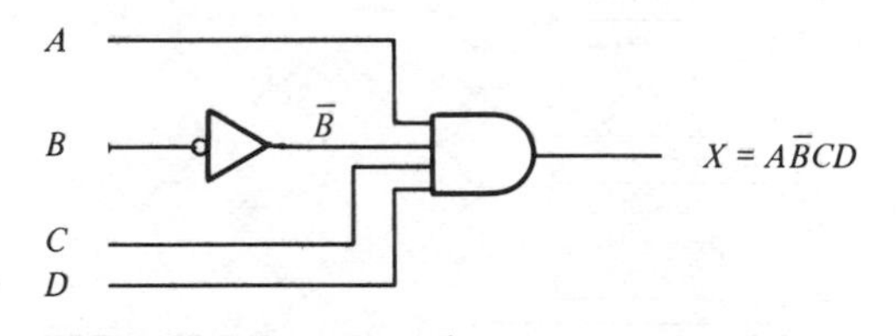

FIGURE 6–34 *Decoding logic for producing a HIGH output when A = 1, B = 0, C = 1, and D = 1.*

Four-Bit Binary Decoder

In order to decode all possible combinations of four bits, 16 decoding gates are required ($2^4 = 16$). This type of decoder is commonly called a *1-of-16* because each of the four-bit code words will activate only one of the 16 total outputs. It is also sometimes referred to as a *4-line-to-16-line* decoder because there are four inputs and 16 outputs. A list of the 16 binary code words and their corresponding decoding functions is given in Table 6–4.

If an active LOW output is desired for each decoded number, the entire decoder can be implemented with NAND gates and inverters as follows:

First, since each variable and its complement are required in the decoder as seen from Table 6–4, they can be generated once and then used for all decoding gates as required, rather than duplicating an inverter each place a complement is used. This arrangement, shown in Figure 6–35, indicates that each variable and its complement are available to be connected to the inputs of the proper decoding gates.

In order to decode each of the 16 binary code words, 16 NAND gates are required (AND gates can be used to produce active HIGH outputs). The decoding gate arrangement is illustrated in Figure 6–35.

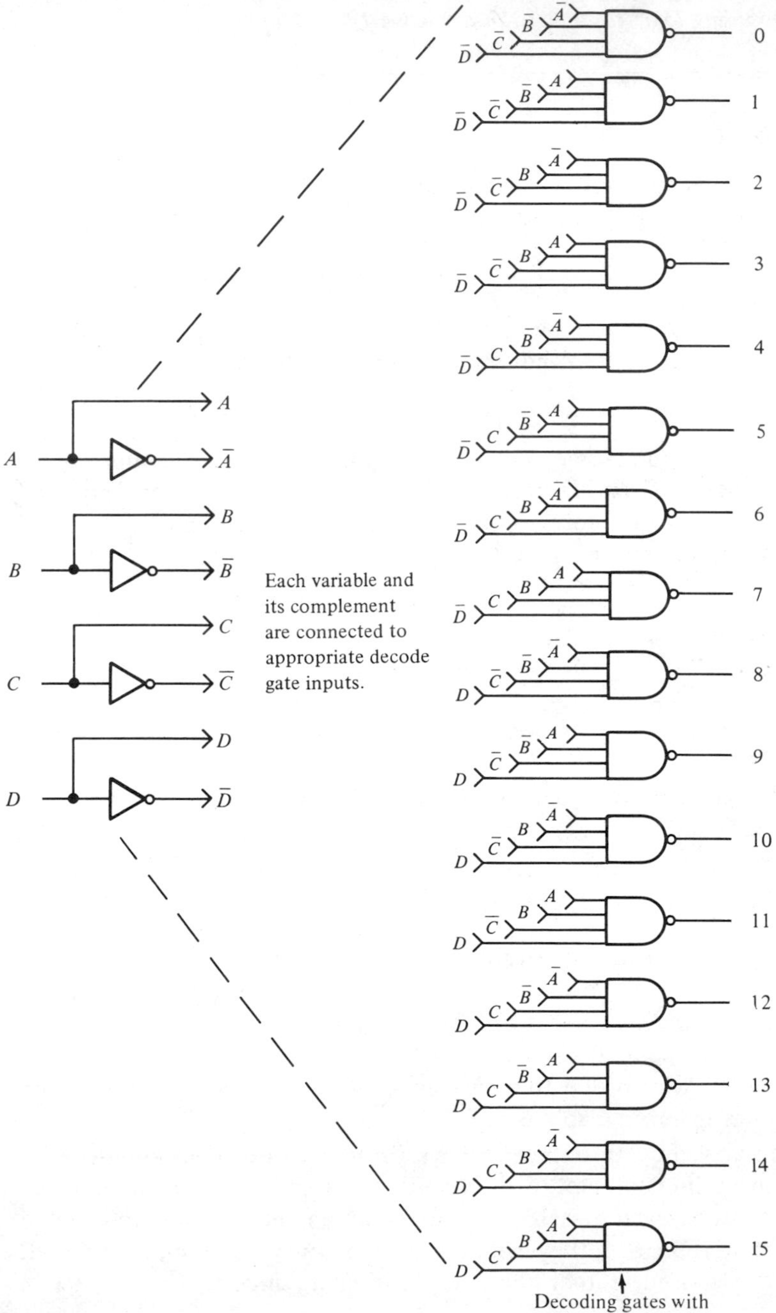

FIGURE 6–35 *Logic for a 1-of-16 decoder.*

Rather than reproduce the complex logic diagram for the decoder each time it is required in a schematic, a simpler representation is normally used. A typical logic symbol for a four-bit binary decoder (1-of-16) is shown in Figure 6–36, and the corresponding truth table is given in Table 6–5.

TABLE 6–4 *Four-bit binary decoding functions.*

	Binary Code				Logic Function
Decimal Digit	D	C	B	A	X
0	0	0	0	0	$\overline{D}\,\overline{C}\,\overline{B}\,\overline{A}$
1	0	0	0	1	$\overline{D}\,\overline{C}\,\overline{B}A$
2	0	0	1	0	$\overline{D}\,\overline{C}B\overline{A}$
3	0	0	1	1	$\overline{D}\,\overline{C}BA$
4	0	1	0	0	$\overline{D}C\overline{B}\,\overline{A}$
5	0	1	0	1	$\overline{D}C\overline{B}A$
6	0	1	1	0	$\overline{D}CB\overline{A}$
7	0	1	1	1	$\overline{D}CBA$
8	1	0	0	0	$D\overline{C}\,\overline{B}\,\overline{A}$
9	1	0	0	1	$D\overline{C}\,\overline{B}A$
10	1	0	1	0	$D\overline{C}B\overline{A}$
11	1	0	1	1	$D\overline{C}BA$
12	1	1	0	0	$DC\overline{B}\,\overline{A}$
13	1	1	0	1	$DC\overline{B}A$
14	1	1	1	0	$DCB\overline{A}$
15	1	1	1	1	$DCBA$

TABLE 6–5 *One-of-16 decoder truth table.*

Inputs				Outputs															
D	C	B	A	0	1	2	3	4	5	6	7	8	9	10	11	12	13	14	15
0	0	0	0	0	1	1	1	1	1	1	1	1	1	1	1	1	1	1	1
0	0	0	1	1	0	1	1	1	1	1	1	1	1	1	1	1	1	1	1
0	0	1	0	1	1	0	1	1	1	1	1	1	1	1	1	1	1	1	1
0	0	1	1	1	1	1	0	1	1	1	1	1	1	1	1	1	1	1	1
0	1	0	0	1	1	1	1	0	1	1	1	1	1	1	1	1	1	1	1
0	1	0	1	1	1	1	1	1	0	1	1	1	1	1	1	1	1	1	1
0	1	1	0	1	1	1	1	1	1	0	1	1	1	1	1	1	1	1	1
0	1	1	1	1	1	1	1	1	1	1	0	1	1	1	1	1	1	1	1
1	0	0	0	1	1	1	1	1	1	1	1	0	1	1	1	1	1	1	1
1	0	0	1	1	1	1	1	1	1	1	1	1	0	1	1	1	1	1	1
1	0	1	0	1	1	1	1	1	1	1	1	1	1	0	1	1	1	1	1
1	0	1	1	1	1	1	1	1	1	1	1	1	1	1	0	1	1	1	1
1	1	0	0	1	1	1	1	1	1	1	1	1	1	1	1	0	1	1	1
1	1	0	1	1	1	1	1	1	1	1	1	1	1	1	1	1	0	1	1
1	1	1	0	1	1	1	1	1	1	1	1	1	1	1	1	1	1	0	1
1	1	1	1	1	1	1	1	1	1	1	1	1	1	1	1	1	1	1	0

1 = HIGH, 0 = LOW

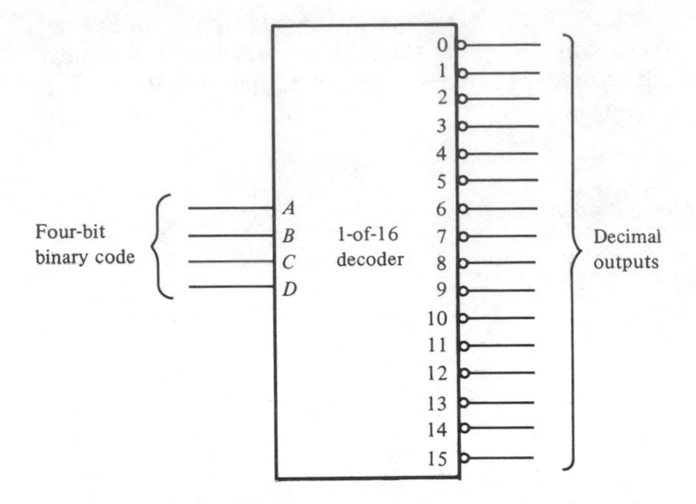

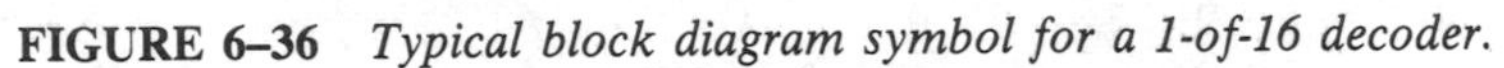
FIGURE 6–36 *Typical block diagram symbol for a 1-of-16 decoder.*

The 74154 Four-Line to 16-Line Decoder

The 74154 is a good example of a TTL MSI decoder. Its logic diagram is shown in Figure 6–37(a) and the block symbol with pin connections appears in Figure 6–37(b). The additional inverters on the inputs are required to prevent excessive loading of the driving source(s). Each input is connected to the input of only one inverter, rather than to the inputs of several NAND gates as in Figure 6–35. There is

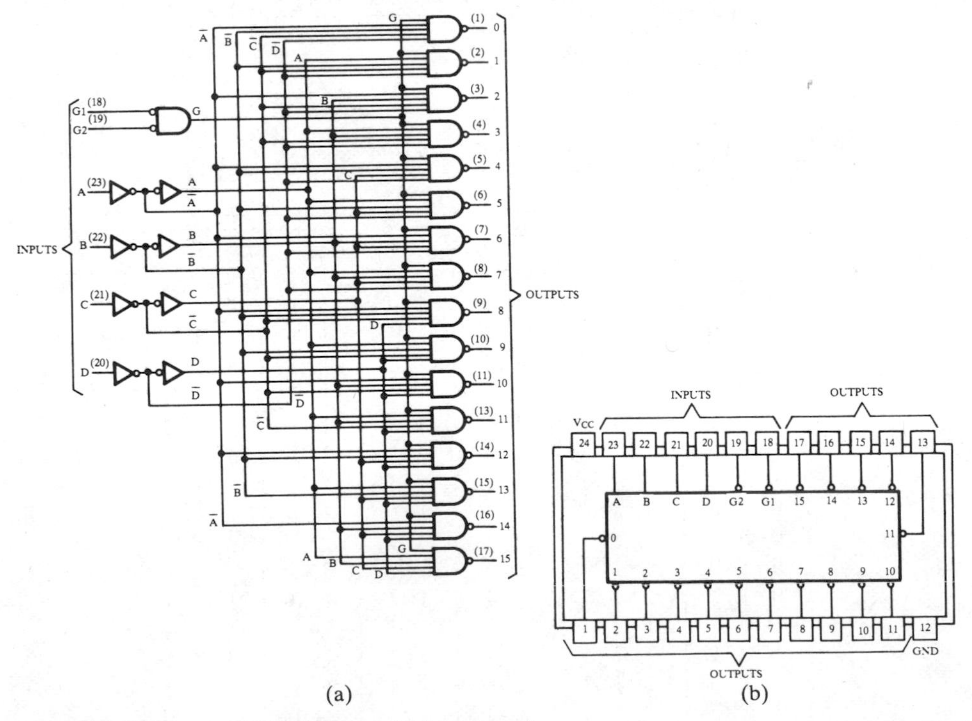

FIGURE 6–37 *A 74154 4-line to 16-line decoder.*

also an enable function provided on this particular device. This is implemented with a NOR gate used as a negative AND. A LOW level on each input, G_1 and G_2, is required in order to make the enable gate ouput *(G)* HIGH. The enable gate output is connected to an input of *each* NAND gate, so it must be HIGH for the gates to be enabled. If the enable gate is not activated (LOW on both inputs), then all 16 decoder outputs will be HIGH *regardless* of the states of the four input variables, *A, B, C,* and *D*.

The BCD Decoder

The BCD decoder converts each BCD code word (8421 code) into one of ten possible decimal digit indications. It is typically referred to as a *1-of-10* or a *4-line-to-10-line* decoder, although other types of decoders also fall into this category (such as an Excess-3 decoder).

The method of implementation is essentially the same as for the 1-of-16 decoder previously discussed, except that only *ten* decoding gates are required because the BCD code represents only the ten decimal digits 0 through 9. A list of the ten BCD code words and their corresponding decoding functions is given in Table 6–6. Each of these decoding functions is implemented with NAND gates to provide active LOW outputs, as shown in Figure 6–38. If an active HIGH output is required, AND gates are used for decoding. Notice that the logic is identical to the first ten decoding gates in the 1-of-16 decoder (Figure 6–35).

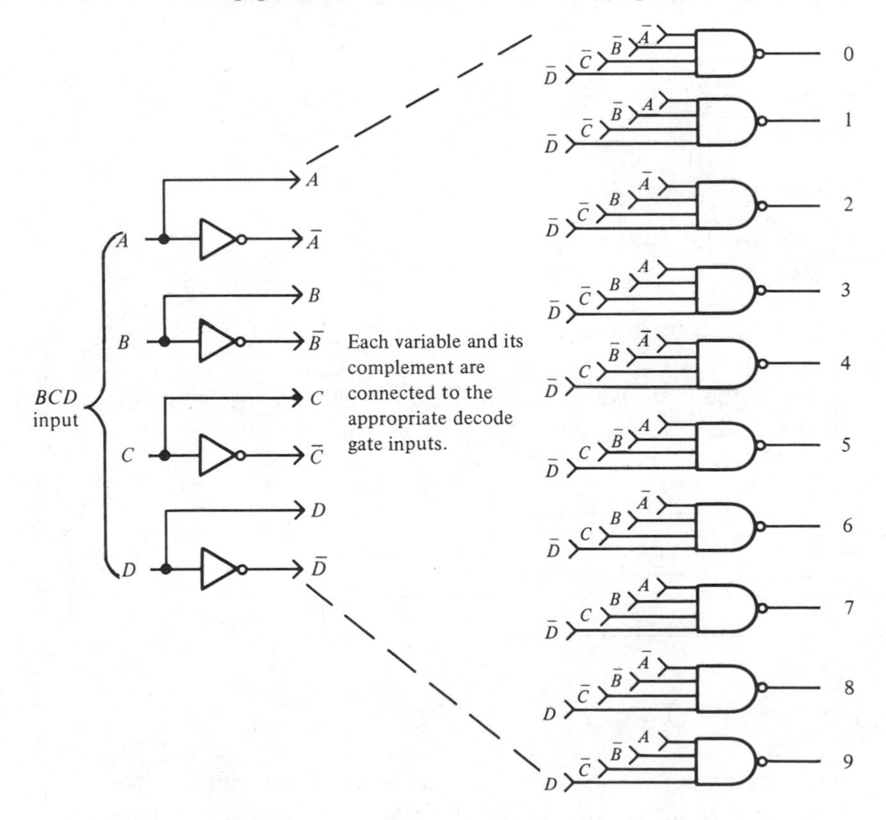

FIGURE 6–38 *Logic for a BCD decoder.*

TABLE 6–6 *BCD decoding functions.*

	BCD Code				Logic Function
Decimal Digit	D	C	B	A	X
0	0	0	0	0	$\overline{D}\,\overline{C}\,\overline{B}\,\overline{A}$
1	0	0	0	1	$\overline{D}\,\overline{C}\,\overline{B}A$
2	0	0	1	0	$\overline{D}\,\overline{C}B\overline{A}$
3	0	0	1	1	$\overline{D}\,\overline{C}BA$
4	0	1	0	0	$\overline{D}C\overline{B}\,\overline{A}$
5	0	1	0	1	$\overline{D}C\overline{B}A$
6	0	1	1	0	$\overline{D}CB\overline{A}$
7	0	1	1	1	$\overline{D}CBA$
8	1	0	0	0	$D\overline{C}\,\overline{B}\,\overline{A}$
9	1	0	0	1	$D\overline{C}\,\overline{B}A$

Example 6–11

The 7442A is an integrated circuit BCD-to-decimal decoder. The block and pin diagram is shown in Figure 6–39. Note that on this device the inputs are A, B, C, and D where A is the least significant bit.

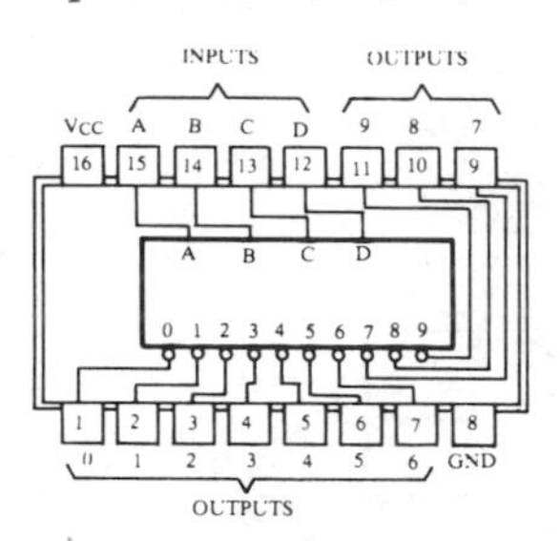

FIGURE 6–39 *A 7442A BCD-to-decimal decoder.*

If the input waveforms in Figure 6–40 are applied to the inputs of the 7442A, sketch the output waveforms.

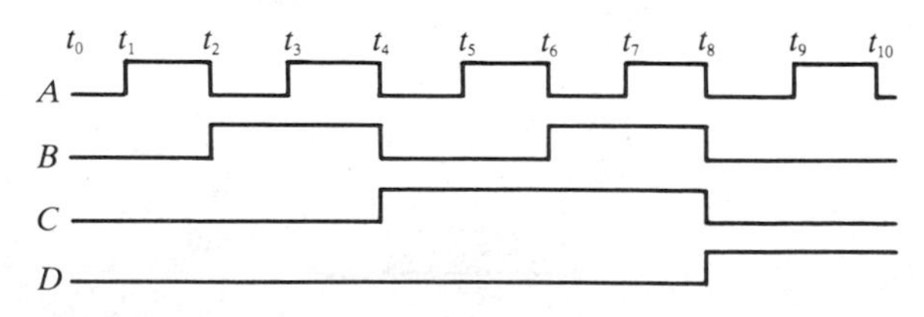

FIGURE 6–40

Solution:

The output waveforms are shown in Figure 6–41. As you can see, the inputs are sequenced through the BCD for digits 0 through 9. The output waveforms indicate that sequence.

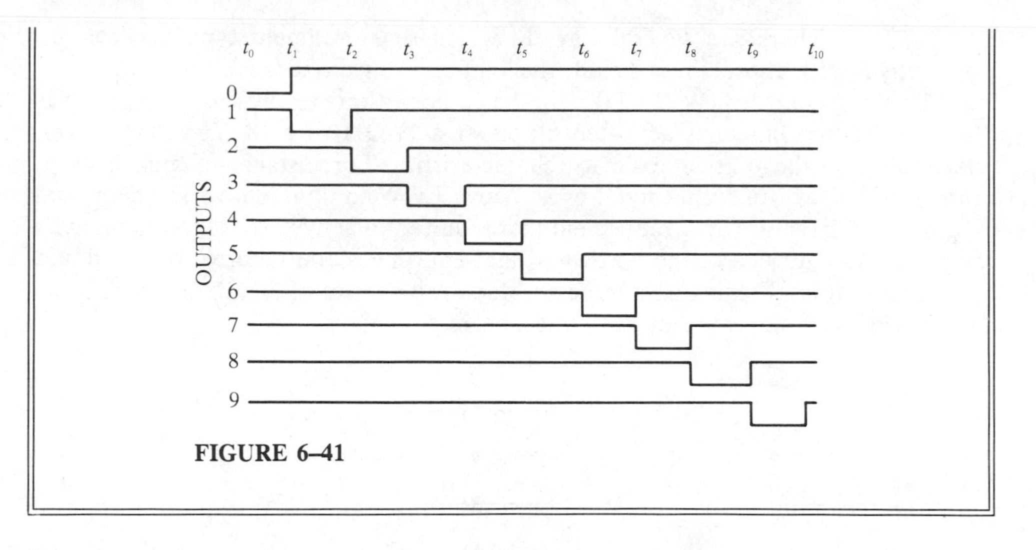

FIGURE 6–41

BCD to Seven-Segment Decoder

This type of decoder accepts the BCD code on its inputs and provides outputs to energize seven-segment display devices in order to produce a digital readout. Before proceeding with a discussion of this decoder, let us examine the basics of a seven-segment display device.

Figure 6–42 shows a common display format composed of seven light-emitting elements or segments. By lighting certain combinations of these segments, each of the ten decimal digits can be produced. Figure 6–43 illustrates this method of digital display for each of the ten digits by using a darker segment to represent one that is illuminated. To produce a 1, segments *f* and *e* are illuminated; to produce a 2, segments *a, b, g, e,* and *d* are used; and so on.

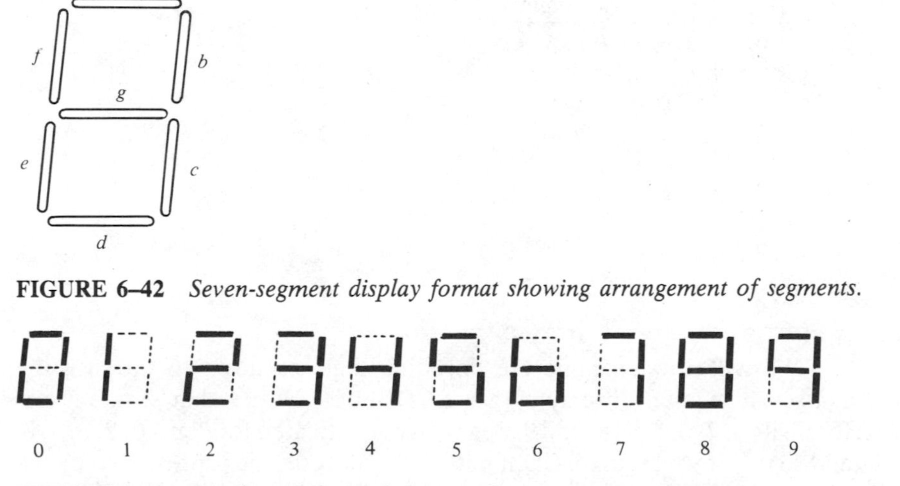

FIGURE 6–42 *Seven-segment display format showing arrangement of segments.*

FIGURE 6–43 *Display of decimal digits with a seven-segment device.*

There are several ways in which seven-segment type displays are currently implemented; these include the light-emitting diode (LED), incandescent, and liquid crystal display (LCD). The basic operation, regardless of type, is as follows. Each segment is activated by either a HIGH or a LOW voltage level, depending on the particular device characteristics. For instance, Figure 6–44(a) shows an LED display that requires an active LOW to illuminate a segment, and Figure 6–44(b) shows an arrangement that requires an active HIGH to illuminate a segment. In either case, the particular light-emitting diode is forward biased and conducts current, causing light to be emitted. The physics of the diode which bring about light emission are beyond the scope of this book.

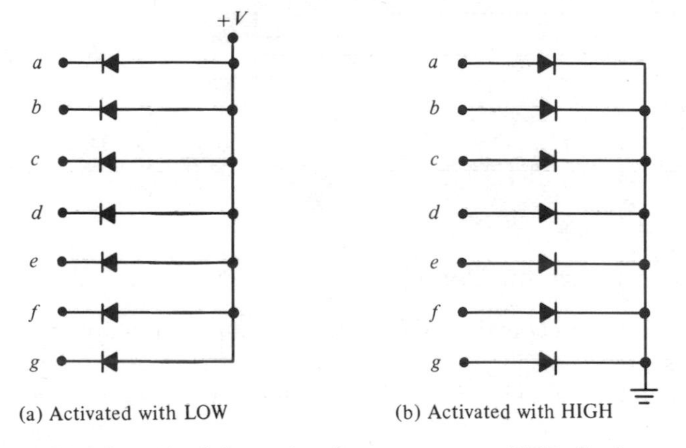

FIGURE 6–44 *Schematic of seven-segment LED displays.*

The activated segments for each of the ten decimal digits are listed in Table 6–7.

TABLE 6–7 *Seven-segment display format.*

Digit	Segments Illuminated
0	*a, b, c, d, e, f*
1	*e, f*
2	*a, b, g, e, d*
3	*a, b, c, d, g*
4	*b, c, f, g*
5	*a, c, d, f, g*
6	*c, d, e, f, g*
7	*a, b, c*
8	*a, b, c, d, e, f, g*
9	*a, b, c, f, g*

We will now examine the decoding logic required to produce the format for a seven-segment display with a BCD input. Notice that segment *a* is activated for digits 0, 2, 3, 5, 7, 8, and 9. Segment *b* is activated for digits 0, 2, 3, 4, 7, 8, and 9, and so on. If we let the BCD inputs to the decoder be represented by the general form *DCBA,* a Boolean expression can be found for each segment in the display, and this will tell us the logic circuitry required to drive or activate each segment. For example, the equation for segment a is as follows:

$$a = \bar{D}\bar{C}\bar{B}\bar{A} + \bar{D}\bar{C}B\bar{A} + \bar{D}\bar{C}BA + \bar{D}C\bar{B}A + \bar{D}CBA + D\bar{C}\bar{B}\bar{A} + D\bar{C}\bar{B}A$$

This equation says that segment a is activated or "true" if the BCD code is 0 "OR 2 OR 3 OR 5 OR 7 OR 8 OR 9." Table 6–8 lists the logic function for each of the seven segments.

TABLE 6–8 *Seven-segment decoding functions.*

Segment	Used in Digits	Function
a	0, 2, 3, 5, 7, 8, 9	$\bar{D}\bar{C}\bar{B}\bar{A} + \bar{D}\bar{C}B\bar{A} + \bar{D}\bar{C}BA + \bar{D}C\bar{B}A + \bar{D}CBA + D\bar{C}\bar{B}\bar{A} + D\bar{C}\bar{B}A$
b	0, 2, 3, 4, 7, 8, 9	$\bar{D}\bar{C}\bar{B}\bar{A} + \bar{D}\bar{C}B\bar{A} + \bar{D}\bar{C}BA + \bar{D}C\bar{B}\bar{A} + \bar{D}CBA + D\bar{C}\bar{B}\bar{A} + D\bar{C}\bar{B}A$
c	0, 3, 4, 5, 6, 7, 8, 9	$\bar{D}\bar{C}\bar{B}\bar{A} + \bar{D}\bar{C}BA + \bar{D}C\bar{B}\bar{A} + \bar{D}C\bar{B}A + \bar{D}CB\bar{A} + \bar{D}CBA + D\bar{C}\bar{B}\bar{A} + D\bar{C}\bar{B}A$
d	0, 2, 3, 5, 6, 8	$\bar{D}\bar{C}\bar{B}\bar{A} + \bar{D}\bar{C}B\bar{A} + \bar{D}\bar{C}BA + \bar{D}C\bar{B}A + \bar{D}CB\bar{A} + D\bar{C}\bar{B}\bar{A}$
e	0, 1, 2, 6, 8	$\bar{D}\bar{C}\bar{B}\bar{A} + \bar{D}\bar{C}\bar{B}A + \bar{D}\bar{C}B\bar{A} + \bar{D}CB\bar{A} + D\bar{C}\bar{B}\bar{A}$
f	0, 1, 4, 5, 6, 8, 9	$\bar{D}\bar{C}\bar{B}\bar{A} + \bar{D}\bar{C}\bar{B}A + \bar{D}C\bar{B}\bar{A} + \bar{D}C\bar{B}A + \bar{D}CB\bar{A} + D\bar{C}\bar{B}\bar{A} + D\bar{C}\bar{B}A$
g	2, 3, 4, 5, 6, 8, 9	$\bar{D}\bar{C}B\bar{A} + \bar{D}\bar{C}BA + \bar{D}C\bar{B}\bar{A} + \bar{D}C\bar{B}A + \bar{D}CB\bar{A} + D\bar{C}\bar{B}\bar{A} + D\bar{C}\bar{B}A$

From the expressions in Table 6–8, the logic for the BCD-to-seven-segment decoder can be implemented. Each of the ten BCD code words is decoded, and then the decoding gates are ORed as dictated by the logic expression for each segment. For instance, segment a requires that the decoded BCD digits 0, 2, 3, 5, 7, 8, and 9 be ORed, as shown in Figure 6–45.

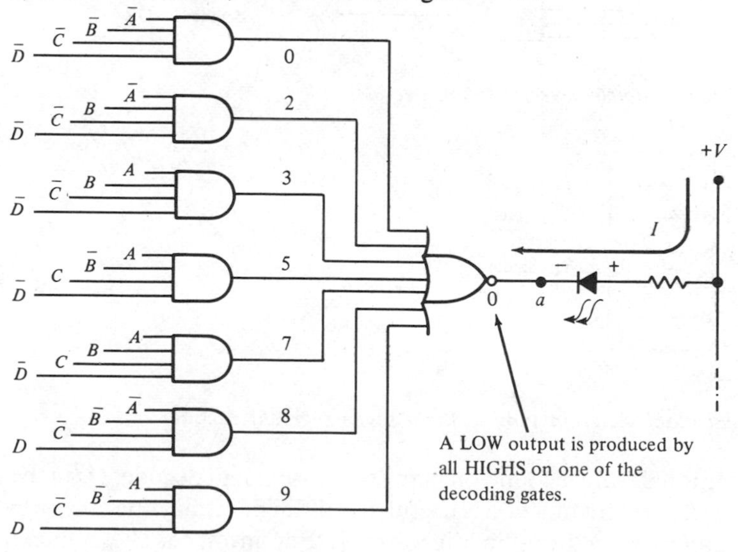

FIGURE 6–45 *Decoding logic for the "a" segment.*

Segment b requires that the decoded BCD digits 0, 2, 3, 4, 7, 8, and 9 be ORed. Segment c is activated by ORing the outputs of the 0, 3, 4, 5, 6, 7, 8, and 9 decode gates, etc. The total logic for the BCD-to-seven-segment decoder is shown in Figure 6–46 and a typical block diagram representation in Figure 6–47.

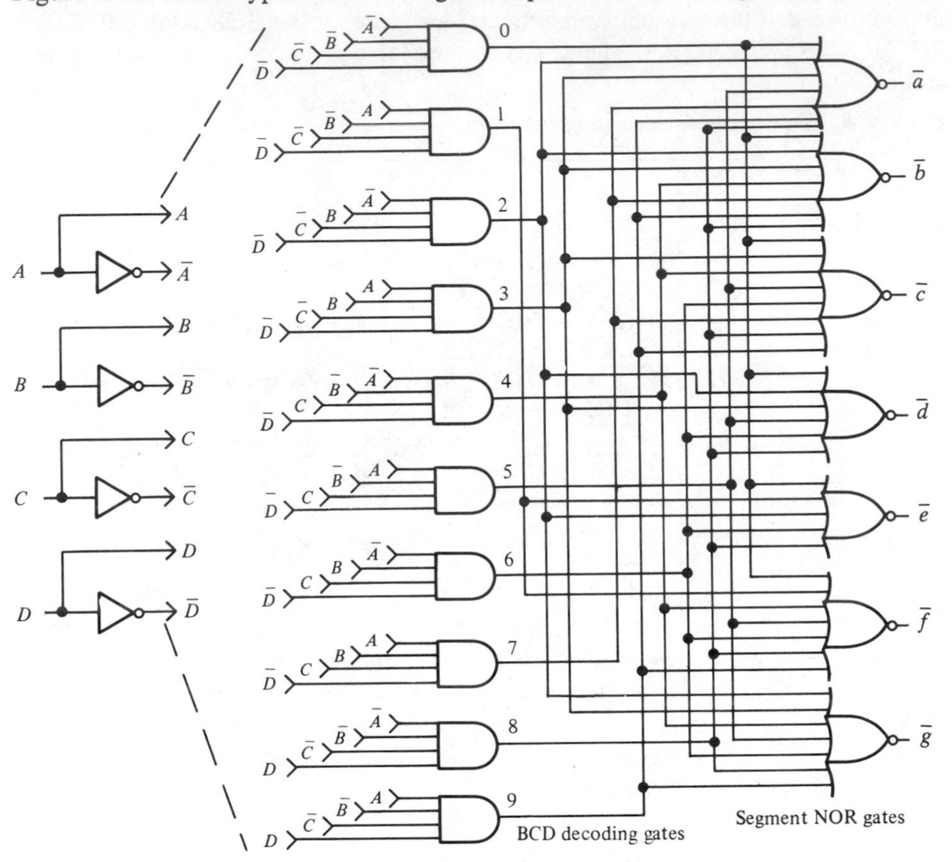

FIGURE 6–46 *Logic for complete seven-segment decoder.*

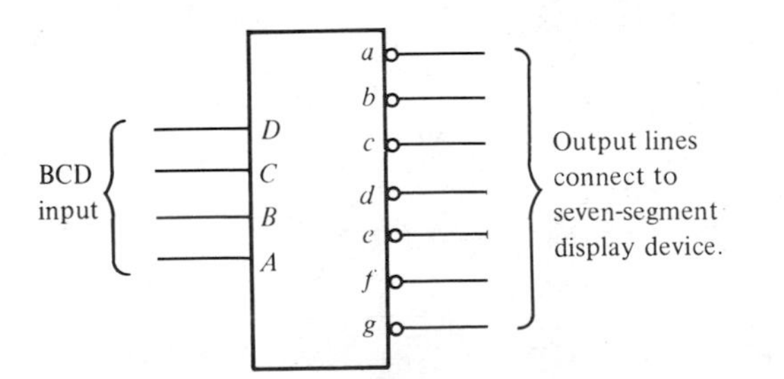

FIGURE 6–47 *Typical block diagram symbol for a seven-segment decoder.*

An additional feature found on many seven-segment decoders is in the *zero suppression logic.* This extra function is useful in multidigit displays because it is used to blank out unnecessary zeros in the display. For instance, the number 0006.400 would be displayed as 6.4, which is read more easily. Blanking of the zeros on the front of the number is called *leading edge zero suppression* and the blanking of the zeros after the number is called *trailing edge zero suppression*.

A block diagram of a four-digit display is shown in Figure 6–48, and will be used to illustrate the requirements for leading edge zero suppression. Notice that two additional functions have been added to each BCD-to-seven-segment decoder, a *ripple blanking input* (RBI) and a *ripple blanking output* (RBO). The highest order digit position is always blanked if a 0 code appears on its BCD inputs *and* the blanking input is HIGH. Each lower order digit position is blanked if a 0 code appears on its BCD inputs, *and* the next higher order digit is a 0 as indicated by a HIGH on its blanking output. The ripple blanking output of any decoder indicates that it has a BCD 0 on its inputs, and *all* higher order digits are also 0. The blanking output of each stage is connected to the blanking input of the next lower order stage, as shown in the diagram. As an example, in Figure 6–48 the highest order digit is 0, which is therefore blanked. Also, the next digit is a 0, and because the highest order digit is 0, it is also blanked. The remaining two digits are displayed.

For the fractional portion of a display (the digits to the right of the decimal point), trailing edge zero suppression is used; that is, the lowest order digit is blanked if it is 0 and each digit that is 0 *and* is followed by 0s in all the lower order positions is blanked. To illustrate, Figure 6–49 shows a block diagram in which there are three digits to the right of the decimal point. In this example, the lowest order digit is blanked because it is a 0. The next digit is also 0, *and* its blanking input is HIGH, so it is blanked. The highest order digit is a 5, and it is displayed. Notice that the blanking output of each decoder stage is connected to the blanking input of the next higher order stage.

One way of implementing the 0 suppression function is shown in Figure 6–50. Here, a NAND gate G_1 has been added to the basic BCD-to-seven-segment decoder, and is shown within the dashed lines. Gate G_1 detects the presence of a BCD 0 on the inputs *and* a HIGH on the blanking input. The complement of the output of this gate is the blanking output, and it is used to inhibit the decoding logic in the BCD-to-seven-segment decoder so that the seven-segment display device will not be activated. In this particular implementation, the output of gate G_1 is connected to an input of each BCD decoding gate so that when a blanking condition occurs (a 0 code on the inputs and BI = 1), each decoding gate is inhibited by the LOW on the output of gate G_1. As a result, the output lines from the decoder that drive or activate the display segments are held HIGH, preventing the segments from lighting. (We are using active LOW outputs in this case).

The 7447 BCD to Seven-Segment Decoder/Driver

As an example of a specific device, let us look at the 7447. The block and pin diagram is shown in Figure 6–51. Notice that it has an RBI (pin 5) and a BI/RBO (pin 4). The BI/RBO pin can be used as a ripple blanking output or as an overriding blanking input to control lamp intensity. The LT input (pin 3) is for *lamp test*. When a low level is applied to this input, all of the seven segment outputs are activated to check all segments of the display. This particular device has open collector outputs that can drive the display segments directly. See Appendix A for a discussion of open-collector TTL circuits.

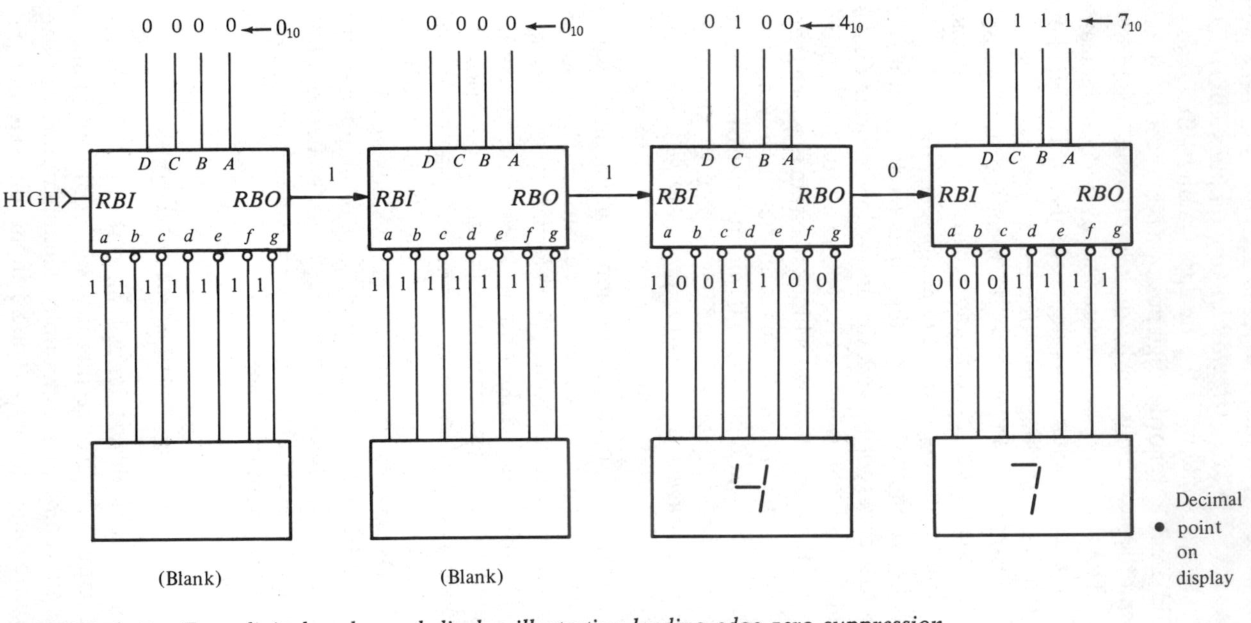

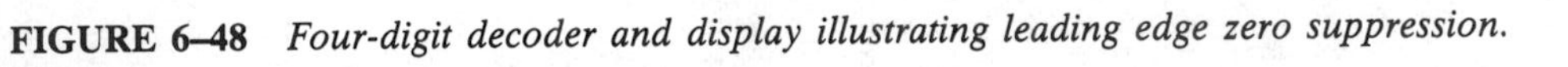

FIGURE 6–48 *Four-digit decoder and display illustrating leading edge zero suppression.*

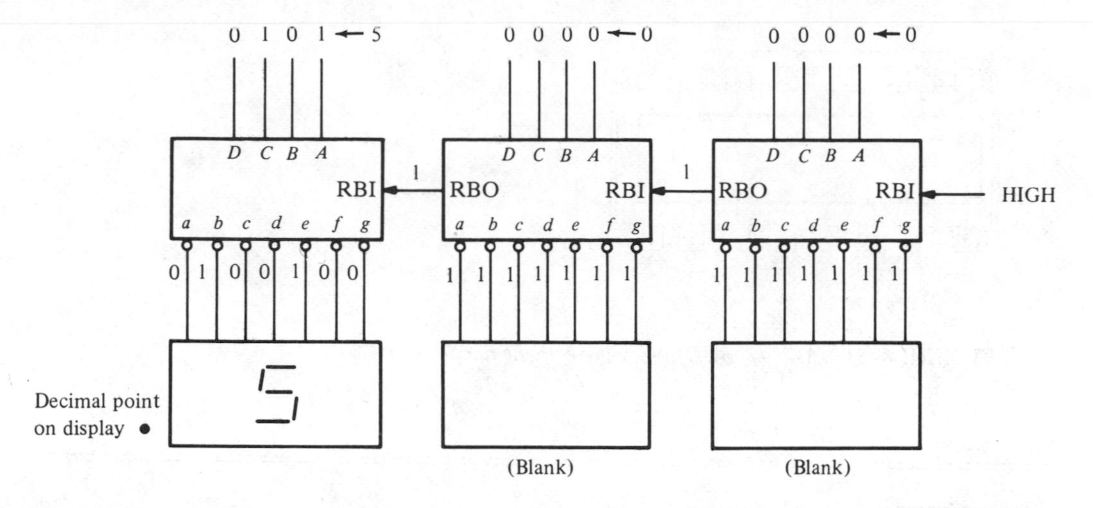

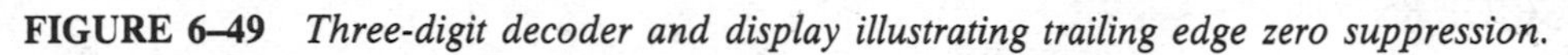

FIGURE 6–49 *Three-digit decoder and display illustrating trailing edge zero suppression.*

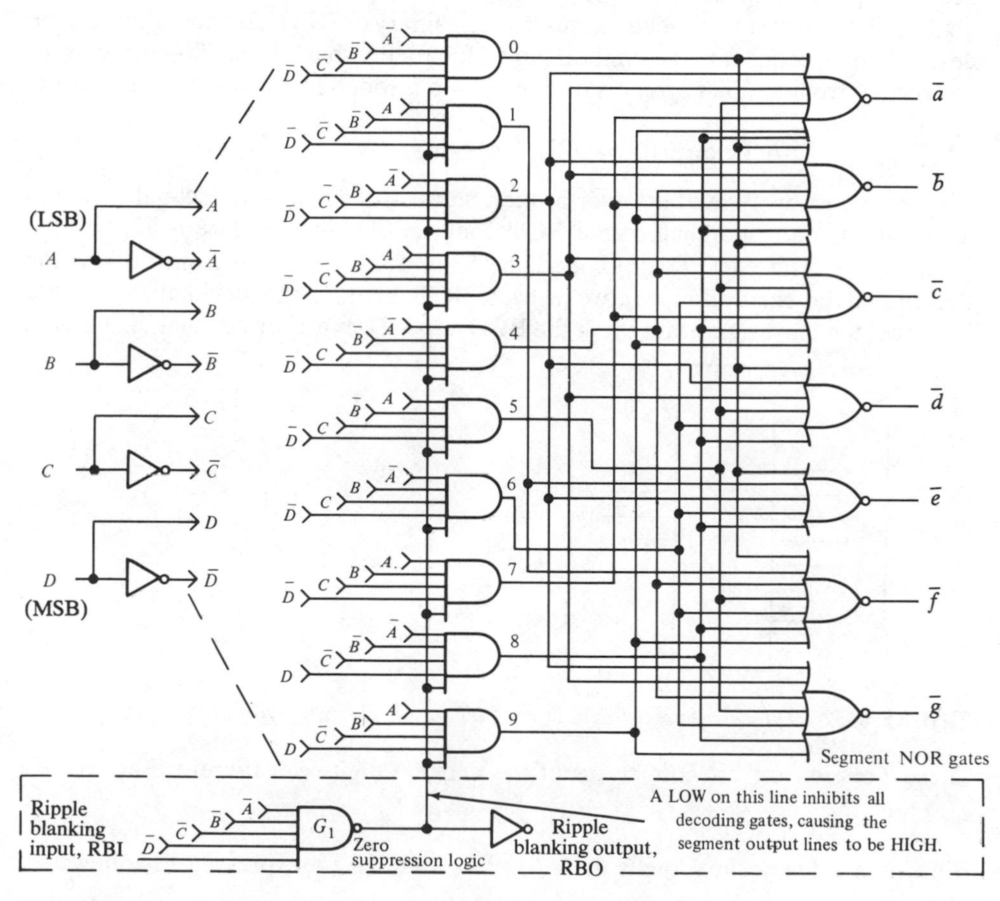

FIGURE 6–50 *Seven-segment decoder with zero suppression logic.*

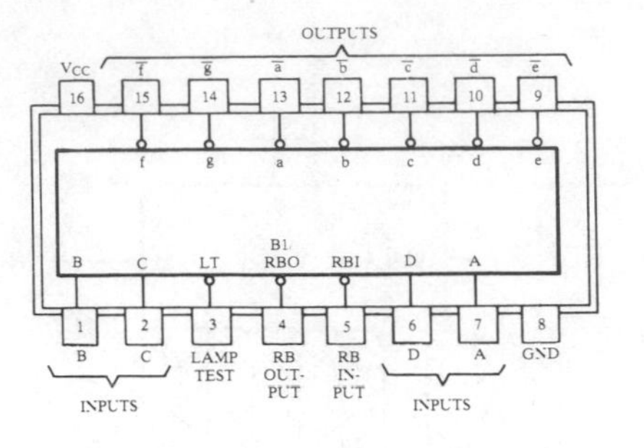

FIGURE 6–51 *A 7447 BCD-to-seven-segment decoder/driver.*

6–6 ENCODERS

An *encoder* is a combinational logic circuit that essentially performs a "reverse" decoder function. An encoder accepts a digit on its inputs, such as a decimal or octal digit, and converts it to a coded output, such as binary or BCD. Encoders can also be devised to encode various symbols and alphabetic characters. This process of converting from familiar symbols or numbers to a coded format is called *encoding*.

The Decimal-to-BCD Encoder

This type of encoder has ten inputs—one for each decimal digit and four outputs corresponding to the BCD code, as shown in Figure 6–52.

The BCD (8421) code is listed in Table 6–9, and from this we can determine the relationship between each BCD bit and the decimal digits. For instance, the most significant bit of the BCD code, D, is a 1 for decimal digit 8 or 9.

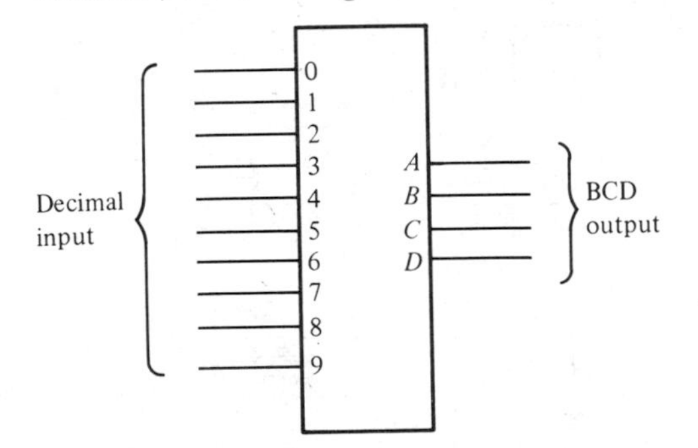

FIGURE 6–52 *Typical decimal-to-BCD encoder block diagram symbol.*

The expression for bit D in terms of the decimal digits can therefore be written

$$D = 8 + 9$$

Bit C is a 1 for decimal digits 4, 5, 6, or 7, and can be expressed as follows:

$$C = 4 + 5 + 6 + 7$$

B is a 1 for decimal digits 2, 3, 6, or 7, and can be expressed as

$$B = 2 + 3 + 6 + 7$$

Finally, A is a 1 for digits 1, 3, 5, 7, or 9. The expression for A is

$$A = 1 + 3 + 5 + 7 + 9$$

TABLE 6–9

Decimal Digit	BCD Code D	C	B	A
0	0	0	0	0
1	0	0	0	1
2	0	0	1	0
3	0	0	1	1
4	0	1	0	0
5	0	1	0	1
6	0	1	1	0
7	0	1	1	1
8	1	0	0	0
9	1	0	0	1

Now we can implement the logic circuitry required for encoding each decimal digit to a BCD code by using the logic expressions just developed. It is simply a matter of ORing the appropriate decimal digit input lines to form each BCD output. The basic encoder logic resulting from these expressions is shown in Figure 6–53.

The basic operation is as follows: when a HIGH appears on one of the decimal digit input lines, the appropriate levels occur on the four BCD output lines. For instance, if input line 9 is HIGH (assuming all other input lines are LOW), this will produce a HIGH on output A and output D and LOWs on outputs B and C, which is the BCD code (1001) for decimal 9.

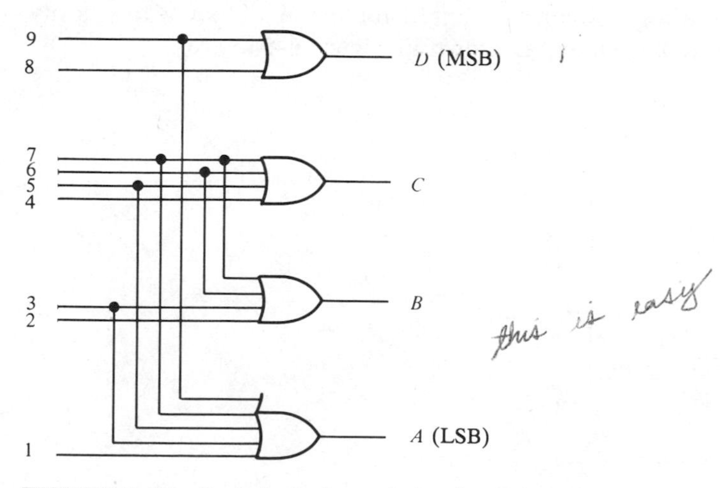

FIGURE 6–53 *Basic logic for a decimal-to-BCD encoder.*

Decimal-to-BCD Priority Encoder

This type of encoder performs the same basic encoding function as that previously discussed. It also offers additional flexibility in that it can be used in applications requiring priority detection. The *priority* function means that the encoder will produce a BCD output corresponding to the *highest order decimal digit* appearing on the inputs, and ignore all others. For instance, if the 6 and the 3 inputs are both HIGH, the BCD output is 0110 (which represents decimal 6).

Now let us look at the requirements for the priority logic. The purpose of this logic circuitry is to prevent a lower order digit input from disrupting the encoding of a higher order digit; this is accomplished by using inhibit gates. We will start by examining each BCD output (beginning with output A). Referring to Figure 6–53, note that A is HIGH when 1, 3, 5, 7, or 9 is HIGH. Digit input 1 is allowed to activate the A output if no higher order digits *other than those that also activate* A are HIGH. This can be stated as follows:

1. A is HIGH if 1 is HIGH and 2, 4, 6, and 8 are LOW. Similar statements can be made for the other digit inputs to the A OR gate.
2. A is HIGH if 3 is HIGH and 4, 6, and 8 are LOW.
3. A is HIGH if 5 is HIGH and 6 and 8 are LOW.
4. A is HIGH if 7 is HIGH and 8 is LOW.
5. A is HIGH if 9 is HIGH.

These five statements describe the priority of encoding for the BCD bit A. The A output is HIGH if any of the conditions listed occur; that is, A is true if statement 1, statement 2, statement 3, statement 4, or statement 5 is true. This can be expressed in the form of the following logic equation:

$$A = 1 \cdot \overline{2} \cdot \overline{4} \cdot \overline{6} \cdot \overline{8} + 3 \cdot \overline{4} \cdot \overline{6} \cdot \overline{8} + 5 \cdot \overline{6} \cdot \overline{8} + 7 \cdot \overline{8} + 9$$

From this expression the logic circuitry required for the A output with priority inhibits can be readily implemented, as shown in Figure 6–54.

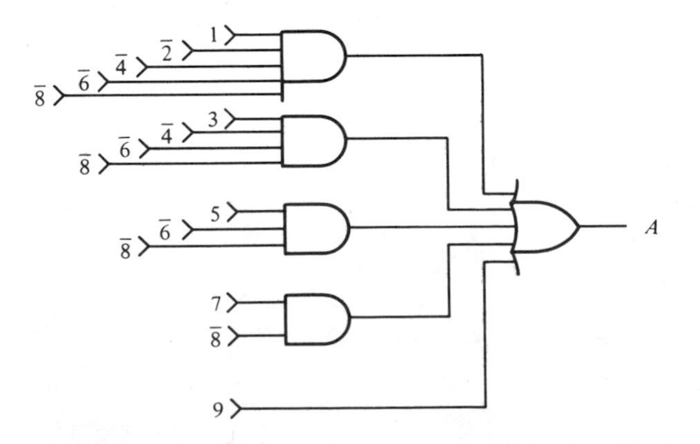

FIGURE 6–54 *Logic for the A-bit output of a decimal-to-BCD priority encoder.*

The same reasoning process can be applied to output B, and the following logical statements can be made:

1. B is HIGH if 2 is HIGH and 4, 5, 8, and 9 are LOW.
2. B is HIGH if 3 is HIGH and 4, 5, 8, and 9 are LOW.
3. B is HIGH if 6 is HIGH and 8 and 9 are LOW.
4. B is HIGH if 7 is HIGH and 8 and 9 are LOW.

These statements are summarized in the following equation and the logic implementation is shown in Figure 6–55:

$$B = 2 \cdot \overline{4} \cdot \overline{5} \cdot \overline{8} \cdot \overline{9} + 3 \cdot \overline{4} \cdot \overline{5} \cdot \overline{8} \cdot \overline{9} + 6 \cdot \overline{8} \cdot \overline{9} + 7 \cdot \overline{8} \cdot \overline{9}$$

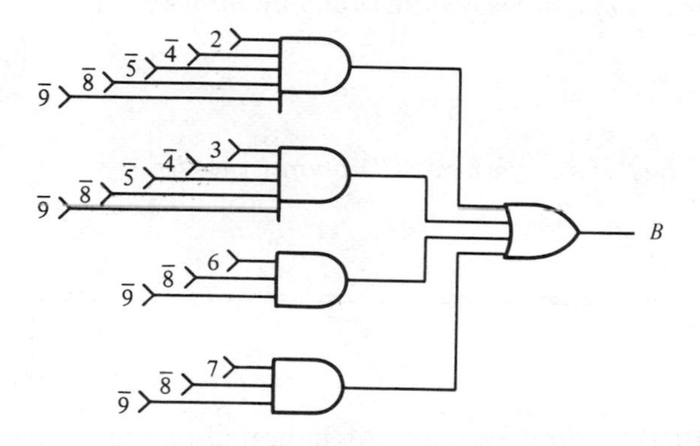

FIGURE 6–55 *Logic for the B-bit output of a decimal-to-BCD priority encoder.*

Output C can be described as follows:

1. C is HIGH if 4 is HIGH and 8 and 9 are LOW.
2. C is HIGH if 5 is HIGH and 8 and 9 are LOW.
3. C is HIGH if 6 is HIGH and 8 and 9 are LOW.
4. C is HIGH if 7 is HIGH and 8 and 9 are LOW.

In equation form, output C is

$$C = 4 \cdot \overline{8} \cdot \overline{9} + 5 \cdot \overline{8} \cdot \overline{9} + 6 \cdot \overline{8} \cdot \overline{9} + 7 \cdot \overline{8} \cdot \overline{9}$$

The logic circuitry for the C output appears in Figure 6–56.

Finally, for the D output,

D is HIGH if 8 is HIGH or if 9 is HIGH.

This statement appears in equation form as follows:

$$D = 8 + 9$$

The logic for this output is in Figure 6–57. No inhibits are required. We now have developed the basic logic for the decimal-to-BCD priority encoder. All of the complements of the input digit variables are realized by inverting the inputs. A complete logic diagram is shown in Figure 6–58.

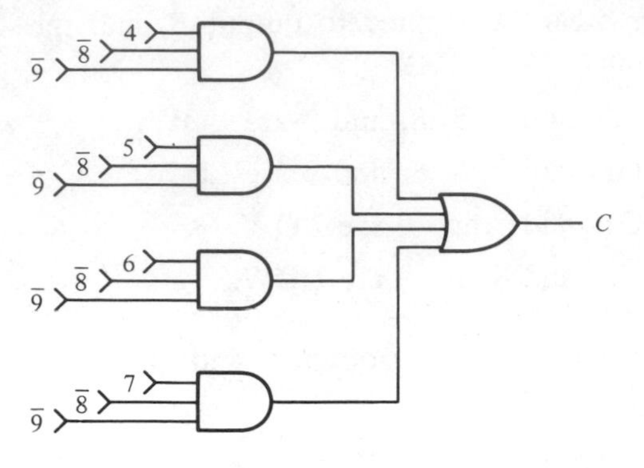

FIGURE 6–56 *Logic for the C-bit output of a decimal-to-BCD priority encoder.*

FIGURE 6–57 *Logic for the D-bit output of a decimal-to-BCD priority encoder.*

Example 6–12

The 74147 is a decimal-to-BCD priority encoder. As indicated on the block and pin diagram in Figure 6–59, its inputs and outputs are all active LOW.

If LOW levels appear on pins 1, 4, and 13, indicate the state of the four outputs. All other inputs are HIGH.

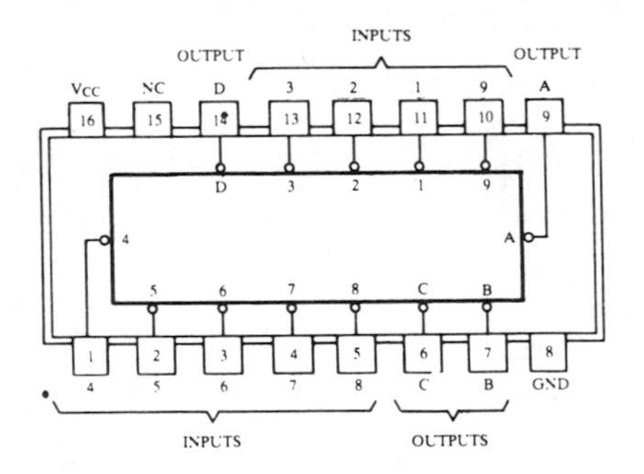

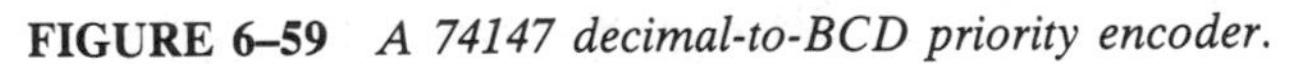

FIGURE 6–59 *A 74147 decimal-to-BCD priority encoder.*

Solution:

Pin 4 is the highest order decimal input having a LOW level and represents decimal 7. Therefore the output levels indicate the BCD code for decimal 7 where A is the LSB and D is the MSB. A is LOW, B is LOW, C is LOW, and D is HIGH.

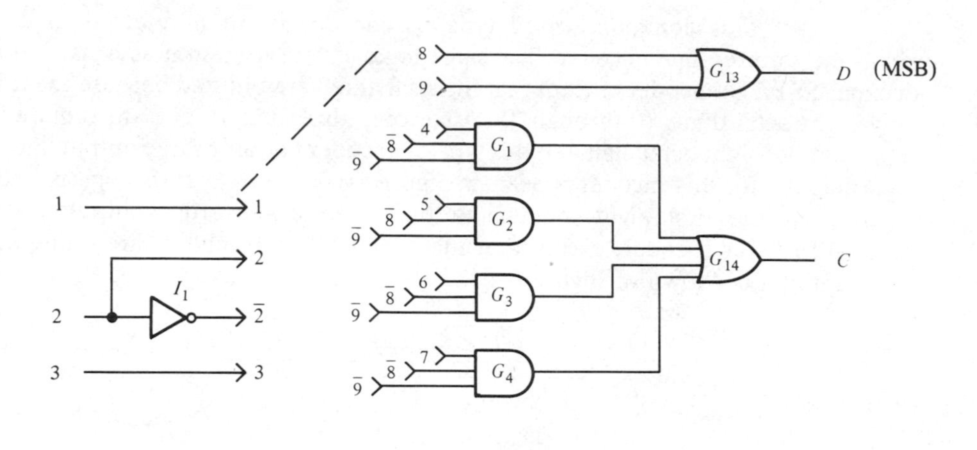

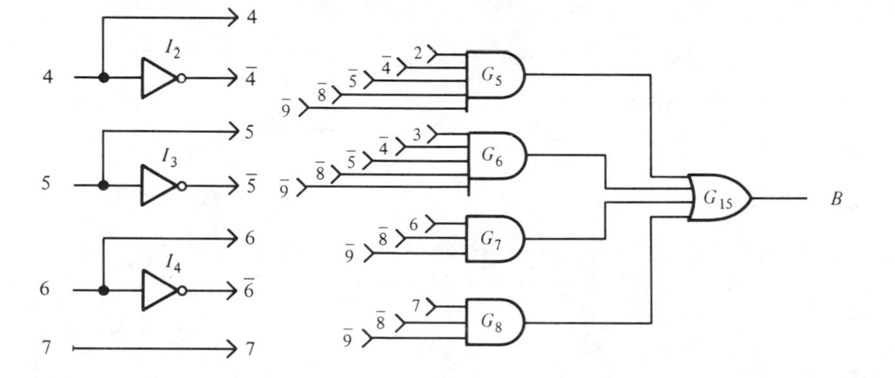

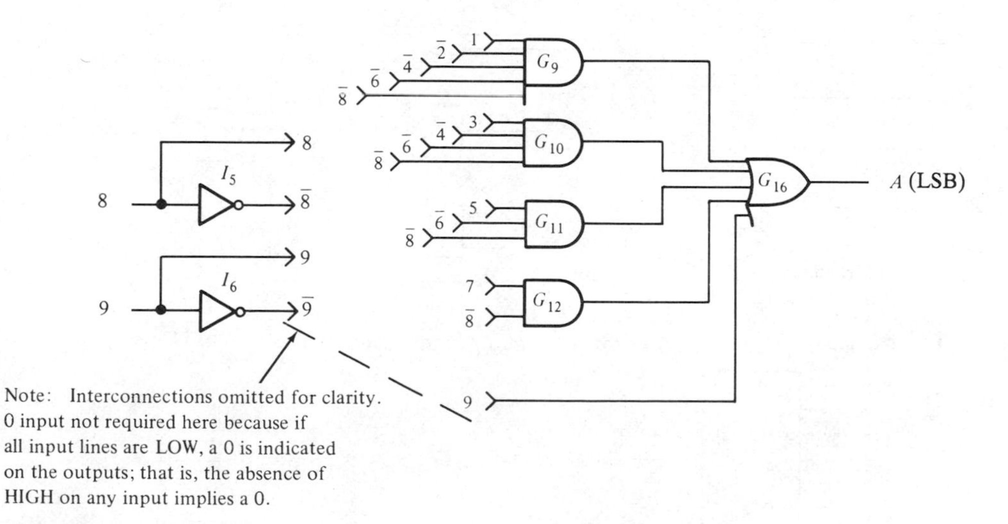

FIGURE 6–58 *Full logic diagram of a decimal-to-BCD priority encoder.*

Octal-to-Binary Encoder

This commonly used type of encoder is often referred to as an eight-line to three-line encoder. The same logic can be used in it as is used in the decimal-to-BCD encoder, except that inputs 8 and 9 are omitted because there are only eight octal digits (0 through 7). Also, only three binary bits are required to represent the eight octal digits, so this type of encoder has only three output lines. A logic diagram for this encoder appears in Figure 6–60. Notice that the logic is similar to that for the first eight inputs and the three lower order outputs of the decimal-to-BCD encoder, except that the decimal 8 or 9 inhibits are omitted. A block symbol is shown in Figure 6–61.

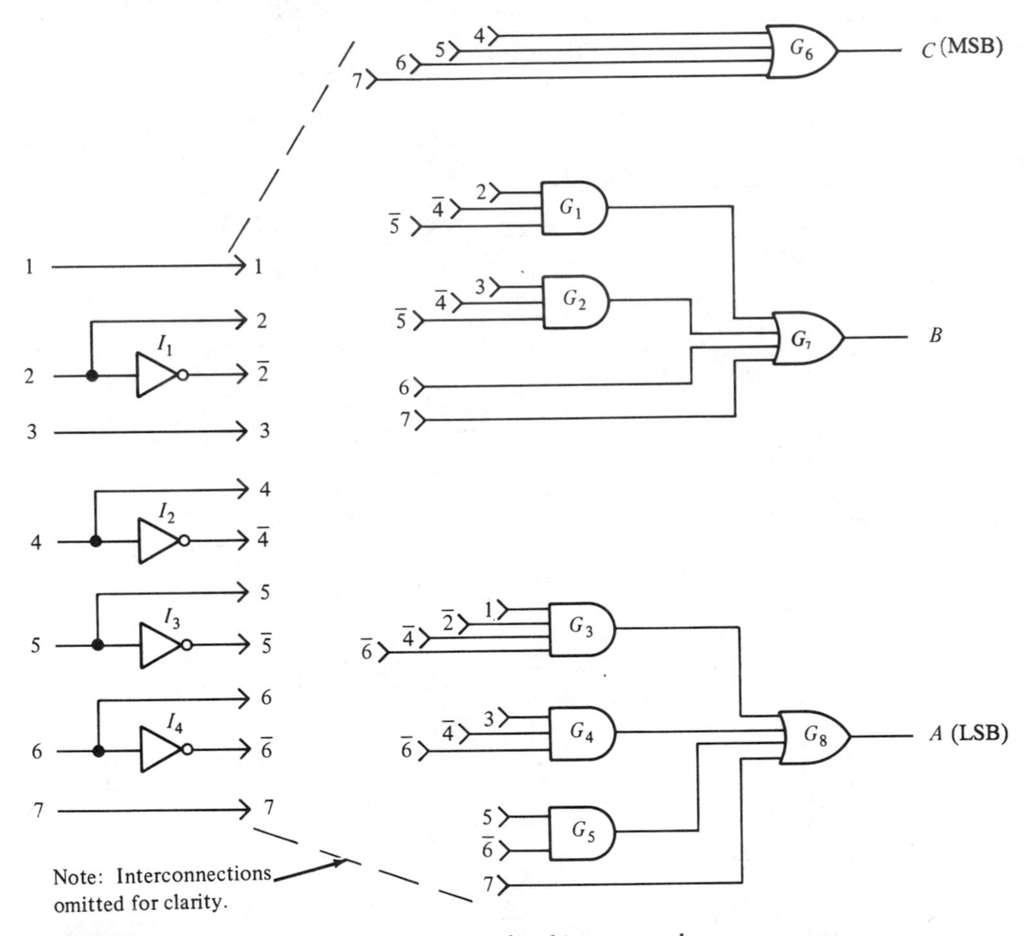

FIGURE 6–60 *Logic diagram for an octal-to-binary encoder.*

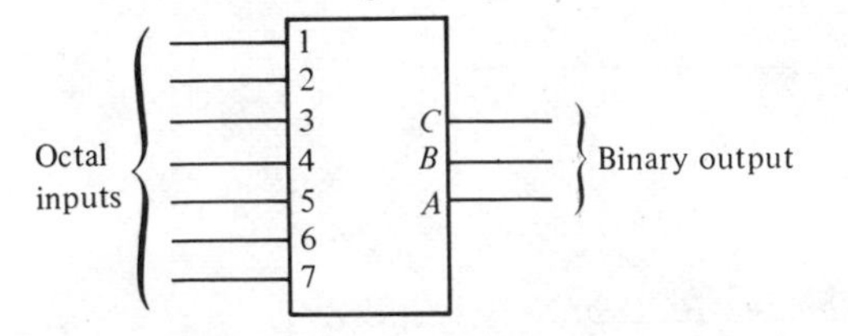

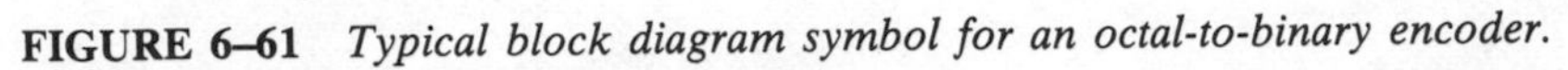
FIGURE 6–61 *Typical block diagram symbol for an octal-to-binary encoder.*

CODE CONVERTERS 6–7

In this section we will examine some combinational circuits that convert from one code to another.

BCD-to-Binary Converter

One method of BCD to binary code conversion involves the use of adder circuits. The basic conversion process is as follows:

1. The *value* of *each* bit in the BCD number is represented by a binary number.
2. All of the binary representations of bits that are 1s in the BCD number are added.
3. The result of this addition is the binary equivalent of the BCD number.

A more concise statement of this operation is that the binary numbers for the weights of the BCD bits are summed to produce the total binary number.

We will examine an eight-bit BCD code (one that represents a two-digit decimal number) in order to understand the relationship between BCD and binary. For instance, we already know the decimal number 87 can be expressed in BCD as

$$\underbrace{1000}_{8}\underbrace{0111}_{7}$$

The left-most four-bit group represents 80 and the right-most four-bit group represents 7; that is, the left-most group has a weight of 10 and the right-most has a weight of 1. Within each group, the binary weight of each bit is as follows:

	Tens digit				*Units digit*			
Weight:	80	40	20	10	8	4	2	1
Bit Designation:	D_1	C_1	B_1	A_1	D_0	C_0	B_0	A_0

The binary equivalent of each BCD bit is a binary number representing the *weight* of that bit within the total BCD number. This representation is given in Table 6–10.

If the binary representations for the weight of each 1 in the BCD number are added, the result is the binary number corresponding to the BCD number. An example will illustrate this.

With this basic procedure in mind, let us determine how the process can be implemented with logic circuits. Once the binary representation for each 1 in the BCD number is determined, adder circuits can be used to add the 1s in each column of the binary representation. The 1s occur in a given column only when the corresponding BCD bit is a 1. The occurrence of a BCD 1 can therefore be used to generate the proper binary 1 in the appropriate column of the adder structure. To handle a two-decimal digit (two-decade) BCD code, eight BCD input lines and seven

TABLE 6-10 *Binary representations of BCD bit weights.*

BCD Bit	Weight	Binary Representation						
		64	32	16	8	4	2	1
A_0	1	0	0	0	0	0	0	1
B_0	2	0	0	0	0	0	1	0
C_0	4	0	0	0	0	1	0	0
D_0	8	0	0	0	1	0	0	0
A_1	10	0	0	0	1	0	1	0
B_1	20	0	0	1	0	1	0	0
C_1	40	0	1	0	1	0	0	0
D_1	80	1	0	1	0	0	0	0

Example 6–13

Convert the BCD numbers 00100111 (decimal 27) and 10011000 (decimal 98) to binary.

Solutions:

Write the binary for the weights of all 1s appearing in the numbers, and then add them together.

```
00100111
  |  |||
  |  ||└──────────► 0000001      1
  |  |└───────────► 0000010      2
  |  └────────────► 0000100      4
  └───────────────►+0010100     20
                   -------
                    0011011     binary number for decimal 27

10011000
|  ||
|  |└─────────────► 0001000      8
|  └──────────────► 0001010     10
└─────────────────►+1010000     80
                   -------
                    1100010     binary number for decimal 98
```

binary outputs are required. (It takes 7 binary bits to represent numbers up through 99.) A block diagram is shown in Figure 6–62.

Referring to Table 6–10, notice that the "1" (LSB) column of the binary representation has only a single 1 and no possibility of an input carry, so that a straight connection from the A_0 bit of the BCD input to the least significant binary output is sufficient. In the "2" column of the binary representation, the possible occurrence of the two 1s can be accommodated by adding the B_0 bit and the A_1 bit of the BCD number. In the "4" column of the binary representation, the possible occurrence of the two 1s is handled by adding the C_0 bit and the B_1 bit of the BCD number. In the "8" column of the binary representation, the possibility of the three 1s is handled by adding the D_0, A_1, and C_1 bits of the BCD number. In the "16" column, the B_1 and the D_1 bits are added. In the "32" column, only a single 1 is possible, so the C_1 bit is added to the carry from the "16" column. In the "64" column, only a single 1 can occur, so the D_1 bit is added only to the carry from the "32" column. A method of implementing these requirements is shown in Figure 6–63.

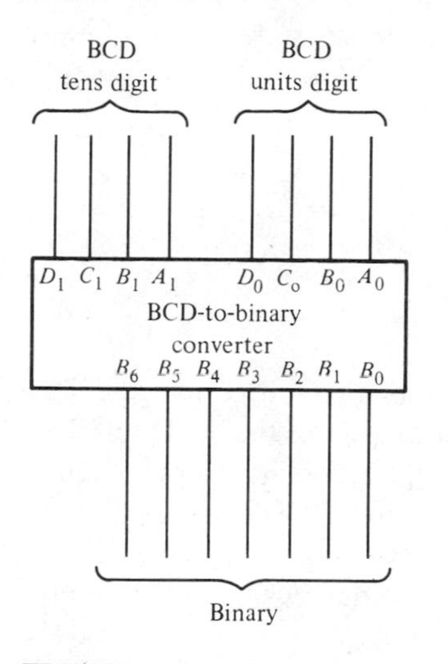

FIGURE 6–62 *Typical block diagram symbol for a two-decade BCD-to-binary converter.*

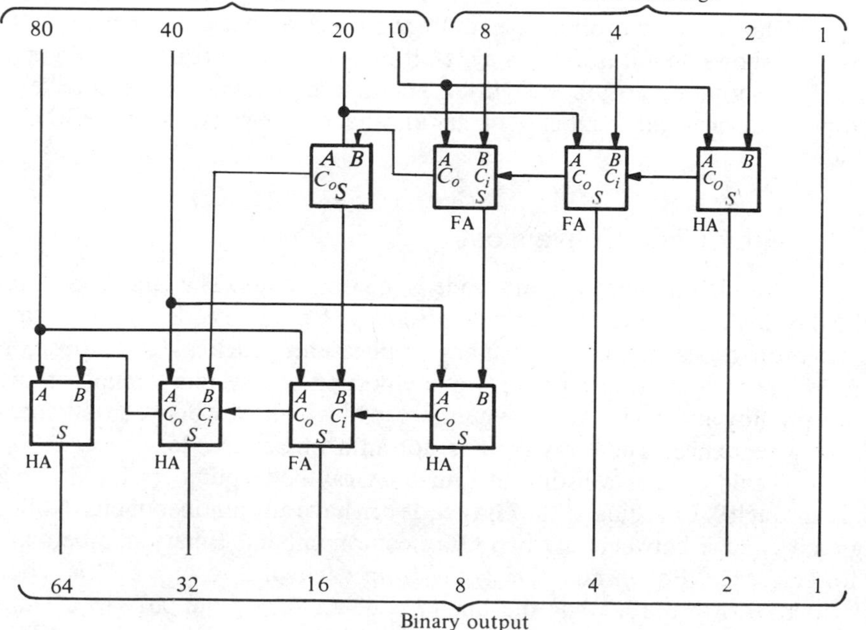

FIGURE 6–63 *Two-digit BCD-to-binary converter.*

74184 BCD-to-Binary Converter

The block and pin diagram for this device is shown in Figure 6–64.

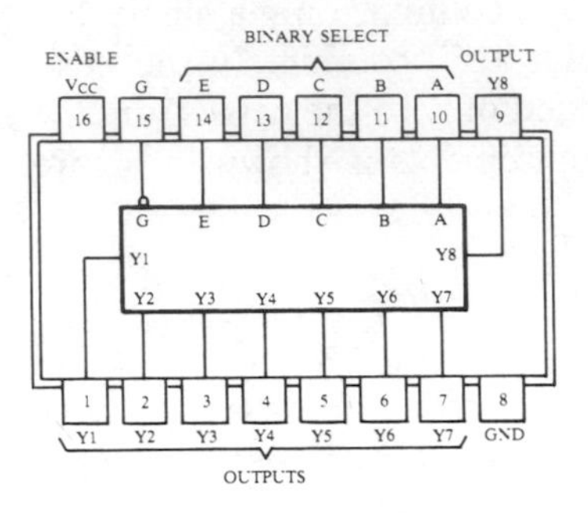

FIGURE 6–64 *A 74184 BCD-to-binary converter.*

Figure 6–65(a) shows how this IC is used as a six-bit converter. The single device can handle four bits of the least significant decade of the BCD number and two bits of the most significant decade. It produces a six-bit binary output. Figure 6–65(b) shows how two 74184s are connected for handling two BCD decades. Figure 6–65(c) shows a configuration for three BCD decades. By using additional devices the number of BCD decades that can be converted to binary can be expanded.

74185 Binary-to-BCD Converter

This device performs essentially the reverse operation as the 74184. Figure 6–66(a) shows how it is used as a six-bit binary-to-BCD converter. Figures 6–66(b) and (c) shows expansion to eight-bit and nine-bit converters, respectively. With additional devices, the number of binary bits that can be converted to BCD can be expanded.

Gray Code Conversions

Another important digital code is the Gray code. The Gray code is an unweighted code that exhibits only a *single-bit change* from one code number to the next. This property is important in many applications, such as shaft position encoders, where data are transmitted from one portion of a system to another and error susceptibility increases with the number of bit changes between adjacent numbers in a sequence. The Gray code is not an arithmetic code.

Table 6–11 is a listing of four-bit Gray code numbers for decimal numbers 0 through 9. Like binary, the Gray code can have any number of bits. Notice the single-bit change between any two Gray code numbers. Binary numbers are shown for reference. For instance, in going from decimal 3 to 4, the Gray code changes from 0010 to 0110, while the binary code changes from 0011 to 0100, a change of three bits. The only bit change is in the third bit from the right in the Gray code; the others remain the same.

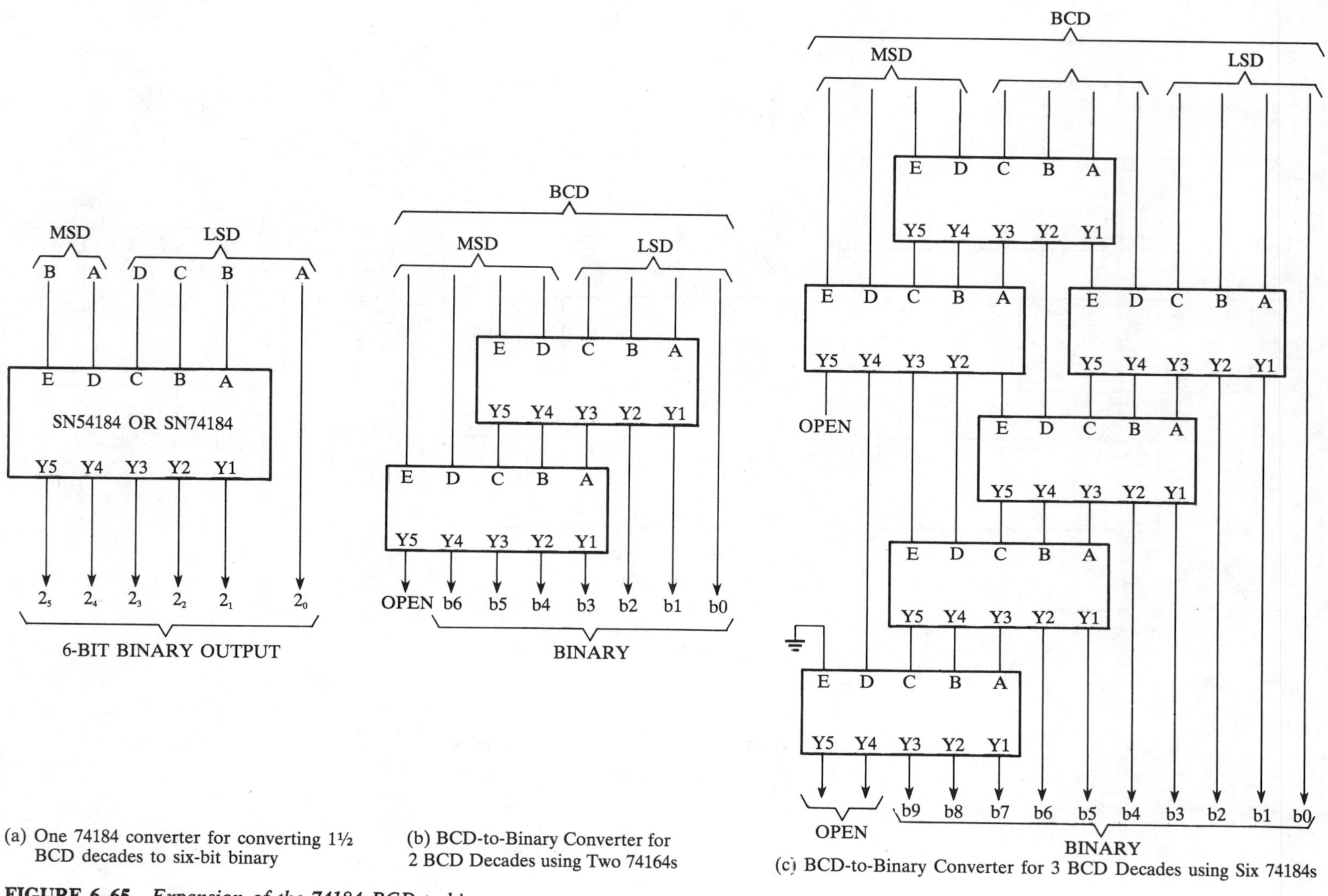

(a) One 74184 converter for converting 1½ BCD decades to six-bit binary

(b) BCD-to-Binary Converter for 2 BCD Decades using Two 74164s

(c) BCD-to-Binary Converter for 3 BCD Decades using Six 74184s

FIGURE 6–65 *Expansion of the 74184 BCD-to-binary converter.*

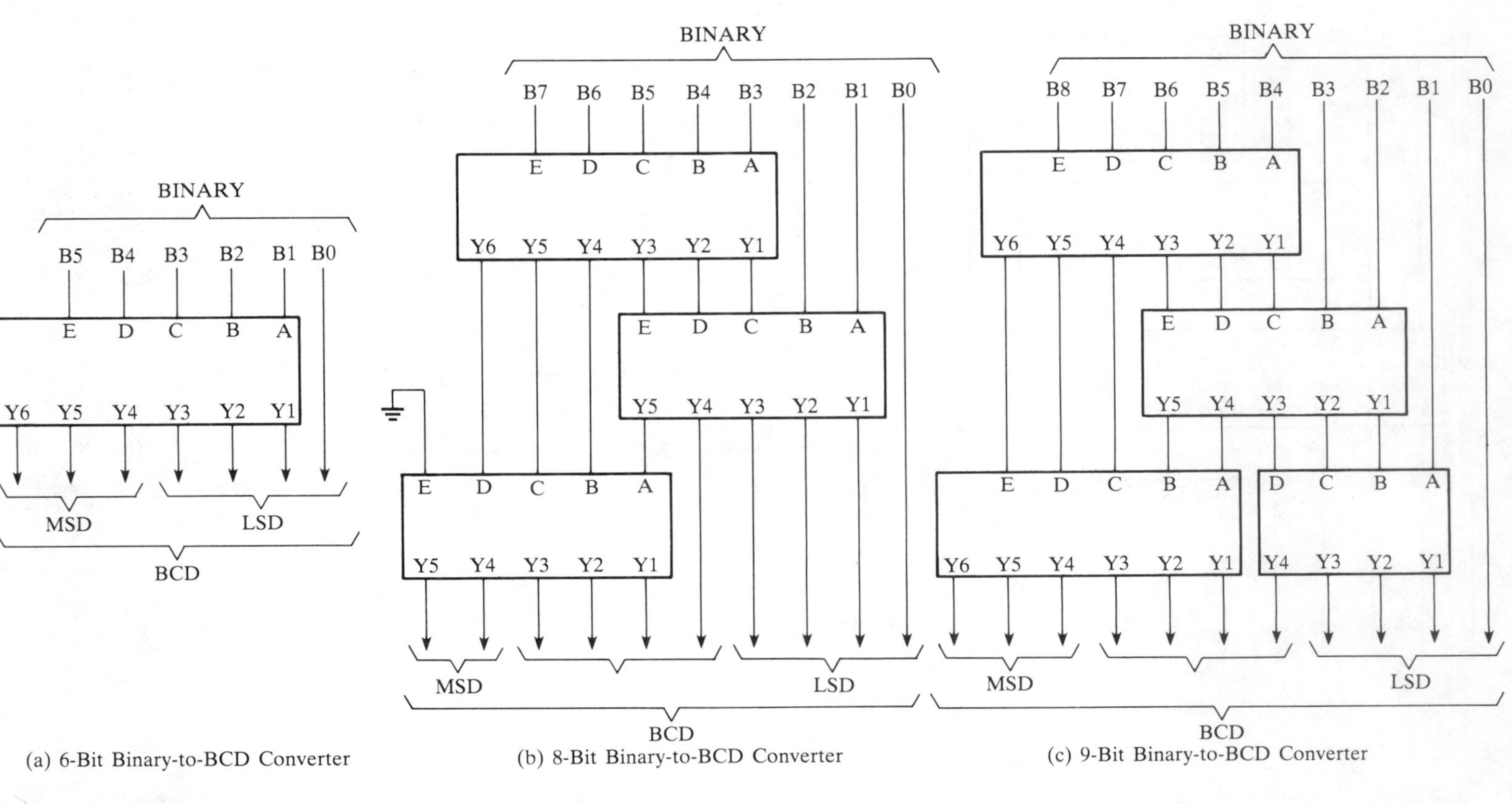

(a) 6-Bit Binary-to-BCD Converter

(b) 8-Bit Binary-to-BCD Converter

(c) 9-Bit Binary-to-BCD Converter

FIGURE 6–66 *Expansion of the 74185 binary-to-BCD converter.*

TABLE 6–11 *Four-bit Gray code.*

Decimal	Binary	Gray
0	0000	0000
1	0001	0001
2	0010	0011
3	0011	0010
4	0100	0110
5	0101	0111
6	0110	0101
7	0111	0100
8	1000	1100
9	1001	1101

Binary to Gray Conversion. By representing the ten decimal digits with a four-bit Gray code, we have another form of BCD code. The Gray code, however, can be extended to any number of bits, and conversion between binary code and Gray code is sometimes useful. First, we will discuss how to convert from a binary number to a Gray code number. The following rules apply:

1. The most significant digit (left-most) in the Gray code is the same as the corresponding digit in the binary number.
2. Going from left to right, add each adjacent pair of binary digits to get the next Gray code digit. Disregard carries. This is an exclusive-OR addition.

For example, let us convert the binary number 10110 to Gray code.

Step 1. The left-most Gray digit is the same as the left-most binary digit.

$$\begin{array}{ccccccl} 1 & 0 & 1 & 1 & 0 & & \text{binary} \\ \downarrow & & & & & & \\ 1 & & & & & & \text{Gray} \end{array}$$

Step 2. Add the left-most binary digit to the adjacent one.

$$\begin{array}{ccccccl} 1 \oplus & 0 & 1 & 1 & 0 & & \text{binary} \\ & \downarrow & & & & & \\ 1 & 1 & & & & & \text{Gray} \end{array}$$

Step 3. Add the next adjacent pair.

$$\begin{array}{ccccccl} 1 & 0 \oplus & 1 & 1 & 0 & & \text{binary} \\ & & \downarrow & & & & \\ 1 & 1 & 1 & & & & \text{Gray} \end{array}$$

Step 4. Add the next adjacent pair and discard carry.

$$\begin{array}{ccccccl} 1 & 0 & 1 \oplus & 1 & 0 & & \text{binary} \\ & & & \downarrow & & & \\ 1 & 1 & 1 & 0 & & & \text{Gray} \end{array}$$

Step 5. Add the last adjacent pair.

$$\begin{array}{ccccccl} 1 & 0 & 1 & 1 \oplus 0 & & \text{binary} \\ & & & \downarrow & & \\ 1 & 1 & 1 & 0 & 1 & \text{Gray} \end{array}$$

The conversion is now complete and the Gray code is 11101.

In order to convert from binary to Gray, the summing operations are performed by exclusive-OR gates connected as shown in Figure 6–67.

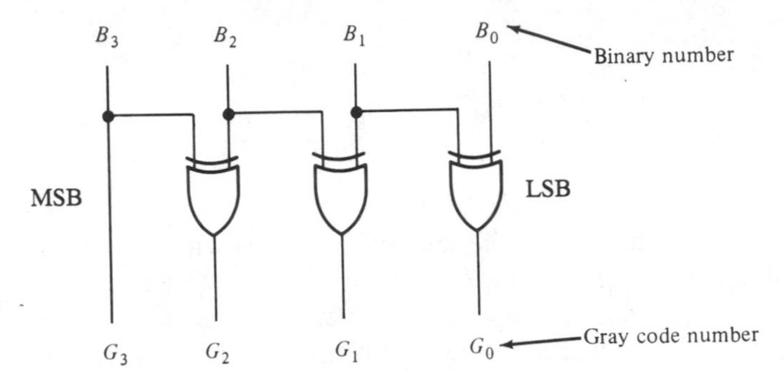

FIGURE 6–67 *Four-bit binary-to-Gray converter.*

Example 6–14

Convert the binary codes 0101, 00111, and 101011 to Gray code, using exclusive-OR gates.

Solutions:

Binary 0101 is 0111 Gray, 00111 is 00100 Gray, and 101011 is 111110 Gray. See Figure 6–68.

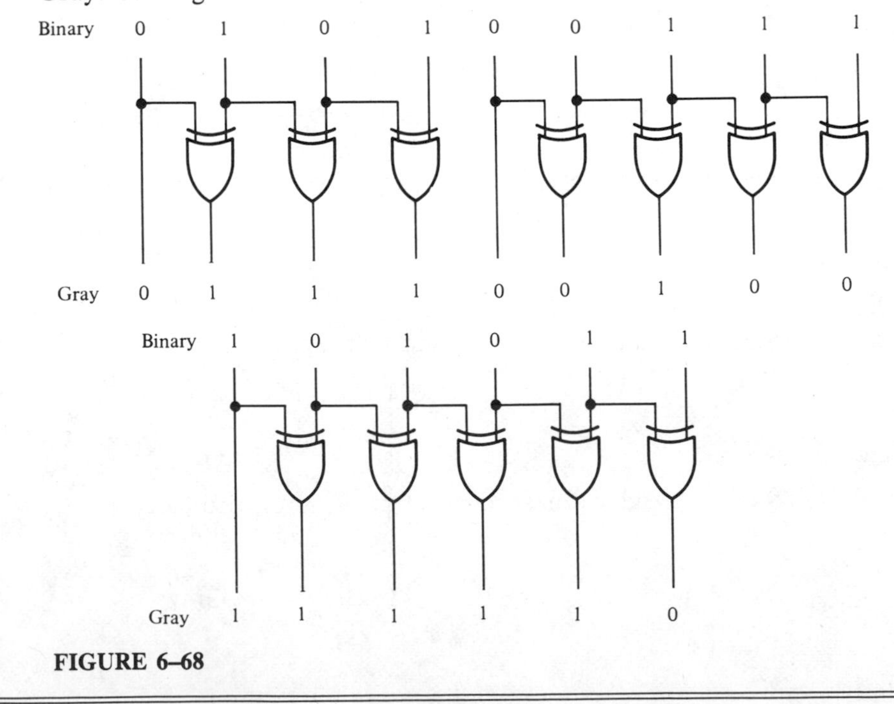

FIGURE 6–68

Gray to Binary Conversion. To convert from Gray code to binary, a similar method is used, but there are some differences. The following rules apply:

1. The most significant digit (left-most) in the binary code is the same as the corresponding digit in the Gray code.
2. Add each binary digit generated to the Gray digit in the next adjacent position. Disregard carries. Again, this is an exclusive-OR addition.

For example, the conversion of the Gray code number 11011 to binary is as follows:

Step 1. The left-most digits are the same.

```
1  1  0  1  1     Gray
↓
1                 binary
```

Step 2. Add the last binary digit just generated to the Gray digit in the next position. Discard carry.

```
1  1  0  1  1     Gray
  ⊕↓
1  0              binary
```

Step 3. Add the last binary digit generated to the next Gray digit.

```
1  1  0  1  1     Gray
     ⊕↓
1  0  0           binary
```

Step 4. Add the last binary digit generated to the next Gray digit.

```
1  1  0  1  1     Gray
        ⊕↓
1  0  0  1        binary
```

Step 5. Add the last binary digit generated to the next Gray digit. Discard carry.

```
1  1  0  1  1     Gray
           ⊕↓
1  0  0  1  0     binary
```

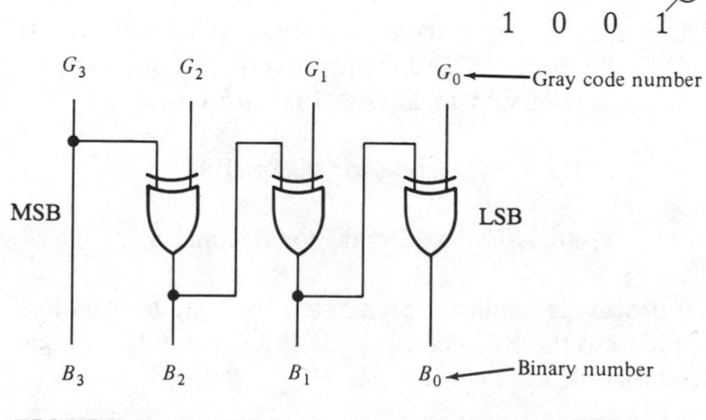

FIGURE 6–69 *Four-bit Gray-to-binary converter.*

This completes the conversion. The final binary number is 10010. Figure 6–69 shows a four-bit Gray-to-binary converter using exclusive-OR gates.

Example 6–15

Convert the Gray codes 1011, 11000, and 1001011 to binary, using exclusive-OR gates.

Solutions:
Gray code 1011 is 1101 binary, 11000 is 10000, and 1001011 is 1110010. See Figure 6–70.

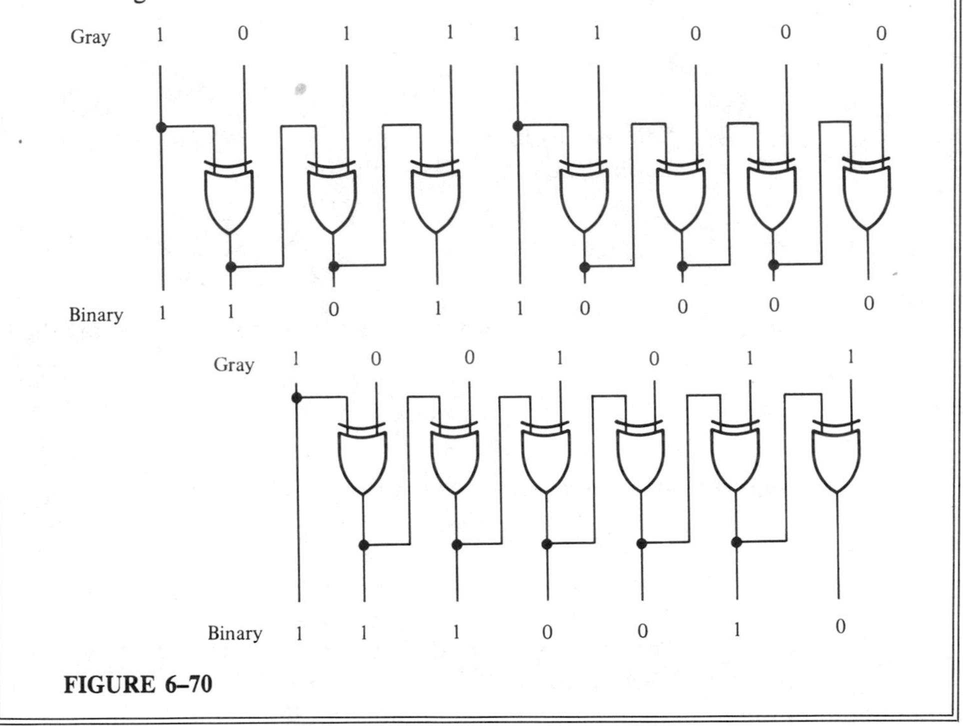

FIGURE 6–70

BCD to Excess-3 Converter

Recall from Chapter 2 that the Excess-3 code for a decimal digit is formed by adding three (0011_2) to the BCD (8421) code for that digit. For example, given the BCD (8421) code 1001, conversion to Excess-3 is as follows:

1001	8421 code for decimal 9
+0011	add 3
1100	Excess-3 code for decimal 9

This conversion process is readily implemented by using four adders wired such that a binary 3 is added to the BCD code applied to the inputs. A logic diagram of this operation is shown in Figure 6–71.

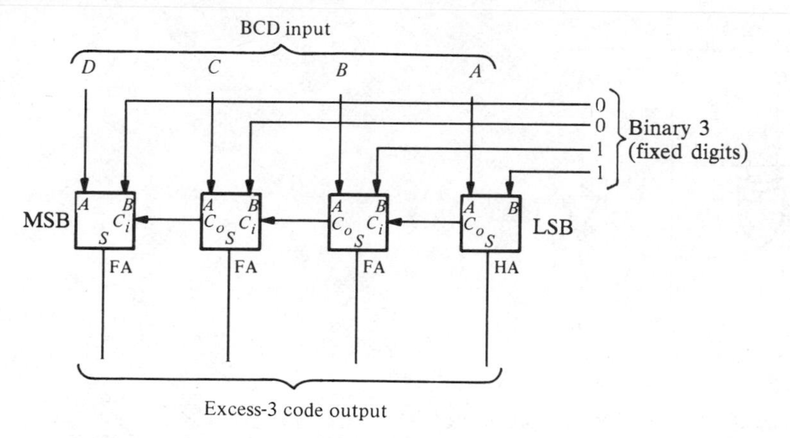

FIGURE 6–71 *BCD-to-Excess-3 converter.*

9's Complement Generation

The conversion of a BCD number to its 9's complement is a form of code conversion that is useful in BCD arithmetic. The 9's complement of a decimal number was discussed in Chapter 2. Table 6–12 shows the binary representation of the 9's complement for each BCD digit.

TABLE 6–12 *Table of 9's complements.*

Decimal	9's Complement	BCD				9's Complement (Binary)			
		D	*C*	*B*	*A*	*D*	*C*	*B*	*A*
0	9	0	0	0	0	1	0	0	1
1	8	0	0	0	1	1	0	0	0
2	7	0	0	1	0	0	1	1	1
3	6	0	0	1	1	0	1	1	0
4	5	0	1	0	0	0	1	0	1
5	4	0	1	0	1	0	1	0	0
6	3	0	1	1	0	0	0	1	1
7	2	0	1	1	1	0	0	1	0
8	1	1	0	0	0	0	0	0	1
9	0	1	0	0	1	0	0	0	0

Notice in Table 6–12 that each *A* bit of the 9's complement code is the complement of the corresponding *A* bit of the BCD code. An inverter can be used to make this conversion.

Also notice that the *B* bits of both codes are identical, and that each *C* bit in the 9's complement code is the sum (exclusive-OR) of the *B* and *C* bits in the corresponding BCD code. Lastly, the *D* bit in the 9's complement code can be generated by NORing the *D, C,* and *B* bits of the corresponding BCD code. This implementation is illustrated in Figure 6–72.

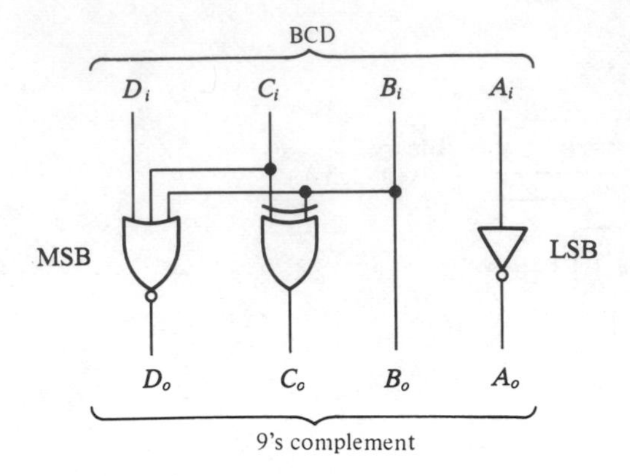

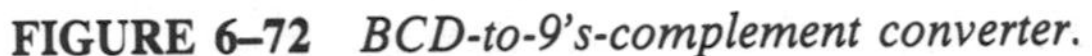
FIGURE 6–72 *BCD-to-9's-complement converter.*

The 74184 as a BCD-to-9's Complement Converter

As you have already seen, the 74184 can be used as a BCD-to-binary converter (Figure 6-64). It can also function as a 9's complement converter when connected as shown in Figure 6–73. Notice that the Y_7, Y_8, and Y_9 outputs that were not employed in the BCD-to-binary application are used in this case.

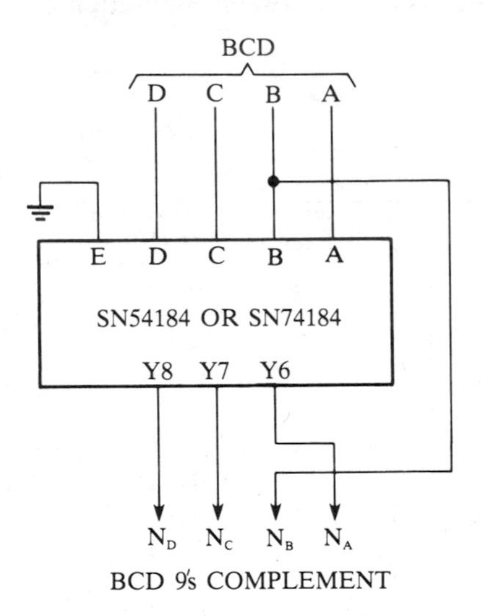

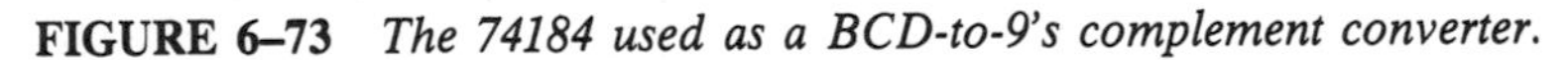
FIGURE 6–73 *The 74184 used as a BCD-to-9's complement converter.*

10's Complement Generation

Recall that the 10's complement of a decimal number is found by adding 1 to the 9's complement. In order to convert BCD to 10's complement, the 9's complement can first be produced as previously discussed and then 0001_2 added to it. This can be implemented as shown in Figure 6–74 with the use of a four-bit full adder such as the 7483A.

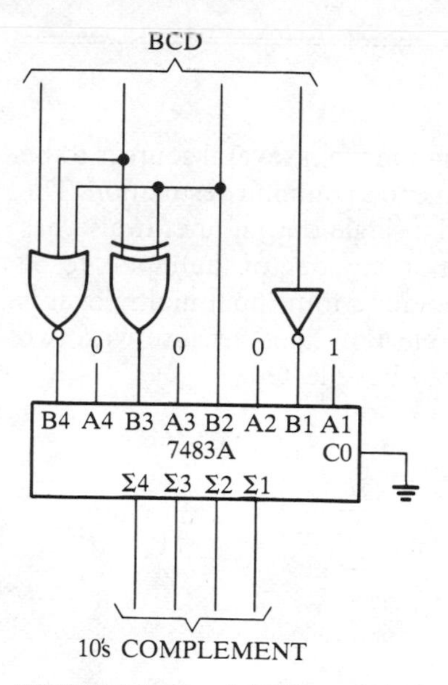

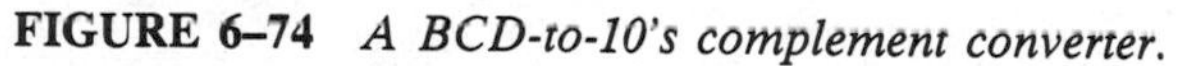
FIGURE 6–74 *A BCD-to-10's complement converter.*

The 74184 as a BCD-to-10's Complement Converter

As you have seen, the 74184 can be used as a 9's complement converter. With slightly different connections it can also perform as a BCD-to-10's-complement converter as shown in Figure 6–75. The E input functions as a mode control; when it is LOW the device acts as a 9's complement converter, and when it is HIGH the device is a 10's complement converter. Notice that the same outputs are used in both cases.

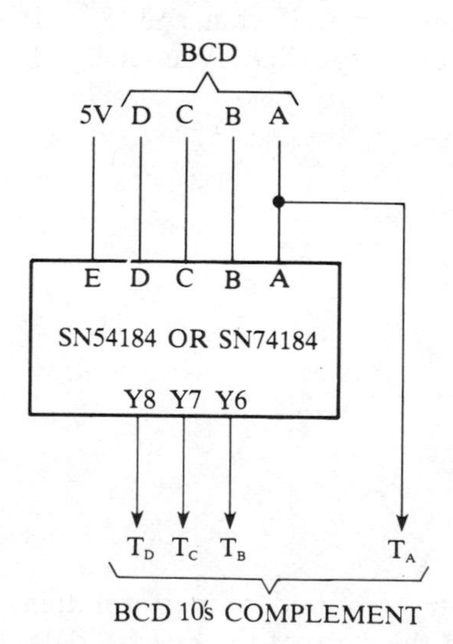

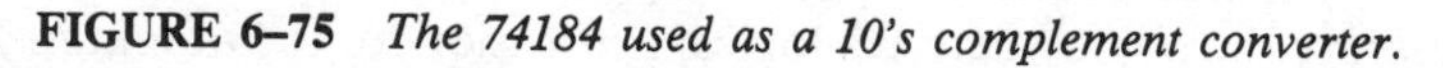
FIGURE 6–75 *The 74184 used as a 10's complement converter.*

6–8 MULTIPLEXERS (DATA SELECTORS)

A multiplexer is a device that allows digital information from several sources to be routed onto a single line for transmission over that line to a common destination. The basic multiplexer, then, has several input lines and a single output line. It also has control or selection inputs that permit digital data on any one of the inputs to be switched to the output line. A block diagram symbol for a four-input multiplexer is shown in Figure 6–76. Notice that there are two selection lines because with two selection bits, each of the four data input lines can be selected.

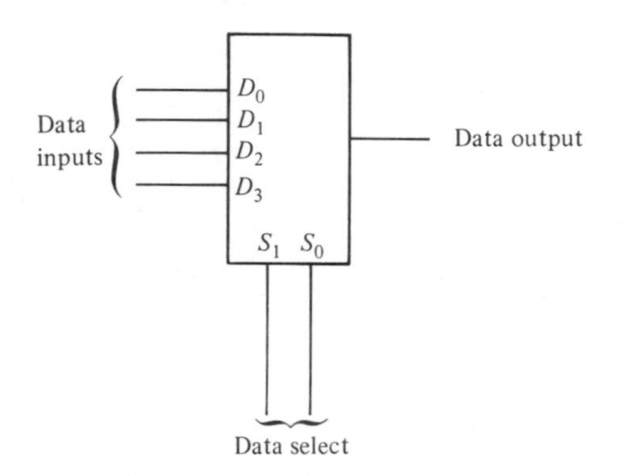

FIGURE 6–76 *Four-line multiplexer block diagram.*

In reference to Figure 6–76, a two-bit binary code on the data select inputs will allow the data on the corresponding data input to pass through to the data output. If a binary 0 ($S_1 = 0$ and $S_0 = 0$) is applied to the data select lines, the data on input D_0 appear on the data output line. If a binary 1 ($S_1 = 0$ and $S_0 = 1$) is applied to the data select lines, the data on input D_1 appear on the data output. If a binary 2 ($S_1 = 1$ and $S_0 = 0$) is applied, the data on D_2 appear on the output. If a binary 3 ($S_1 = 1$ and $S_0 = 1$) is applied, the data on D_3 are switched to the output line. A summary of this operation is given in Table 6–13.

TABLE 6–13 *Data selection for a four-input multiplexer.*

Data Select Inputs		
S_1	S_0	Input Selected
0	0	D_0
0	1	D_1
1	0	D_2
1	1	D_3

Now let us look at the logic circuitry required to perform this multiplexing operation. The data output is equal to the state of the *selected* data input. We should, therefore, be able to derive a logical expression for the output in terms of the data input and the select inputs. This can be done as follows:

The data output is equal to the data input D_0 if and only if $S_1 = 0$ and $S_0 = 0$:

$$\text{Data output} = D_0\overline{S}_1\overline{S}_0$$

The data output is equal to D_1 if and only if $S_1 = 0$ and $S_0 = 1$:

$$\text{Data output} = D_1\overline{S}_1S_0$$

The data output is equal to D_2 if and only if $S_1 = 1$ and $S_0 = 0$:

$$\text{Data output} = D_2S_1\overline{S}_0$$

The data output is equal to D_3 if and only if $S_1 = 1$ and $S_0 = 1$:

$$\text{Data output} = D_3S_1S_0$$

If these terms are ORed, the total expression for the data output is

$$\text{Data output} = D_0\overline{S}_1\overline{S}_0 + D_1\overline{S}_1S_0 + D_2S_1\overline{S}_0 + D_3S_1S_0$$

The implementation of this equation requires four three-input AND gates, a four-input OR gate, and two inverters to generate the complements of S_1 and S_0, as shown in Figure 6–77.

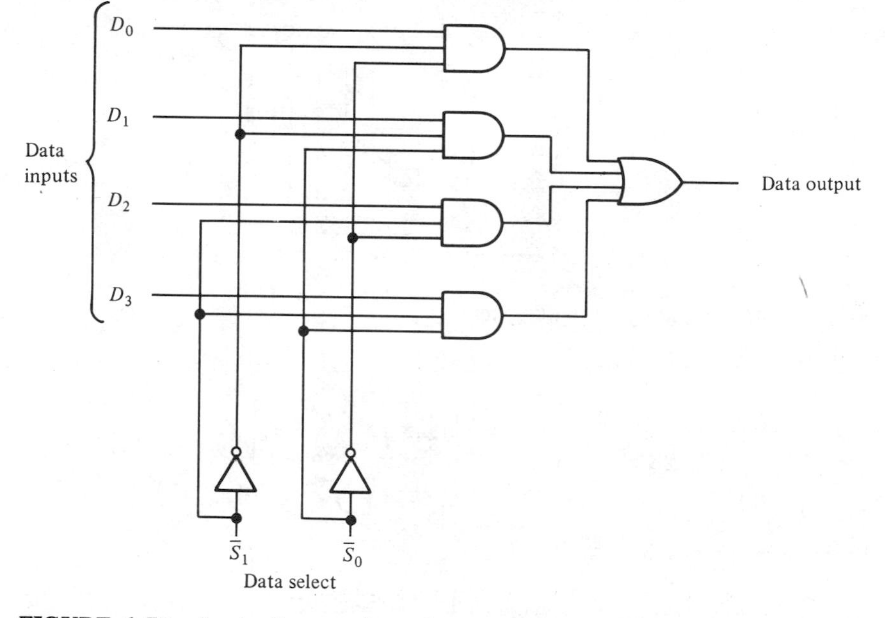

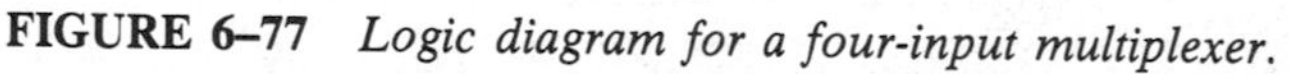

FIGURE 6–77 *Logic diagram for a four-input multiplexer.*

Because data can be selected from any of the input lines, this circuit is sometimes referred to as a *data selector.* It can be implemented with any number of input lines as you will see in the following discussion of specific devices.

Integrated Circuit Multiplexers

Figure 6–78(a) shows the block and pin diagram for the 74150 sixteen-line multiplexer, and Figure 6–78(b) shows the logic diagram. Notice there are four *data select* lines for selecting one of sixteen input lines. The *strobe* input enables the selected gate when a LOW is applied. This allows sampling of data inputs at stable intervals to avoid transitions.

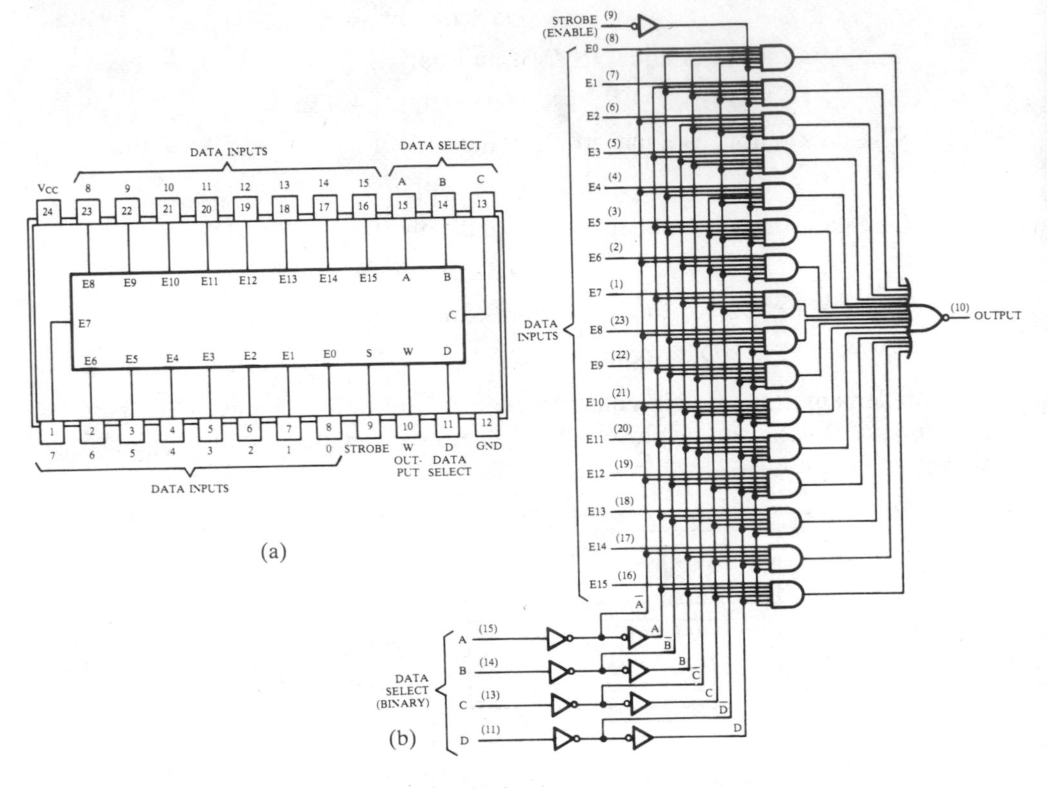

FIGURE 6–78 *A 74150 sixteen-line multiplexer.*

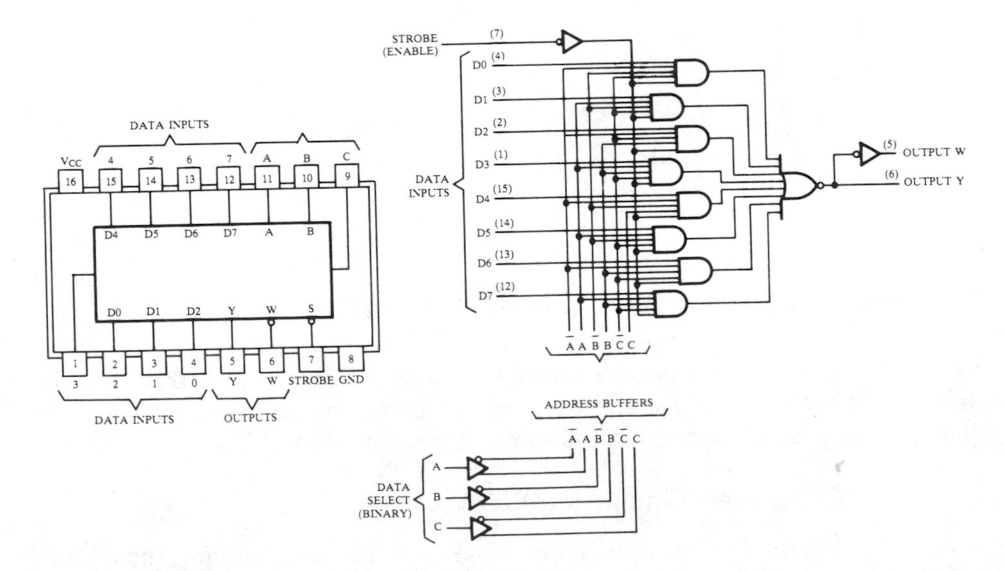

FIGURE 6–79 *A 74151A eight-line multiplexer.*

Another example of an IC multiplexer is the 74151A, an eight-line device as shown in Figure 6–79. Notice that the output is available in both true and complement form.

DEMULTIPLEXERS 6–9

A demultiplexer basically reverses the multiplexing function. It takes data from one line and distributes them to a given number of output lines. Figure 6–80 shows a one-line-to-four-line demultiplexer circuit. The input data line goes to all of the AND gates. The two select lines enable only one gate at a time and the data appearing on the input line will pass through the selected gate to the associated output line.

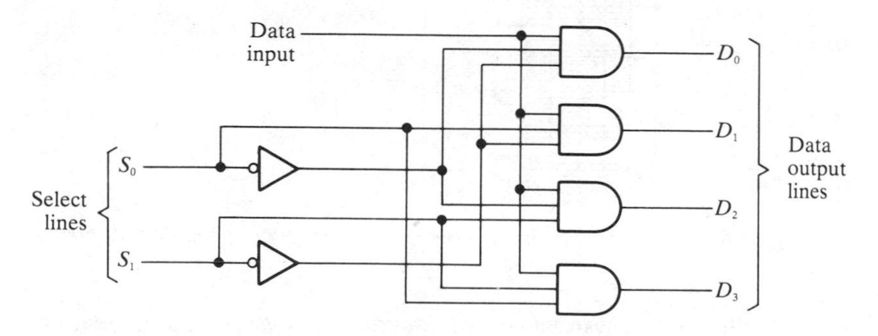

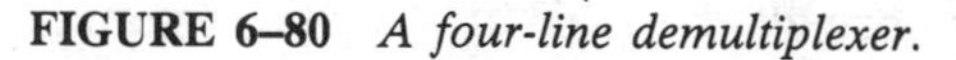
FIGURE 6–80 *A four-line demultiplexer.*

The 74154 as a Demultiplexer

We have already discussed the 74154 in its application as a 1-of-16 or binary-to-decimal decoder (Figure 6–37). This device and other decoders can also be used in demultiplexing applications. The logic diagram for this device is shown in Figure 6–81.

In demultiplexer applications the input lines *A, B, C,* and *D* are used as the data select lines. One of the enable inputs, *G*1 or *G*2, is used as the data input line with the other enable input held LOW to enable the gate.

PARITY GENERATORS/CHECKERS 6–10

Parity

Errors can occur as digital codes are being transferred from one point to another within a digital system or while codes are being transmitted from one system to another. The errors take the form of undesired changes in the bits that make up the coded information; that is, a 1 can change to a 0, or a 0 to 1, due to component manlfunctions or electrical noise. In most digital systems the probability that even a single-bit error will occur is very small and the likelihood of more than one

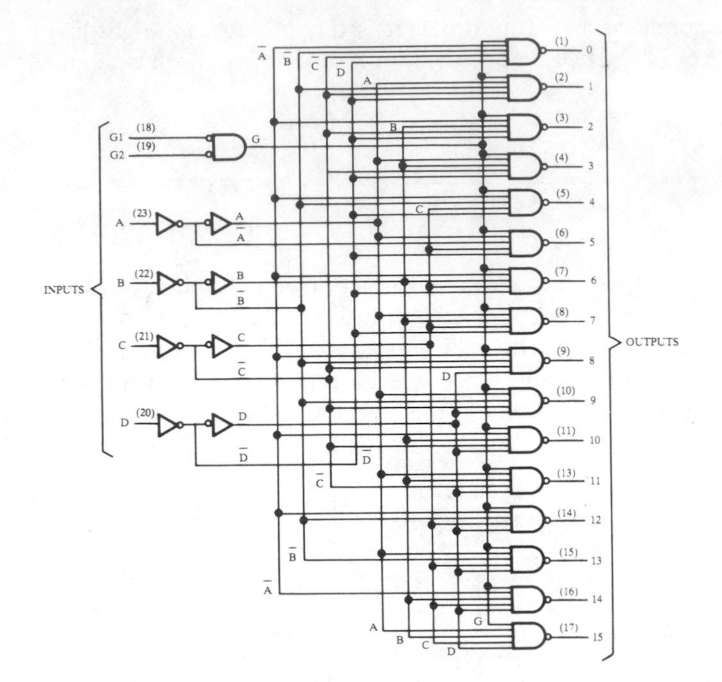

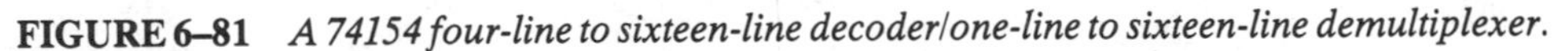
FIGURE 6–81 *A 74154 four-line to sixteen-line decoder/one-line to sixteen-line demultiplexer.*

occurring is even smaller. Many systems, however, employ a *parity bit* as a means of detecting a bit error. Binary information is normally handled by a digital system in groups of bits called *words.* A word always contains either an even or an odd number of 1s. A parity bit is attached to the group of information bits in order to make the *total* number of 1s *always even* or *always odd.* An even parity bit makes the total number of 1s even, and an odd parity bit makes the total odd.

A given system operates with even or odd parity, but not both. For instance, if a system operates with even parity, a check is made on each group of bits received to make sure the total number of 1s in that group is even. If there is an odd number of 1s, an error has occurred.

As an illustration of how parity bits are attached to a code word, Table 6–14 lists the parity bits for each 8421 BCD code number for both even and odd parity. The parity bit for each BCD number is in the P column.

The parity bit can be attached to the code group at either the beginning or the end, depending on system design. Notice that the total number of 1s, *including the parity bit,* is always even for even parity and always odd for odd parity.

Detecting an Error

A parity bit provides for the detection of a *single* error (or any odd number of errors, which is very unlikely) but cannot check for two errors. For instance, let us assume that we wish to transmit the BCD code 0101. (Parity can be used with any number of bits; we are using four for illustration.) The total code transmitted, including the parity bit, is

TABLE 6–14 *8421 BCD code with parity bits.*

Even Parity	Odd Parity
8421P	8421P
00000	00001
00011	00010
00101	00100
00110	00111
01001	01000
01010	01011
01100	01101
01111	01110
10001	10000
10010	10011

even parity bit

01010

BCD code

Now, let us assume an error occurs in the second bit from the left (the 1 becomes a 0), as follows:

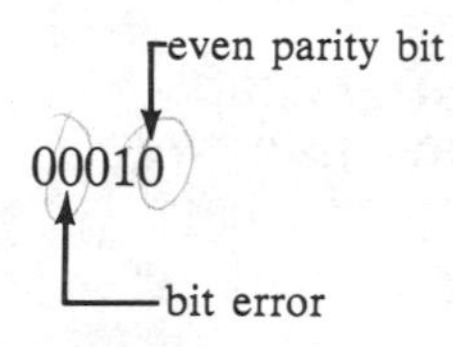

When this code is received, the parity check circuitry determines that there is only a single 1 (odd number), when there should be an even number of 1s. Because an even number of 1s does not appear in the code when it is received, an error is indicated.

Let us now consider what happens if two bit errors occur as follows:

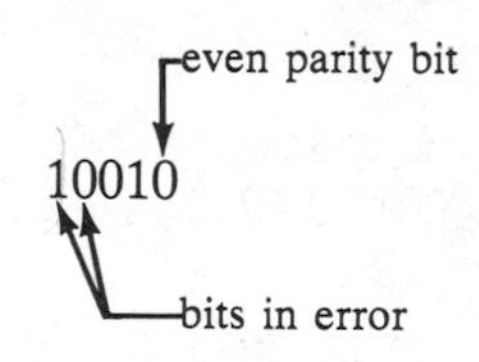

When a check is made, an even number of 1s appears, and although there are two errors, the parity check indicates a *correct* code.

An odd parity bit also provides in a similar manner for the detection of a single error in a given group of bits.

Example 6–16

Assign the proper even parity bit to the following code words: 1010, 111000, 101101, 100011100101, and 101101011111.

Example 6–16, continued

Solutions:

Make the parity bits either 1 or 0 to make the total number of 1s even. The parity will be the right-most bit.

(a) 10100
(b) 1110001
(c) 1011010
(d) 1000111001010
(e) 1011010111111

Example 6–17

Assign the proper odd parity bit to the following code words: 1011, 101010, 1110001, and 110011100.

Solutions:

Make the parity bits either 1 or 0 in order to make the total number of 1s odd. The parity will be the right-most bit.

(a) 10110
(b) 1010100
(c) 11100011
(d) 1100111000

Example 6–18

An odd parity system receives the following code words: 10110, 11010, 110011, 110101110100, and 1100010101010. Determine which ones, if any, are in error.

Solutions:

Since odd parity is required, any code with an even number of 1s is incorrect. The following codes are in error: 110011 and 1100010101010.

Several specific codes also provide inherent error detection; a few of the most important, as well as the Hamming error *correcting* code, are discussed in Appendix C.

Parity Logic

In order to check for or generate the proper parity in a given code word, a very basic principle can be used: *The sum* (disregarding carries) *of an even number of 1s is always 0, and the sum of an odd number of 1s is always 1.* Therefore, in order to determine if a given code word is even or odd parity, all of the bits in that code word are summed. As we know, the sum of two bits can be generated by an exclusive-OR gate, as shown in Figure 6–82(a); the sum of three bits can be formed by two exclusive-OR gates connected as shown in Figure 6–82(b), and so on.

The parity generation/detection logic for a four-bit code (including parity) is shown in Figure 6–83(a), and that for an eight-bit code is in Figure 6–83(b).

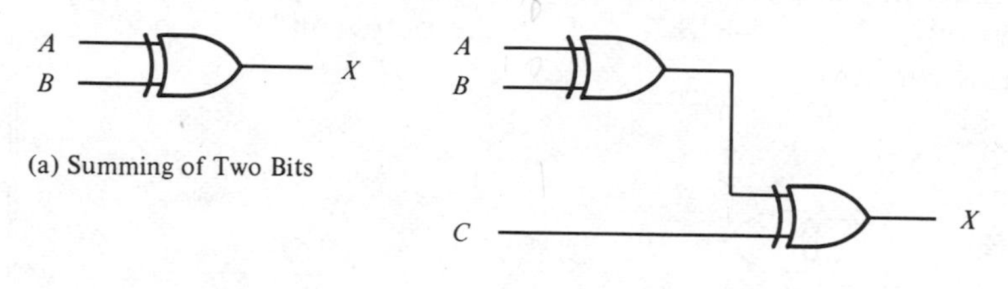

(a) Summing of Two Bits

(b) Summing of Three Bits

FIGURE 6–82

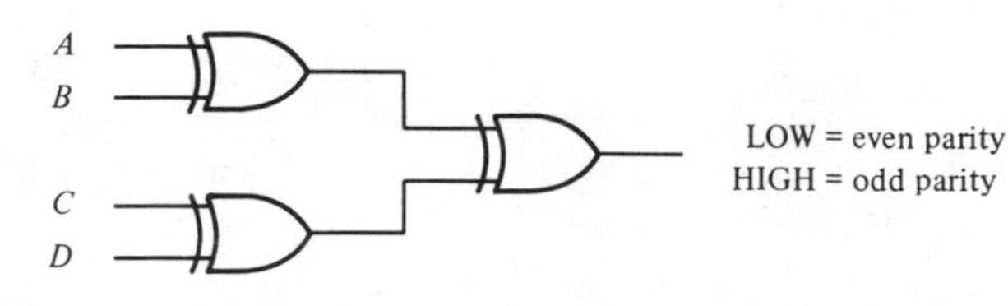

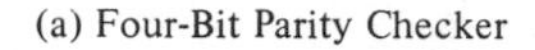

(a) Four-Bit Parity Checker

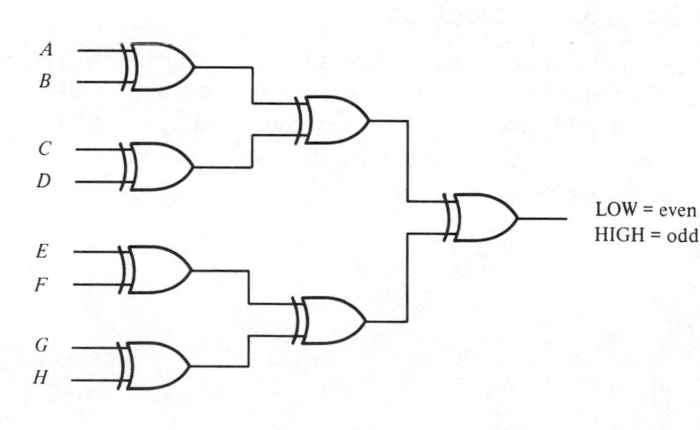

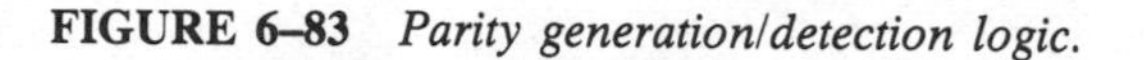

(b) Eight-Bit Parity Checker

FIGURE 6–83 *Parity generation/detection logic.*

Example 6–19

Verify that the parity detector circuit of Figure 6–84(a) produces the proper indication for the code words 1010 and 1101.

Solutions:

First we will take the first code word listed and follow the logic states of each gate through to the output, as shown in Figure 6–84(a). A 0 output results which indicates even parity. This is the proper output since there are two 1s in the code word. Next, the second code word is applied to the inputs, as shown in Figure 6–84(b). A 1 on the output indicates odd parity, which is correct.

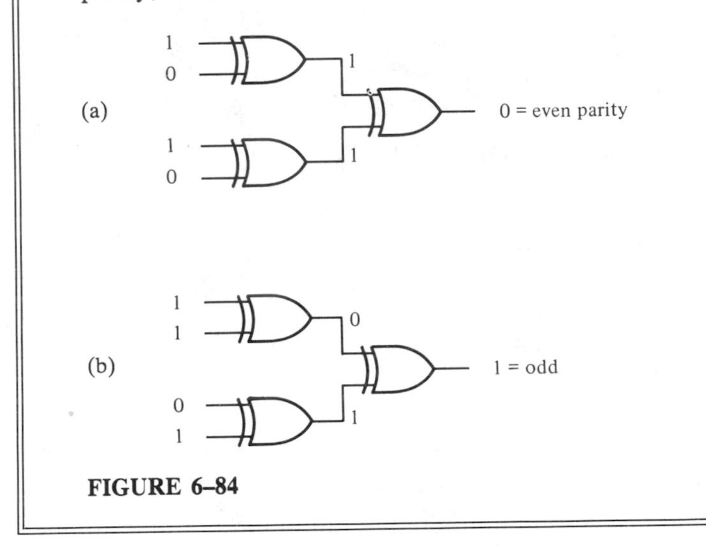

FIGURE 6–84

The 74180 Parity Generator/Checker

The pin diagram for this integrated circuit is shown in Figure 6–85. It is a nine-bit TTL parity generator/checker. The odd and even outputs and control inputs allow operation in either odd or even parity applications. Depending on whether even or odd parity is being generated or checked, the even or odd inputs can be used as the parity or ninth bit input.

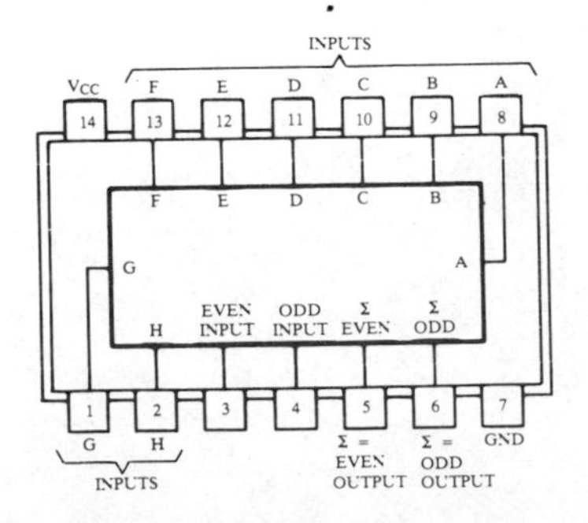

FIGURE 6–85 *A 74180 parity generator/checker.*

Problems

Section 6–1

6–1 Describe the difference between a half-adder and a full-adder.

6–2 For the full-adder of Figure 6–4, determine the logic state (1 or 0) at each gate output for the following inputs:

(a) $A = 1, B = 1, C_i = 1$ (b) $A = 0, B = 1, C_i = 1$
(c) $A = 0, B = 1, C_i = 0$

6–3 Repeat Problem 6–2 for the following inputs:

(a) $A = 0, B = 0, C_i = 0$ (b) $A = 1, B = 0, C_i = 0$
(c) $A = 1, B = 0, C_i = 1$

6–4 Simplify, if possible, the full-adder circuit of Example 6–1, using Karnaugh map simplification.

Section 6–2

6–5 For the parallel adder in Figure 6–86, determine the sum by analysis of the logical operation of the circuit. Verify your result by long hand addition of the two input numbers.

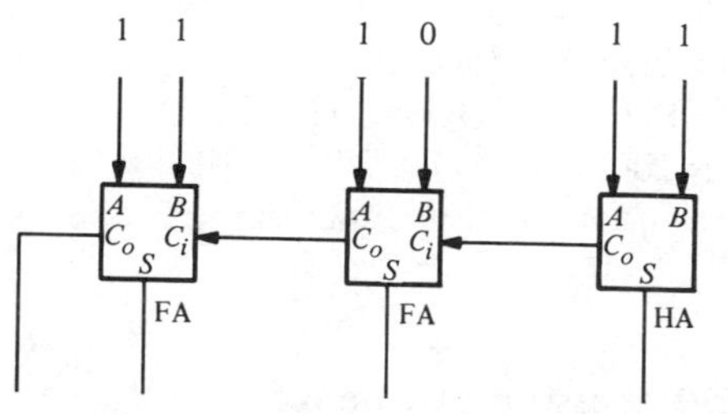

FIGURE 6–86

6–6 Repeat Problem 6–5 for the circuit and input conditions in Figure 6–87.

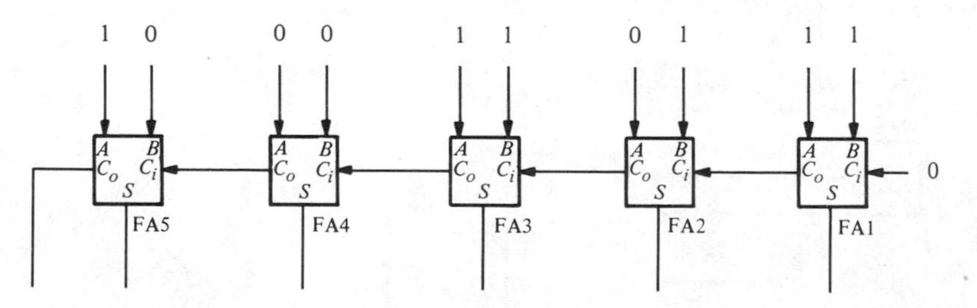

FIGURE 6–87

6–7 The input waveforms in Figure 6–88 are applied to a 7482 two-bit adder. Determine the waveforms for the sum and carry outputs in relation to the inputs.

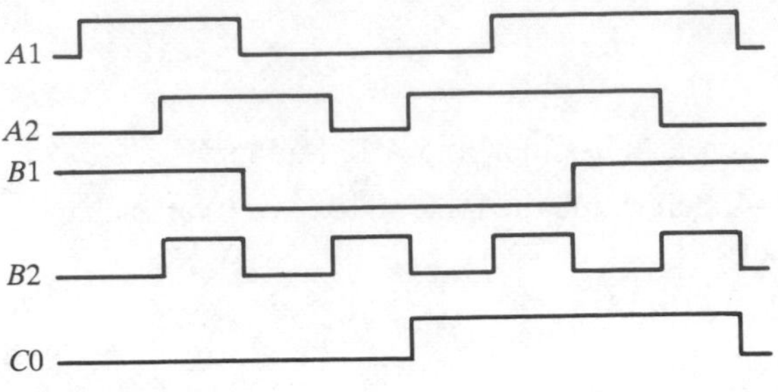

FIGURE 6–88

6–8 The following sequences of bits appear on the inputs to the adder developed in Example 6–3. Determine the resulting sequence of bits on each sum output.
Adder 1: $A1$- 1001011010110100 $C0$- 0001000011010011
$A2$- 1110100010110010
$B1$- 0000101011110010
$B2$- 1011101011111001
Adder 2: $A1$- 1111100001111000
$A2$- 1100110011001100
$B1$- 1010101010101010
$B2$- 0010010010010010

6–9 In the process of checking a 7483A four-bit full-adder, the following voltage levels are observed on its pins: 1-LOW, 2-HIGH, 3-HIGH, 4-HIGH, 6-HIGH, 7-HIGH, 8-LOW, 9-LOW, 10-LOW, 11-LOW, 13-LOW, 14-HIGH, 15-LOW, and 16-HIGH. Determine if the IC is functioning properly.

Section 6–3

6–10 Show the additional logic circuitry necessary to make the four-bit carry-look-ahead adder in Figure 6–19 into a five-bit adder.

Section 6–4

6–11 The waveforms in Figure 6–89 are applied to the comparator as shown. Determine the output $(A = B)$ waveform. Assume a HIGH on the output indicates equality of the inputs.

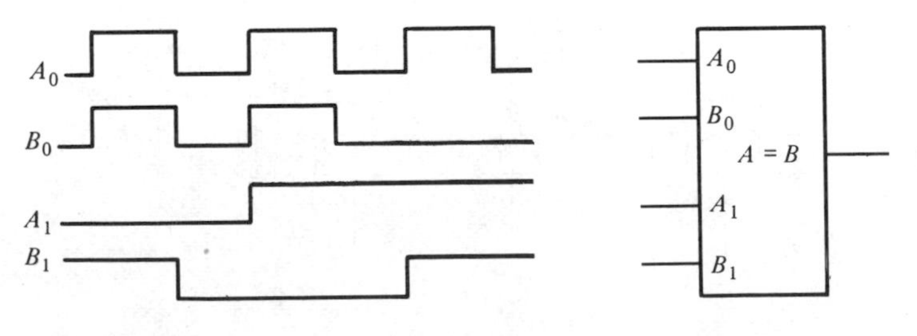

FIGURE 6–89

6–12 For the four-bit comparator in Figure 6–90, plot each output waveform for the inputs shown. The outputs are active HIGH.

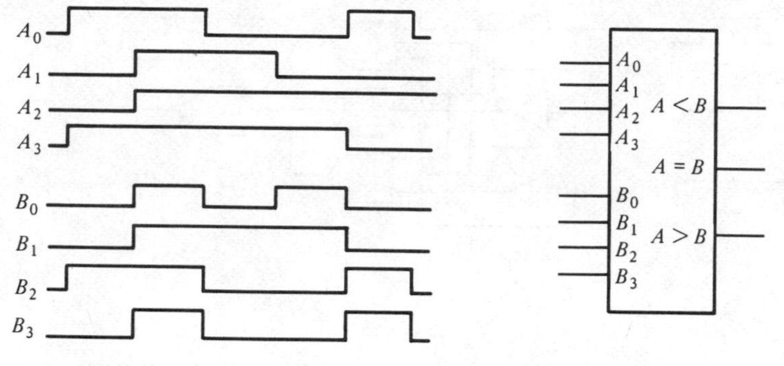

FIGURE 6–90

6–13 For each following set of binary numbers, determine the logic states at each point in the comparator circuit of Figure 6–27, and verify that the output indications are correct:

(a) $A_3\ A_2\ A_1\ A_0 = 1100$
$B_3\ B_2\ B_1\ B_0 = 1001$

(b) $A_3\ A_2\ A_1\ A_0 = 1000$
$B_3\ B_2\ B_1\ B_0 = 1011$

(c) $A_3\ A_2\ A_1\ A_0 = 0100$
$B_3\ B_2\ B_1\ B_0 = 0100$

Section 6–5

6–14 When a HIGH is on the output of each of the decoding gates in Figure 6–91, what is the binary code appearing on the inputs? D is the MSB.

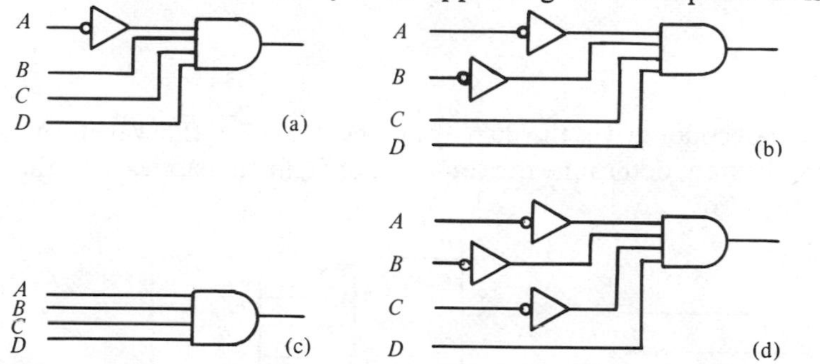

FIGURE 6–91

6–15 Show the decoding logic for each of the following code words if an active HIGH indication is required:

(a) 1101 **(b)** 1000 **(c)** 11011 **(d)** 11100
(e) 101010 **(f)** 111110 **(g)** 000101 **(h)** 1110110

6–16 Repeat Problem 6–15 given that an active LOW output is required.

6–17 We wish to detect only the presence of the code words 1010, 1100, 0001, and 1011. An active HIGH output is required to indicate their presence. Develop the complete decoding logic with a single output that will tell us when any one of these codes is on the inputs. For any other code, the output must be LOW.

6–18 If the input waveforms are applied to the decoding logic as indicated in Figure 6–92, sketch the output waveform in proper relation to the inputs.

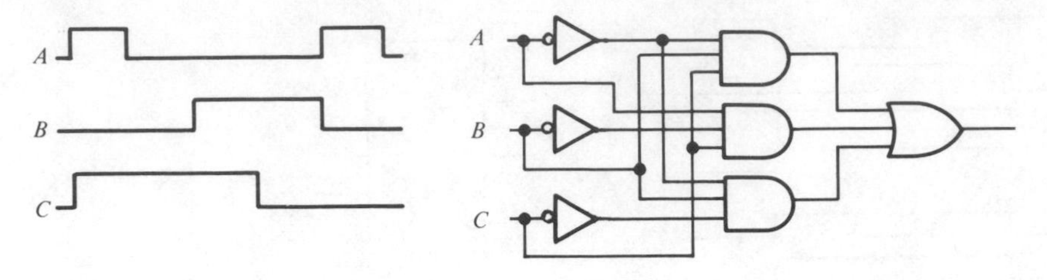

FIGURE 6–92

6–19 BCD numbers are applied sequentially to the BCD-to-decimal decoder in Figure 6–93. Draw the ten output waveforms, showing each in the proper relationship to the others and to the inputs. *D* is the MSB.

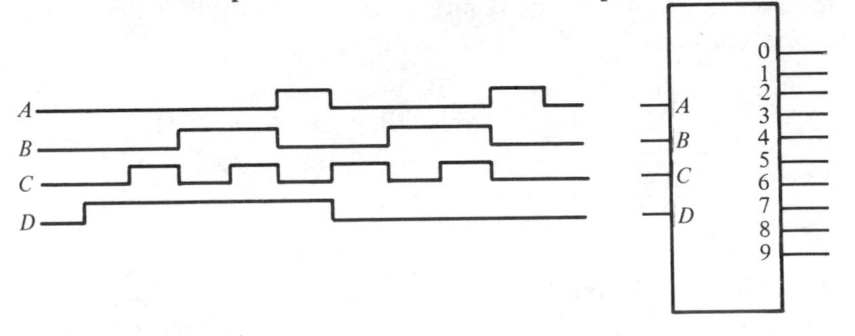

FIGURE 6–93

6–20 A seven-segment decoder drives the display in Figure 6–94. If the waveforms are applied as indicated, determine the sequence of digits that appears on the display. *D* is the MSB.

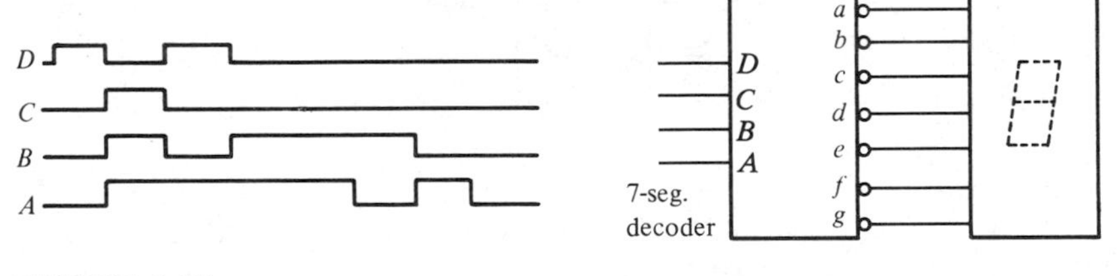

FIGURE 6–94

Section 6–6

6–21 In the decimal-to-BCD encoder of Figure 6–53, assume the 9 input and the 3 input are both HIGH. What is the output code? Is it a valid BCD (8421) code?

6–22 For the priority encoder of Figure 6–58, analyze the operation and verify that the proper output occurs when a HIGH is simultaneously applied to both the 2 and the 7 inputs.

6–23 Repeat Problem 6–22 for inputs 4, 6, and 9 being HIGH at the same time.

6–24 Explain why an input line for the decimal digit 0 is not required in Figure 6–58.

6–25 A 74147 encoder has LOW levels on pins 2, 5, and 12. What BCD code appears on the outputs if all the other inputs are HIGH?

Section 6–7

6–26 Show the logic required to convert a ten-bit binary number to Gray code, and use that logic to convert the following binary code words to Gray code:

(a) 1010101010 (b) 1111100000
(c) 0000001110 (d) 1111111111

6–27 Show the logic required to convert a ten-bit Gray code to binary, and use that logic to convert the following Gray code words to binary:

(a) 1010000000 (b) 0011001100
(c) 1111000111 (d) 0000000001

6–28 Convert the following decimal numbers first to BCD and then, using the logical operation of the BCD-to-binary converter of Figure 6–63, convert the BCD to binary. Verify the result in each case.

(a) 2 (b) 8 (c) 13 (d) 26 (e) 33
(f) 45 (g) 61 (h) 70 (i) 84 (j) 99

Section 6–8

6–29 For the four-input multiplexer in Figure 6–95, determine the output state. Refer to logic diagram in Figure 6–77 if necessary.

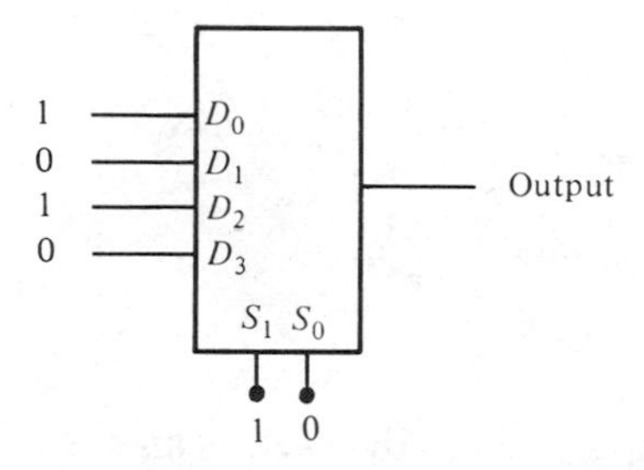

FIGURE 6–95

6–30 For the multiplexer of Problem 6–29, determine the output for the following input states: $D_0 = 0$, $D_1 = 1$, $D_2 = 1$, $D_3 = 0$, $S_0 = 1$, $S_1 = 0$.

6–31 If the data select inputs to the multiplexer of Problem 6–29 are sequenced as shown by the waveforms in Figure 6–96, determine the output waveform with the data inputs as shown in Figure 6–95.

6–32 The waveforms in Figure 6–97 are observed on the inputs of a 74151A eight-line multiplexer. Sketch the true output waveform.

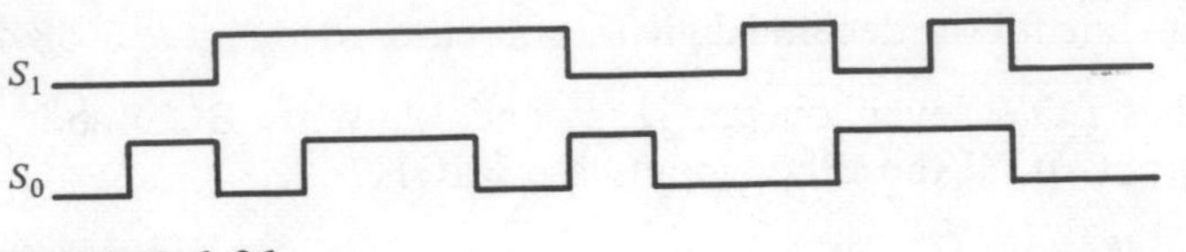

FIGURE 6–96

Section 6–10

6–33 The waveforms in Figure 6–98 are applied to the four-bit parity checker. Determine the output waveform in proper relation to the inputs. How many times does even parity occur?

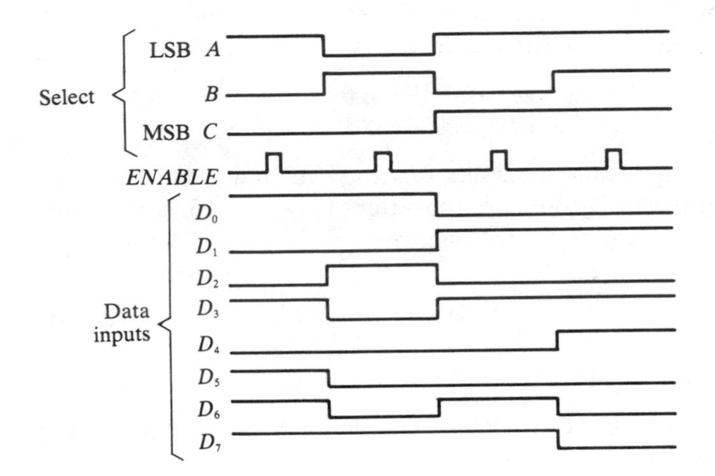

FIGURE 6–97

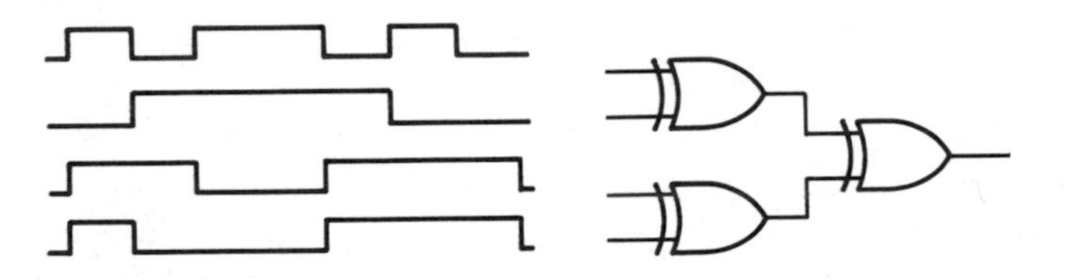

FIGURE 6–98

6–34 The following bit patterns are applied to the inputs of a 74180 used as an even parity generator. Determine the output bit in each case.
(a) 10101001 **(b)** 11101011 **(c)** 11001011

The flip-flop belongs to a category of digital circuits known as *multivibrators.* There are three basic types of multivibrators in common use: the *bistable multivibrator,* the *monostable multivibrator,* and the *astable multivibrator.* The bistable multivibrator is commonly called a flip-flop, and will be emphasized in this chapter. The other two types are also discussed.

The flip-flop has two stable states. It is capable of being in either a HIGH state (logic 1) or a LOW state (logic 0) indefinitely. Since it can retain either state, it is useful as a storage or memory device. The flip-flop finds wide application in digital systems as a "building block" for counters, registers, memories, control logic, and other functions. It is the most widely applied type of multivibrator and therefore, in general, ranks highest in importance. In this chapter we will study several important types of flip-flops.

The emphasis will be on the functional operation and characteristics of the flip-flop as a logic device rather than on the detailed circuitry. The advent of integrated circuits and the number of circuit technologies with which a given type of function can be implemented have made the internal circuitry less important *from an application point of view* than the characteristics associated with the inputs, outputs, and functional or logical operation. For application pruposes, it is generally more important to know what logic function the circuit performs, and what the limitations are on his performance, than it is to know the details of the circuit design.

7

FLIP-FLOPS AND OTHER MULTIVIBRATORS

7–1 LATCHES

Latches are a type of bistable multivibrator that are normally placed in a unique category separate from that of flip-flops. They are basically similar to flip-flops, however, because they are bistable devices that can reside in either of two states by virtue of a feedback arrangement. The main difference is in the method used for changing their state.

The S–R Latch

An *S–R* (set-reset) latch is formed with two cross-coupled NOR gates as shown in Figure 7–1(a) or with two cross-coupled NAND gates as in Figure 7–1(b). Notice that the output of each gate is connected to an input of the opposite gate. This produces the *feedback* that is characteristic of all multivibrators.

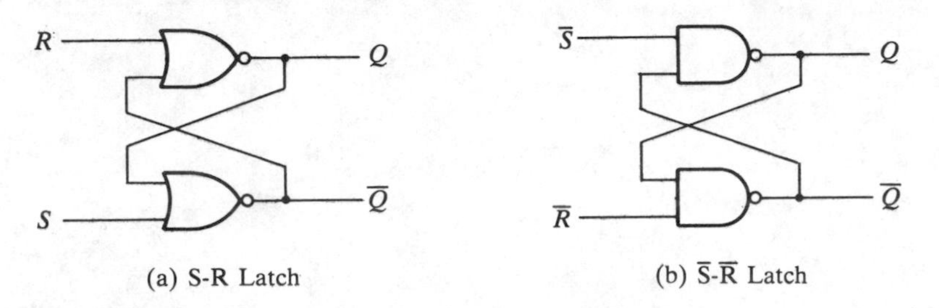

FIGURE 7–1 *Two versions of set-reset latches.*

To understand the operation of the latch, we will take the NAND gate $\overline{S}$–$\overline{R}$ latch in Figure 7–1(b) and use it as an example. This latch is redrawn in Figure 7–2 using the negative-OR gate equivalents in place of the NAND gates. This is done because LOWs on the $\overline{R}$ and $\overline{S}$ lines are the activating inputs.

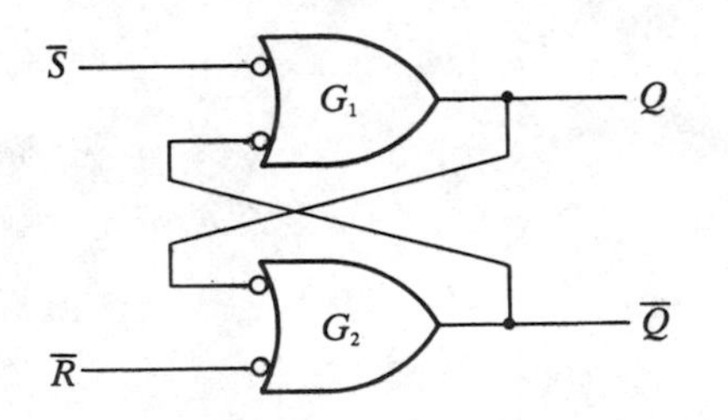

FIGURE 7–2

The latch has two inputs, $\overline{S}$ and $\overline{R}$, and two outputs, Q and $\overline{Q}$. We will start by assuming that both inputs and the Q output are HIGH. Since the Q output is connected back to an input of gate G_2, and the $\overline{R}$ input is HIGH, the output of G_2 must be LOW. This LOW output is coupled back to an input of gate G_1, insuring that its output is HIGH.

When the Q output is HIGH, the latch is said to be in the SET state. It will remain in this state indefinitely until a LOW is temporarily applied to the $\overline{R}$

input. With a LOW on the $\overline{R}$ input, the output of gate G_2 is forced HIGH. This HIGH on the $\overline{Q}$ output is coupled back to an input of G_1, and since the $\overline{S}$ input is HIGH, the output of G_1 goes LOW. This LOW on the Q output is coupled back to an input of G_2, insuring that the $\overline{Q}$ output remains HIGH even when the LOW on the $\overline{R}$ input is removed. When the Q output is LOW, the latch is said to be in the RESET state. Now the latch remains indefinitely in the RESET state until a LOW is applied to the $\overline{S}$ input.

Notice that the outputs are always complements of each other: when Q is HIGH, $\overline{Q}$ is LOW, and when Q is LOW, $\overline{Q}$ is HIGH.

A condition that is not allowed in the operation of an $\overline{S}$–$\overline{R}$ latch occurs when LOWs are applied to both $\overline{S}$ and $\overline{R}$ at the same time. As long as the LOW levels are simultaneously held on the inputs, both the Q and $\overline{Q}$ outputs are forced HIGH thus violating the basic complementary operation of the outputs. Also, if the LOWs are released simultaneously, both outputs will attempt to go LOW. Since there is always some small difference in the propagation delay of the gates, one of the gates will dominate in its transition to the LOW output state. This, in turn, forces the output of the slower gate to remain HIGH. This creates an unpredictable situation because we cannot reliably predict the next state of the latch.

Figure 7–3 illustrates the $\overline{S}$–$\overline{R}$ latch operation for each of the four possible combinations of levels on the inputs. (The first three combinations are allowed, but the last is not.) Table 7–1 summarizes the logical operation in truth table form. Operation of the NOR implemented latch in Figure 7–1(a) is similar but requires the use of opposite logic levels. You should analyze its operation as an exercise.

TABLE 7–1 *Truth table for an $\overline{S}$–$\overline{R}$ latch.*

Inputs		Outputs		Comments
$\overline{S}$	$\overline{R}$	Q	$\overline{Q}$	
HIGH	HIGH	NC	NC	No change. Latch remains in present state.
LOW	HIGH	HIGH	LOW	Latch SETS
HIGH	LOW	LOW	HIGH	Latch RESETS
LOW	LOW	HIGH	HIGH	Next state is not predictable.

In many cases, a simplified symbol is used to represent a latch. Figure 7–4 shows a typical logic symbol that represents the $\overline{S}$–$\overline{R}$ latch of Figure 7–2.

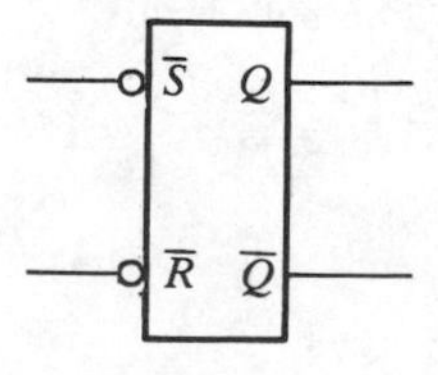

FIGURE 7–4 *Logic symbol for the $\overline{S}$–$\overline{R}$ latch.*

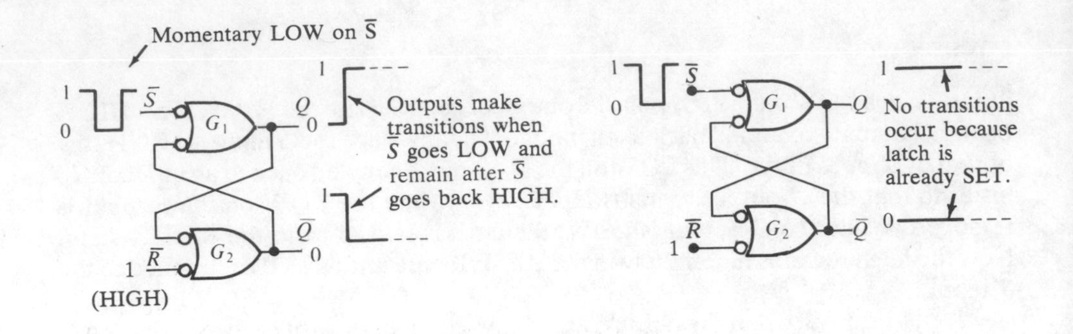

Latch starts out RESET.

Latch starts out SET.

(a) Two possibilities for the SET operation

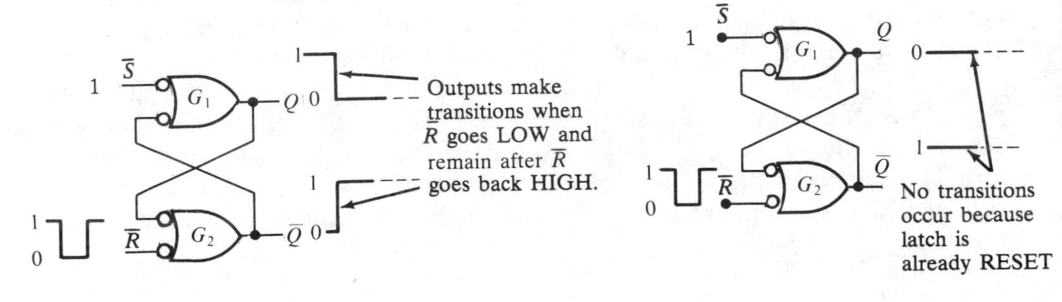

Latch starts out SET.

Latch starts out RESET.

(b) Two possibilities for the RESET operation

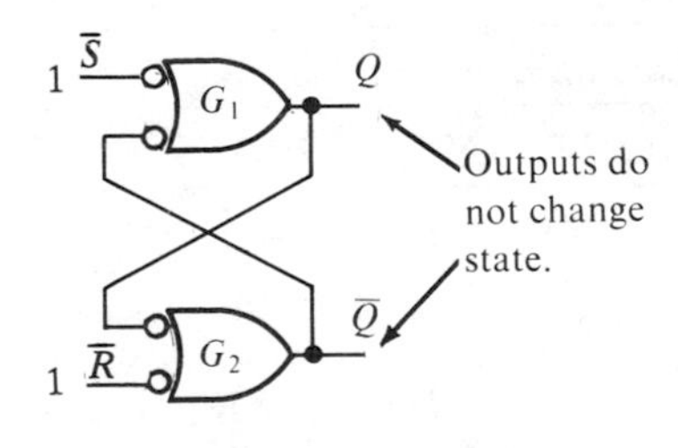

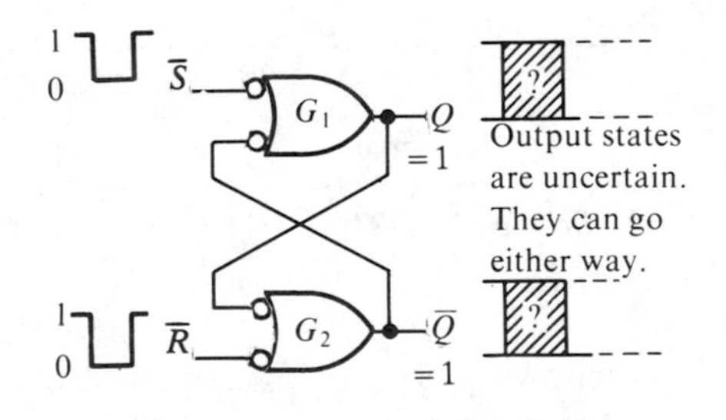

HIGHS on both inputs.

Simultaneous LOWS on both inputs.

(c) NO CHANGE condition

(d) Indeterminate condition

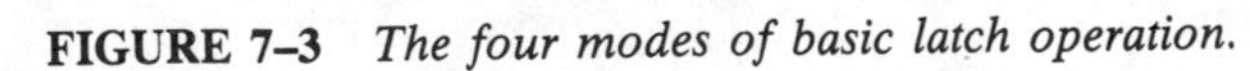

FIGURE 7–3 *The four modes of basic latch operation.*

The following example will illustrate how an $\overline{S}$–$\overline{R}$ latch responds to conditions on its inputs. We will pulse LOW levels on each input in a certain sequence and observe the resulting Q output waveform. The indeterminate condition is avoided because it results in an undesirable mode of operation and is a major drawback of any set-reset type of latch.

Example 7–1

If the waveforms in Figure 7–5(a) are applied to the inputs of the flip-flop of Figure 7–4, determine the waveform that would be observed on the Q output. Assume that Q is initially LOW.

Solution:

See Figure 7–5(b).

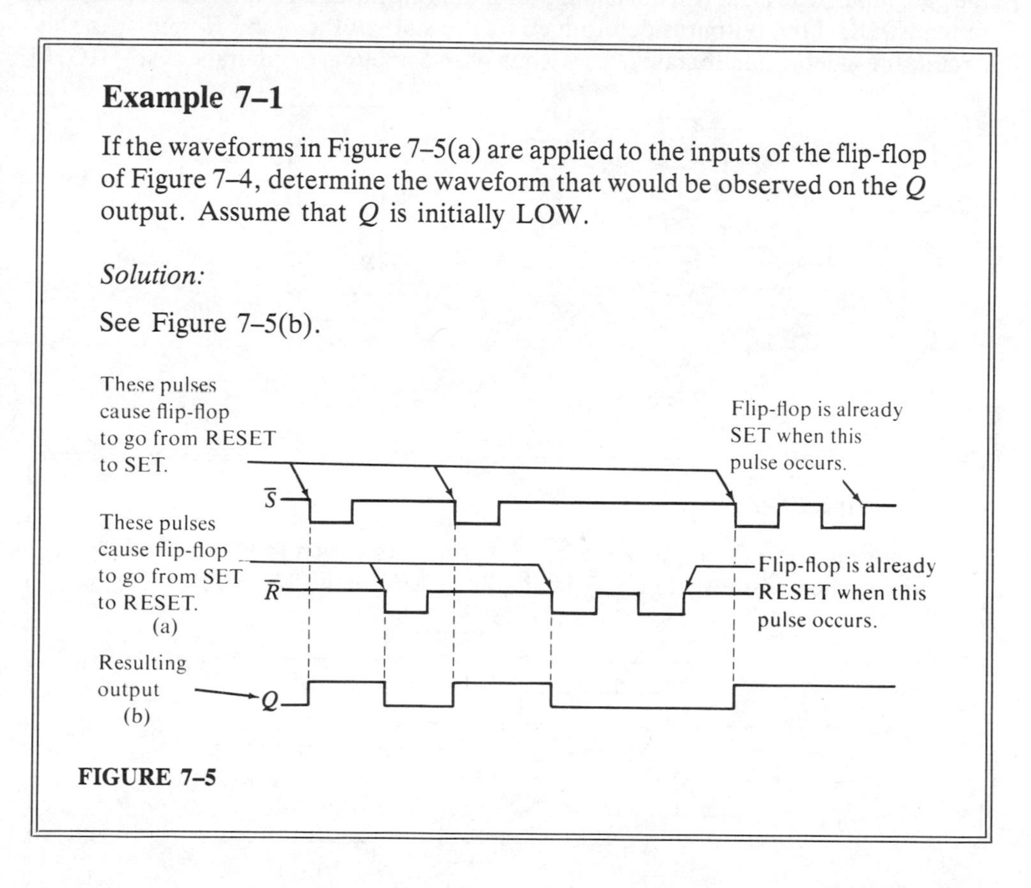

FIGURE 7–5

An example of an integrated circuit $\overline{S}$–$\overline{R}$ latch is the 74279, a block and pin diagram, which is shown in Figure 7-6. As you can see, there are four latches in a single package.

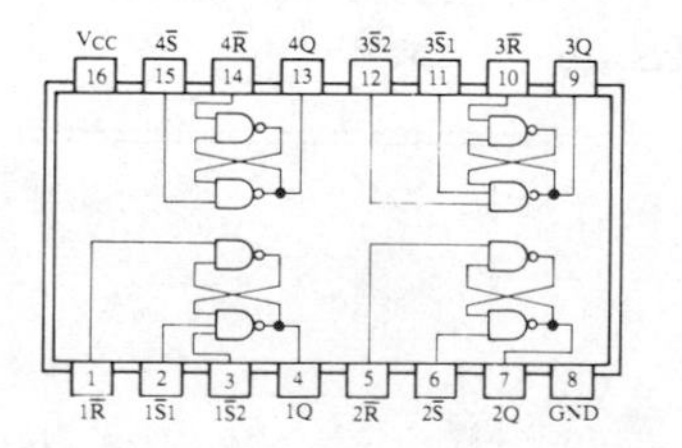

FIGURE 7–6 *A 74279 quad $\overline{S}$–$\overline{R}$ latch.*

Gated S–R Latch

A gated latch requires an additional input called the gate, *G*. The logic diagram and block symbol for a gated S–*R* latch is shown in Figure 7–7. The *S* and *R* inputs control the state to which the latch will go upon application of a HIGH level on the gate input. The latch will not change until the gate input is HIGH, but as long as it remains HIGH the output is determined by the state of the *S* and *R* inputs. In this circuit, the indeterminate state occurs when both *S* and *R* are simultaneously HIGH.

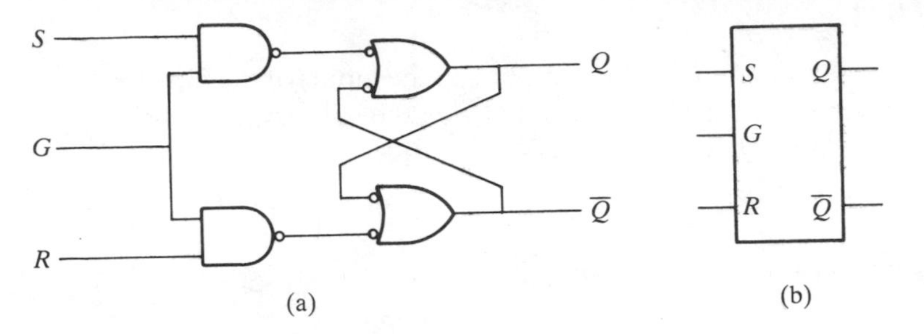

FIGURE 7–7 *A gated S–R latch.*

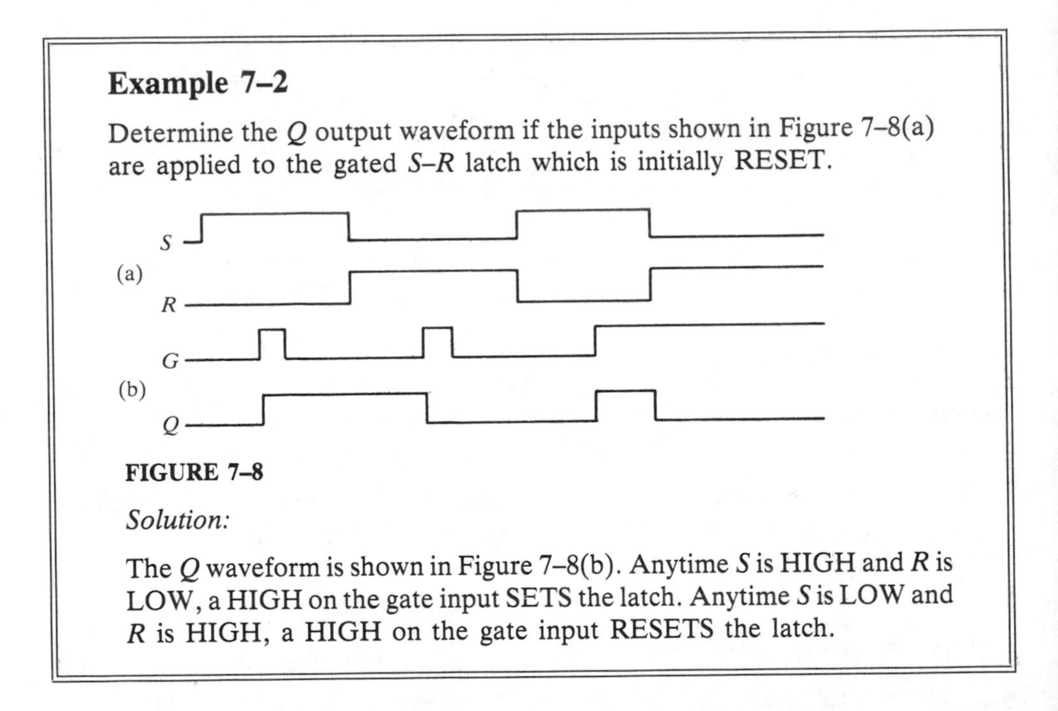

Example 7–2

Determine the *Q* output waveform if the inputs shown in Figure 7–8(a) are applied to the gated *S–R* latch which is initially RESET.

FIGURE 7–8

Solution:

The *Q* waveform is shown in Figure 7–8(b). Anytime *S* is HIGH and *R* is LOW, a HIGH on the gate input SETS the latch. Anytime *S* is LOW and *R* is HIGH, a HIGH on the gate input RESETS the latch.

Gated D Latch

Another type of gated latch is called the *D* latch. It differs from the *S–R* latch in that it has only one input in addition to the gate. This input is called the *D* input. Figure 7–9 is a logic diagram and block symbol of a *D* latch. When the *D* input is HIGH and the gate input is HIGH, the latch will SET. When the *D* input is LOW and the gate is HIGH, the latch will RESET. Stated another way, the output *Q* follows the input *D* when the gate is HIGH.

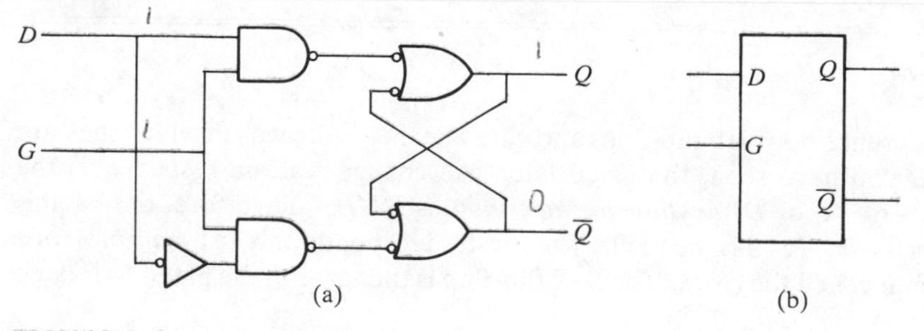

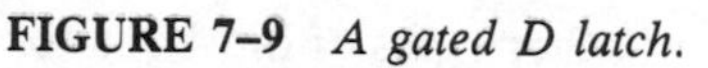
FIGURE 7–9 *A gated D latch.*

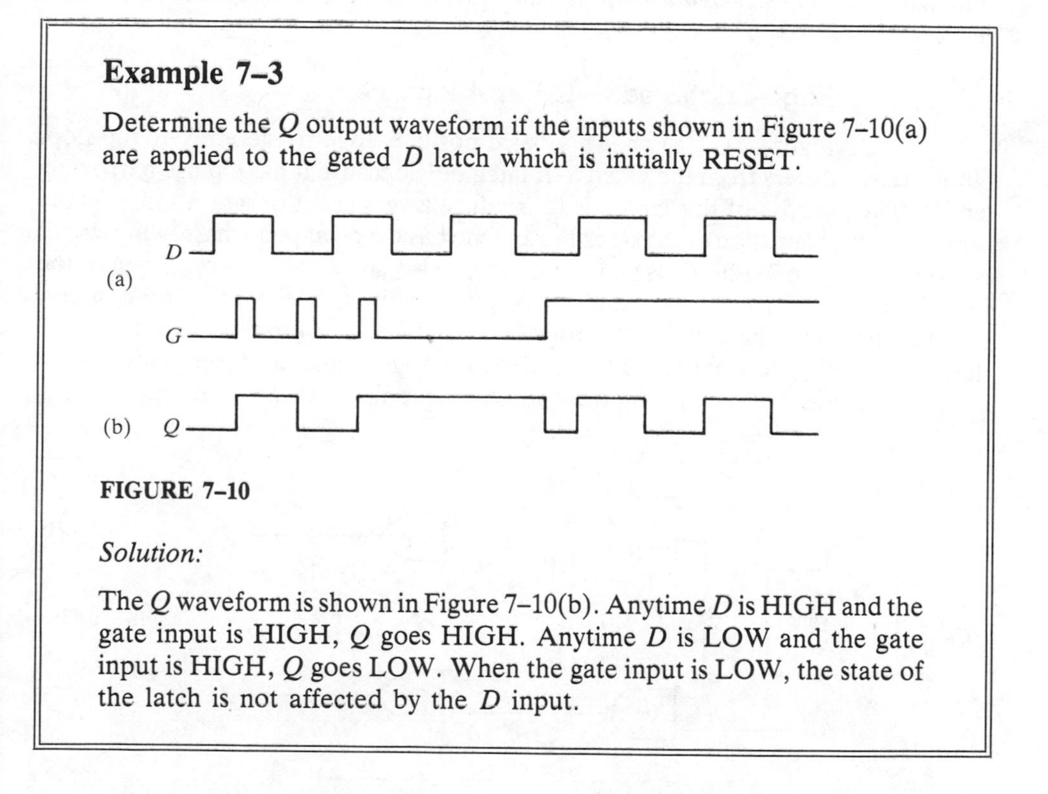

Example 7–3

Determine the Q output waveform if the inputs shown in Figure 7–10(a) are applied to the gated D latch which is initially RESET.

FIGURE 7–10

Solution:

The Q waveform is shown in Figure 7–10(b). Anytime D is HIGH and the gate input is HIGH, Q goes HIGH. Anytime D is LOW and the gate input is HIGH, Q goes LOW. When the gate input is LOW, the state of the latch is not affected by the D input.

An example of an integrated D latch is the 7475 four-bit latch in Figure 7–11. Four latches are contained in the single IC package. Notice that each gate input is shared by two latches.

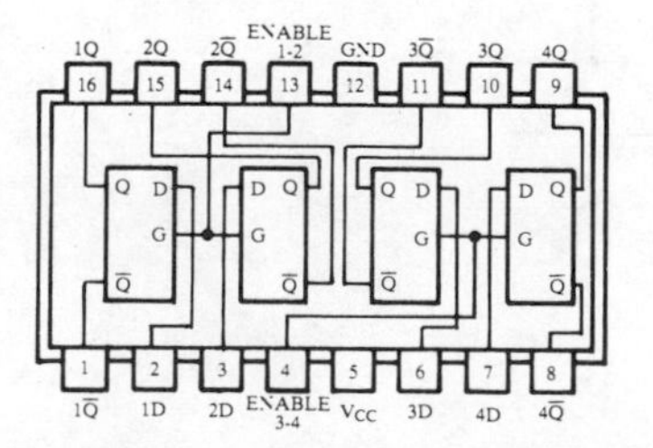

FIGURE 7–11 *A 7475 quad D gated latch.*

7–2 S–R FLIP-FLOPS

The basic difference between flip-flops and gated latches is the way in which they are *triggered.* As you have seen, the gated latch can change state in response to the control inputs *(S, R,* or *D) anytime the gate input is HIGH.* Therefore, the latch is controlled by the *level* of its gate input. The flip-flop responds only to a *transition* of a triggering input called the *clock.* The *S–R* flip-flop is the same as the gated *S–R* latch in all other ways.

There are two types of *S–R* flip-flops (and other types of flip-flops for that matter): *edge-triggered* and *master-slave.* Both of these types are covered in this section.

Edge-Triggered S–R Flip-Flop

Figure 7–12(a) shows a basic positive edge-triggered *S–R* flip-flop. Notice that it differs from the gated *S–R* latch only because it has a pulse narrowing circuit. The purpose of this circuit is to produce a very short duration spike on the positive-going transition of the clock pulse. One basic type of pulse-narrowing circuit is shown in Figure 7–12(b). As you can see, there is a small delay on one input to the NAND gate so that the inverted clock pulse arrives at the gate input a few nanoseconds after the true clock pulse. This produces an output spike with a time duration of only a few nanoseconds. A negative edge-triggered flip-flop inverts the clock pulse thus producing a narrow spike on the negative-going edge.

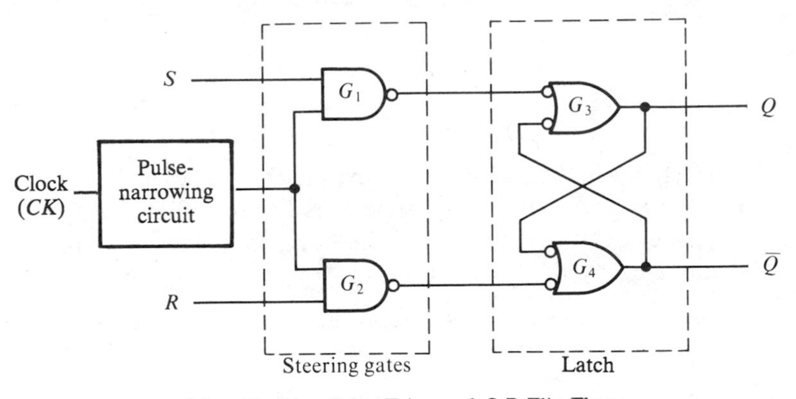

(a) A Positive Edge-Triggered *S-R* Flip-Flop

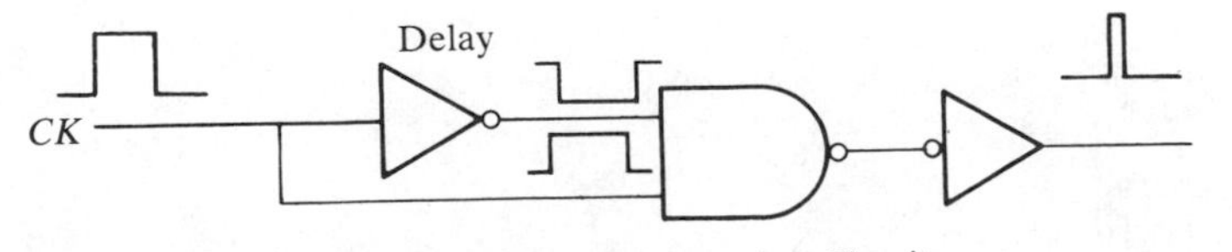

(b) A type of Pulse-Narrowing Circuit

FIGURE 7–12

Notice that the circuit in Figure 7–12 is partitioned into two sections, one labeled *steering gates* and the other *latch*. The steering gates direct or "steer" the clock spike to either the input to gate G_3 or the input to gate G_4, depending on the state of the S and R inputs. In order to understand the operation of this flip-flop, let us begin with the assumption that it is in the RESET state ($Q = 0$) and the S, R, and clock inputs are all LOW. For this condition, the outputs of gate G_1 and gate G_2 are both HIGH. The LOW on the Q output is coupled back into one input of gate G_4, making the $\overline{Q}$ output HIGH. Because $\overline{Q}$ is HIGH, both inputs to gate G_3 are HIGH (remember, the output of gate G_1 is HIGH), holding the Q output LOW. If a positive-going edge of the clock is applied to the clock input, the outputs of gates G_1 and G_2 remain HIGH because they are disabled by the LOWs on the S input and the R input; therefore, there is no change in the state of the flip-flop—it remains RESET.

We will now make S HIGH, leave R LOW, and apply a clock pulse. Because the S input to gate G_1 is now HIGH, the output of gate G_1 goes LOW for a very short time (spike) when the clock input goes HIGH, causing the Q output to go HIGH. Both inputs to gate G_4 are now HIGH (remember, gate G_2 output is HIGH because R is LOW), forcing the $\overline{Q}$ output LOW. This LOW on $\overline{Q}$ is coupled back into one input of gate G_3, insuring that the Q output will remain HIGH. The flip-flop is now in the SET state. Figure 7–13 illustrates the logic level transitions that take place within the flip-flop for this condition.

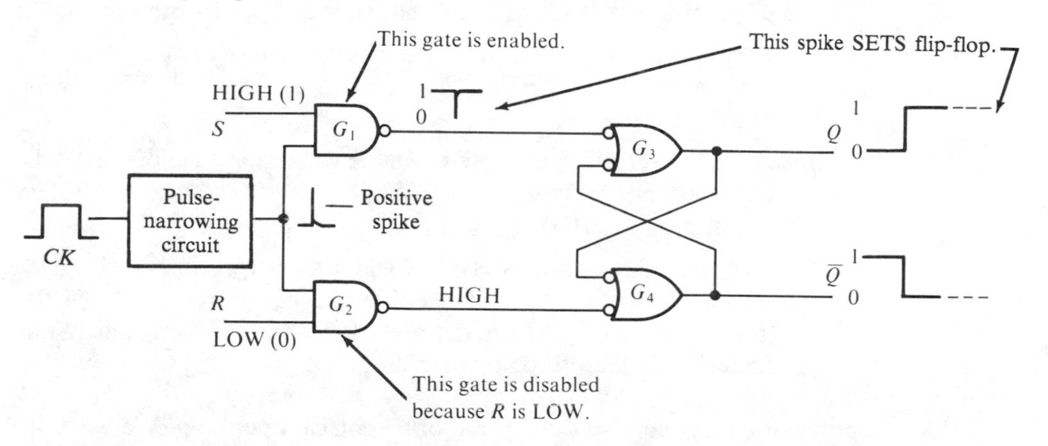

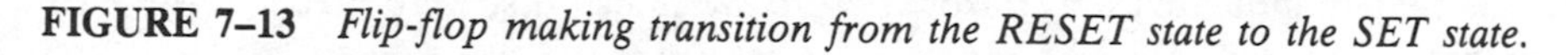
FIGURE 7–13 *Flip-flop making transition from the RESET state to the SET state.*

Next, we will make S LOW, R HIGH, and apply a clock pulse. Because the R input is now HIGH, the positive-going edge of the clock produces a negative going spike on the output of gate G_2, *causing the* $\overline{Q}$ output to go HIGH. Because of this HIGH on $\overline{Q}$, both inputs to gate G_3 are now HIGH (remember, the output of gate G_1 is HIGH because of the LOW on S), *forcing the* Q output to go LOW. This LOW on Q is coupled back into one input of gate G_4, insuring that $\overline{Q}$ will remain HIGH. The flip-flop is now in the RESET state. Figure 7–14 illustrates the logic level transitions that occur within the flip-flop for this condition.

As with the gated latch, an indeterminate condition exists when both S and R are HIGH at the same time. Because we are trying to SET and RESET the

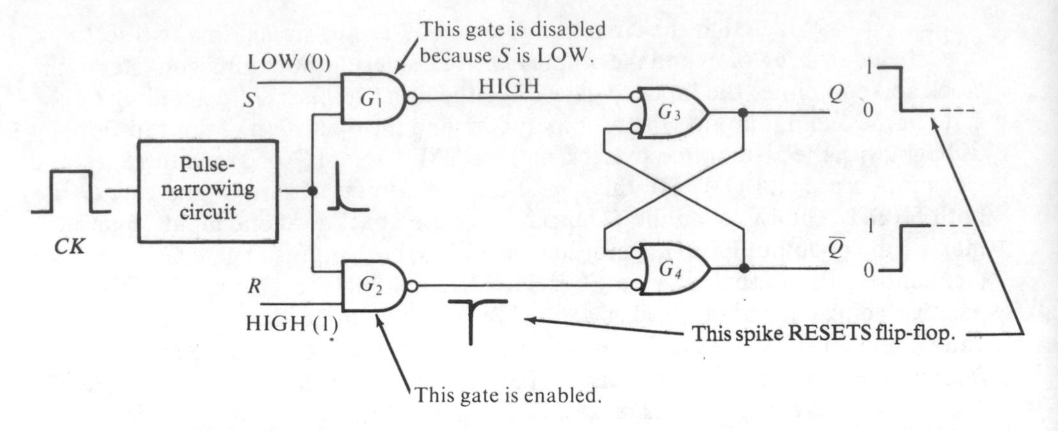

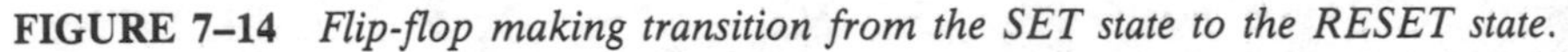

FIGURE 7–14 *Flip-flop making transition from the SET state to the RESET state.*

flip-flop simultaneously, a race condition occurs when the clock pulse is removed, and the final state depends on the delays through the gates. The four possible input conditions for the edge-triggered S–R flip-flop are summarized below:

1. If S is LOW and R is LOW when the triggering edge of the clock pulse occurs, the flip-flop will not change from its present state.
2. If S is LOW and R is HIGH when the triggering clock edge occurs, the flip-flop will RESET ($Q = 0$). If Q is already LOW, it will remain LOW.
3. If S is HIGH and R is LOW when the triggering edge of the clock pulse occurs, the flip-flop will SET ($Q = 1$). If Q is already HIGH, it will remain HIGH.
4. If S is HIGH and R is HIGH, when the triggering clock edge occurs, the next state of the flip-flop is unpredictable. This imposes a restriction on the use of the RS flip-flop and is an undesirable feature of this flip-flop.

Table 7–2 provides a more concise description of the logical operation. We will use the notation Q_n to indicate the Q output *before* the clock pulse is applied and Q_{n+1} to indicate the Q output after the clock pulse is applied.

TABLE 7–2 *Truth table for an S–R flip-flop.*

Inputs		Output	Comments
S	R	Q_{n+1}	
LOW	LOW	Q_n	NO CHANGE. The output remains in its previous state.
LOW	HIGH	LOW	RESET condition.
HIGH	LOW	HIGH	SET condition.
HIGH	HIGH	?	Indeterminate.

A logic symbol for a positive edge-triggered *S–R* flip-flop is shown in Figure 7–15(a). Notice the triangle shaped symbol on the clock *(CK)* input. This is called a *dynamic input indicator,* and it means that triggering occurs on a *transition* rather than a level. As mentioned before, a negative edge-triggered flip-flop operates the same as the one just discussed, except that it is triggered on negative going transitions of the clock. This is shown in the logic symbol in Figure 7–15(b) where the bubble indicates an active low input.

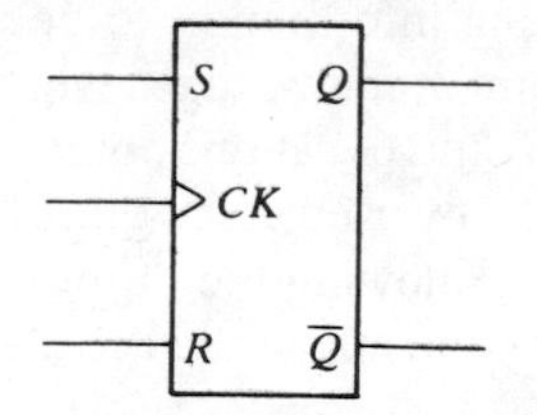

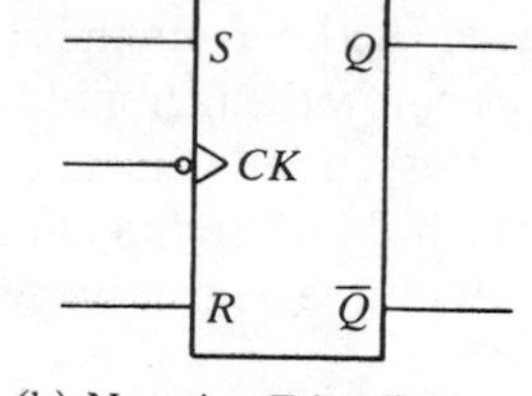

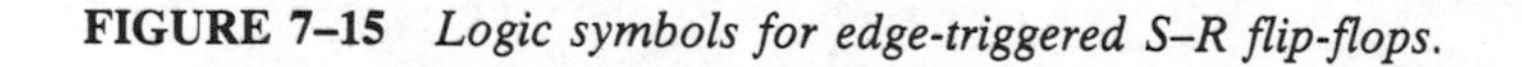

FIGURE 7–15 *Logic symbols for edge-triggered S–R flip-flops.*

Master-Slave S–R Flip-Flop

Another important type of S–*R* flip-flop is commonly referred to as a *master-slave* type and is shown in Figure 7–16. The truth table operation is identical to that of an edge-triggered *S–R* flip-flop. The primary difference in operation is the manner in which the flip-flop is clocked; it is level-triggered rather than edge-triggered.

This type of flip-flop is composed of two sections, the *master* section and the *slave* section. The master section is basically a gated *S–R* latch, and the slave section is the same except that it is triggered or clocked on the inverted clock pulse and is controlled by the outputs of the master section rather than by external inputs.

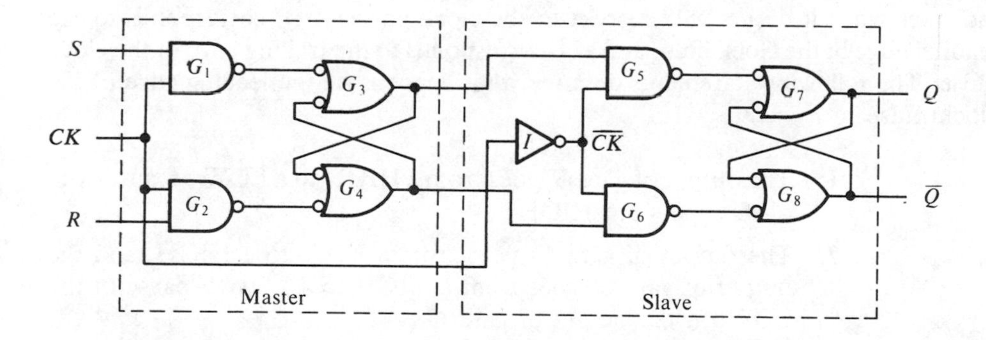

FIGURE 7–16 *Logic diagram for a master-slave S–R flip-flop.*

The master section will assume the state determined by the *R* and *S* inputs at the positive-going edge of the clock pulse. The state of the master section is then transferred into the slave section on the negative-going edge of the clock pulse because the outputs of the master are applied to the inputs of the slave, and the clock pulse to the slave is inverted. The state of the slave then immediately appears on the Q and $\overline{Q}$ outputs.

Figure 7–17 illustrates the master-slave timing operation, which is described as follows. The clock pulse shown is nonideal since it has a rise time and fall time other than 0 (as is always the situation, in practice). At point 1 on the rising edge, the voltage level on the inverted clock pulse reaches the threshold point below which the G_5 and G_6 gates of Figure 7–16 are disabled. At point 2, the voltage level on the true (noninverted) clock enables gates G_1 and G_2, allowing information to be entered into the master section from the R and S inputs. At point 3, the voltage level on the true clock goes below the threshold and disables gates G_1 and G_2; at this point, the S and R inputs are essentially "locked out." At point 4, the inverted clock pulse $(\overline{CK})$ reaches the threshold level and enables gates G_5 and G_6, allowing the output levels of the master section to be transferred to the slave section and appear on the Q and $\overline{Q}$ outputs.

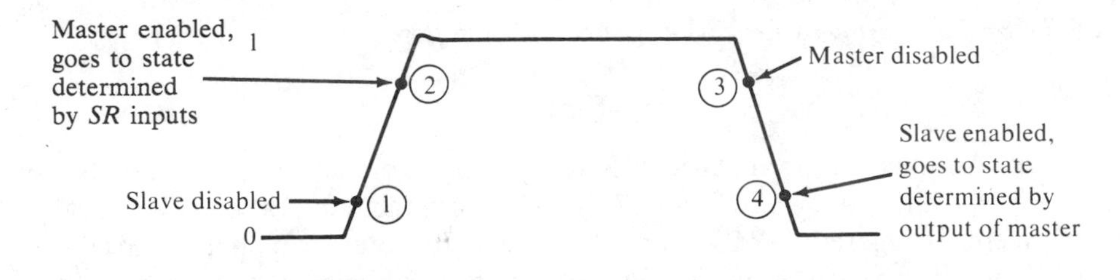

FIGURE 7–17 *Clock pulse showing master-slave operation.*

Let us now further examine the operation of this circuit. To begin, we will assume that the flip-flop is RESET and that S is HIGH and R is LOW. Figure 7–18 shows the logic level transitions within the flip-flop when it goes from the RESET state to the SET state. The circled number associated with each transition indicates when it occurs with respect to the clock pulse. A ① corresponds to the leading edge of the clock pulse and a ② corresponds to the trailing edge of the clock pulse. The following statements describe what happens on the leading edge of the clock pulse:

1. The output of gate G_1 goes from a HIGH to a LOW because both of its inputs are HIGH.
2. The output of gate G_3 goes from a LOW to a HIGH, and the output of gate G_4 goes from a HIGH to a LOW because of the LOW on the output of gate G_1.
3. The inverted clock $(\overline{CK})$ input to gates G_5 and G_6 goes LOW because the true clock goes HIGH; this disables the G_5 and G_6 gates and forces their outputs HIGH.

Let us review what has taken place on the leading edge of the clock pulse. The master section has been SET because the S input is HIGH and the R input is LOW. The slave section has not changed state, and therefore the Q and $\overline{Q}$ outputs remain in the initial RESET state.

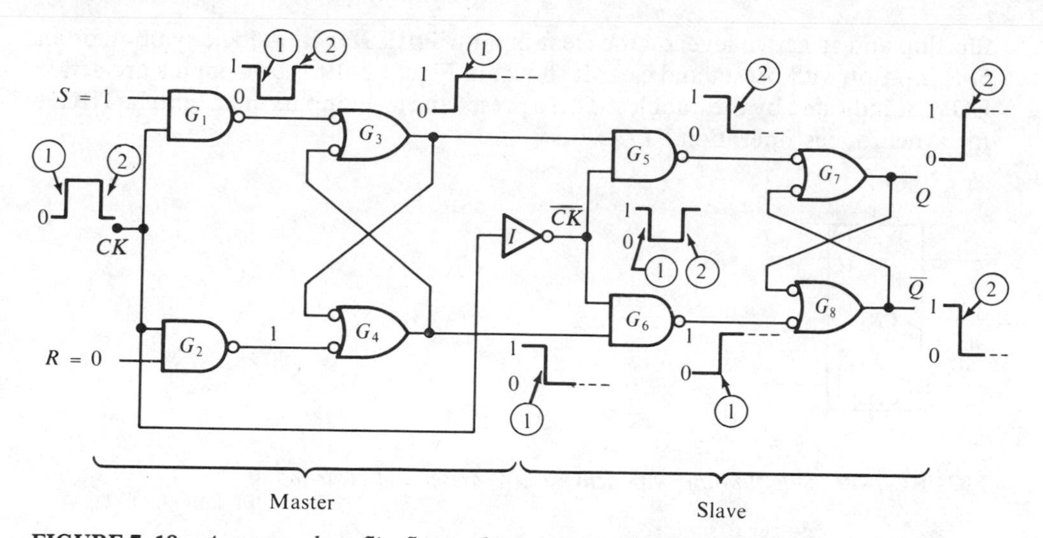

FIGURE 7–18 *A master-slave flip-flop making transition from a RESET state to a SET state.*

Now let us look at what happens on the trailing edge of the clock pulse.

1. The master section remains in the SET condition.
2. The output of gate G_5 goes from a HIGH to a LOW because both of its inputs are now HIGH.
3. The output of gate G_7 goes from a LOW to a HIGH, and the output of gate G_8 goes from a HIGH to a LOW because the output of gate G_5 is LOW.

The flip-flop is now in the SET state, and therefore the Q output is HIGH. It did not become SET until the trailing edge of the clock pulse, although the master section was SET on the leading edge. You should verify the operations for the RESET and no-change conditions.

The logic symbol for master-slave *S–R* flip-flops is the same as those for the edge-triggered type shown in Figure 7–15. The principal advantage of the master-slave flip-flop over the edge-triggered type is that the master-slave permits transitions of the *S* and *R* inputs during and after the leading edge of the clock pulse. The inputs of the edge-triggered flip-flop, however, must be stabilized at the triggering edge of the clock. The master-slave feature can be used where *race* conditions might cause input transitions at or near the clock edge.

Synchronous and Asynchronous Inputs

The *S* and *R* inputs are called *synchronous inputs* because data on these inputs are transferred to the flip-flop's output only on the triggering edge of the clock pulse; that is, the data are transferred synchronous with the clock.

Most integrated circuit flip-flops also have *asynchronous inputs*. These are inputs that affect the state of the flip-flop independent of the clock. They are normally labeled *preset* and *clear*. An active level on the preset input will SET the flip-flop and an active level on the clear input will RESET it. A logic symbol for an S–R flip-flop with preset and clear is shown in Figure 7–19. These inputs are active LOW as indicated by the bubbles. These preset and clear inputs must both be HIGH for synchronous operation.

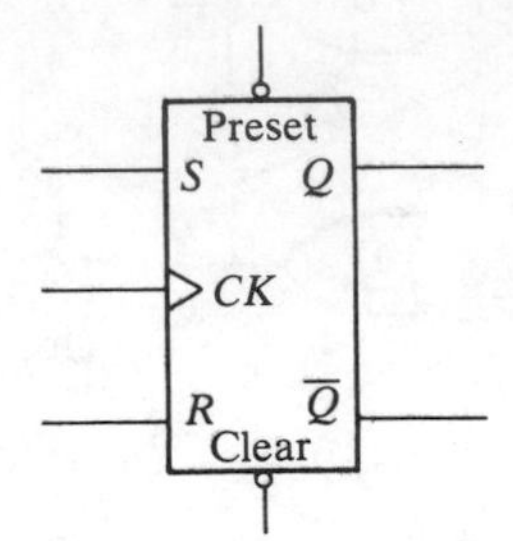

FIGURE 7–19 *S–R flip-flop with active-LOW preset and clear inputs.*

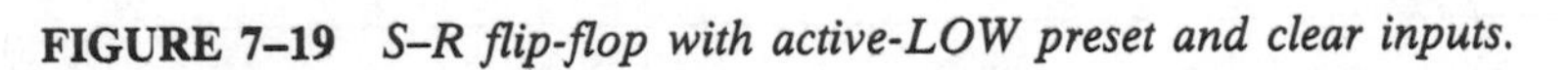

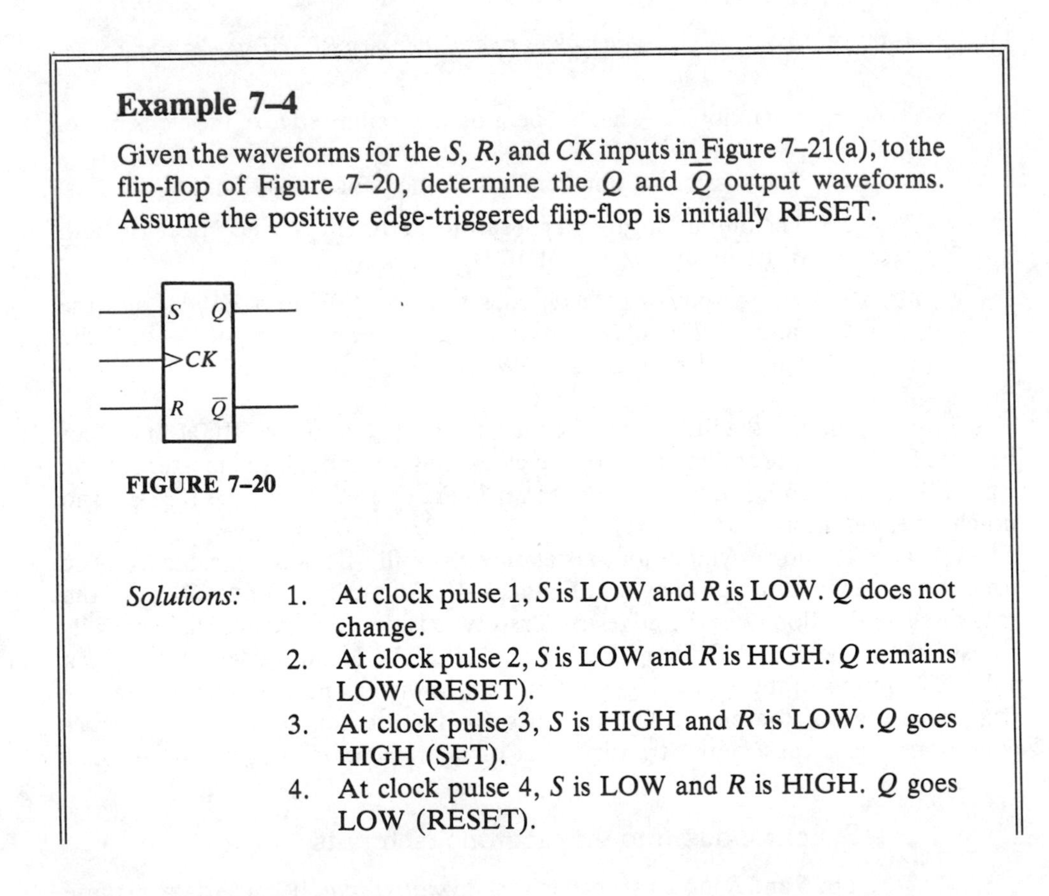

Example 7–4

Given the waveforms for the S, R, and CK inputs in Figure 7–21(a), to the flip-flop of Figure 7–20, determine the Q and $\overline{Q}$ output waveforms. Assume the positive edge-triggered flip-flop is initially RESET.

FIGURE 7–20

Solutions:

1. At clock pulse 1, S is LOW and R is LOW. Q does not change.
2. At clock pulse 2, S is LOW and R is HIGH. Q remains LOW (RESET).
3. At clock pulse 3, S is HIGH and R is LOW. Q goes HIGH (SET).
4. At clock pulse 4, S is LOW and R is HIGH. Q goes LOW (RESET).

5. At clock pulse 5, S is HIGH and R is LOW. Q goes HIGH (SET).
6. At clock pulse 6, S is HIGH and R is LOW. Q stays HIGH.

Once Q is determined, $\overline{Q}$ is easily found since it is simply the complement of Q. The resulting waveforms for Q and $\overline{Q}$ are shown in Figure 7–21(b) for the input waveforms in (a).

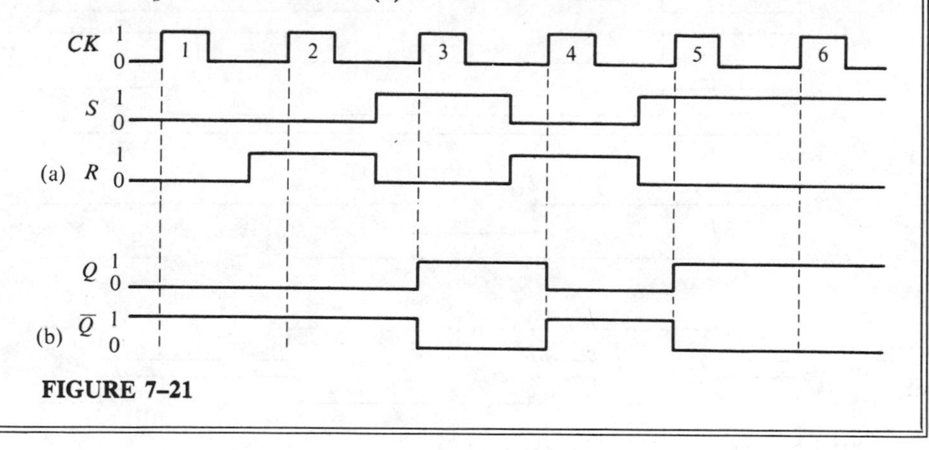

FIGURE 7–21

Example 7–5

A 74L71 S–R flip-flop is shown in Figure 7–22. Notice that this particular device has AND gates on the S and R inputs. All inputs to the associated gate must be HIGH in order to make the S or R input to the flip-flop HIGH. This device also has active LOW clear (CLR) and preset (PR) inputs. It is a master-slave type of flip-flop.

From the input waveforms in Figure 7–23(a), determine the Q and $\overline{Q}$ output waveforms. Assume it starts RESET.

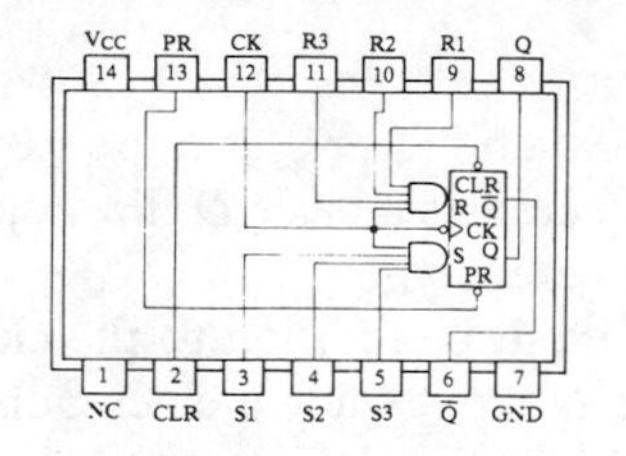

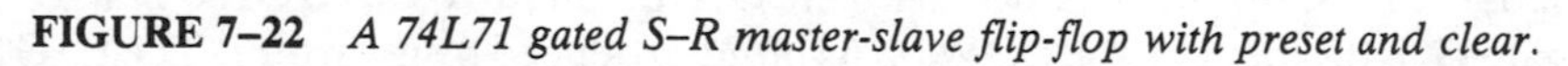

FIGURE 7–22 *A 74L71 gated S–R master-slave flip-flop with preset and clear.*

Example 7–5, continued

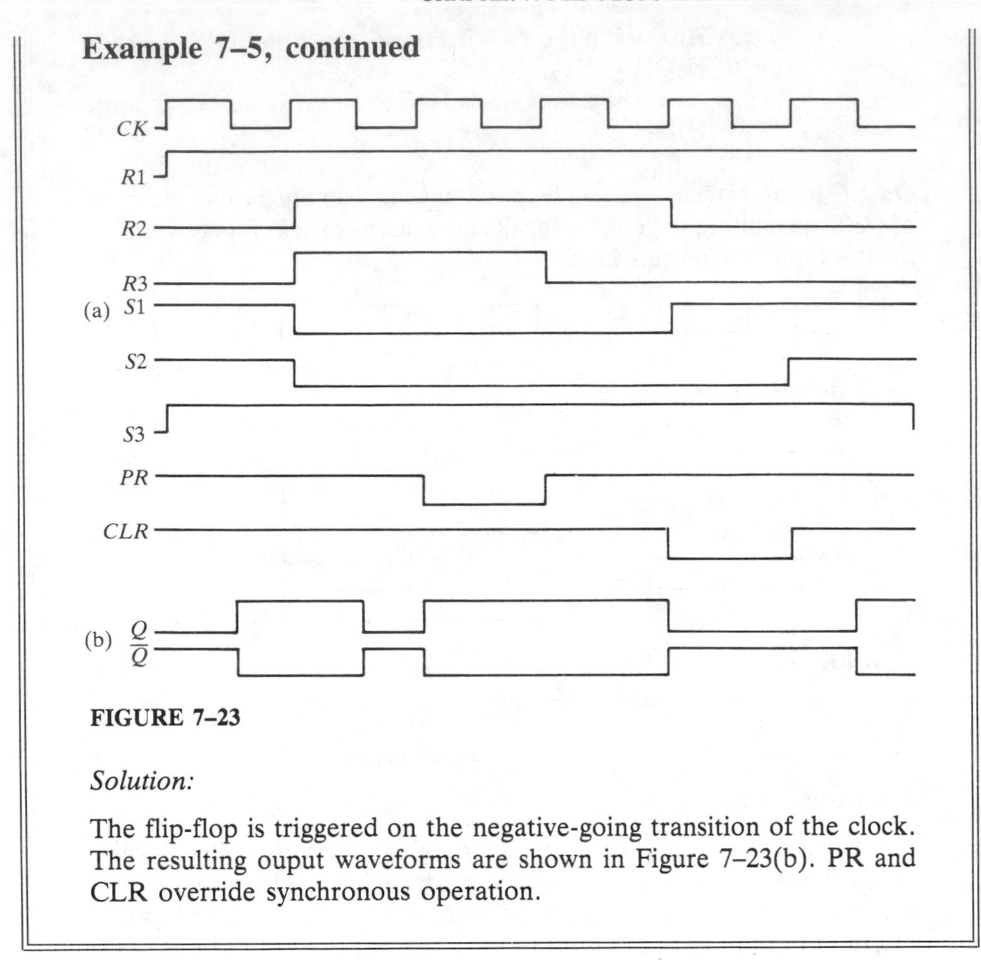

FIGURE 7–23

Solution:

The flip-flop is triggered on the negative-going transition of the clock. The resulting ouput waveforms are shown in Figure 7–23(b). PR and CLR override synchronous operation.

7–3 D FLIP-FLOPS

The *D* flip-flop is very useful in cases where a single data bit (1 or 0) is to be stored. The simple addition of an inverter to an *S–R* flip-flop creates a basic *D* flip-flop, as shown in Figure 7–24, which is a positive edge-triggered type.

Notice that this flip-flop has only one input in addition to the clock. This is called the *D* input. If there is a HIGH on the *D* input when a clock pulse is applied, the flip-flop will SET because gate G_1 is enabled by the HIGH on the *D* input and gate G_2 is disabled by the LOW on its input produced by inverter *I*. The HIGH on the *D* input is "stored" by the flip-flop on the leading edge of the clock pulse. If there is a LOW on the *D* input when the clock pulse is applied, the flip-flop will RESET because gate G_1 is disabled by the LOW on the *D* input and gate G_2 is enabled by the HIGH produced by the inverter *I*. The LOW on the *D* input is thus stored by the flip-flop on the leading edge of the clock pulse. In the SET state, the flip-flop can be said to be storing a logic 1, and in the RESET state it is storing a logic 0.

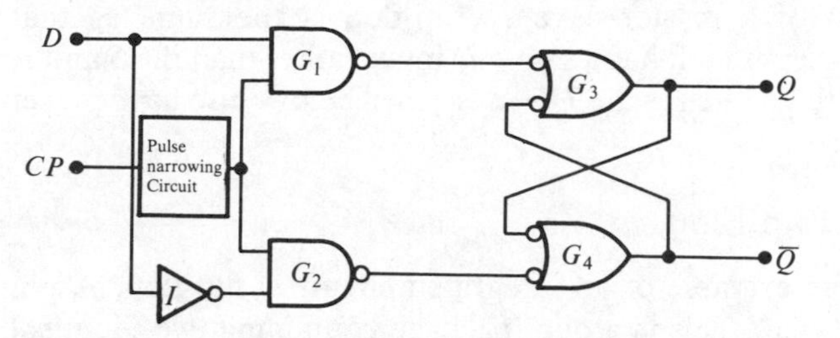

FIGURE 7–24 *Positive edge-triggered D flip-flop.*

As we have seen, the state of the D input (HIGH or LOW) is transferred to the Q output on the leading edge of the clock pulse (negative edge-triggering is also possible). Figure 7–25 illustrates the transitions within the flip-flop when a HIGH is entered on the D input, followed by a LOW. The HIGH is transferred to the Q output on the first clock pulse, and the LOW is transferred to the Q output on the second clock pulse. Q was initially LOW.

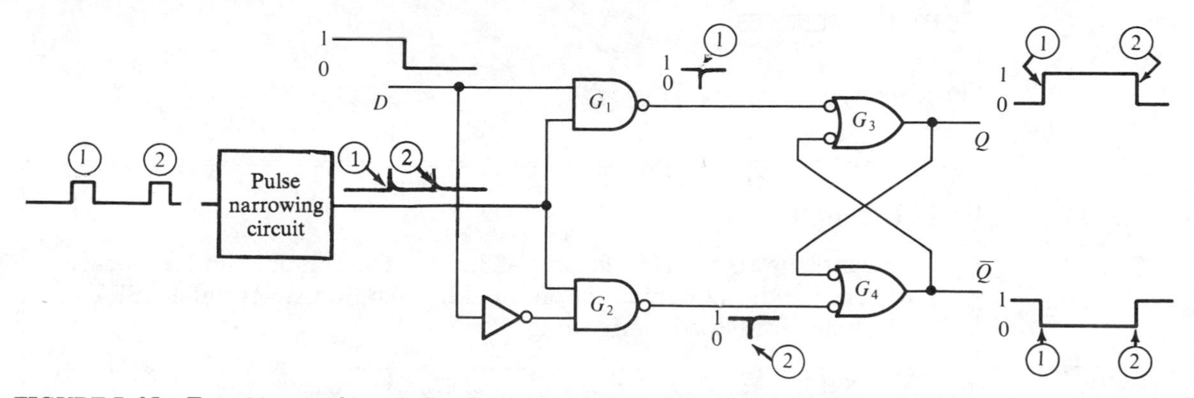

FIGURE 7–25 *Transitions within a D flip-flop when a HIGH followed by a LOW is clocked into the flip-flop.*

A typical logic symbol for the D flip-flop appears in Figure 7–26, and the truth table is given in Table 7–3.

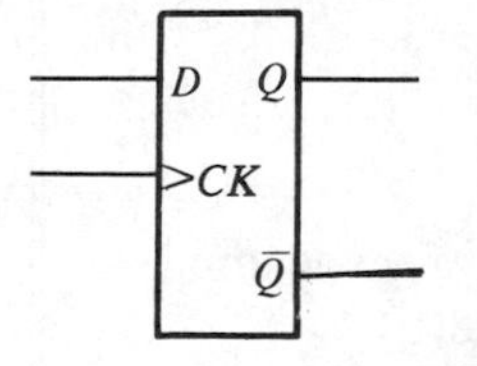

FIGURE 7–26 *Logic symbol for a positive edge-triggered D flip-flop.*

TABLE 7–3 *Truth table for a D flip-flop.*

D	Q_{n+1}	Comments
LOW HIGH	LOW HIGH	The Q output assumes the state of D at each clock.

The operation of a master-slave *D* flip-flop is the same as that discussed for an *S–R* type, except that it has a single *D* input rather than the *S* and *R* inputs. Many IC *D*-type flip-flops, such as the 7474 described below, also have preset and clear inputs.

The 7474 D Flip-Flop

The 7474 is one example of an integrated circuit *D* flip-flop. A pin diagram is shown in Figure 7–27. This is a dual package containing two identical flip-flops. The flip-flops are positive edge-triggered and have active LOW preset and clear inputs.

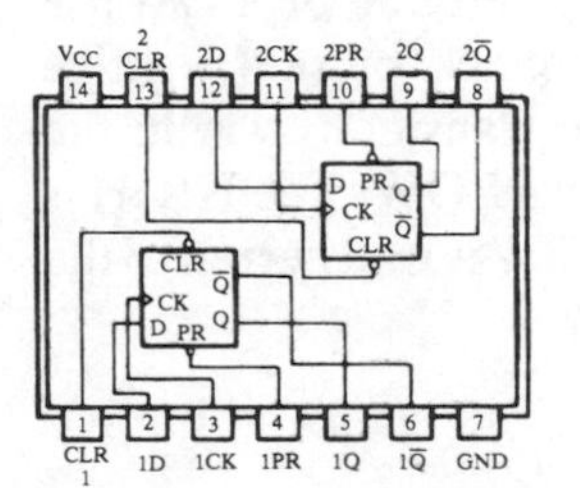

FIGURE 7–27 *7474 dual D flip-flops.*

Example 7–6

Given the waveforms in Figure 7–28(a) for the *D* input and the clock, determine the *Q* output waveform if the flip-flop starts out RESET. Assume positive edge-triggering.

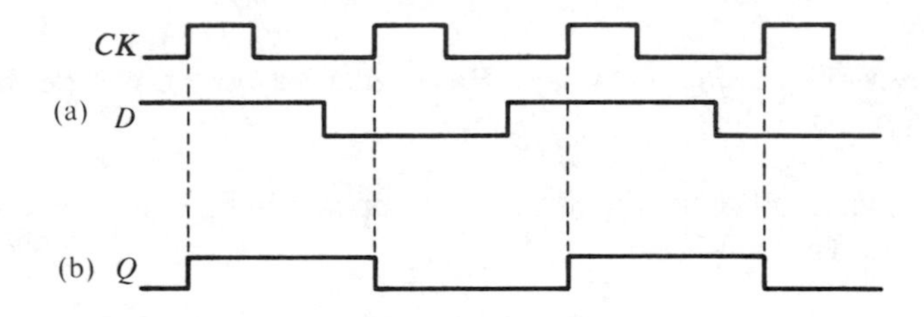

FIGURE 7–28

Solution:

The *Q* output assumes the state of the *D* input at the time of the leading clock edge. The resultant output is shown in Figure 7–28(b).

7–4 J–K FLIP-FLOPS

The *J–K* flip-flop is very versatile and is perhaps the most widely used type of flip-flop. The *J* and *K* designations for the inputs have no known significance except that they are adjacent letters in the alphabet.

The functioning of the *J–K* flip-flop is identical to that of the *S–R* flip-flop in SET, RESET, and *no change* conditions of operation. The difference is that the *J–K* flip-flop has no indeterminate state as does the *S–R* flip-flop. Therefore, the *J–K* flip-flop is a very flexible device that finds wide application in digital systems.

Two types of clocked *J–K* flip-flops will be considered: the edge-triggered type in this section and the master-slave type in the next section. The truth table operation for both of these types is identical; the primary difference is the clocking operation. For an edge-triggered type, the data on the inputs are entered into the flip-flop and appear on the outputs on the same edge of the clock pulse. For a master-slave type, the data on the *J* and *K* inputs are entered on the leading edge of the clock but do not appear on the outputs until the trailing edge of the clock.

Edge-Triggered J–K Flip-Flop

Figure 7–29 shows a basic positive edge-triggered *J–K* flip-flop. Notice that it differs from the *S–R* edge-triggered flip-flop in that the Q output is connected back to the input of gate G_2, and the $\overline{Q}$ output is connected back to the input of gate G_1. The two control inputs are labeled *J* and *K*. A *J–K* flip-flop can also be of the negative edge-triggered type in which case the clock input is inverted.

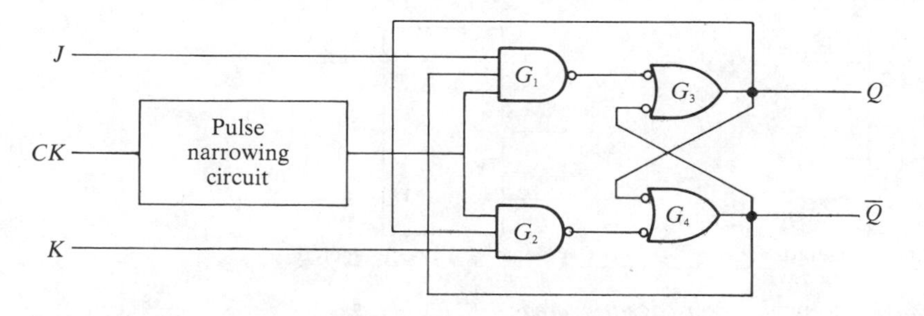

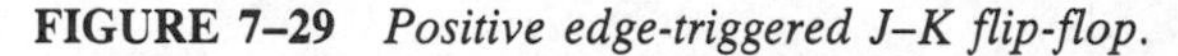

FIGURE 7–29 *Positive edge-triggered J-K flip-flop.*

We will start by assuming the flip-flop is RESET and that the *J* input is HIGH and the *K* input is LOW. When a clock spike occurs, it passes through gate G_1 because $\overline{Q}$ is HIGH and *J* is HIGH. This will cause the latch portion of the flip-flop to change to the SET state. It is important that there be sufficient delay from the occurrence of the clock spike to the change in Q and $\overline{Q}$ outputs so that the clock spike has time to go LOW before the change in the output state is felt on the steering gate input.

The flip-flop is now SET. If we now make *J* LOW and *K* HIGH, the next clock spike will pass through gate G_2 because Q is HIGH and *K* is HIGH. This will cause the latch portion of the flip-flop to change to the RESET state.

Now if a LOW is applied to both the *J* and *K* inputs, the flip-flop will stay in its present state when a clock pulse occurs. So, a LOW on both *J* and *K* results in a *no-change* condition.

So far, the basic operation of the *J–K* flip-flop is the same as for the *S–R* type in the SET, RESET, and *no-change* modes. The difference in operation occurs when both the J and K inputs are HIGH. To see this, assume the flip-flop is RESET. The HIGH on the $\overline{Q}$ enables gate G_1 so that the clock spike passes through to SET the flip-flop. Now, there is a HIGH on Q which allows the next clock spike to pass through gate G_2 and RESET the flip-flop. As you can see, on each successive clock, the flip-flop changes to the opposite state. This is called *toggle* operation. Figure 7–30 illustrates the transitions when the flip-flop is in the toggle mode.

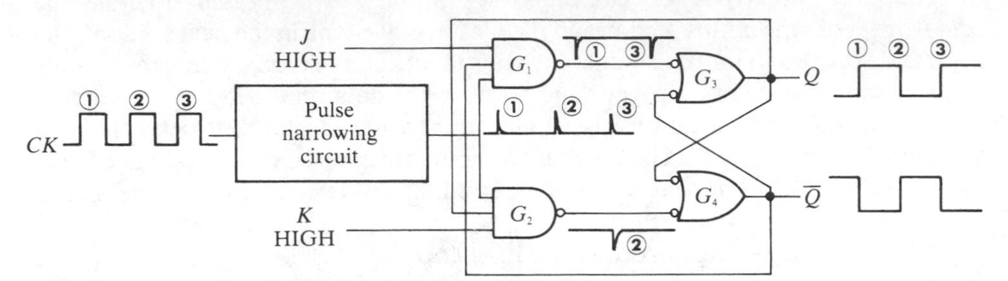

FIGURE 7–30 *Transitions illustrating the toggle operation.*

Table 7–4 is the truth table for a J–K flip-flop. It is the same for both edge-triggered types and master-slave types. Figure 7–31 shows the logic symbols for both the positive edge-triggered and negative edge-triggered J–K flip-flops.

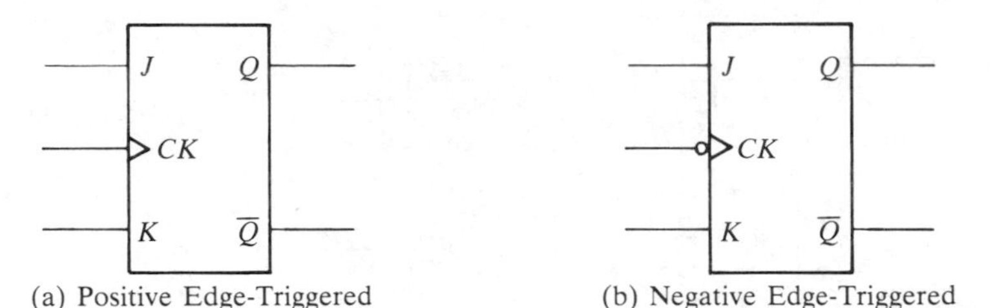

(a) Positive Edge-Triggered (b) Negative Edge-Triggered

FIGURE 7–31 *Logic symbols for edge-triggered J–K flip-flops.*

TABLE 7–4 *Truth table for a J–K flip-flop.*

Inputs		Output	Comments
J	K	Q_{n+1}	
LOW	LOW	Q_n	NO CHANGE
LOW	HIGH	LOW	RESET
HIGH	LOW	HIGH	SET
HIGH	HIGH	$\overline{Q}_n$	TOGGLE

Master-Slave J–K Flip-Flop

Figure 7–32 is the logic diagram for a master-slave J–K flip-flop. A brief inspection of this diagram shows that this circuit is very similar to the master-slave S–R flip-flop. Notice the two main differences: first, the Q output is connected back into the input of gate G_2 and the $\overline{Q}$ output is connected back into the input of gate G_1; and second, one input is now designated J and the other input is now designated K. We will now examine the operation of this circuit in detail.

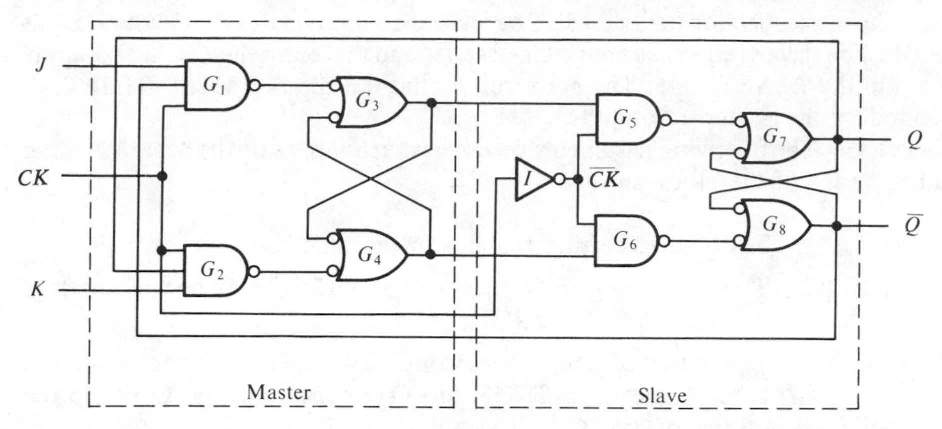

FIGURE 7–32 *Logic diagram for a master-slave J–K flip-flop.*

To begin, let us assume that the flip-flop is RESET. We will also let J be HIGH and K be LOW, creating a SET condition on the inputs. Figure 7–33 shows the transitions for the flip-flop going from the RESET state to the SET state upon application of a clock pulse. The circled number associated with each transition indicates when that transition occurs with respect to the clock pulse. A ① corresponds to the leading edge of the clock pulse and a ② corresponds to the trailing edge. The following statements describe what happens on the leading edge of the clock pulse:

1. The output of gate G_1 goes from HIGH to LOW because all of its inputs are HIGH.
2. The output of gate G_3 goes from LOW to HIGH and the output of gate G_4 goes from HIGH to LOW because of the LOW on the output of gate G_1.
3. The inverted clock ($\overline{CK}$) goes LOW, disabling both the G_5 and G_6 gates. This insures that their outputs remain HIGH.

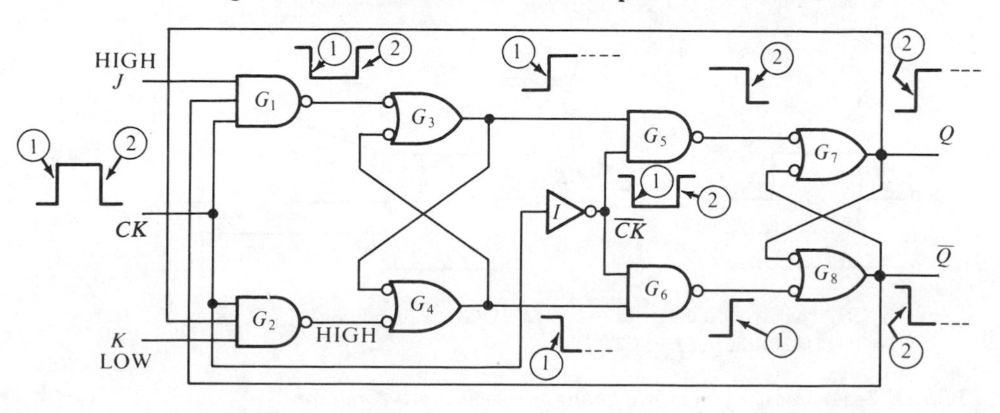

FIGURE 7–33 *Flip-flop making transition from the RESET state to the SET state.*

Let us review what has taken place on the positive-going (leading) edge of the clock pulse. The master section has been SET because the J input is HIGH and the K input is LOW. The slave section has not changed state, and therefore the Q and $\overline{Q}$ outputs remain in the RESET state. The net result is that the flip-flop is still RESET, as indicated by the Q and $\overline{Q}$ outputs.

The following statements describe what happens on the negative-going (trailing) edge of the clock pulse:

1. The master section remains in the SET state.
2. The output of gate G_5 goes from HIGH to LOW because both of its inputs are now HIGH.
3. The output of gate G_7 goes from LOW to HIGH and the output of gate G_8 goes from HIGH to LOW because of the LOW on the output of gate G_5.

The flip-flop is now in the SET state because the Q output is HIGH. It did not become SET until the trailing (negative-going) edge of the clock pulse, although the master section was SET on the leading (positive-going) edge. Figure 7–34 is a timing diagram illustrating the relationship between the clock pulse and the Q output for each condition on the J and K inputs.

The transitions for the SET condition have been examined. You should verify the operation for the other conditions in a similar manner.

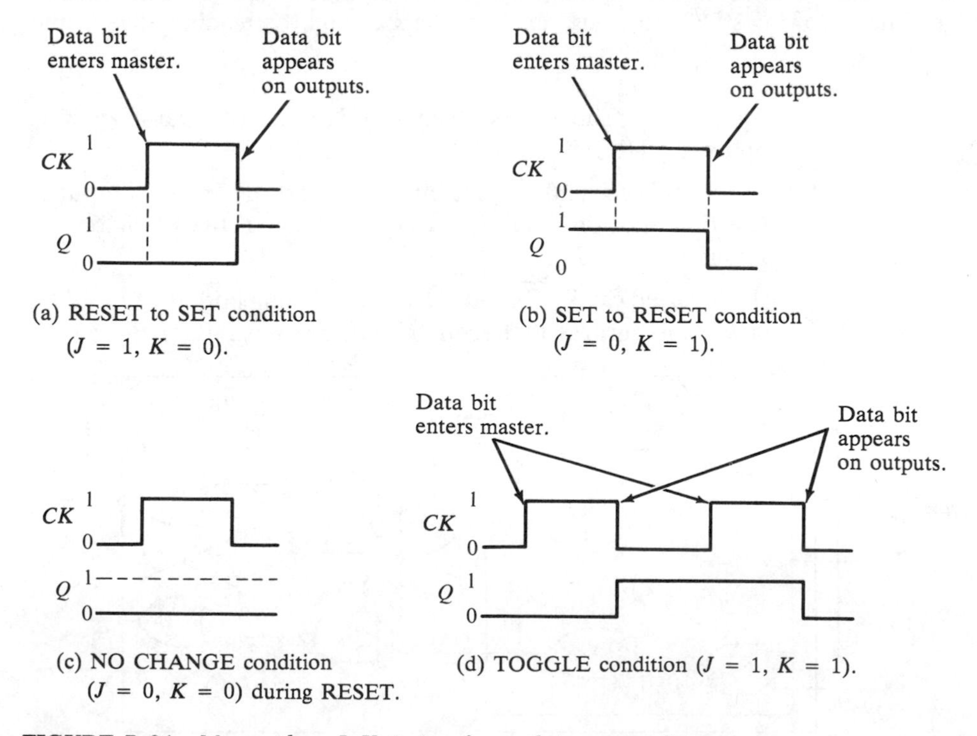

FIGURE 7–34 *Master-slave J–K timing for each input condition.*

The logic symbols for the master-slave *J–K* flip-flop are the same as those for the edge-triggered type in Figure 7–31. As with the *S–R* flip-flop, most IC *J–K* flip-flops have preset and clear inputs.

J–K Flip-Flop with Preset and Clear

Figure 7–35 shows the logic diagram and block symbol for an edge-triggered *J–K* flip-flop with *preset (PR)* and *clear (CLR)* inputs. This illustrates basically how these inputs work. They can be either active LOW or active HIGH. As you can see, they are connected directly into the latch portion of the flip-flop so that they override the effect of the synchronous inputs, *J, K,* and the clock.

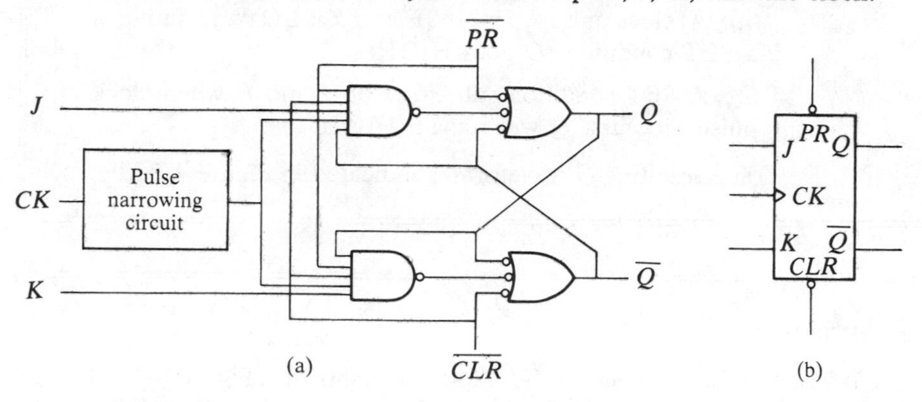

FIGURE 7–35 *J–K flip-flops with active LOW preset and clear.*

Example 7–7

The waveforms in Figure 7–36(a) are applied to the *J, K,* and clock inputs as indicated. Determine the *Q* output, assuming the flip-flop is initially RESET and that it is a negative edge-triggered type.

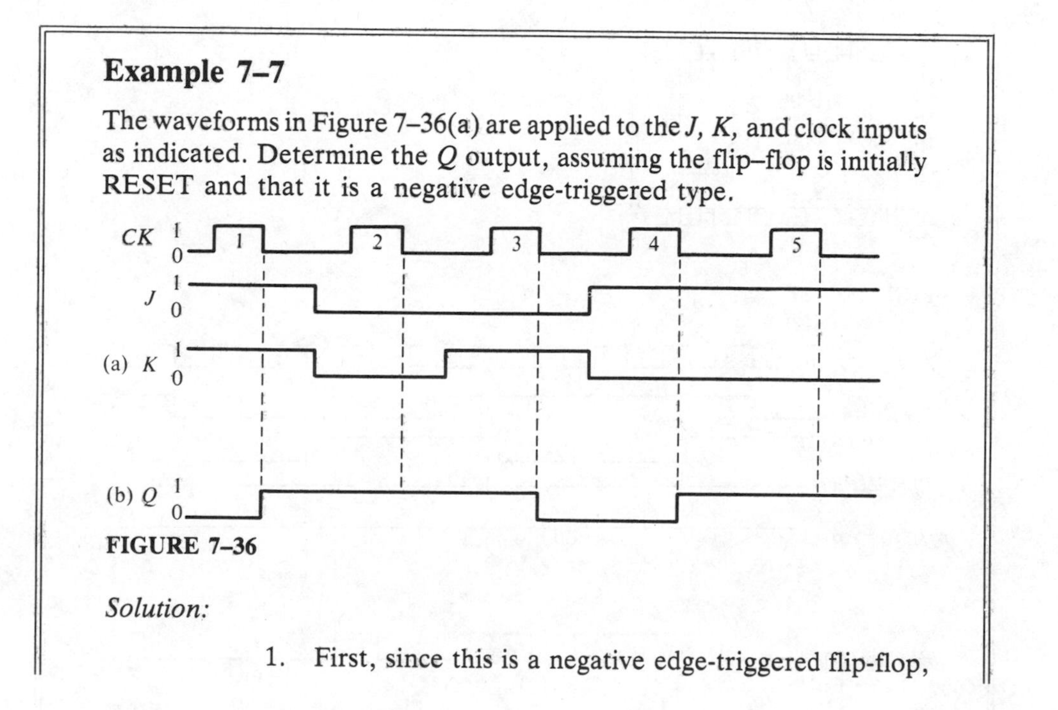

FIGURE 7–36

Solution:

1. First, since this is a negative edge-triggered flip-flop,

Example 7–7, continued

we know that the Q output will change only on the negative-going edge of the clock pulse.

2. At the first clock pulse both J and K are HIGH, and because this is a toggle condition, Q goes HIGH.

3. At clock pulse 2, a *no change* condition exists on the inputs, keeping Q at a HIGH level.

4. When clock pulse 3 occurs, J is LOW and K is HIGH, resulting in a RESET condition. Q goes LOW.

5. At clock pulse 4, J is HIGH and K is LOW, resulting in a SET condition, Q goes HIGH.

6. A SET condition still exists on J and K when clock pulse 5 occurs. Q will remain HIGH.

The resulting Q waveform is indicated in Figure 7–36(b).

Example 7–8

The 7476 is an IC containing two J–K flip-flops as shown in Figure 7–37. If the waveforms in Figure 7–38(a) are applied to the pins indicated, what is the Q output waveform?

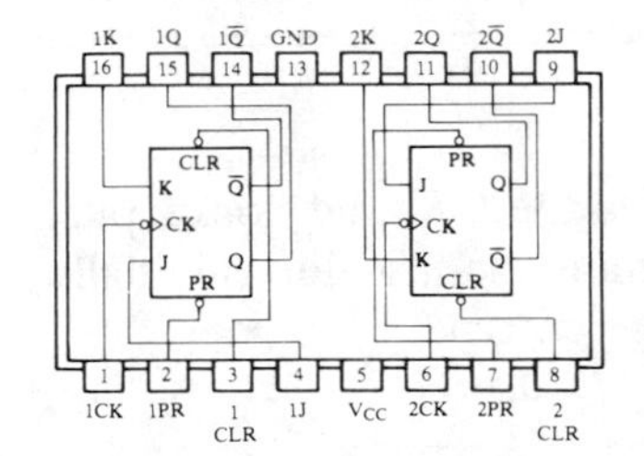

FIGURE 7–37

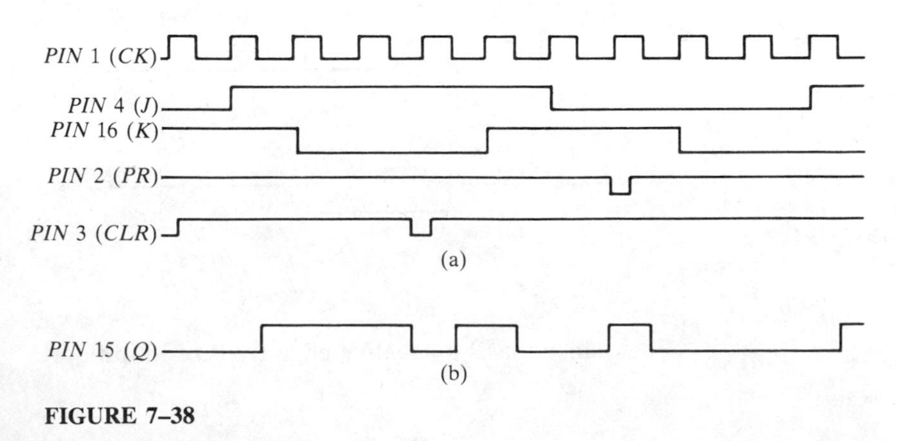

FIGURE 7–38

Solution:

The resulting Q output is shown in Figure 7–38(b)

Example 7–9

For the edge-triggered *J-K* flip-flop with *preset* and *clear* inputs in Figure 7–39(a), determine the Q output for the inputs shown in the timing diagram. Q is initially LOW.

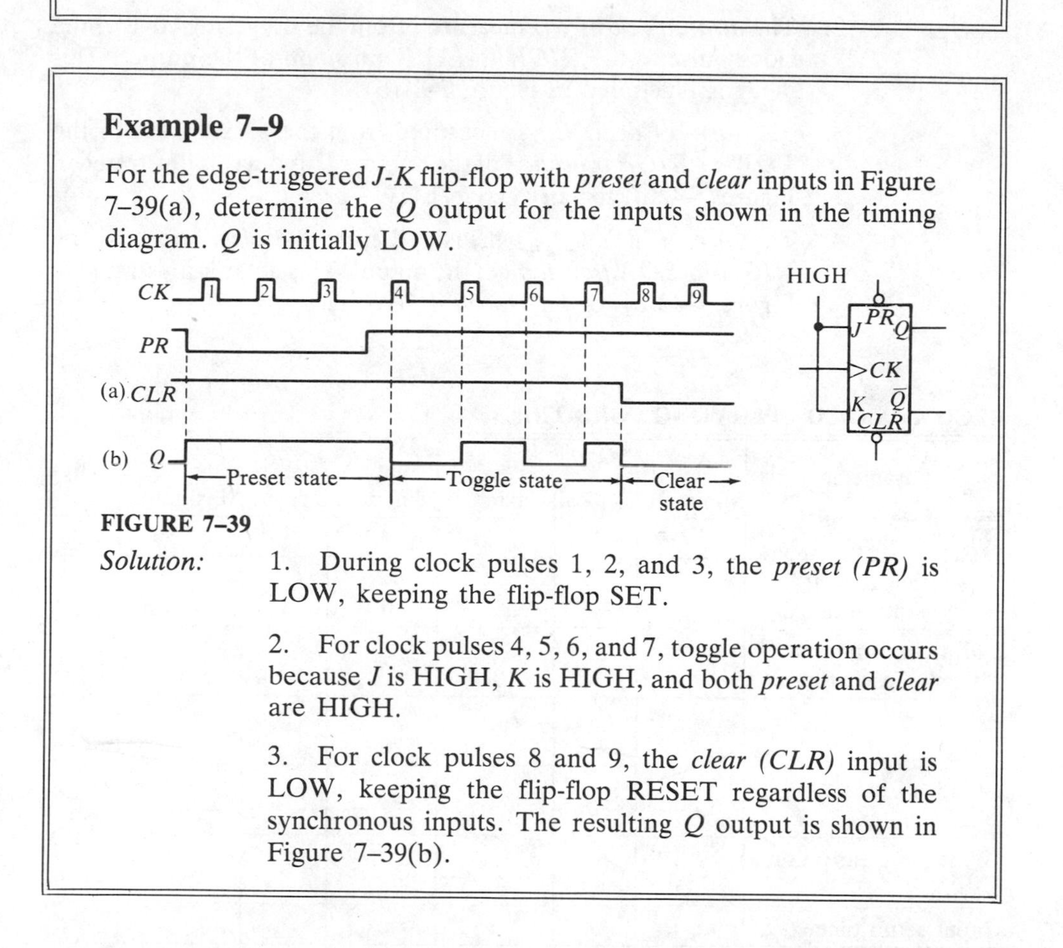

FIGURE 7–39

Solution:

1. During clock pulses 1, 2, and 3, the *preset (PR)* is LOW, keeping the flip-flop SET.

2. For clock pulses 4, 5, 6, and 7, toggle operation occurs because J is HIGH, K is HIGH, and both *preset* and *clear* are HIGH.

3. For clock pulses 8 and 9, the *clear (CLR)* input is LOW, keeping the flip-flop RESET regardless of the synchronous inputs. The resulting Q output is shown in Figure 7–39(b).

7–5 OPERATING CHARACTERISTICS

Several operating characteristics or parameters important in the application of flip-flops specify the performance, operating requirements, and operating limitations of the circuit. They are typically found in the data sheets for integrated circuits, and are applicable to all flip-flops regardless of the particular form of the circuit. Figure 7–40 is a typical flip-flop data sheet.

Propagation Delays

A propagation delay is the interval of time required after the input signal has been applied for the resulting output change to occur.

There are several categories of propagation delay that are important in the operation of a flip-flop.

1. The *turn-off delay* (t_{pLH}) measured from the triggering edge of the clock pulse to the *LOW-to-HIGH transition* of the output. This delay is illustrated in Figure 7–41(a).
2. The *turn-on delay* (t_{pHL}) measured from the triggering edge of the clock pulse to the *HIGH-to-LOW transition* of the output. This delay is illustrated in Figure 7–41(b).
3. The *turn-off delay* (t_{pLH}) measured from the *preset* input to the *LOW-to-HIGH transition* of the output. This delay is illustrated in Figure 7–42(a) for an "active LOW" *preset*.
4. The *turn-on delay* (t_{pHL}) measured from the *clear* input to the *HIGH-to-LOW transition* of the output. This delay is illustrated in Figure 7–42(b) for an "active LOW" *clear*.

RECOMMENDED OPERATING CONDITIONS

Parameter	5476			7476			Units
	Minimum	Typical	Maximum	Minimum	Typical	Maximum	
Supply voltage (V_{CC})	4.5	5.0	5.5	4.75	5.0	5.25	Volts
Operating free-air temperature range	−55	25	125	0	25	70	°C
Normalized fan out from each output			10			10	UL
Width of clock pulse $t_{p(clock)}$	20			20			ns
Width of preset pulse, $t_{p(preset)}$	25			25			ns
Width of clear pulse, $t_{p(clear)}$	25			25			ns
Input setup time, t_{setup}	$\geq t_{p(clock)}$			$\geq t_{p(clock)}$			
Input hold time, t_{hold}	0			0			

ELECTRICAL CHARACTERISTICS OVER OPERATING TEMPERATURE RANGE (unless otherwise noted)

Symbol	Parameter	Limits			Units	Test Conditions (note 1)
		Minimum	Typical (note 2)	Maximum		
V_{IH}	Input HIGH voltage	2.0			Volts	Guaranteed input HIGH
V_{IL}	Input LOW voltage			0.8	Volts	Guaranteed input LOW
V_{OH}	Output HIGH voltage	2.4	3.5		Volts	V_{CC} = Min., $I_{OH} = -0.4$ mA

ELECTRICAL CHARACTERISTICS OVER OPERATING TEMPERATURE RANGE (unless otherwise noted)

V_{OL}	Output LOW voltage		0.22	0.4	Volts	V_{CC} = Min., I_{OL} = 16 mA
I_{IH}	Input HIGH current at J or K			40	μA	V_{CC} = Max., V_{IN} = 2.4V
				1.0	mA	V_{CC} = Max., V_{IN} = 5.5 V
	Input HIGH current at clear, preset, or clock			80	μA	V_{CC} = Max., V_{IN} = 2.4 V
				1.0	mA	V_{CC} = Max., V_{IN} = 5.5 V
I_{IL}	Input LOW current at J or K			−1.6	mA	V_{CC} = Max., V_{IN} = 0.4V
	Input LOW current at clear, preset, or clock			−3.2	mA	V_{CC} = Max., V_{IN} = 0.4 V
I_{OS}	Output short circuit current (note 3)	−20		−57	mA	5476 V_{CC} = Max.
		−18		−57	mA	7476 V_{IN} = 0 V
I_{CC}	Supply current		20	40	mA	V_{CC} = Max.

SWITCHING CHARACTERISTICS (T_A = 25°C)

Symbol	Parameter	Limits			Units	Test Conditions
		Minimum	Typical	Maximum		
f_{max}	Maximum clock frequency	15	20		MHz	
t_{PLH}	Turn-off delay clear or preset to output		16	25	ns	V_{CC} = 5.0 V
t_{PHL}	Turn-on delay clear or preset to output		25	40	ns	C_L = 15 pF
t_{PLH}	Turn-off delay clock to output	10	16	25	ns	R_L = 400 Ω
t_{PHL}	Turn-on delay clock to output	10	25	40	ns	

NOTES:
1. For conditions shown as Min. or Max., use the appropriate value specified under recommended operating conditions for the applicable device type.
2. Typical limits are at V_{CC} = 5.0 V, 25°C.

FIGURE 7–40 *Typical J–K flip-flop data sheet.*

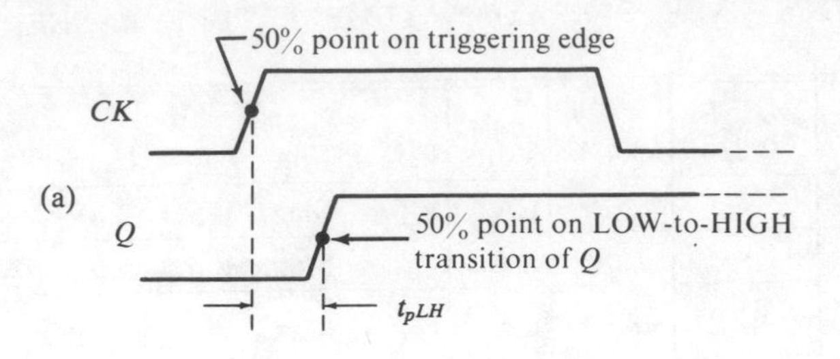

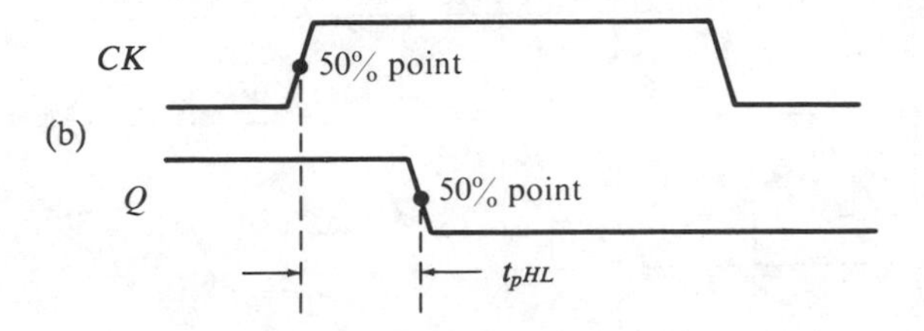

FIGURE 7–41 *Propagation delays, clock to output.*

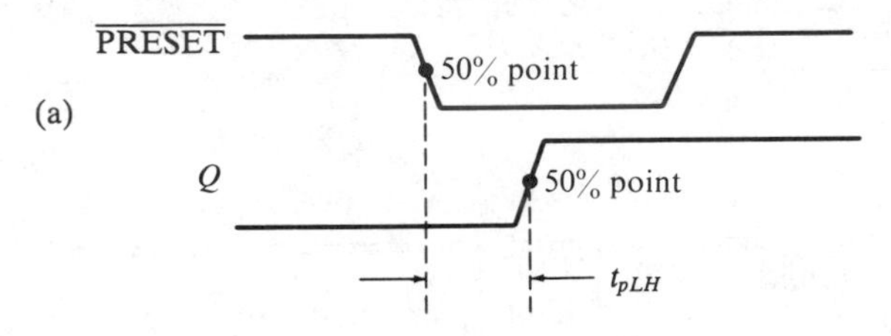

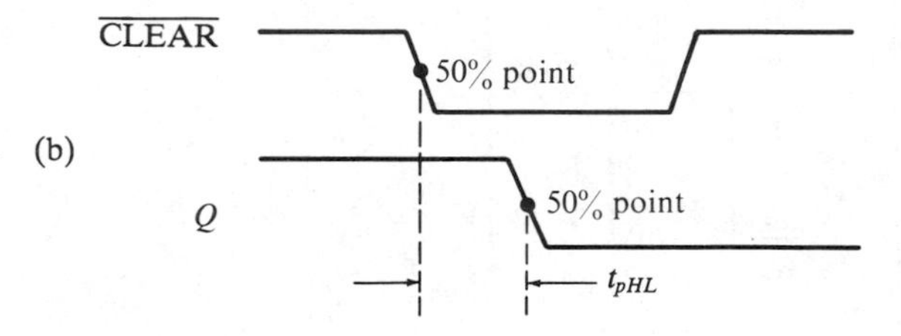

FIGURE 7–42 *Propagation delays, preset and clear to output.*

Set-Up Time

The *set-up time* is the minimum interval required for the control levels to be on the inputs (J and K, or S and R, or D) *prior* to the triggering edge of the clock pulse in order for the levels to be reliably clocked into the flip-flop. This is illustrated in Figure 7–43 for a D flip-flop.

Hold Time

The *hold time* is the minimum interval required for the control levels to remain on the inputs *after* the triggering edge of the clock pulse in order for the levels

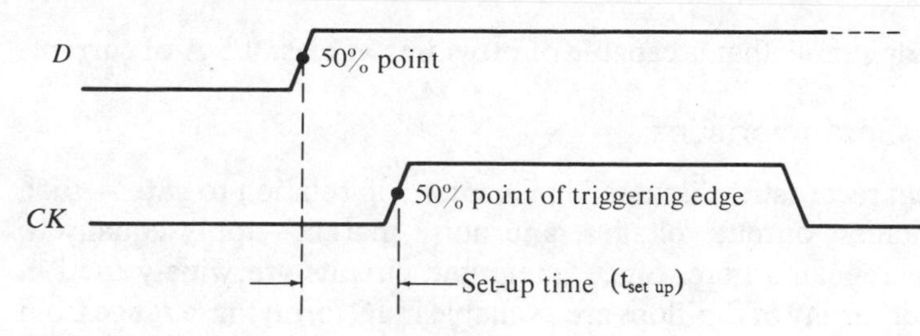

FIGURE 7–43 *Set-up time (t_{setup}).*

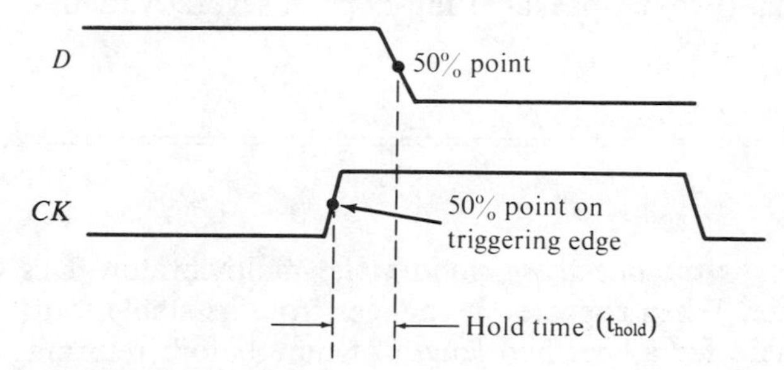

FIGURE 7–44 *Hold time (t_{hold}).*

to be reliably clocked into the flip-flop. This is illustrated in Figure 7–44 for a *D* flip-flop.

Maximum Clock Frequency

The *maximum clock frequency* (f_{max}) is the highest rate at which a flip-flop can be reliably triggered. At clock frequencies above the maximum, the flip-flop would be unable to respond quickly enough and its operation would be impaired.

Power Dissipation

The power dissipation of a flip-flop is the total power consumption of the device. For example, if the flip-flop operates on a +5 V dc source and draws 50 mA of current, the power dissipation is

$$P = V_{CC} \times I_{CC} = 5\text{ V} \times 50\text{ mA} = 250\text{ mW}$$

The power dissipation is very important in most applications where the capacity of the dc supply is a concern. As an example, let us assume we have a digital system that requires a total of ten flip-flops, and each flip-flop dissipates 250 mW of power. The total power requirement is

$$P_{TOT} = 10 \times 250\text{ mW} = 2500\text{ mW} = 2.5\text{ W}$$

This tells us the type of dc supply that is required as far as output capacity is concerned. If the flip-flops operate on +5 V dc, then the amount of current that the supply must provide is as follows:

$$I = \frac{2.5\text{ W}}{5\text{V}} = 0.5\text{ A}$$

We must use a +5 V dc supply that is capable of providing at least 0.5 A of current.

Other Characteristics

Many characteristics discussed in Chapter 3 in relation to gates—such as fan-out, input voltages, output voltages, and noise margin—apply equally to flip-flops and are not repeated here. Since integrated circuits are widely used in digital systems today, a variety of flip-flops are available in IC form; these range from single (one-per-package) flip-flops with various types of gated inputs and different operating characteristics to dual (two-per-package) flip-flops in several varieties.

7–6 ONE-SHOTS

A second type of multivibrator is the *one-shot (monostable)* multivibrator. This device has only one stable state. When triggered it changes from its stable to its unstable state and remains there for a specified length of time before returning automatically to its stable state.

Figure 7–45 shows a basic one-shot circuit composed of a logic gate and an inverter.

When a pulse is applied to the *trigger* input, the output of gate G_1 goes LOW. This HIGH-to-LOW transition is coupled through the capacitor to the input to inverter, G_2. The apparent LOW on G_2 makes its output go HIGH. This HIGH is connected back into G_1, keeping its output LOW. Up to this point, the trigger pulse has caused the output of the one-shot, Q, to go HIGH.

The capacitor immediately begins to charge through R toward the high voltage level. The rate at which it charges is determined by the RC time constant. When the capacitor charges to a certain level which appears as a HIGH to G_2, the output goes back LOW.

To summarize, the output of inverter G_2 goes high in response to the trigger input. It remains high for a time set by the RC time constant. At the end of this time it goes LOW. So, a single narrow trigger pulse produces a *single* output pulse whose time duration is controlled by the RC time constant. This operation is illustrated in Figure 7–45. A typical one-shot logic symbol is shown in Figure 7–46(a) and the same symbol with an external R and C is shown in Figure 7–46(b).

Integrated Circuit One-Shots

The two basic types of IC one-shots are non-retriggerable and retriggerable. A non-retriggerable type will not respond to any additional trigger pulses once it is triggered into its unstable state (fired) until it returns to its stable state (times out). In other words, the period of the trigger pulses must be greater than the time the one-shot remains in its unstable state. The time that it remains in its unstable state is the *pulse width* of the output.

Figure 7–47 shows the one-shot being triggered at intervals greater than its pulse width and at intervals less than the pulse width. Notice that in the second case, the additional pulses are ignored.

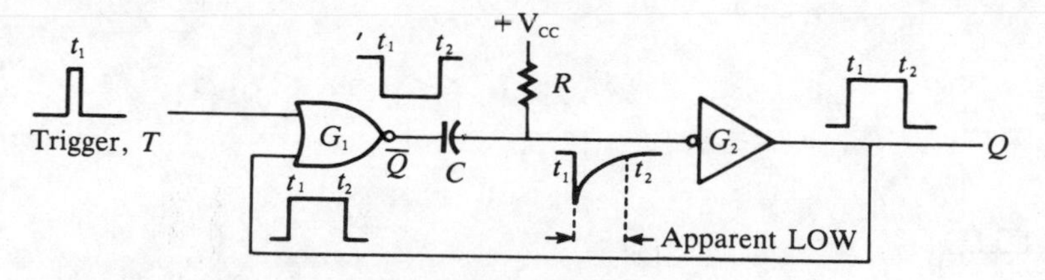

FIGURE 7–45 *A simple one-shot circuit.*

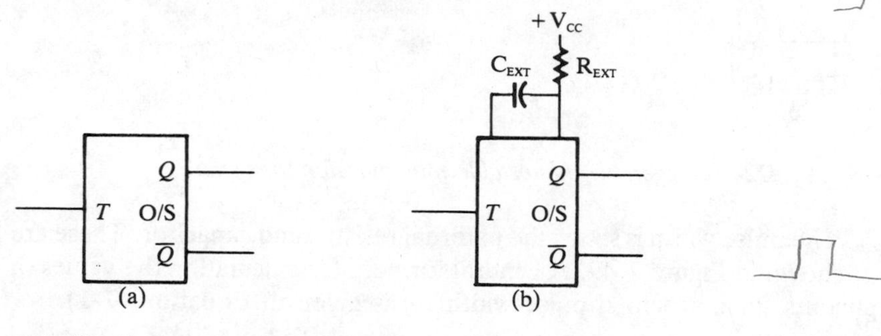

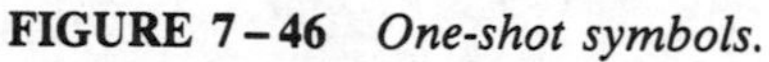
FIGURE 7–46 *One-shot symbols.*

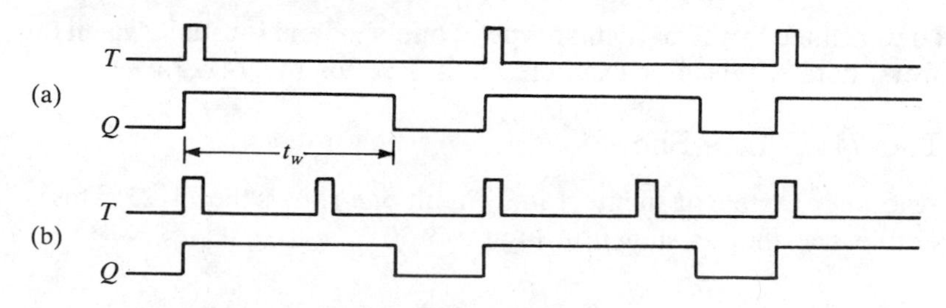

FIGURE 7–47 *Nonretriggerable one-shot action.*

A retriggerable one-shot can be triggered before it times out. The result of retriggering is an extension of the pulse width as illustrated in Figure 7–48.

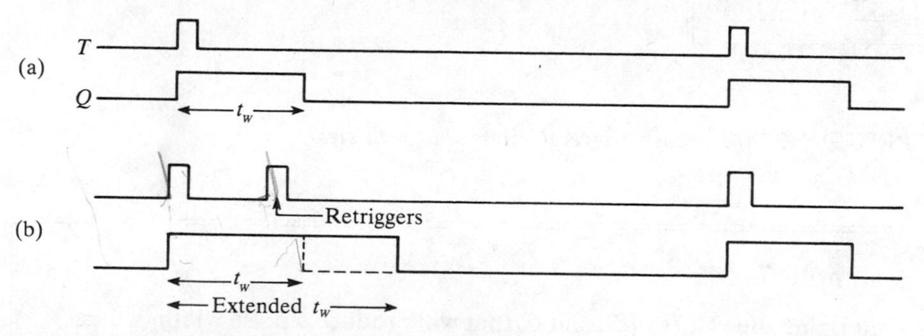

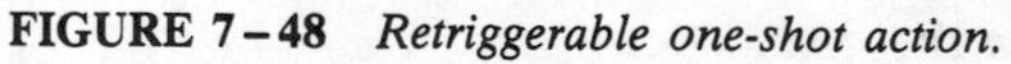
FIGURE 7–48 *Retriggerable one-shot action.*

The 74122 One-Shot

The 74122 is an example of an IC one-shot with a gated trigger and a clear input. It has provisions for external R and C as shown in Figure 7–49.

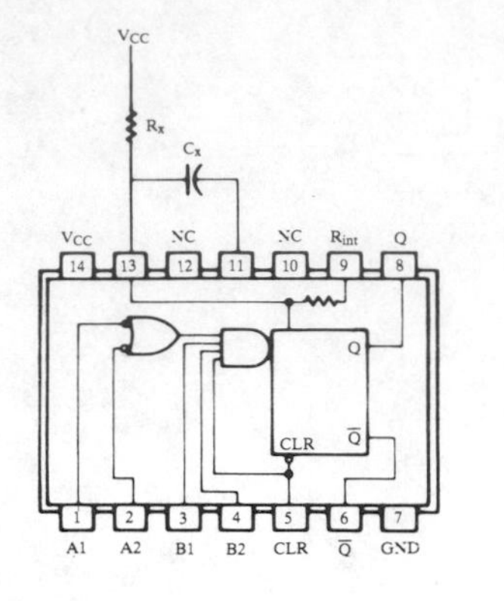

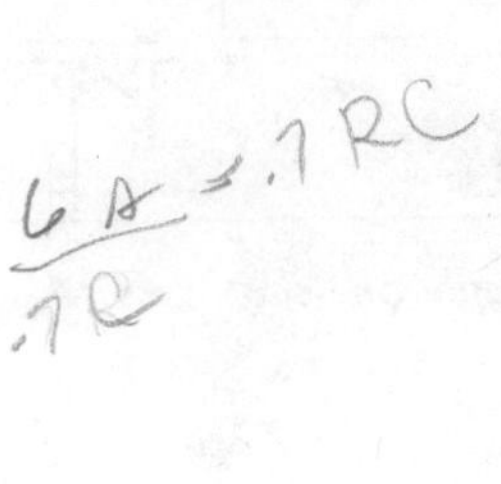

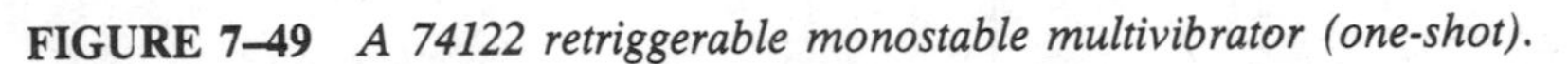

FIGURE 7–49 *A 74122 retriggerable monostable multivibrator (one-shot).*

The pulse width is set by the external resistor and capacitor. These are connected as shown in Figure 7–49. A general formula for calculating the values of these components for a specified pulse width(t_W) is given in Equation (7–1).

$$t_W = KR_xC_x(1 + 0.7/R_x) \qquad \textbf{(7–1)}$$

K is a constant determined by the particular type of one-shot and is usually given on the manufacturers' data sheets. For example, K is 0.32 for the 74122.

The 74123 One-Shot

Another example of an integrated circuit one-shot is the 74123. This is a dual retriggerable one-shot as shown in Figure 7–50.

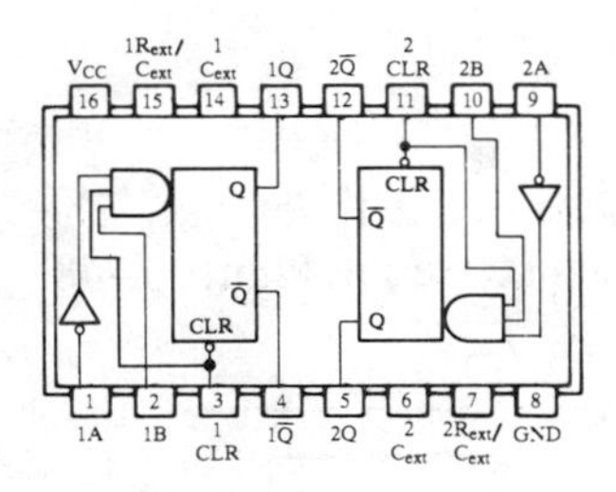

FIGURE 7–50 *A 74123 dual retriggerable one shot with clear.*

Example 7–10

Determine the values of R_x and C_x that will produce a pulse width of 1 μs when connected to a 74123. Show the connections.

Solution:

The manufacturer's data sheet gives $K = 0.28$ for this device. R_x is in kΩ. C_x is in pF. t_W is in ns.

$$t_W = KR_xC_x(1 + 0.7/R_x)$$

Assume a value of $C_x = 500$ pF and then solve for R_x.

$$t_W = KR_xC_x + 0.7\ (KR_xC_x)/R_x$$
$$t_W = KR_xC_x + 0.7KC_x$$
$$R_x = \frac{t_W - 0.7KC_x}{KC_x} = \frac{t_W}{KC_x} - 0.7$$
$$= \frac{1000\text{ ns}}{(0.28)500\text{pf}} - 0.7$$
$$= 7.14\text{ k}\Omega$$

Connections are shown in Figure 7–51.

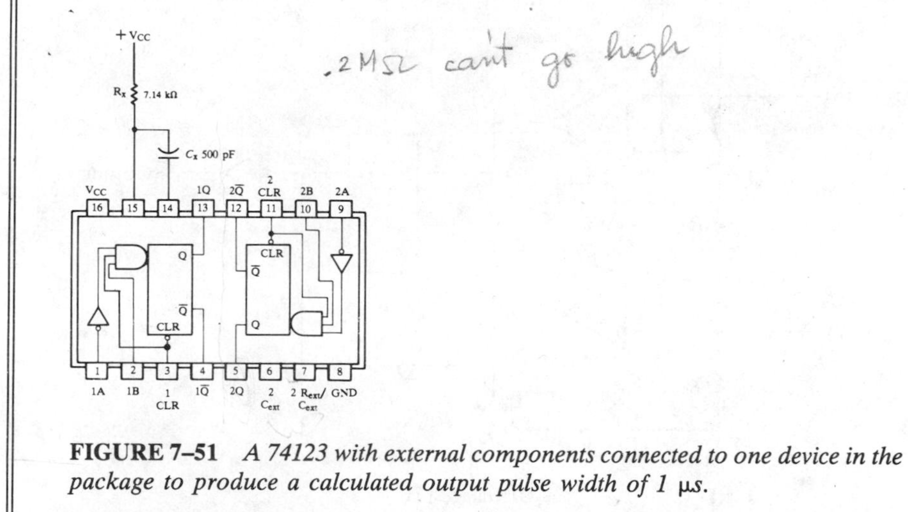

FIGURE 7–51 *A 74123 with external components connected to one device in the package to produce a calculated output pulse width of 1 μs.*

ASTABLE MULTIVIBRATORS 7–7

A third type of multivibrator is the *astable* (free-running) type. This device has no stable states but it switches back and forth between two unstable states. The astable multivibrator is used as an oscillator to provide clock signals for timing purposes.

Figure 7–52 shows a basic astable using two inverters. Notice the capacitive coupling from the output of each inverter to the input of the other. The operation is similar to that of the one-shot in its unstable state but the two capacitive coupling networks prevent either inverter from having a stable state. If the circuit is designed properly, it will start oscillating on its own and requires no initial input trigger.

The 555 Timer

The 555 timer is an integrated circuit that can be used as an astable or monostable multivibrator and for many other applications.

Figure 7–53 shows a functional diagram of a 555 timer.

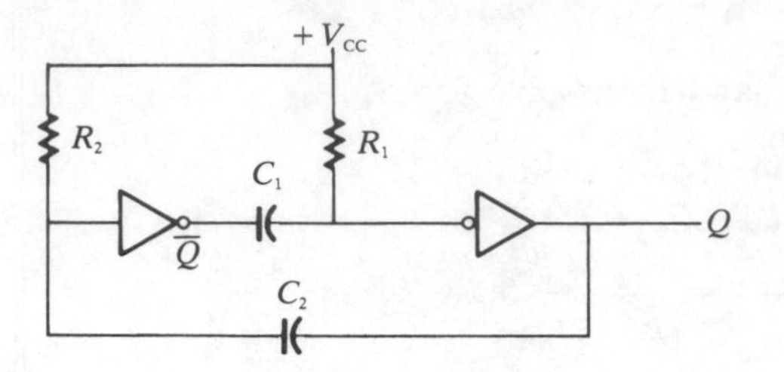

FIGURE 7–52 *An example of an astable multivibrator.*

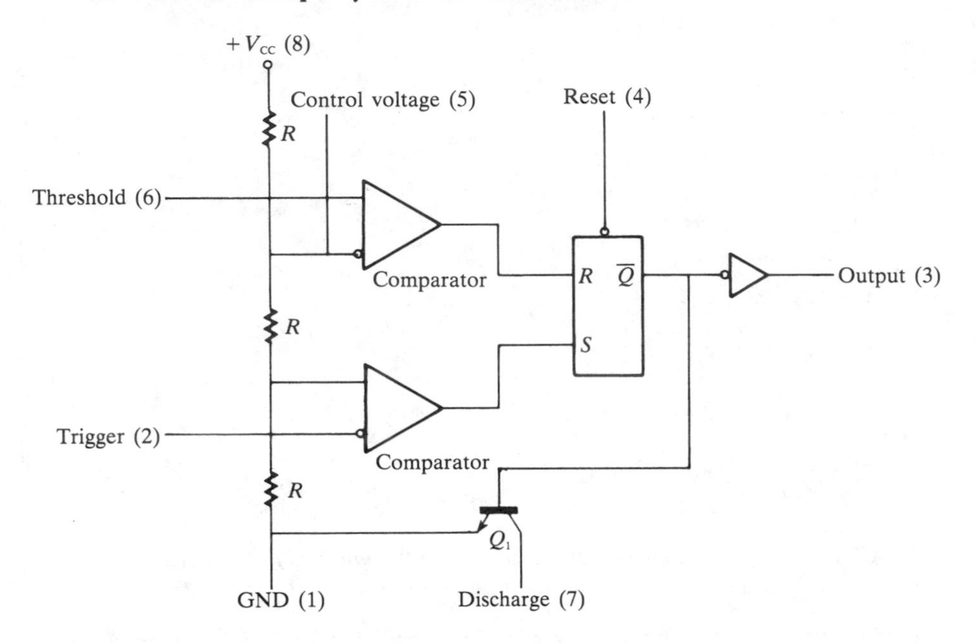

FIGURE 7–53 *A 555 timer.*

The *threshold* and *trigger* levels are normally (2/3) V_{CC} and (1/3) V_{CC} respectively. These levels can be altered with external connections to the *control voltage* terminal. When the *trigger* input goes below the trigger level, the internal flip-flop is SET and the output goes HIGH. When the *threshold* input goes above the threshold level, the flip-flop is RESET and the output goes LOW.

The RESET inputs can override the other inputs. Here is what happens when the RESET is LOW. The flip-flop is RESET causing the output to go LOW. This turns on Q_1 which provides a low impedance path from the *discharge* terminal to ground.

Some IC Timer Applications

Astable Operation A 555 timer is shown in Figure 7–54 connected as a *free-running multivibrator*. Notice that the trigger input and threshold inputs are connected together. The capacitor C charges through R_1 and R_2 and discharges through R_2. Therefore, the frequency and duty cycle of the output waveform can be set by selecting proper values for these two resistors. Figure 7–55 shows typical waveforms produced by this device. It can be used as a pulse signal source in various applications.

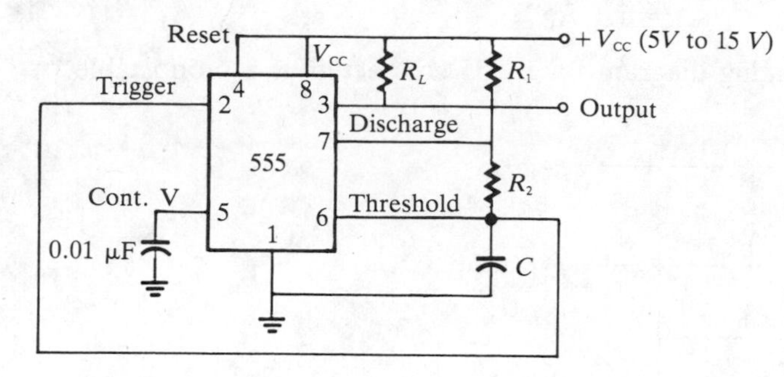

FIGURE 7–54 *A 555 timer connected for astable operation.*

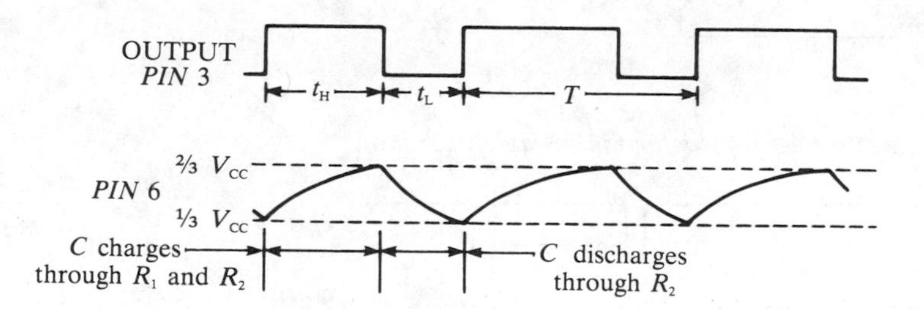

FIGURE 7–55 *Astable waveforms.*

The following formula can be used to calculate the frequency of the 555 timer for astable operation.

$$f = \frac{1}{T} = \frac{1.44}{(R_1 + 2R_2)C} \tag{7–2}$$

The time that the output is in the HIGH state is

$$t_H = 0.693(R_1 + R_2)C \tag{7–3}$$

The time that the output is in the LOW state is

$$t_L = 0.693R_2C \tag{7–4}$$

The period of the output waveform is the sum of the HIGH time and the LOW time.

$$T = t_H + t_L = 0.693(R_1 + 2R_2)C \tag{7–5}$$

Finally, the duty cycle is developed as follows:

$$\text{duty cycle} = \frac{t_H}{T} = \frac{t_H}{t_H + t_L} = \frac{R_1 + R_2}{R_1 + 2R_2} \tag{7–6}$$

Monostable Operation The 555 timer can also be used as a one-shot when connected as shown in Figure 7–56.

One-shot action is initiated by a negative-going input pulse on the *trigger* input. Once initiated, the timing interval will complete even if retriggering occurs before the end of the output pulse. The output pulse width can be determined by the following equation.

$$t_W = 1.1\ R_1C \tag{7–7}$$

Figure 7–57 is a typical timing diagram for a 555 connected as a monostable.

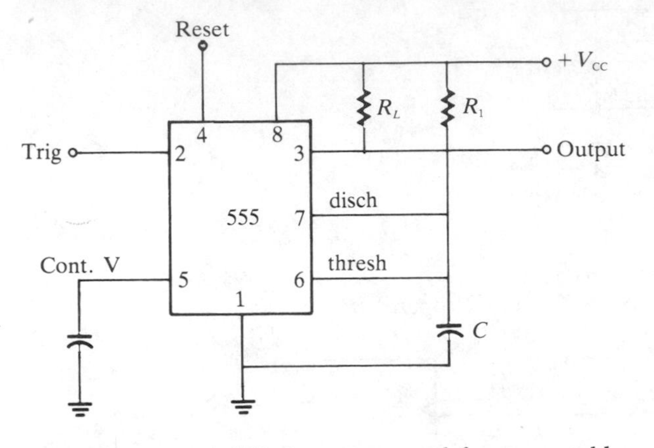

FIGURE 7–56 *A 555 timer connected for monostable operation.*

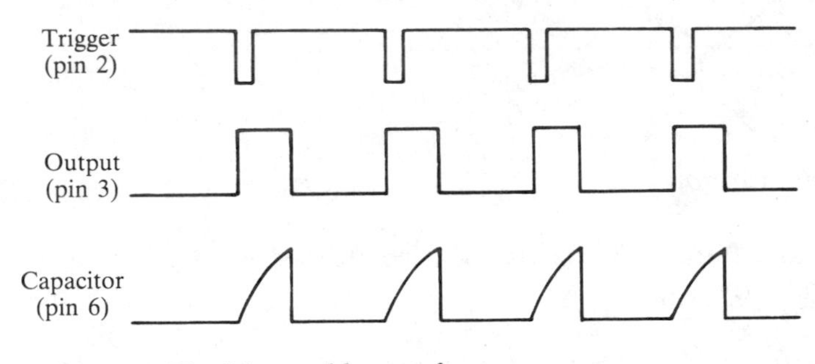

FIGURE 7–57 *Monostable waveforms.*

Example 7–11

A 555 timer is to be used as an astable multivibrator. Determine the values for R_1 and R_2 if the output frequency is to be 10 kHz with a 50 percent duty cycle. The capacitor C is 0.1 μF.

Solution:

For a 50 percent duty cycle, $t_H = t_L$; therefore $R_1 + R_2 \cong R_2$. To satisfy this requirement, $R_2 >> R_1$. So, make $R_1 = 10\ \Omega$.

$$f = \frac{1.44}{(R_1 + 2R_2)C} \cong \frac{1.44}{2R_2C}$$

$$R_2 = \frac{1.44}{2fC} = \frac{1.44}{2(10\text{kHz})(0.1\ \mu F)}$$

$$= 720\ \Omega$$

Problems

Section 7–1

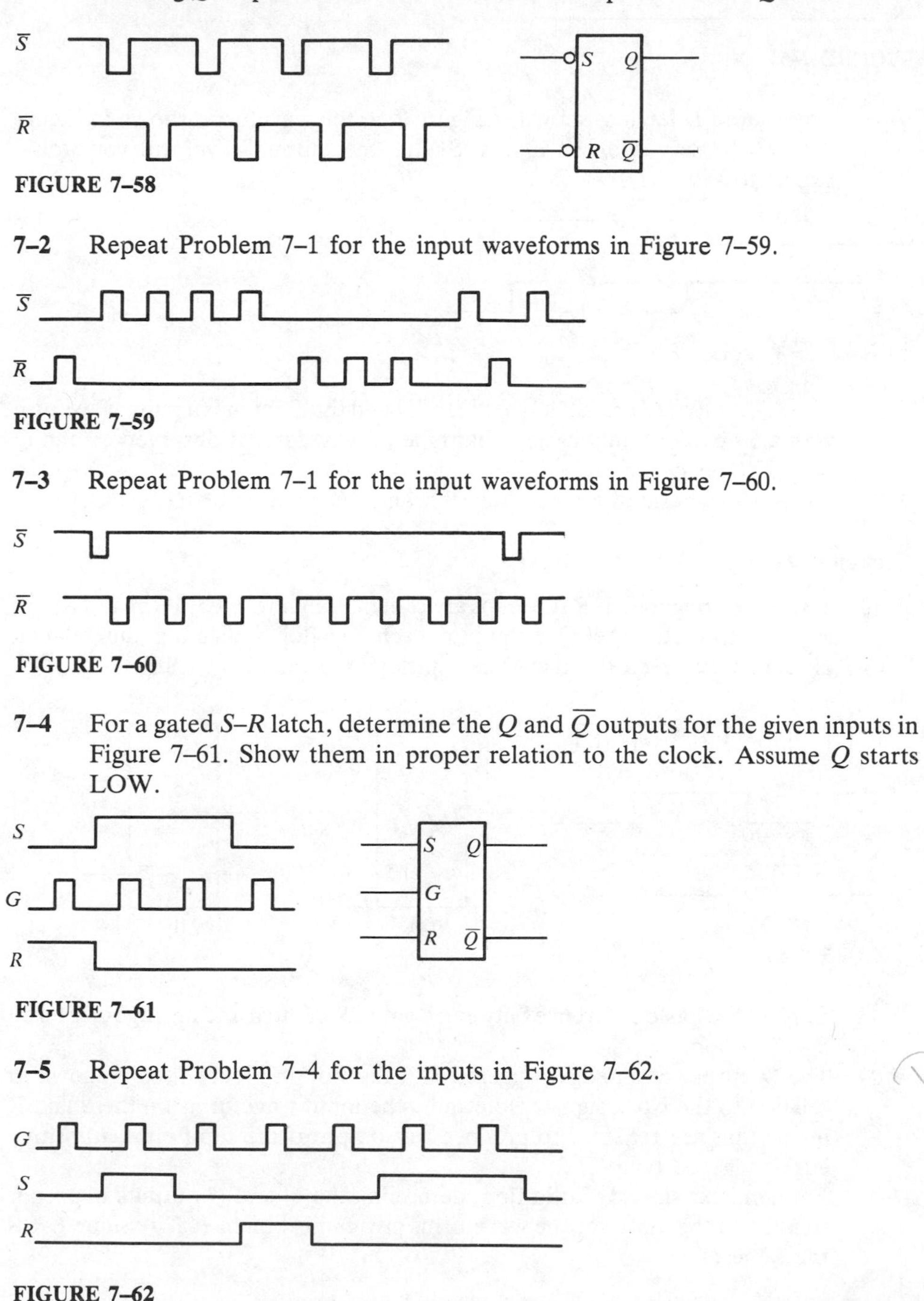

7–1 If the waveforms in Figure 7–58 are applied to an $\overline{S}$–$\overline{R}$ latch, sketch the resulting Q output waveform in relation to the inputs. Assume Q starts LOW.

FIGURE 7–58

7–2 Repeat Problem 7–1 for the input waveforms in Figure 7–59.

FIGURE 7–59

7–3 Repeat Problem 7–1 for the input waveforms in Figure 7–60.

FIGURE 7–60

7–4 For a gated S–R latch, determine the Q and $\overline{Q}$ outputs for the given inputs in Figure 7–61. Show them in proper relation to the clock. Assume Q starts LOW.

FIGURE 7–61

7–5 Repeat Problem 7–4 for the inputs in Figure 7–62.

FIGURE 7–62

7–6 Repeat Problem 7–4 for the inputs in Figure 7–63.

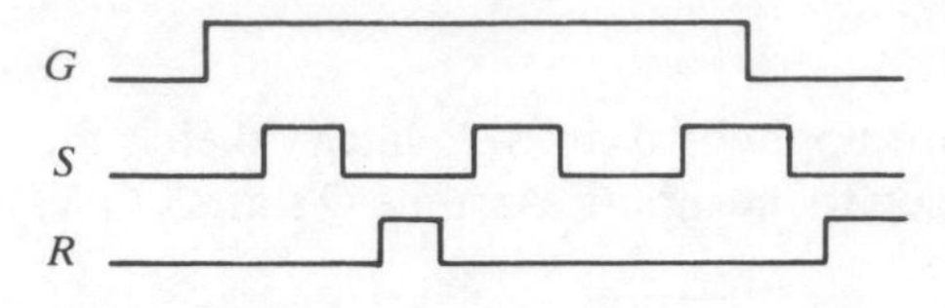

FIGURE 7–63

7–7 For a gated D latch as shown in Figure 7–9, the waveforms shown in Figure 7–64 are observed on its inputs. Sketch the output waveform you would expect to see at Q.

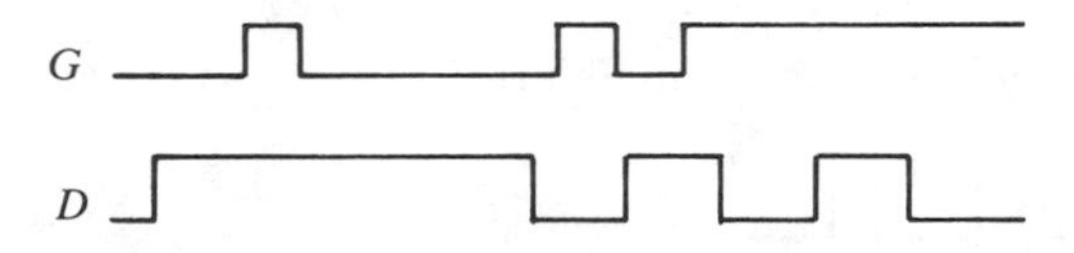

FIGURE 7–64

7–8 If the gate input of a D latch is held HIGH and the D input is a pulse waveform with a 50 percent duty cycle, what type of waveform is observed on the Q output? The $\overline{Q}$ output?

7–9 Explain the basic difference between an S–R latch and a D latch.

Section 7–2

7–10 Two edge-triggered S–R flip-flops are shown in Figure 7–65. If the inputs are as shown, sketch the Q output of each flip-flop indicating the relative difference between the two. The flip-flops are initially RESET.

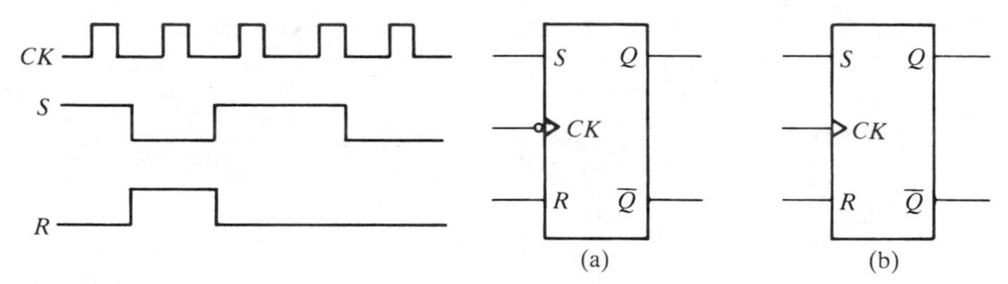

FIGURE 7–65

7–11 Explain the basic difference between a gated S–R latch and an edge-triggered S–R flip-flop.

7–12 The Q output of an edge-triggered S–R flip-flop in Figure 7–66 is shown in relation to the clock signal. Determine the input waveforms on the S and R inputs that are required to produce this output if the flip-flop is a positive edge-triggered type.

7–13 For a master-slave S–R flip-flop, determine the Q and $\overline{Q}$ outputs in proper relation to the clock for the waveforms given in Problem 7–5. Assume CK is the same as G.

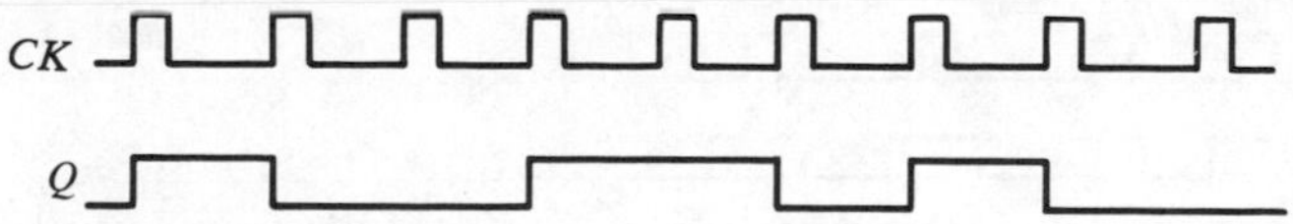

CK

Q

FIGURE 7–66

Section 7–3

7–14 Sketch the Q output for a D flip-flop, with the inputs as shown in Figure 7–67. Assume positive edge-triggering and Q initially LOW.

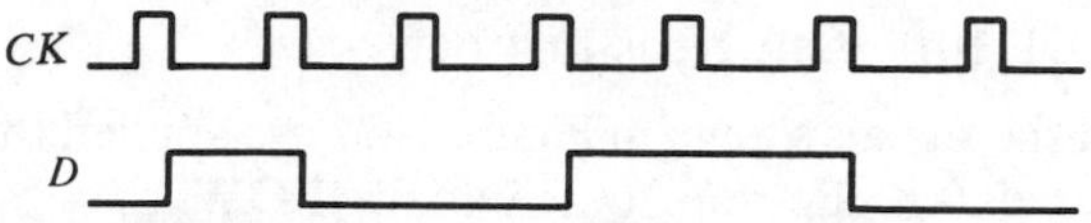

CK

D

FIGURE 7–67

7–15 Repeat Problem 7–14 for the inputs in Figure 7–68.

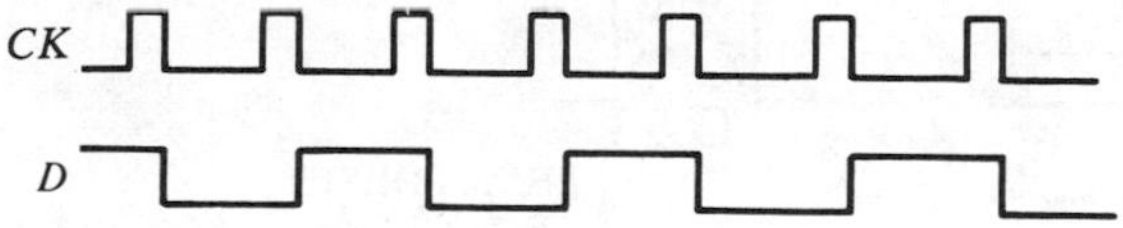

CK

D

FIGURE 7–68

Section 7–4

7–16 For a positive edge-triggered J–K flip-flop with inputs as shown in Figure 7–69, determine the Q output. Assume Q starts LOW.

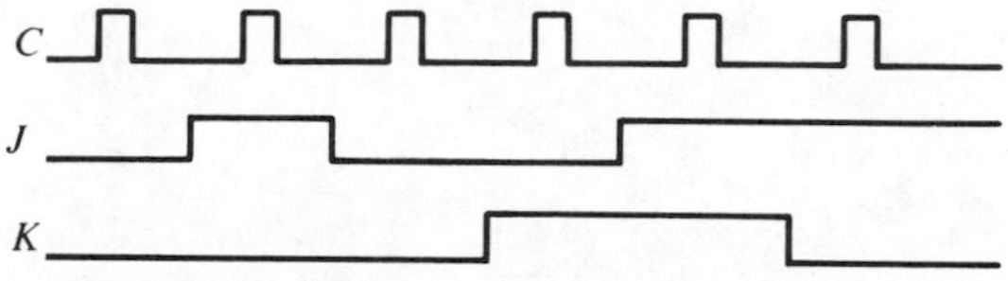

C

J

K

FIGURE 7–69

7–17 Repeat Problem 7–16 for the inputs in Figure 7–70.

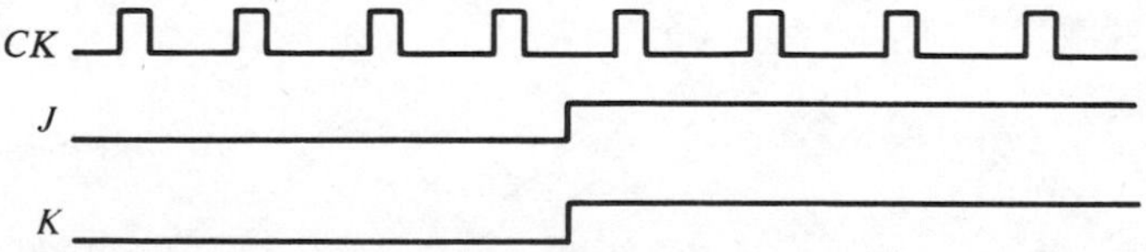

CK

J

K

FIGURE 7–70

7–18 For a master-slave J–K flip-flop with the inputs in Figure 7–71, sketch the Q output waveform. Assume Q is initially LOW. Assume the flip-flop accepts data at the positive-going edge of the clock pulse.

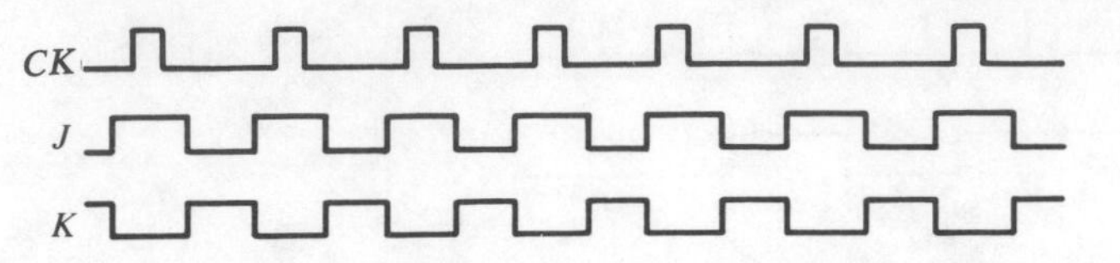

FIGURE 7–71

7–19 The following serial data stream is to be generated using a J–K positive edge-triggered flip-flop. Determine the inputs required. The data are to be in NRZ format.

1001011000110100001111101 Left bit is first out.

7–20 Determine the Q waveform if the signals shown in Figure 7–72 are applied to the inputs of the edge-triggered J–K flip-flop. Q is initially LOW.

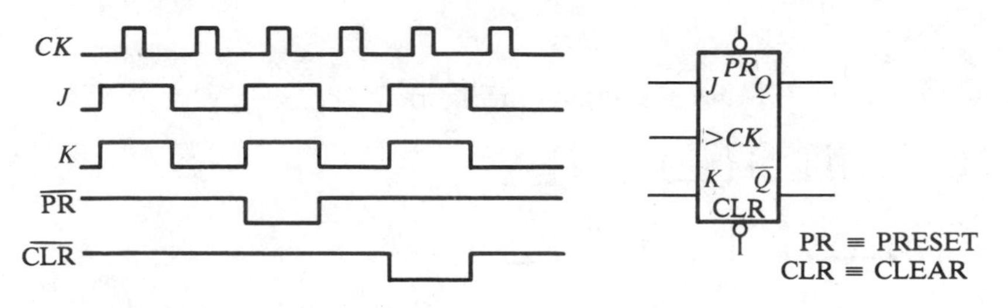

FIGURE 7–72

7–21 The following serial data are applied to the flip-flop as indicated in Figure 7–73. Determine the resulting serial data that appear on the Q output. There is one clock pulse for each bit time. Assume Q is initially 0.

J_1: 1010011
J_2: 0111010
J_3: 1111000
K_1: 0001110
K_2: 1101100
K_3: 1010101

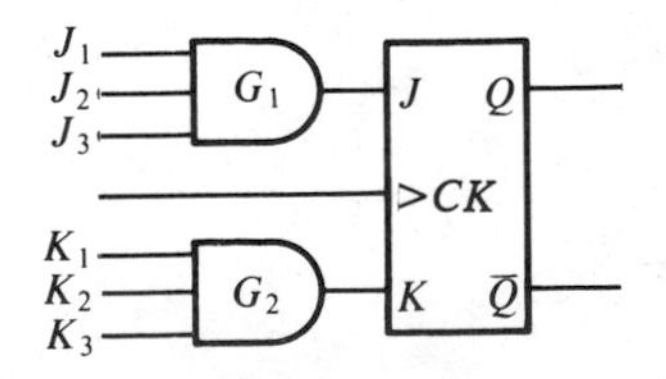

FIGURE 7–73

7–22 The 7472 is an AND-gated master-slave J–K flip-flop with a pin connection diagram shown in Figure 7–74(a). Complete the timing diagram in Figure 7–74(b) by sketching the Q output (which is initially LOW).

7–23 Sketch the Q output of flip-flop B in Figure 7–75 in proper relation to the clock. The flip-flops are initially RESET.

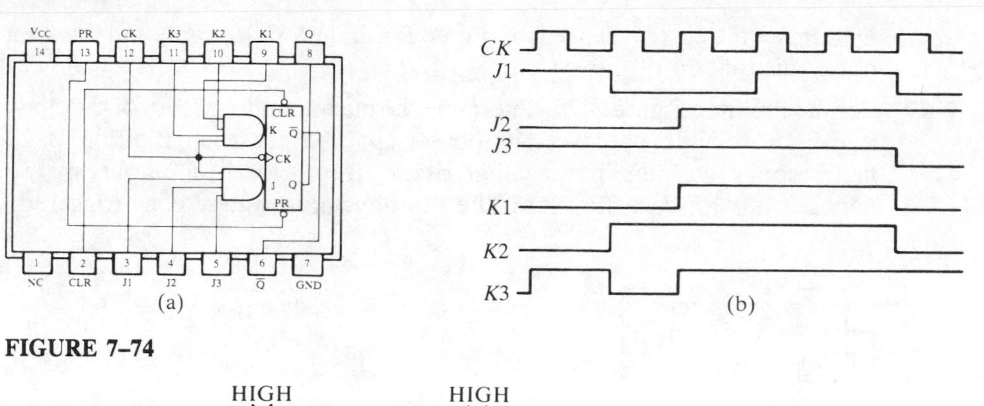

FIGURE 7–74

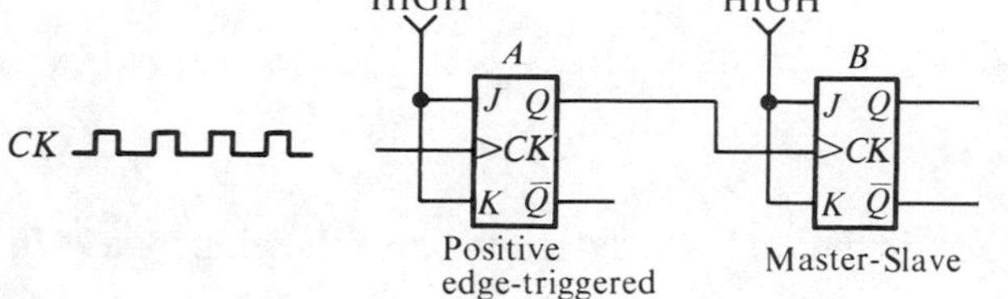

FIGURE 7–75

7–24 Sketch the Q output of flip-flop B in Figure 7–76 in proper relation to the clock. The flip-flops are initially RESET.

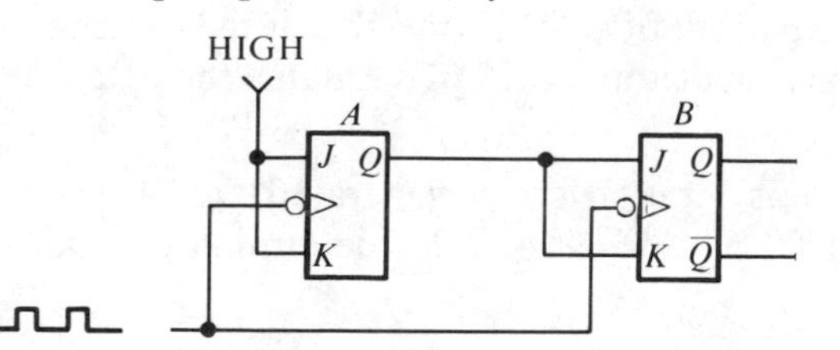

FIGURE 7–76

7–25 The flip-flop in Figure 7–77 is initially RESET. Show the relation between the Q output and the clock pulse if the propagation delay is 8 nanosecond. The flip-flop is initially RESET.

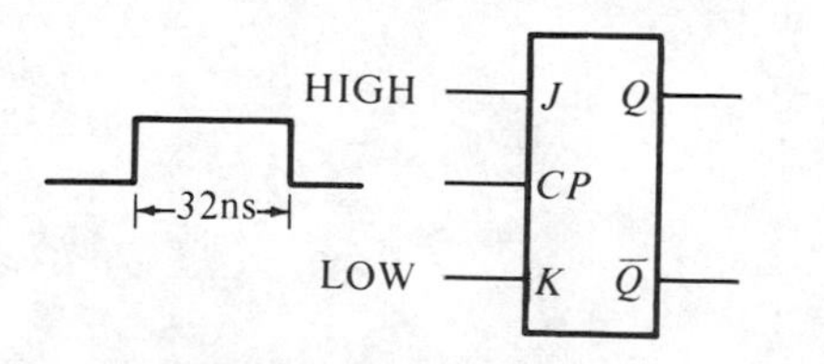

FIGURE 7–77

Section 7–5

7–26 The dc current required by a particular flip-flop that operates on +5 V dc is found to be 10 mA. A certain digital system uses 15 of these flip-flops.

Determine the current capacity required for the +5 V dc supply and the total power dissipation of the system.

7–27 For the circuit in Figure 7–78, determine the maximum frequency of the clock signal for reliable operation if the set-up time for each flip-flop is 20 nanoseconds and the propagation delays from clock to output are 50 nanoseconds for each flip-flop. The flip-flops are positive edge-triggered.

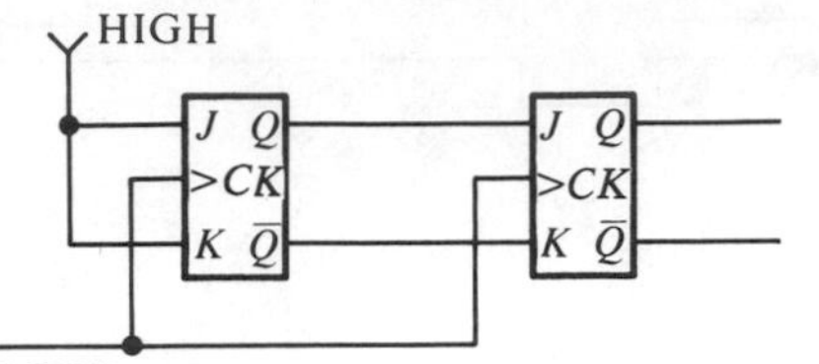

FIGURE 7–78

7–28 From the sample data sheet in Figure 7–40, determine the following for a 7476 flip-flop:
(a) Maximum dc supply voltage
(b) Maximum operating temperature
(c) The number of other TTL inputs that the flip-flop output can drive.
(d) Typical power dissipated with V_{CC} = 5.25 V.

7–29 Determine the pulse width of a 74123 one-shot if the external resistor is 3.3 kΩ, and the external capacitor is 2000 pF. Assume the data sheet gives K = 0.28.

7–30 An output pulse of 5 μs duration is to be generated by a 74123 one-shot. Using a capacitor of 10,000 pF and K = 0.28, determine the value of external resistance required.

7–31 Explain the difference in operating characteristics of a monostable multivibrator and an astable multivibrator.

7–32 A 555 timer is configured to run as an astable multivibrator as shown in Figure 7–79. Determine its frequency.

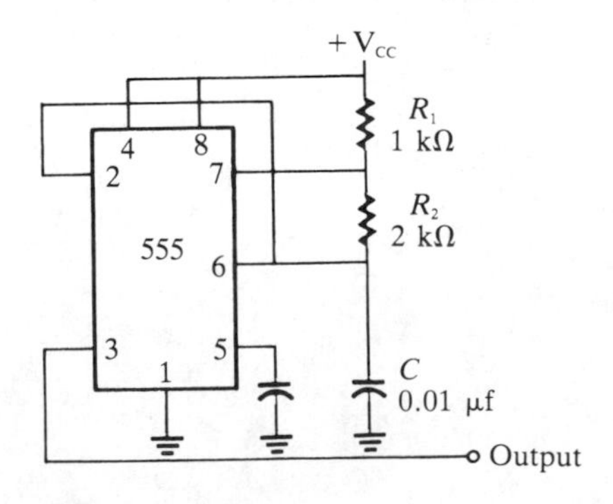

FIGURE 7–79

7–33 Determine the values of R_1 and R_2 for a 555 timer used as an astable multivibrator with an output frequency of 20 kHz, if the external capacitor C is 0.001 μF and the duty cycle is to be 50 percent.

7–34 Repeat Problem 7–33 for a duty cycle of 25 percent. (The output is HIGH for 75 percent of the period.)

When we count a sequence of numbers, two basic factors are involved: we must know where we are in the sequence at any given time, and we must know the next number in the sequence. A digital *counter* is a logic circuit that can progress through a sequence of numbers or states when activated by a clock pulse input. The outputs of a counter indicate the binary number contained within the counter at any given time. Counters are all characterized by a storage or memory capability because they must be able to retain their present state until the clock forces a change to the next state in the sequence. The clock signal is normally a series of pulses (pulse train) of a specified frequency.

Digital counters have many wide-ranging applications in digital systems, but all have one thing in common—they utilize the counter's ability to progress through a certain sequence of states. The number of states through which a counter progresses before it goes back to the original state (recycles) is called the *modulus* of the counter.

Another category of sequential circuits known as *registers* is very important in applications involving the *storage* and *transfer* of data in a digital system. The basic difference between a register and a counter is that a register has no specified sequence of states except in certain very specialized applications. A register, in general, is used solely for the purpose of *storing* and *shifting* data (1s and 0s) entered into it from an external source and possesses no characteristic internal sequence of states.

As you will learn later, both counters and registers are important elements in microprocessor and microcomputer systems.

8

COUNTERS AND REGISTERS

8–1 BINARY COUNTERS

Using Flip-Flops to Count

Since the flip-flop has memory (storage capability), it is the primary logic element in digital counters. In fact, a single flip-flop can be viewed as a simple counter. Let's look at Figure 8–1 and see how this works. First, assume the flip-flop is RESET ($Q = 0$) and a clock pulse occurs. The flip-flop is connected for toggle operation ($J = 1, K = 1$), so the first clock pulse (CK_1) causes it to SET ($Q = 1$). The fact that Q has gone from a 0 to a 1 state tells us that a single clock pulse has occurred—that is, the flip-flop has essentially "counted" one clock pulse. At CK_2, the flip-flop RESETS. If we know that the flip-flop has been SET only *once* and is then RESET, we know that two pulses have occurred. The flip-flop has essentially counted two pulses.

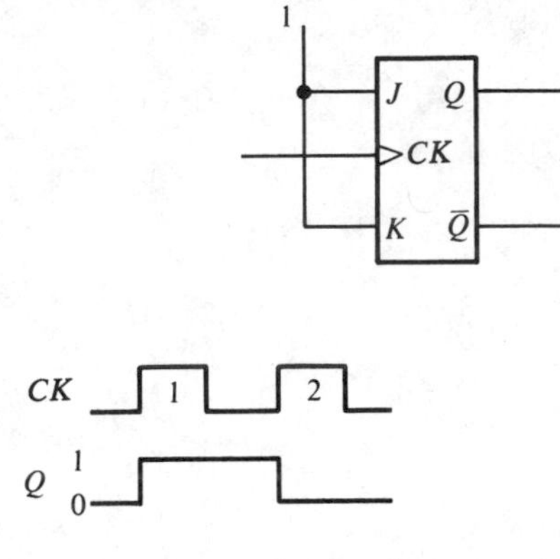

FIGURE 8–1 *The flip-flop as a simple counter.*

A limitation of the "counting" ability of a single flip-flop should be apparent at this time. Assuming RESET is the beginning state, a series of clock pulses (a square wave is shown in this case) will produce the output indicated in Figure 8-2.

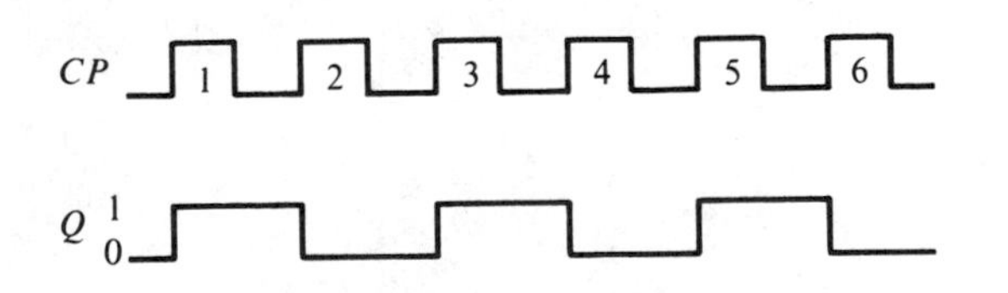

FIGURE 8–2

Notice that the Q output is HIGH (1) after the start of each odd clock pulse, and it is LOW (0) initially and after each even clock pulse. Therefore, if $Q = 0$ at any time, either no clock pulses or an even number of clock pulses have occurred. If $Q = 1$ at any time, this tells us that an odd number of clock pulses have occurred, but not how many.

If there is a situation where either *none* or *only one* clock pulse can possibly occur, then the single flip-flop counter can give us a precise indication. Assuming that it is initially RESET, then $Q = 0$ indicates that no clock pulses have occurred and $Q = 1$ indicates that one has occurred. As you can see, the single flip-flop is limited to counting two occurrences—none or one.

In order to increase the counting capacity, additional flip-flops are used. For instance, let us look at the counting capacity of two flip-flops. At this point, we are not concerned with the actual connections of the flip-flops. As indicated in Figure 8–3, the two flip-flops together have a total of four possible states and can therefore be used to count the occurrence of none, one, two, or three clock pulses when connected properly.

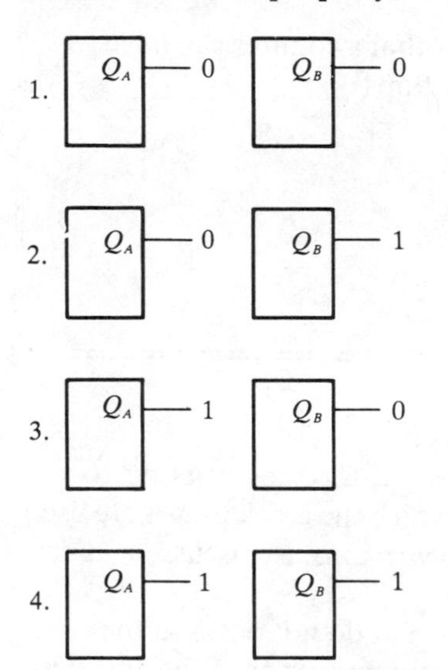

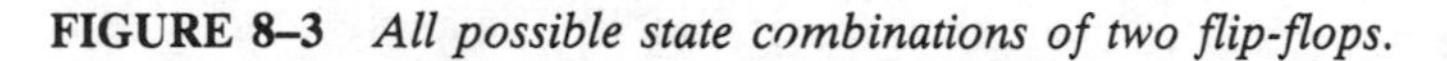

FIGURE 8–3 *All possible state combinations of two flip-flops.*

When three flip-flops are used together, there are eight possible combinations of states, as shown in Table 8-1.

TABLE 8–1 *All possible state combinations of a three flip-flop counter.*

Q_C	Q_B	Q_A
0	0	0
0	0	1
0	1	0
0	1	1
1	0	0
1	0	1
1	1	0
1	1	1

As you can see, the three flip-flops together can be used to count from zero to seven clock pulses when connected properly. It becomes apparent that the maximum number of binary states in which a counter can exist is dependent on the number of flip-flops used to construct the counter; this can be expressed by the following relationship, where N = maximum number of counter states and n = number of flip-flops in the counter:

$$N = 2^n \tag{8–1}$$

Example 8–1

Determine the maximum number of binary states that a counter can have if it is constructed with four, five, and six flip-flops.

Solutions:

Four flip-flops: $2^4 = 16$ states
Five flip-flops: $2^5 = 32$ states
Six flip-flops: $2^6 = 64$ states

Asynchronous Binary Counters

Binary counters can be placed into two categories relating to the method by which they are *clocked,* that is, the way in which the clock pulses are used to sequence the counter. The category called *asynchronous* is discussed in this section.

The term *asynchronous* refers to events that do not occur at the same time. With respect to counter operation, asynchronous means that the flip-flops within the counter are not made to change states at exactly the same time; this is because the clock pulses are not connected directly to the *CK* input of each flip-flop in the counter. Figure 8–4 shows a two-stage counter connected for asynchronous operation. Each flip-flop in a counter is commonly referred to as a *stage* of the counter.

Notice in Figure 8–4 that the clock line is connected to the clock *(CK)* input of only the first stage, FF*A*. The second stage, FF*B*, is triggered by the $\overline{Q}_A$ output of *FFA*. FF*A* changes state at the positive-going edge of each clock pulse, but FF*B* changes only when triggered by a positive-going transition of the $\overline{Q}_A$ output of

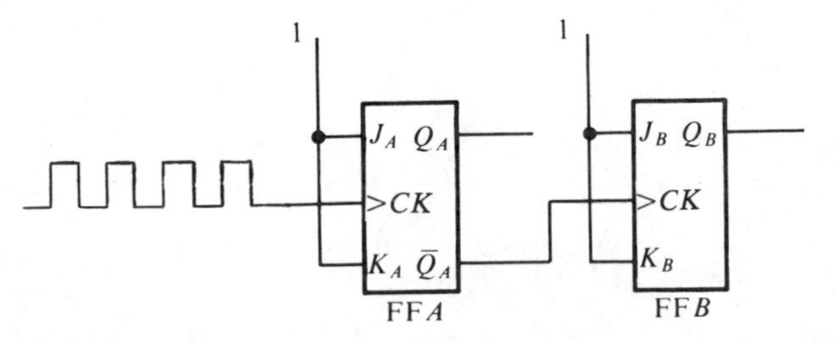

FIGURE 8-4 *A two-stage asynchronous binary counter.*

FF*A*. Because of the inherent propagation delay through a flip-flop, a transition of the input clock pulse and a transition of the $\overline{Q}_A$ output of FF*A* can never occur at exactly the same time. Therefore, the two flip-flops are *never simultaneously triggered,* which results in *asynchronous* counter operation.

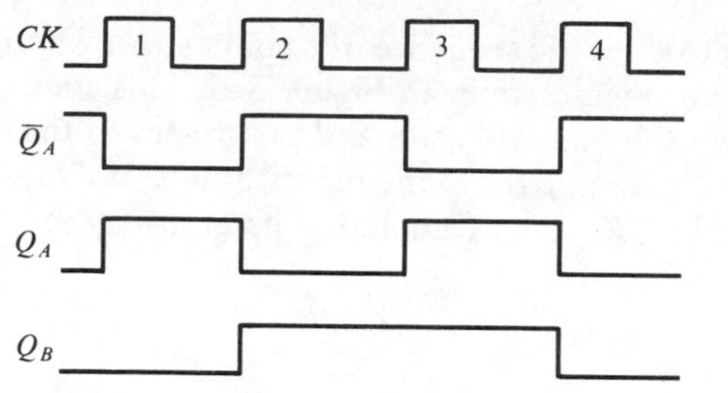

FIGURE 8–5 *Timing diagram for the counter of Figure 8-4.*

Let us examine the basic operation of the counter of Figure 8–4 by applying four clock pulses to FF*A* and observing the *Q* output of each flip-flop; Figure 8–5 illustrates the changes in the state of the flip-flop outputs in response to the clock pulses. Also, both flip-flops are connected for toggle operation ($J = 1, K = 1$) and are initially RESET.

The positive-going edge of CK_1 (clock pulse 1) causes the Q_A output of FF*A* to go HIGH. The $\overline{Q}_A$ output at the same time goes LOW, but has no effect on FF*B* because a *positive-going* transition must occur to trigger the flip-flop. After the leading edge of CK_1, $Q_A = 1$ and $Q_B = 0$. The positive-going edge of CK_2 causes Q_A to go LOW. $\overline{Q}_A$ goes HIGH and triggers FF*B*, causing Q_B to go HIGH. After the leading edge of CK_2, $Q_A = 0$ and $Q_B = 1$. The positive-going edge of CK_3 causes Q_A to go HIGH again. $\overline{Q}_A$ goes LOW and has no effect on FF*B*. Thus, after the leading edge of CK_3, $Q_A = 1$ and $Q_B = 1$. The positive-going edge of CK_4 causes Q_A to go LOW. $\overline{Q}_A$ goes HIGH and triggers FF*B*, causing Q_B to go LOW. After the leading edge of CK_4, $Q_A = 0$ and $Q_B = 0$. The counter is back in its original state (both flip-flops RESET).

The waveforms of the Q_A and Q_B outputs are shown relative to the clock pulses in Figure 8–5. This graphic waveform relationship is called a *timing diagram.* It should be pointed out that, for simplicity, the transitions of Q_A, Q_B, and the clock pulses are shown simultaneous even though this is an asynchronous counter. There is, of course, some small delay between the Q_A and the Q_B transitions.

Notice that the two-flip-flop counter exhibits four different states, as you would expect with two flip-flops ($N = 2^2 = 4$). Also, notice that if Q_A represents the least significant bit (LSB) and Q_B represents the most significant bit (MSB), the sequence of counter states is actually a sequence of binary numbers as shown in Table 8-2, which is a state table or truth table for a two-stage binary counter.

TABLE 8–2

Clock Pulse	Q_B	Q_A
0	0	0
1	0	1
2	1	0
3	1	1

Since it goes through a binary sequence, the counter in Figure 8–4 is a form of *binary counter*. It actually counts the number of clock pulses up to three, and on the fourth pulse it recycles to its original state ($Q_A = 0$, $Q_B = 0$). The term *recycle* is commonly applied to counter operation, and refers to the transition of the counter from its final state back to its original state.

Sometimes it is convenient to show the sequence of counter states with a *state diagram,* as illustrated for the two-stage counter in Figure 8–6. This state diagram tells us that the counter starts in the binary 0 state and progresses to the binary 1 state at CK_1, goes to the binary 2 state at CK_2, to the binary 3 state at CK_3, and then recycles to the binary 0 state at CK_4, thus completing its entire cycle.

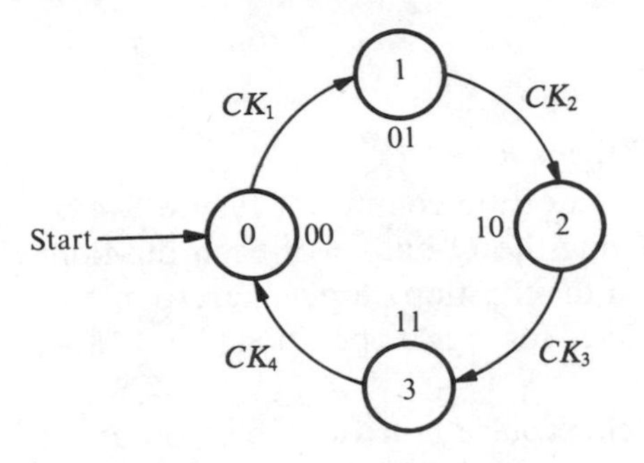

FIGURE 8–6 *State diagram for a two-stage binary counter.*

A three-stage asynchronous binary counter is shown in Figure 8–7(a). The basic operation, of course, is the same as that of the two-stage counter just discussed, except that it has *eight* states due to its three stages. A timing diagram appears in Figure 8–7(b) for eight clock pulses.

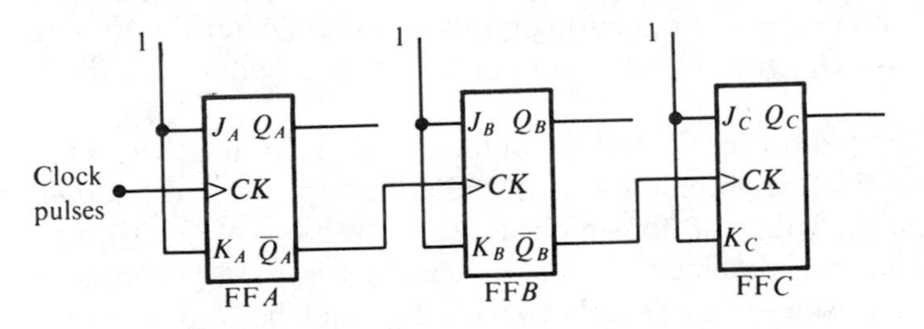

(a) Three-Stage Binary Counter (Asynchronous or Ripple Type)

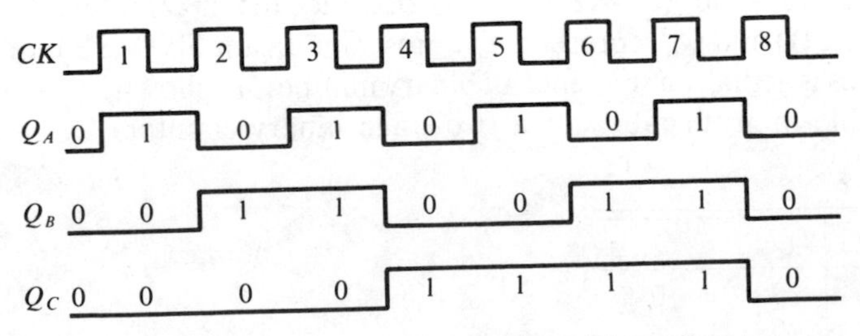

(b) Timing Diagram for the Three-Stage Binary Counter

FIGURE 8–7

Notice that the counter progresses through a binary count of 0 to 7 and then recycles to the 0 state. This counter sequence is presented in the state diagram of Figure 8–8 and the state table of Table 8–3.

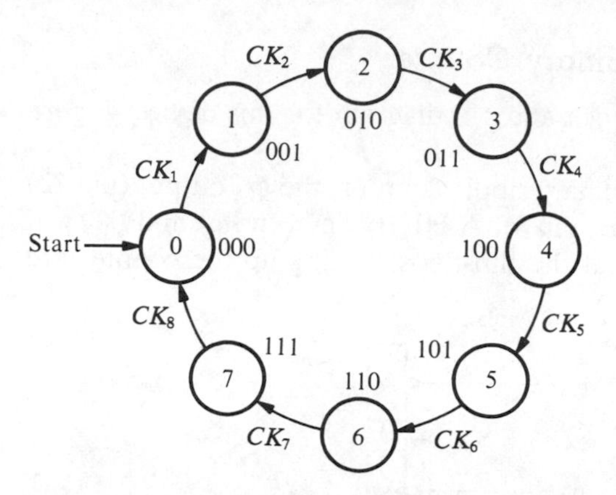

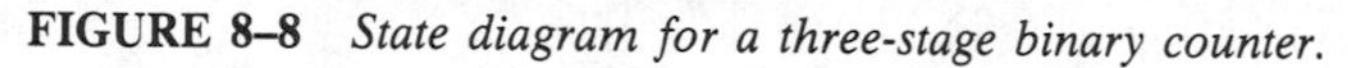

FIGURE 8–8 *State diagram for a three-stage binary counter.*

TABLE 8–3. *State table for a three-stage binary counter.*

Clock Pulse	Q_C	Q_B	Q_A
0	0	0	0
1	0	0	1
2	0	1	0
3	0	1	1
4	1	0	0
5	1	0	1
6	1	1	0
7	1	1	1

Asynchronous counters are commonly referred to as *ripple* counters for the following reason. The effect of the input clock pulse is first "felt" by FF*A*. This effect cannot get to FF*B* immediately due to the propagation delay through FF*A*. Then there is the propagation delay through FF*B* before FF*C* can be triggered. Thus, the effect of an input clock pulse "ripples" through the counter, taking some time due to propagation delays to reach the last flip-flop. To illustrate, all three flip-flops in the counter of Figure 8–7 change state as a result of CK_4. The HIGH-to-LOW transition of Q_A occurs one delay time after the positive-going transition of the clock pulse. The HIGH-to-LOW transition of Q_B occurs one delay time after the positive-going transition of $\overline{Q}_A$. The LOW-to-HIGH transition of Q_C occurs one delay time after the positive-going transition of $\overline{Q}_B$. As you can see, FF*C* is not triggered until two delay times after the positive-going edge of the clock pulse, CK_4. In other words, it takes two flip-flop delay times for the effect of the clock pulse to "ripple" through the counter and trigger FF*C*.

This cumulative delay of an asynchronous counter is a major disadvantage in many applications because it limits the rate at which the counter can be clocked and creates decoding problems.

The 7493A Four-Bit Binary Counter

Figure 8–9(a) shows a block and pin diagram for this device. Figure 8–9(b) is the logic diagram.

To operate as a four-bit binary ripple counter, the Q_A output (pin 12) must be externally connected to *input B* (pin 8). A HIGH on both $R_{0(1)}$ and $R_{0(2)}$ will clear (RESET) the counter. Notice that the flip-flops making up the counter are negative edge-triggered.

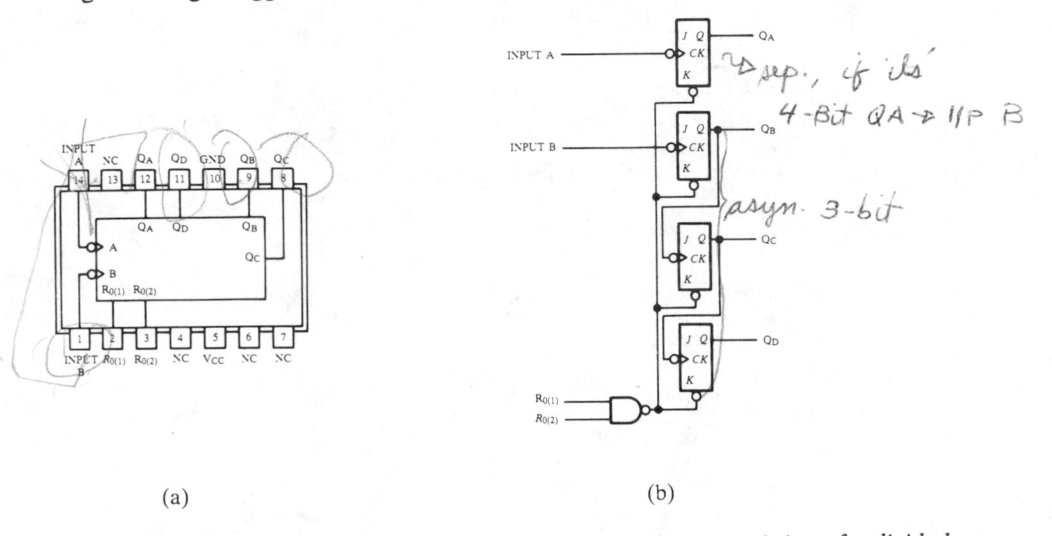

FIGURE 8–9 *A 7493A four-bit binary asynchronous counter consisting of a divide-by-two and a divide-by-eight counter.*

Example 8–2

A four-stage asynchronous binary counter is shown in Figure 8–10(a). Each flip-flop is positive edge-triggered and has a propagation delay of 10 nanoseconds (ns). Draw a timing diagram showing the Q output of each stage, and determine the total delay time from the leading edge of a clock pulse until a corresponding change can occur in the state of Q_D.

Solutions:

The timing diagram with delays omitted is as shown in Figure 8–10(b). To determine the total delay time, the effect of CK_8 or CK_{16} must propagate through four flip-flops before Q_D changes.

$$t_D = 4 \times 10 \text{ ns} = 40 \text{ ns total delay}$$

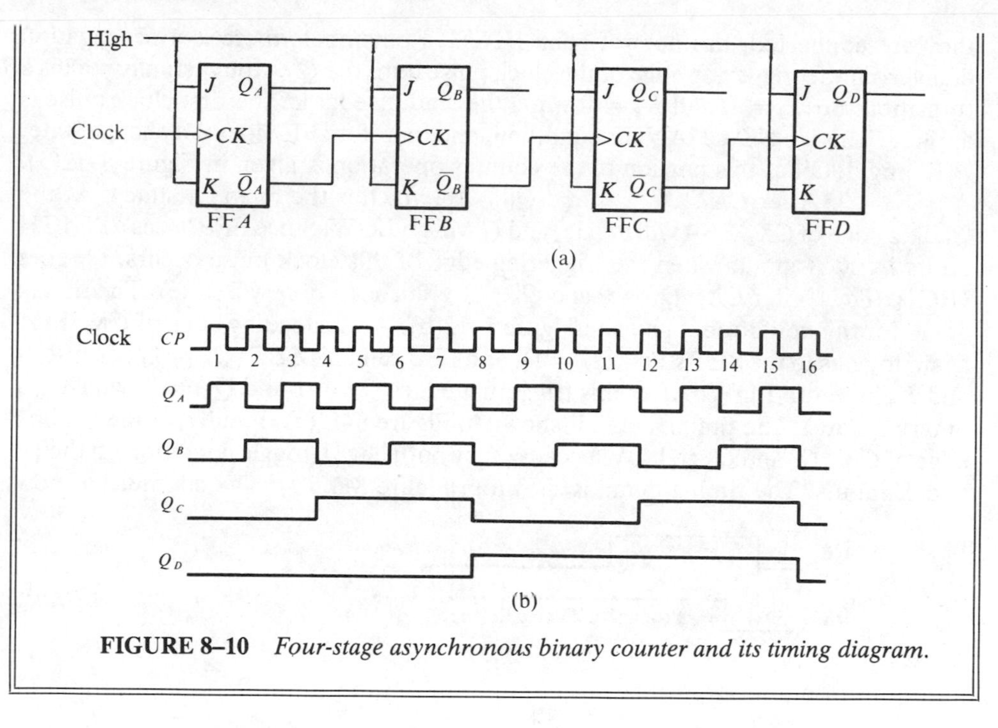

FIGURE 8–10 *Four-stage asynchronous binary counter and its timing diagram.*

Synchronous Binary Counters - long duration

The term *synchronous* as applied to counter operation means that the counter is clocked such that each flip-flop in the counter is *triggered at the same time.* This is accomplished by connecting the clock line to *each* stage of the counter, as shown in Figure 8–11 for a two-stage counter. The synchronous counter is also called a *parallel* counter because the clock line is connected in parallel to each flip-flop.

Notice that an arrangement different from that for the asynchronous counter must be used for the J and K inputs of FFB in order to achieve a binary sequence. The operation of this counter is as follows. First, we will assume the

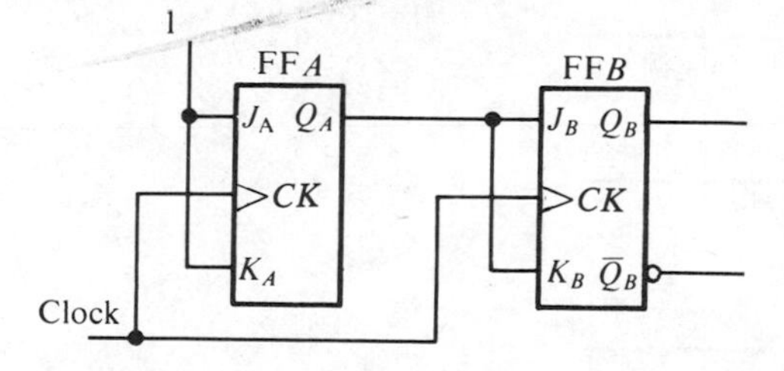

FIGURE 8–11 *A two-stage synchronous binary counter.*

counter is initially in the binary 0 state—both flip-flops RESET. When the positive edge of the first clock pulse is applied, FFA will toggle, and Q_A will therefore go HIGH. What happens to FFB at the positive-going edge of CK_1? To find out, let us look at the input conditions of FFB. J_B and K_B are both LOW because Q_A, to which they are connected, has not yet gone HIGH. Remember, there is a propagation delay from the triggering edge of the clock pulse until the Q output actually makes a transition. So, $J_B = 0$ and $K_B = 0$ when the leading edge of the first clock pulse is applied. This is a NO CHANGE condition, and therefore FF_B does not change state. A timing detail of this portion of the counter operation is given in Figure 8–12(a).

After CK_1, $Q_A = 1$ and $Q_B = 0$ (which is the binary 1 state). At the leading edge of CK_2, FFA will toggle, and Q will go LOW. Since FFB "sees" a HIGH on its J and K inputs when the triggering edge of this clock pulse occurs, Q_B goes HIGH. Thus, after CK_2, $Q_A = 0$ and $Q_B = 1$ (which is a binary 2 state). The timing detail for this condition is given in Figure 8–12(b). At the leading edge of CK_3, FFA again toggles to the SET state ($Q_A = 1$), and FFB remains SET ($Q_B = 1$) because J_B and K_B are both LOW. After this triggering edge, $Q_A = 1$ and $Q_B = 1$ (which is a binary 3 state). The timing detail is shown in Figure 8–12(c). Finally, at the leading edge of CK_4, Q_A and Q_B go LOW because they both have a toggle condition on their J and K inputs. The timing detail is shown in Figure 8–12(d). The counter has now

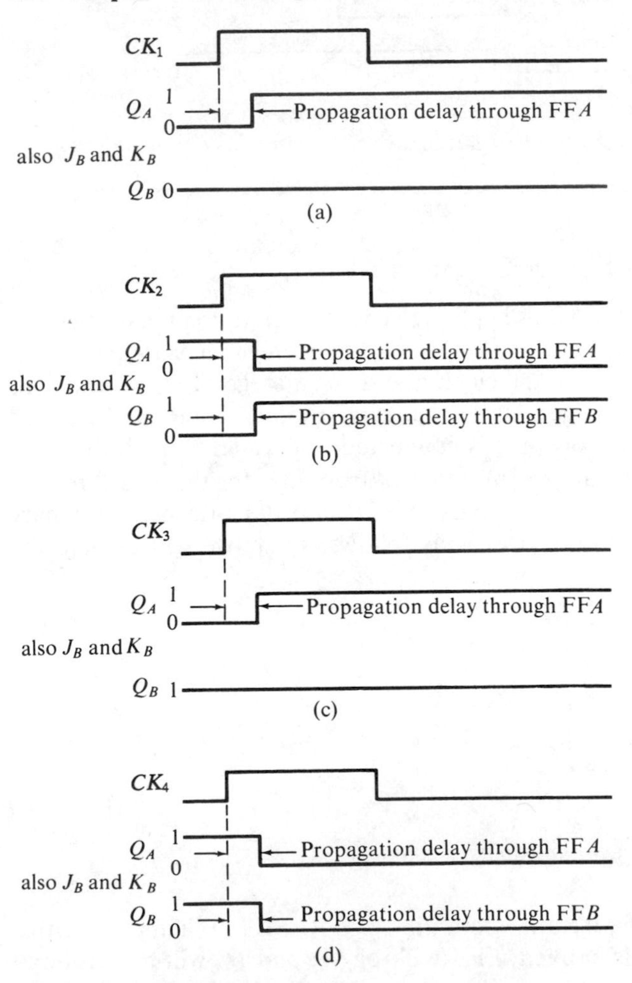

FIGURE 8–12 *Timing details for the synchronous counter operation.*

recycled back to its original state, binary 0. The complete timing diagram with the delays is shown in Figure 8–13.

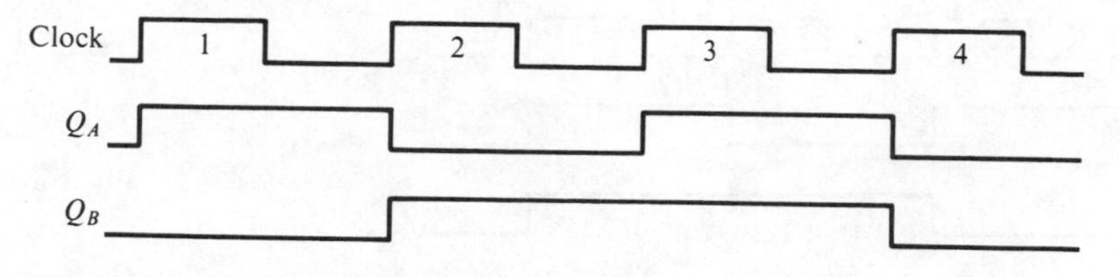

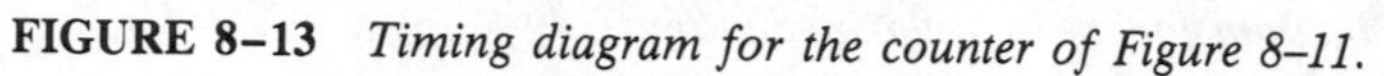

FIGURE 8–13 *Timing diagram for the counter of Figure 8–11.*

Notice that all of the waveform transitions appear coincident; that is, the delays are not indicated. Although the delays are a very important factor in the counter operation, as we have seen in the preceding discussion, in an overall timing diagram they are normally omitted for simplicity. Major waveform relationships resulting from the logical operation of a circuit can be conveyed completely without showing minute delay and timing differences.

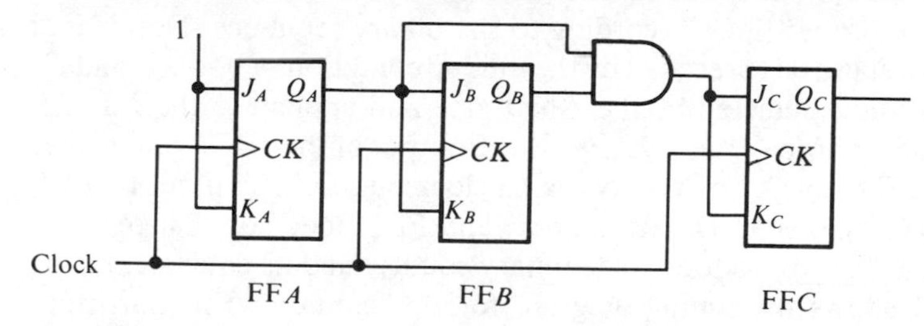

FIGURE 8–14 *A three-stage synchronous binary counter.*

A three-stage synchronous binary counter is shown in Figure 8–14 and its timing diagram in Figure 8–15. An understanding of this counter can be achieved by a careful examination of its sequence of states as shown in Table 8–4.

TABLE 8–4 *State table for a three-stage binary counter.*

CK	Q_C	Q_B	Q_A
0	0	0	0
1	0	0	1
2	0	1	0
3	0	1	1
4	1	0	0
5	1	0	1
6	1	1	0
7	1	1	1

First, let us look at Q_A. Notice that Q_A changes on each clock pulse as we progress from its original state to its final state and then back to its original state.

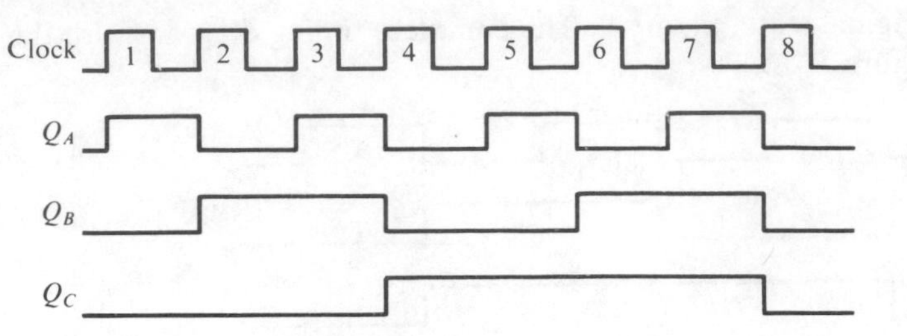

FIGURE 8–15 *Timing diagram for the counter of Figure 8–14.*

To produce this operation FF*A* must be held in the toggle mode by constant HIGHs on its *J* and *K* inputs. Now let us see what Q_B does. Notice that it goes to the opposite state following each time Q_A is a 1. This occurs at CK_2, CK_4, CK_6, and the clock pulse after CK_7 that causes the counter to recycle. To produce this operation, Q_A is connected to the *J* and *K* inputs of FF*B*. When Q_A is a 1 and a clock pulse occurs, FF*B* is in the toggle mode and will change state. The other times when Q_A is a 0, FF*B* is in the *no change* mode and reamins in its present state. Next, let us see how FF*C* is made to change at the proper times according to the binary sequence. Notice that both times Q_C changes state, it is preceded by the unique condition of both Q_A and Q_B being 1s. This condition is detected by the AND gate and applied to the *J* and *K* inputs of FF*C*. Whenever both Q_A and Q_B are 1s, the output of the AND gate makes J_C and K_C HIGH and FF*C* toggles on the following clock pulse. At all other times J_C and K_C are held LOW by the AND gate output, and FF*C* does not change state.

Figure 8–16 shows a four-stage binary counter and its equivalent logic symbol. Figure 8–17 shows the timing diagram for this counter. This particular counter is implemented with master-slave flip-flops rather than the edge-triggered types, and the counter makes its transitions on the trailing (negative-going) edge of the clock pulses. The reasoning behind the *J* and *K* input control for the first three flip-flops is the same as presented previously for the three-stage counter. The fourth stage, FF*D,* changes only twice in the sequence. Notice that both of these transitions occur following the times Q_A, Q_B, and Q_C are all 1s. This condition is detected by AND gate G_2, so that when a clock pulse occurs, FF*D* will change state. For all other times FF*D*'s *J* and *K* inputs are LOW, and it is in a *no change* condition.

The 74163 Synchronous Four-Bit Binary Counter

The 74163 is an example of an integrated circuit synchronous binary counter. A block and pin diagram is shown in Figure 8–18. This counter has several features in addition to basic functions previously discussed for the general synchronous binary counter.

First, the counter can be *preset* to any four-bit binary number by applying the proper levels to the *data inputs.* When a LOW is applied to the *load input,* the counter will assume the state of the data inputs on the next clock pulse. This, of course, allows the counter sequence to be started with any four-bit binary number.

Also, there is an active LOW *clear* input which resets all four flip-flops in the counter. There are two *enable inputs, P* and *T.* These inputs must be high for the counter to sequence through its binary states. When at least one is LOW, the

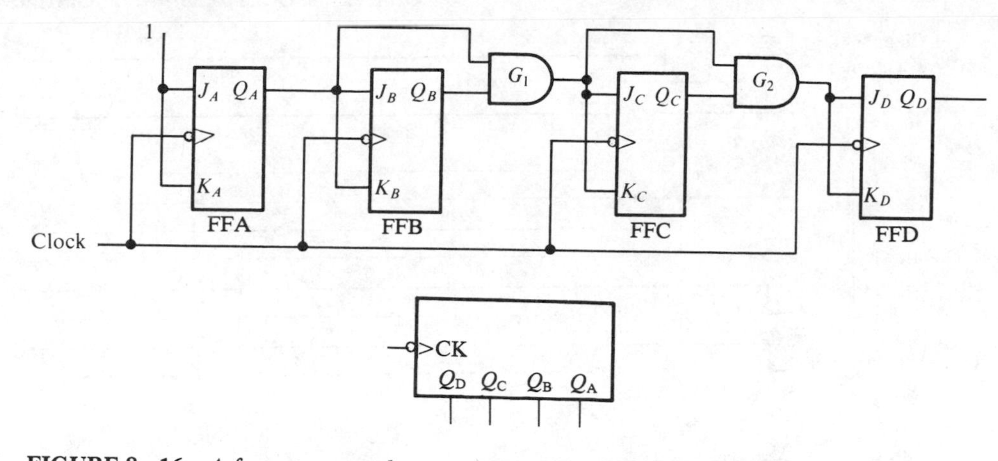

FIGURE 8–16 *A four-stage synchronous binary counter and its typical logic block symbol.*

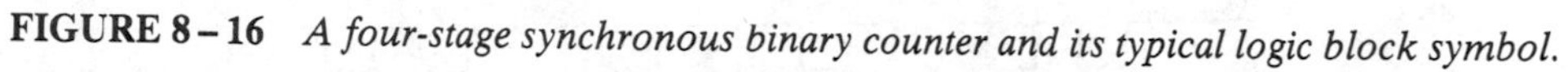

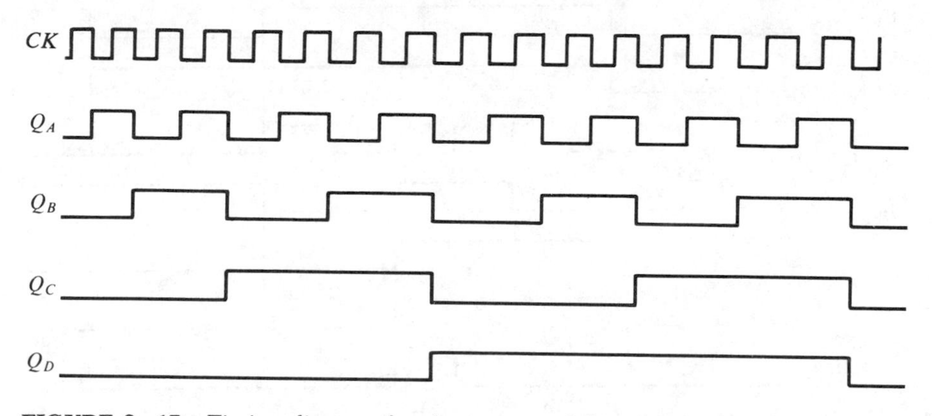

FIGURE 8–17 *Timing diagram for the counter of Figure 8–16.*

counter is disabled. The *carry output* goes HIGH when the counter reaches the last state in the sequence, binary 15. This output, in conjunction with the enable inputs, allows these counters to be cascaded for higher count sequences as will be discussed later. Figure 8–19 shows a timing diagram of this counter being preset to binary 12 (1100) and then counting up to its terminal count, binary 15(1111). Notice particularly the carry and enable waveforms.

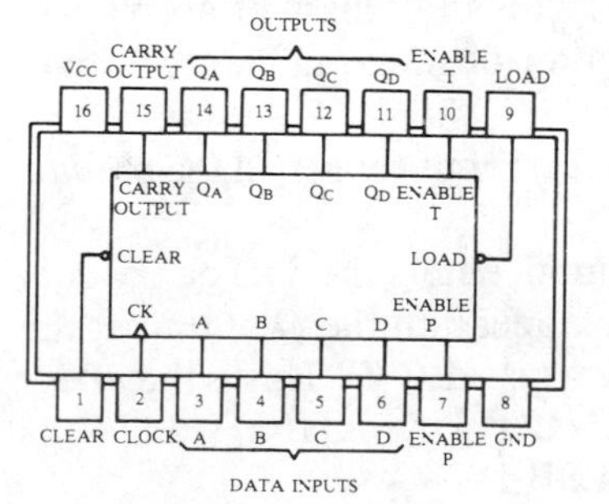

FIGURE8–18 *A 74163 binary synchronous counter.*

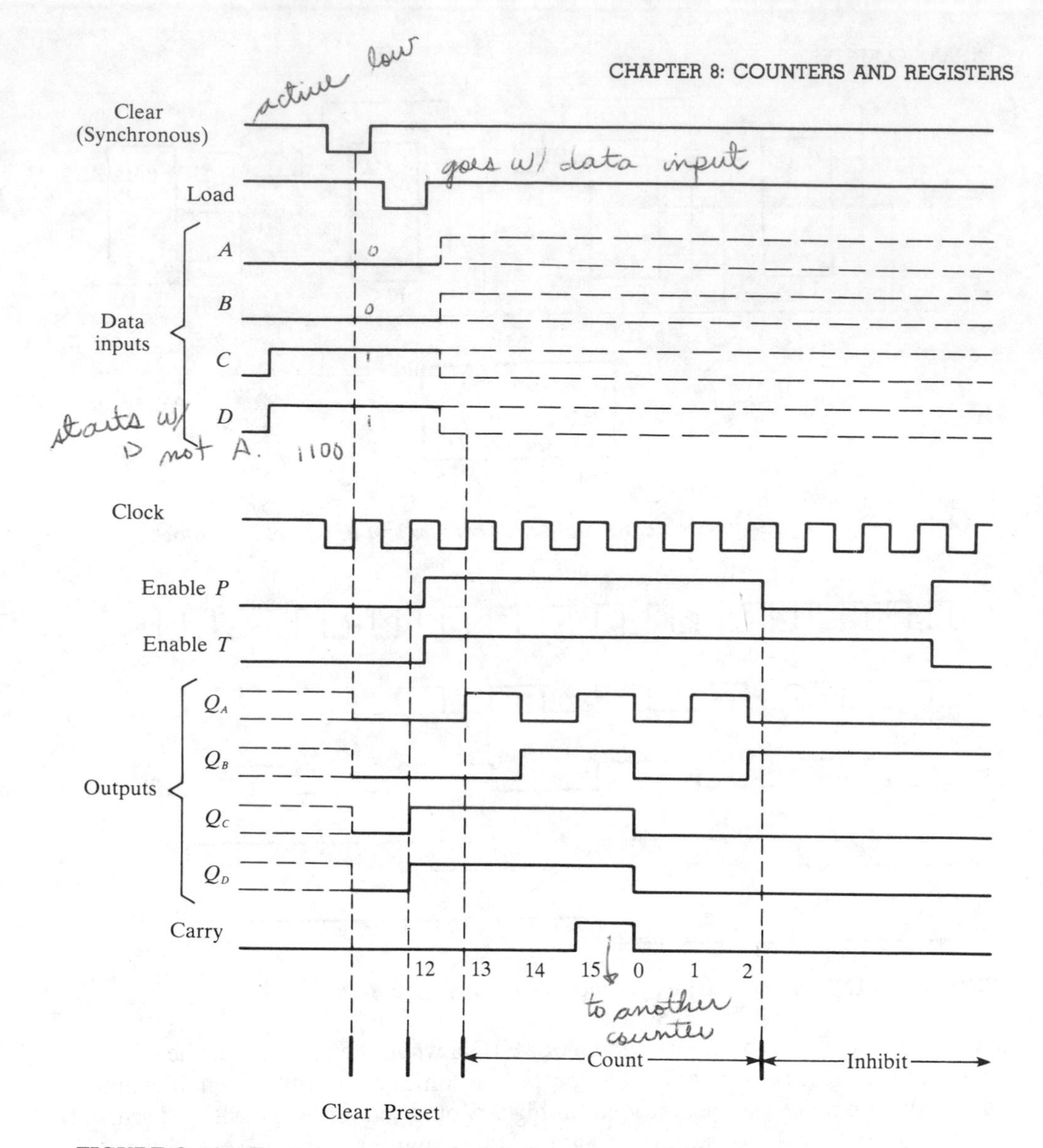

FIGURE 8–19 *Timing example for a 74163.*

Let us examine this timing diagram in detail. This will aid you in interpreting timing diagrams found later in this chapter or in manufacturer's data sheets.

To begin, the LOW level pulse on the *clear* input causes all the *outputs* *(Q_A, Q_B, Q_C,* and *Q_D)* to go LOW.

Next, the LOW level pulse on the *load* input enters the data on the *data inputs (A, B, C,* and *D)* into the counter. These data appear on the *Q* output at the time of the first positive-going *clock* edge after the *load* goes LOW. This is the *preset* operation. In this particular example, Q_A is LOW, Q_B is LOW, Q_C is HIGH, and Q_D is HIGH. This, of course, is a binary 12 (Q_A is the LSB).

The counter now advances through binary counts 13, 14, 15 on the next three positive-going clock edges. It then recycles to 0, 1, 2 on the following clock pulses. Notice that both *enable P* and *enable T* inputs are HIGH during the count sequence.

When *enable P* goes LOW, the count is inhibited, and the counter remains in the binary 2 state.

DECADE COUNTERS 8–2

Decade counters are a very important category of digital counter because of their wide application. A decade counter has *ten* states in its sequence—that is, it has a *modulus of ten.* A logic symbol for a typical decade counter is shown in Figure 8–20. It consists of four stages and can have any given sequence of states as long as there are ten. A very common type of decade counter is the BCD (8421) counter, which exhibits a Binary-Coded-Decimal sequence as shown in Table 8–5. A state diagram is shown in Figure 8–21.

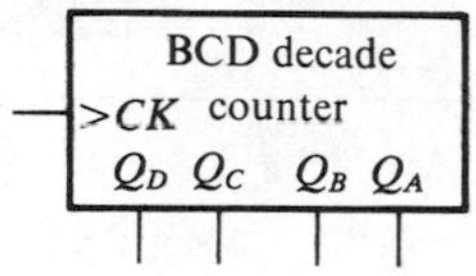

BCD decade
>CK counter
Q_D Q_C Q_B Q_A

FIGURE 8–20 *A typical logic block symbol for a BCD decade counter.*

TABLE 8–5 *States of a BCD decade counter.*

CK	Q_D	Q_C	Q_B	Q_A
0	0	0	0	0
1	0	0	0	1
2	0	0	1	0
3	0	0	1	1
4	0	1	0	0
5	0	1	0	1
6	0	1	1	0
7	0	1	1	1
8	1	0	0	0
9	1	0	0	1

As you can see, the BCD decade counter goes through a straight binary sequence through the binary 9 state. Rather than going to the binary 10 state, it recycles to the 0 state. A synchronous BCD decade counter is shown in Figure 8–22.

The counter operation can be understood by examining the sequence of states in Table 8–5. First, notice that the FF*A* toggles on each clock pulse, so the logic equation for its J and K inputs is

$$J_A = K_A = 1$$

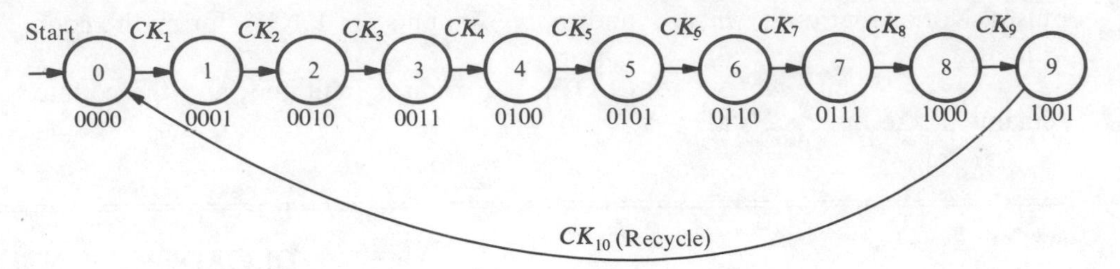

FIGURE 8–21 *State diagram for a BCD decade counter.*

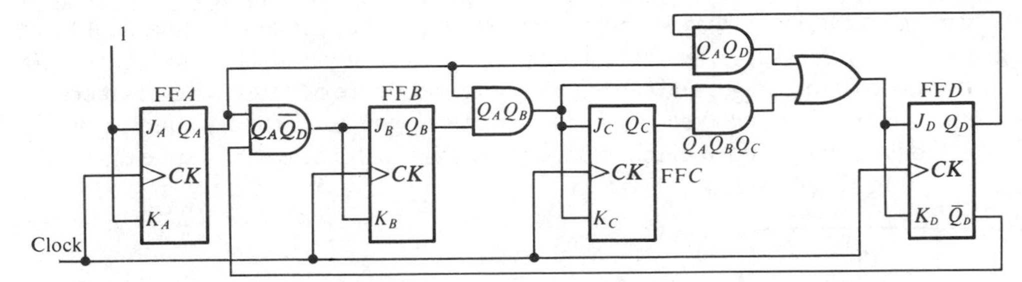

FIGURE 8–22 *A synchronous BCD decade counter.*

This is implemented by connecting these inputs to a constant HIGH level. Next notice that FFB changes on the next clock pulse each time $Q_A = 1$ and $Q_D = 0$, so the logic equation for its J and K inputs is

$$J_B = K_B = Q_A\overline{Q}_D$$

This is implemented by connecting Q_A to the J_B and K_B inputs. FFC changes on the next clock pulse each time both $Q_A = 1$ and $Q_B = 1$. This requires an input logic equation as follows:

$$J_C = K_C = Q_AQ_B$$

The control is implemented by ANDing Q_A and Q_B and connecting the gate output to J_C and K_C. Finally, FFD changes to the opposite state on the next clock pulse each time $Q_A = 1$, $Q_B = 1$, and $Q_C = 1$ (state 7), or when $Q_A = 1$ and $Q_D = 1$ (state 9). The equation for this is as follows:

$$J_D = K_D = Q_AQ_BQ_C + Q_AQ_D$$

This function is implemented with the AND/OR logic as shown in the logic diagram in Figure 8–22. Notice that the only difference between this decade counter and a four-stage binary counter is the Q_AQ_D AND gate and the OR gate; this essentially detects the occurrence of the 9 state and causes the counter to recycle properly on the next clock pulse. The timing diagram for the decade counter is given in Figure 8–23.

We will now look at examples of integrated circuit decade counters both of the asynchronous and synchronous types.

The 7490A Asynchronous Decade Counter

A block and pin connection diagram for the 7490A is shown in Figure 8–24. In addition to the four Q outputs, there are two gated RESET inputs, $R_{0(1)}$ and

CK 1 2 3 4 5 6 7 8 9 10

Q_A 0 1 0 1 0 1 0 1 0 1 0

Q_B 0 0 1 1 0 0 1 1 0 0 0

Q_C 0 0 0 0 1 1 1 1 0 0 0

Q_D 0 0 0 0 0 0 0 0 1 1 0

FIGURE 8–23 *Timing diagram for the BCD decade counter.*

$R_{0(2)}$. When HIGHs are applied to both of these pins, the counter is RESET to its zero count. Also, there are two *set-to-nine* gated inputs, $R_{9\,(1)}$ and $R_{9\,(2)}$. When HIGHs are applied to both of these pins, the counter is present to count 9 (1001).

To use the device as a BCD counter, the Q_A output must be connected to the B input, and the clock must be applied to the A input. This configuration results in a 0 through 9 BCD sequence as indicated in Table 8–6.

This device can also be used to produce a symetrical divide-by-ten count by connecting the Q_D output to the A input and using the B input as the clock. In this configuration, the Q_A output provides a *square wave* which is one tenth the frequency of the clock as indicated in Table 8–7. Notice that although Q_A is 1 for five clock pulses, this HIGH level occurs only once during ten clock pulses.

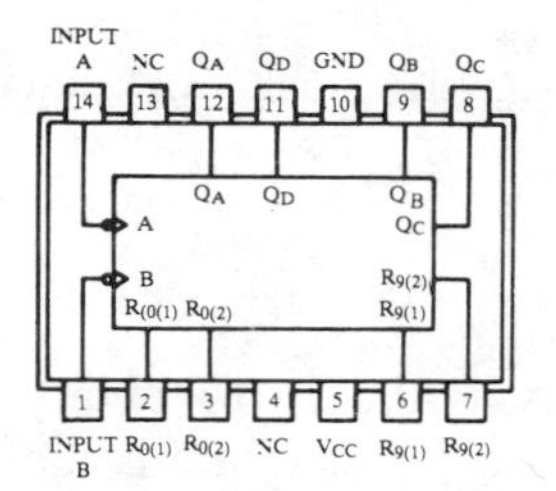

FIGURE 8–24 *A 7490A asynchronous decade counter.*

TABLE 8–6 *BCD count sequence.*

Clock	Q_D	Q_C	Q_B	Q_A
0	0	0	0	0
1	0	0	0	1
2	0	0	1	0
3	0	0	1	1
4	0	1	0	0
5	0	1	0	1
6	0	1	1	0
7	0	1	1	1
8	1	0	0	0
9	1	0	0	1

TABLE 8–7 *Symmetrical divide-by-10 count sequence.*

Clock	Q_D	Q_C	Q_B	Q_A
0	0	0	0	0
1	0	0	1	0
2	0	1	0	0
3	0	1	1	0
4	1	0	0	0
5	0	0	0	1
6	0	0	1	1
7	0	1	0	1
8	0	1	1	1
9	1	0	0	1

Example 8–3

Show a 7490A configured to operate as a symmetrical divide-by-ten counter and show the output waveform with respect to the clock.

Solution:

The connections are shown in Figure 8–25(a), and the timing diagram for the Q_A output is shown in 8–25(b).

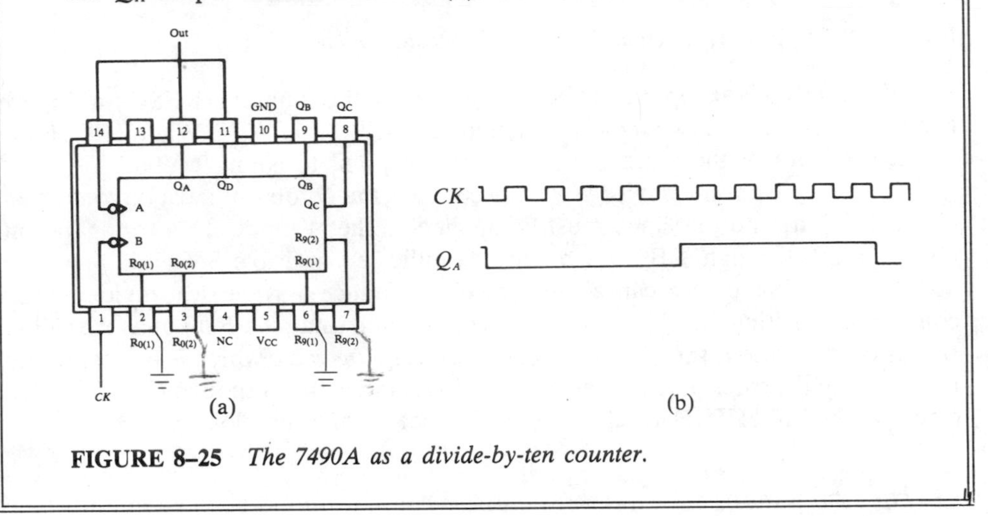

FIGURE 8–25 *The 7490A as a divide-by-ten counter.*

The 74160 Synchronous Decade Counter

This device has the same inputs and outputs as the 74163 binary counter previously discussed (Figure 8–18), and its block and pin diagram is the same. As with the 74163, the 74160 may be preset to any BCD count using the data inputs and an active LOW on the *load* input. An active LOW on the *clear* will RESET the counter. The count enable inputs *P* and *T* must both be high for the counter to advance through its sequence of states in response to a negative transition on the *CK* input. Also, like the 74163, the enable inputs in conjunction with the carry out (terminal count of 1001) provide for cascading several decade counters. Cascaded counters will be discussed later in this chapter.

Figure 8–26 shows the block and pin diagram, and Figure 8–27 is a timing diagram showing the counter being preset to BCD seven (0111).

8–3 UP-DOWN COUNTERS

An up-down counter is one that is capable of progressing in *either direction* through a certain sequence. An up-down counter is sometimes called a *bidirectional* counter and can have any specified sequence of states. For example, a BCD decade counter

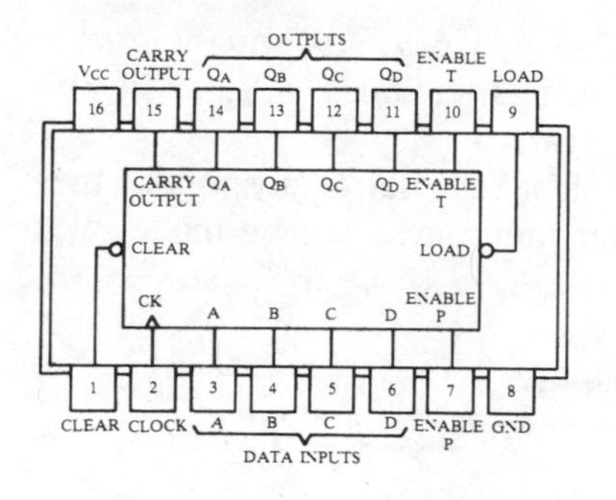

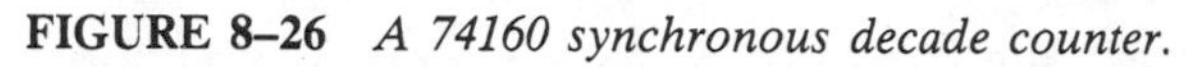
FIGURE 8–26 *A 74160 synchronous decade counter.*

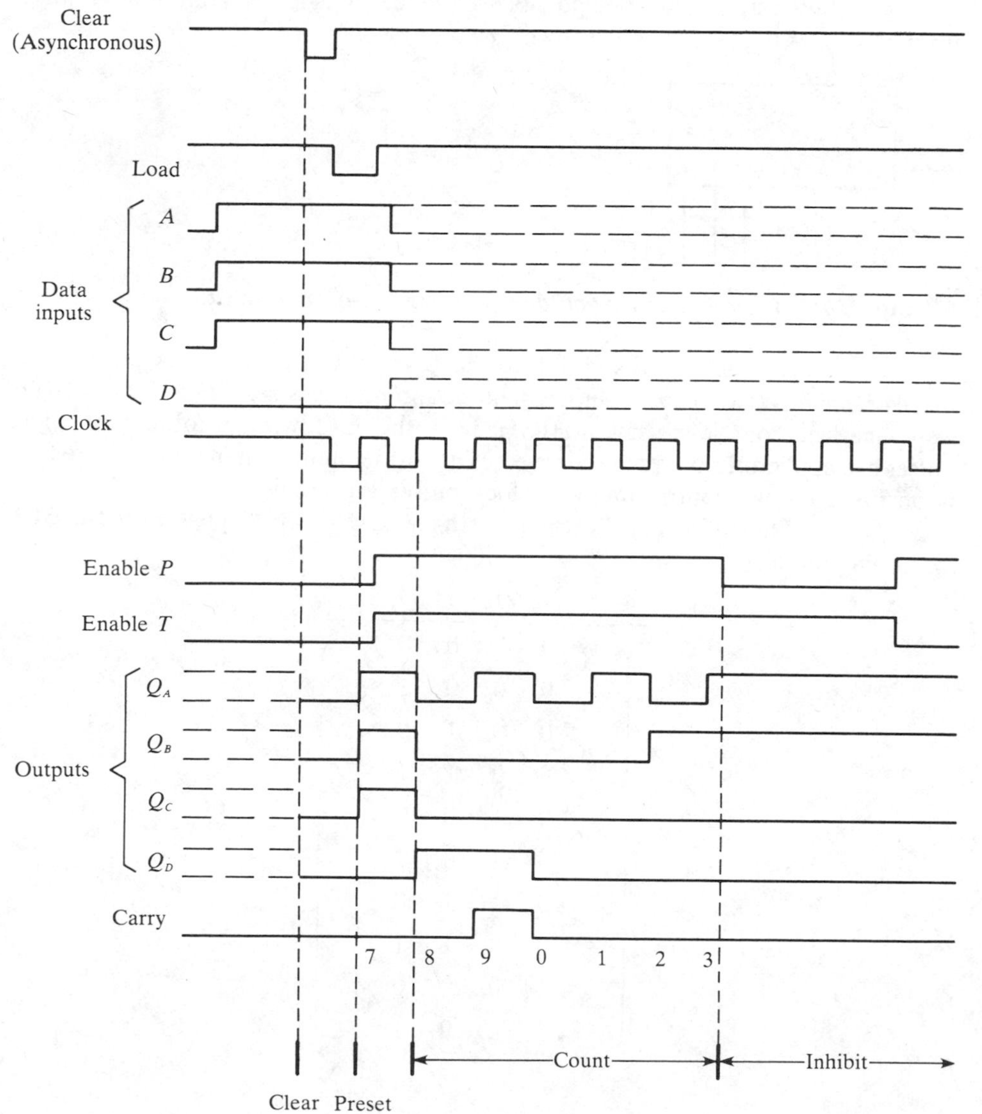

FIGURE 8–27 *Timing example for a 74160. A represents the LSB and D the MSB.*

that advances upwards through its sequence (0, 1, 2, 3, 4, 5, 6, 7, 8, 9) and then can be reversed so it goes through the sequence in the opposite direction (9, 8, 7, 6, 5, 4, 3, 2, 1, 0) is an illustration of up-down sequential operation.

In general, most up-down counters can be reversed at any point in their sequence. For instance, the BCD decade counter mentioned can be made to go through the following sequence:

$$\overbrace{0,\ 1,\ 2,\ 3,\ 4,\ 5,}^{\text{up}}\ \underbrace{4,\ 3,\ 2,}_{\text{down}}\ \overbrace{3,\ 4,\ 5,\ 6,\ 7,\ 8,}^{\text{up}}\ \underbrace{7,\ 6,\ 5,}_{\text{down}}\ \text{etc.}$$

This is an illustration of just one possible sequence. A block diagram symbol for an up-down four-stage counter is shown in Figure 8–28.

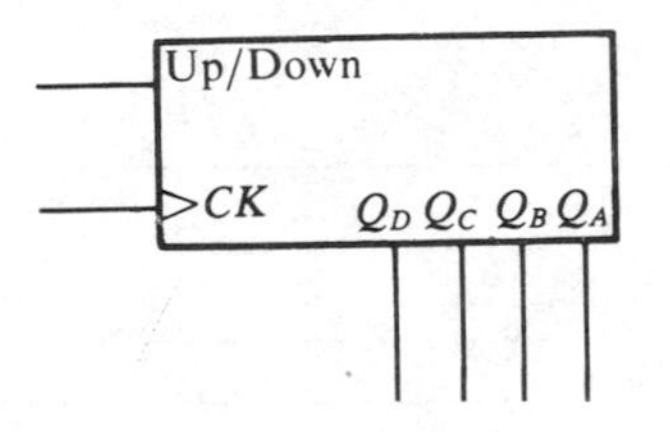

FIGURE 8–28 *Typical logic symbol for a four-stage up-down counter.*

Notice that in addition to the Q outputs and the clock pulse input, there is an *up-down control* input. The counter progresses in its forward (up) sequence when one level, say, a HIGH, is on the up-down control line and clock pulses are applied. It moves in the reverse (down) sequence when the opposite level is on the up-down control line and clock pulses are applied.

The following illustrates both up and down sequences for a BCD up-down counter:

Q_D	Q_C	Q_B	Q_A
0	0	0	0
0	0	0	1
0	0	1	0
0	0	1	1
0	1	0	0
0	1	0	1
0	1	1	0
0	1	1	1
1	0	0	0
1	0	0	1

up down

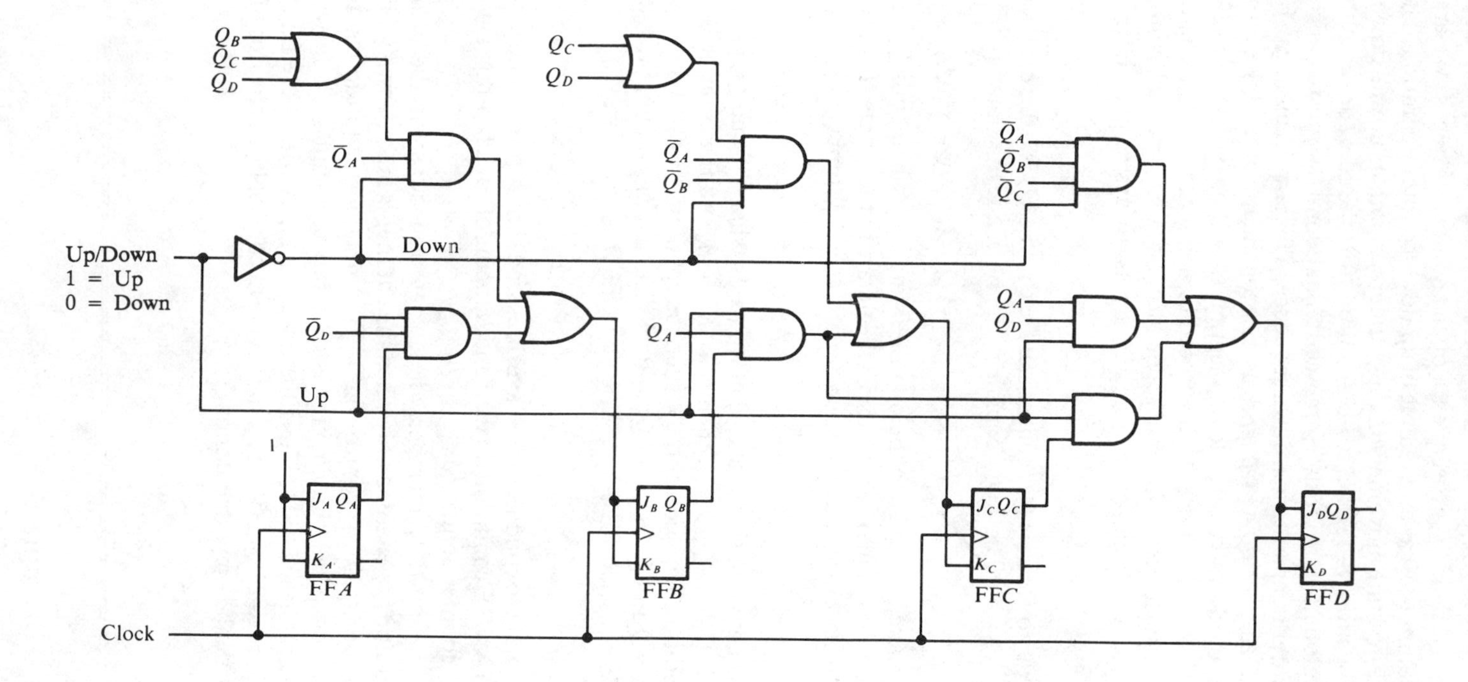

FIGURE 8–29 *A BCD up-down counter.*

Keep in mind that, in general, an up-down counter can be characterized by any sequence or any number of stages. We are using the BCD type of counter just as an example of up-down counter operation. The arrows indicate the state-to-state movement of the counter for both its up and down modes of operation. In Section 8–2, the logic required for the decade counter up sequence was developed. Obviously, some additional controls must be provided to enable the counter to operate in a down sequence. This requirement can best be understood by analyzing the *down sequence* as was done in the case of the forward BCD sequence. The up-down counter logic diagram is shown in Figure 8–29.

Notice in the preceding sequence that Q_A changes state each time a clock pulse occurs for either sequence. So, FF*A* operates continuously in the toggle mode, which means its *J* and *K* inputs always remain HIGH.

$$J_A = K_A = 1$$

Next, notice that Q_B changes state each time Q_A is a 0 *except* between the 0 state and the 9 state. The logic equation for the *J* and *K* inputs of FF*B* is therefore

$$J_B = K_B = \overline{Q}_A(Q_B + Q_C + Q_D)$$

Q_C makes a change to the opposite state only when both $Q_A = 0$ and $Q_B = 0$ *except* when the counter recycles from the 0 state to the 9 state. In order to make FF*C* toggle at these times, the equation for its *J* and *K* inputs is

$$J_C = K_C = \overline{Q}_A\overline{Q}_B(Q_C + Q_D)$$

This equation says that J_C and K_C are 1s when Q_A and Q_B are 0 *and* Q_C or Q_D is 1. If both Q_C and Q_D are 0, then $J_C = K_C = 0$, and the counter is in the NO CHANGE condition. This occurs only when the counter is in the 0 state. Q_D changes state when $Q_A = 0$ and $Q_B = 0$ and $Q_C = 0$. Notice that this occurs twice in the down sequence. The equation for *J* and *K* logic of FF*D* is therefore

$$J_D = K_D = \overline{Q}_A\overline{Q}_B\overline{Q}_C$$

Each of the conditions we have discussed simply produces a toggle condition or a NO CHANGE condition on the *J* and *K* inputs of each flip-flop at the appropriate time in the counter sequence.

Remember, these logic functions for the *J* and *K* inputs to each flip-flop are enabled only when the counter is in the down mode of operation (LOW on the up-down control line). The functions required for the up mode must be enabled when the counter is required to count in a forward sequence. This is accomplished by ANDing the up-down input with these functions, as illustrated in Figure 8–29. A HIGH on the up-down control line activates the up sequence for this particular implementation.

Example 8–4

Determine the sequence of the four-stage *binary* up-down counter if the clock and up-down control inputs have waveforms as shown in Figure 8–30(a). Counter starts in all 0s state. Q_A is the LSB.

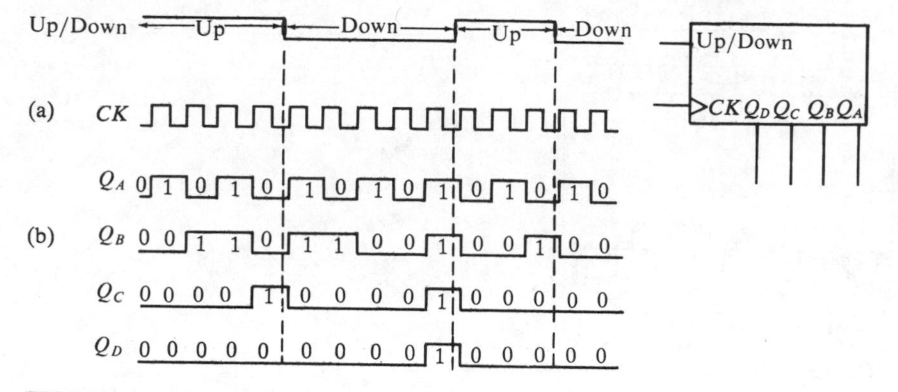

FIGURE 8–30

Solution:

From the waveforms in Figure 8–30(b), we can see that the counter sequence is as follows:

Q_D	Q_C	Q_B	Q_A	
0	0	0	0	up
0	0	0	1	
0	0	1	0	
0	0	1	1	
0	1	0	0	
0	0	1	1	down
0	0	1	0	
0	0	0	1	
0	0	0	0	
1	1	1	1	
0	0	0	0	up
0	0	0	1	
0	0	1	0	
0	0	0	1	down
0	0	0	0	

The 74190 Up-Down Decade Counter

Figure 8–31 is a block and pin diagram for the 74190 counter, a good example of an integrated circuit up-down counter.

The direction of the count is determined by the level of the *up-down input.* When this input is HIGH the counter counts down, and when it is LOW the counter counts up. Also, this device can be preset to any desired BCD digit determined by the states of the *data inputs* when the *load input* is LOW.

The *max-min output* produces a HIGH pulse when the terminal count 9 (1001) is reached in the up mode or when the terminal count 0 (0000) is reached in the down mode. This output along with the *ripple clock output* and the *enable input* are used when cascading counters. Again, the topic of cascading counters is discussed later in this chapter.

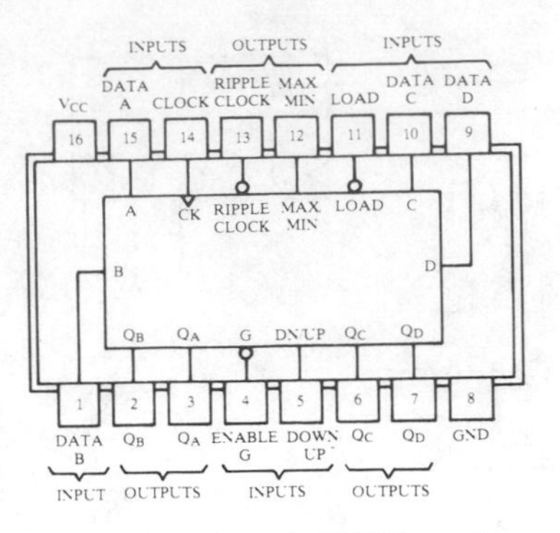

FIGURE 8–31 *A 74190 up-down decade counter.*

Figure 8–32 is an example timing diagram showing the 74190 counter preset to 7 (0111) and then going through a count up sequence followed by a count down sequence. The *max/min* output is HIGH when the counter is in either the all 0s state (min) or the 1001 state (max).

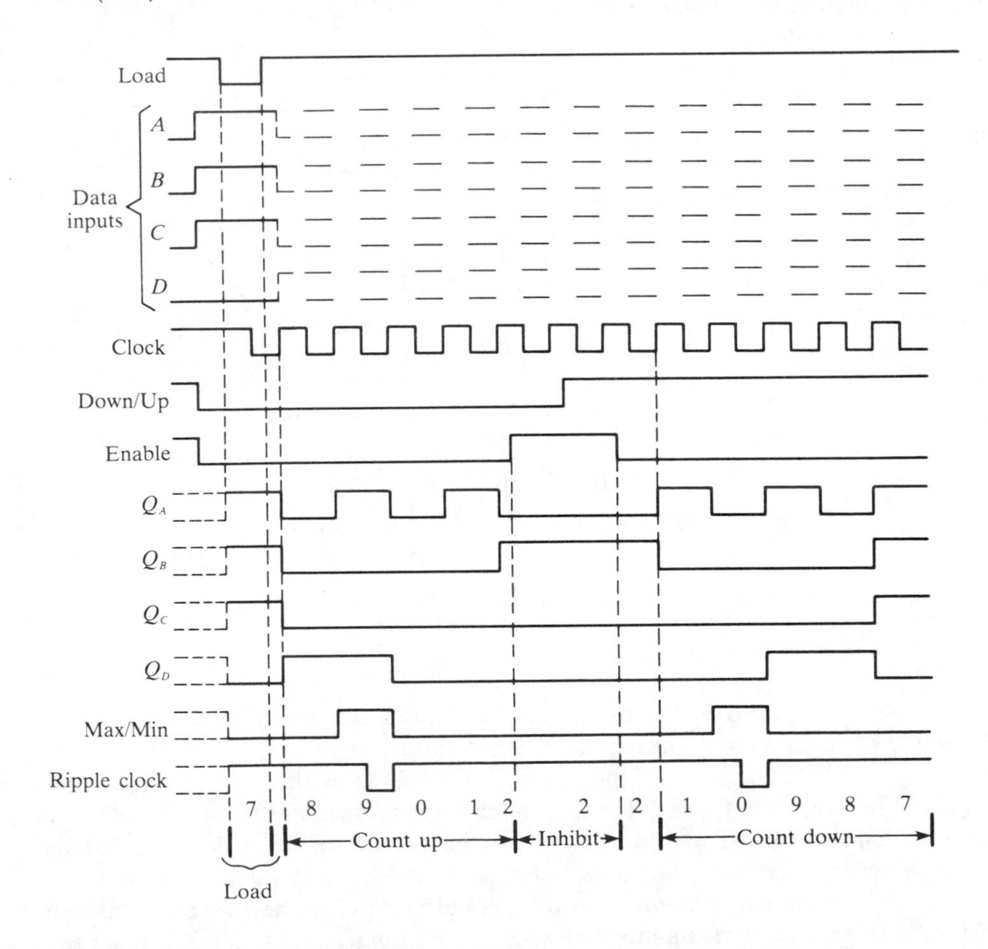

FIGURE 8–32 *Timing example for a 74190.*

SPECIAL PURPOSE COUNTERS 8–4

The Johnson Counter

The Johnson counter is sometimes known as a *shift counter* and produces a special sequence; the sequence for a four-stage counter is shown in Table 8–8(a) and that for a five-stage counter appears in Table 8–8(b).

TABLE 8–8

(a) Four-bit Johnson sequence.

CK	Q_A	Q_B	Q_C	Q_D
0	0	0	0	0
1	1	0	0	0
2	1	1	0	0
3	1	1	1	0
4	1	1	1	1
5	0	1	1	1
6	0	0	1	1
7	0	0	0	1
	(back to 0)			

(b) Five-bit Johnson sequence.

CK	Q_A	Q_B	Q_C	Q_D	Q_E
0	0	0	0	0	0
1	1	0	0	0	0
2	1	1	0	0	0
3	1	1	1	0	0
4	1	1	1	1	0
5	1	1	1	1	1
6	0	1	1	1	1
7	0	0	1	1	1
8	0	0	0	1	1
9	0	0	0	0	1
	(back to 0)				

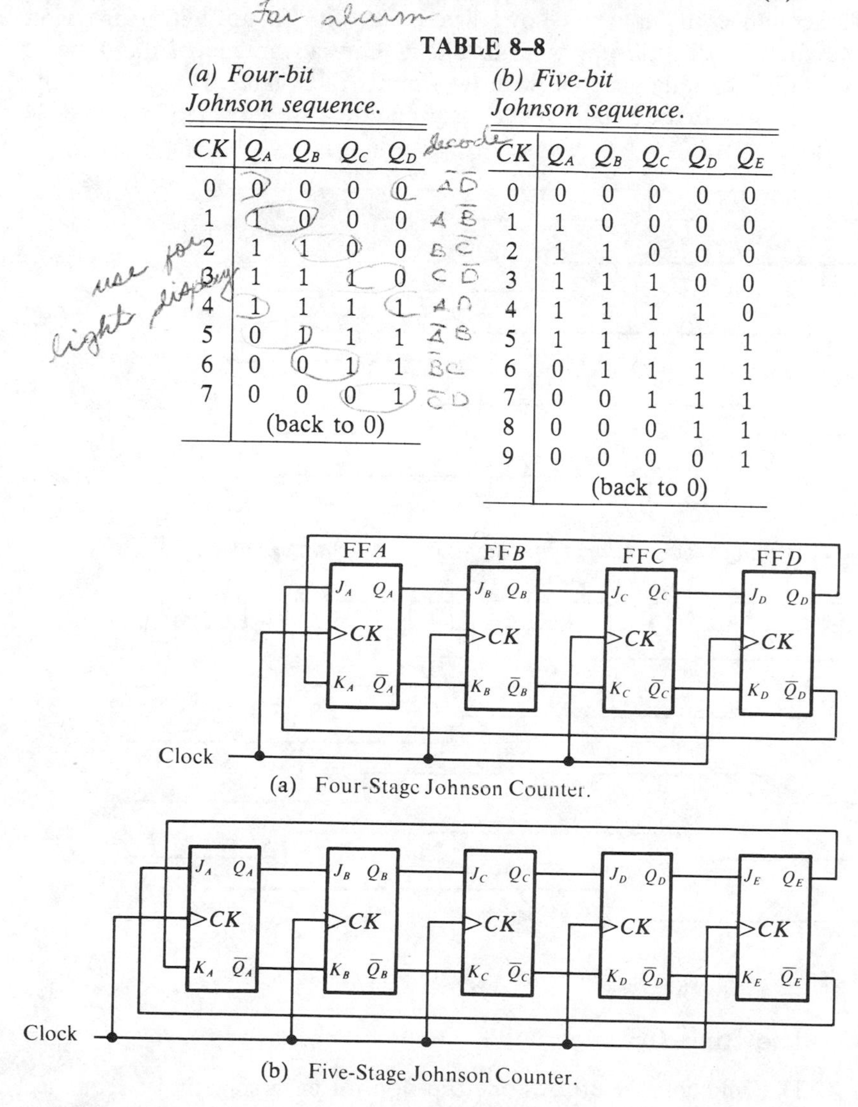

(a) Four-Stage Johnson Counter.

(b) Five-Stage Johnson Counter.

FIGURE 8–33

Notice that the four-bit sequence has a total of *eight* states and that the five-bit sequence has a total of *ten* states. In general, an *n*-stage Johnson counter will produce a modulus of 2*n*, where *n* is the number of stages in the counter. The implementations for the four- and five-stage Johnson counters are shown in Figure 8–33.

The implementation of a Johnson counter is very straightforward and is the same regardless of the number of stages. The Q output of each stage is connected to the J input of the next stage (assuming J–K flip-flops are used), and the $\overline{Q}$ output of each stage is connected to the K input of the following stage. The single exception is that the last stage Q ouput is connected back to the K input of the first stage, and the $\overline{Q}$ output of the last stage is connected back to the J input of the first stage. As the sequences in Table 8–8 show, the counter will "fill up" with 1s from left to right, and then it will "fill up" with 0s again. One advantage of this type of sequence is that it is readily decoded with two-input AND gates. Diagrams of the timing operations of both the four- and five-stage counters are shown in Figures 8–34 and 8–35, respectively.

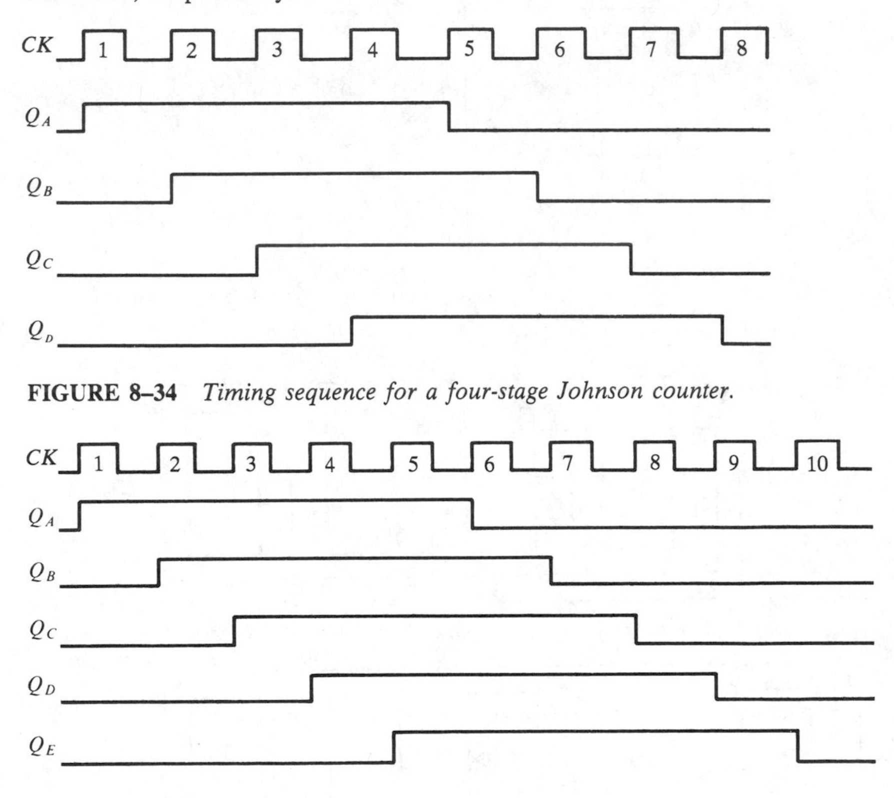

FIGURE 8–34 *Timing sequence for a four-stage Johnson counter.*

FIGURE 8–35 *Timing sequence for a five-stage Johnson counter.*

The Ring Counter

The ring counter utilizes one flip-flop for each stage in its sequence, and is therefore the most wasteful of flip-flops. It does have the advantage that decoding is not required for decimal conversion. A logic diagram for a ten-stage ring counter is shown in Figure 8–36. The sequence for this ring counter is given in Table 8–9.

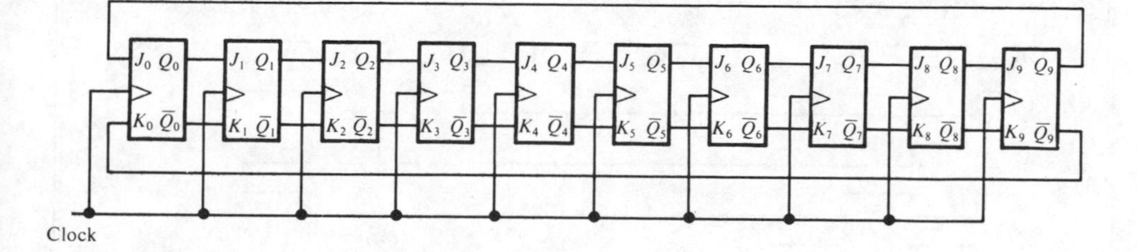

FIGURE 8–36 *A ten-stage ring counter.*

TABLE 8–9 *Ring counter sequence (10 bits).*

CK	Q_0	Q_1	Q_2	Q_3	Q_4	Q_5	Q_6	Q_7	Q_8	Q_9
0	1	0	0	0	0	0	0	0	0	0
1	0	1	0	0	0	0	0	0	0	0
2	0	0	1	0	0	0	0	0	0	0
3	0	0	0	1	0	0	0	0	0	0
4	0	0	0	0	1	0	0	0	0	0
5	0	0	0	0	0	1	0	0	0	0
6	0	0	0	0	0	0	1	0	0	0
7	0	0	0	0	0	0	0	1	0	0
8	0	0	0	0	0	0	0	0	1	0
9	0	0	0	0	0	0	0	0	0	1
	(back to 0)									

Actually, the ring counter can be thought of as a form of shift register since the 1 is simply shifted from one stage to the next. However, since a particular sequence is characteristic of this device, it is normally classified as a counter. Notice that the interstage connections are the same as for a Johnson counter, except that the feedback connections from the last stage to the first are not reversed. The ten outputs of the counter indicate directly the decimal count of the clock pulse. For instance, a 1 on Q_0 is a zero, a 1 on Q_1 is a one, a 1 on Q_2 is a two, a 1 on Q_3 is a three, and so on. You should verify for yourself that the 1 is always retained in the counter and simply shifted "around the ring," advancing one stage for each clock pulse.

Modified sequences can be achieved by having more than a single 1 in the counter, as will be illustrated in the following example.

Example 8–5

If the ten-stage ring counter of Figure 8–36 has the initial state 1010000000, determine the waveforms for each of the Q outputs.

Solutions:

See Figure 8–37.

Example 8–5, continued

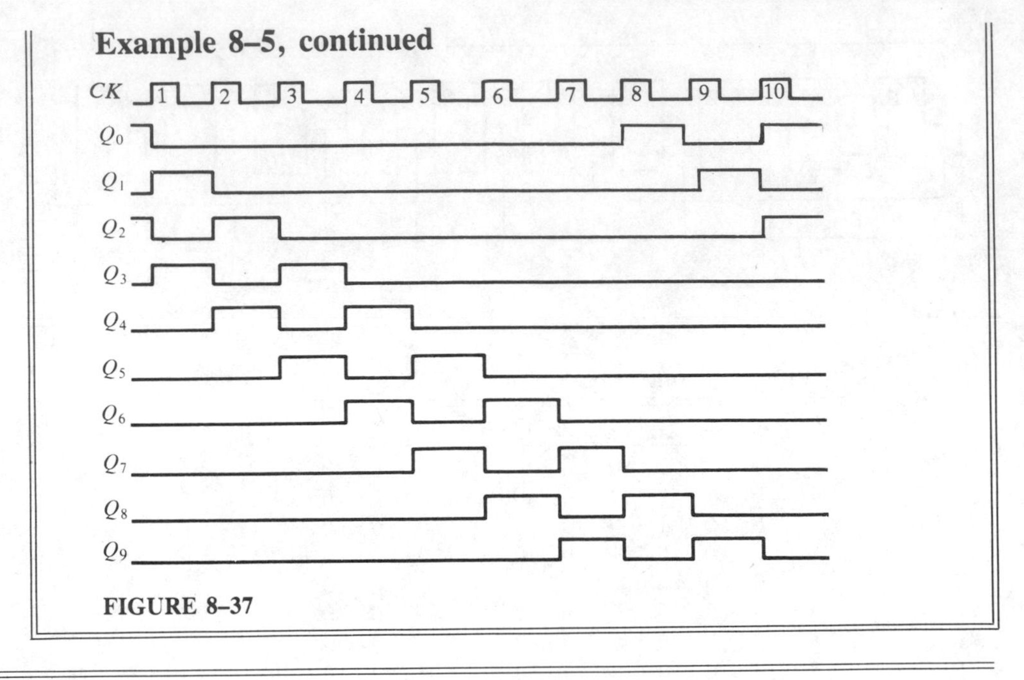

FIGURE 8–37

8–5 CASCADED COUNTERS

Counters can be connected in series or *cascaded* in order to achieve higher modulus operation. In essence, cascading means that the last stage of one counter drives the input of the next counter. An example of two counters connected in cascade is shown for a two-stage and a three-stage ripple counter in Figure 8–38. The timing diagram is in Figure 8–39.

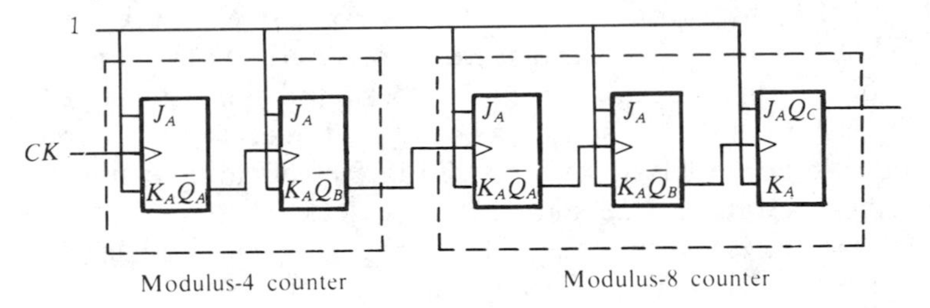

FIGURE 8–38 *Two cascaded counters.*

Notice in the timing diagram of Figure 8–39 that the final output of the modulus-8 counter Q_C occurs once for every 32 input clock pulses applied to the modulus-4 counter. The overall modulus of the cascaded counters is 32—that is, it acts as a divide-by-32 counter.

In general, the overall modulus of cascaded counters is equal to the product of each individual modulus. For instance, for the counter in Figure 8–38, $4 \times 8 = 32$.

When operating synchronous counters in a cascaded configuration, it is convenient to use the *count enable* and the *terminal count* functions to achieve

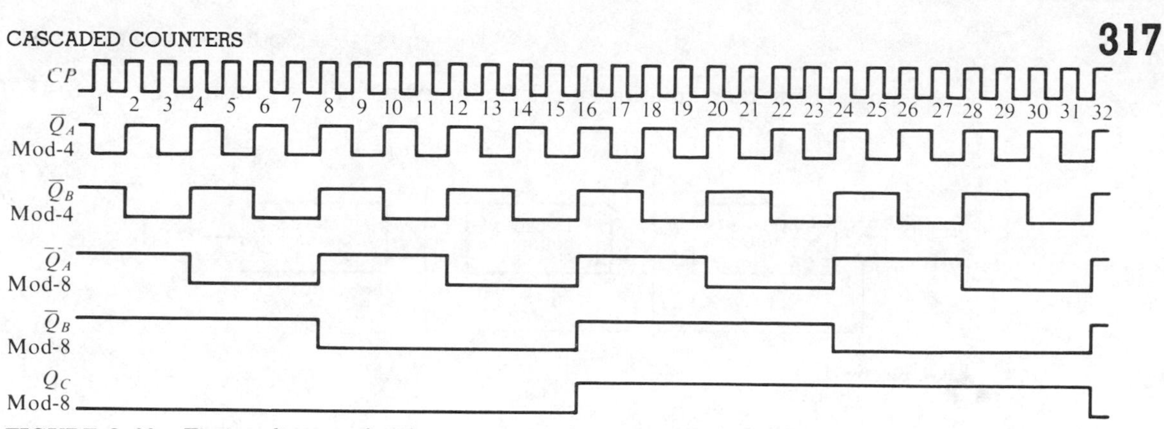

FIGURE 8–39 *Timing diagram for the counter configuration of Figure 8–38.*

higher modulus operation. On some devices the count enable is labeled simply *enable* or some other similar designation and terminal count is analogous to *ripple clock* or *carry output* on some IC counters.

Figure 8–40 shows two decade counters connected in cascade. The terminal count *(TC)* output of counter 1 is connected to the count enable *(CE)* input of counter 2. Counter 2 is inhibited by the LOW on its *CE* input until counter 1 reaches its last or terminal state and its terminal count output goes HIGH. This HIGH now enables counter 2, so that on the first clock pulse after counter 1 reaches its terminal count (CK_{10}), counter 2 goes from its initial state to its second state. Upon completion of the entire second cycle of counter 1 (when counter 1 reaches terminal count the second time), counter 2 is again enabled and advances to its next state. This sequence continues. Since these are decade counters, counter 1 must go through ten complete cycles before counter 2 complete its first cycle. In other words, for every ten cycles of counter 1, counter 2 goes through one cycle. This means that counter 2 will complete one cycle after 100 clock pulses. The overall modulus of these two cascaded counters is $10 \times 10 = 100$.

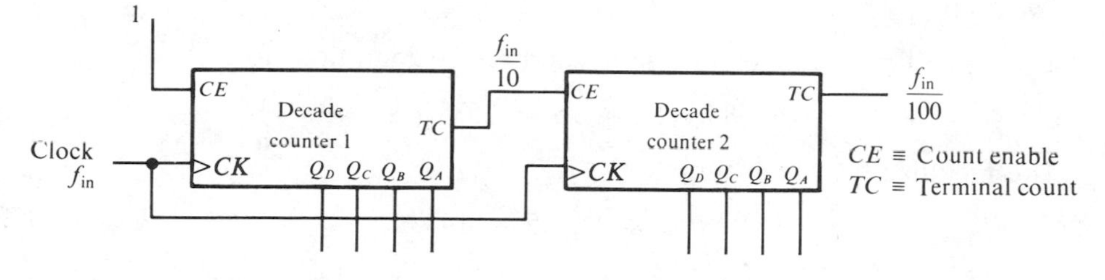

FIGURE 8–40 *A Modulus-100 cascaded counter.*

When viewed as a frequency divider, the circuit of Figure 8–40 divides the input clock frequency by 100. Cascaded counters are often used to divide a high-frequency clock signal to obtain highly accurate pulse frequencies. Cascaded counter configurations used for such purposes are sometimes called *countdown chains.*

For example, suppose we have a basic clock frequency of 1 MHz and we wish to obtain 100 kHz, 10 kHz, and 1 kHz; a series of cascaded counters can be used. If the 1-MHz signal is divided by ten, we get 100 kHz. Then if the 100-kHz signal is divided by ten, we get 10 kHz. Another division by ten yields the 1-kHz frequency. The implementation of this countdown chain is shown in Figure 8–41.

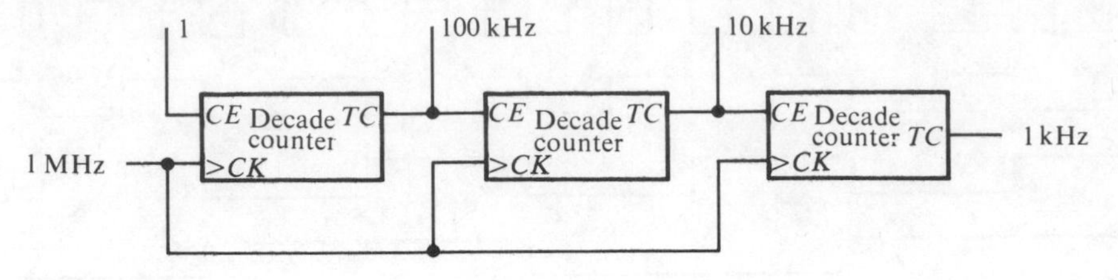

FIGURE 8–41

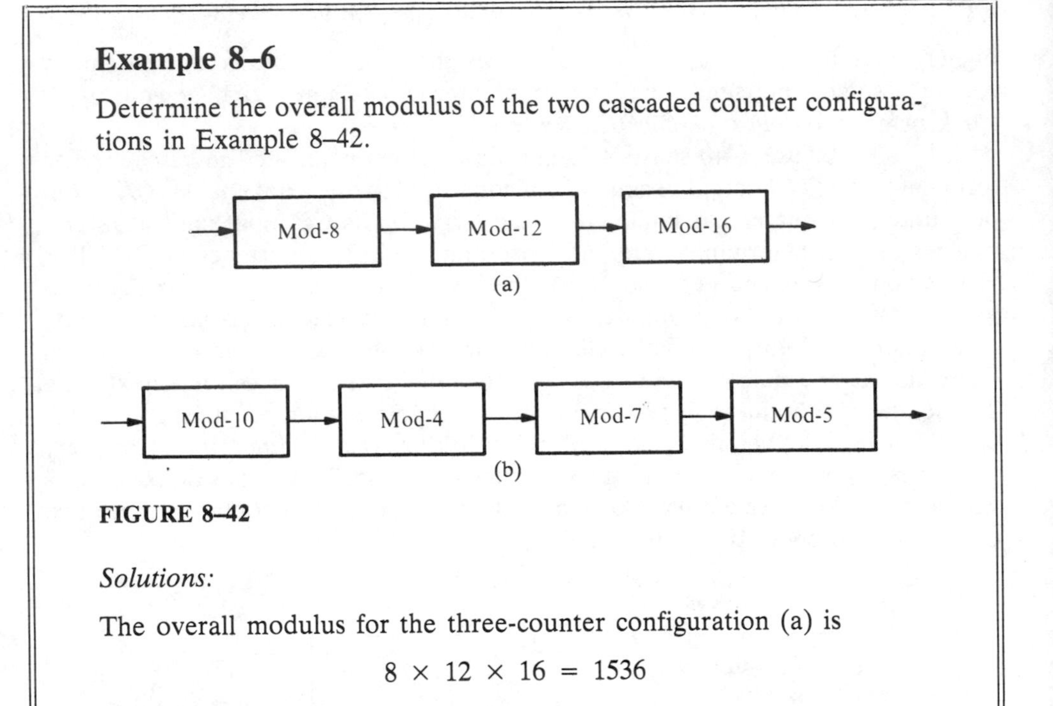

Example 8–6

Determine the overall modulus of the two cascaded counter configurations in Example 8–42.

FIGURE 8–42

Solutions:

The overall modulus for the three-counter configuration (a) is

$$8 \times 12 \times 16 = 1536$$

The modulus for four counters (b) is

$$10 \times 4 \times 7 \times 5 = 1400$$

Example 8–7

Use 7490A counters to obtain a 10-kHz square wave from a 1-MHz clock. Show the connection diagram.

Solution:

To obtain 10 KHz from a 1-MHz clock requires a division factor of 100.

Two 7490A counters in the divide-by-ten configuration must be cascaded as shown in Figure 8–43.

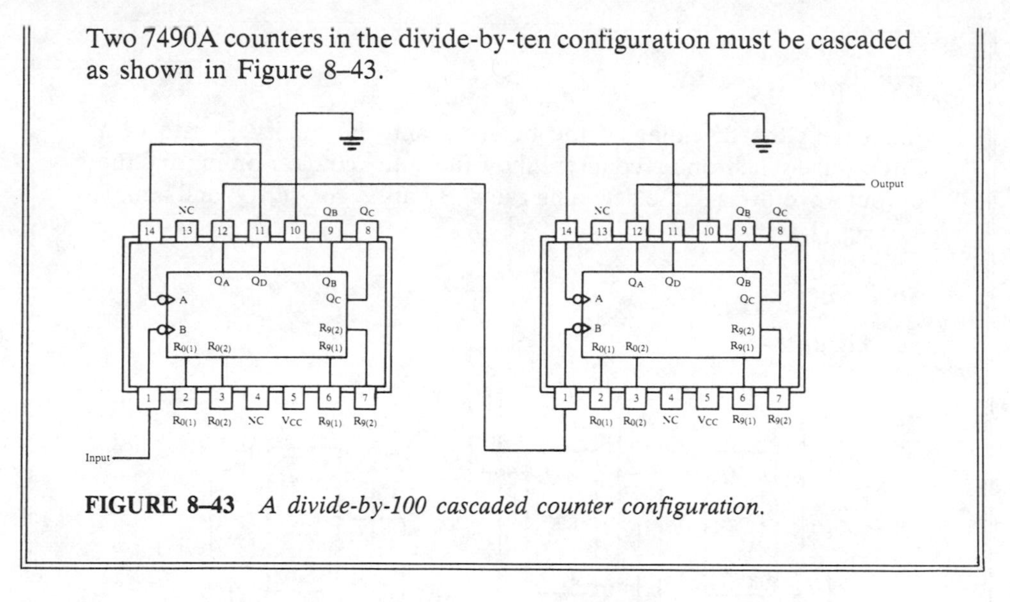

FIGURE 8–43 *A divide-by-100 cascaded counter configuration.*

COUNTER DECODING 8–6

In many digital applications it is necessary that some or all of the counter states be decoded. The decoding of a counter involves using decoders or logic gates to determine when the counter is in a certain state or states in its sequence. For instance, the terminal count function previously discussed is a case of a single state (the last state) in the counter sequence. Any state in a counter sequence can be decoded in a similar manner.

For example, let us say we wish to decode the binary 6 state of a three-stage binary counter. This can be done as shown in Figure 8–44. When $Q_C = 1$, $Q_B = 1$, and $Q_A = 0$, a HIGH appears on the output of the decoding gate indicating that the counter is in state 6.

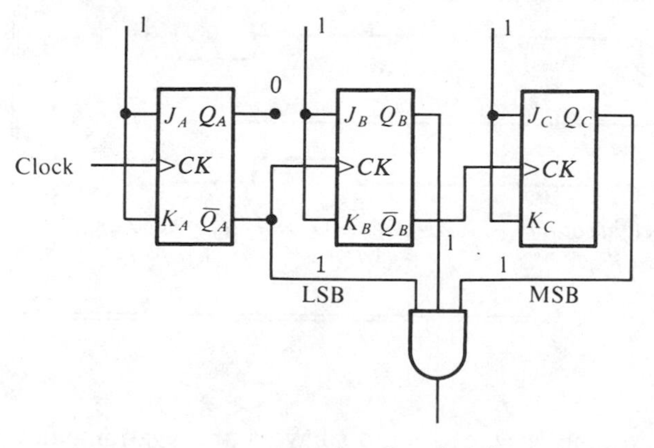

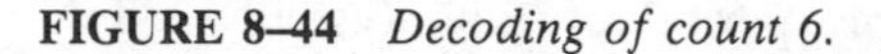
FIGURE 8–44 *Decoding of count 6.*

Example 8–8

Implement the decoding of the binary 2 and the binary 7 state of a three-stage synchronous counter. Show the entire counter timing and the output waveforms of the decoding gates. Binary 2 = $\overline{Q}_C Q_B \overline{Q}_A$ and binary 7 = $Q_C Q_B Q_A$.

Solutions:

See Figure 8–45.

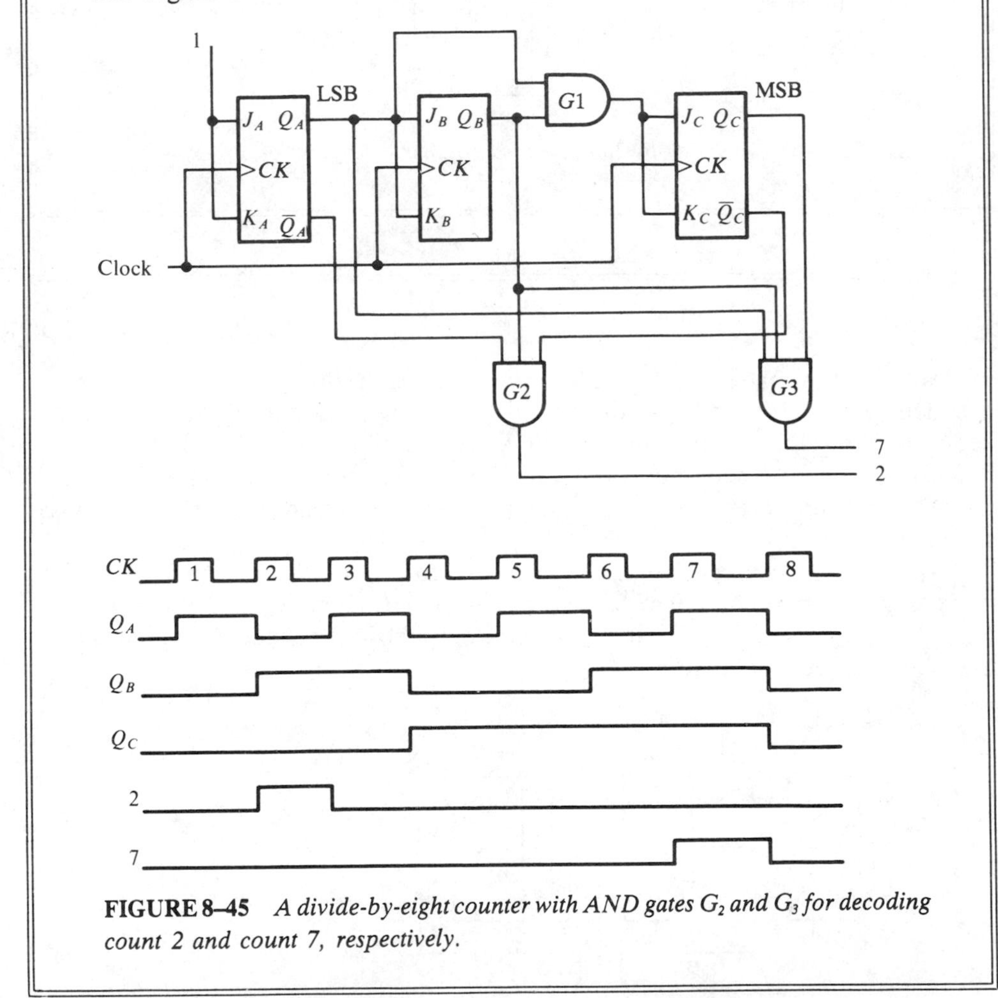

FIGURE 8–45 *A divide-by-eight counter with AND gates G_2 and G_3 for decoding count 2 and count 7, respectively.*

Decoding Glitches

Recall that asynchronous (ripple) counters have a propagation delay from the triggering transition on the output of one flip-flop to the resulting change in the output of the next flip-flop. For a short period of time (usually a few nanoseconds) the counter is in a transition state as the flip-flops change state. A decoder can respond to these transition states and produce undesired short duration spikes or *glitches* on the output which may cause problems in certain applications.

As an example of the decoding glitch problem, let us look at an asynchronous BCD decade counter with ouputs connected to a 4-line-to-10-line decoder as in Figure 8–46.

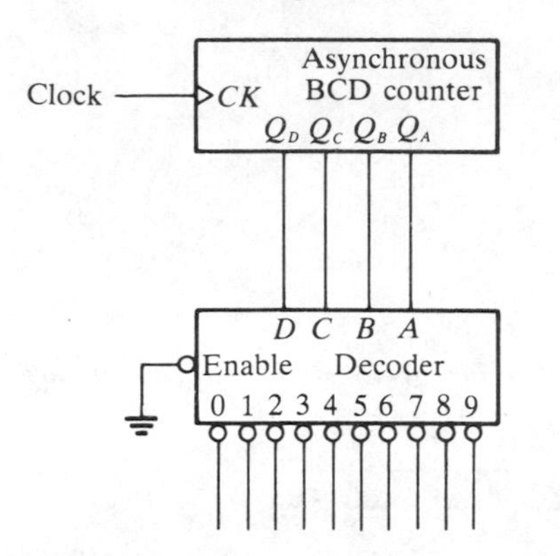

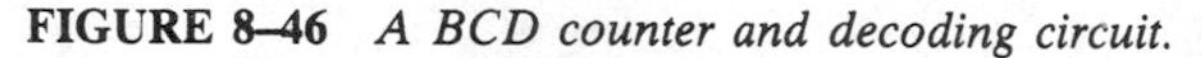
FIGURE 8–46 *A BCD counter and decoding circuit.*

To see what happens in this case, we will look at a timing diagram where the clocking delays are taken into account as in Figure 8–47. Notice how the transition delays cause false states of short duration. The decimal value of the false binary state at each critical transition is indicated on the diagram. The resulting glitches can be seen on the decoder outputs.

One way to eliminate the glitches is to sample the decoded outputs at a time *after* the glitches have had time to disappear. This method is known as *strobing* and can be accomplished in this case by using the LOW level of the clock to enable the decoder, as shown in Figure 8–48. The resulting improved timing diagram is shown in Figure 8–49.

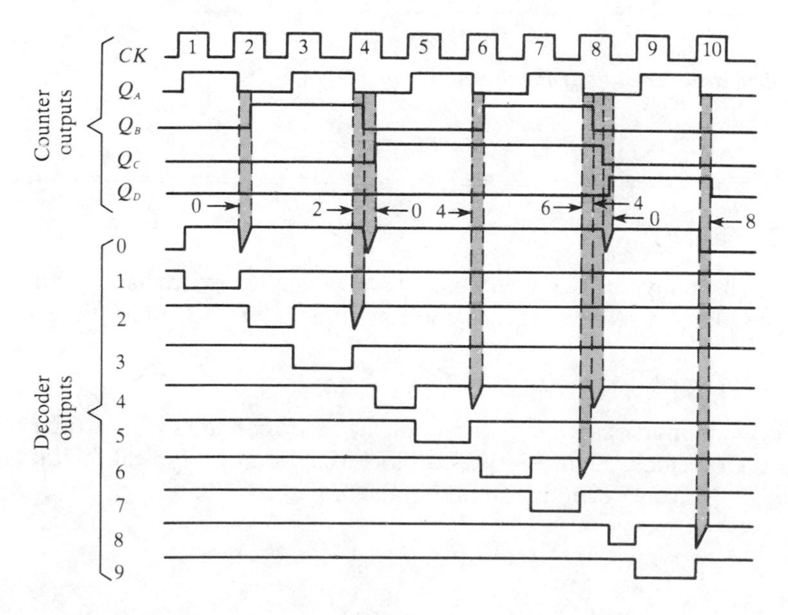

FIGURE 8–47 *Outputs with glitches from the decoder in Figure 8–46.*

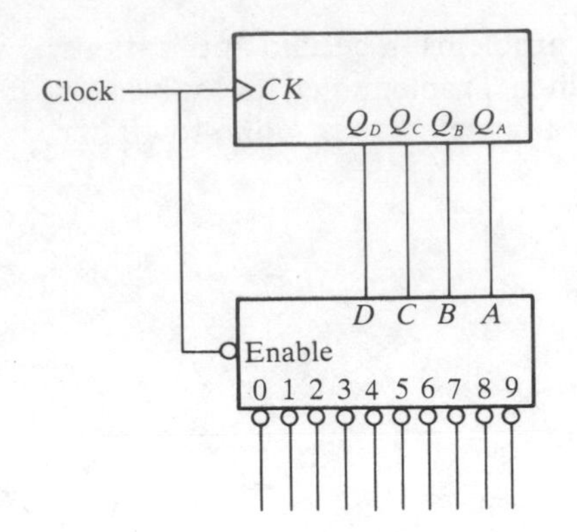

FIGURE 8–48 *A BCD counter and decoder circuit of Figure 8–46 modified to eliminate glitches.*

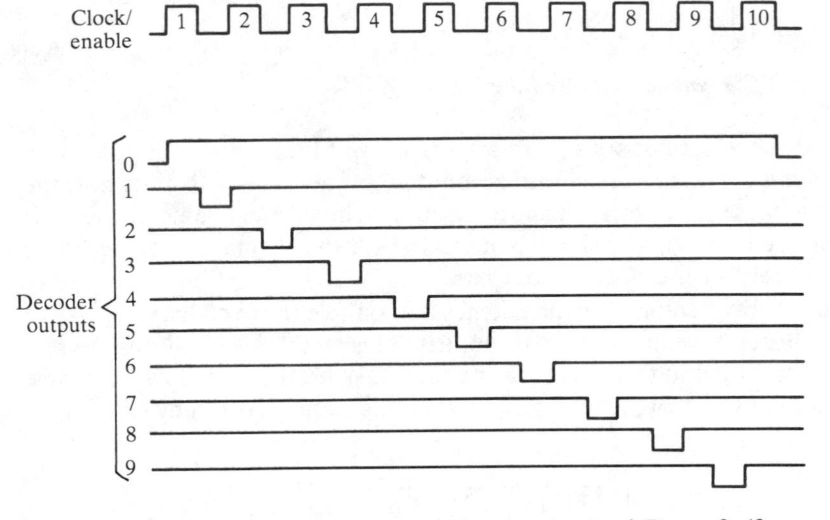

FIGURE 8–49 *Strobed decoder outputs for the circuit of Figure 8–48.*

8–7 COUNTER APPLICATIONS

The digital counter in its many forms is a very useful and versatile device that is used in many applications. In this section we will look at some example applications.

The Digital Clock

A very common example of the application of counters is in a time keeping device or digital clock. Figure 8–50 is a block diagram of a typical digital clock arrangement to indicate seconds, minutes, and hours.

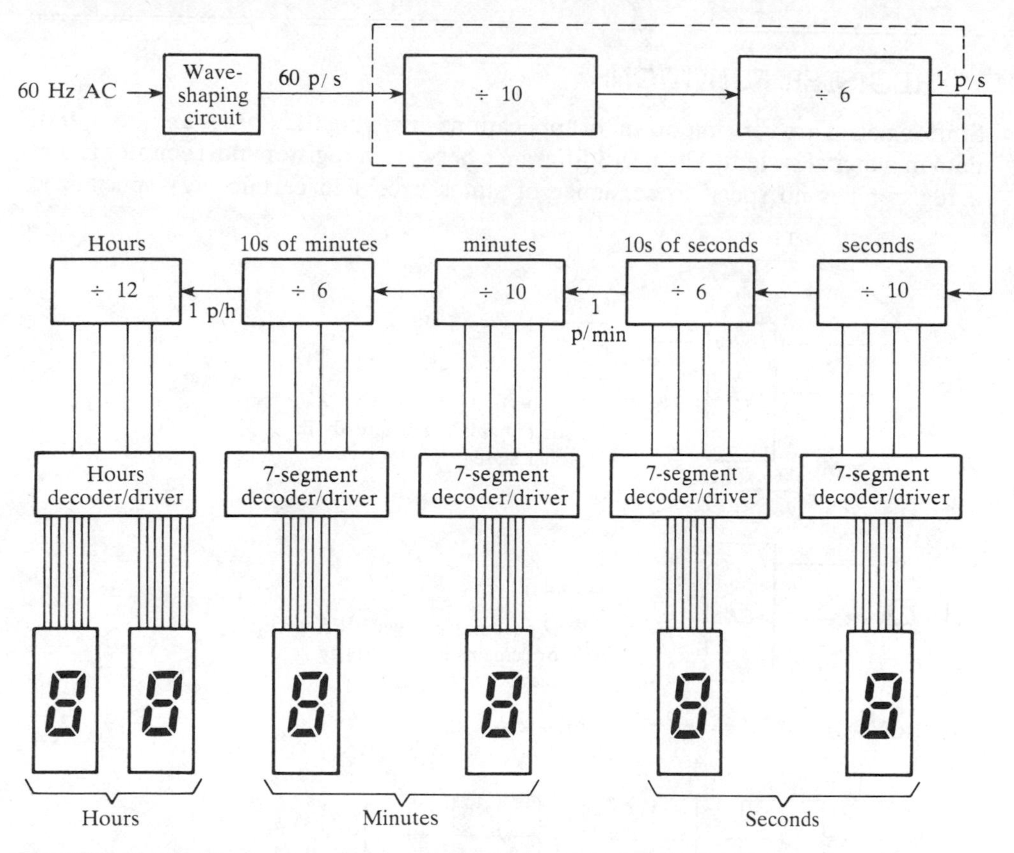

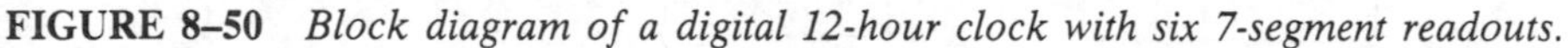

FIGURE 8–50 *Block diagram of a digital 12-hour clock with six 7-segment readouts.*

The input frequency of 60 Hz is first shaped to produce a square-wave signal and then counted down by a mod-60 counter to obtain a 1-Hz pulse waveform. This can be implemented with a decade counter followed by a divide-by-six as shown.

The *seconds count* is produced by a BCD counter followed by a divide-by-six that counts from 0 to 59 and then recycles to 0. A seven-segment decoder/driver translates the counter's state into a seven-segment display. The terminal count of the seconds divide-by-six counter is a 1-pulse/minute signal used to enable the *minutes counter,* which is also a BCD counter followed by a divide-by-six. The output of the minutes counter is a 1-pulse/hour signal used as input to the divide-by-twelve hours counter. The hours counter must be decoded somewhat differently from the preceeding counters because it goes through states 1 to 12 and then recycles to 1.

Actually the digital clock function with additional functions such as sleep timer, snooze alarm, AM/PM circuitry, etc. is contained in a single integrated circuit package available from several manufacturers. We have used the digital clock as only one example of how counters can be used.

8–8 SHIFT REGISTER FUNCTIONS

Shift registers are very important in applications involving the *storage* and *transfer* of data in a digital system. The basic difference between a register and a counter is that a register has no specified sequence of states except in certain very specialized

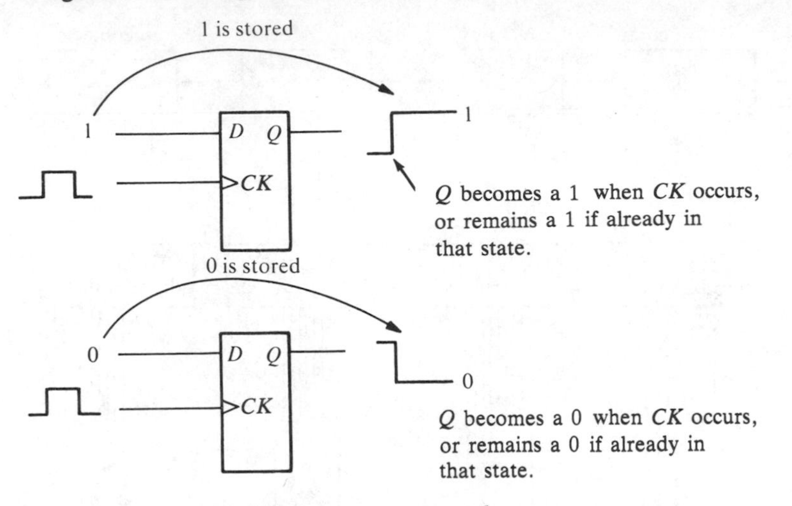

FIGURE 8–51 *The flip-flop as a storage element.*

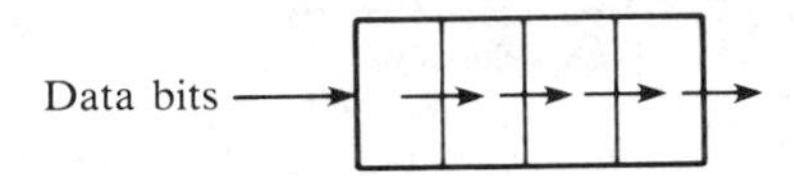

(a) Serial shift in and right, then out.

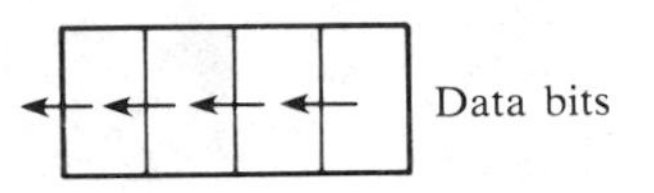

(b) Serial shift left, then out.

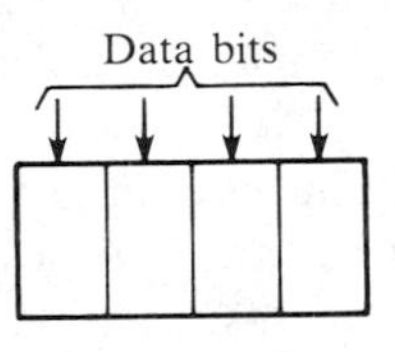

(c) Parallel shift in.

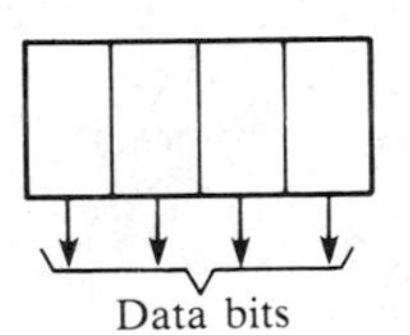

(d) Parallel shift out.

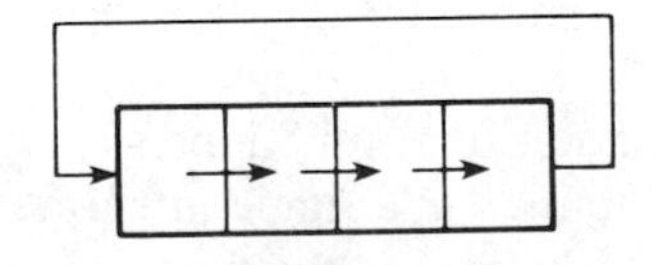

(e) Rotate right.

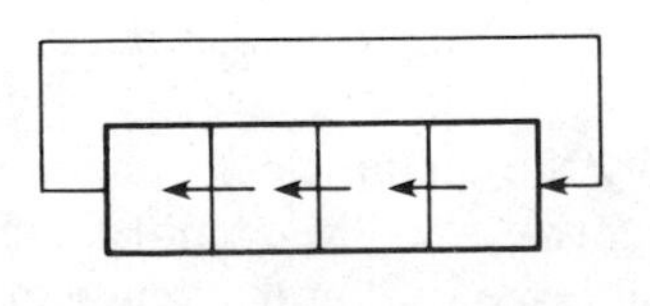

(f) Rotate left.

FIGURE 8–52 *Basic data movement in registers.*

applications. A register, in general, is used solely for the purpose of *storing* and *shifting* data (1s and 0s) entered into it from an external source and possesses no characteristic internal sequence of states.

The storage capability of a register is one of its two basic functional characteristics and makes it an important type of *memory* device. Figure 8–51 illustrates the concept of "storing" a 1 or a 0 in a flip-flop. A 1 is applied to the input as shown, and a clock pulse is applied that stores the 1 by setting the flip-flop. When the 1 on the input is removed, the flip-flop remains in the SET state thereby storing the 1. The same procedure applies to the storage of a 0, as also illustrated in Figure 8–51.

The *storage capacity* of a register is the number of bits (1s and 0s) of digital data it can retain. Each stage of a shift register represents one bit of storage capacity, and therefore the number of stages in a register determines its total storage capacity.

Registers are commonly used for the *temporary* storage of data within a digital system. Registers are implemented with a flip-flop for each stage or by utilizing the inherent capacitance in MOSFET devices for the storage elements. The *shift* capability of a register permits the movement of data stored from stage to stage within the register or into or out of the register upon application of clock pulses. Figure 8–52 shows symbolically the types of data movement in shift register operations. The block represents any arbitrary register, and the arrow indicates direction and type of data movement.

8–9 SERIAL IN–SERIAL OUT SHIFT REGISTERS

This type of shift register accepts digital data serially—that is, one bit at the time on one line. It produces the stored information on its output also in serial form. Let us first look at the serial entry of data into a typical shift register with the aid of Figure 8–53, which shows a four-bit device implemented with *S–R* flip-flops.

With four stages, this register can store up to four bits of digital data; its *storage capacity* is four bits. We will illustrate the entry of the four-bit binary number 1010 into the register, beginning with the right-most bit. The 0 is put onto the data input line, making $S_A = 0$ and $R_A = 1$; when the first clock is applied, FF*A* is RESET, thus "storing" the 0. Next the 1 is applied to the data input, making $S_A = 1$ and $R_A = 0$; $S_B = 0$ and $R_B = 1$, because they are connected to the Q_A and $\overline{Q}_A$ outputs, respectively. When the second clock pulse occurs, the 1 on the data input is "shifted" into FF*A* because FF*A* SETS, and the 0 that was in FF*A* is "shifted" into FF*B*. The

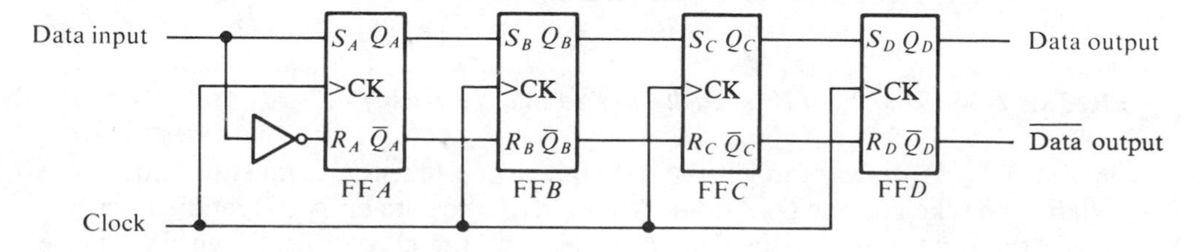

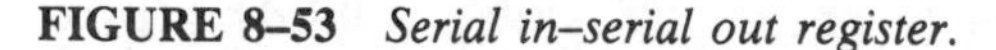
FIGURE 8–53 *Serial in–serial out register.*

next 0 in the binary number is now put onto the data input line and a clock pulse is applied. The 0 is entered into FF*A*, the 1 stored in FF*A* is shifted into FF*B*, and the 0 stored in FF*B* is shifted into FF*C*. Examination of the *S* and *R* inputs of each of the flip-flops will verify this operation. The last bit in the binary number, a 1, is now applied to the data input, and a clock pulse is applied to the *CK* line. This time the 1 is entered into FF*A*, the 0 stored in FF*A* is shifted into FF*B*, the 1 stored in FF*B* is shifted into FF*C*, and the 0 stored in FF*C* is shifted into FF*D*. This completes the serial entry of the four-bit number into the shift register, where it can be retained for any length of time. Figure 8–54 illustrates each step in the shifting of the four bits into the register.

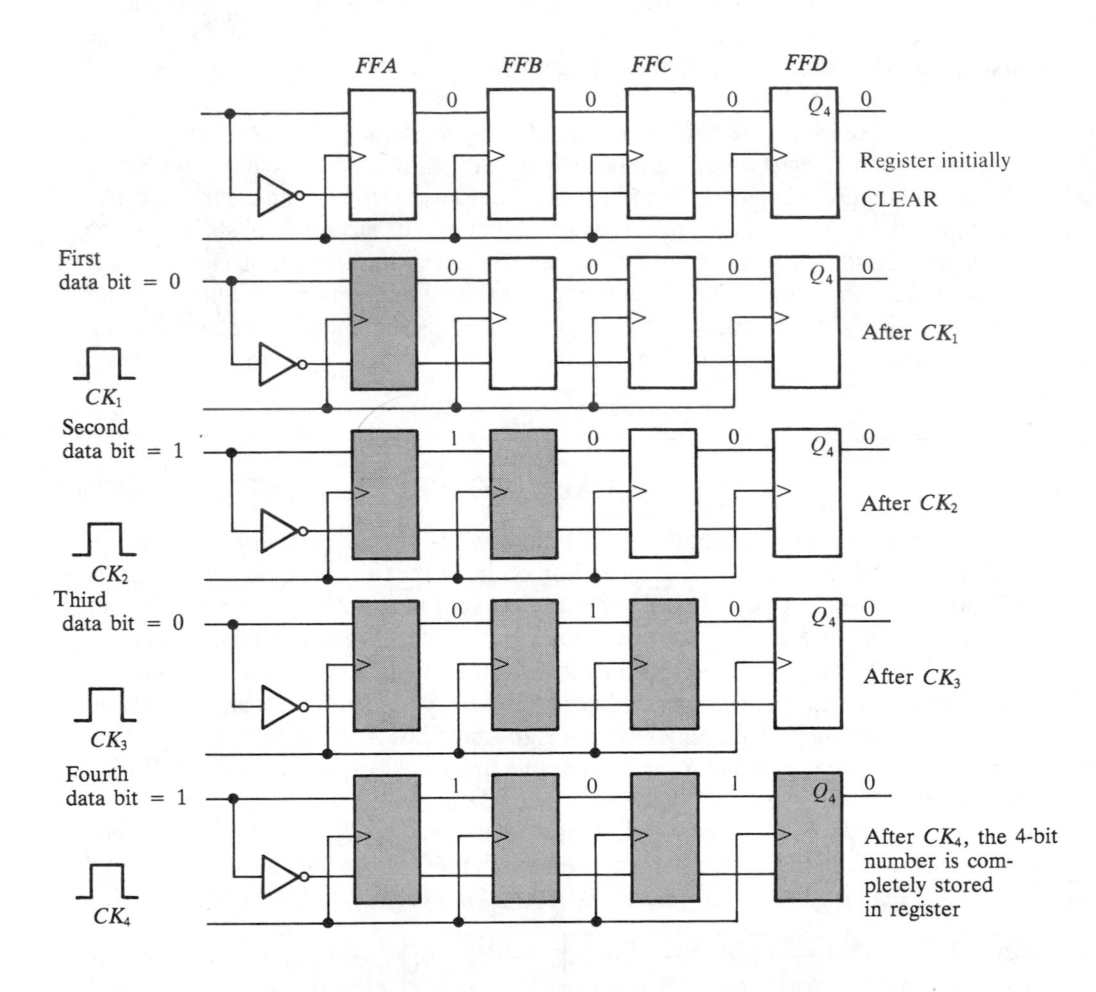

FIGURE 8–54 *Four bits being serially entered into the register.*

If we want to get the data out of the register, it must be shifted out serially and taken off the Q_D output. After CK_4 in the data entry operation described above, the right-most 0 in the number appears on the Q_D output. When clock pulse CK_5 is applied, the second bit appears on the Q_D output. CK_6 shifts the third bit to the output, and CK_8 shifts the fourth bit to thc output, as illustrated in Figure 8–55. Notice that while the original four bits are being shifted out, a new four-bit number can be shifted in.

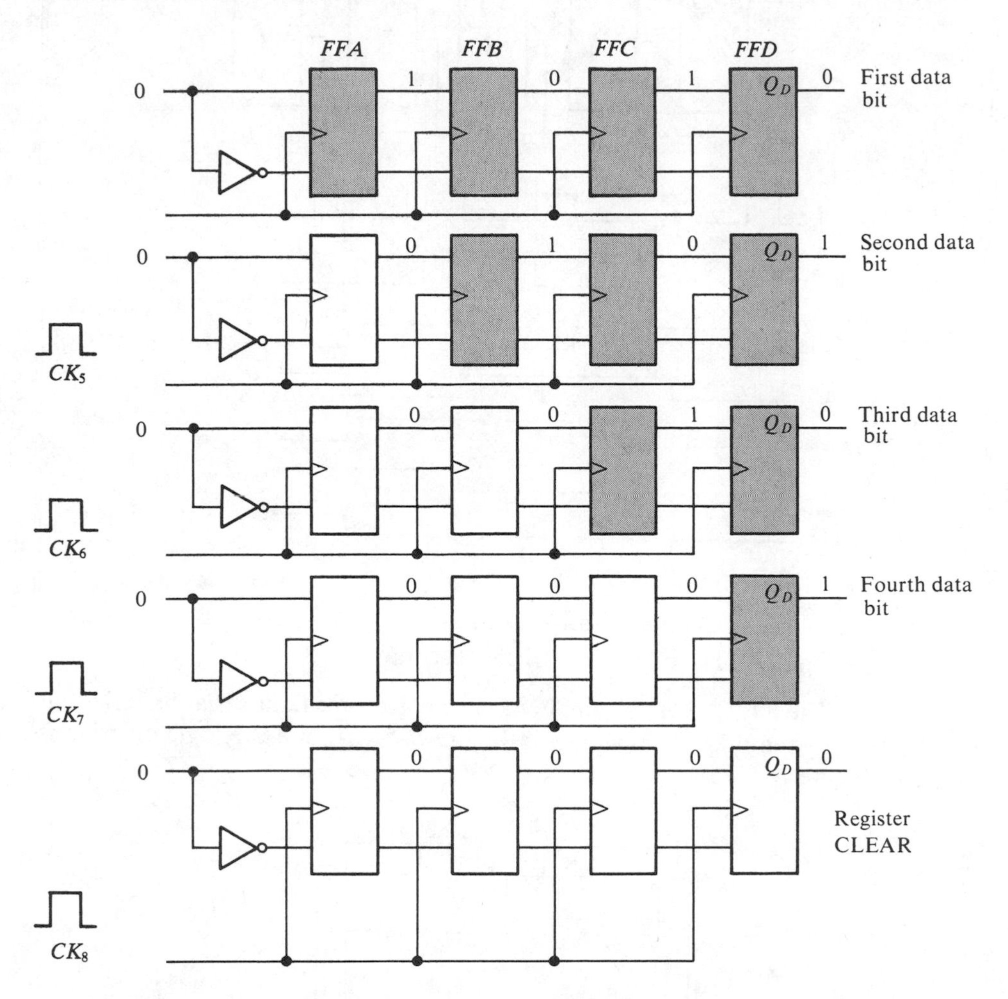

FIGURE 8–55 *Four bits being serially shifted out of the register.*

Example 8–9

Show the states of the five-bit register in Figure 8–56(a) for the specified data input and clock waveforms. Assume the register is initially cleared (all 0s). D flip-flops are used for this particular implementation.

Solutions:

See Figure 8–56(b).

Example 8–9, continued

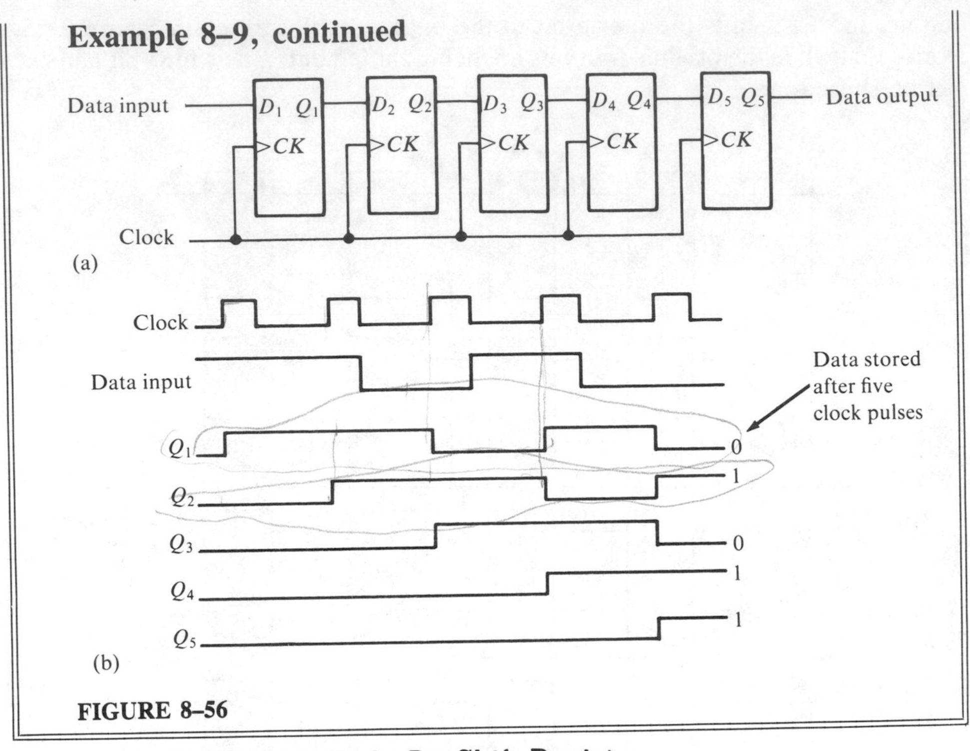

FIGURE 8–56

The 7491A Eight-Bit Shift Register

The 7491A is an example of an integrated circuit serial in-serial out shift register. A pin diagram is shown in Figure 8–57(a), and the logic diagram is shown in Figure 8–57(b).

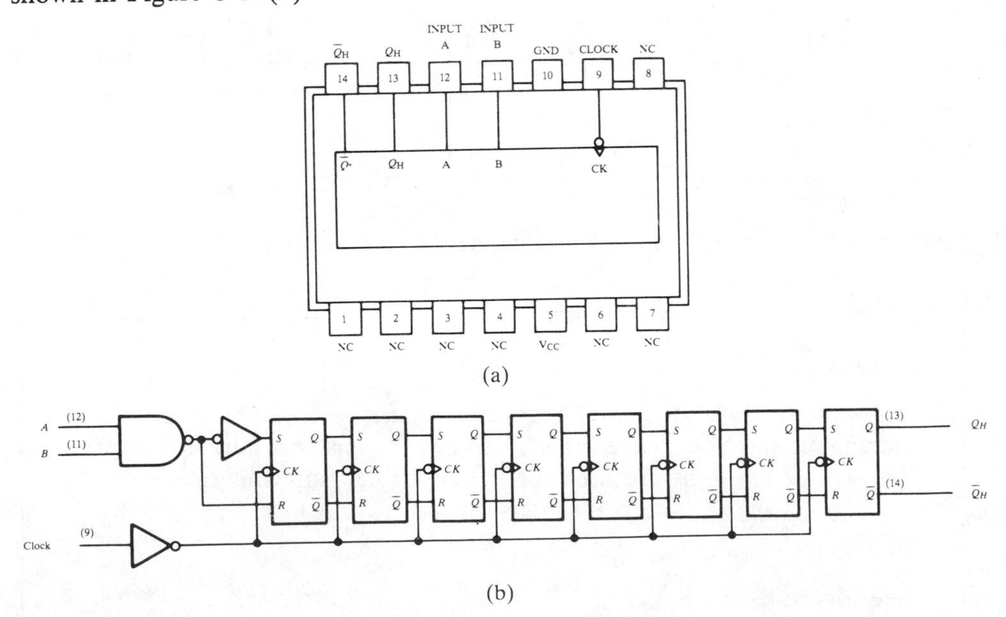

FIGURE 8–57 *A 7491A eight-bit serial in–serial out shift register.*

There are two gated input lines for serial data entry. If data are entered on *A*, then *B* must be HIGH and vice versa. The data output is Q_H and the data complement output is $\overline{Q}_H$.

SERIAL IN–PARALLEL OUT SHIFT REGISTERS 8–10

Data are entered into this type of register in the same manner as discussed in the last section—serially. The difference is the way in which the data are taken out of the register; in the parallel output register, the output of each stage is available. Once the data are stored, each bit appears on its respective output line and all bits are available simultaneously, rather than on a bit-by-bit basis as with the serial output. Figure 8–58 shows a four-bit serial in-parallel out register and its equivalent logic block symbol.

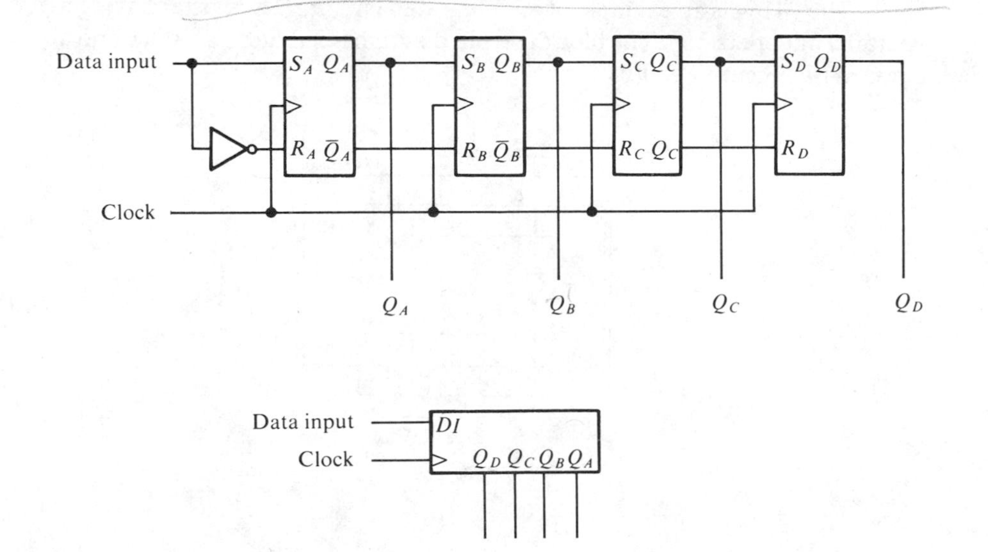

FIGURE 8–58 *A serial in–parallel out register.*

Example 8–10

Show the states of the four-bit register for the data input and clock waveforms in Figure 8–59(a). The register initially contains all 1s.

Solutions:

The counter contains 0110 after four clock pulses. See Figure 8–59(b).

Example 8–10, continued

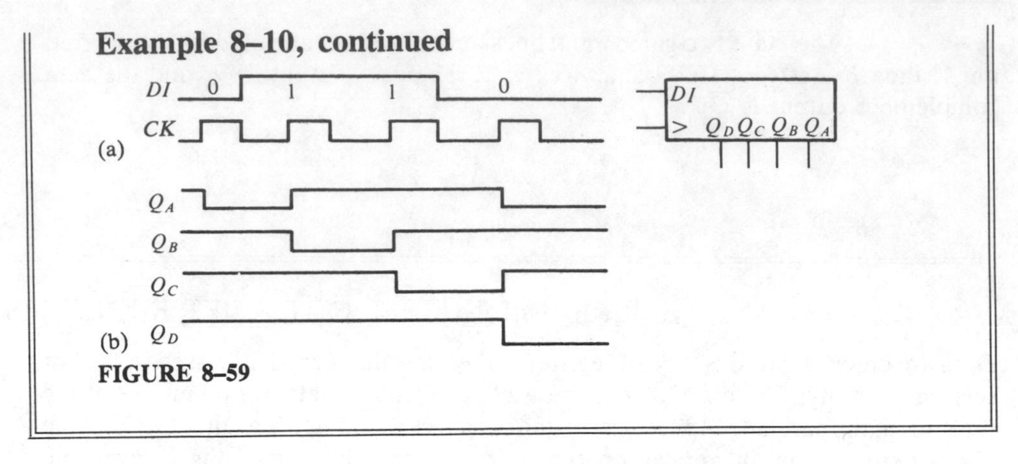

FIGURE 8–59

The 74164 Eight-Bit Parallel Out Serial Shift Register

The 74164 is an example of an integrated circuit shift register having a serial in-parallel out operation. The block and pin diagram is in Figure 8–60(a), and a logic diagram is in Figure 8–60(b).

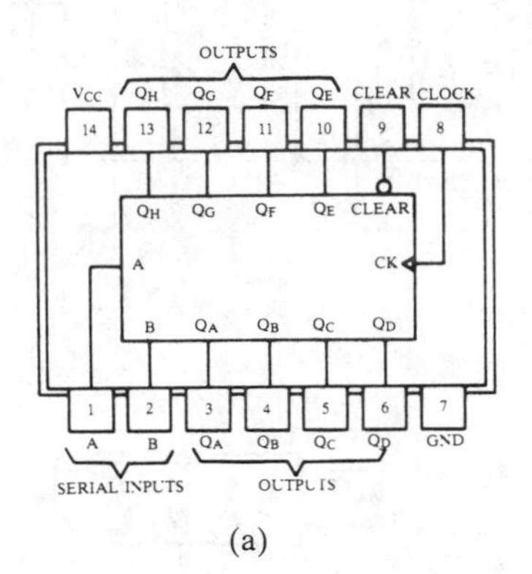

(a)

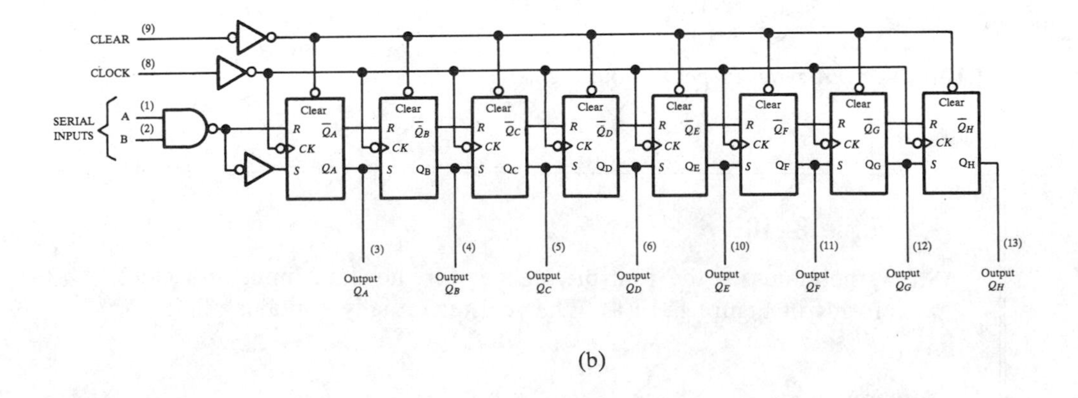

(b)

FIGURE 8–60 *A 74164 eight-bit parallel out serial shift register.*

Notice that this device has two gated serial inputs, A and B, and a *clear* input which is active LOW. The parallel outputs are Q_A through Q_H. Figure 8–61 is an example timing diagram for this register.

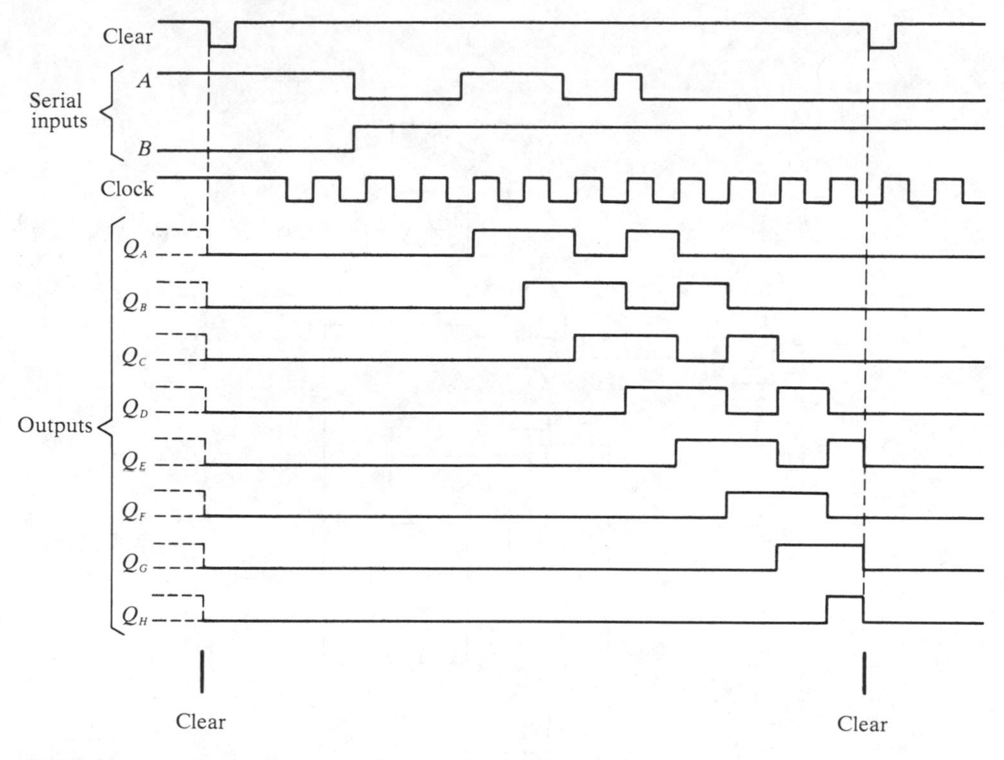

FIGURE 8–61 *Example timing diagram for a 74164 shift register.*

PARALLEL IN–SERIAL OUT SHIFT REGISTERS 8–11

For a register with parallel data inputs, the bits are entered simultaneously into their respective stages on parallel lines rather than on a bit-by-bit basis on one line as with serial data inputs. The serial output is executed as described in Section 8–9 once the data are completely stored in the register. Figure 8–62 illustrates a four-bit parallel in-serial out register. Notice that there are four data input lines, D_A, D_B, D_C, and D_D, and a parallel enable line, PE, that allows four bits of data to be entered in parallel into the register. When the PE line is HIGH, gates G_1 through G_4 are enabled, allowing each data bit to be applied to the S input of its respective flip-flop and the data bit complement to be applied to the R input. When a clock pulse is applied, the flip-flops with a 1 data bit will SET and those with a 0 data bit will RESET, thereby storing the data word on one clock pulse.

When the PE line is LOW, parallel data input gates G_1 through G_4 are disabled and shift enable gates G_5 through G_7 are enabled, allowing the data bits to shift from one stage to the next. The OR gates allow either the normal shifting operation to be carried out or the parallel data entry operation to be accomplished, depending on which AND gates are enabled by the level on the PE line.

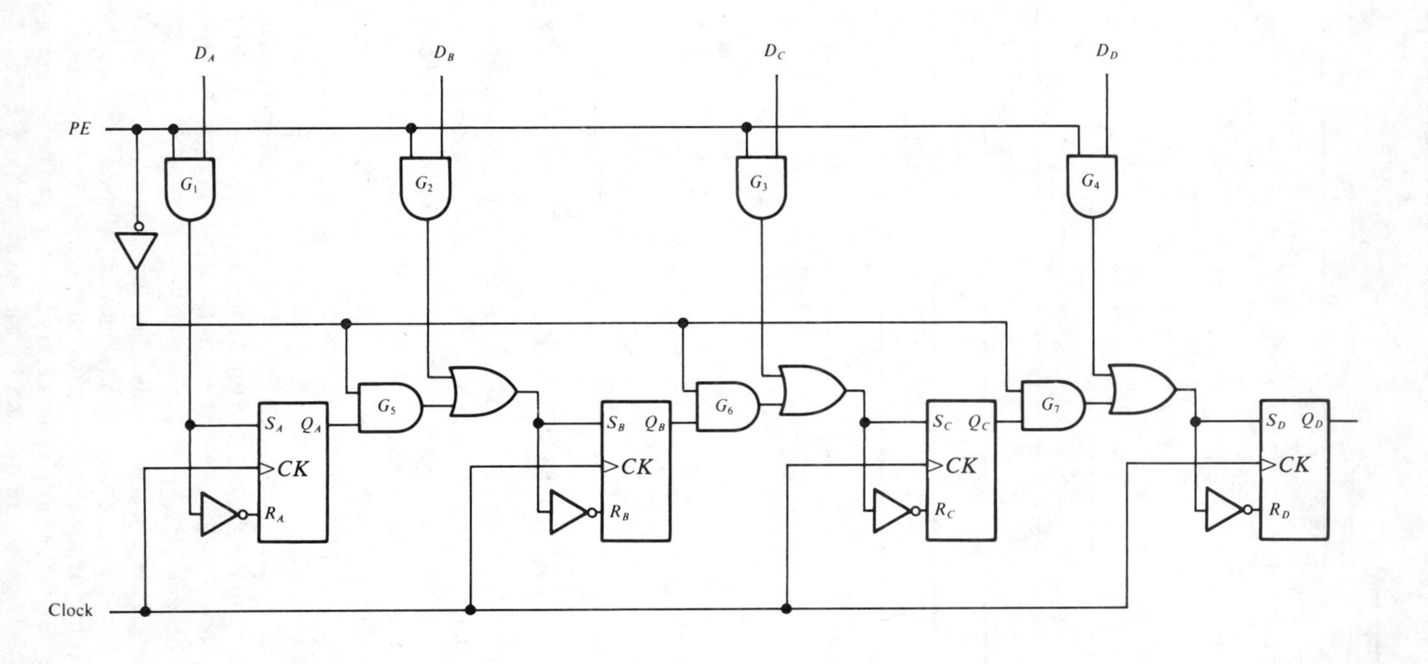

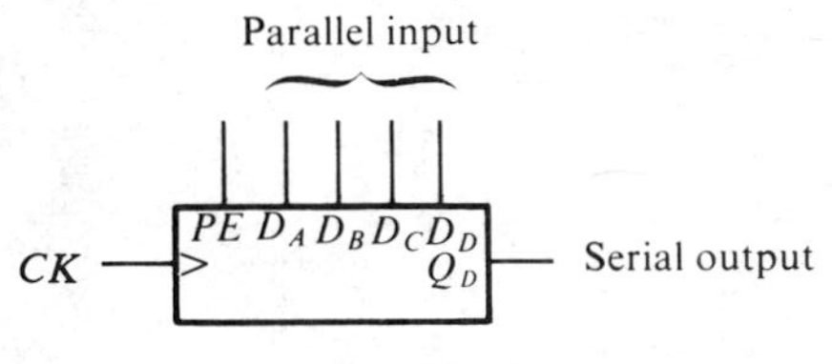

FIGURE 8–62 *Parallel in–serial out register.*

Example 8–11

Show the Q_D waveform for a four-bit register with the input data, *PE,* and clock waveform as given in Figure 8–63(a). Refer to Figure 8–62 for the logic diagram. The register is initially cleared.

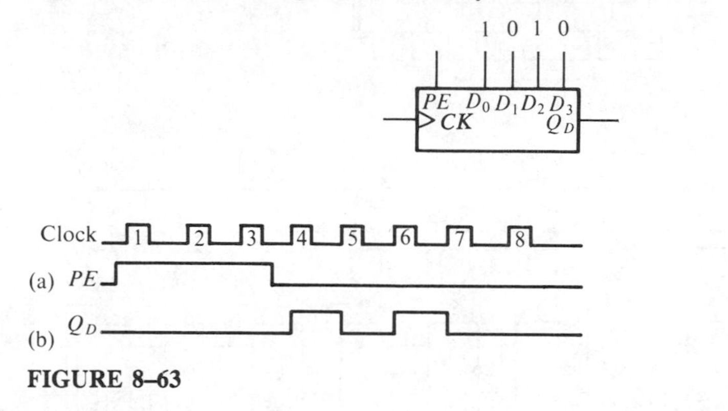

FIGURE 8–63

Solution:

On clock pulses 1, 2, and 3, the same parallel data are respectively loaded into the register, keeping Q_D a 0. On clock pulse 4, the 1 from Q_C is shifted onto Q_D; on clock pulse 5, the 0 is shifted onto Q_D; on clock pulse 6, the next 1 is shifted onto Q_D; and on clock pulses 7 and 8, all data bits have been shifted out and only 0s remain in register because no new data have been entered. See Figure 8–63(b).

The 74165 Eight-Bit Parallel Load Shift Register

The 74165 is an example of an integrated circuit device that has a parallel in-serial out operation. The block and pin diagram is shown in Figure 8–64(a), and the logic diagram is in Figure 8–64(b).

The *shift-load* input performs basically the same function as the *parallel enable (PE)* input of the generalized register previously discussed. Notice the method in which the parallel data bits are loaded. The asynchronous *preset* and *clear* inputs are used to store each data bit in the appropriate flip-flop. This approach is different from the synchronous method of parallel loading used in the generalized register and points up the fact that there are usually several ways to accomplish the same function.

Also, notice that the 74165 has provisions for serial data entry (pin 10). The Q_H line (pin 7) is the serial data output.

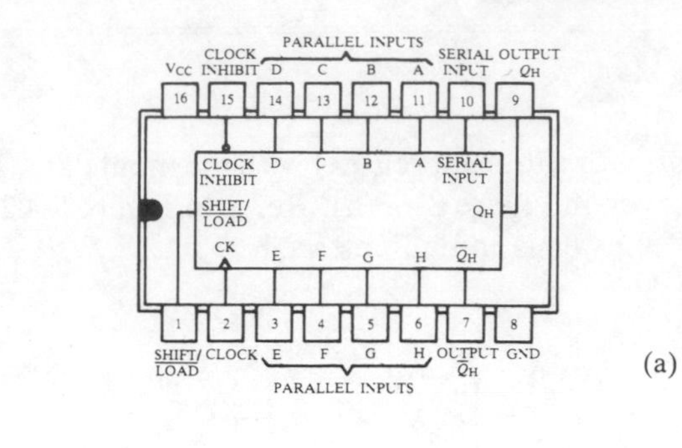

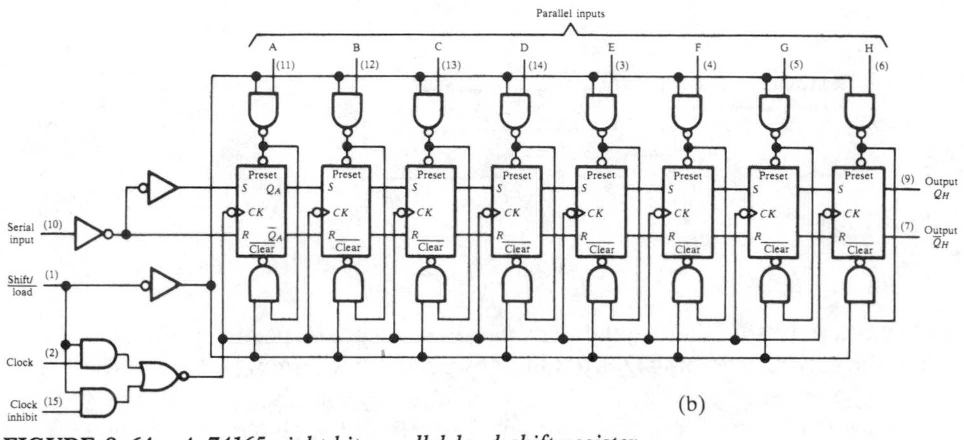

FIGURE 8–64 *A 74165 eight-bit parallel load shift register.*

8–12 PARALLEL IN–PARALLEL OUT SHIFT REGISTERS

Parallel entry of data was described in Section 8–11, and parallel output of data was also previously discussed. The parallel in-parallel out register employs both methods—immediately following the simultaneous entry of all data bits, the bits appear on the parallel outputs. This type of register is shown in Figure 8–65.

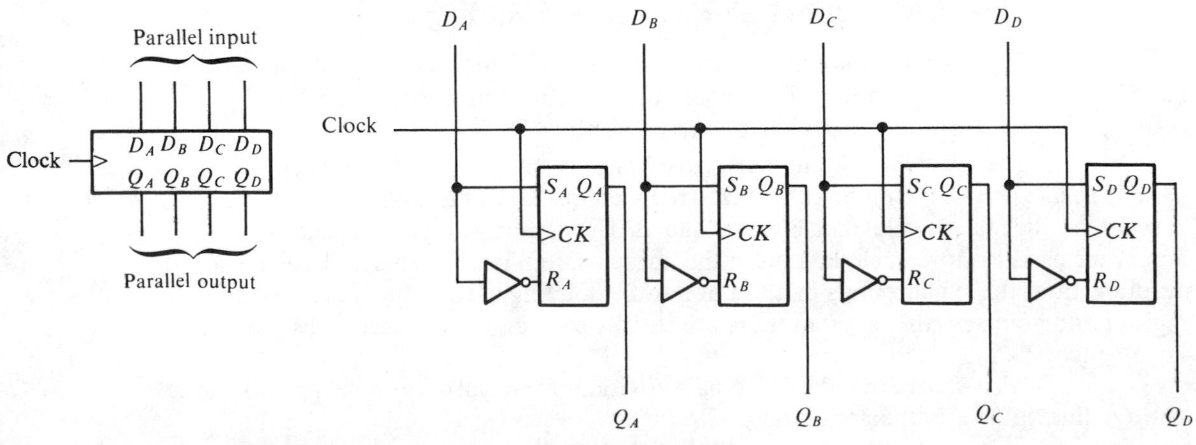

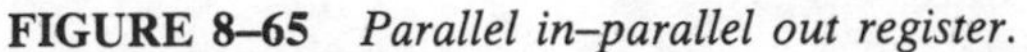
FIGURE 8–65 *Parallel in–parallel out register.*

The 74195 Four-Bit Shift Register

An example of an integrated circuit shift register having a parallel in-parallel out arrangement is the 74195. As shown in Figure 8–66, this device has serial inputs in addition to the parallel inputs. It also has an active LOW *clear* and a *shift-load* input. To load parallel data, the *shift-load* input must be LOW. When a positive clock transition occurs, the data are stored in the flip-flops and appear on the parallel outputs. When the *shift-load* input is HIGH data stored in the register will shift in a direction from Q_A toward Q_D. An example timing diagram is shown in Figure 8–67.

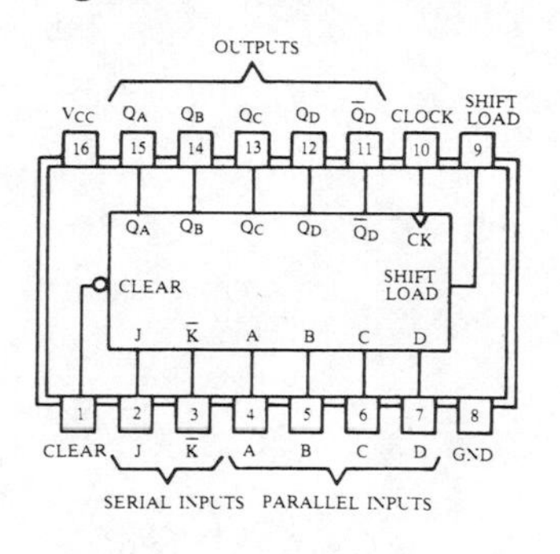

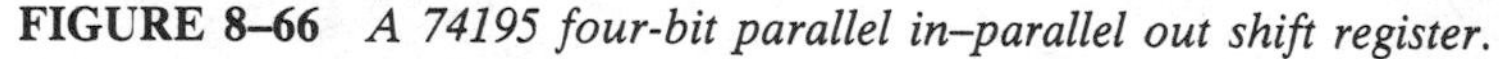
FIGURE 8–66 *A 74195 four-bit parallel in–parallel out shift register.*

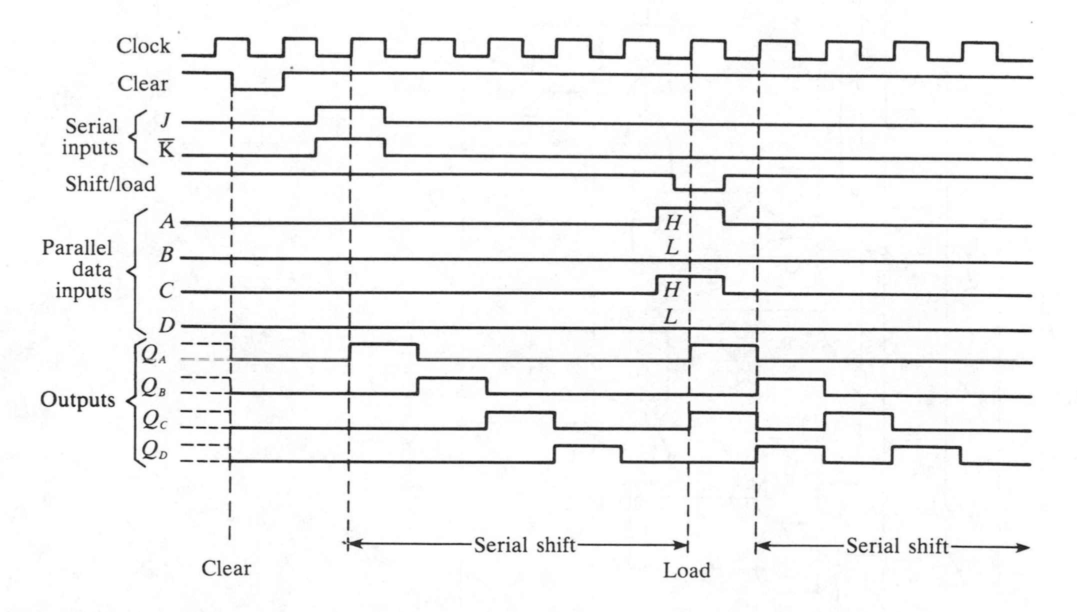

FIGURE 8–67 *Example timing diagram for a 74195.*

BIDIRECTIONAL SHIFT REGISTERS 8–13

A bidirectional shift register is one in which the data can be shifted either left or right. This can be implemented by using gating logic that enables the transfer of a data bit

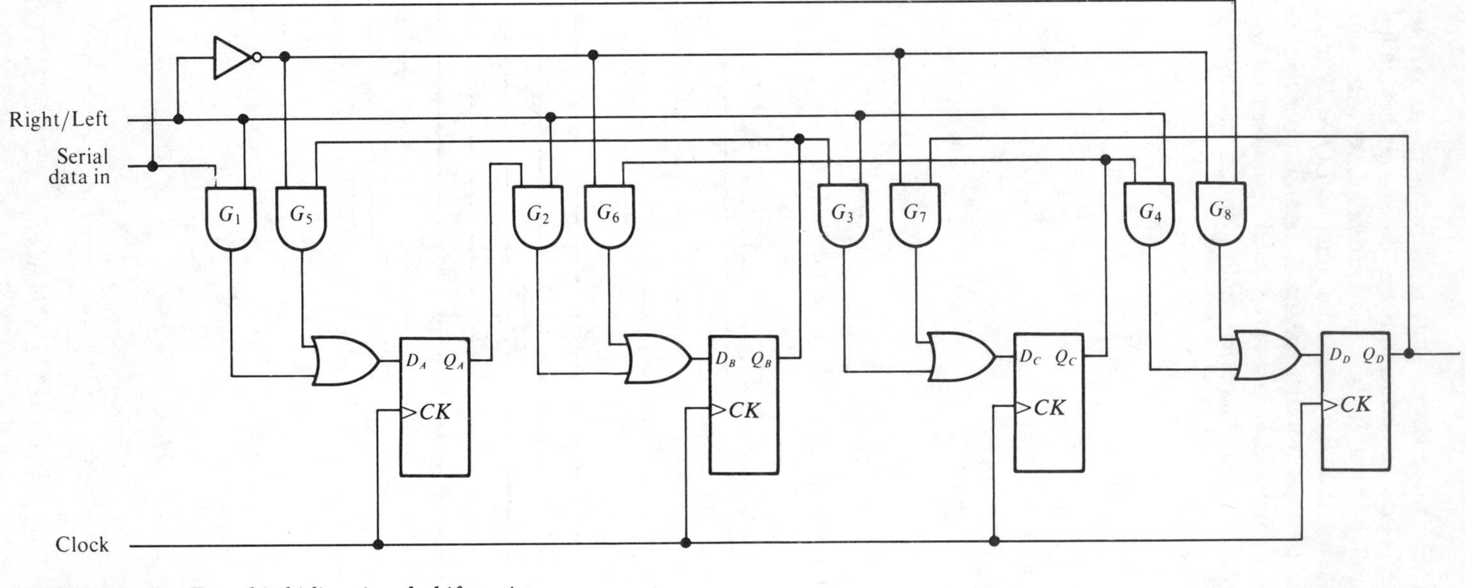

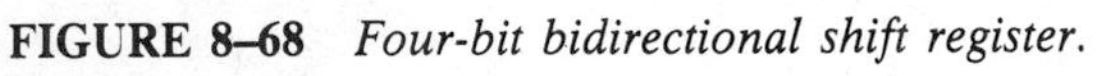

FIGURE 8–68 *Four-bit bidirectional shift register.*

from one stage to the next stage to the right or to the stage preceding it, depending on the level of a control line. A four-stage implementation is shown in Figure 8–68 for illustration purposes. A HIGH on the right/left control input allows data to be shifted to the right and a LOW enables a left shift of data. An examination of the gating logic should make the operation apparent. When the right/left control is HIGH, gates G_1 through G_4 are enabled and the state of the Q output of each flip-flop is passed through to the D input of the *following* flip-flop. When a clock pulse occurs, the data are then effectively shifted one place to the *right.* When the right/left control is LOW, gates G_5 through G_8 are enabled and the Q output of each flip-flop is passed through to the D input of the *preceding* flip-flop. When a clock pulse occurs, the data are then effectively shifted one place to the *left.*

Example 8–12

Determine the state of the shift register of Figure 8–68 after each clock pulse for the given right/left control input waveform in Figure 8–69(a). Assume $Q_A = 1$, $Q_B = 1$, $Q_C = 0$, $Q_D = 1$, and serial data in line is LOW.

Solution:

See Figure 8–69(b).

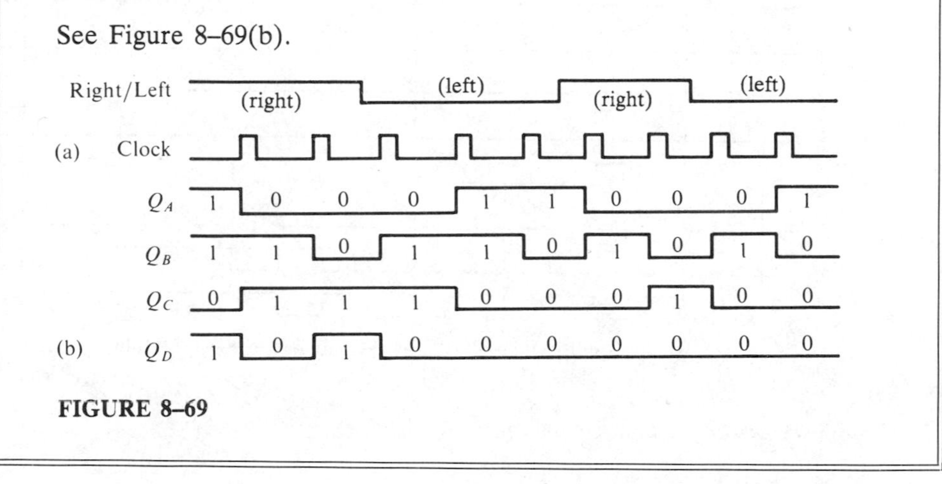

FIGURE 8–69

The 74194 Four-Bit Bidirectional Universal Shift Register

The 74194 is an example of a bidirectional shift register in integrated circuit form. A block and pin diagram is shown in Figure 8–70.

Parallel loading, which is synchronous with a positive transition of the clock, is accomplished by applying the four bits of data to the parallel inputs and a HIGH to the S_0 and S_1 inputs.

Shift right is accomplished synchronously with the positive edge of the clock when S_0 is HIGH and S_1 is LOW. Serial data in this mode is entered at the *shift-right serial input.* When S_0 is LOW and S_1 is HIGH, data bits shift left synchronous with the clock and new data is entered at the *shift-left serial input.* An example timing diagram is shown in Figure 8–71.

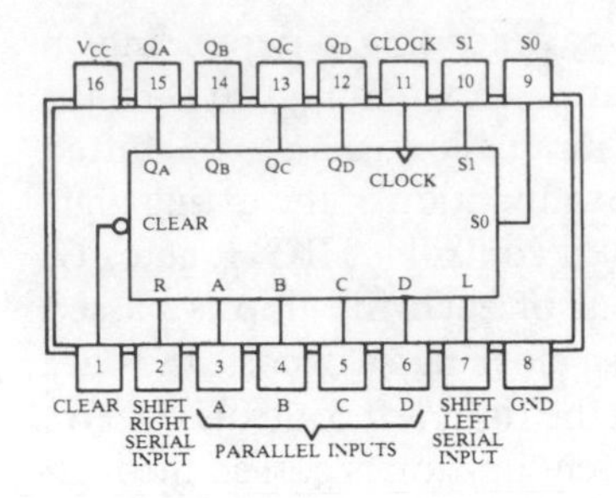

FIGURE 8–70 *A 74194 bidirectional shift register.*

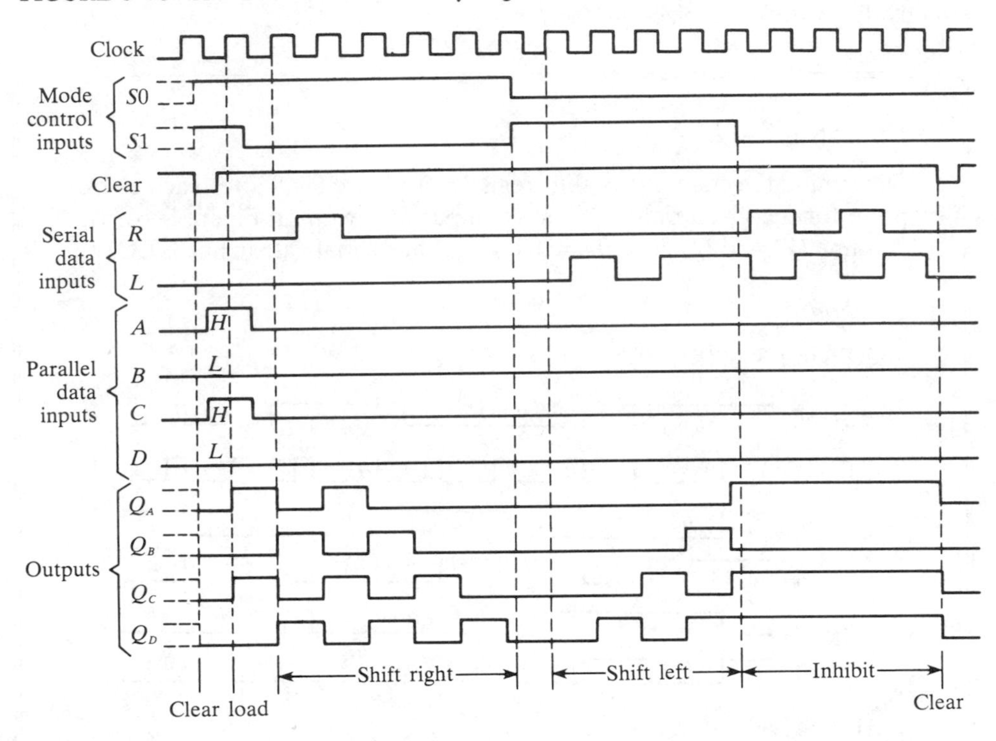

FIGURE 8–71 *Example timing diagram for a 74194.*

8–14 STATIC AND DYNAMIC REGISTERS

Two important categories of shift registers according to their circuit technologies are *static* and *dynamic*. The registers in the 54/74 TTL family such as those that have been presented as example devices are all *static* type registers. A static register is one that uses the flip-flop as the storage element.

There are basically two types of static registers: *MOS* and *bipolar*. The TTL devices are bipolar static registers. Functionally, MOS static registers are the same as bipolar except that the flip-flops are implemented with MOSFETs (metal oxide semiconductor field effect transistors).

A 1024-Bit Static Shift Register

Figure 8–72(a) is a block diagram of a 1024-bit static register made with PMOS circuit technology which is discussed in Appendix A. Figure 8–72(b) is the logic symbol, and Figure 8–72(c) is the connection diagram. D_1 and D_2 are the data inputs, *SEL* is the data select input, *CP* is the clock input, and *Q* is the data output. As you might recognize, this is a serial in-serial out shift register.

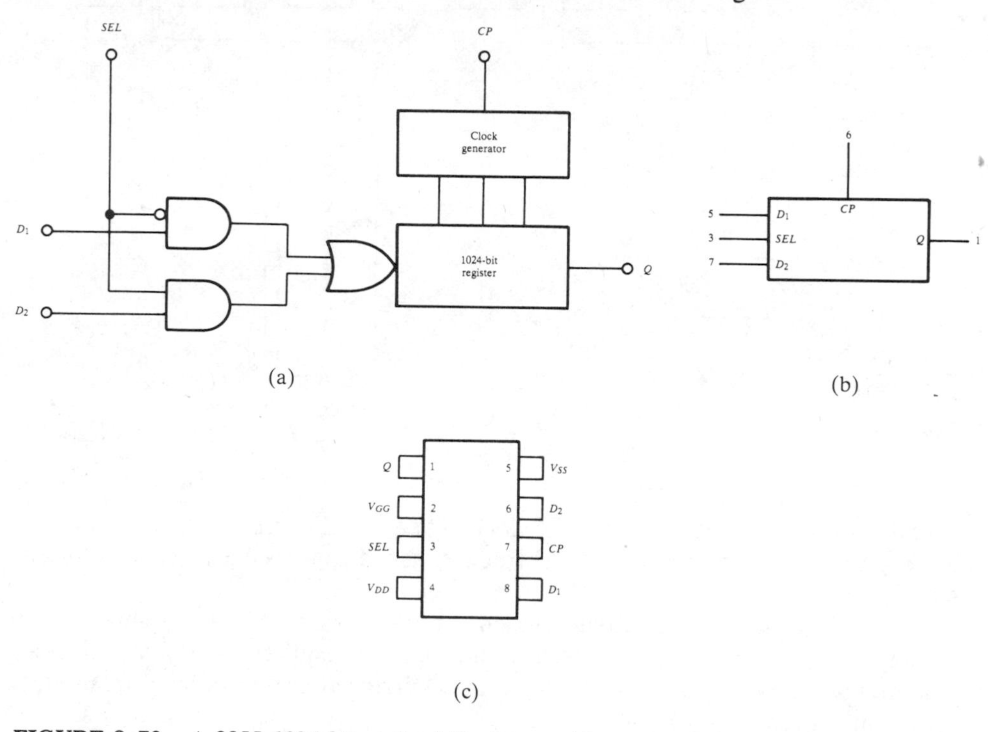

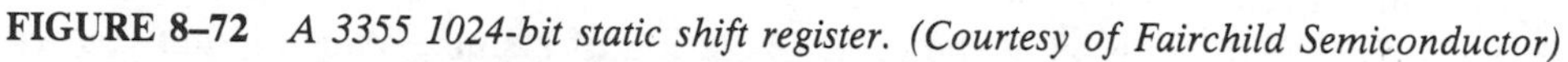

FIGURE 8–72 *A 3355 1024-bit static shift register. (Courtesy of Fairchild Semiconductor)*

A 64-Stage Static Shift Register

Figure 8–73 shows an example of a static shift register made with CMOS circuit technology. The block and logic diagrams are shown along with a pin diagram.

This is an edge-triggered device with two serial data inputs (D_0 and D_1), a data select input *(S)*, a clock input *(CP)*, a clock output *(CO)*, and data outputs from the 64th stage (Q_{63} and $\overline{Q}_{63}$).

Dynamic Shift Registers

The term *dynamic* refers to a type of MOS register that uses the inherent gate *capacitance* of a MOSFET as the basic storage element and takes advantage of the very high impedance of the MOSFET. The capacitance cannot retain its charge indefinitely and must be replenished periodically so that stored information will not be lost; this restriction requires that a shift register implemented

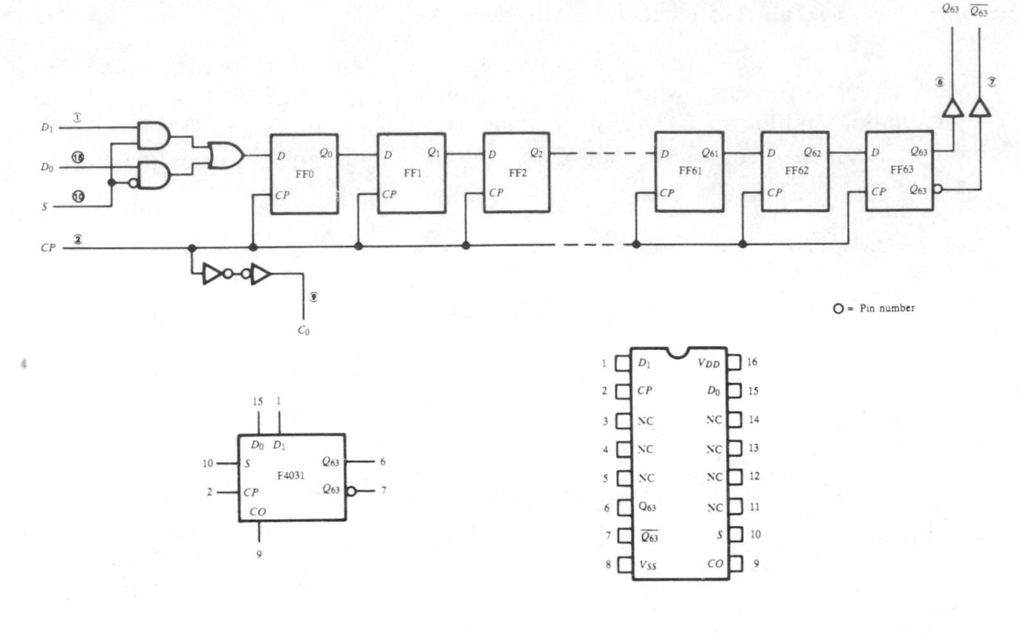

FIGURE 8–73 *A F4031 64-stage static shift register. (Courtesy of Fairchild Semiconductor)*

with dynamic storage devices must be clocked at a specified minimum rate or higher to keep the data in the register from being lost as the charges on the capacitors leak off.

A single stage circuit implementation for a MOS dyamic shift register is shown in Figure 8–74(a). This type of shift register requires a *two-phase* clocking system; that is, two clock signals with pulses occurring at different times are used, as shown in Figure 8–74(b).

We will look at the basic operation of a single stage so that you will understand how a bit of data is entered into one stage and shifted out to another in a dynamic shift register. The operation is basically the same regardless of the number of stages, and therefore the following discussion applies to all stages in a register of any *length* (number of stages).

When a logic 1 (near 0 V) is applied to the stage input (the gate of Q_1) and a *phase-one* (ø 1) clock pulse occurs, Q_1 is OFF because of the logic 1, and Q_2 and Q_3 are turned ON by the clock. The gate capacitance of Q_4 (C_1) charges to near $-V_{DD}$ through Q_2 and Q_3, which represents a logic 1 being stored. When the *phase-two* clock pulse occurs, Q_4 is ON because of the negative charge on its gate, and both Q_5 and Q_6 are turned ON by the clock pulse. The gate capacitance of the input to the next stage discharges to near 0 V through Q_4 and Q_6, thereby shifting the logic 1 that was stored in this stage onto the input of the next stage.

Now let us look at a condition for a logic 0 (negative voltage) on the gate of Q_1. When a *phase-one* clock pulse occurs, both Q_2 and Q_3 turn on. Q_1 is ON because of the logic 0 on its gate. The gate capacitance of Q_4 (C_1) then discharges to near ground potential through Q_1 and Q_3, which represents the storing of the logic 0. When a *phase-two* clock pulse occurs, both Q_5 and Q_6 are turned ON. The logic 0 on the gate of Q_4 keeps it OFF, and allows the gate capacitance of the input to the next stage to charge to near $-V_{DD}$. At this time the logic 0 that was stored in this stage appears on the input to the next stage.

Figure 8–75 is a logic block symbol for an *n*-bit dynamic MOS shift register. Notice that it has one data input, *two* clock inputs, and a data output. Remember, the important thing in the operation of a *dynamic* shift register is that *it cannot be operated below a specified clock frequency or the data will be lost* due to the short-term capacitive storage.

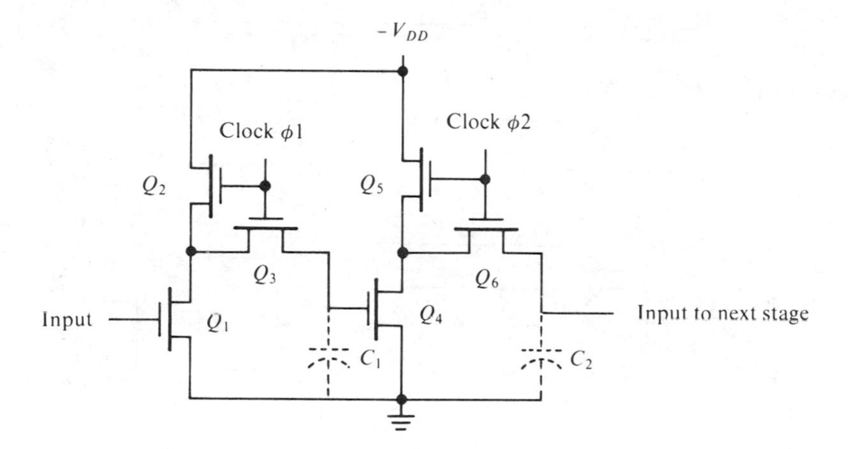

(a) Typical MOS Dynamic Shift Register Stage

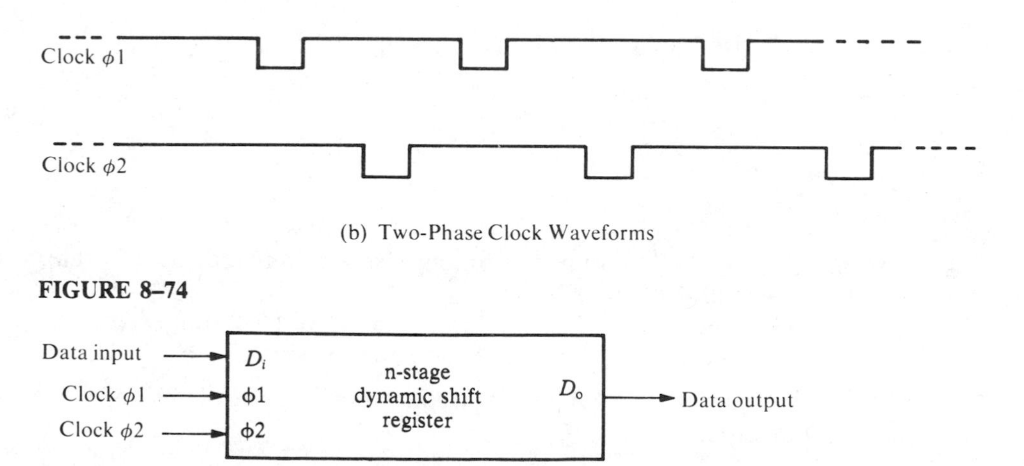

(b) Two-Phase Clock Waveforms

FIGURE 8–74

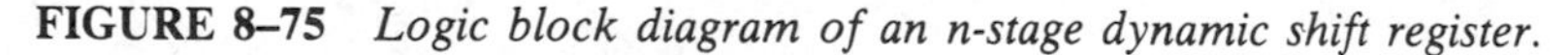

FIGURE 8–75 *Logic block diagram of an n-stage dynamic shift register.*

CCD Registers

Charge-coupled devices (CCDs) are a class of integrated circuits based on the principle of discrete charge-packet transfer. Although they share the same technological base with transistors, they are functional devices that manipulate information in charge packets rather than by electrical currents.

Charge coupling is the collective transfer of all the mobile electric charge within a semiconductor storage element to a similar adjacent storage element. These devices are characterized by very low power dissipation and high density so that large functions can be implemented on a small chip.

The CCD450 is a device containing nine 1024-bit serial dynamic shift registers as shown in Figure 8–76.

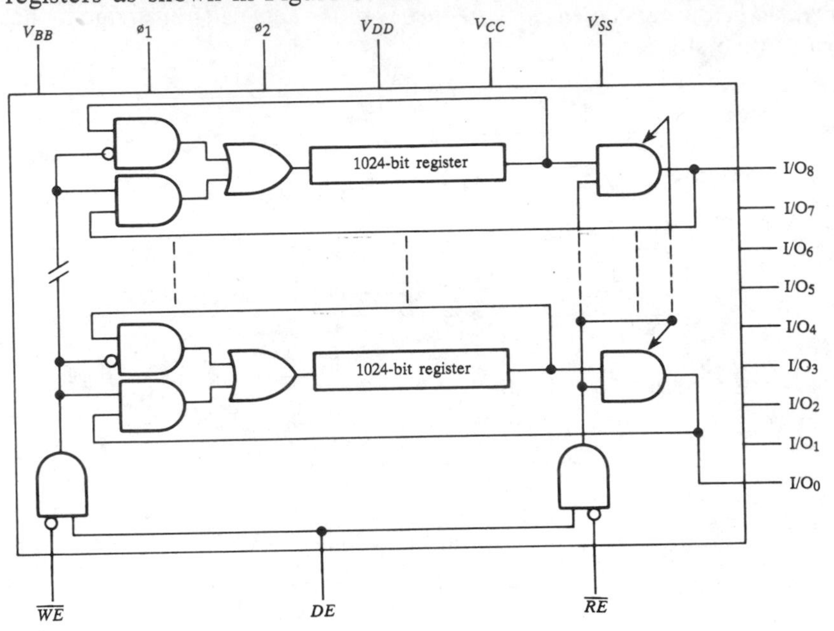

FIGURE 8–76 *A CCD450 dynamic shift register memory.*

Problems

Section 8–1

8–1 Assuming a *J–K* flip-flop connected to toggle starts out in the RESET state, in what state will it be after
(a) two clock pulses **(b)** five clock pulses
(c) ten clock pulses **(d)** twenty-one clock pulses

8–2 What is the maximum modulus for a counter with each of the following numbers of flip-flops?
(a) 2 **(b)** 4 **(c)** 5 **(d)** 6
(e) 7 **(f)** 8 **(g)** 9 **(h)** 10

8–3 Determine the number of flip-flops required to implement each of the following counters:
(a) modulus-4 **(b)** modulus-3 **(c)** modulus-5
(d) modulus-9 **(e)** modulus-12 **(f)** modulus-17
(g) modulus-36 **(h)** modulus-65

8–4 Repeat Problem 8–3 for the following:

(a) modulus-32 (b) modulus-39 (c) modulus-50
(d) modulus-64 (e) modulus-75 (f) modulus-144
(g) modulus-257 (h) modulus-512

8–5 For the ripple counter shown in Figure 8–77, draw the complete timing diagram for eight clock pulses showing the clock, Q_A, and Q_B waveforms.

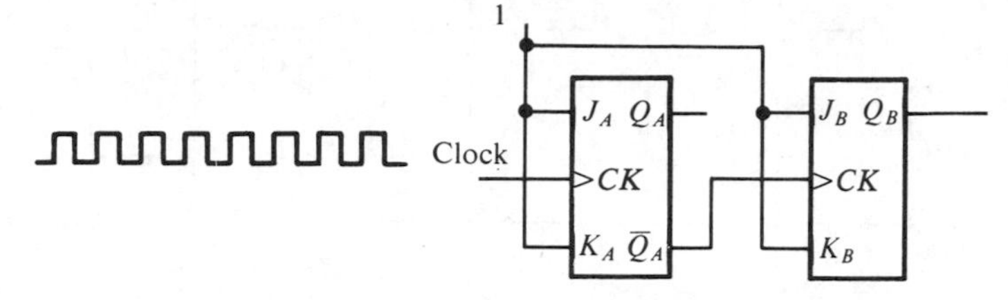

FIGURE 8–77

8–6 For the ripple counter in Figure 8–78, draw the complete timing diagram for 16 clock pulses. Show the clock, Q_A, Q_B, and Q_C waveforms.

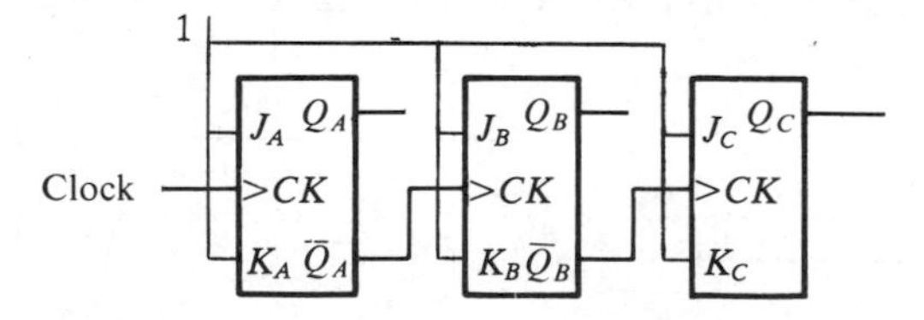

FIGURE 8–78

8–7 Repeat Problem 8–6 using master-slave flip-flops in which data enter the master on the positive edge of the clock pulse.

8–8 In the counter of Problem 8–6, assume each flip-flop has a propagation delay from the triggering edge of the clock to a change in the Q output of 12 nanoseconds. Determine the worst-case (longest) delay time from a clock pulse to the arrival of the counter in a given state. Specify the state or states for which this worst-case delay occurs.

8–9 The waveforms in Figure 8–79 are applied to the count enable, clear, and clock inputs as indicated. Sketch the counter output waveforms in proper relation to these inputs. Changes in the outputs occur on the LOW-to-HIGH transition of the clock. Count enable is active HIGH and clear is active LOW.

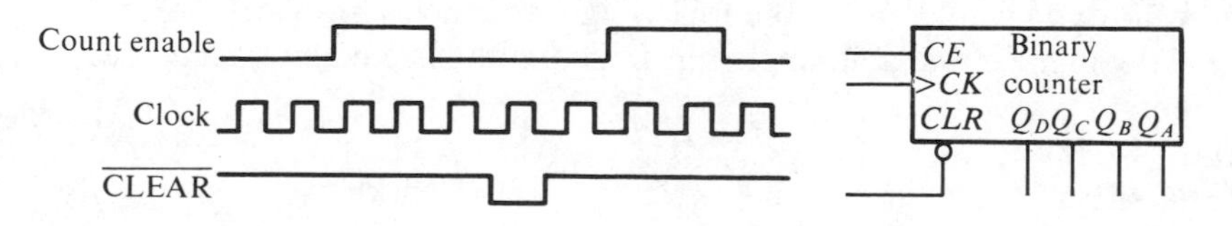

FIGURE 8–79

8–10 If the counter of Problem 8–8 were synchronous rather than asynchronous, what would be the longest delay time?

8–11 Draw the complete timing diagram for the five-stage synchronous binary counter in Figure 8–80. Verify that the waveforms of the Q outputs represent the proper binary number after each clock pulse.

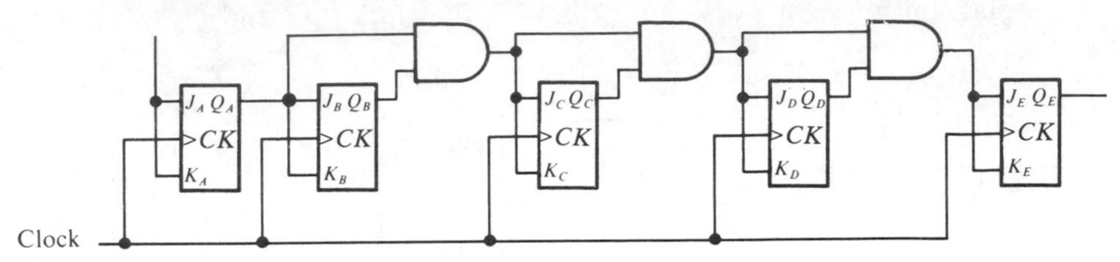

FIGURE 8–80

Section 8–2

8–12 Prove that the decade counter in Figure 8–81 progresses through a BCD sequence by analyzing the J and K inputs to each flip-flop prior to each clock pulse. Explain how these conditions in each case cause the counter to go to the next proper state.

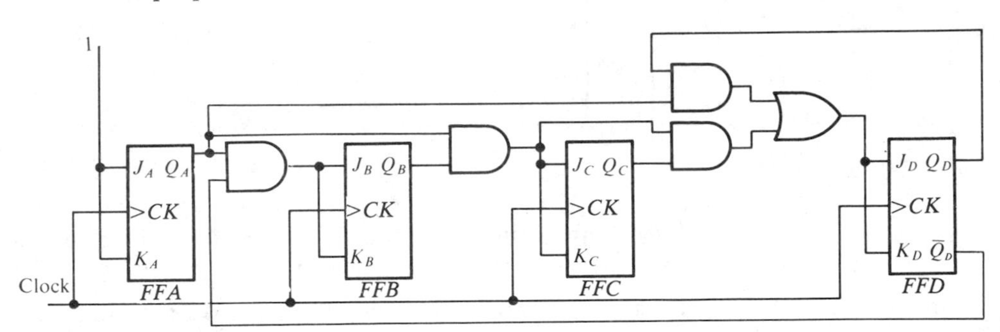

FIGURE 8–81

8–13 Explain how the BCD decade counter acts as a divide-by-10 device, and show this with the proper waveforms.

8–14 List the unallowed states of the BCD decade counter.

8–15 Show how you would obtain the following frequencies from a 10-MHz clock signal if you have available single flip-flops, modulus-5 counters, and decade counters:

(a) 5 MHz **(b)** 2.5 MHz **(c)** 2 MHz
(d) 1 MHz **(e)** 500 kHz **(f)** 250 kHz
(g) 62.5 kHz **(h)** 40 kHz **(i)** 10 kHz
(j) 1 kHz

8–16 The logic symbol for a BCD decade counter is shown in Figure 8–82. The waveforms are applied to the clock and clear inputs as indicated. Determine the waveforms for each of the counter ouputs (Q_A, Q_B, Q_C, and Q_D). Changes in the outputs occur on the LOW-to-HIGH transitions of the clock. The clear is asynchronous.

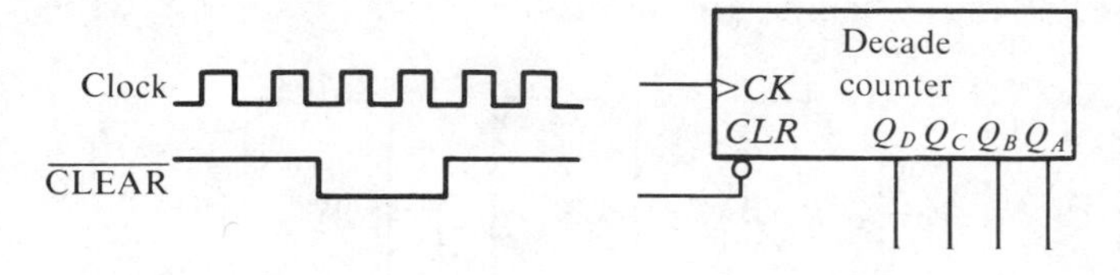

FIGURE 8–82

8–17 The waveforms in Figure 8–83 are applied to the count enable *(CE)* and clock inputs *(CK)* as indicated. Draw the counter output waveforms in proper relation to these inputs. Changes in the outputs occur on the LOW-to-HIGH transitions of the clock. Count enable is active HIGH.

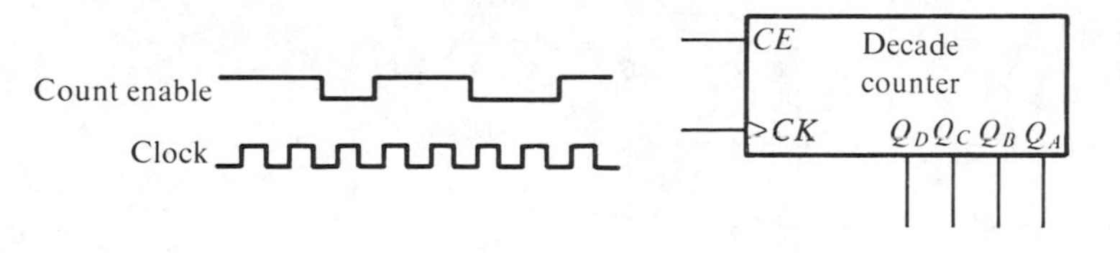

FIGURE 8–83

Section 8–3

8–18 Draw a complete timing diagram for an up-down counter that goes though the following sequence. Indicate when the counter is in the up mode and when it is in the down mode. Assume positive edge-triggering.

0, 1, 2, 3, 2, 1, 2, 3, 4, 5, 6, 5, 4, 3, 2, 1, 0

Section 8–4

8–19 The state diagram for a positive edge-triggered counter is shown in Figure 8–84. Sketch the corresponding timing diagram.

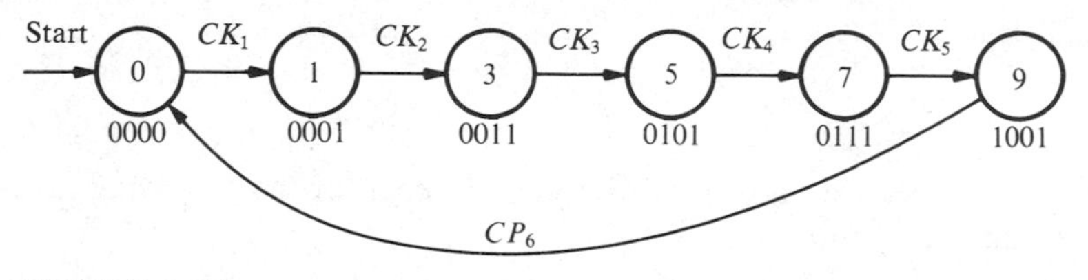

FIGURE 8–84

8–20 Repeat Problem 8–19 for the state diagram in Figure 8–85.

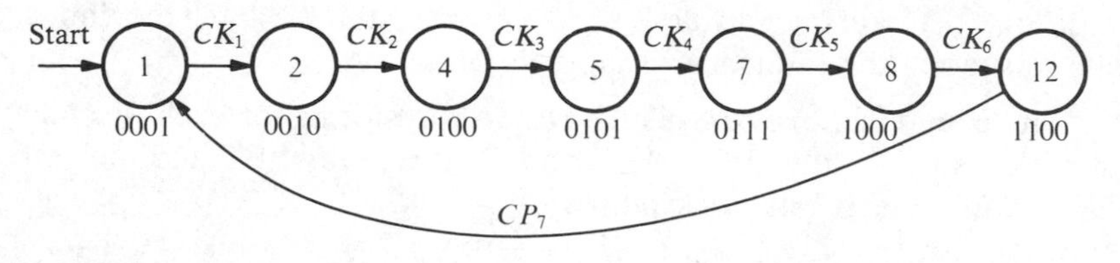

FIGURE 8–85

8–21 For the counter logic in Figure 8–86, properly connect the preset and clear inputs of each stage so that the counter may be initialized to 10110 (left-most bit = Q_A). Assume J and K inputs are all HIGH.

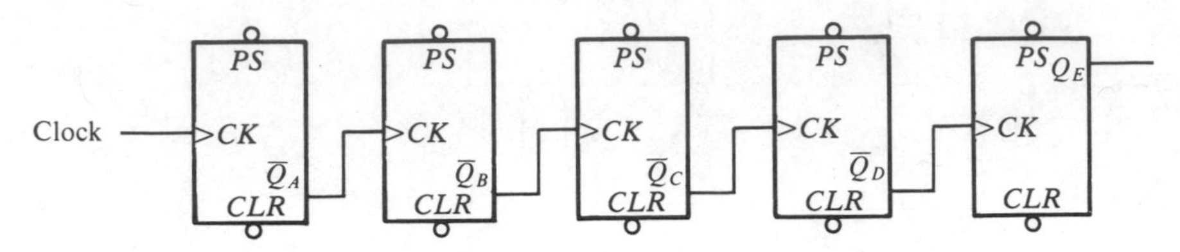

FIGURE 8–86

8–22 Determine the sequence of the counter in Figure 8–87.

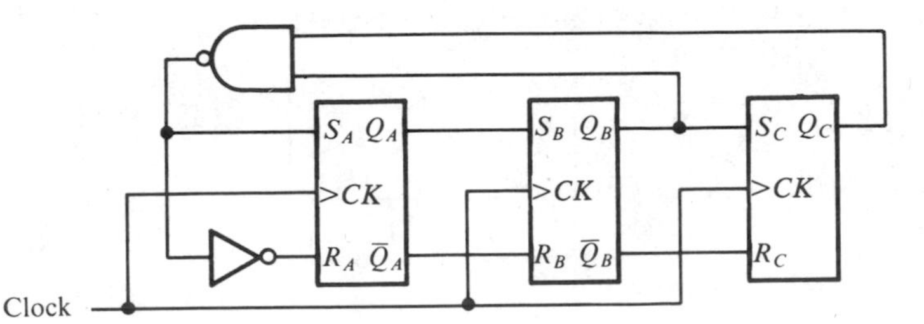

FIGURE 8–87

8–23 Determine the sequence of the counter in Figure 8–88. Begin with the counter cleared.

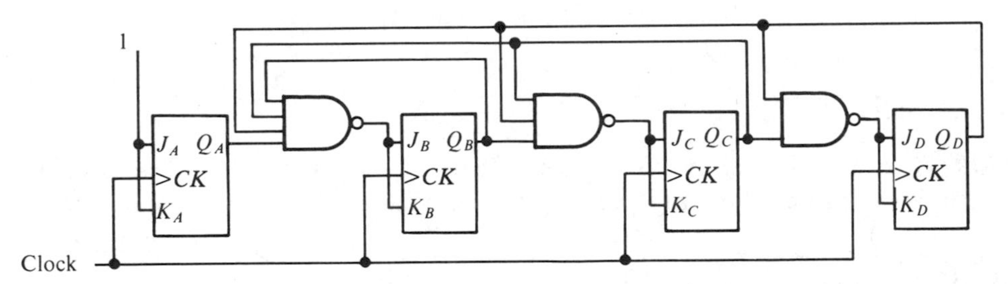

FIGURE 8–88

8–24 How many flip-flops are required to implement each of the following in a Johnson counter configuration?

(a) divide-by-6 **(b)** divide-by-10 **(c)** divide-by-14
(d) divide-by-16 **(e)** divide-by-20 **(f)** divide-by-24
(g) divide-by-36

8–25 Draw the logic diagram for a divide-by-18 Johnson counter. Sketch the timing diagram and write the sequence in tabular form.

8–26 For the ring counter in Figure 8–89, draw the waveforms for each flip-flop output with respect to the clock. Assume FF0 is initially SET and the rest RESET. Show at least ten clock pulses.

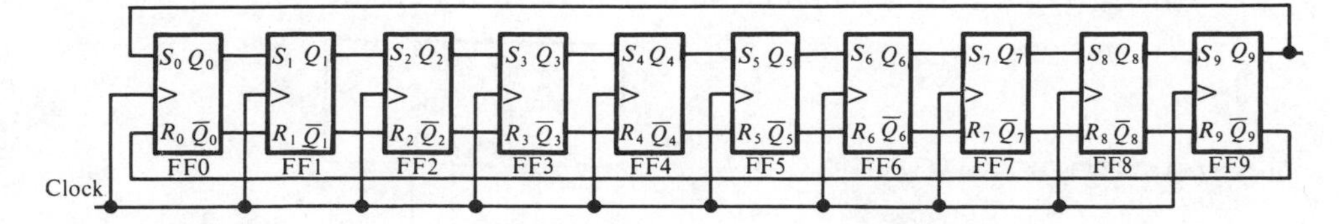

FIGURE 8–89

8–27 The waveform pattern in Figure 8–90 is required. Show a ring counter and indicate how it can be preset to produce this waveform.

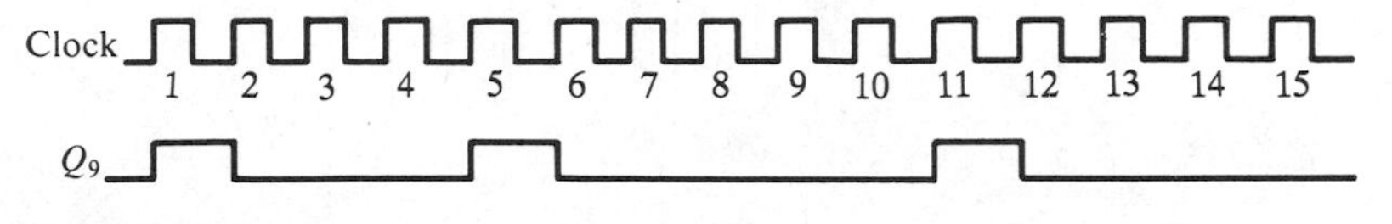

FIGURE 8–90

Section 8–5

8–28 For each of the cascaded counter configurations in Figure 8–91, determine the frequency of the waveform at each point indicated by circled numbers and determine the overall modulus.

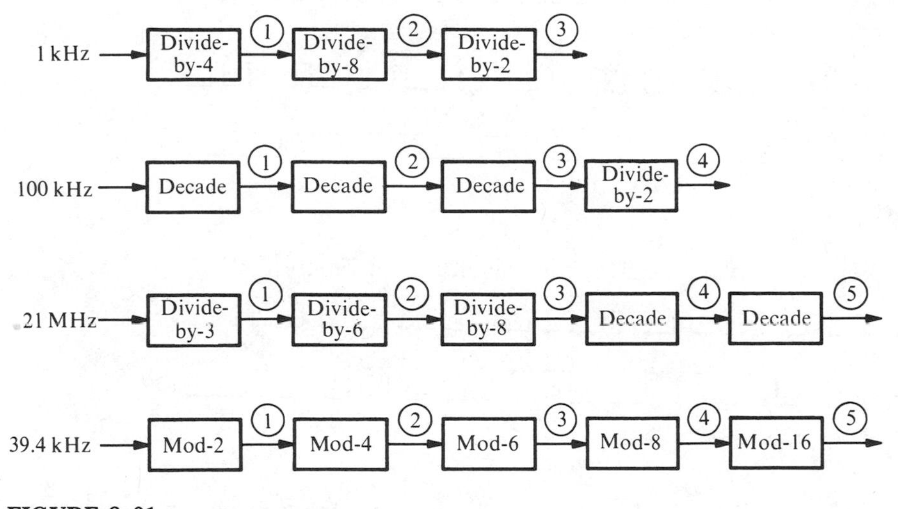

FIGURE 8–91

Section 8–6

8–29 Given a BCD decade counter, show the decoding logic required to decode each of the following states and how it should be connected to the counter. A HIGH output indication is required for each decoded state. MSB is to the left.
(a) 0001 **(b)** 0011 **(c)** 0101 **(d)** 0111 **(e)** 1000

8–30 For the four-stage binary counter connected to the 1-of-10 decoder in Figure

8–92, determine each of the decoder output waveforms in relation to the clock pulses. Q_A is LSB of counter, and A is LSB of decoder.

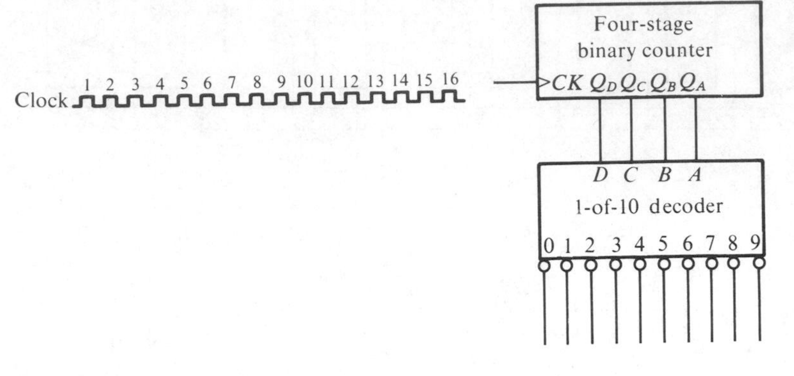

FIGURE 8–92

Section 8–8

8–31 Explain how a flip-flop can store a data bit.

Section 8–9

8–32 For the data input and clock timing diagram in Figure 8–93, determine the state of each flip-flop in the shift register of Figure 8–53. Assume the register contains all 1s initially.

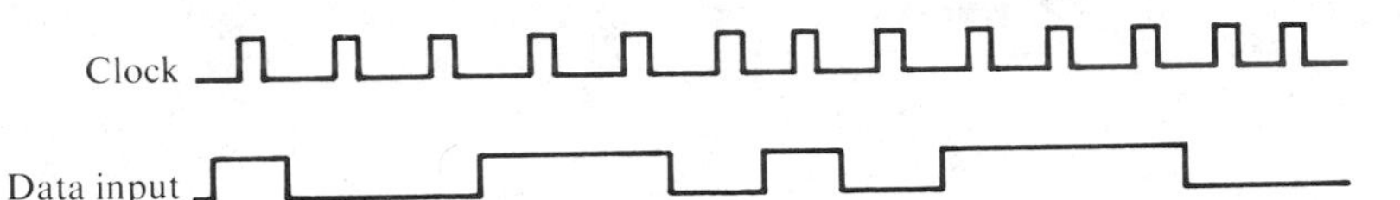

FIGURE 8–93

8–33 Repeat Problem 8–32 for the waveforms in Figure 8–94.

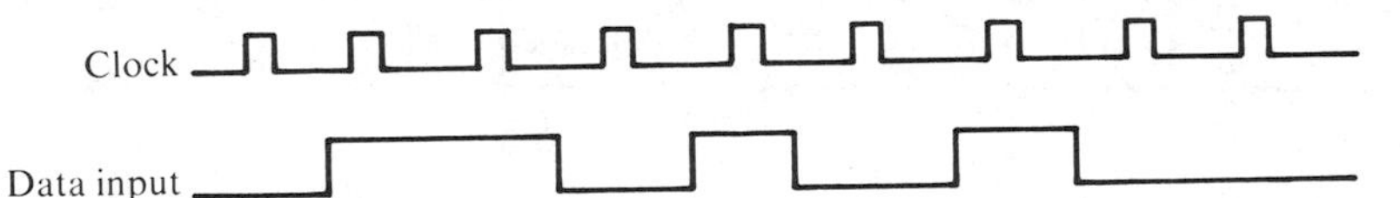

FIGURE 8–94

8–34 What is the state of the register in Figure 8–95 after each clock pulse if it starts in the 101001111000 state?

8–35 For the serial in-serial out shift register, determine the data output waveform for the data input and clock waveforms in Figure 8–96. Assume the register is initially cleared and clocks on the positive edge.

8–36 Repeat Problem 8–35 for the waveforms in Figure 8–97.

8–37 The data output waveform in Figure 8–98 is related to the clock as indicated. What binary number was stored in the register if the first bit out is the LSB?

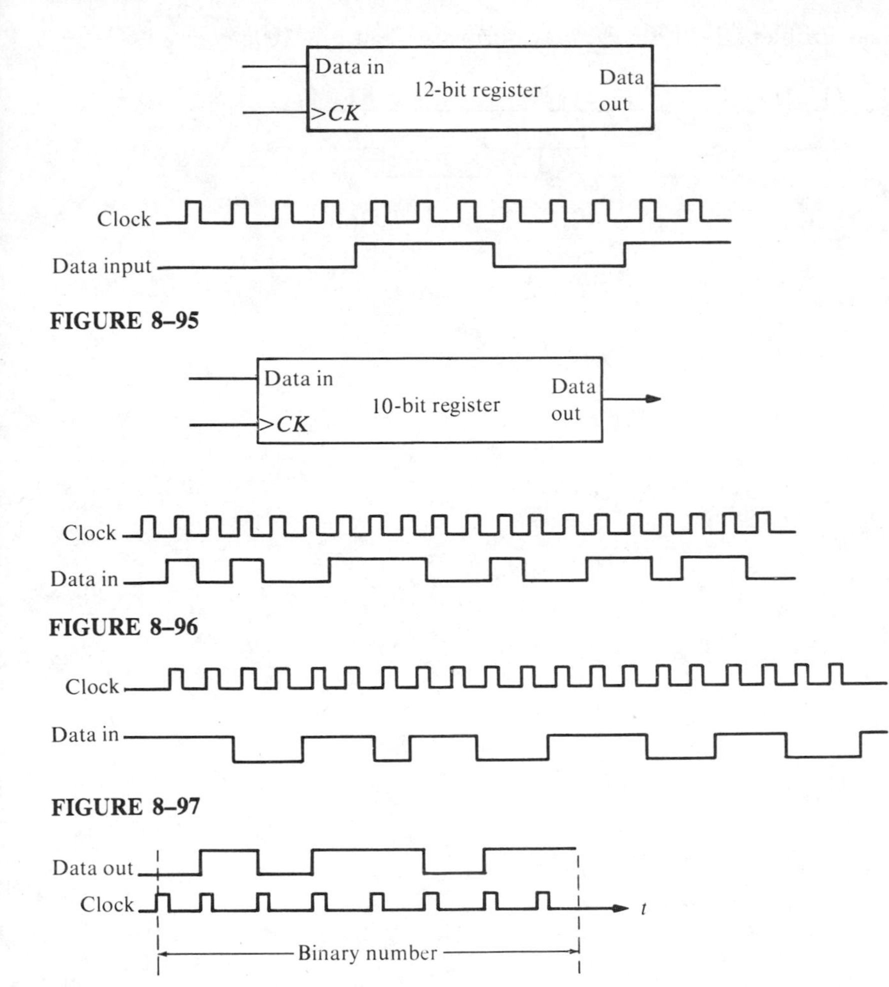

FIGURE 8–95

FIGURE 8–96

FIGURE 8–97

FIGURE 8–98

Section 8–13

8–38 For the eight-stage bidirectional register in Figure 8–99, determine the state of the register after each clock pulse for the right/left control waveform given. A 1 on this input enables a shift to the right and a 0 enables a shift to the left. Assume the register is initially storing a "76" in binary, with the right-most position being the LSB.

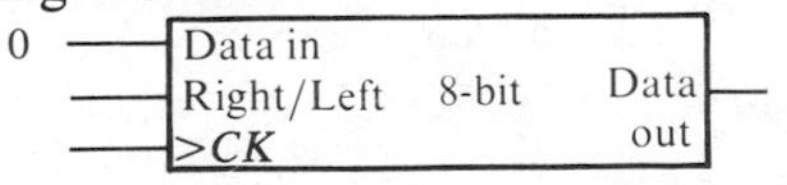

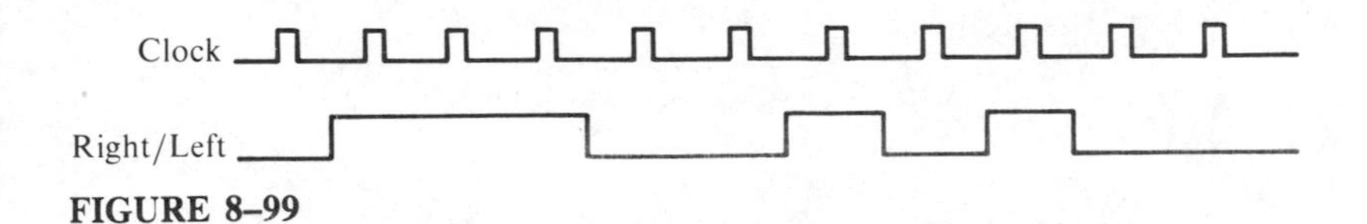

FIGURE 8–99

8–39 Repeat Problem 8–38 for the waveforms in Figure 8–100.

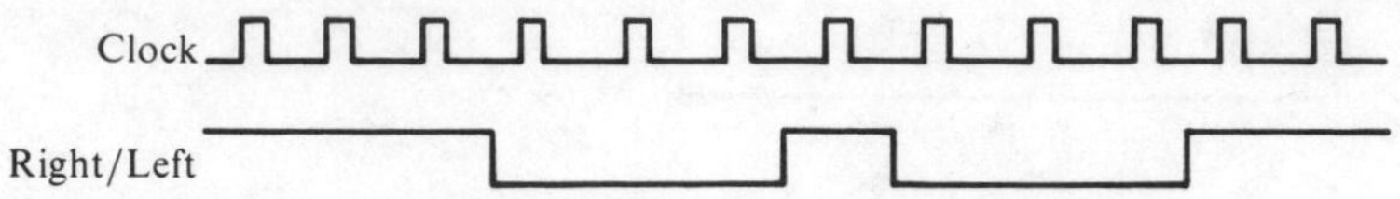

Clock

Right/Left

FIGURE 8–100

The last chapter covered registers, a type of storage device. Registers are normally considered temporary storage devices, whereas memories are typically used for longer term storage of binary data. Also, a typical memory can store a much larger amount of data than a typical register.

Modern applications of data processing systems require that huge amounts of information be permanently stored and readily accessible. Banking, inventory control, the census, and social security are just a few examples where a great deal of information must be kept and processed.

In this chapter we will discuss the two categories of memories in current use, *semiconductor* and *magnetic*. Within each category are a variety of memory types. Generally, the semiconductor memories are used for smaller capacity and faster access applications. The various types of magnetic memories are used for larger capacity bulk storage, but it generally takes much longer to access information.

9 MEMORIES

9–1 REGISTER FILES

We will lead into our coverage of memory devices by looking at a device that uses a number of registers for storage of data in groups of bits (words). This register oriented device is known as a *register file.*

Figure 9–1 shows an arrangement of four registers as an example of a register file. In this device, we will assume that each register contains four stages or cells. It is therefore a 4 × 4 register file for sixteen bits.

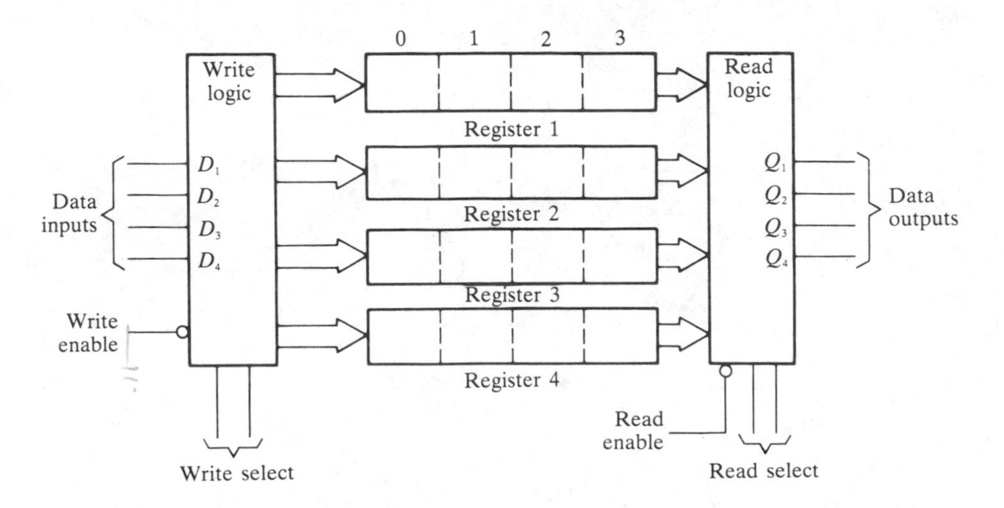

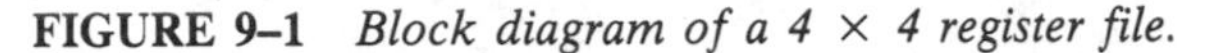

FIGURE 9–1 *Block diagram of a 4 × 4 register file.*

Two basic memory operations are demonstrated in this diagram, the *write* and the *read* operations. The write operation is the process of entering binary data into the memory for storage. The read operation is the process of retrieving binary data from the memory.

In Figure 9–1, when the *write enable* line is active (LOW), the four bits on the *data inputs* ($D1$ $D2$ $D3$ $D4$) are routed to the location in each register selected by a binary code on the *write select* lines and stored. For example, assume that the input bits are 1010_2 ($D1$ $D2$ $D3$ $D4$) and a 10_2 is applied to the *write select* inputs with the *write enable* LOW. The four bits of input data will be stored in cell 2 of each register in the following manner: bit $D1$ (1) is stored in cell 2 of register 1; bit $D2$ (0) is stored in cell 2 of register 2: bit $D3$ (1) is stored in cell 2 of register 3; and bit $D4$ (0) is stored in cell 2 of register 4. This result is illustrated in Figure 9–2(a). The register file can, of course, be filled up with four bits in each register for a total of 16 bits which can be stored in any order by the binary code on the *write select* lines.

To retrieve or recall the stored data, the *read enable* must be activated and the register cell selected by a two-bit code on the *read select* lines. The four data bits will appear on the data output lines ($Q1$ $Q2$ $Q3$ $Q4$). For example, if we wish to recall the data just written, a 10_2 must be applied to the *read select* and a LOW applied to the *read enable.* The 1010_2 will be routed to the data outputs by the read logic. This is illustrated in Figure 9–2(b).

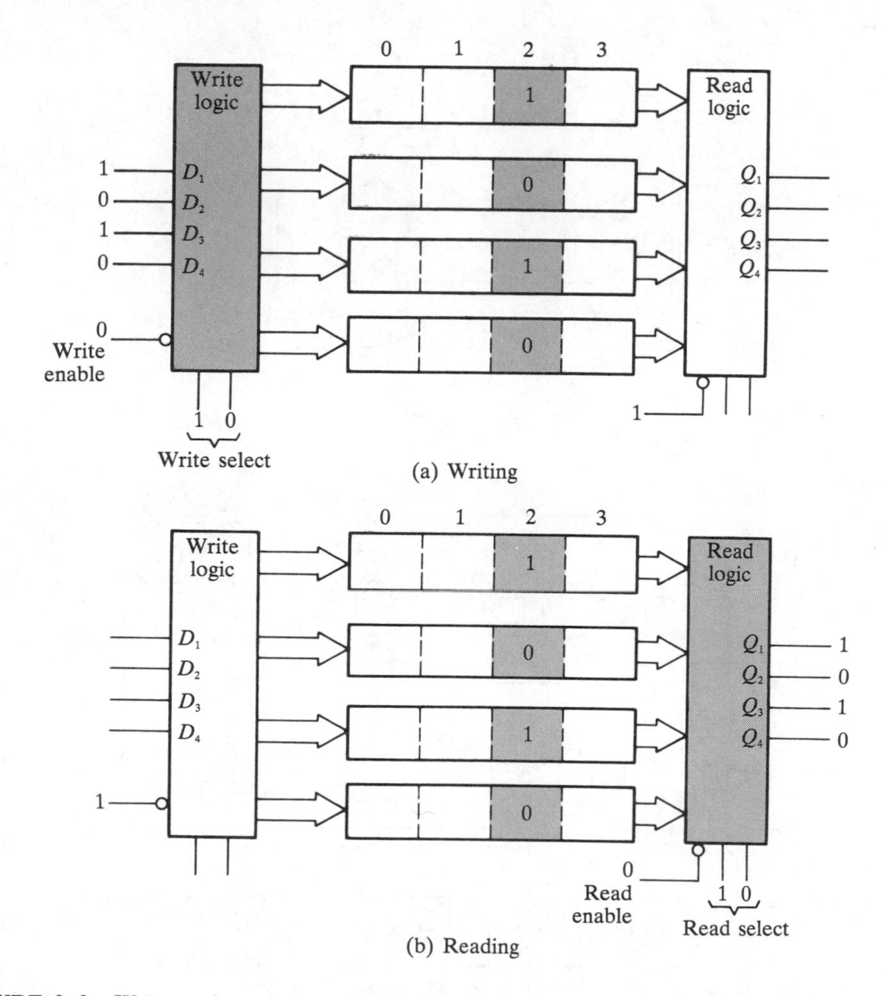

FIGURE 9–2 *Write and read operations in the register file.*

As a specific example of a register file, Figure 9–3 shows the block and pin diagram and the logic diagram for the 74170 4 × 4 register file. Notice that the write enable and read enables are labeled G_W and G_R respectively. The write selects are W_A and W_B, and the read selects are R_A and R_B.

The registers in the file are composed of gated D latches.

RANDOM ACCESS MEMORIES (RAMs) 9–2

A random access memory (RAM) is a read/write memory in which any given cell or groups of cells can be selected in any sequence (randomly). The location of a given cell is called its *address*. The arrangement of cells is called a memory *array* and there are many ways in which the cells can be arranged or *organized*.

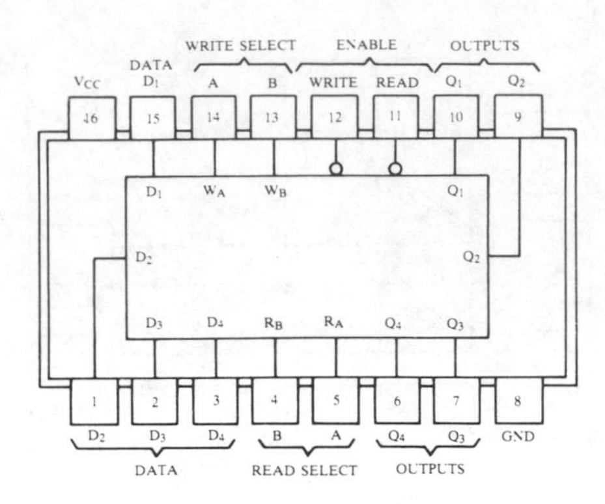

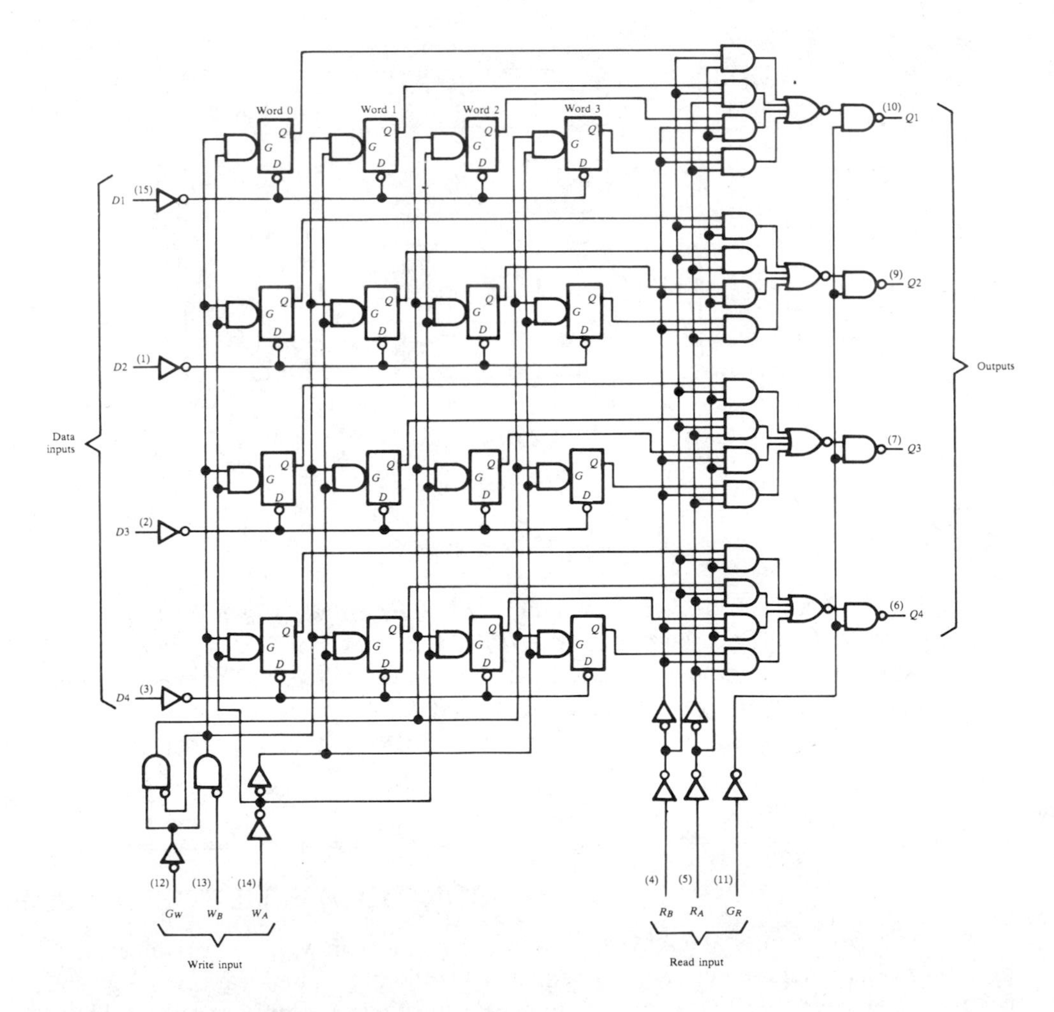

FIGURE 9–3 *The 74170 4 × 4 register file.*

Figure 9–4 is a representation of a memory array containing 256 cells organized into 16 rows of 16 columns each. Each of the 256 cells can be selected (addressed) by activating the appropriate *row select* line and the appropriate *column select* line. These lines are activated by the appropriate *address code* inputs to the *row address decoder* and the *column address decoder*.

In addition to the address decoders, the memory has a data input and data output as well as a read/write enable input.

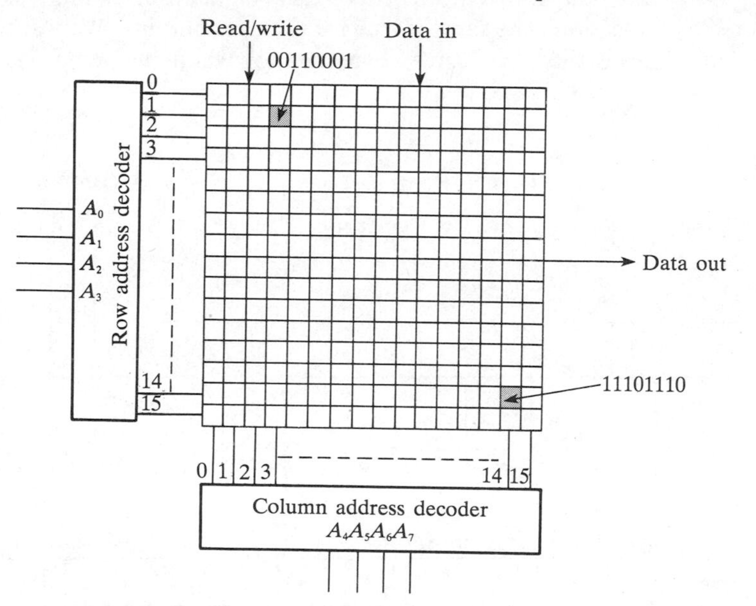

FIGURE 9–4 *Representation of a 256-bit RAM.*

Eight bits are required to address the memory because $2^8 = 256$. There are four row address bits $(A_3A_2A_1A_0)$ and four column address bits $(A_7A_6A_5A_4)$. The four row address bits can select one of 16 rows, and the four column address bits can select one of 16 columns. One row address and one column address uniquely specify one of the 256 memory locations.

For example, an address code of 00110001_2 $(A_7 - A_0)$ specifies memory address 49_{10} (31_{16}) as indicated in Figure 9–4. Notice that the row address is $0001_2 = 1_{10}$ and the column address is $0011_2 = 3_{10}$. This is cell 49_{10} counting from top to bottom beginning in the upper left corner.

An address code of 11101110_2 $(A_7 - A_0)$ specifies memory address 238_{10} (EE_{16}) as also indicated in Figure 9–4.

Memory Expansion

In many applications, data are handled in groups of eight bits. These eight-bit groups are called *bytes*. Often, it is desirable to have a byte-organized rather than a bit-organized memory. Expansion of the 256-bit RAM just discussed can accomplish this objective. The memory array that we have examined can be thought of as a 256×1 array; that is, it can store 256 one-bit words. If we wish to store 256 eight-bit words (256 bytes), eight of the memories can be connected together, as shown in Figure 9–5.

The eight address lines are connected to each of the eight memory arrays so that the same corresponding cell in each memory array can be selected at the same time. A single *read/write* line goes to each array so that they are all being written or all being read at the same time.

A separate data input goes to each array so that eight different bits can be stored in a selected byte location. The same is true for the data outputs. We want to be able to have eight different bits from a given byte location when the memory is read.

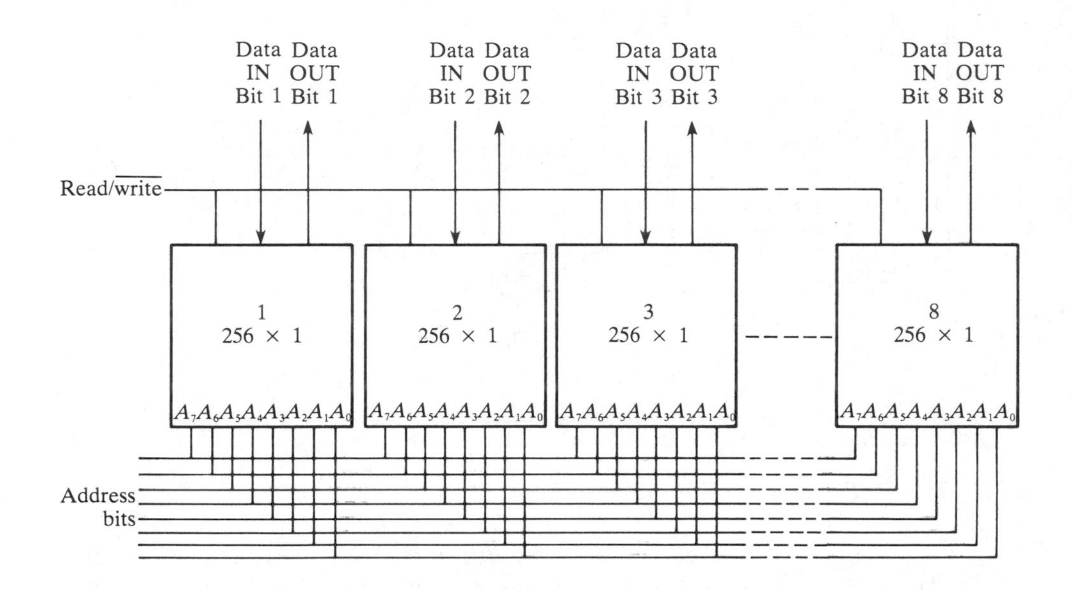

FIGURE 9–5 *Eight 256-bit memory arrays expanded to form a 256-byte RAM.*

A TTL RAM

The 74200 is one example of a RAM constructed with TTL technology. Although it is a 256-bit read/write memory, its organization is somewhat different from the generalized memory previously discussed. Whereas the other memory was organized into a 16 × 16 cell array, this device has a 32 × 8 cell array as shown in the block diagram of Figure 9–6. Of course, both 16 × 16 and 32 × 8 memories have the same number of cells (256).

To enable the memory, all of the *memory enable inputs* must be LOW. The DEFGH address inputs select one of the 32 eight-bit (byte) locations. The ABC address inputs select one of the eight-bit locations in the selected row. When the *write enable (WE)* input is LOW the bit on the *data input* is stored in the selected address. When the *WE* input is HIGH, the memory is in the read mode and the selected data bit appears on the output. The 74200 pin diagram is shown in Figure 9–7.

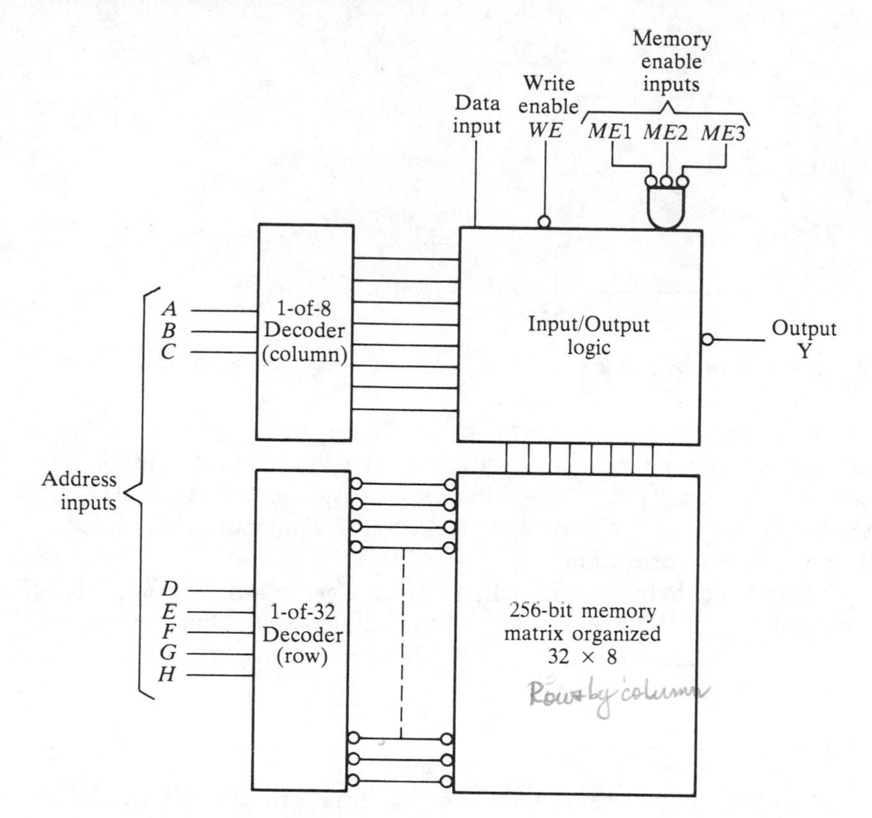

FIGURE 9–6 *Block diagram of 74200 TTL RAM.*

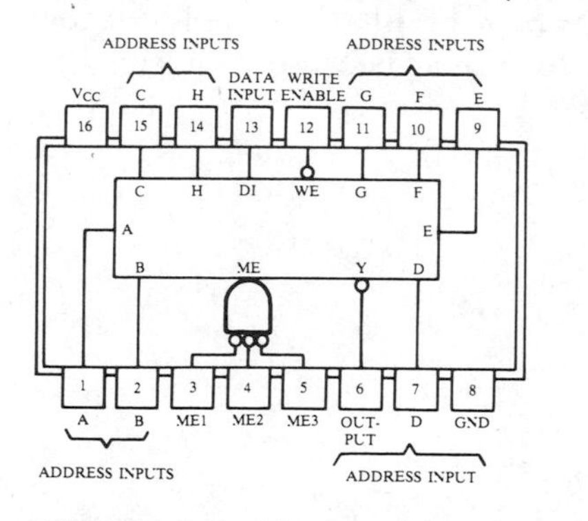

FIGURE 9–7 *Block and pin diagram for 74200 RAM.*

The TTL Memory Cell

The basic latch storage cell is shown in Figure 9–8 and, as you can see, uses multiple-emitter bipolar transistors. There are 256 of these in the 74200 memory, for example.

Prior to addressing, the row and column select lines are normally LOW, conducting current from the associated emitters of the *on* transistor. Since this

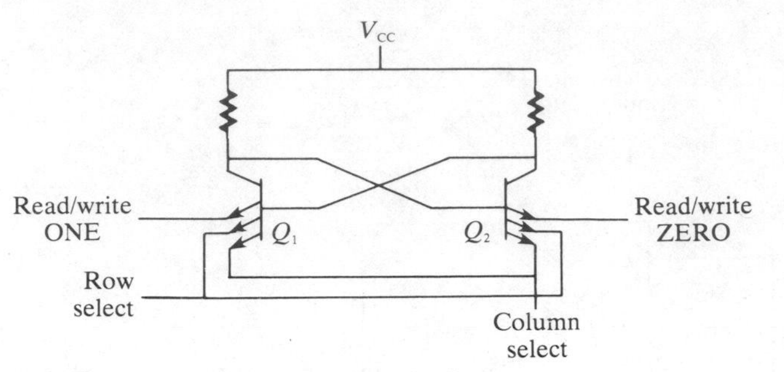

FIGURE 9–8 *A TTL memory cell.*

is a latch, one of the transistors is on and the other off, depending on the bit being stored. To address or select the cell, HIGHs are applied to both the row and column select lines, causing the emitter current of the *on* transistor to be diverted to the associated read/write line where it is *sensed* by a circuit in the memory called a sense amplifier. This is the read operation.

To write a data bit into the cell, a HIGH is placed on the appropriate read/write line and one of the row and one of the column select lines are taken HIGH. This causes the transistor which has the HIGH on its *read/write* line to turn *off* and the other transistor to turn *on*.

An MOS Static RAM

MOS (metal oxide semiconductor) memories are found in two basic types, *static* and *dynamic*.

The memory cells of a static RAM are MOSFET latches. A MOS static memory cell is shown in Figure 9–9. It is referred to as *static* because, as a latch, its binary state is retained as long as power is applied.

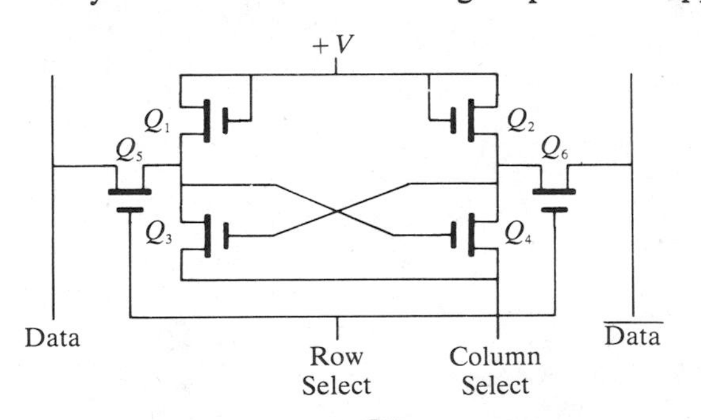

FIGURE 9–9 *An MOS static memory cell.*

Transistors Q_1 and Q_2 are used only as load resistors. Transistors Q_3 and Q_4 form the latch in which a data bit is written or read via transistor Q_5 and Q_6 from the *data* and $\overline{\textit{data}}$ lines when the cell is selected by the row and column select lines.

As an example of an MOS static RAM, Figure 9–10 shows a block diagram and logic symbol for the Intel 2147. This device has a 4096 × 1 organization in a 64-row–by–64-column array.

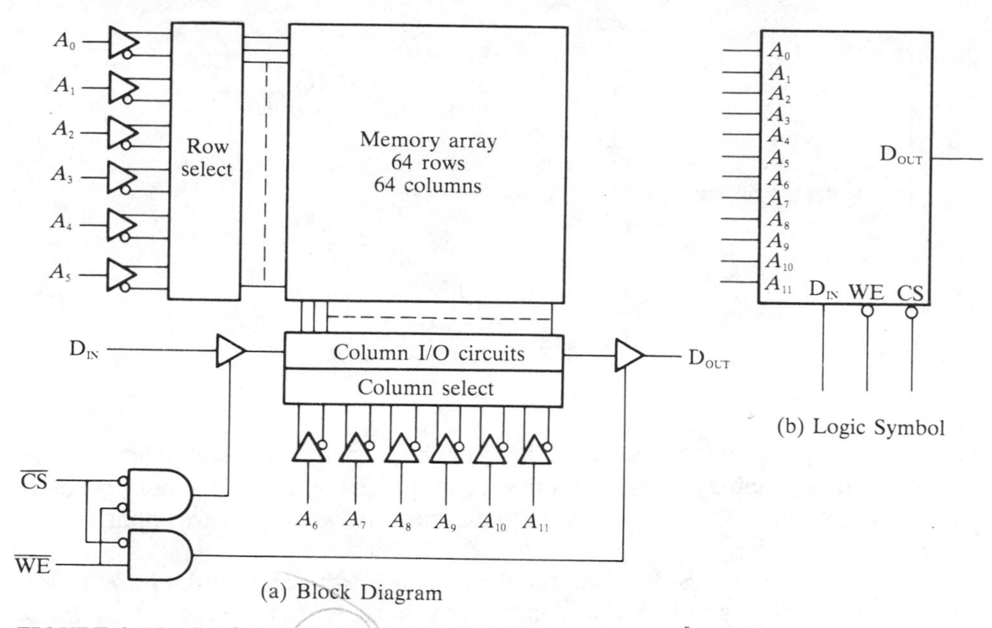

FIGURE 9–10 *Intel MOS 2147 static RAM. (Courtesy of Intel Corporation.)*

Notice that there are 12 address lines ($A_0 - A_{11}$) required to select any one of the 4096 memory locations ($2^{12} = 4096$). A 4096 address memory is often referred to simply as a *4K* memory. When the *chip select* ($\overline{CS}$) input is LOW, the memory is enabled and a LOW on the *write enable* ($\overline{WE}$) allows a bit on the Data Input (D_{IN}) line to be stored at the selected location. A HIGH on the $\overline{WE}$ input causes a bit to be read from the selected location to the Data Output (D_{OUT}) line.

The major advantages of MOS memories over bipolar (TTL) memories are that the MOS devices require much less power to operate and that larger memories can be fabricated on a given chip area. The main disadvantage is that the MOS devices are slower than TTL.

An MOS Dynamic RAM

Dynamic RAMs are the second major category. Dynamic memory cells store a data bit in a small *capacitor* rather than in a latch as does the static memory. A great advantage of this type of cell is that it is very simple thus allowing very large memory arrays to be constructed on a chip. Its disadvantage is that the storage capacitor cannot hold its charge over an extended period of time and will lose the stored data bit unless its charge is *refreshed* periodically. This process of refreshing requires additional circuitry and complicates the operation of the dynamic memory. Figure 9–11 shows a typical dynamic cell that is constructed of a single MOS transistor and a capacitor.

In this type of cell, Q acts as a switch. When the row select line is activated, Q is turned on and connects the storage capacitor to the data/sense line. A write amplifier and a read (sense) amplifier are connected to the data/sense line of

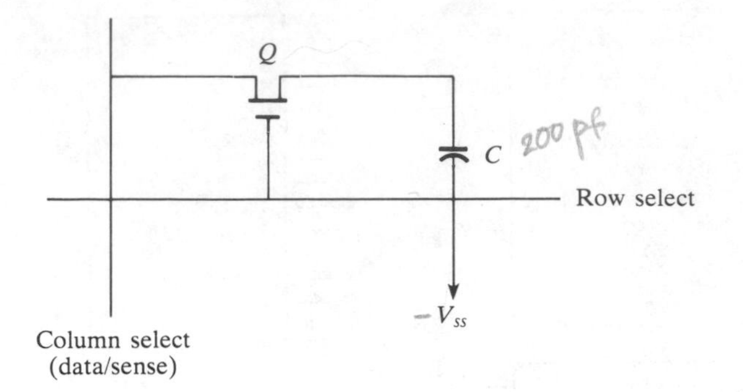

FIGURE 9–11 *MOS dynamic memory cell.*

each column in the memory array. When the memory is being read, the data bit stored by the capacitor is sensed by the read amplifier; when the memory is being written into, a data bit is placed on the data/sense line by the write amplifier and stored by the capacitor.

As mentioned before, the dynamic memory cell must be *refreshed* at certain intervals or the data bit will be lost due to charge leakage. Basically, a refresh operation consists of a read operation in which the data bit is sensed and then applied back to the write amplifier followed by a write operation which restores the same data bit into the cell. The refresh cycle time is typically 2 to 3 ms.

As an example of an MOS dynamic RAM, Figure 9–12 shows a block diagram, logic diagram and pin diagram for the Intel 2104A 4096 × 1 bit memory. The 4096 addresses require twelve address bits ($2^{12} = 4096$). Notice, however, that there are only six address lines ($A_0 - A_5$) due to the pin limitations of the package size. Therefore, a 12-bit address is multiplexed into the *row* and *column* latches on the address bus six bits at a time. The address bus is simply six parallel lines that connect the address inputs to both latches.

The *row address strobe* ($\overline{RAS}$) and the *column address strobe* ($\overline{CAS}$) are used to enter the two sets of six address bits into the latches. First, the $\overline{RAS}$ pulse strobes in the lower order address bits which select one of the 64 rows. Next, the $\overline{CAS}$ pulse strobes in the higher order address bits which select one of the 64 columns. The $\overline{CAS}$ pulse also clocks in the *chip select* ($\overline{CS}$) to bit 7 of the column latch, which in turn enables the data output buffer.

Unlike a static memory, a dynamic memory read cycle removes a data bit when it is read from a cell because the process requires that the capacitor be discharged. This is called *destructive read.* In a static memory, the information is stored in a latch cell and is not destroyed when read. This is called *nondestructive read.*

A dynamic memory requires that each read cycle be followed by a write cycle so that the data bit is restored back into the cell. If a *new* data bit is to be written into a cell, the memory goes through a *read-modify-write* cycle in which the present bit is read from the cell and a new bit on the data input is stored in its place.

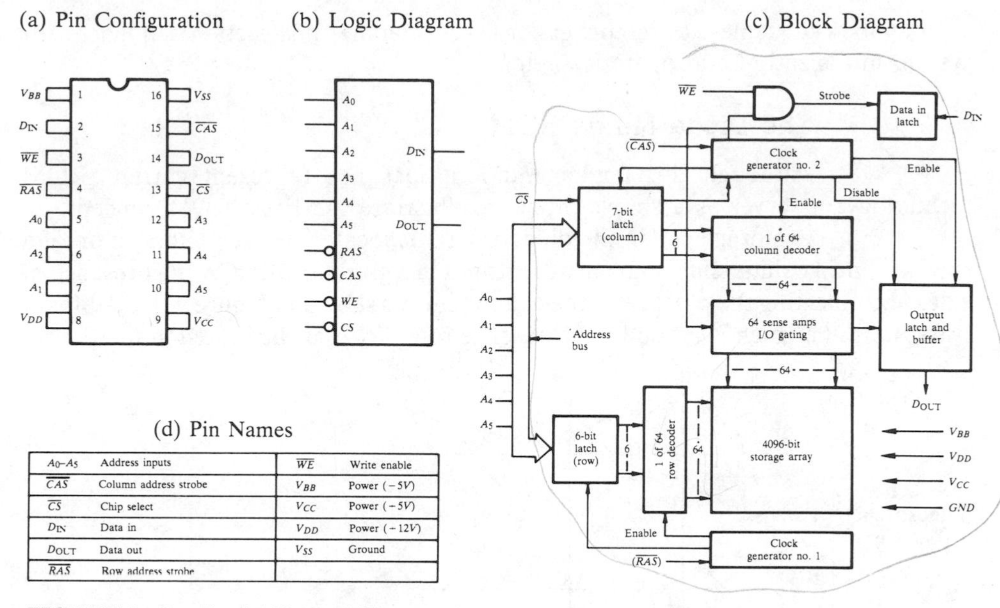

A_0–A_5	Address inputs	$\overline{WE}$	Write enable
$\overline{CAS}$	Column address strobe	V_{BB}	Power (−5V)
$\overline{CS}$	Chip select	V_{CC}	Power (−5V)
D_{IN}	Data in	V_{DD}	Power (−12V)
D_{OUT}	Data out	V_{SS}	Ground
$\overline{RAS}$	Row address strobe		

FIGURE 9–12 *Intel 2104A 4096 × 1 dynamic RAM. (Courtesy of Intel Corporation.)*

The *write enable* ($\overline{WE}$) is taken HIGH for a read cycle and LOW for a write cycle in the 2104A RAM. Chip selection or enable is done by applying a LOW to the *chip select* ($\overline{CS}$) input during a read, write, or read-modify-write cycle. Deselection occurs when the $\overline{CS}$ input is HIGH during any of these cycles. The basic purpose of the chip select function is to allow connection or disconnection from a common data line that may be shared with other devices.

Each of the 64 rows of cells in the memory array must be refreshed every 2 ms to keep from losing the stored data. This is accomplished by going through the $\overline{RAS}$-only sequence for the 64 row addresses with the chip deselected.

At the present state-of-the-art technology, dynamic RAMs are available with up to 64K bits of storage capacity; 64K actually means 65,536 bits in currently accepted terminology.

9–3 READ ONLY MEMORIES (ROMs)

A *read only memory* (ROM) contains permanently or semipermanently stored binary data which can be *read* from the memory but which either cannot be changed at all or cannot be changed without a relatively involved process.

A ROM differs from a RAM in that the ROM has no normal *write* cycle. ROM addresses can be selected in any given sequence just as in a RAM. ROMs are used to store data that are used over and over again in a system application such as look-up tables, code conversions, programmed instruction sequences for system initialization and operation, etc.

The Diode Matrix ROM

Although this implementation does not represent current ROM technology, it serves as a simple model to illustrate the basic ROM concept.

A storage cell in the diode matrix is located at each intersection of a row line and a column line. A binary 0 is stored in a given location by the presence of a diode connecting the row line to the column line as shown in Figure 9–13. A binary 1 is stored in a given location by the absence of a diode as illustrated in the figure.

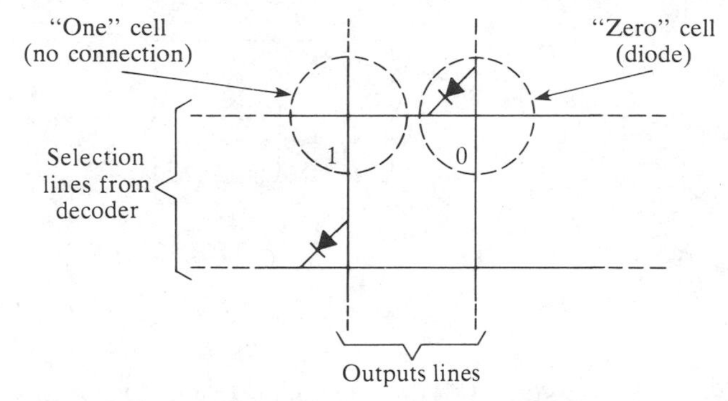

FIGURE 9–13 *Storage in a diode ROM.*

As an example of a very simple diode matrix ROM, we will produce the squares of decimal numbers 0 through 7. A three-bit address is required to select one of the eight locations. When one of the locations representing a given decimal number is addressed, the binary number representing the square of that number should appear on the output lines. Since the highest value possible on the outputs is

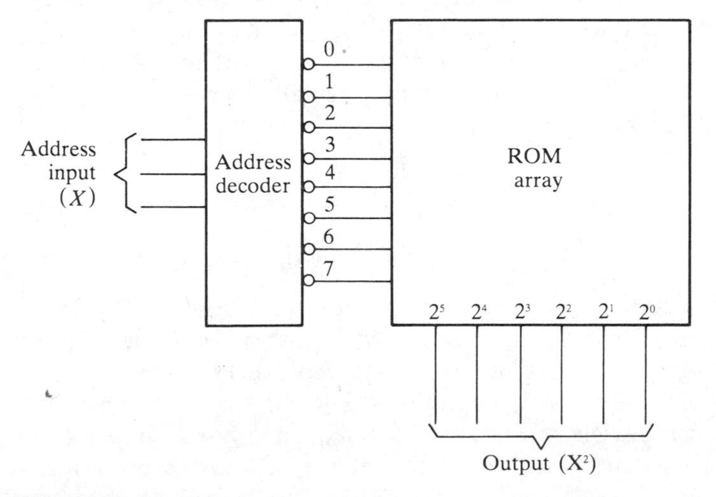

FIGURE 9–14 *Block diagram of x^2 look-up table ROM.*

$7^2 = 49$, six bits are necessary. Figure 9–14 illustrates this simple ROM in block diagram form.

A diode matrix implementation of the x^2 ROM is shown in Figure 9–15. The address decoder produces a LOW on one of the row lines corresponding to the binary inputs. Each row represents a decimal digit and is connected through diodes to the appropriate columns where 0s should appear in the output code for x^2. A 1-of-8 decoder with open collector outputs or a 1-of-10 (BCD-to-Decimal) decoder such as the 7445 with the most significant input bit grounded (0) can be used.

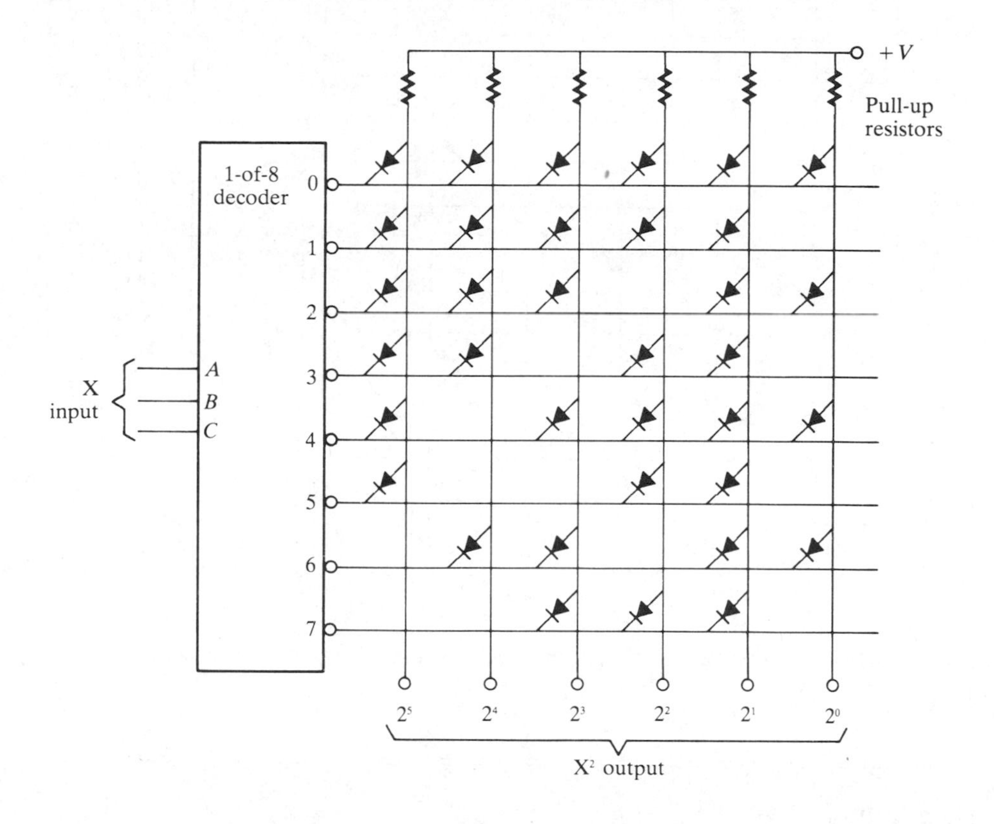

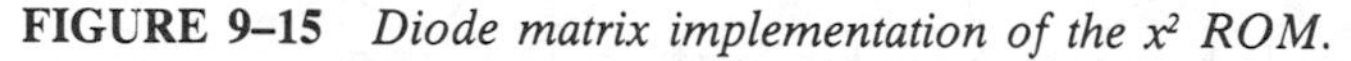
FIGURE 9–15 *Diode matrix implementation of the x^2 ROM.*

If a binary 011 (3_{10}) is applied to the x input, a LOW level is produced on the "3" output of the decoder. This forward biases each of the diodes connected to this row and pulls the corresponding columns LOW. The two columns which have no diode connected to the "3" row remain HIGH because of the pull-up resistors. The resulting code on the x^2 output is 001001 (9_{10}). You can verify that the output code represents the square of any of the other digits when the appropriate three-bit code is applied to the input.

Another example of a simple ROM application is shown in Figure 9–16. This ROM is programmed as a four-bit binary-to-gray code converter. An

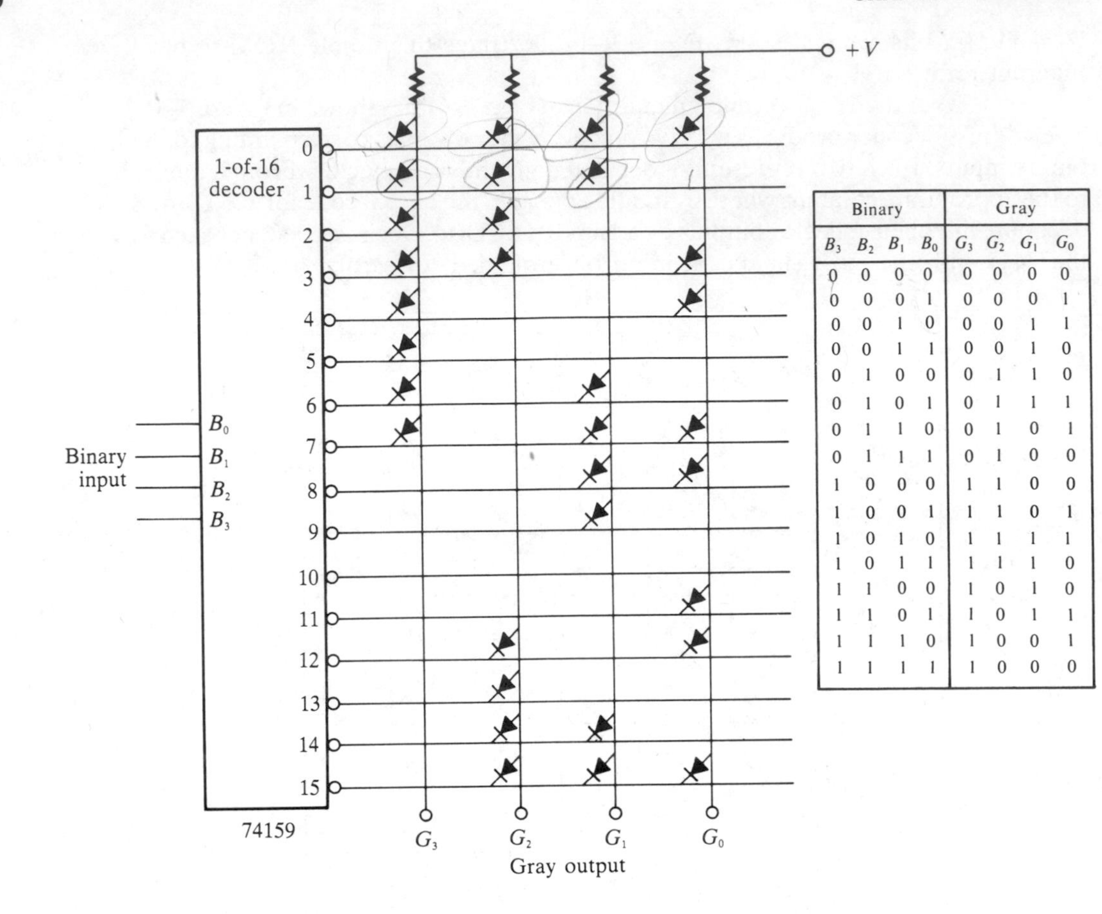

Binary				Gray			
B_3	B_2	B_1	B_0	G_3	G_2	G_1	G_0
0	0	0	0	0	0	0	0
0	0	0	1	0	0	0	1
0	0	1	0	0	0	1	1
0	0	1	1	0	0	1	0
0	1	0	0	0	1	1	0
0	1	0	1	0	1	1	1
0	1	1	0	0	1	0	1
0	1	1	1	0	1	0	0
1	0	0	0	1	1	0	0
1	0	0	1	1	1	0	1
1	0	1	0	1	1	1	1
1	0	1	1	1	1	1	0
1	1	0	0	1	0	1	0
1	1	0	1	1	0	1	1
1	1	1	0	1	0	0	1
1	1	1	1	1	0	0	0

FIGURE 9–16 *Diode matrix ROM programmed for binary-to-gray conversion.*

Example 9–1

Show how a ROM can be programmed to perform the following logic functions (A is the LSB):

$$X = \overline{A}B\overline{C} + A\overline{B}C$$
$$Y = AB\overline{C} + \overline{A}BC + \overline{A}\,\overline{B}\,\overline{C}$$
$$Z = A\overline{B}C + \overline{A}\,\overline{B}\,\overline{C} + \overline{A}\,\overline{B}C + ABC$$

Solution:

See Figure 9–17.

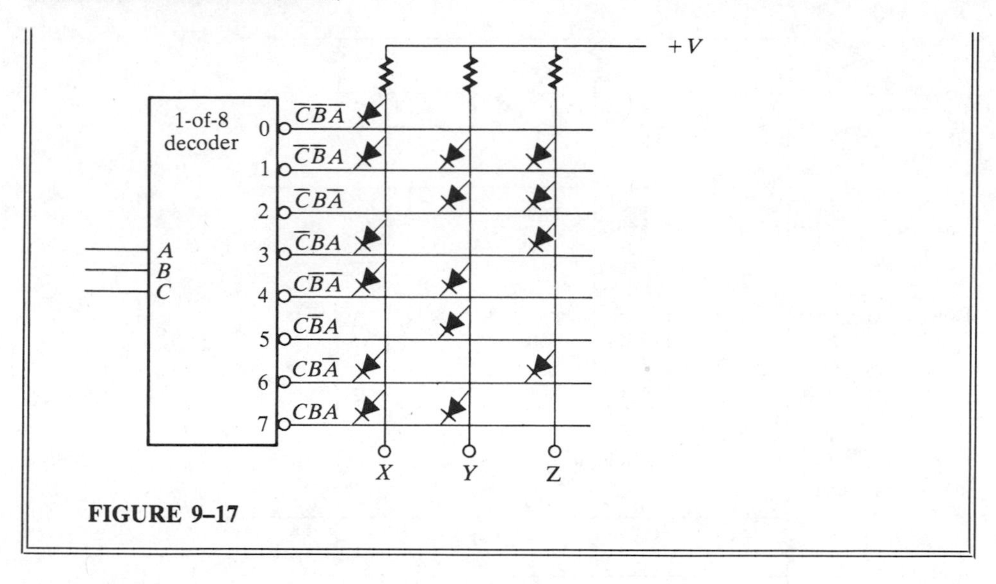

FIGURE 9–17

inspection of this matrix will show that any four-bit binary code applied to the address inputs will produce the corresponding four-bit gray code on the outputs as shown in the table.

Mask Programmed ROM

Although the diode matrix serves to demonstrate the ROM concept, ROMs are actually manufactured using transistors. The mask programmed ROM is a type that is permanently programmed during the manufacturing process to customer specified functions.

Once the memory is programmed, it cannot be changed. Most integrated circuit ROMs utilize the presence or absence of a transistor at a row/column junction to represent a 1 or 0. ROMs can be either bipolar or MOS. Figure 9–18(a) shows a bipolar TTL ROM array. The presence of a connection from a row line to the base of a transistor represents a "0" at that location because when the row line is taken HIGH, all transistors with a base connection turn on pulling the associated column lines LOW.

At row/column junctions where there is no base connection the column line remains HIGH when the row is addressed. This represents a "1" at that location.

Figure 9–18(b) illustrates an MOS ROM array. It is basically the same concept as the bipolar device, but uses MOSFETs with the presence or absence of a gate connection at a junction to store a 1 or a 0.

A ROM is *nonvolatile* memory because the stored data are retained if power is lost. In fact, the device can be removed from the circuit without loss of the stored bits. In contrast to this, a RAM is a *volatile* memory. If the V_{CC} is removed, all stored data are lost.

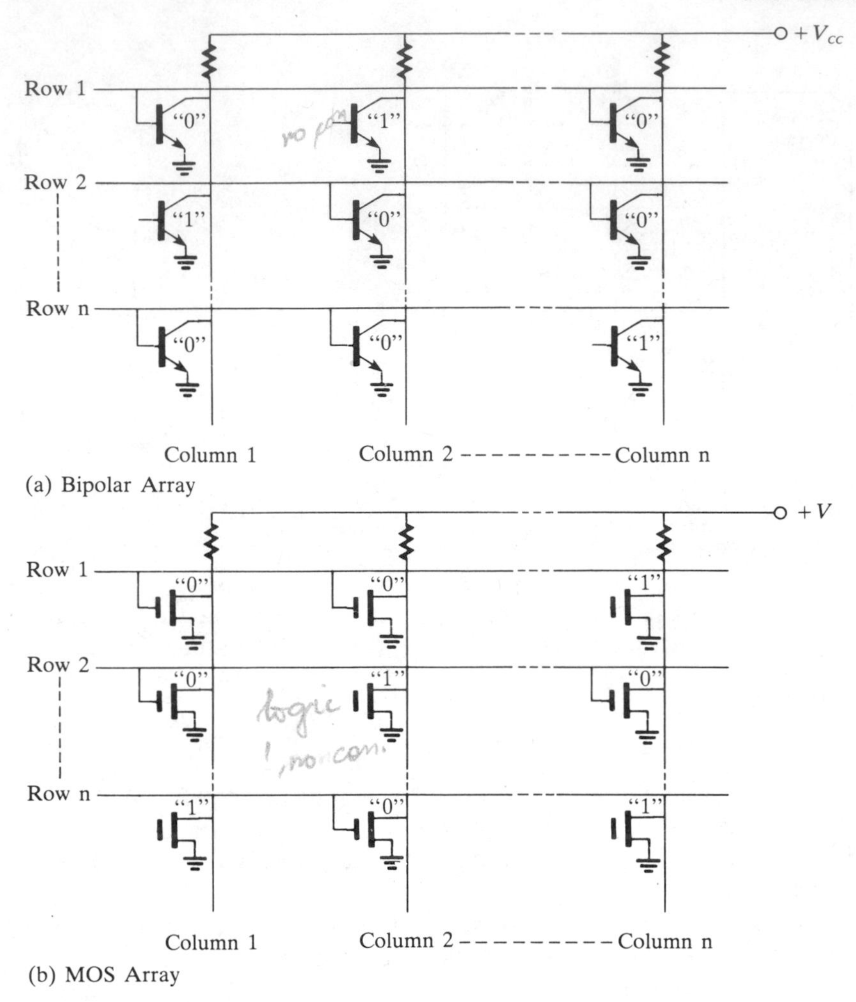

FIGURE 9–18 *Semiconductor ROM arrays.*

9–4 PROGRAMMABLE READ ONLY MEMORIES (PROMs)

There are several categories of programmable read only memories: PROMs, EAROMs, EPROMs, UVEPROMs, and EEPROMs. They differ from the ROM in that they can be programmed by the user to meet specific requirements after they leave the manufacturer. In the last four categories mentioned above, the programs stored in the memory can be *erased* and the device reprogrammed.

PROMs

These are nonerasable memories that can be programmed by the user, but once programmed they cannot be changed. The process of programming involves storing a specified pattern of 1s and 0s in the memory array so that each address corresponds to a given bit pattern.

PROMs are memory arrays with address decoding and buffering as shown in Figure 9–19. Programming of the memory arrays requires special equipment to apply certain voltages and logic conditions at various pins on the device.

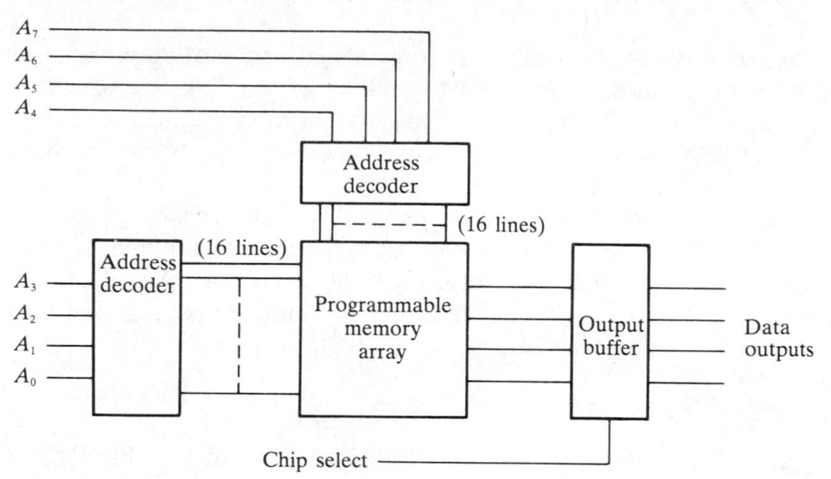

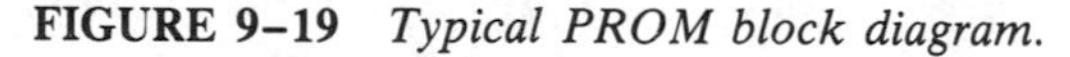
FIGURE 9–19 *Typical PROM block diagram.*

PROMs are available in both bipolar and MOS technology and basically have four-bit and eight-bit output word formats with bit capacities up to 64K. The nonerasable PROMs have some form of fusing process where a memory link is fused open or closed to represent a 1 or a 0 at a given memory location. The fusing process involves an electochemical action that is irreversible so that once a PROM is programmed, it cannot be changed.

Figure 9–20 illustrates a bipolar PROM array with fusible links. There are three types of fuse technologies used in PROMs: *metal links, silicon links,* and *shorted junctions.*

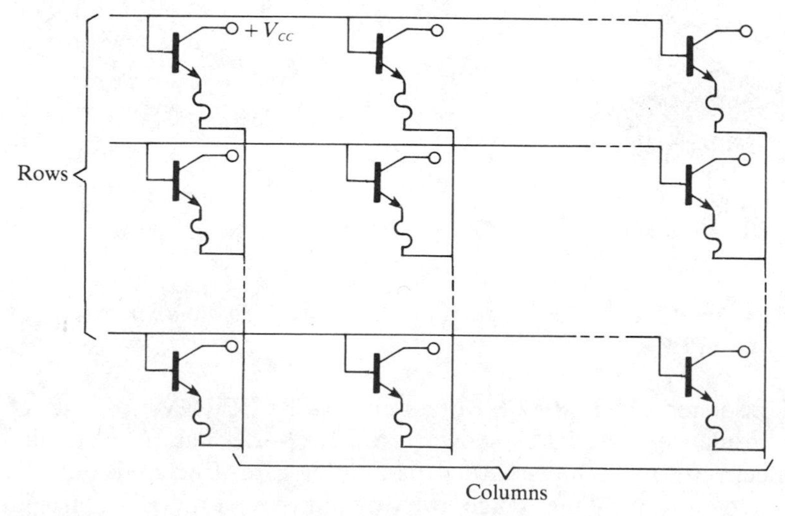

FIGURE 9–20 *Bipolar PROM array with fusible links. (All collectors are commonly connected to V_{CC}.)*

Metal links are made of a material such as nichrome. Each bit in the memory array is represented by a separate link. During programming, the link is either "blown" open or left intact. This is basically done by first addressing a given cell and then forcing a sufficient amount of current through the link to cause it to open.

Silicon links are formed by narrow notched strips of polycrystalline silicon. Programming of these fuses requires melting of the links by passing a sufficient amount of current through them. This amount of current causes a high temperature at the fuse location that oxidizes the silicon and forms an insulation around the now open link.

Shorted junction or *avalanche-induced migration* technology consists basically of two PN junctions arranged back to back. During programming, one of the diode junctions is avalanched and the resulting voltage and heat cause aluminum ions to migrate and short the junction. The remaining junction is then used as a forward biased diode to represent a data bit.

EPROMs

An EPROM is an *erasable* PROM. Unlike an ordinary PROM, an EPROM can be reprogrammed by first erasing an existing program in the memory array. These devices use an NMOS FET array with an *isolated gate* structure.

The isolated transistor gate has no electrical connections and can store an electrical charge for indefinite periods of time. The data bits in this type of an array are represented by the presence or absence of a stored gate charge. Erasure of a data bit is a process that removes the gate charge.

There are two basic types of erasable PROMs, the *ultra-violet light erasable* PROM (UV EPROM) and the *electrically erasable* PROM (EEPROM).

UV EPROMs

You can recognize this UV EPROM device by the quartz lid on the package as in Figure 9–21.

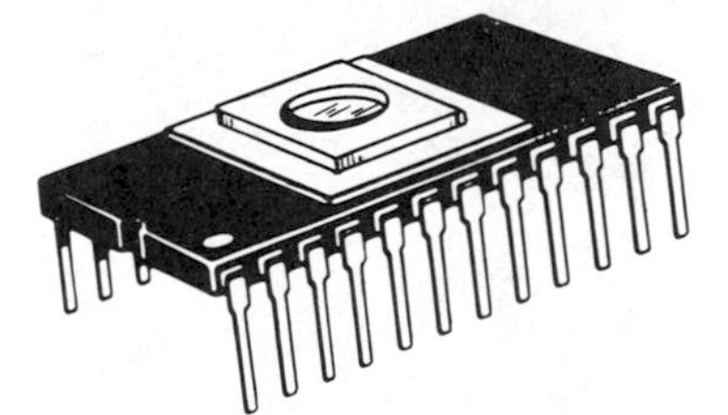

FIGURE 9–21 *Ultra-violet erasable PROM package. (Courtesy of Motorola Semiconductor Products)*

The isolated gate in the FET of an ultra-violet EPROM is "floating" within an oxide insulating material as shown in Figure 9–22. The programming process causes electrons to be removed from the floating gate. Erasure is done by exposure of the memory array chip to high intensity ultra-violet radiation through the quartz window. This neutralizes the positive charge stored on the gate after several minutes to an hour of exposure time.

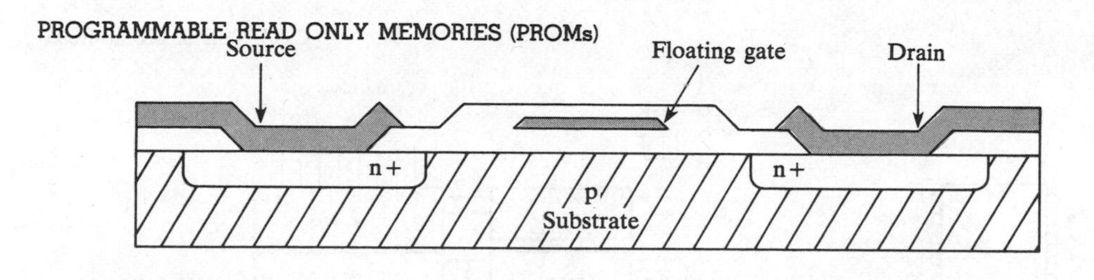

FIGURE 9–22 *UV EPROM MOS memory cell.*

EEPROMs

Electrically erasable PROMs can be both erased and programmed with electrical pulses. This type of device is also known as an *electrically alterable* ROM (EAROM). Since it can be both electrically written into and electrically erased, the EEPROM can be programmed and erased in-circuit for reprogramming.

There are two types of EEPROMs, floating gate MOS and metal-nitride-oxide-silicon (MNOS). Both structures are shown in Figure 9–23. The application of a voltage on the control gate in the floating gate structure permits the storage and removal of charge from the floating gate. In the MNOS structure, the isolated gate exists on the boundary between the SiO_2 layer and the Si_3N_4 layer.

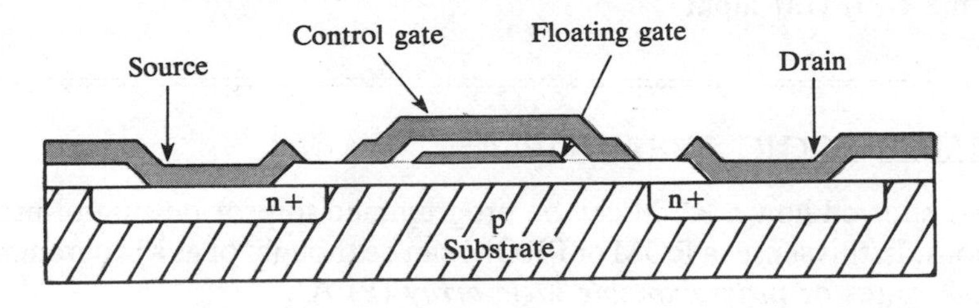

(a) Floating Gate EEPROM Cell

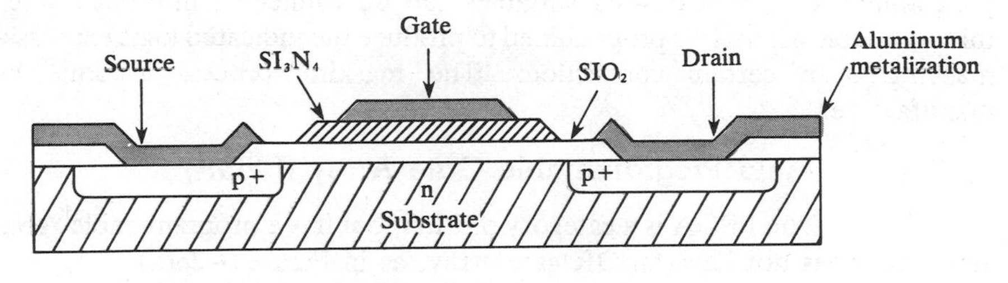

(b) MNOS EEPROM Cell

FIGURE 9–23 *EEPROM cells.*

An EPROM Example

As a specific example of an EPROM, we will look at the Intel 2716 MOS UV EPROM which is a 16K-bit device (actually 16,384 bits). The block and pin diagrams are shown in Figure 9–24.

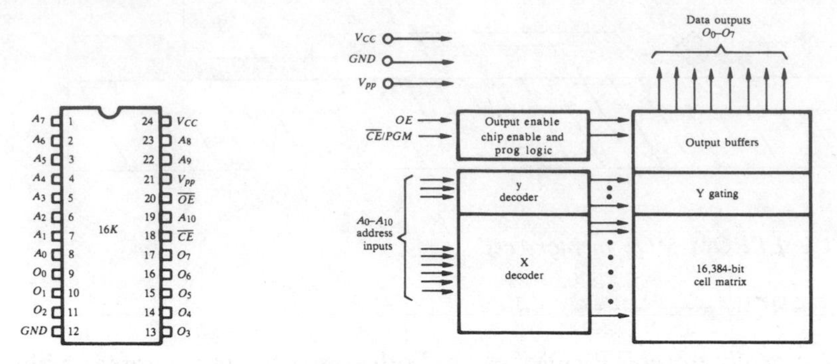

FIGURE 9–24 *A 2716 MOS UV EPROM (Courtesy of Intel Corporation).*

After an erasure by exposure to ultra-violet light for a specified amount of time, all bits in the array are in the "1" state. During programming, only 0s are written into the array at selected locations.

The 2716 is programmed by setting the V_{pp} power supply to 25V and taking the *OE* input HIGH. The data to be programmed are applied eight bits at a time to the data output pins. The address into which the eight bits are to be entered is then applied to the address inputs and a high level pulse of a specified duration is applied to the $\overline{CE}/PGM$ input.

9–5 PROGRAMMABLE LOGIC ARRAYS (PLAs)

Example 9–1 showed how a ROM can be programmed to provide sum-of-product logic functions. In this sense, a ROM or PROM can be thought of as a programmable array of logic gates or *programmable logic array* (PLA).

Basically a PLA consists of an array of AND-OR logic with inverters that can be programmed to produce desired logic functions on the outputs. An example of a small PLA is shown in Figure 9–25. Each connection is mask programmable so that desired variables can be connected into each gate. This three-variable example is programmed to produce the indicated logic expressions by masking open certain connections. The masking process is done by the manufacturer.

Field Programmable Gate Array (FPGA)

The FPGA is a category of PLA that has a programmable AND gate array but does not have an OR gate array, as in Figure 9–26(a).

Programmable Array Logic (PAL)

The PAL is a type of PLA having a programmable AND aray and a *fixed* OR gate array, as in Figure 9–26(b).

Field Programmable Logic Array (FPLA)

This is a PLA that can be user programmed rather than mask programmed by the manufacturer. Figure 9–26(c) shows a block diagram for this type of logic array.

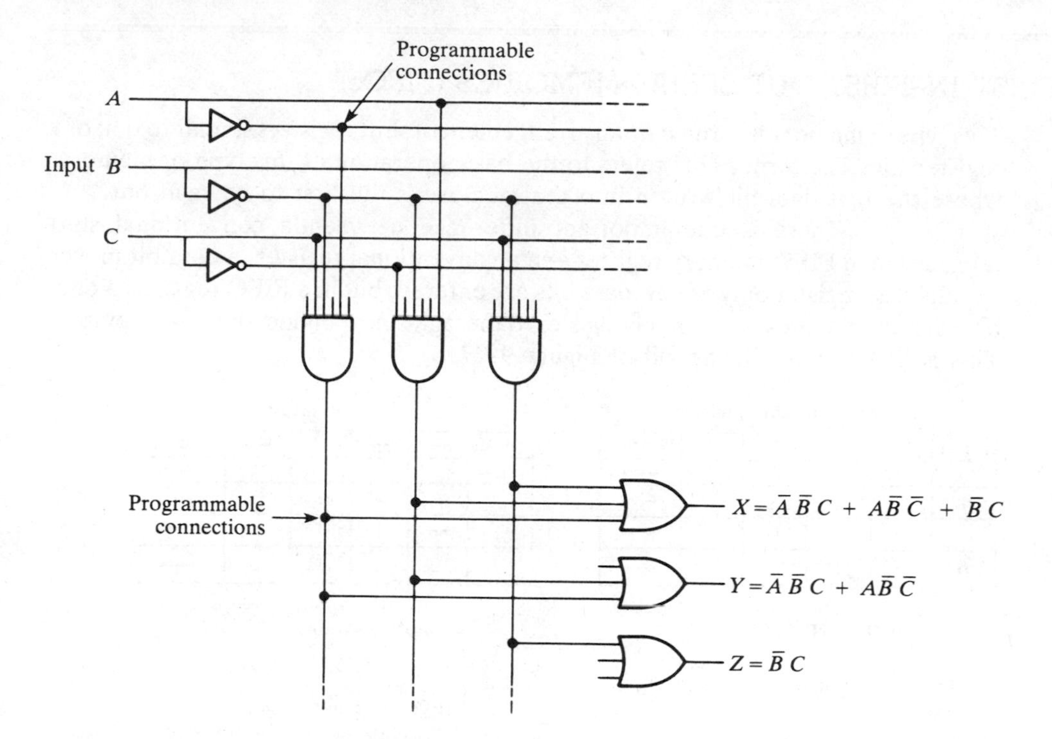

FIGURE 9–25 *Simple example of a PLA.*

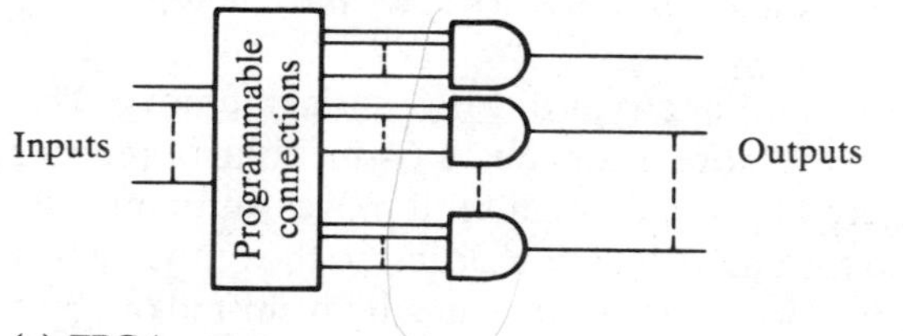

(a) FPGA

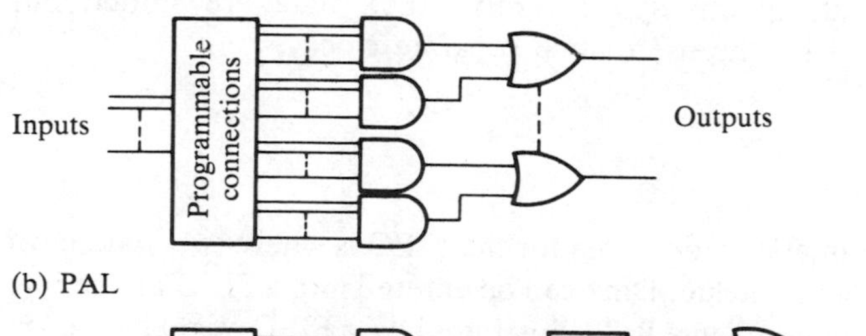

(b) PAL

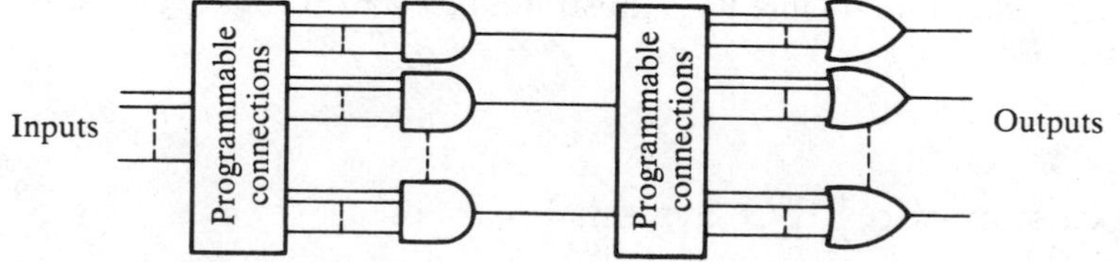

(c) FPLA

FIGURE 9–26 *FPGA, PAL, and FPLA block diagrams.*

9–6 FIRST IN–FIRST OUT SERIAL MEMORIES (FIFOs)

This type of memory is formed by an arrangement of shift registers similar to that of a register file. The term *FIFO* refers to the basic operation of this type of memory, where the first data bit written into the memory is the first to be read out.

There is one important difference between a conventional shift register and a FIFO memory register—in a conventional register, a data bit moves through the register only as new data bits are entered; but in a FIFO register, a data bit immediately goes through the register to the right-most bit location that is empty. This is illustrated with the aid of Figure 9–27.

Conventional shift register

Input	X	X	X	X	Output
0	0	X	X	X	→
1	1	0	X	X	→
1	1	1	0	X	→
0	0	1	1	0	→

X = unknown data bits

In a conventional shift register, data stay to left until "forced" through by additional data.

FIFO shift register

Input	—	—	—	—	Output
0	—	—	—	0	→
1	—	—	1	0	→
1	—	1	1	0	→
0	0	1	1	0	→

— = empty positions

In a FIFO shift register, data "fall" through—go right.

FIGURE 9–27 *Comparison of conventional and FIFO register operation.*

Figure 9–28 is a block diagram of a typical FIFO serial memory. This particular memory has four serial 64-bit data registers and a 64-bit control register (marker register). When data are entered by a shift-in pulse, it moves automatically under control of the marker register to the empty location closest to the output. Data cannot advance into occupied positions. However, when a data bit is shifted out by a shift-out pulse, the data bits remaining in the registers automatically move to the next position toward the output. In an *asynchronous* FIFO, data are shifted out independent of data entry with the use of two separate clocks.

Applications

One important application area for the FIFO is where two systems of differing data rates must communicate. Data can be entered into a FIFO at one rate and be outputted at another rate. Figure 9–29 illustrates how a FIFO might be used in these situations.

Expansion of a FIFO System

In order to handle large numbers of data bits in situations such as those illustrated, expansion may be done by connecting a number of individual FIFO

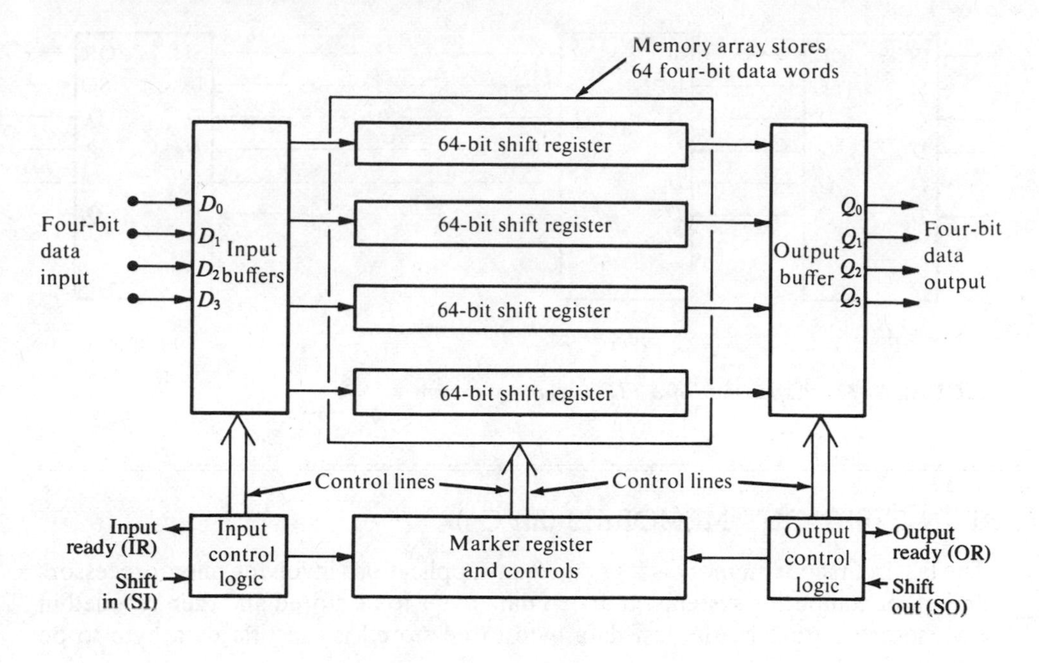

FIGURE 9–28 *Block diagram of a typical FIFO serial memory.*

FIFO

(a) Irregular telemetry data can be stored and retransmitted at a constant rate.

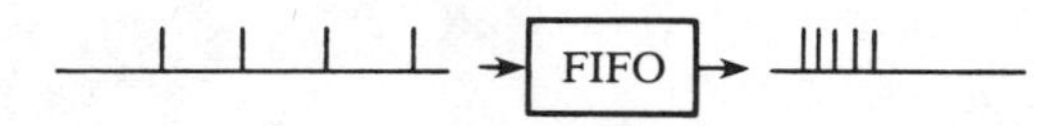

(b) Data input at a slow keyboard rate can be stored and then transferred at a higher rate for processing.

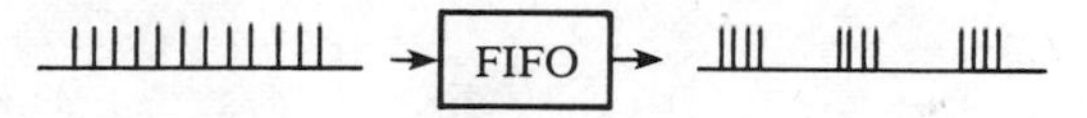

(c) Data input at a steady rate can be stored and then output in even bursts.

(d) Data in bursts can be stored and reformatted into a steady-rate output.

FIGURE 9–29 *The FIFO in data rate buffering applications.*

devices together as shown in Figure 9–30. Assuming the FIFO contains 64–bit registers, the capacity is increased by 64 bits for each device connected as shown.

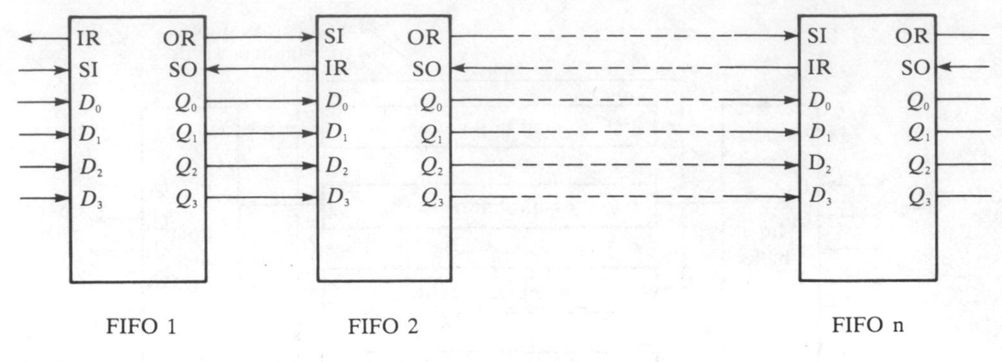

FIGURE 9–30 *Expansion of a FIFO memory system.*

9–7 LAST IN–FIRST OUT MEMORIES (LIFOs)

The last in/first out memory is found in many applications involving microprocessors and other computing systems. It allows data bytes to be stored and then recalled in reverse order; that is, the last data byte to be stored is the first data byte to be retrieved.

Stacks

A LIFO memory is commonly referred to as a *push-down stack.* In many microprocessor systems, it is implemented with a group of registers as shown in Figure 9–31. A stack can consist of any number of registers but the register at the top is called the *top of the stack.*

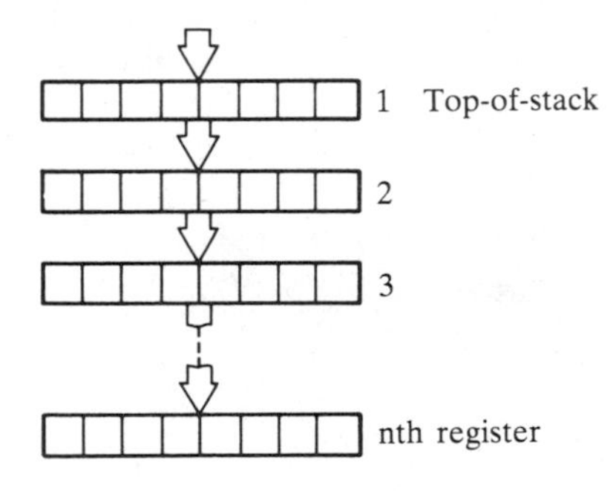

FIGURE 9–31 *Register stack.*

A byte of data (eight bits) is loaded in parallel onto the top of the stack. Each successive byte "pushes" the previous one down into the next register. This process is illustrated in Figure 9–32. Notice that the new data byte is always loaded into the top register and the previously stored bytes are pushed deeper into the stack. This is where the name *push down* stack comes from.

Data bytes are retrieved in the reverse order. The last byte entered is always at the top of the stack so that when it is "pulled" from the stack the other bytes "pop" up into the next higher locations. This process is illustrated in Figure 9–33.

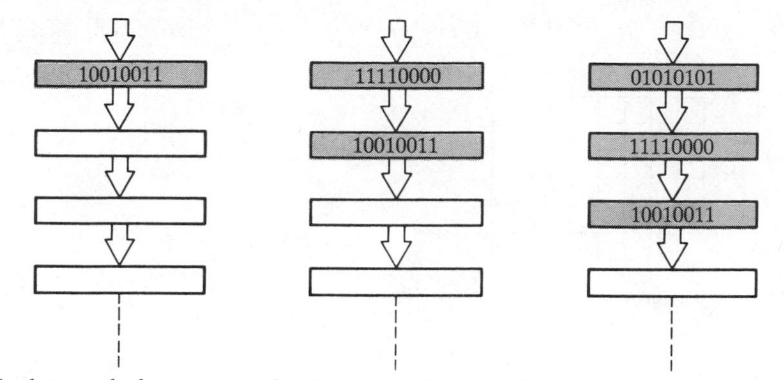

1st byte pushed onto stack. 2nd byte pushed onto stack. 3rd byte pushed onto stack.

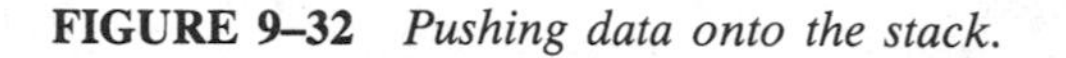
FIGURE 9–32 *Pushing data onto the stack.*

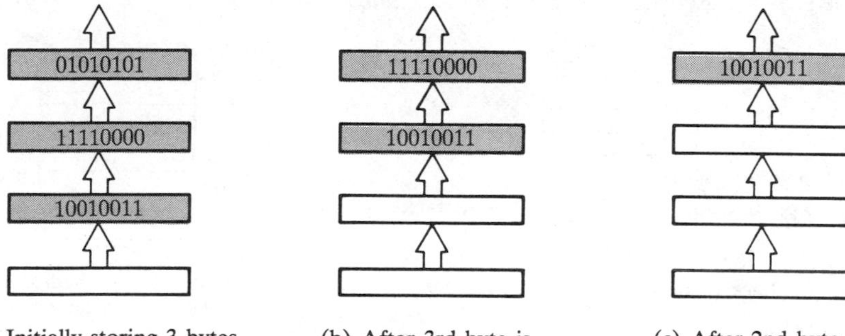

(a) Initially storing 3 bytes.

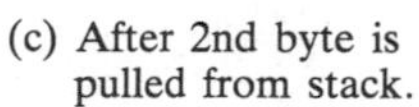
(b) After 3rd byte is pulled from stack.

(c) After 2nd byte is pulled from stack.

FIGURE 9–33 *Pulling data from the stack.*

RAM Stack

Another approach used in some microprocessors including the 6800 (described in Chapter 12) is the allocation of a section of RAM as the stack rather than dedicating a set of registers.

Consider a random access memory that is *byte* oriented—that is, each address contains eight bits as illustrated in Figure 9–34.

Next, consider a section of RAM set aside for use as a stack. A special separate register called a *stack pointer* contains the address of the top of the stack, as illustrated in Figure 9–35. A four-digit hexadecimal representation is used for the binary addresses (we are assuming a 16-bit address code). In the figure, the addresses are chosen arbitrarily for purposes of illustration.

Now let us see how data are pushed onto the stack. A data byte is stored by a normal memory *write* operation at address 00FF which is the top of the stack. The stack pointer is then decremented (decreased by 1) to 00FE. This moves the top of the stack to the next lower memory address as shown in Figure 9–35(b).

Notice that the top of the stack is not stationary as in the fixed register stack but moves *downward* (to lower addresses) in the RAM as data bytes are stored.

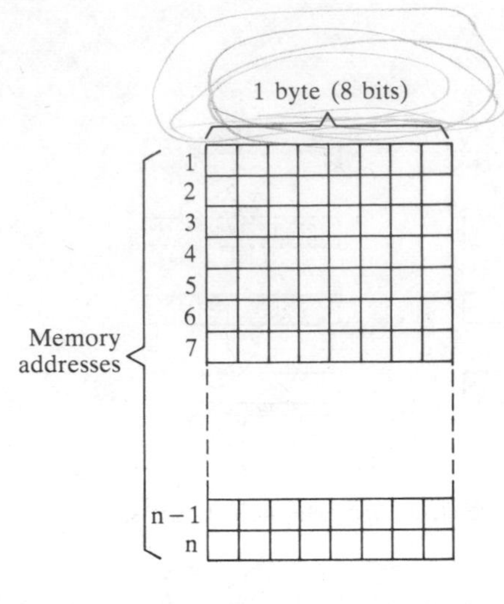

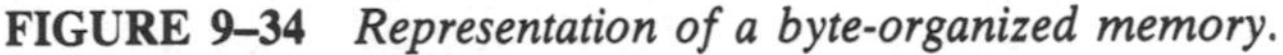

FIGURE 9–34 *Representation of a byte-organized memory.*

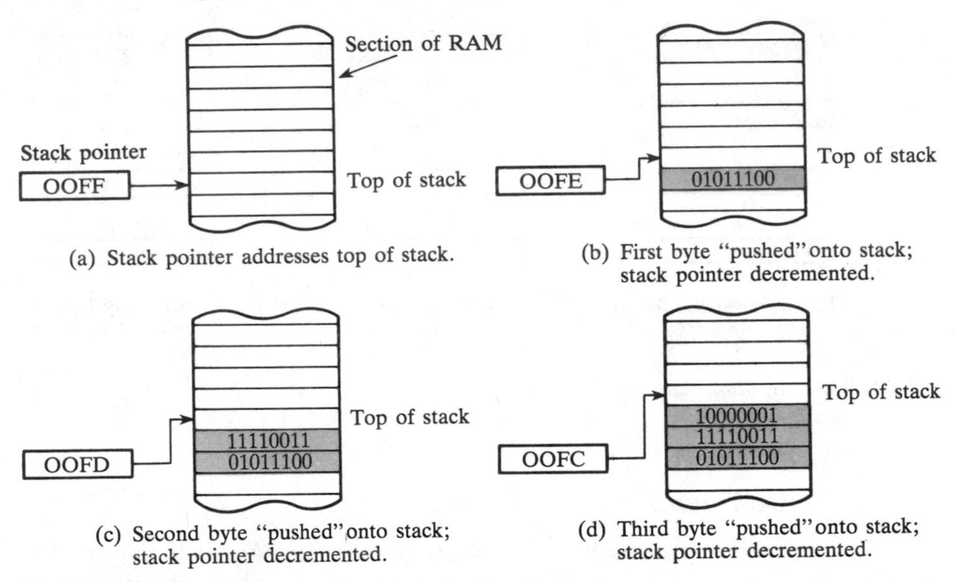

FIGURE 9-35 *Illustration of pushing data onto a RAM stack.*

Figure 9–35(c) and (d) show two more bytes being pushed onto the stack. After the third byte is stored the top of the stack is at 00FC.

To retrieve data from the stack, the stack pointer is first incremented (increased by 1) and then a RAM *read* operation is performed. The first data byte to be pulled from the stack is at address 00FD. When this byte is read, the stack pointer is again incrementeed to 00FE. Another read operation pulls the next byte and so on as shown in Figure 9–36. Keep in mind that most RAMs are nondestructive when read so that the data byte still remains in the memory after a read operation. A data byte is destroyed only when a new byte is written over it.

A RAM stack can be of any depth depending on the amount of continuous memory addresses available that are not being or will not be used for anything else.

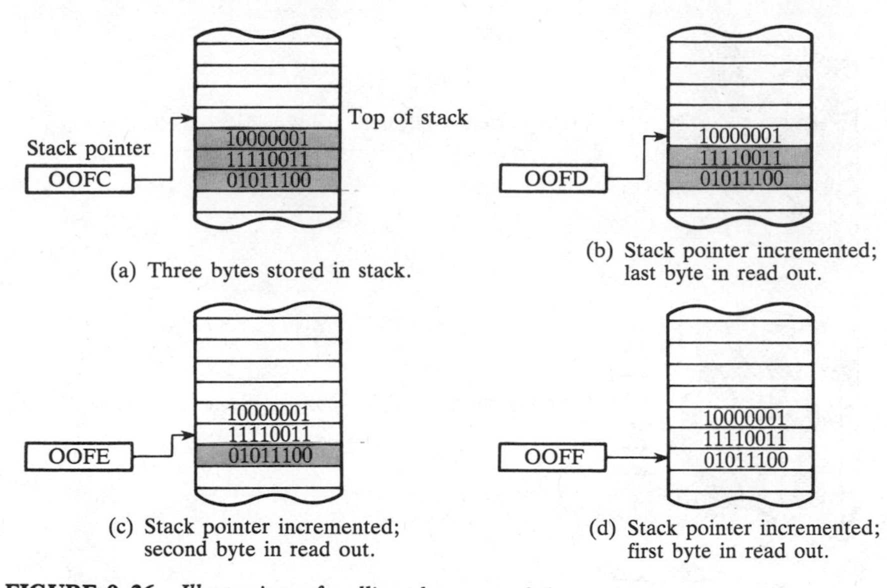

(a) Three bytes stored in stack.

(b) Stack pointer incremented; last byte in read out.

(c) Stack pointer incremented; second byte in read out.

(d) Stack pointer incremented; first byte in read out.

FIGURE 9–36 *Illustration of pulling data out of the stack.*

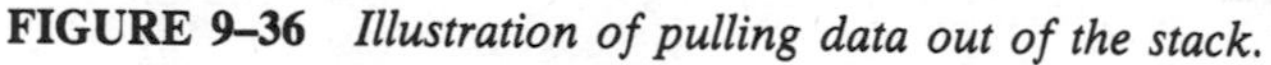

MAGNETIC BUBBLE MEMORIES (MBMs) 9–8

Magnetic bubble memories can be considered in some ways analogous to magnetic disk memories. In both types, data are stored as states of magnetization in a thin magnetic film.

In an MBM data bits are stored in the form of magnetic "bubbles" moving in thin films of magnetic material. The bubbles are actually cylindrical magnetic domains whose polarization is opposite to that of the thin magnetic film in which they are embedded.

The Magnetic Film

When a thin wafer of magnetic *garnet* is viewed by polarized light through a microscope, a pattern of "wavy" strips of magnetic domains can be seen. In one set of strips, the tiny internal magnets point up and in the other areas they point down. As a result, one set of strips appear bright and the other dark when exposed to the polarized light. This is illustrated graphically in Figure 9–37(a).

Now, if an *external magnetic field* is applied perpendicular to the wafer and slowly increased in strength, the wavy domain strips whose magnetization is opposite to that of the external field begin to narrow. This is illustrated in Figure 9–37(b).

At a certain magnitude of external field strength, all these domains suddenly contract into small circular areas called "bubbles." This is shown in Figure 9–37(c).

These bubbles typically are only a few micrometers in diameter and act as tiny magnets floating in the external field. The bubbles can be easily moved and

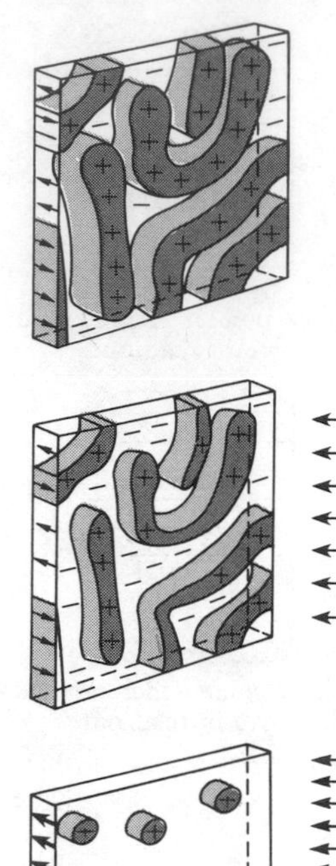

(a) No external magnetic field—wide magnetic domains.

(b) Small magnetic field—positive magnetic domains shrink.

(c) Large magnetic field—bubbles form.

FIGURE 9–37 *Creation of magnetic bubbles in a thin magnetic wafer by application of an external magnetic field.*

controlled within the wafer by rotating magnetic fields in the plane of the wafer or by current carrying conductive elements.

The method that uses external driving fields *in the plane* of the wafer to move the bubbles is called the *field-access* method. The bubbles are moved from point to point by a rotating magnetic field acting in conjunction with a pattern of thin-film permalloy geometric shapes on the wafer.

For example, Figure 9–38 shows a pattern of permalloy Ts and vertical bars deposited on the magnetic garnet wafer. Various other patterns are also possible. Imagine a magnetic field rotating clockwise *in the plane* of the wafer. At successive points in its rotation, it will be pointing right, down, left, and up as illustrated.

The segments in the pattern that are *parallel* with the field at a particular time are magnetized in the direction of the field. The segments that are perpendicular to the field are neutral.

If a bubble is introduced at the left end of the T-bar pattern at the level of the top of the T, it will move step by step to the right as the field rotates clockwise. In one complete rotation of the field, a bubble will move from the center of one T to the center of the next one.

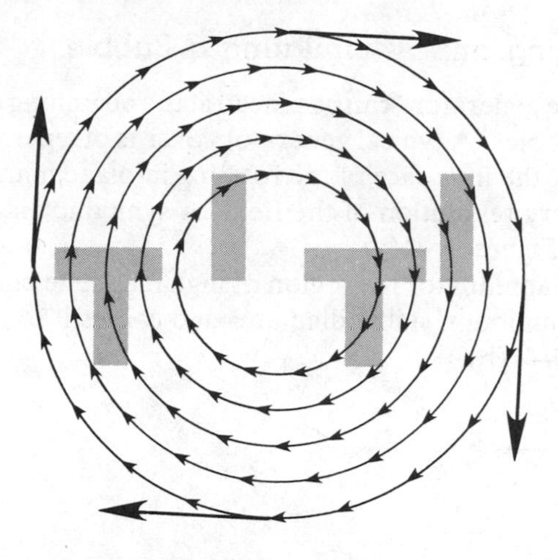

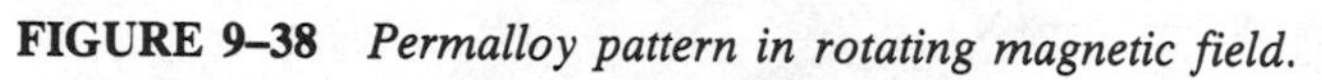

FIGURE 9–38 *Permalloy pattern in rotating magnetic field.*

This process is illustrated in Figure 9–39 where the top of the bubble is assumed to have a negative polarization. A shift to the right can be represented symbolically as shown in Figure 9–40.

Direction of external field.

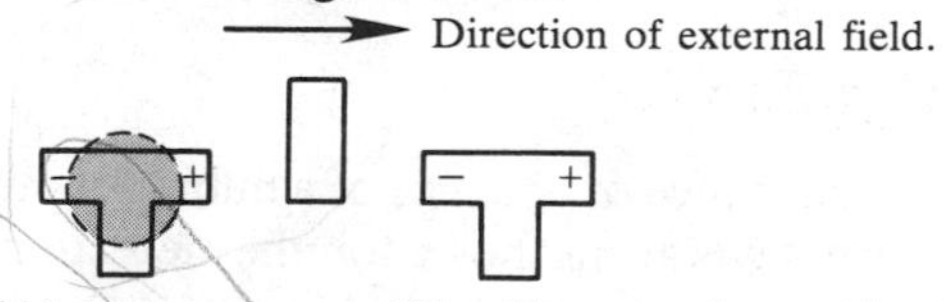

(a) Bubble begins at center of T and is attracted to positive right end.

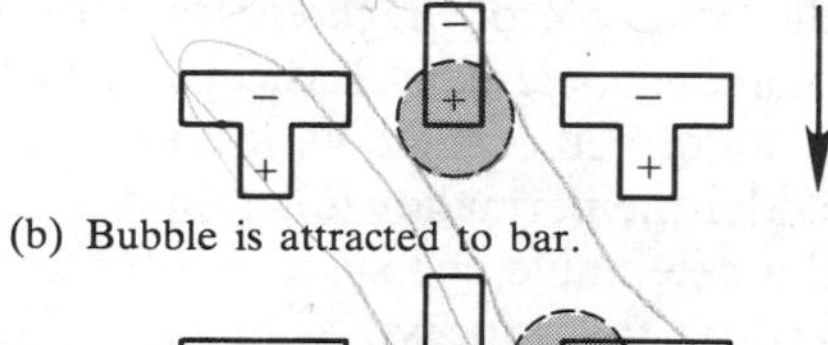

(b) Bubble is attracted to bar.

(c) Bubble is attracted to positive left end of T.

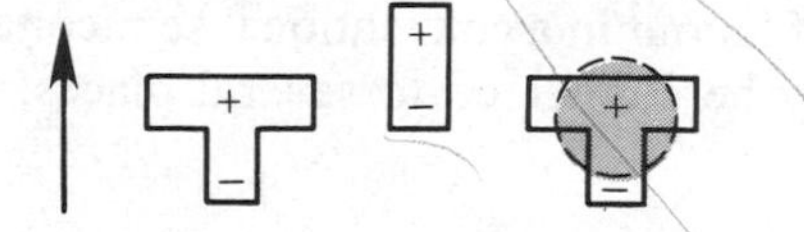

(d) Bubble is attracted to center of T.

FIGURE 9–39 *How the bubbles move in a T-bar pattern.*

FIGURE 9–40 *Bubbles shifting to the right.*

Generating and Annihilating a Bubble

A "bubble generator" can be placed at the beginning of a row of Ts and bars to produce new bubbles. A typical generator consists of a permalloy disk with a small projection. Under the influence of the rotating in-plane magnetic field, a new bubble emerges for every revolution of the field. A symbolic diagram of a bubble generator is shown in Figure 9–41(a).

A bubble annihilator is a region of high magnetic bias that shrinks the bubble to less than its minimum stable diameter and causes it to disappear. This is symbolized in Figure 9–41(b).

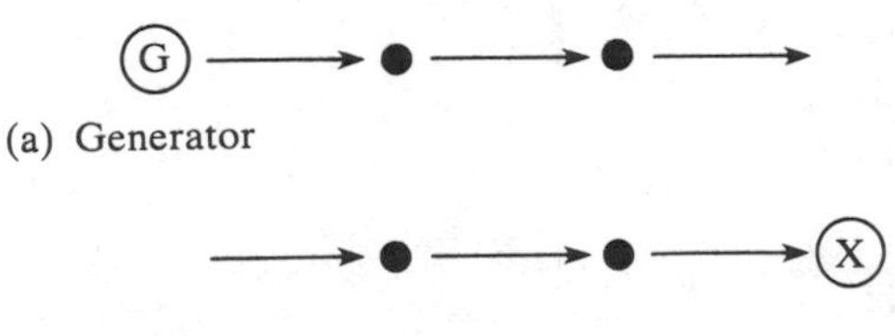

FIGURE 9–41 *Symbols representing bubble generator and annihilator.*

Detection of a Bubble

In an MBM, the presence or absence of a bubble represents a binary 1 or a 0 respectively. There are several methods for the detection or sensing the presence of a bubble.

One method involves the principle of *magnetoresistance* whereby a magnetic field changes the resistance of a sensor element. Basically, the detector is a permalloy junction whose resistance to direct current varies with the rotating magnetic field. When a bubble lands under the permalloy junction, the resistance is reduced by an additional amount and a detectable change in the sensor current is produced. The bubble is not destroyed by this process.

Replication

Replication is the duplication of existing binary states represented by bubbles. It is similar to the concept of fanout in a conventional semiconductor logic circuit where a given output bit may be connected to several places.

Memory Organization

In the simplest MBM arrangement, bubbles can be generated, shifted, and detected in an *endless-loop shift register* formed by T-bar or other similar patterns on the magnetic garnet film. These endless-loop shift registers provide the basis for mass data storage.

Basically, the data bits are stored in several *minor loops* and transferred into a single *major loop* for read out or alteration. Bubble generation and erasure are also required in an MBM.

Data are entered by selectively transferring bits in response to a write command from a special minor loop equipped with a generator at one end and an annihilator at the other.

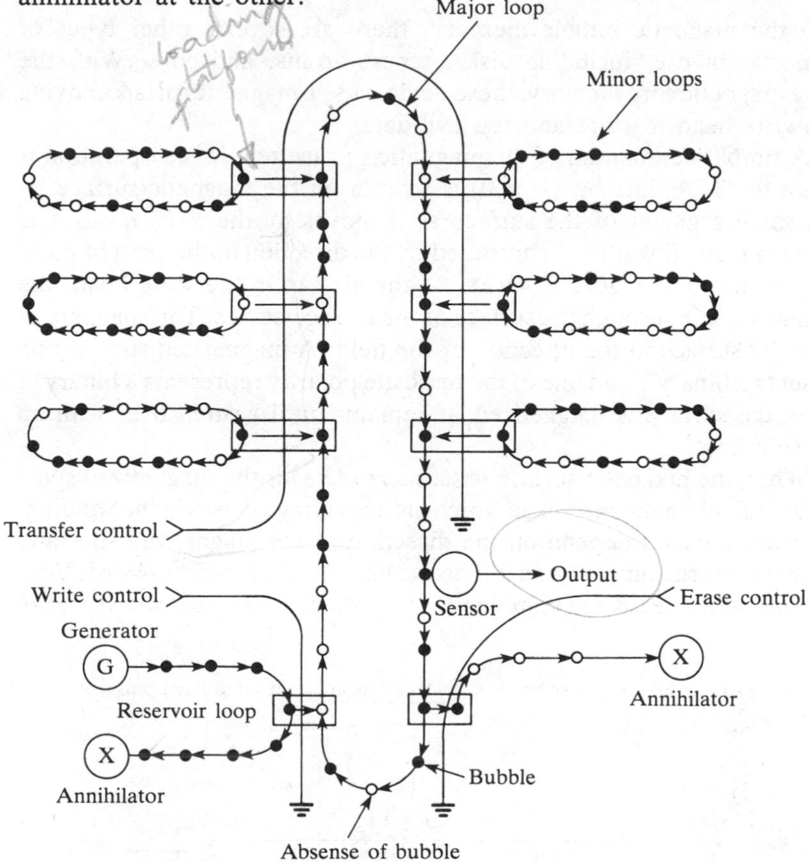

FIGURE 9–42 *Diagram of a basic MBM concept.*

Data are erased by transferring bits into a special minor loop that has an annihilator at the end.

Figure 9–42 shows a simplified diagram of the major/minor loop organization. To write data, bubbles are entered into the major loop from the reservoir loop in the lower right under control of a write mechanism. Then, from the major loop, data bits can be transferred into a minor loop for storage. Keep in mind that the in-plane rotating magnetic field keeps the bubbles moving in the loops at all times. Selective field application with the various control inputs is used to produce transfers from loop to loop.

To read, data bits are transfered from a minor loop to the major loop. Once in the major loop, the bubbles are sensed by the detector element and the result becomes the output data bits. Data can also be erased by transfer from the major loop into the annihilator loop under control of the erase mechanism.

Presently, MBMs having a storage capacity of one million bits (1 Mbit) are available.

9–9 MAGNETIC SURFACE STORAGE DEVICES

In addition to the magnetic bubble memory, there are several other types of magnetic memories in use, including disks, tapes, drums, and core. With the exception of the magnetic core memory, these devices use a magnetic surface moving passed a read/write head to store and retrieve data.

A simplified diagram of the magnetic surface read/write operation is shown in Figure 9–43. A data bit (1 or 0) is written on the magnetic surface by magnetizing a small segment of the surface as it moves by the *write head.* The direction of the magnetic flux lines is controlled by the direction of the current pulse in the winding as shown in Figure 9–43(a). At the air gap in the write head, the magnetic flux takes a path through the surface of the storage device. This magnetizes a small spot on the surface in the direction of the field. A magnetized spot of one polarity represents a binary 1 and one of the opposite polarity represents a binary 0. Once a spot on the surface is magnetized, it remains until written over with an opposite magnetic field.

When the magnetic surface passes a *read head,* the magnetized spots produce magnetics fields in the read head which induce voltage pulses in the winding. The polarity of these pulses depend on the direction of the magnetized spot and indicate whether the stored bit is a 1 or a 0. This is illustrated in Figure 9–43(b). Very often the read and write heads are combined into a single unit as shown in Figure 9–43(c).

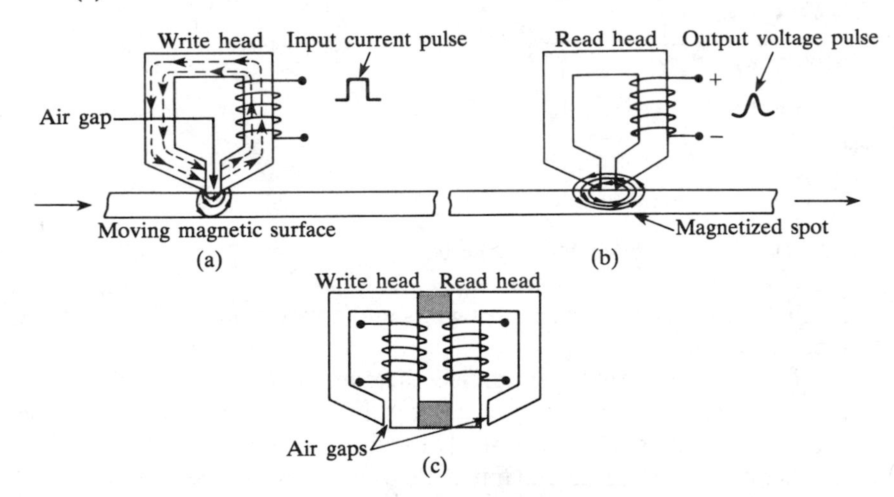

FIGURE 9–43 *Read/write function on a magnetic surface.*

Magnetic Recording Formats

Several ways in which digital data can be respresented for purposes of magnetic surface recording are called *return-to-zero* (RZ), *non-return-to-zero* (NRZ), *biphase, Manchester,* and the *Kansas City standard.* These waveform representations are separated into *bit times,* the intervals during which the level or frequency of the waveform indicates a 1 or 0 bit. These bit times are definable by their relation to a basic system timing signal or *clock.*

Figure 9–44 shows an example of a *return-to-zero* (RZ) waveform. In this case a fixed-width pulse occurring during a bit time represents a 1, and no pulse during a bit time is a 0. There is always a return to the 0 level after a 1 occurs. The period of the clock waveform determines the bit time interval.

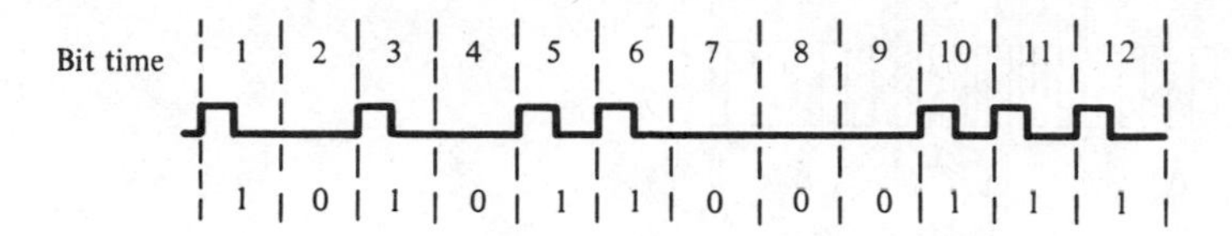

FIGURE 9–44 *An RZ waveform representing 10101100011.*

Figure 9–45 illustrates a *non-return-to-zero* (NRZ) waveform. In this case a 1 or 0 level remains during the entire bit time. If two or more 1s occur in succession, the waveform does not return to the 0 level until a 0 occurs.

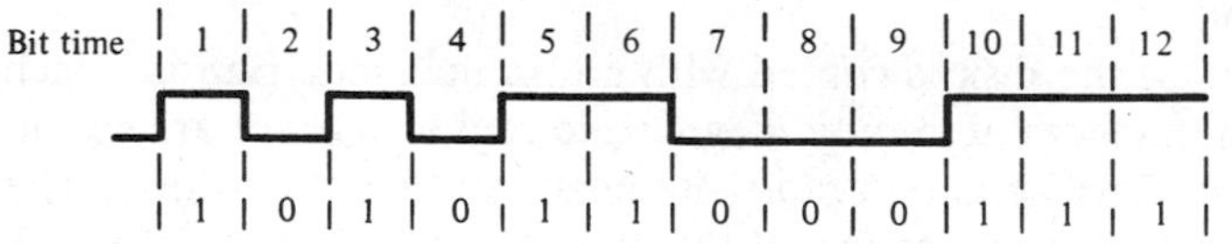

FIGURE 9–45 *An NRZ waveform representing 101011000111.*

Figure 9–46 is an illustration of a *biphase* waveform. In this type, a 1 is a HIGH level for the first half of a bit time and a LOW level for the second half, so a *high-to-low transition occurring in the middle of a bit time is interpreted as a 1.* A 0 is represented by a LOW level during the first half of a bit time followed by a HIGH level during the second half, so a *low-to-high transition in the middle of a bit time is interpreted as a 0.*

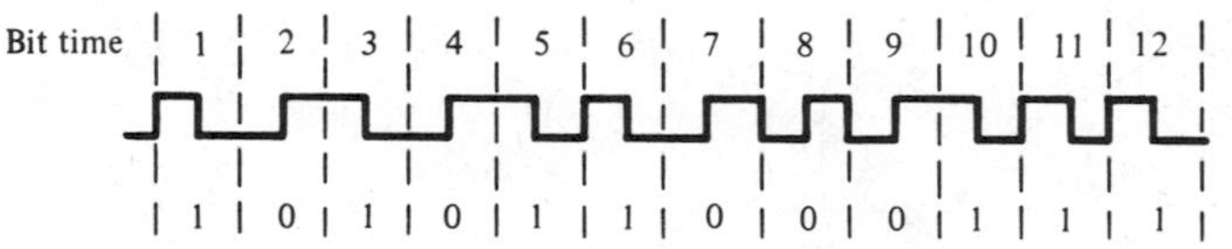

FIGURE 9–46 *A biphase waveform representing 101011000111.*

Manchester is another type of phase encoding where a high-to-low transition at the start of a bit time represents a 0 and no transition represents a 1. Figure 9–47 illustrates a Manchester waveform.

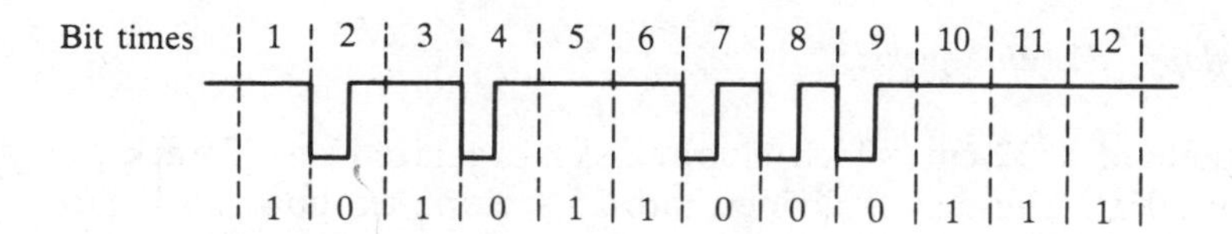

FIGURE 9–47 *A Manchester waveform for 101011000111.*

The *Kansas City* method uses two different frequencies to represent 1s and 0s. The standard 300 baud (*baud* means bits per second) version uses eight cycles of 2400 Hz to represent a 1 and four cycles of 1200 Hz to represent a 0. This is illustrated in Figure 9–48.

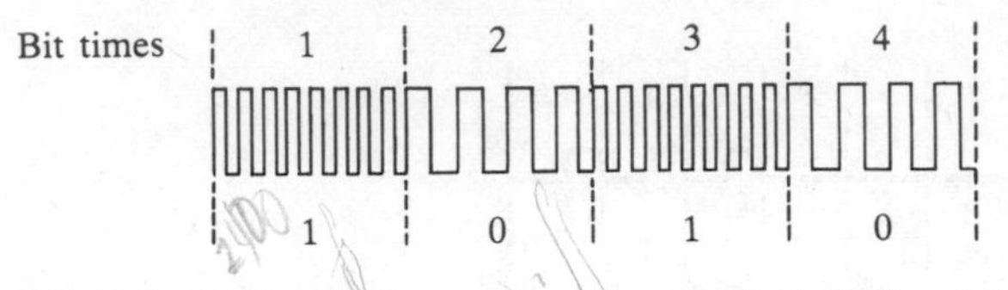

FIGURE 9–48 *A Kansas City Standard Code for 1010.*

The Floppy Disk

The floppy disk (diskette) is a small, flexible, mylar disk with a magnetic surface. It is *permanently* housed in a square jacket for protective purposes as shown in Figure 9–49.

The surface of the disk is coated with a thin magnetic film in which binary data are stored in the form of minute magnetized regions. There are cutout areas in the jacket for the drive spindle, read/write head, and index position. The index hole establishes a reference point for all the tracks on the disk. As the disk rotates at 360 rpm within the stationary jacket, the read/write head makes contact through the access window.

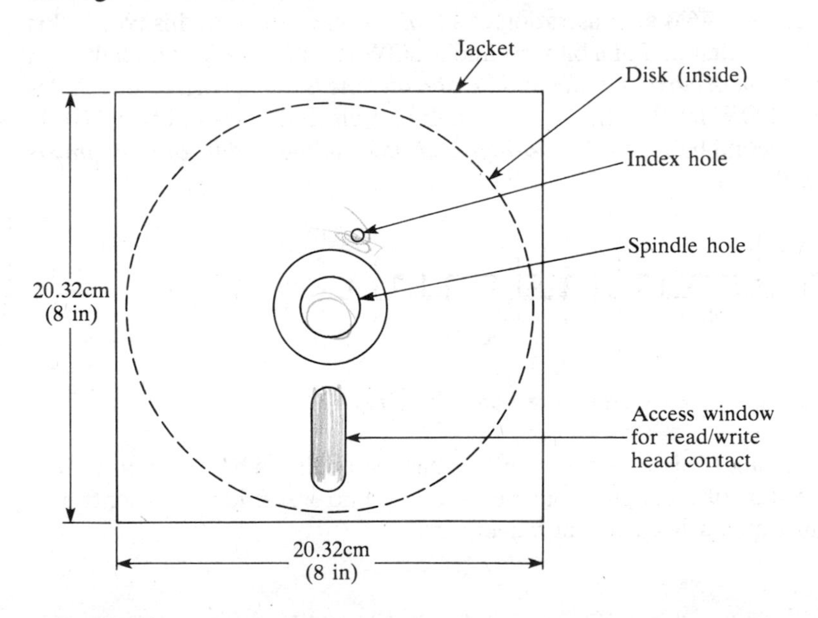

FIGURE 9–49 *A floppy disk in a jacket.*

A standard 20.32-cm (8-inch) floppy disk is organized into 77 tracks, as shown in Figure 9–50(a). The disk is divided into 26 sectors, as shown in Figure 9–50(b) so that each of the 77 tracks are also divided into 26 equal-sized sectors. The longer outside track has the same number of sectors and the same sector length as does the shorter inside track. There is just more unused space between sectors in the longer track. Notice that the index hole appears between the first and the last sectors.

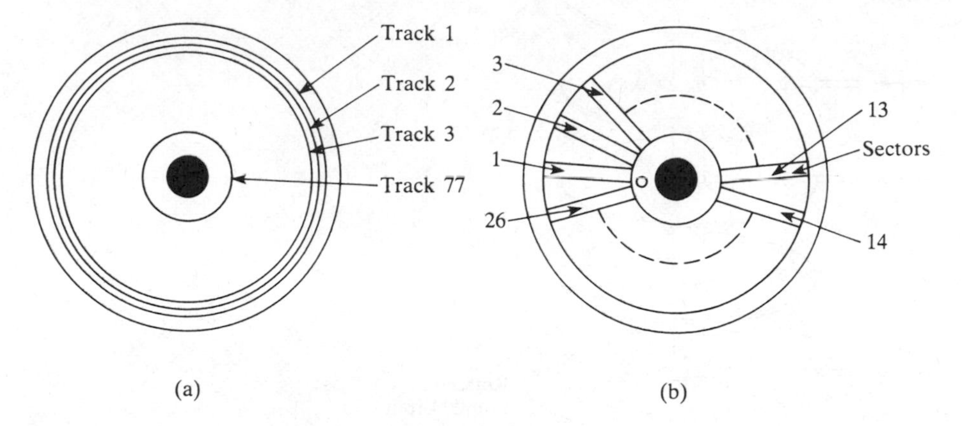

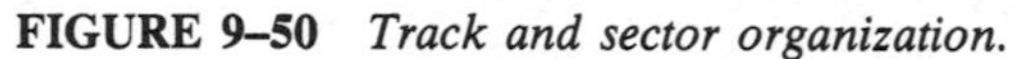
FIGURE 9–50 *Track and sector organization.*

Each sector can store 128 bytes of data. (Typically, 1 byte represents one character or numeral.) The total storage capacity of the disk is therefore

$$(128 \text{ bytes/sector})\ (26 \text{ sectors/track})(77 \text{ tracks}) = 256{,}000 \text{ bytes}$$

A *double density* diskette can store 256 bytes in a sector.

A typical sector format is shown in Figure 9–51 where each sector is divided into fields. The address mark passes the read/write head first and identifies the upcoming areas of the sector as the ID field. The ID field identifies the data field by sector and track number. The data mark indicates whether the upcoming data field contains a good record or a deleted record. The data field is the portion of the sector that contains the 128 data bytes. The average access time to a given sector is about 500 ms. This is much faster than for magnetic tape but much slower than the semiconductor memories.

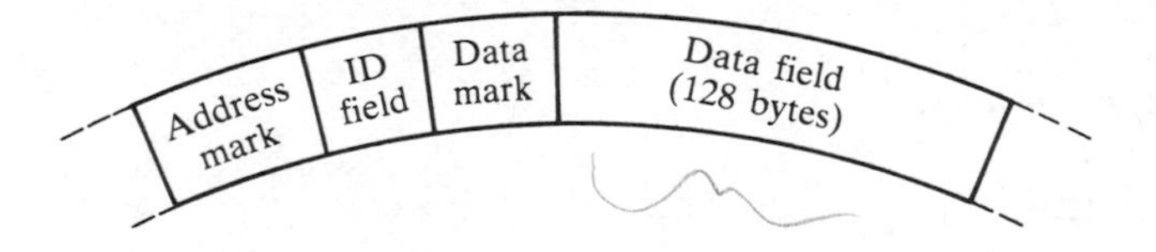

FIGURE 9–51 *A typical sector of format for one track of a floppy disk.*

Rigid Disks

This type of magnetic disk is larger (diameter up to about 40 inches) and much more rigid than the floppy disk. Stacks of these disks are used as mass storage devices in large computer systems. Data are recorded on concentric tracks by read/write heads that move in and out in the spaces between the disks as illustrated in the simplified diagram in Figure 9–52.

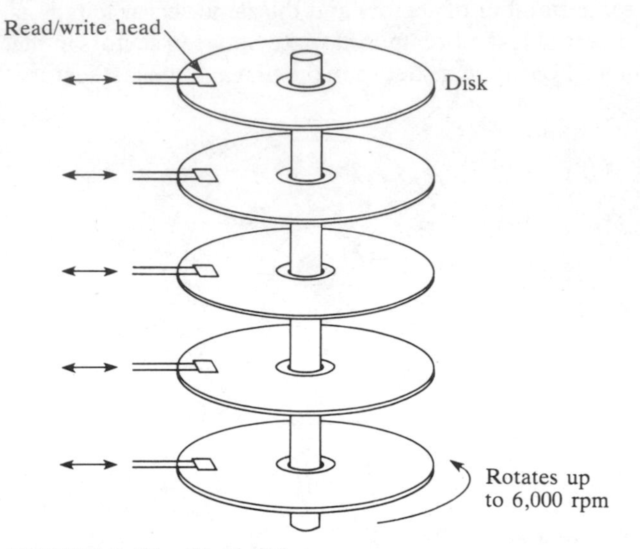

FIGURE 9–52 *Rigid disk memory.*

Drum Memory

The drum memory is a cylinder with a magnetic material on its surface. Data are stored in tracks around the drum as illustrated in Figure 9–53 with a read/write head for each track. As the drum rotates, data can be written or read from each track simultaneously.

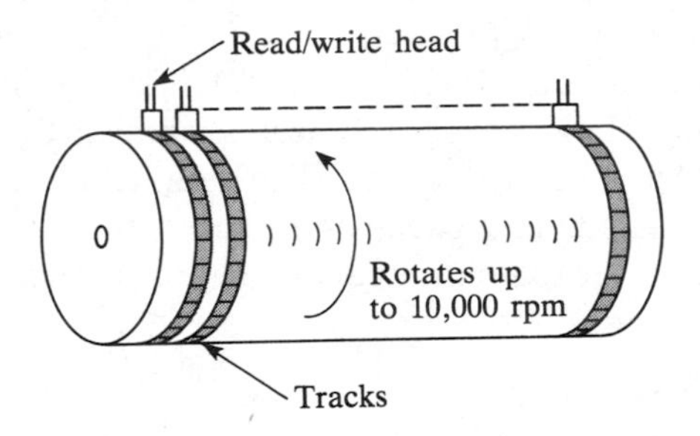

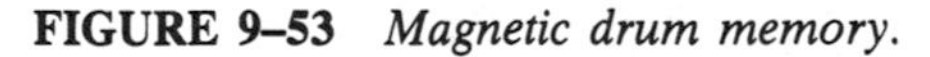
FIGURE 9–53 *Magnetic drum memory.*

Magnetic Tape

We have all seen the tape cassettes used in small microcomputer systems to store data. Larger reel type systems are used in many large computer systems.

A typical tape format with seven tracks is shown in Figure 9–54(a). There is a separate read/write head for each track as indicated. The tape is divided in records with gaps between records for the starting and stopping of the tape. A record may be organized into a typical format such as shown in Figure 9–54(b).

In reference to the format, the marker indicates the beginning of a record, the ID specifically identifies the record, the stored data then appears followed by a checksum.

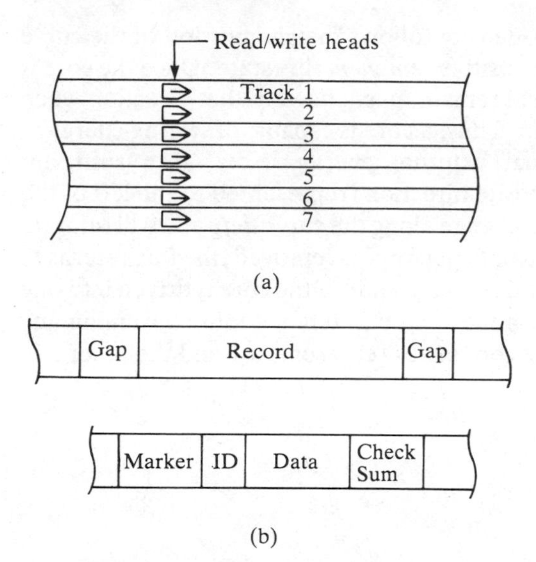

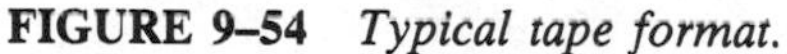

FIGURE 9–54 *Typical tape format.*

The *checksum* is the arithmetic sum of all the bytes in a record and is used by the computer to check for errors in the data as it is being taken from the tape. This is basically done by adding the data bytes as they are read from the tape and comparing this figure with the checksum.

MAGNETIC CORE MEMORY 9–10

The magnetic core memory is one of the older memory technologies but still has several advantages over some of the other memories. It is much faster than the other magnetic memories and it is nonvolatile. This, of course, means that data can be written into and read from the memory in a very short time, and data are retained even when power is lost.

The Magnetic Core

The typical magnetic core is torodial (doughnut-shaped) and is composed of a ferromagnetic composition material. Memory cores are typically very small, with diameters ranging down to a few tenths of mils (thousandths of an inch). Figure 9–55 is a pictorial representation of a magnetic core.

The magnetic properties of this type of core exhibit a two-state characteristic; that is, a core can be magnetized in two directions, making it ideal for the storage of binary information. This magnetic property can be described graphically by what is called a *hysteresis curve,* which is a plot of the *magnetic flux density B* versus the *magnetic force H,* as shown in Figure 9–56.

If sufficient magnetizing force is applied in a given direction, the magnetic flux density increases along the curve *abc*. The core is in a *saturated* condition from point *b* to point *c,* meaning that any further increase in the magnetizing force will produce very little change in the flux density. When the magnetizing force is removed, the flux density follows the *cbd* portion of the curve and stays at point *d,* which is called the positive *remanent* flux state. Once the core is saturated in the positive direction, it will remain magnetized in that direction when the magnetizing force is reduced to zero. Thus, a core is capable of storing energy in the magnetic field without additional external energy. Now, if a sufficient magnetizing force is applied in the opposite direction (represented to the left of the origin of the graph), the flux density will change along the curve *defg* and will saturate in the opposite direction. When this magnetizing force is removed, the flux returns to point *h,* which is the negative remanent flux state. Thus, if the core is driven into one of its two magnetic states, it can remain there indefinitely without consuming any energy. In a digital memory, one of the core states represents a 1 and the other a 0.

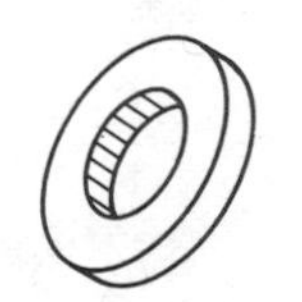

FIGURE 9–55 *Toroidal magnetic core.*

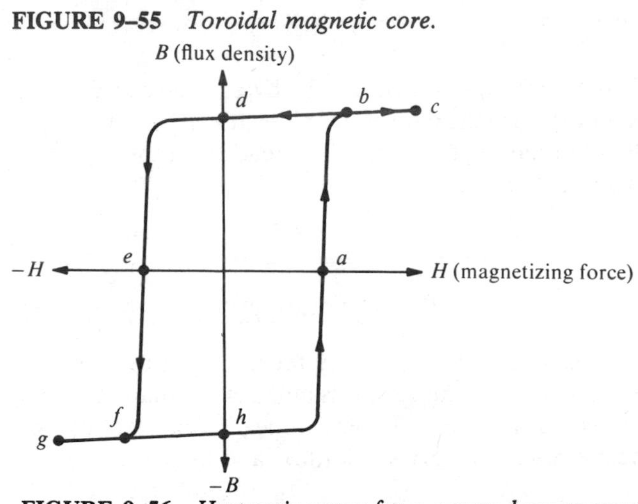

FIGURE 9–56 *Hysteresis curve for a square loop magnetic core.*

How can a magnetic core be driven into either of its two states? Simply run a wire through the core and pass a sufficient amount of current through the wire to create a magnetizing force by virtue of the magnetic field around the conductor. Figure 9–57 shows the direction of the magnetic field within the core for each direction of current through the wire. This direction can be remembered by using the "right-hand rule," which says that if the thumb of the right hand points in the direction of the conventional current, the fingers point in the direction of the magnetic field.

The value of current required to switch a core into either of its two states is known as the *critical value* or the *full-select value* of magnetizing current, which we will designate I_m. As you can see, if the full-select current is made to flow momentarily in one direction, the core will switch to the corresponding magnetic state. If a full-select value of current momentarily flows in the opposite direction, the core will go to its other magnetic state. By this basic method a 1 or a 0 can be stored, depending on the direction in which the current flows through the wire; entering or storing a 1 or a 0 in a memory core is termed *writing* into the memory.

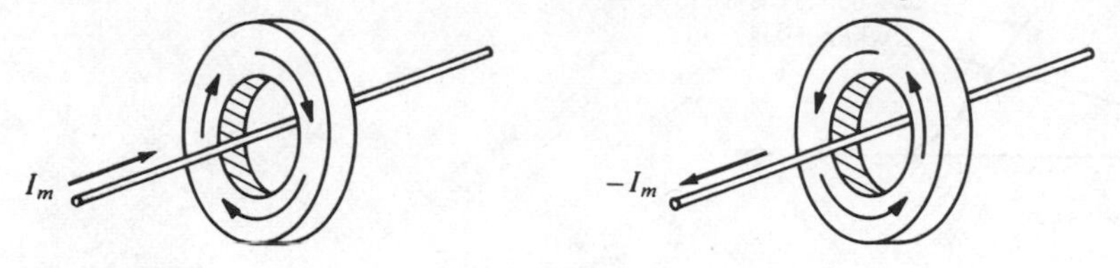

FIGURE 9–57 *Direction of the magnetizing current determines the direction of the magnetic field within the core.*

Once a bit of information is written into a core, how do we detect what is stored? Notice that on the hysteresis curve, a large transition of magnetic flux density occurs when the core is driven from one of its remanent flux states to the other. This change in the magnetic field direction will induce a voltage in a wire passing through the core. Therefore, when the core is driven from one state to the other, the transition can be detected by measuring this induced voltage. If the positive remanent state is selected as the 1 state, the negative remanent state is then the 0 state. If a current is made to flow in a direction through the wire such that the core will always be driven to the 0 state, an induced voltage will occur on another wire if the core is in the 1 state and makes the transition to the 0 state. If the core is already in the 0 state, a much smaller induced voltage will occur due to a small transition from the 0 remanent state to a saturated state and back. By this basic method we can detect whether a 1 or a 0 is stored in the core; this process is called *reading* information from the core. Basically, two wires through the core are required—one to carry the *read* or *write* currents and the other to detect or sense the induced voltage. The latter is called the *sense line*. This is illustrated in Figure 9–58.

Figure 9–59(a) shows a typical curve for the voltage on the sense line for both a 1 and a 0 being read from the curve. The voltage induced by a stored 1 is

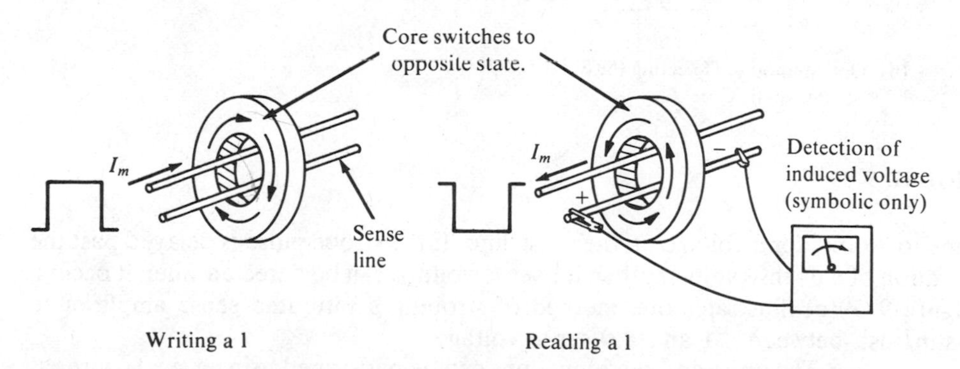

FIGURE 9–58 *Basic illustration of read and write operations with a magnetic core.*

considerably longer in duration and normally greater in amplitude than that for a stored 0 and can be readily detected by a strobed amplifier connected to the sense line. "Strobed" means that the sense amplifier circuit is enabled only at a specified

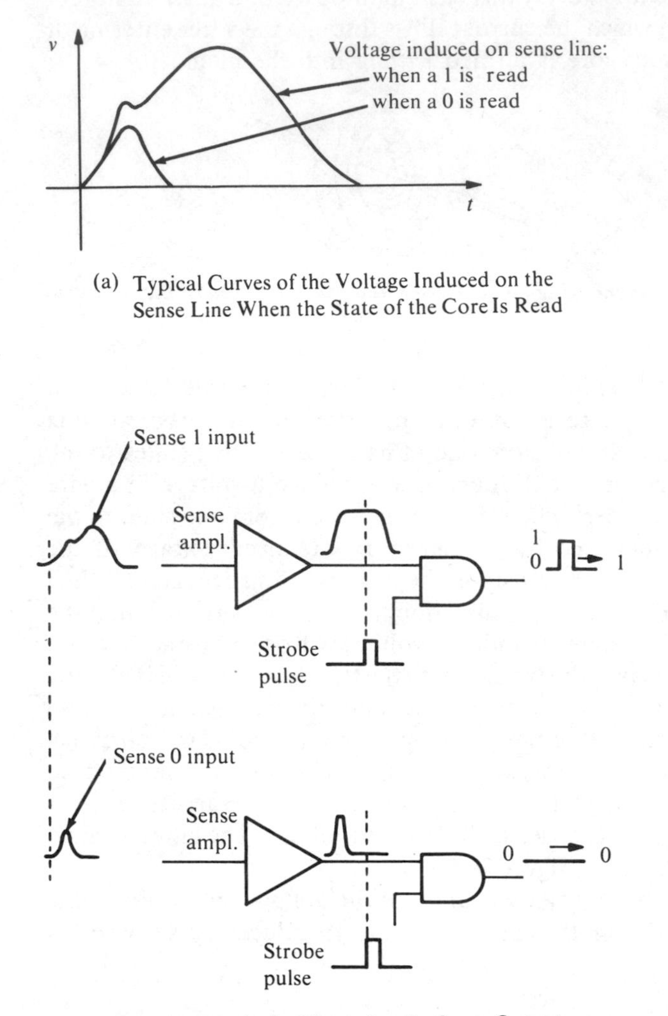

(a) Typical Curves of the Voltage Induced on the Sense Line When the State of the Core Is Read

(b) One Method of Detecting the Sense Output of a Magnetic Core

FIGURE 9–59

time to "look" for a voltage on the sense line. If the strobe pulse is delayed past the duration of a 0 sense voltage, then a 1 sense voltage can be detected when it occurs. Figure 9–59(b) illustrates one method of strobing a saturated sense amplifier to distinguish between a 1 and a 0 sense voltage.

The write and read functions can be performed using a single wire, as previously discussed, by properly timing each current pulse so that the time at which a write pulse occurs is distinct from the time of a read pulse. This operation is illustrated in Figure 9–60. As shown, a bit is stored in the core by a pulse of write current. Later a pulse of read current is applied in the opposite direction, and at an appropriate time the sense line is strobed.

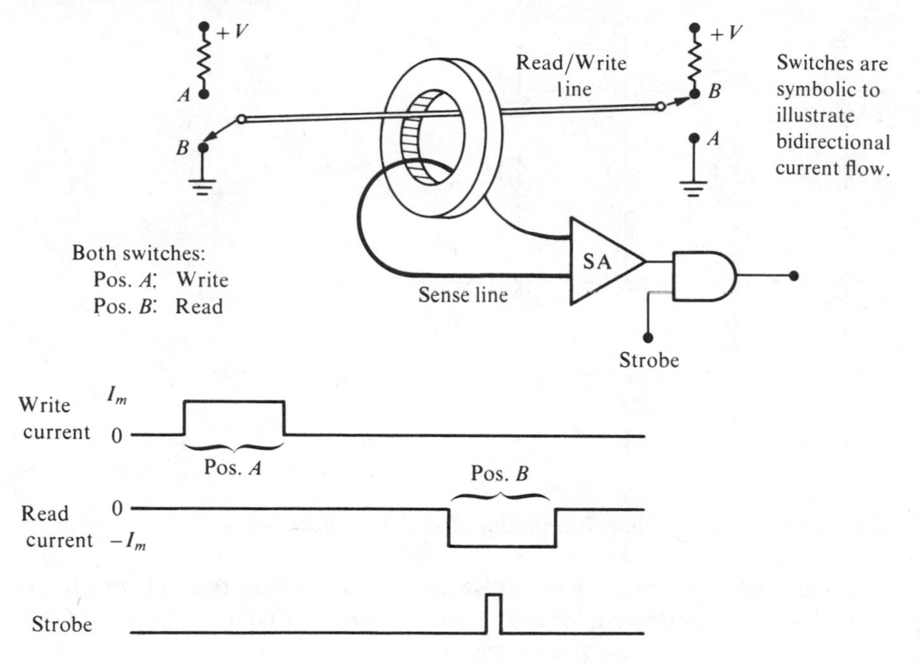

FIGURE 9–60 *Simplifed illustration of read/write implementation and timing.*

Magnetic Core Memory Arrays

In most applications it is necessary to store large numbers of bits (1s and 0s), and therefore a magnetic core memory is made up of many individual cores. We will discuss the basic configuration of an array of cores in this section, using small arrays for clarity and purposes of illustration. The same principles apply to larger arrays.

Each core in the simplified array of Figure 9–61 (sometimes called a *core plane)* occupies a unique location that is identified by X and Y (row and column) coordinates; the location is called the *address* of the core within the array. For instance, the coordinates X_1 Y_1 identify the core in the lower left corner of the array; the coordinates X_4 Y_4 identify the core in the upper right corner; and so on.

As previously discussed, a full-select current, I_m, is required to put the core into one of its magnetic states. If one-half of the current (the *half-select* current or $I_m/2$) is made to flow through one of the X lines, and the same value is made to flow through one of the Y lines in a corresponding direction, then the currents will intersect at a particular core dependent on the X and Y lines selected. When a half-select current passes through each of the select lines in a particular core, the two currents are additive and the core is switched to either the 1 or the 0 state, depending on the direction of the two currents. In this manner, information can be written into or read out of the core array at specified locations or addresses. The basic operation is illustrated in Figure 9–62 where the X_3Y_2 core is selected and the half-select currents through the X_3 line and the Y_2 line cause a 1 to be written into the core. Notice that a half-select current passes through each core "threaded" on the X_3 line and each core on the Y_2 line. Since these cores have only one-half of the critical value of current

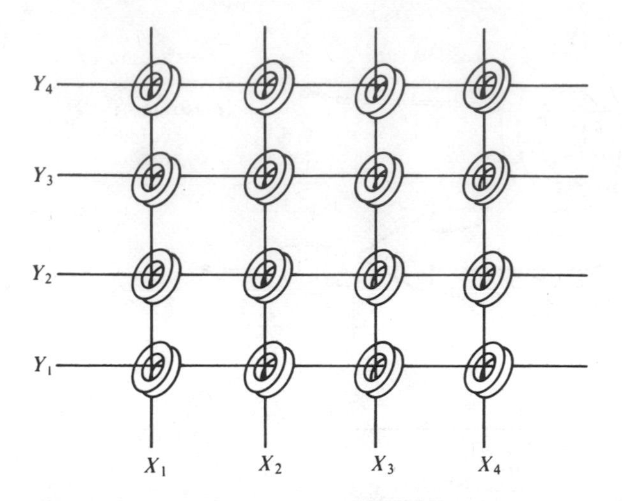

FIGURE 9–61 *An array of cores showing the X and Y coordinates.*

passing through each of them, they do not switch. Only the core through which the currents are *coincident* is affected; for this reason, a memory that uses this type of cell selection is called a *coincident-current* memory.

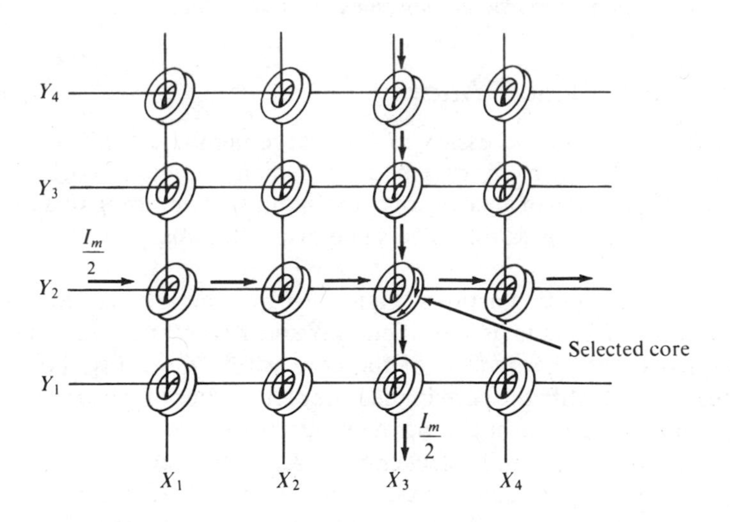

FIGURE 9–62 *A core array showing the selection of a specific core by coincident current on the X and Y lines.*

In addition to the X and Y select lines, a sense line is required to detect an induced voltage when the memory core array is being "read." A single wire is used for the sense line and threaded through *each* core as shown in Figure 9–63. Because only one core is selected at a time, only one sense line is required. Any induced voltage appearing on the sense line is a result of the particular core selected.

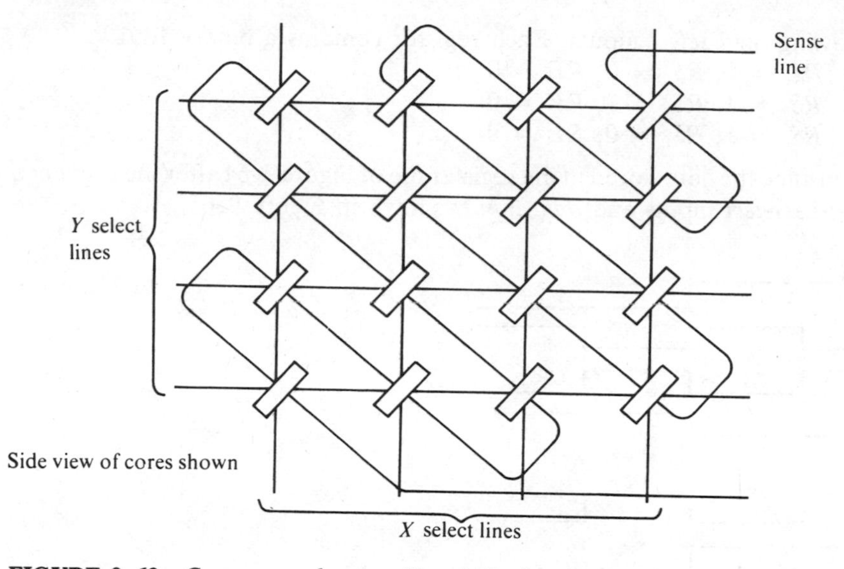

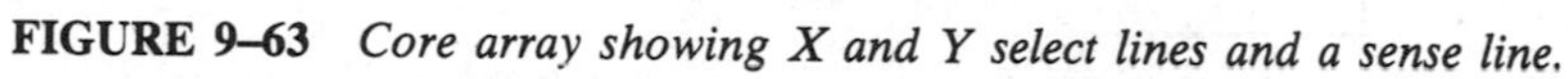
FIGURE 9–63 *Core array showing X and Y select lines and a sense line.*

Problems

Section 9–1

9–1 For the 4 × 4 register file in Figure 9–64, determine the contents of the registers for each of the following input conditions. Assume all registers are initially clear.

(a) $D_0 = 0$, $D_1 = 1$, $D_2 = 1$, $D_3 = 0$, $WS_0 = 0$, $WS_1 = 1$, $WE = 0$
(b) $D_0 = 1$, $D_1 = 0$, $D_2 = 1$, $D_3 = 1$, $WS_0 = 1$, $WS_1 = 0$, $WE = 0$
(c) $D_0 = 0$, $D_1 = 0$, $D_2 = 1$, $D_3 = 1$, $WS_0 = 1$, $WS_1 = 1$, $WE = 0$

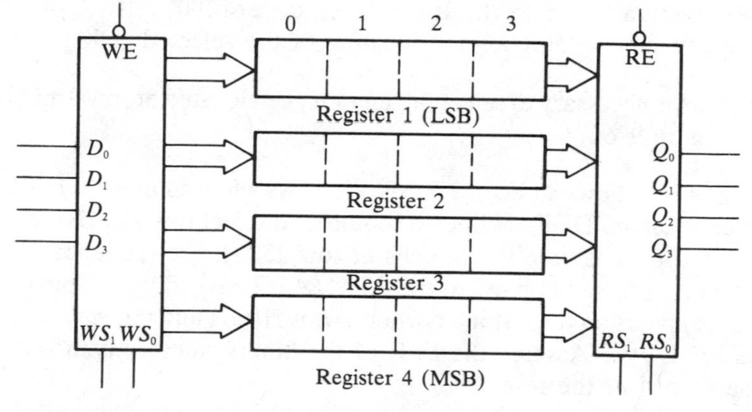

FIGURE 9–64

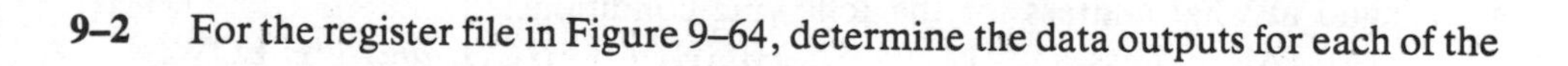
9–2 For the register file in Figure 9–64, determine the data outputs for each of the

following *read select* inputs. Each register contains a binary 1001.
(a) $RS_0 = 0$, $RS_1 = 0$, $RE = 0$
(b) $RS_0 = 1$, $RS_1 = 1$, $RE = 0$
(c) $RS_0 = 1$, $RS_1 = 0$, $RE = 0$

9–3 Determine the data stored in the register file of Figure 9–64 after the sequence of *write select* inputs and *data inputs* shown in Figure 9–65.

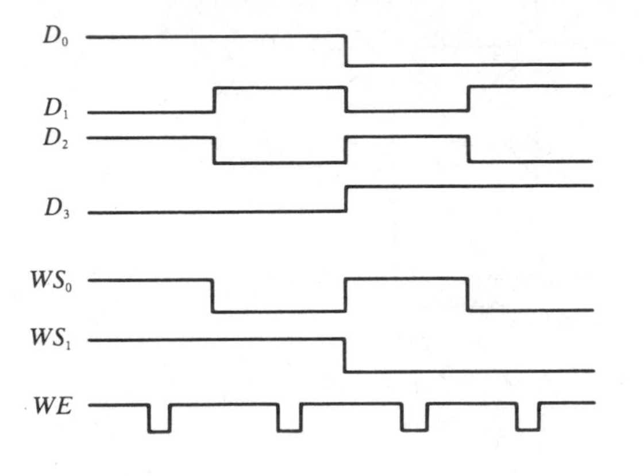

FIGURE 9–65

9–4 In Figure 9–3, the following inputs are applied to the 74170 pins: pin 1 = 1, pin 2 = 0, pin 3 = 1, pin 4 = 0, pin 5 = 0, pin 11 = 1, pin 12 = 0, pin 13 = 1, pin 14 = 0, and pin 15 = 1. What are the states of the Q outputs if pin 12 goes HIGH and pin 11 goes LOW?

Section 9–2

9–5 Figure 9–66 represents a 256-bit RAM. If an address code of 10010010 (A_7-A_0) is applied, what is the row and column location of the selected cell?

9–6 What address code is necessary to select the memory cell located at row 9 and column 12 in Figure 9–66?

9–7 The 74200 TTL RAM shown in Figure 9–6 has 32 rows and 8 columns. First, what address code on the DEFGH lines is required to select row 25_{10}? Next, assume that 10110101_2 is stored in the cells of row 25_{10}. If the sequence of addresses in Figure 9–67 are applied to the ABC (A is LSB) address inputs, what bit pattern appears on the output Y when WE is HIGH and the *memory enable* inputs are LOW? Assume the LSB of the binary number is in the lowest numbered cell of the row.

9–8 For the 2147 static RAM in Figure 9–10, what operation is being performed and at what address for the following conditions:
(a) All address lines LOW, D_{in} HIGH, $\overline{CS}$ *HIGH, and* $\overline{WE}$ LOW.
(b) All address lines HIGH, D_{in} LOW, $\overline{CS}$ LOW, and $\overline{WE}$ LOW.
(c) $A_0 - A_5$ LOW, $A_6 - A_{11}$ HIGH, D_{in} HIGH, $\overline{CS}$ LOW, and $\overline{WE}$ HIGH.

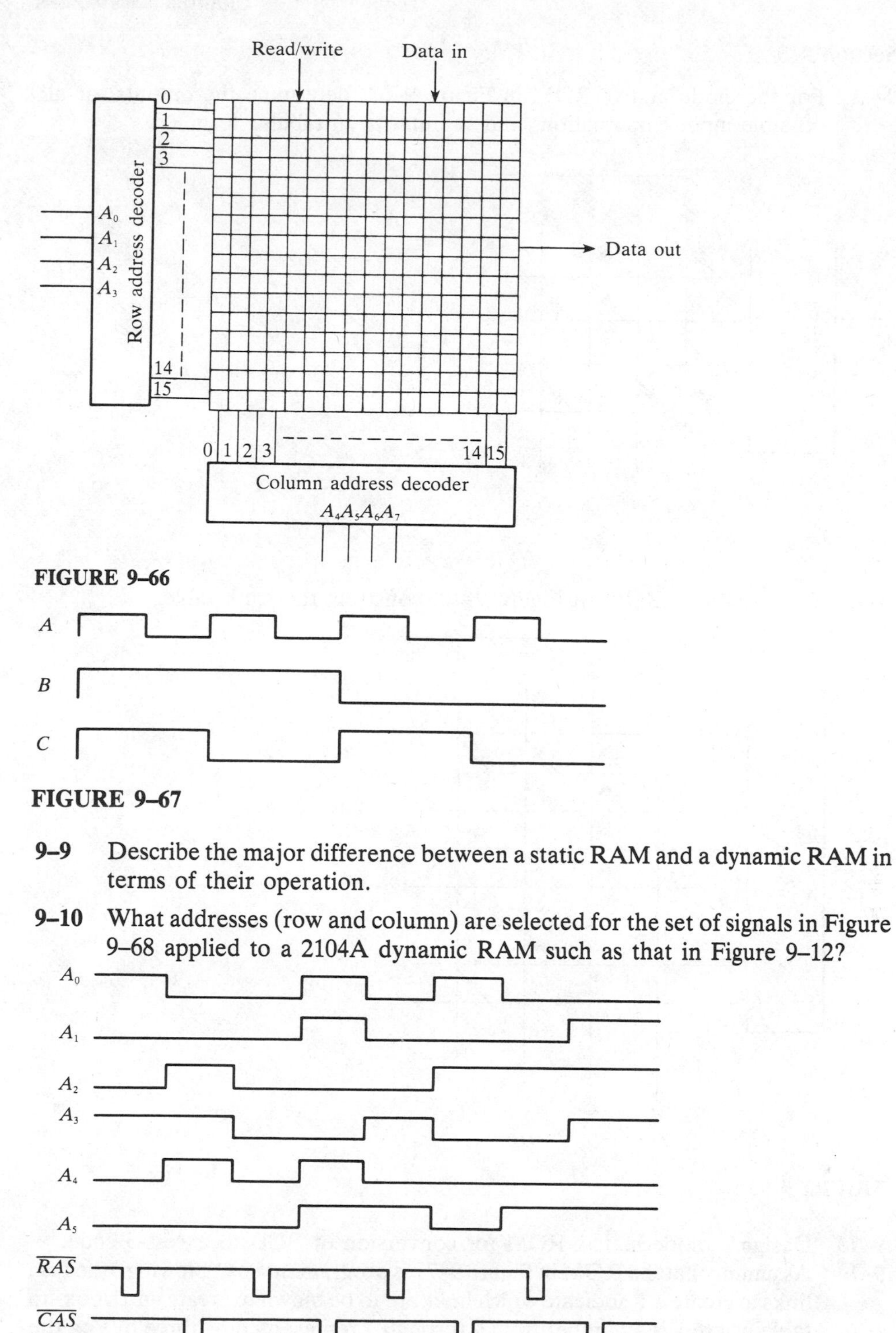

FIGURE 9–66

FIGURE 9–67

9–9 Describe the major difference between a static RAM and a dynamic RAM in terms of their operation.

9–10 What addresses (row and column) are selected for the set of signals in Figure 9–68 applied to a 2104A dynamic RAM such as that in Figure 9–12?

FIGURE 9–68

Section 9–3

9–11 For the diode matrix ROM in Figure 9–69, determine the outputs for all possible input combinations and summarize in tabular form.

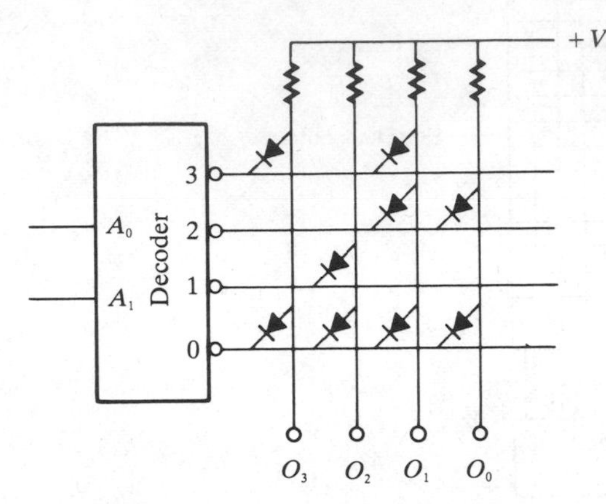

FIGURE 9–69

9–12 For the diode ROM in Figure 9–70, construct the truth table.

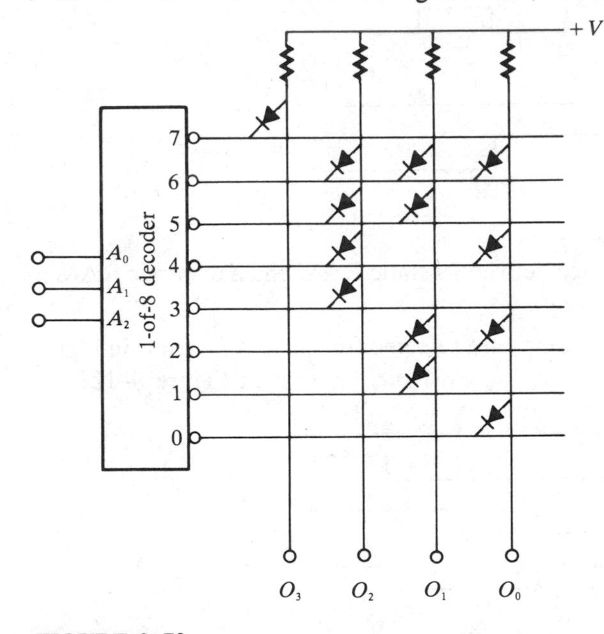

FIGURE 9–70

9–13 Design a diode matrix ROM for conversion of BCD to excess-3 code.

9–14 Assuming that the ROM in Figure 9–71 is programmed by "blowing" the fuse links to create a 1, indicate which links are to be blown to create an x^3 look-up table where x is a number from 0 through 7 represented by three bits on the inputs.

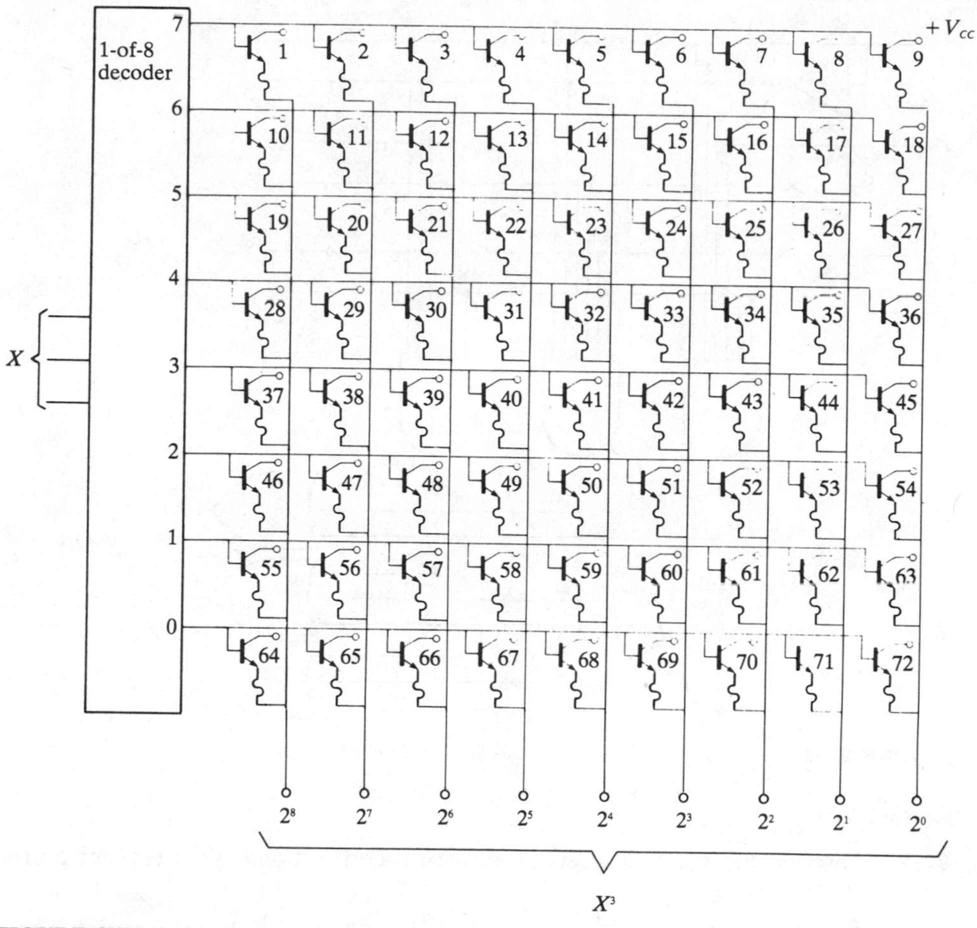

FIGURE 9–71

9–15 Implement the following Boolean expressions with a diode matrix ROM. Assume A is MSB.

(a) $X = AB + A(\overline{B} + C)$

(b) $Y = B(BC + \underline{CD})$

(c) $Z = ABC + \overline{AB}\overline{C}D + \overline{B}C + BC\overline{D}$

9–16 In the simplified PLA in Figure 9–72 determine the points to be opened by mask programming in order to produce the following logic functions

(a) $A\overline{B}C + \overline{A}\overline{B}\overline{C} + ABC$ (c) $\overline{A} + \overline{B} + \overline{C}$

(b) $\overline{A} + BC$

Section 9–5

9–17 Identify each of the following:

(a) A PLA having a programmable AND array but no OR array.

(b) A PLA having a programmable AND array and a fixed OR array.

(c) A PLA that is completely programmable by the user in the field.

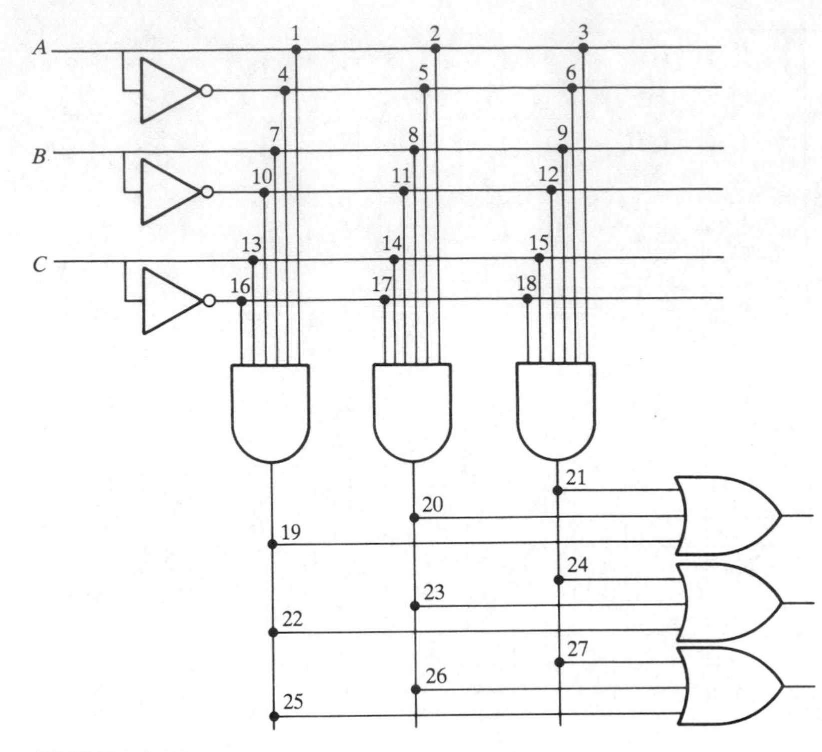

FIGURE 9–72

Section 9–6

9–18 Describe the basic difference between a conventional shift register and a FIFO register.

9–19 Complete the timing diagram in Figure 9–73 by showing the output wave forms for a FIFO serial memory like that shown in Figure 9–28.

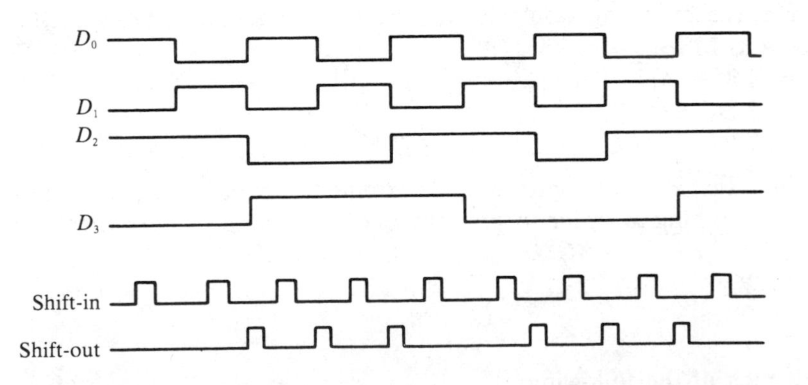

FIGURE 9–73

9–20 Sketch the output waveform for a FIFO eight-bit register having the data input and the input and output clocks shown in Figure 9–74.

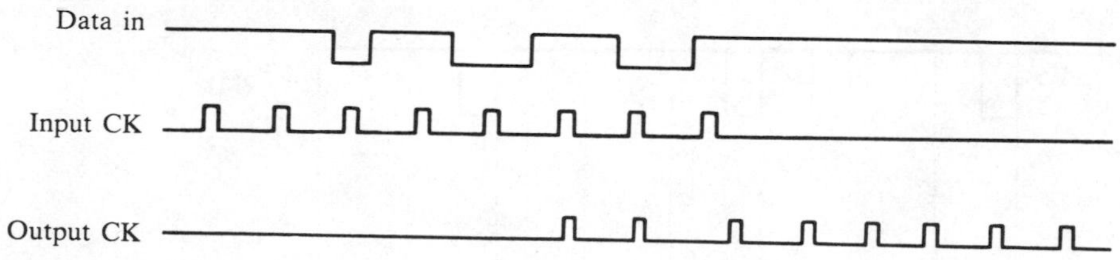

Data in

Input CK

Output CK

FIGURE 9–74

Section 9–7

9–21 Five data bytes are pushed into a RAM stack beginning at address 11BF. What is the top-of-stack address after the data are loaded?

9–22 Data are pushed into a RAM stack beginning at address 05F8. If the top-of-stack address is 05E1 after the data are loaded, how many bytes are stored?

9–23 The top-of-stack address of a given RAM stack is 00CD. Sixteen bytes of data are pulled from the stack. What is the new top-of-stack address?

9–24 Addresses 4ABB to 4AFF are assigned as a stack in a given RAM. How many bytes of storage does this represent?

Section 9–8

9–25 What type of magnetic material is normally used in bubble memories?

9–26 Name two types of storage "loops" in a typical bubble memory.

Section 9–9

9–27 Determine the sequence of bits represented by each of the waveforms in Figure 9–75.

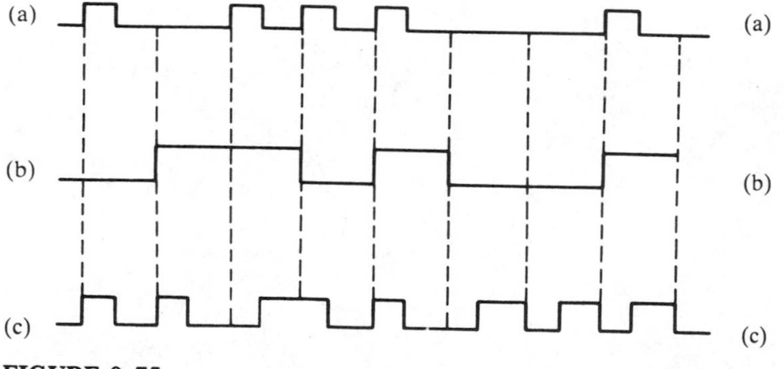

FIGURE 9–75

9–28 Repeat Problem 9–27 for the waveforms in Figure 9–76.

9–29 How many data bytes can be stored in a double density diskette?

9–30 List four types of magnetic media storage.

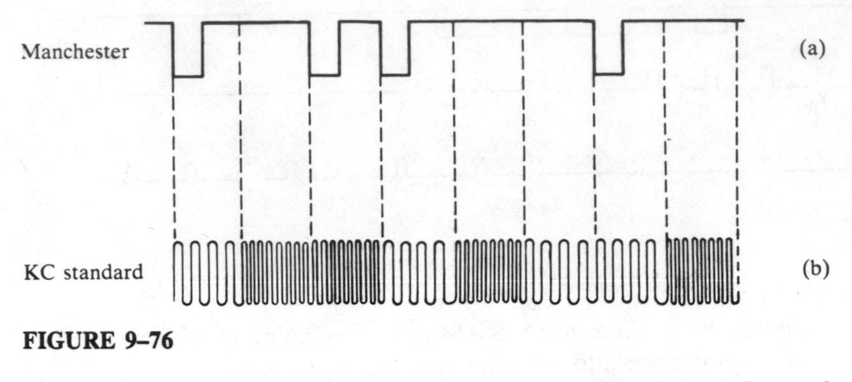

FIGURE 9–76

9–31 What is the purpose of *checksum* in a magnetic tape format?

Interfacing is the proper design and interconnection of two or more electronic devices so that they are operationally compatible. Interfacing involves such considerations as input and output voltages and currents, loading, signal characteristics, and data formats.

10

INTERFACING AND DATA TRANSFER

10–1 BUS STRUCTURES

Physically, a *bus* is a set of conductive paths that serves to interconnect two or more functional components of a system or several diverse systems together. Electrically, a bus is a collection of voltages levels and signals that allow the various devices connected to the bus to work properly together.

For example, a microprocessor is connected to memories and input/output devices by certain bus structures. This is illustrated by the block diagram in Figure 10–1; an *address bus* allows the microprocessor to address the memories, the *data bus* provides for transfer of data between the microprocessor, the memory, and the input/output devices, and the *control bus* allows the microprocessor to control data flow and timing for the various components.

The physical bus is symbolically represented by wide lines with arrow heads indicating direction of data movement. Buses can also be used to interconnect various test instruments or other types of electronic systems.

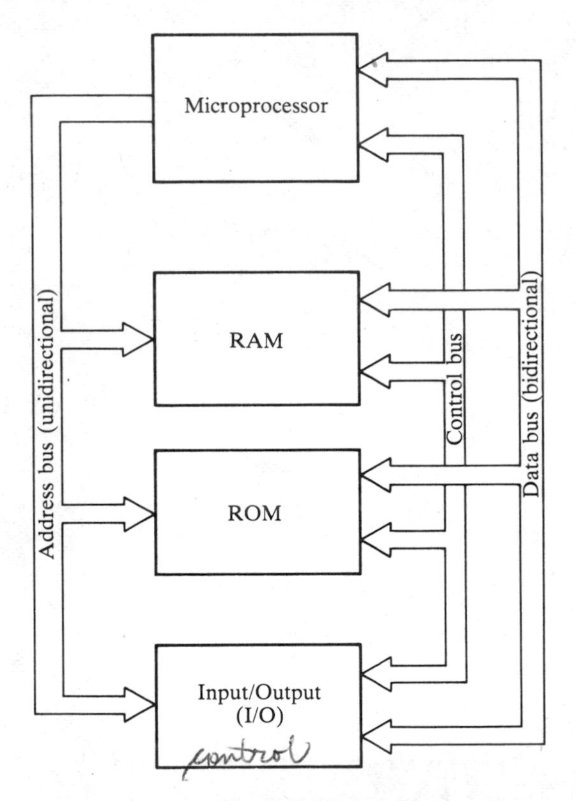

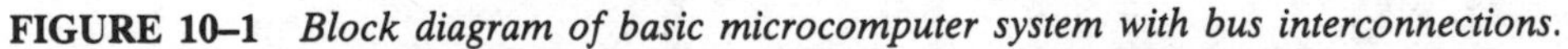
FIGURE 10–1 *Block diagram of basic microcomputer system with bus interconnections.*

Three-State Interface to the Bus

In a typical application, several devices are connected to one bus. For example, a microprocessor, a RAM, and a ROM. For this reason, *three-state* logic circuits are used to interface digital devices to a bus. Figure 10–2(a) shows the logic symbol for a noninverting three-state buffer with an active HIGH *enable.* Part (b) of the figure shows one with an active LOW *enable.*

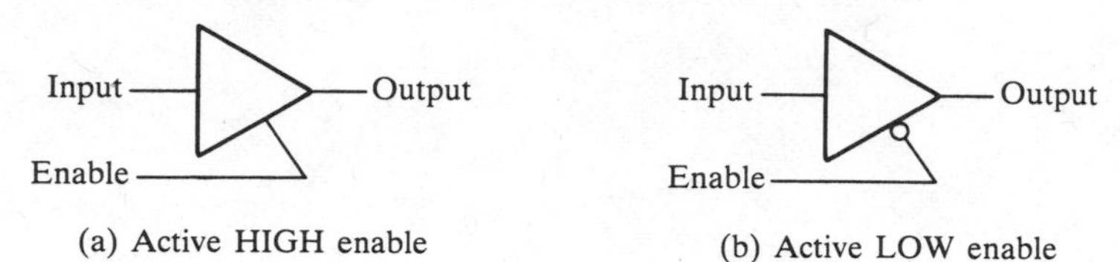

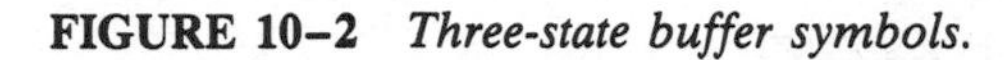

FIGURE 10–2 *Three-state buffer symbols.*

The basic operation of a three-state buffer can be understood in terms of switching action as illustrated in Figure 10–3. When the enable input is active, the gate operates as a normal noninverting circuit. That is, the output is HIGH when the input is HIGH and LOW when the input is LOW as shown in parts (a) and (b). The HIGH and LOW levels represent two of the states. The buffer operates in its *third* state *when the enable input is not active.* In this state, the circuit acts as an open switch and the output is completely disconnected from the input as shown in part (c).

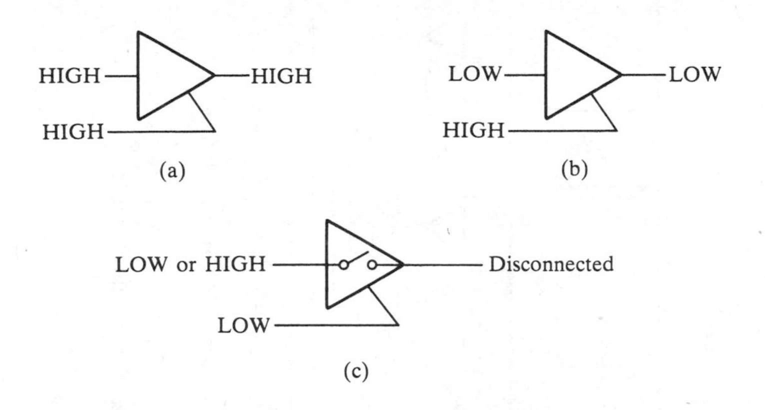

FIGURE 10–3 *Three-state buffer operation.*

Many microprocessors, memories, and other integrated circuit functions have three-state buffers that serve to interface with the buses. This is necessary when two or more devices are connected to a common bus. To prevent the devices from interfering with each other, the three-state buffers are used to disconnect all devices except the ones that are communicating at any given time. Figure 10–4(a) shows four devices connected to a four-bit *unidirectional* (one way) bus with three-state buffers. Data can flow only from device 1 or device 4 to devices 2 and 3. When device 2 is receiving, device 3 can be disconnected from the bus with the three-state buffers; when device 4 is sending, device 1 is disconnected from the bus.

Figure 10–4(b) illustrates a *bidirectional* bus connecting two devices. Data can be transferred back and forth among any of the devices. When not sending or receiving data, a given device can be disconnected from the bus by disabling the three-state buffers.

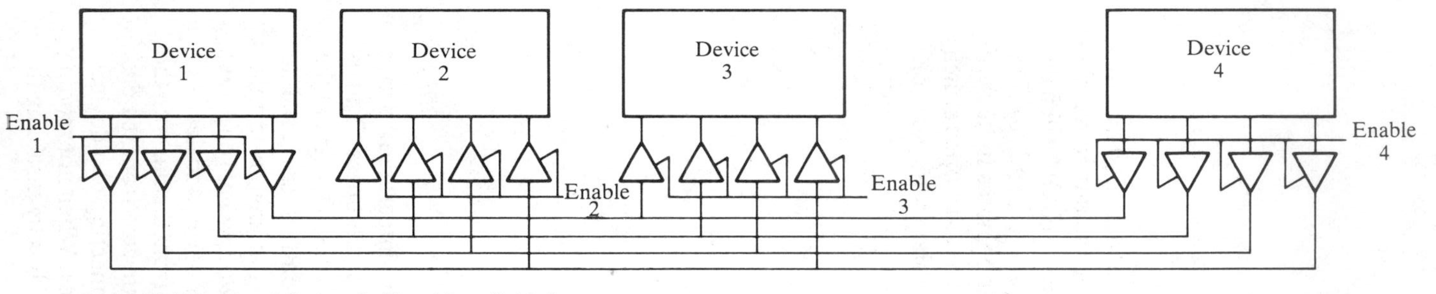

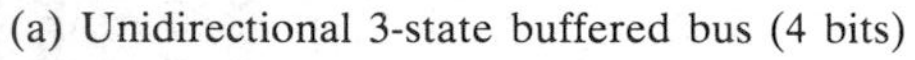
(a) Unidirectional 3-state buffered bus (4 bits)

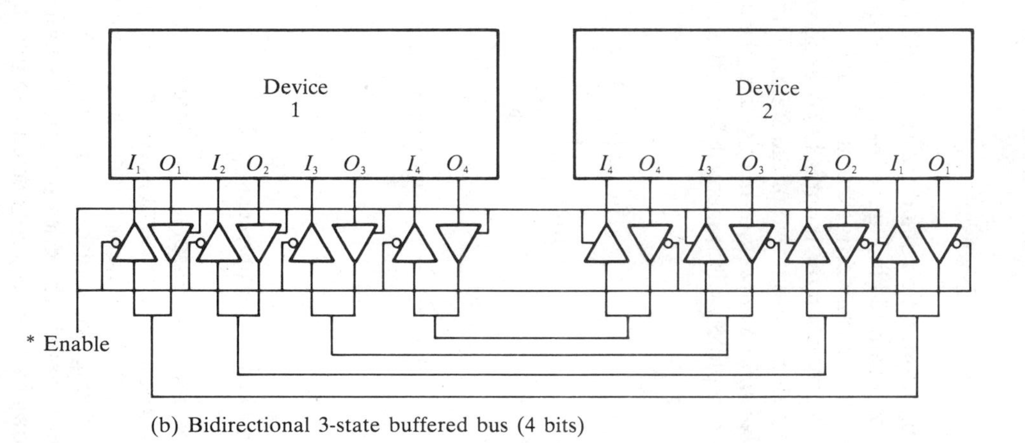

(b) Bidirectional 3-state buffered bus (4 bits)

* Enable LOW: 1 receives, 2 sends. HIGH: 1 sends, 2 receives.

FIGURE 10–4 *Illustration of unidirectional and bidirectional buses.*

Notice that each device in Figure 10–4(b) has a pair of three-state buffers on each input/output line. The reason for this is as follows: When a device is outputting data, the *output* three-state buffers are *enabled,* and the *input* three-state buffers are *disabled.* When a device is inputting data, the *input* three-state buffers are *enabled,* and the *output* three-state buffers are *disabled.* This is illustrated in Figure 10–5 with switch representations of the three-state buffers.

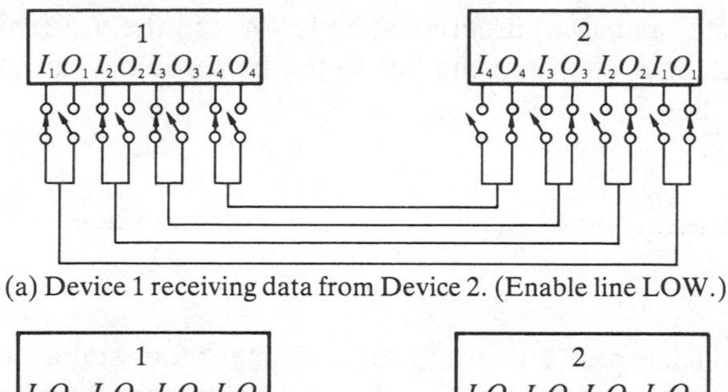

(a) Device 1 receiving data from Device 2. (Enable line LOW.)

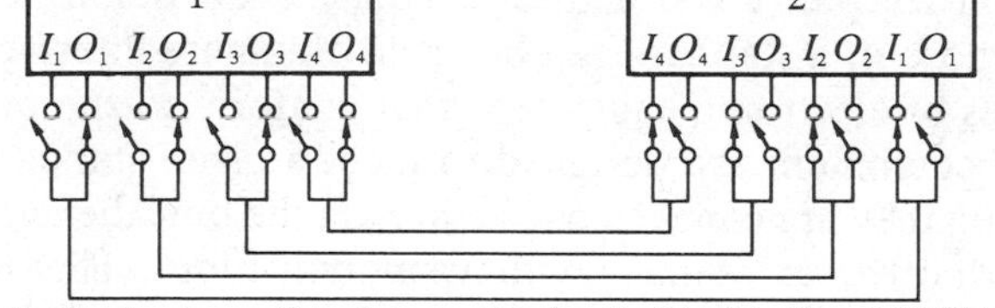

(b Device 1 sending data to Device 2. (Enable line HIGH.)

FIGURE 10–5 *Switch representations of three-stated bus operations.*

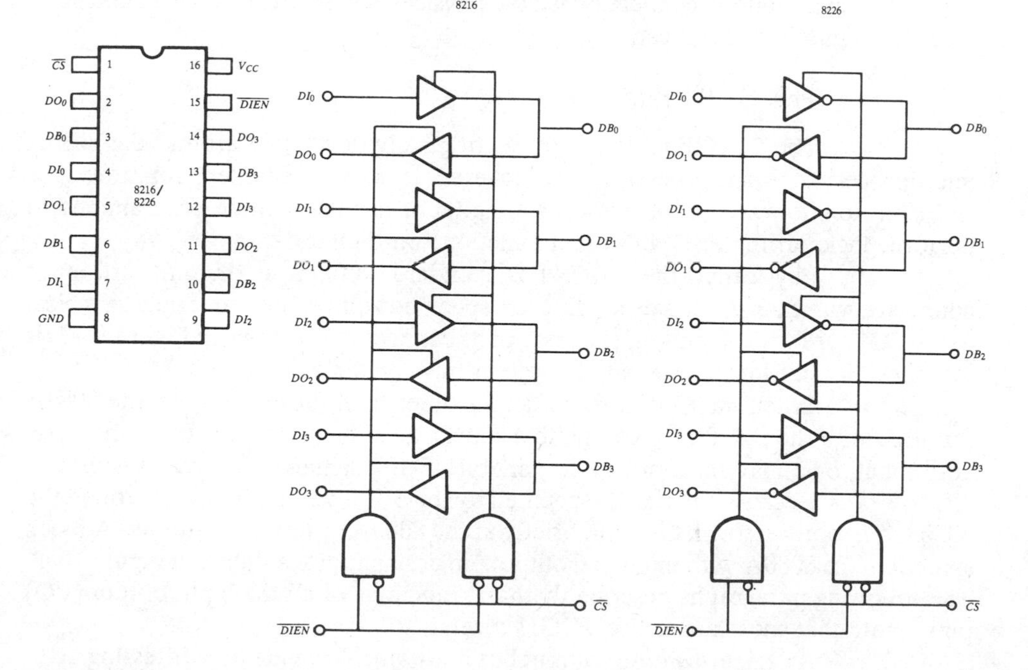

FIGURE 10–6 *The Intel 8216 and 8226 bidirectional three-state bus drivers. (Courtesy of Intel Corporation)*

A Bidirectional Bus Driver

The Intel 8216 is an example of an IC package containing three-state buffers for interfacing a four-bit input to a bus. An inverting version is available in the 8226. The logic diagrams for these devices are shown in Figure 10–6. The DO and DI pins are connected to the data outputs and inputs of the device to be buffered. The DD pins are connected to the bus lines. In order to switch data bits from the buffered device to the bus, the $\overline{DIEN}$ (data in enable) input must be LOW and the $\overline{CS}$ (chip select) must be LOW. To switch data bits from the bus to the buffered device, the $\overline{DIEN}$ line must be HIGH and the $\overline{CS}$ LOW.

10–2 STANDARD BUSES

There are several widely used and accepted bus structures that have been established as so-called *standards*. The purpose of a standardized bus is to eliminate interface problems between various types of electronic equipment that conform to the bus specifications. If two pieces of equipment are designed to meet a given standard interface requirement then, when they are connected together via the bus, they will operate in conjunction with each other as intended without modifications. That is, there should be a perfect fit when the systems conform to the bus standards.

In the general sense, a bus is a set of specifications for voltage levels, signal characteristics, and drive and loading requirements which, if met by a particular device, allows connection via the physical bus structure between units that meet the same bus specifications.

The S-100 Bus

The S-100 bus standard was originally developed around the 8080A microprocessor. Microprocessors are covered in a later chapter; for now it is sufficient to know a microprocessor is a digital device that can be programmed to perform logic, arithmetic, and other functions on digital data.

Physically, the S-100 bus standard defines a 100-pin card edge connector with each pin corresponding to a specified voltage level or signal. A typical S-100 PC board with 50 pins on each side of its connector (as shown in Figure 10–7) is intended to plug into a compatible system bus connector.

When microprocessor systems other than those based on the 8080A are used with the S-100 bus, compatible signals must be produced. In many cases, additional logic may be required to generate S-100 signals in a given system.

Because the signals specified by the S-100 bus are oriented around the 8080A microprocessor, let us look briefly at the 8080A inputs and outputs. A block symbol of the 8080A with input and output pin designations is shown in Figure 10–8. The following paragraphs describe the basic functions of all the input/output (I/O) pins. Note that most are active HIGH level.

A_{15} - A_0 are *address* output bus lines which provide for addressing up to 64K bytes of memory. A_0 is the LSB. These outputs are three-stated.

D_7 - D_0 are bidirectional (input and output) *data bus* lines which

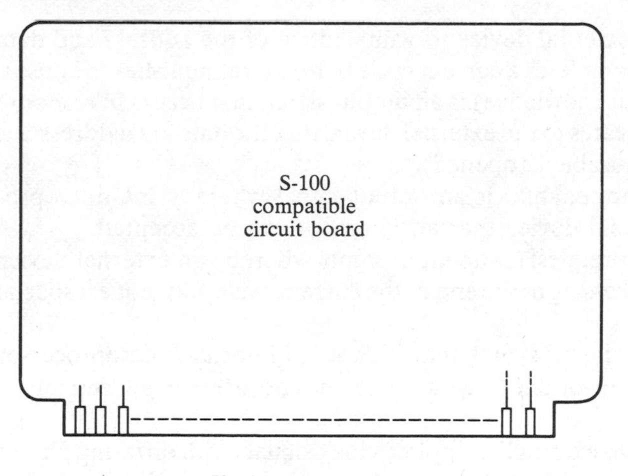

50 pins on each side

FIGURE 10–7 *Typical S-100 card edge board.*

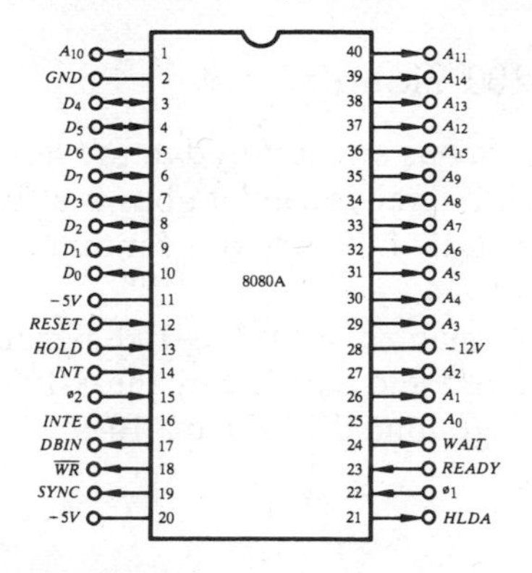

FIGURE 10–8 *8080A microprocessor inputs and outputs.*

provide for transfers of data bytes between the microprocessor, memory, and I/O devices. Both data bits and program instructions are input to the microprocessor on these lines and the microprocessor sends data to the memory or output devices on these lines. These data lines are three-state buffered.

Sync is an output synchronizing signal which indicates the start of a microprocessor cycle.

DBIN (data bus in) is an output signal that tells external circuits that the bidirectional data bus is in the *input* mode.

READY is an input signal that tells the 8080A microprocessor that valid data are available on the data bus. If the microprocessor does not receive a *ready,* it goes into a *wait* state.

WAIT is an output that indicates when the microprocessor is in the *wait* state.

$\overline{WR}$ (write) is an active LOW output signal for memory write operation or I/O control.

HOLD is an input signal that causes the microprocessor to go into a *hold* state which allows an external device to gain control of the address and data buses. The microprocessor completes a current cycle before it relinquishes the buses.

HLDA (hold acknowledge) is an output signal that occurs in response to the *HOLD* signal and indicates to the external device that the data and address bus three-state buffers will be disabled (opened).

INTE (interrupt enable) is an output signal whereby the microprocessor indicates to an external device that an interrupt can be accepted.

INT (interrupt request) is an input signal whereby an external device can interrupt the normal processing at the end of the current cycle and request special service.

RESET is an input signal that clears an internal microprocessor counter (program counter) in order to return the microprocessor addressing to memory location zero.

ϕ_1 and ϕ_2 are two externally supplied clock signals with differing phases and are used for the microprocessor's internal timing.

V_{SS}, V_{DD}, V_{CC} and V_{BB} are the ground; +12 V, +5 V, and −5 V dc supply voltages respectively.

Relation of 8080A Signals to S-100 Signals

Both the 8080A address bus and the data bus are buffered to provide greater drive capability on the S-100 bus. The 8080A bidirectional data bus is also converted to two unidirectional buses—one for input (DI7–DI0) and one for output (DO7–DO0).

The data output lines are three-state enabled by the S-100 signal $\overline{\text{DODBS}}$. The address lines are also three-state buffered and enabled by the S-100 signal $\overline{\text{ADDRDSBL}}$. This set of S-100 signals is diagrammed in Figure 10–9.

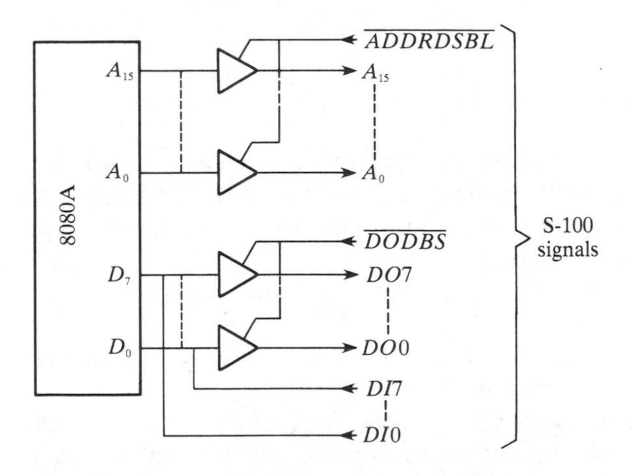

FIGURE 10–9 *S-100 data and address lines.*

The 8080A command and control signals are three-state buffered in the S-100 bus and are enabled by the $\overline{\text{C/C DBS}}$ signal. These 8080A signals are renamed for the S-100 bus. SYNC is PSYNC, DBIN is PDBIN, WAIT is PWAIT, $\overline{\text{WR}}$ is $\overline{\text{PWR}}$, HLDA is PHLDA, and INTE is PINTE. This set of S-100 signals is shown in Figure 10–10.

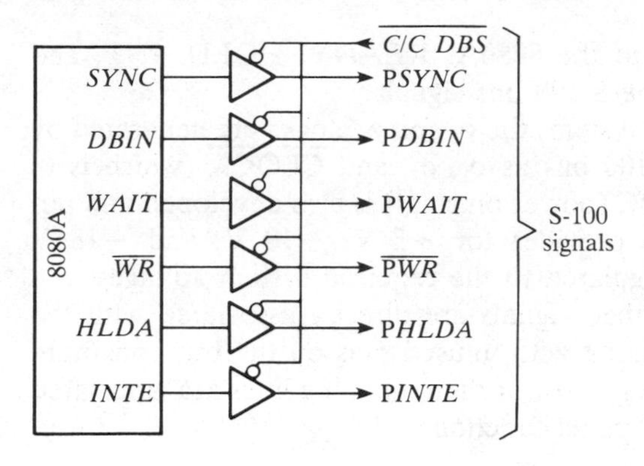

FIGURE 10–10 *S-100 command and control lines.*

The 8080A also outputs *status bits* on the data bus at the beginning of each cycle which describe the particular operation that is underway. These status bits are described in Table 10–1.

TABLE 10–1 *8080A status bits.*

Data bus bit	Signal	Description
D_0	INTA	Interrupt acknowledge
D_1	$\overline{\text{WO}}$	A write or output is about to occur
D_2	STACK	The address bus holds stack address
D_3	HLTA	HALT acknowledge
D_4	OUT	Output device address is on address bus; data bus holds output data when $\overline{\text{WR}}$ active.
D_5	M1	CPU in fetch cycle for 1st instruction byte.
D_6	INP	Input device address on address bus; data bus will accept input data when DBIN active.
D_7	MEMR	Data bus used for memory read.

These status bits are latched by the PSYNC signal and become S-100 signals as indicated in Figure 10–11. The status bits can be disabled by an S-100 signal called STATUS DSBL.

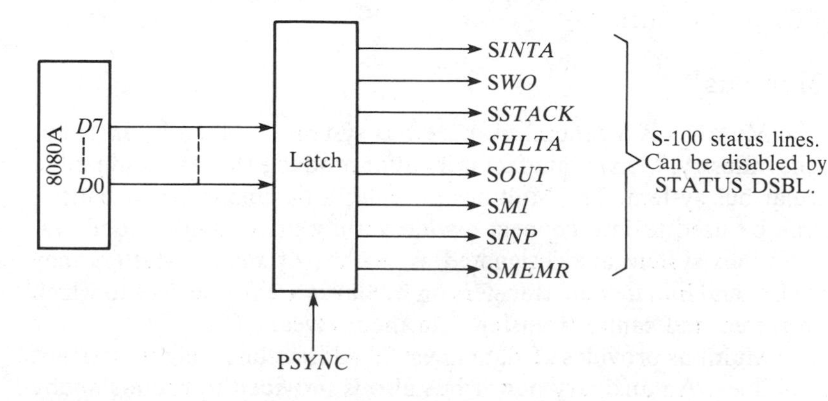

FIGURE 10–11 *S-100 status bits.*

There are four inputs to the 8080A: READY, HOLD, INT, and RESET. These are buffered from the S-100 bus signals.

In an S-100 compatible system, the ϕ_1 and ϕ_2 clocks are generated by S-100 circuitry and put onto the S-100 bus as ϕ_1, ϕ_2, and $\overline{\text{CLOCK}}$ (which is ϕ_2 inverted). The S-100 signal called POC (power on clear) is also developed and put onto the S-100 bus. The S-100 bus provides for +8 V, +18 V, and −18 V unregulated lines. These must be regulated to the required system voltages.

All of the above described signals are directly associated with the 8080A microprocessor. These pins along with unused pins on the bus constitute about 80 percent of the S-100 bus lines. Most of the remaining lines are associated with vectored interrupts and control panel functions.

The General Purpose Interface Bus (GPIB)

This bus system is defined in the IEEE standard 488–1975 and is therefore sometimes known as the IEEE-488 bus. An important application of this bus is that it allows a computer to be connected to several test instruments to form an automated test system.

The IEEE-488 standard defines three basic types of devices that can be connected to the GPIB. These are classified as *talker, listener,* and *controller*. There can also be devices that are combinations of these basic types.

Examples of *talkers* are devices that *produce* information such as digital multimeters and frequency counters. Examples of *listeners* are display devices and programmable instruments such as signal generators, multimeters, or power supplies. A *programmable multimeter,* for example, can be both a talker and a listener. It functions as a talker when outputting voltage measurements and as a listener when receiving program instructions. Also, devices such as the Intel 8291 GPIB Talker/Listener are available to interface microprocessors to the GPIB.

A *controller* is a device that determines when the other devices can use the bus. A microprocessor system is an example of a controller.

The GPIB structure is shown in Figure 10–12.

There are eight bidirectional *data lines* (DI01-DI08). There are three *data byte transfer control lines:* DAV (data available), NRFD (not ready for data), and NDAC (not data accepted). There are five *general interface management lines:* IFCC (interface clear), ATN (attention), SRQ (service request), REN (remote enable), and EOI (end or identify).

Multibus™

The Multibus is a general purpose bus system designed by Intel but widely used in the industry. Many manufacturers offer products that are compatible with this particular bus system. The Multibus provides a flexible communications interface that can be used to interconnect a wide variety of computing modules. Modules in a Multibus system are designated as *masters* or *slaves*. Masters may obtain use of the bus and initiate data transfers on it. Slaves are the devices to which the data is transferred and cannot transfer data themselves.

The Multibus provides 16 data lines, 20 address lines, eight interrupt lines, and control lines. An auxiliary power bus also is provided to route standby

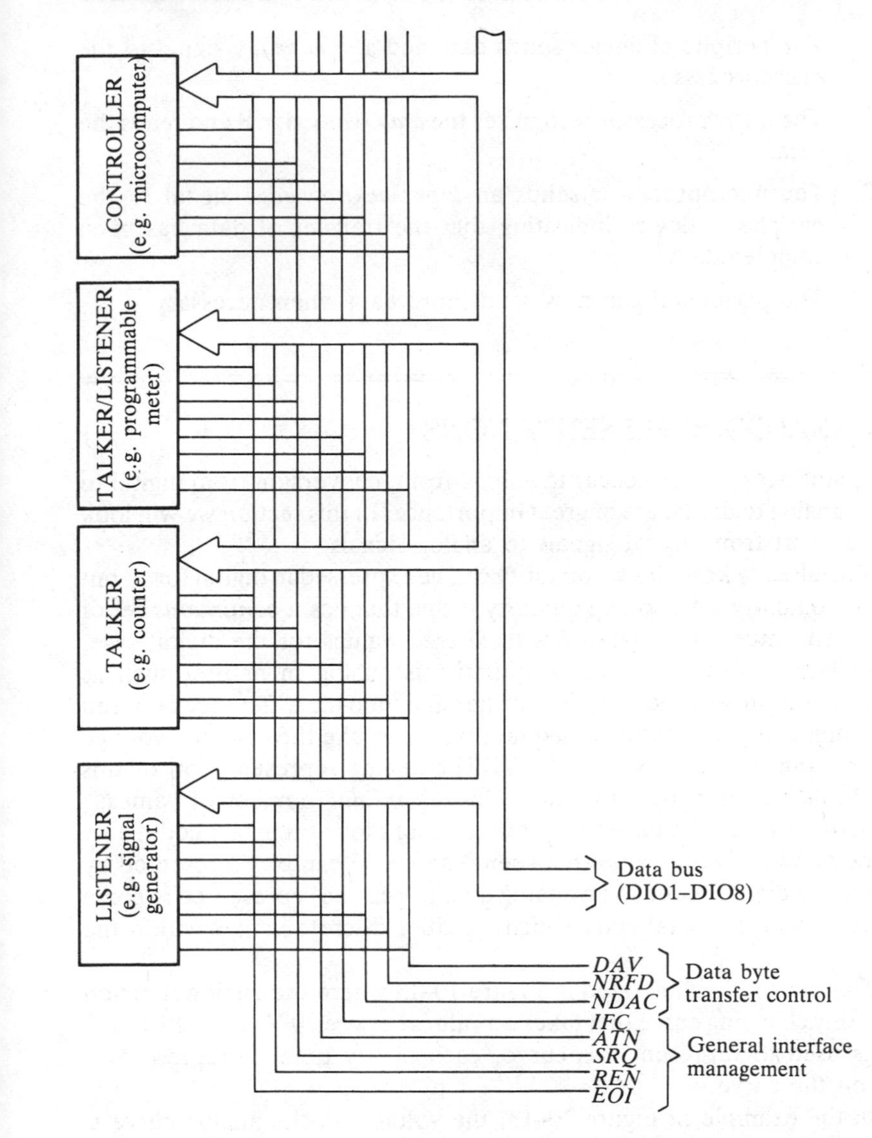

FIGURE 10–12 *The General Purpose Interface Bus (GPIB or IEEE-488).*

power to memories if normal power fails. The Multibus standard defines the form and physical requirements of devices that communicate on the bus.

Handshaking

A widely used interfacing term in digital systems is *handshaking.* It is basically the method or procedure by which two devices establish communication with each other and normally consists of a specified sequence of signals that are transferred between the two devices in a prescribed manner.

For example, when a microprocessor is receiving input data from some peripheral device, the following handshaking procedure occurs:

1. The peripheral device sends data and a *data ready* signal to the microprocessor.
2. The microprocessor recognizes the *data ready* signal and reads the data.
3. The microprocessor sends an *input acknowledge* signal to the peripheral device indicating that the transfer of data has been completed.
4. The peripheral can now send more data when necessary.

10–3 DIGITAL AND ANALOG REPRESENTATIONS

Because most quantities in nature occur in *analog* form, conversion from digital to analog and from analog to digital are of great importance. In this section we will look at methods to convert from digital signals to analog signals.

You already know how a quantity can be expressed in digital form, but what is an *analog* quantity? An analog quantity is one that has a *continuous* set of values over a given range, as contrasted with *discrete* values for the digital case.

Practically any measureable quantity is analog in nature, such as temperature, pressure, speed, and time. To further illustrate the difference between an analog and a digital representation of a quantity, let us take the case of a voltage that varies over a range from 0 V to +15 V. The analog representation of this quantity takes in *all* values between 0 and +15—an *infinite* number of values.

In the case of a digital representation using a four-bit binary code, only 16 values can be defined. More values between 0 and +15 can be represented by using more bits in the digital code. So an analog quantity can be represented to some degree of accuracy with a digital code which specifies discrete values within the range.

This concept is illustrated in Figure 10–13 where the analog function shown is a smoothly changing curve that takes on values between 0 V and +15 V. If a four-bit code is used to represent this curve, each binary number represents a discrete point on the curve.

In the example in Figure 10–13, the voltage on the analog curve is measured or *sampled* at each of 35 equal intervals. The voltage at each of these intervals is represented by a four-bit code as indicated. At this point, we have a series of binary numbers representing various voltage values along the analog curve. This is the basic idea of *analog-to-digital* conversion (A/D).

An approximation of the analog function in Figure 10–13 can be reconstructed from the sequence of digital numbers that have been generated. Obviously, there is going to be some error in the reconstruction because only certain values are represented (35 in this example) and not the continuous set of values. If the digital values at each of the 35 intervals are graphed as shown in Figure 10–14, we have a reconstructed function. As you can see, the graph only approximates the original curve because values between each two points are not known. This is the basic idea behind *digital-to-analog* conversion (D/A).

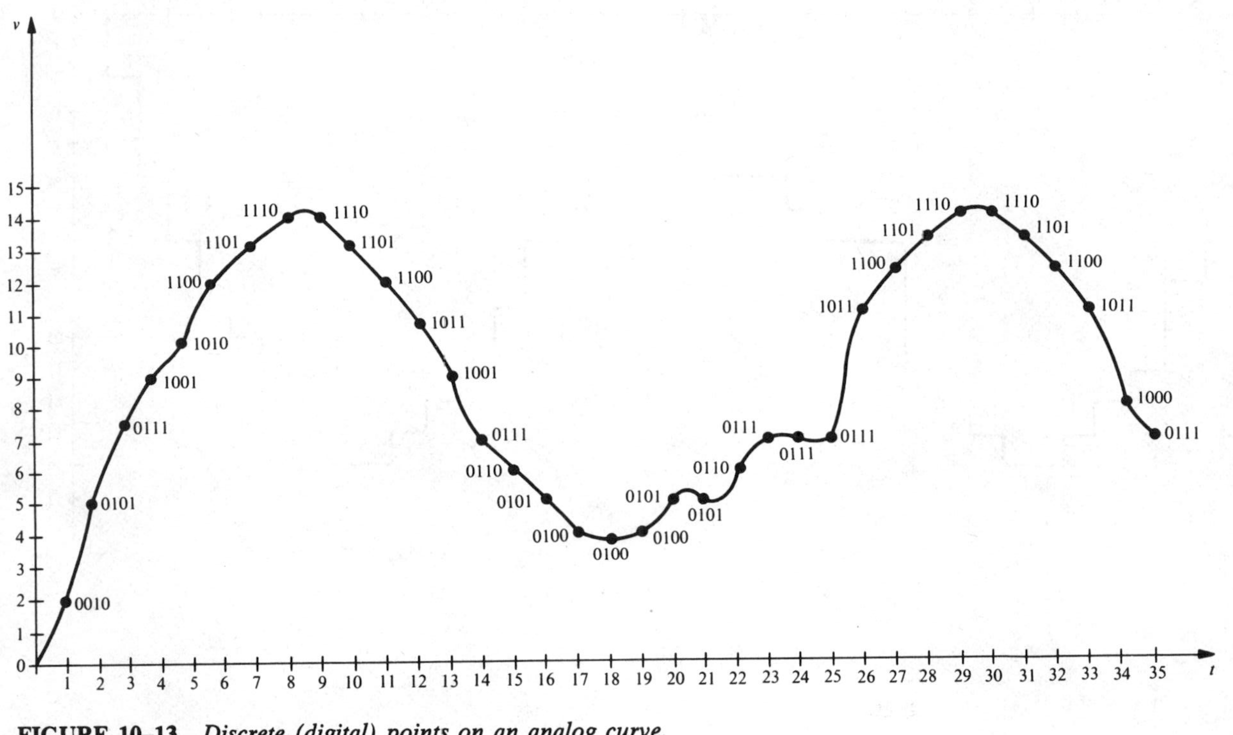

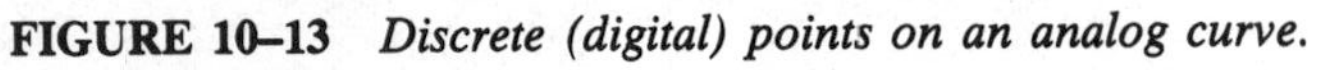
FIGURE 10–13 *Discrete (digital) points on an analog curve.*

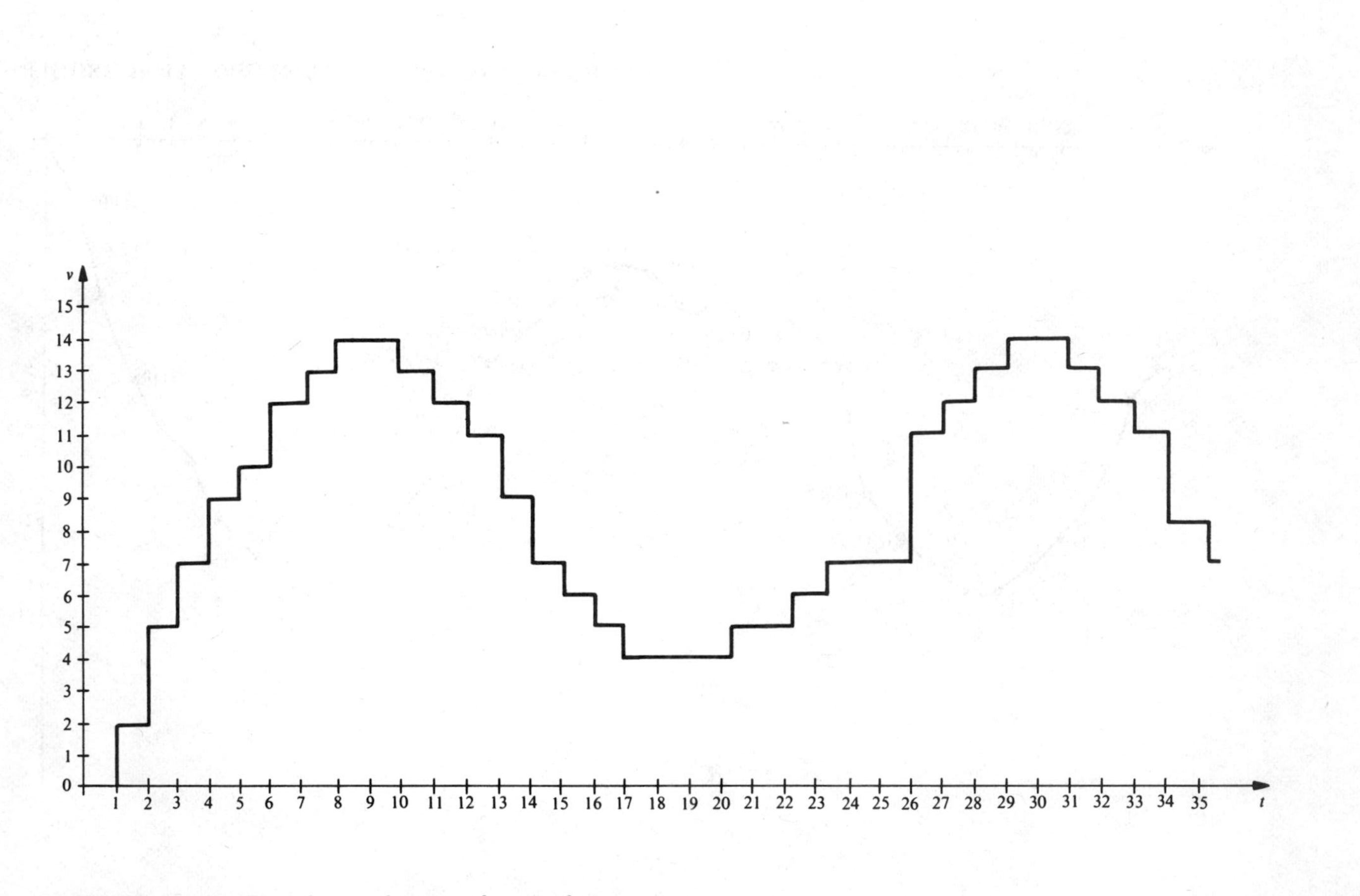

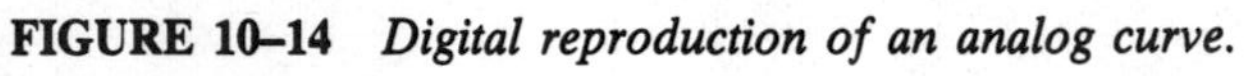
FIGURE 10–14 *Digital reproduction of an analog curve.*

10–4 DIGITAL-TO-ANALOG CONVERSION (D/A)

D/A conversion is an important interface process for I/O operations in many applications. An example is where a voice signal has been digitized for processing or transmission and must be changed back into an approximation of the original signal to ultimately drive a speaker.

Binary Weighted Input D/A Converter

One method of D/A conversion uses a resistor network with values that represent the binary weights of the input bits of the digital code. Figure 10–15 shows a four-bit D/A converter of this type. The switch symbols represent transistor switches for inputting each of the four bits. The operational amplifier provides a very high impedance load to the resistor network, and its inverting input looks like "virtual" ground so that the output is proportional to the current through the feedback resistor R_F (the sum of the input currents.) Practically all the current is through R_F and into the low impedance output of the op amp. The inverting input is approximately at 0 V.

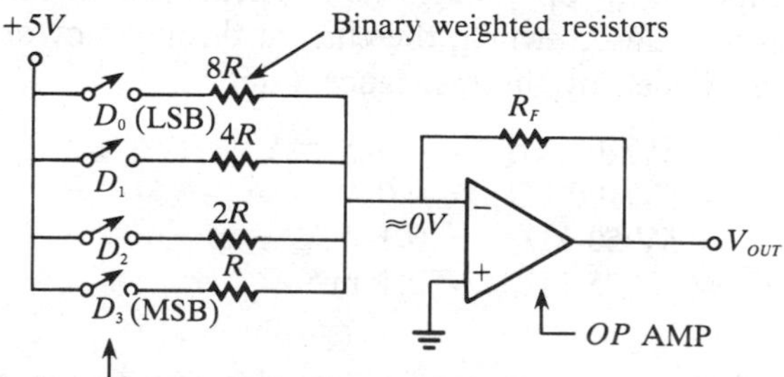

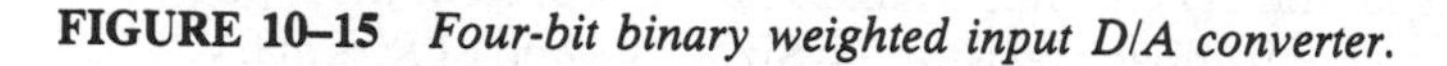

FIGURE 10–15 *Four-bit binary weighted input D/A converter.*

The lowest value resistor R corresponds to the highest binary weighted input 2^3. Each of the other resistors are multiples of R—$2R$, $4R$, and $8R$ corresponding to the binary weights 2^2, 2^1, and 2^0, respectively. One of the disadvantages of this type of D/A converter is the number of different resistor values. For example an eight-bit converter requires eight resistors ranging from some value R to $128R$.

Example 10–1

Determine the output of the D/A converter in Figure 10–16(a) if the sequence of four-bit numbers in Figure 10–16(b) are applied to the inputs. D_0 is the LSB.

Example 10–1, continued

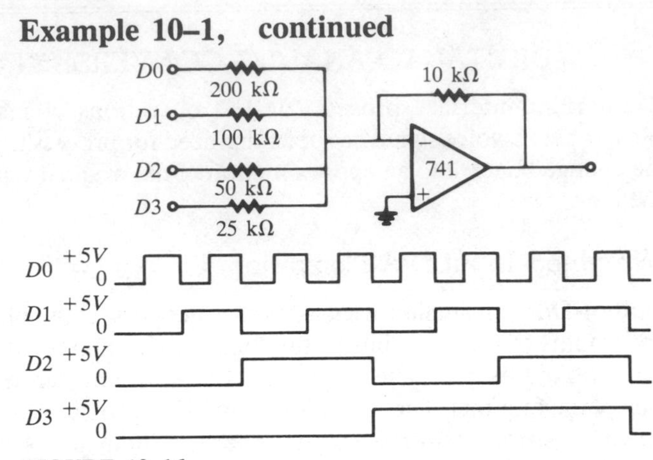

FIGURE 10–16

Solution:

First, let us determine the output voltage for each of the weighted inputs. Since the inverting input of the op amp is at 0V (virtual ground) and a binary 1 corresponds to a closed switch, the current through any of the input resistors is 5 V divided by the resistance value:

$$I_0 = 5V/200\ k\Omega = 0.025\ mA$$
$$I_1 = 5V/100\ k\Omega = 0.05\ mA$$
$$I_2 = 5V/50\ k\Omega = 0.1\ mA$$
$$I_3 = 5V/25\ k\Omega = 0.2\ mA$$

None of the input current goes into the inverting op amp input because of its extremely high impedance. Therefore, all of the input current goes through the feedback resistor R_F. Since one end of R_F is at 0V (virtual ground), the drop across R_F equals the output voltage.

$$V_{OUT(D0)} = (10\ k\Omega)\,(-0.025\ mA) = -0.25\ V$$
$$V_{OUT(D1)} = (10\ k\Omega)\,(-0.05\ mA) = -0.5\ V$$
$$V_{OUT(D2)} = (10\ k\Omega)\,(-0.1\ mA) = -1\ V$$
$$V_{OUT(D3)} = (10\ k\Omega)\,(-0.2\ mA) = -2\ V$$

From Figure 10–16(b), the first input code is 0001 (binary 1). For this, the output voltage is −0.25 V. The next code is 0010 which produces an output voltage of −0.5 V. The next code is 0011 which produces an output voltage of −0.25 V + −0.5 V = −0.75 V. Each successive binary code increases the output voltage by −0.25 V so for this particular straight binary sequence on the inputs, the output is a *stairstep* waveform going from 0 V to −3.75 V in −0.25 V steps. This is shown in Figure 10–17.

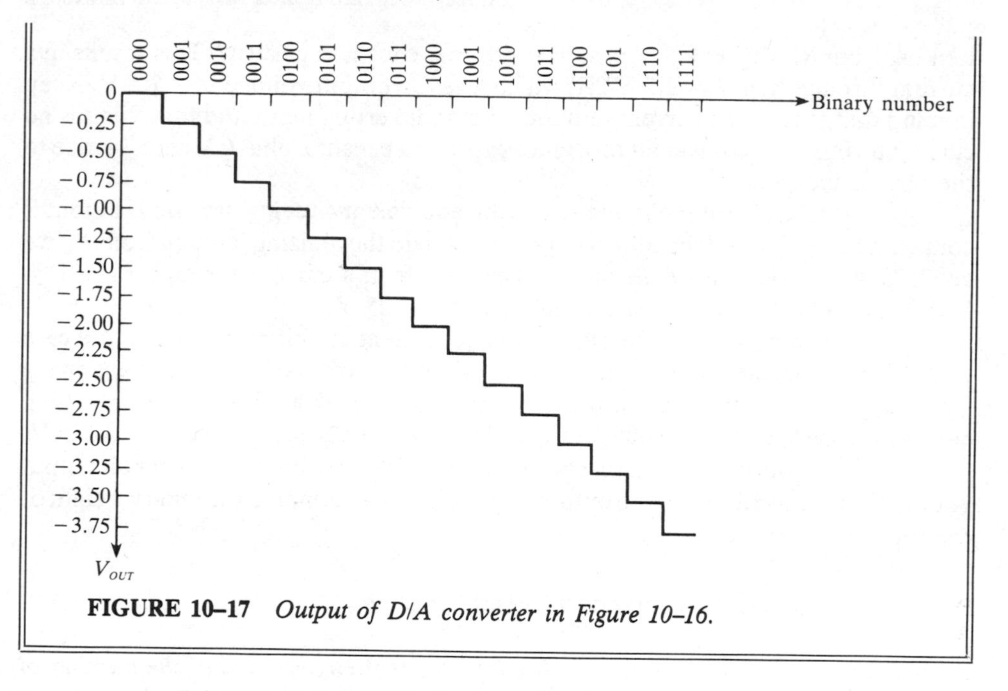

FIGURE 10–17 *Output of D/A converter in Figure 10–16.*

R/2R Ladder D/A Converter

Another method of D/A conversion is the so-called *R/2R ladder,* as shown in Figure 10–18 for four bits. It overcomes one of the problems in the previous type in that it requires only two resistor values. Again, the switch symbols represent transistor switches.

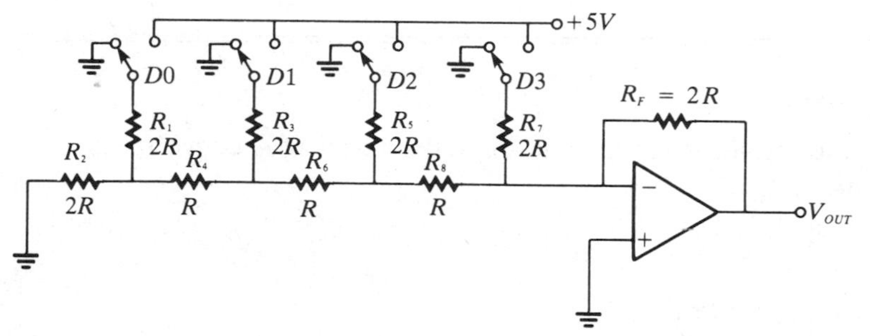

FIGURE 10–18 *An R/2R ladder D/A converter.*

Let us start by assuming the *D*3 switch is connected to +5 V and the others to ground. This represents 1000_2. A little analysis will show you that this reduces to the equivalent form shown in Figure 10–19(a). There is essentially no current through the 2R *equivalent* resistance. This means that all of the current (I = 5V/2R) through R_7 also goes through R_F and the output voltage is −5V.

Part (b) of the figure shows the equivalent circuit when the *D*2 switch is in the +5 V position and the others at ground. This represents 0100_2. If we thevenize looking from R_8, we get 2.5 V in series with R as shown in part (b). This results in a current through R_F of $I = 2.5\ \text{V}/2R$, which gives an output voltage of -2.5 V. Keep in mind that there is no current into the op amp inverting input and that there is no current through the equivalent resistance to ground because it has 0 V across it due to the virtual ground.

Part (c) of the figure shows the equivalent circuit when the $D1$ input is connected to +5 V and the others to ground. Again thevenizing looking from R_8, we get 1.25 V in series with R as shown. This results in a current through R_F of $I = 1.25\ \text{V}/2R$, which gives an output voltage of -1.25 V.

In part (d) of the figure, the equivalent circuit representing the case where $D0$ is connected to +5 V and the other inputs to ground is shown. Thevenizing from R_8 gives an equivalent of 0.625 V in series with R as shown. The resulting current through R_F is $I = 0.625\ \text{V}/2R$, which gives an output voltage of -0.625 V.

Notice that each successively lower weighted input produces an output voltage that is halved so that the output voltage is proportional to the binary weight of the input bits.

D/A Performance Characteristics

Resolution. The resolution of a D/A converter is the *reciprocal of the number of discrete steps* in the D/A output. This, of course, is dependent on the number of input bits. For example, a four-bit D/A converter has a resolution of 1 part in 2^4 (1 part in 16). Expressed as a percentage, this is (1/16)100 = 6.25 percent. Sometimes the resolution is expressed simply as the number of bits: for example, this particular D/A converter has a *four-bit* resolution.

Example 10–2

Determine the resolution of an eight-bit and a 12-bit D/A converter in terms of percentage.

Solution:

For the eight-bit converter:

$$\text{percent resolution} = (1/2^8)100 = (1/256)100 = 0.39\ \%$$

For the 12-bit converter:

$$\text{percent resolution} = (1/2^{12})100 = (1/4096)100 = 0.0244\%$$

Accuracy. Accuracy is a comparison of the *actual* output of a D/A converter with the *expected* output. It is expressed as a percentage of full scale or maximum output voltage. For example, if a converter has a full-scale output of 10 V and the accuracy is ± 0.1 percent, then the maximum error for any output voltage is (10 V) (0.001) = 10 mV. Ideally, the accuracy should be at most ± ½ of an LSB. For an eight-bit converter, 1 LSB is 1/256 = 0.0039 (0.39 percent of full scale). The accuracy should be approximately ± 0.2 percent.

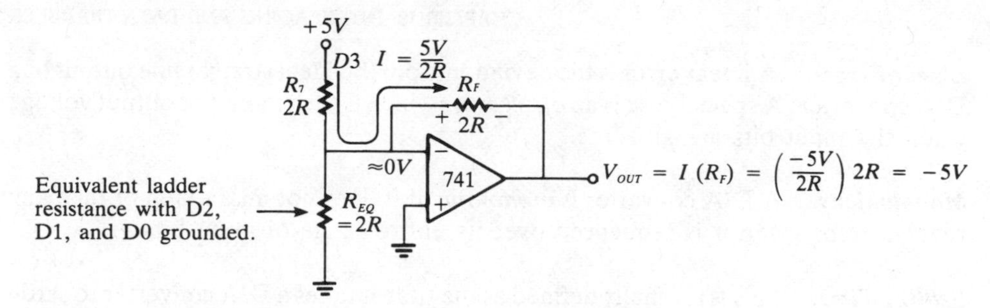

(a) Equivalent circuit for D3 = 1, D2 = 0, D1 = 0, D0 = 0.

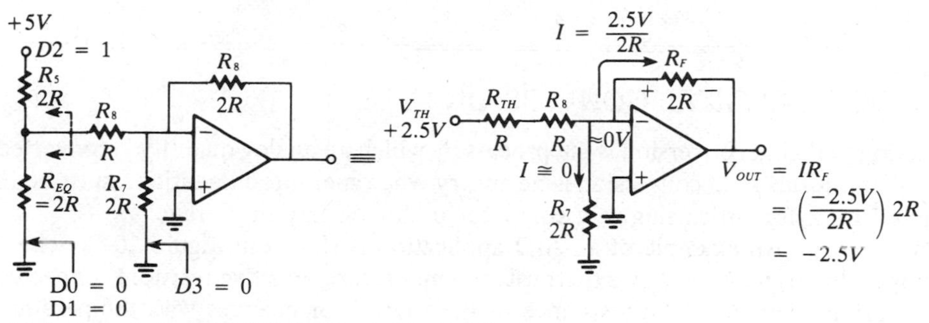

(b) Equivalent circuit for D3 = 0, D2 = 1, D1 = 0, D0 = 0.

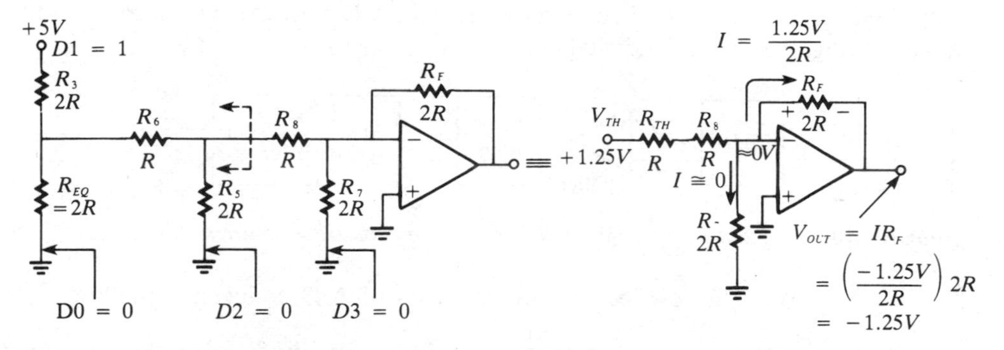

(c) Equivalent circuit for D3 = 0, D2 = 0, D1 = 1, D0 = 0.

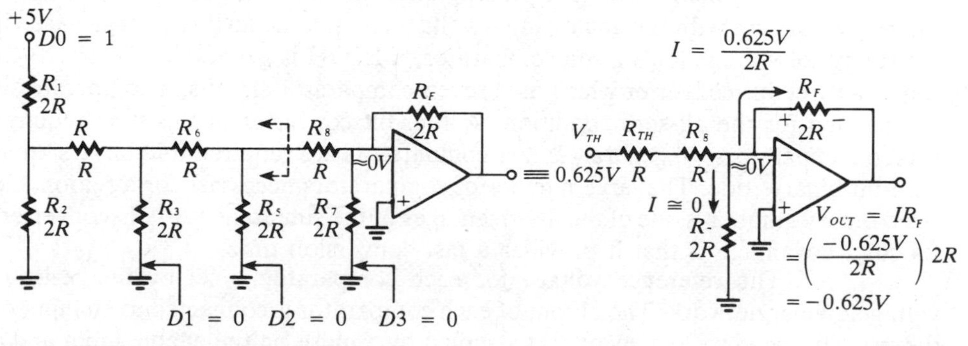

(d) Equivalent circuit for D3 = 0, D2 = 0, D1 = 0, D0 = 1.

FIGURE 10–19 *Analysis of the R/2R ladder D/A converter.*

Linear Errors. A linear error is the deviation from the ideal straight line output of a D/A converter. A special case is an *offset error* which is the amount of output voltage when the input bits are all zeros.

Monotonicity. A D/A converter is *monotonic* if it does not miss a step or take any reverse steps when it is sequenced over its entire range of input bits.

Settling Time. This is normally defined as the time it takes a D/A converter to settle within ± ½ LSB of its final value when a change occurs in the input code.

10–5 ANALOG-TO-DIGITAL CONVERSION (A/D)

Analog to digital conversion is the process by which an analog quantity is converted to digital form. A/D conversion is necessary when measured quantities must be in digital form for processing in a computer or for display or storage.

An example of an A/D application is shown in Figure 10–20 where temperature is measured by a thermister (temperature sensitive resistor) connected in a bridge network. The resistance of the thermistor changes with temperature producing a proportional bridge output. The output voltage of the bridge is amplified, adjusted, and applied to the input of an A/D converter which converts the analog bridge voltage into digital form.

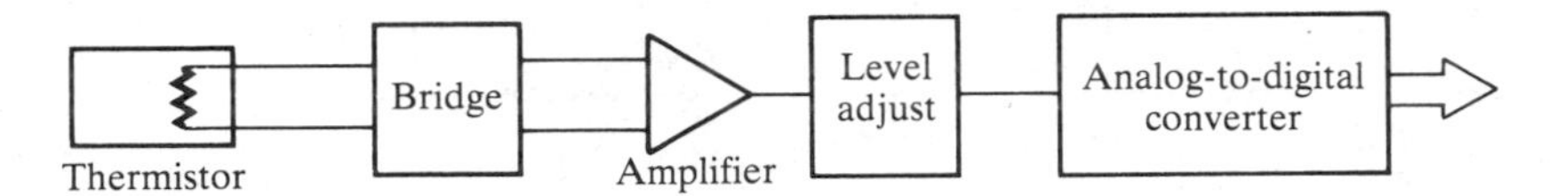

FIGURE 10–20 *Example of A/D application in temperature measurement.*

We will now look at several types of A/D conversion methods.

Simultaneous A/D Converter

This method utilizes parallel differential comparators that compare reference voltages with the analog input voltage. When the analog voltage exceeds the reference voltage for a given comparator, a HIGH is generated. Figure 10–21 shows a three-bit converter which uses seven comparator circuits; a comparator is not needed for the all-zero condition. A four-bit converter of this type requires fifteen comparators. In general, $2^n - 1$ comparators are required for conversion to an n-bit binary code. The large number of comparators necessary for a reasonable sized binary number is one of the disadvantages of the simultaneous A/D converter. Its chief advantage is that it provides a fast conversion time.

The reference voltage for each comparator is set by the resistive voltage divider network. The output of each comparator is connected to an input of the *priority encoder*. The encoder is sampled by a pulse on the enable input and a three-bit binary code proportional to the value of the analog input appears on the encoder's outputs. The binary code is determined by the highest order input having a HIGH level. The three output bits can be transferred into a storage device.

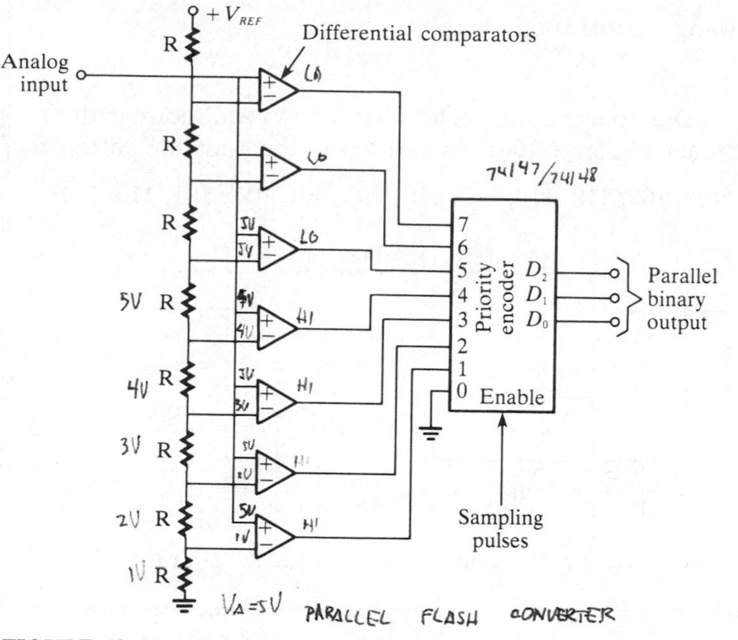

FIGURE 10–21 *A three-bit simultaneous A/D converter.*

The *sampling rate* determines the accuracy with which the sequence of digital codes represents the analog input of the A/D converter. The more samples taken in a given unit of time, the more accurately the analog signal is represented in digital form.

The following example illustrates the basic operation of the simultaneous A/D converter in Figure 10–21.

Example 10–3

Determine the binary code output of the three-bit simultaneous A/D converter for the analog input signal in Figure 10–22 and the sampling pulses (encoder enable) shown.

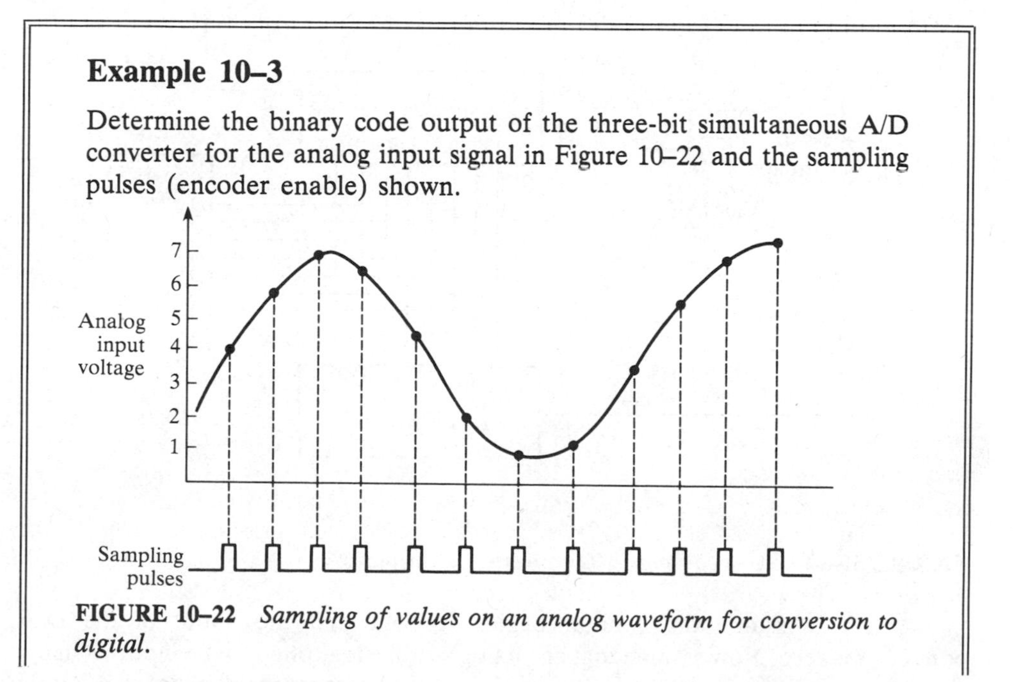

FIGURE 10–22 *Sampling of values on an analog waveform for conversion to digital.*

Example 10–3, continued

Solution:

The resulting A/D output sequence is listed as follows and shown in the waveform diagram of Figure 10–23 in relation to the sampling pulses.

100, 101, 110, 110, 100, 010, 000, 001, 011, 101, 110, 111

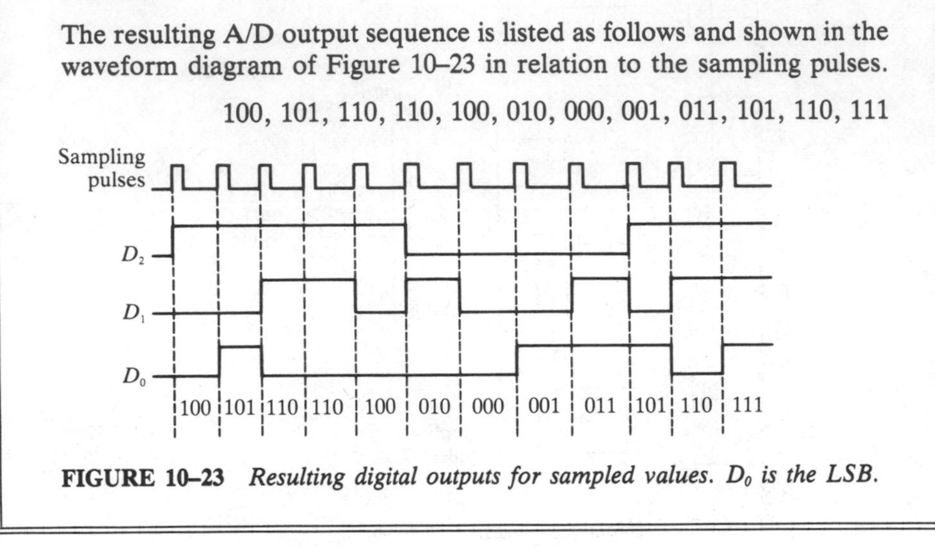

FIGURE 10–23 *Resulting digital outputs for sampled values. D_0 is the LSB.*

Stairstep-Ramp A/D Converter

This method of A/D conversion is also known as *digital-ramp* or *counter* method. It employs a D/A converter and a binary counter to generate the digital value of an analog input. Figure 10–24 shows a diagram of this type of converter.

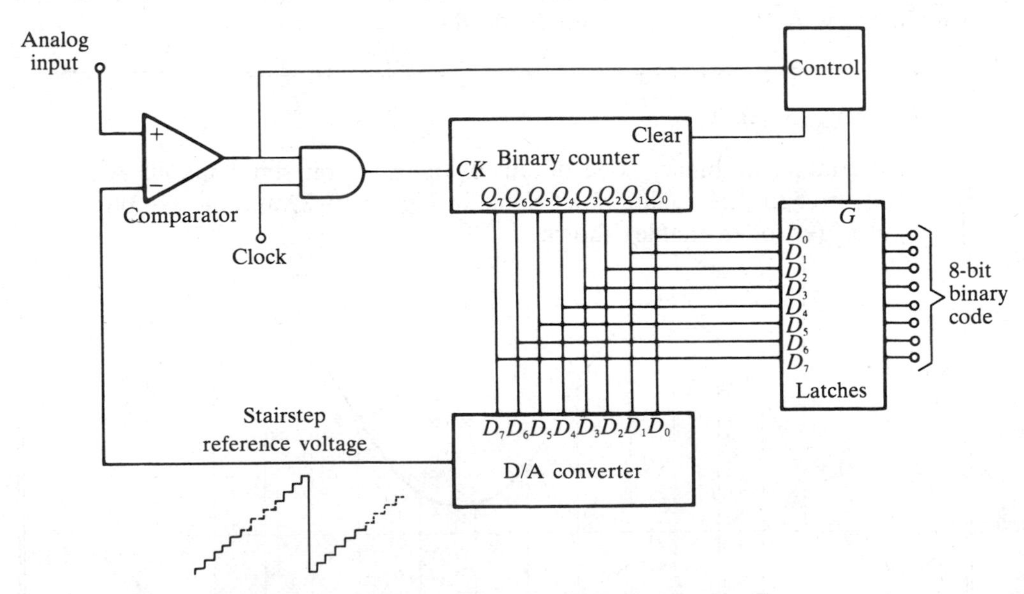

FIGURE 10–24 *Stairstep-ramp A/D converter (eight bits).*

Assume the counter begins RESET and the ouput of the D/A converter is zero. Now assume that an analog voltage is applied to the input. When it exceeds the reference voltage (output of D/A), the comparator switches to a HIGH output state and enables the AND gate. The clock pulses begin advancing the counter through its binary states, producing a stairstep reference voltage from the D/A converter. The counter continues to advance from one binary state to the next, producing successively higher steps in the reference voltage. When the stairstep reference voltage reaches the analog input voltage, the comparator output will go LOW and disable the AND gate, thus cutting off the clock pulses to stop the counter. The state of the counter at this point equals the number of steps in the reference voltage at which the comparison occurs. This binary number, of course, represents the value of the analog input. The control logic loads the binary count into the latches and resets the counter, thus beginning another count sequence to sample the input value.

This method is slower than the simultaneous method because, in the worst case of maximum input, the counter must sequence through its maximum number of states before a comparison occurs. For an eight-bit conversion, this means a maximum of 256 counter states. Figure 10–25 illustrates a conversion sequence for a four-bit conversion. Notice that for each sample, the counter must count from *zero* up to the point where the stairstep reference voltage reaches the analog input voltage. The conversion time varies depending upon the analog voltage.

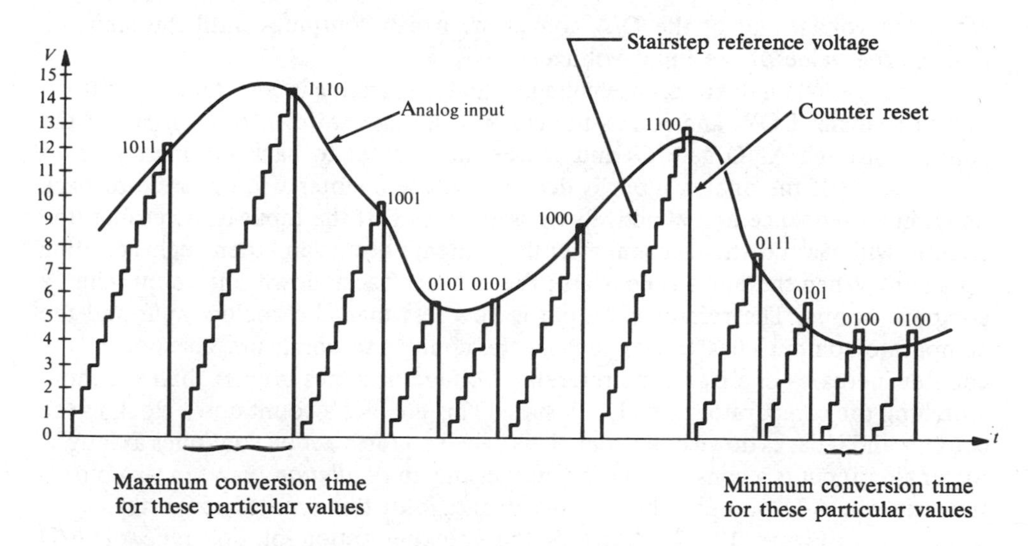

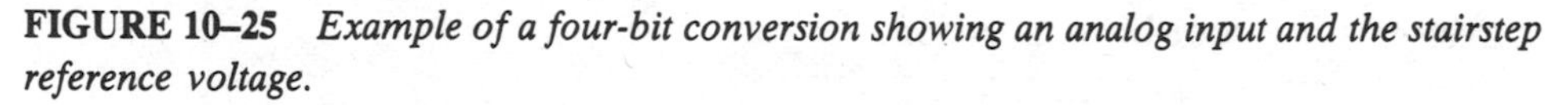

FIGURE 10–25 *Example of a four-bit conversion showing an analog input and the stairstep reference voltage.*

Tracking A/D Converter

This method uses an *up/down counter* and is faster than the stairstep-ramp method because the counter is not reset after each sample but rather tends to *track* the analog input. Figure 10–26 shows a typical eight-bit tracking A/D converter.

As long as the D/A output reference voltage is less than the analog input, the comparator output is HIGH and enables the AND gate G_1. G_1 applies the clock pulses to the *count up (CU)* input of the up/down counter, causing the counter to produce an up sequence of binary counts. This causes an *increasing* stairstep

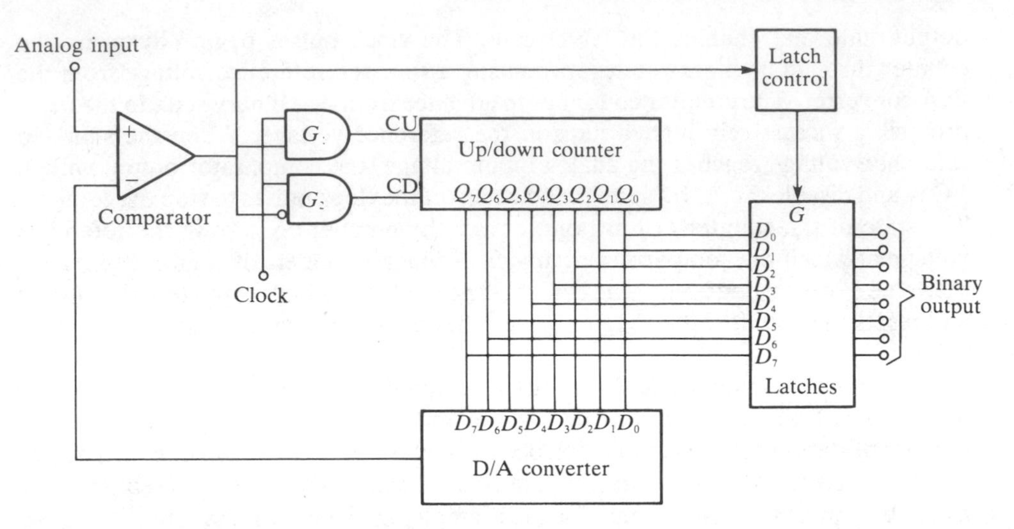

FIGURE 10–26 *An eight-bit tracking A/D converter.*

reference voltage out of the D/A converter, which continues until the stairstep reaches the value of the input voltage.

When the reference voltage equals the analog input, the comparator's output switches LOW and routes the clock to the *count down (CD)* input of the counter through AND gate G_2 and causes the counter to back up one count.

If the analog input is decreasing, the counter will continue to back down in its sequence and effectively *track* the input. If the input is increasing, the counter will back down one count after the compare occurs and then begin counting up again. When the input is constant, the counter backs down one count when a compare occurs. The reference output is now less than the analog input and the comparator output goes HIGH causing the counter to count up. As soon as the counter increases one state, the reference voltage becomes greater than the input switching the comparator to its LOW state. This enables a count down clock to the counter and it backs down one count. This back and forth action continues as long as the analog input is a constant value thus causing an *oscillation* between two binary states in the A/D output. This is a disadvantage of this type of converter.

Figure 10–27 illustrates the tracking action of this type of A/D converter for a four-bit conversion.

Single-Slope A/D Converter

Unlike the previous two methods, this type of converter does not require a D/A converter. It uses a linear ramp generator to produce a constant slope reference voltage. A diagram is shown in Figure 10–28.

At the beginning of a conversion cycle, the counter is RESET and the ramp generator output is 0V. The analog input is greater than the reference voltage at this point and therefore produces a HIGH output from the comparator. This HIGH enables the clock to the counter and starts the ramp generator.

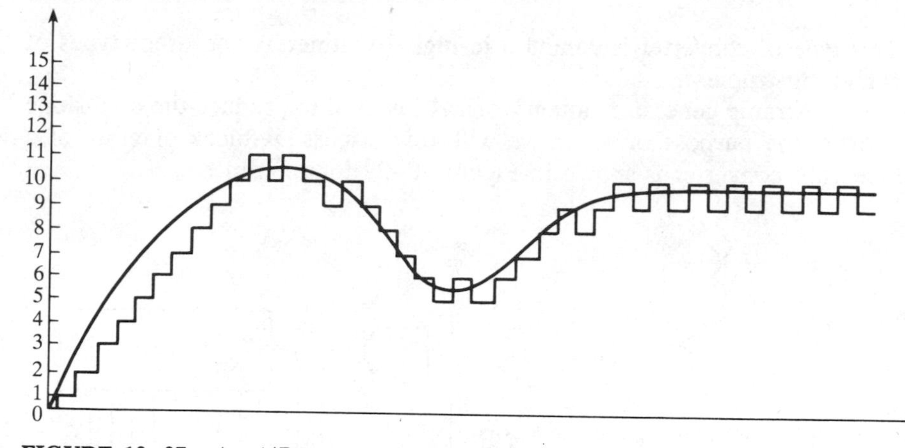

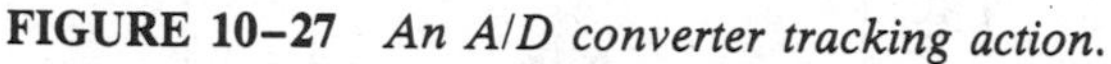
FIGURE 10–27 *An A/D converter tracking action.*

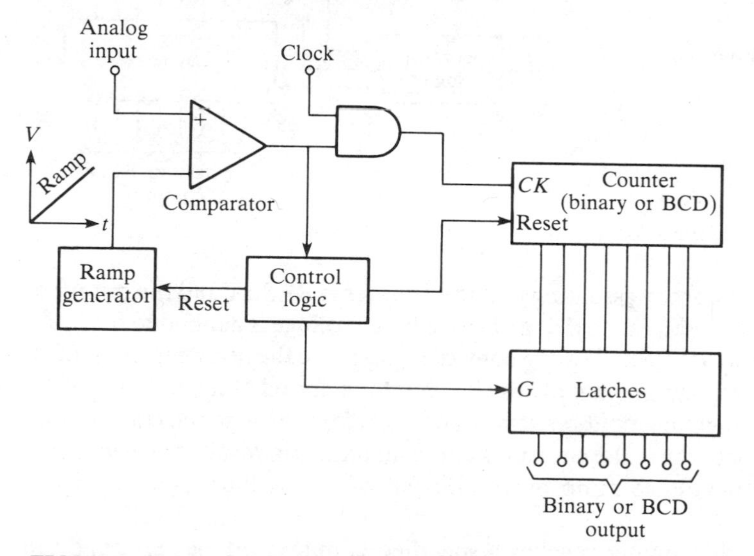

FIGURE 10–28 *Single-slope A/D converter.*

Assume the slope of the ramp is 1 V/ms. It will increase until it equals the analog input; at this point the ramp is RESET and the binary or BCD count is stored in the latches by the control logic. Let us assume that the analog input is 3 V at the point of comparison. This means that the ramp is also 3 V and has been running for 3 ms. Since the comparator output has been HIGH for 3 ms, 300 clock pulses have been allowed to pass through the gate to the counter (assuming a clock frequency of 100 kHz). At the point of comparison, the counter is in the binary state representing decimal 300. With proper scaling and decoding, this binary number can be displayed as 3.00 volts. This basic concept is used in some digital voltmeters.

Dual-Slope A/D Converter

The operation of this type of A/D converter is similar to the single-slope type except that a variable-slope ramp and a fixed-slope ramp are both used. This type of converter is common in digital voltmeters and other types of measurement instruments.

A ramp generator (integrator), A_1, is used to produce the dual-slope characteristic, the purpose of which we will now discuss. A block diagram of a dual-slope A/D converter is shown in Figure 10–29 for reference.

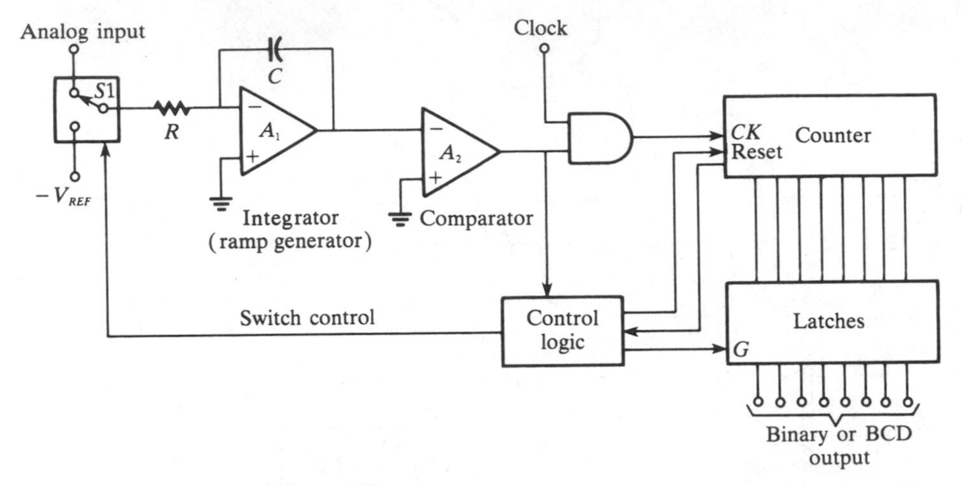

FIGURE 10–29 *Dual-slope A/D converter.*

We will start by assuming that the counter is RESET and the output of the integrator is zero. Now assume that a positive input voltage is applied to the input through the switch (S_1) as selected by the control logic. Since the inverting input of A_1 is at virtual ground and assuming that V_{in} is constant for a period of time, there will be constant current through the input resistor R and therefore through the capacitor C. C will charge linearly because the current is constant and, as a result, there will be a negative going linear voltage ramp on the output of A_1, as illustrated in Figure 10–30(a).

When the counter reaches a specified count, it will be reset and the control logic will switch the negative reference voltage ($-V_{REF}$) to input A_1 as shown in Figure 10–30(b). At this point, the capacitor is charged to a negative voltage ($-$V) *proportional* to the input analog voltage.

Now the capacitor discharges linearly due to the constant current from the $-V_{REF}$ as shown in Figure 10–30(c). This produces a positive-going ramp on the A_1 output, starting at $-V$ and with a *constant slope* that is independent of the charge voltage.

As the capacitor discharges, the counter advances from its reset state. The time it takes the capacitor to discharge to zero depends on the initial voltage $-$V (proportional to V_{in}) because the discharge rate (slope) is constant. When the integrator (A_1) output voltage reaches zero, the comparator (A_2) switches to the LOW state and disables the clock to the counter. The binary count is latched thus completing one conversion cycle.

The binary count is proportional to V_{in} because the time it takes the capacitor to discharge depends only on $-$V and the counter records this interval of time.

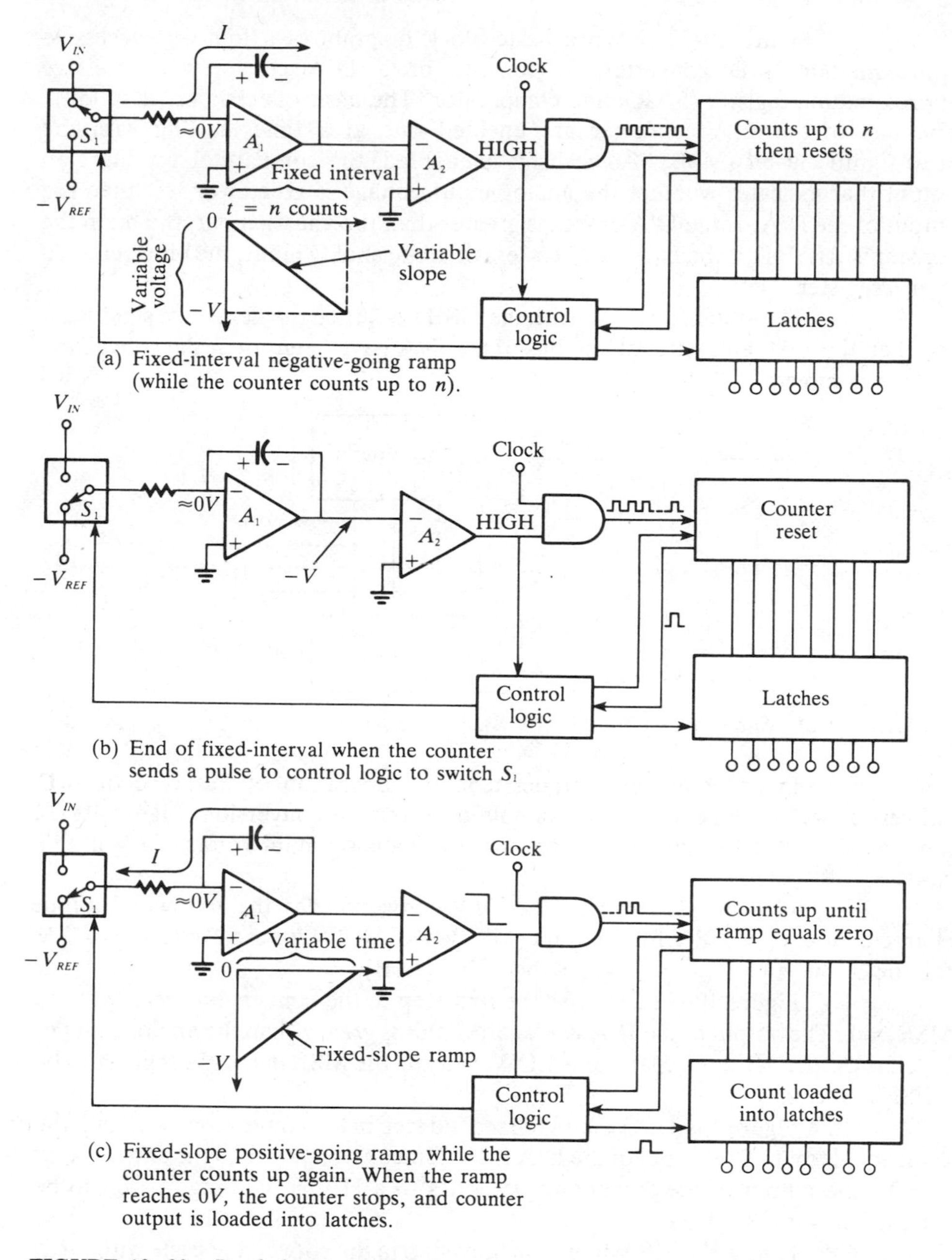

(a) Fixed-interval negative-going ramp (while the counter counts up to n).

(b) End of fixed-interval when the counter sends a pulse to control logic to switch S_1

(c) Fixed-slope positive-going ramp while the counter counts up again. When the ramp reaches $0V$, the counter stops, and counter output is loaded into latches.

FIGURE 10–30 *Dual-slope conversion.*

Successive Approximation A/D Converter

This is perhaps the most widely used method of A/D conversion. It has a much shorter conversion time than the other methods with the exception of the simultaneous method. It also has a fixed conversion time that is the same for any value of the analog input.

Figure 10–31 shows a basic block diagram of a four-bit successive approximation A/D converter. It consists of a D/A converter, successive approximation register (SAR), and comparator. The basic operation is as follows: The bits of the D/A converter are enabled one at a time starting with the most-significant-bit (MSB). As each bit is enabled, the comparator produces an output that indicates whether the analog input voltage is greater or less than the output of the D/A. If the D/A output is greater than the analog input, the bit in the register is RESET. If the D/A output is less than the analog input, the bit is retained in the register.

The system does this with the MSB first, then the next most significant bit, then the next, etc. After all the bits of the D/A have been tried, the conversion cycle is complete.

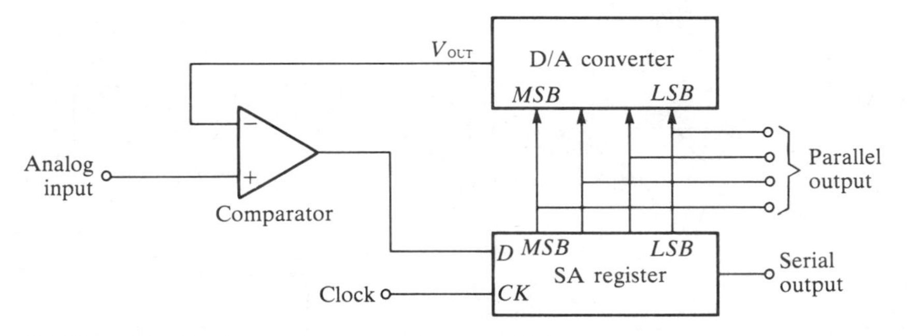

FIGURE 10–31 *Successive approximation A/D converter.*

In order to better understand the operation of this type of A/D converter, we will take a specific example of a four-bit conversion. Figure 10–32 illustrates the step-by-step conversion of a given analog input voltage (5 V in this case).

We will assume that the D/A converter has the following output characteristic: $V_{OUT} = 8$ V for the 2^3 bit (MSB), $V_{OUT} = 4$ V for the 2^2 bit, $V_{OUT} = 2$ V for the 2^1 bit, and $V_{OUT} = 1$ V for the 2^0 bit (LSB).

Figure 10–32(a) shows the first step in the conversion cycle with the MSB = 1. The output of the D/A is 8 V. Since this is *greater* than the analog input of 5 V, the output of the comparator is LOW, causing the MSB in the *SA* register to be RESET to a 0.

Figure 10–32(b) shows the second step in the conversion cycle with the 2^2 bit equal to a 1. The output of the D/A is 4 V. Since this is *less* than the analog input of 5 V, the output of the comparator switches to a HIGH, causing this bit to be retained in the *SA* register.

Figure 10–32(c) shows the third step in the conversion cycle with the 2^1 bit equal to a 1. The output of the D/A is 6 V because there is a 1 on the 2^2 bit input and on the 2^1 bit input, so 4 V + 2 V = 6 V. Since this is *greater* than the analog input of 5 V, the output of the comparator switches to a LOW, causing this bit to be RESET to a 0.

Figure 10–32(d) shows the fourth and final step in the conversion cycle with the 2^0 bit equal to a 1. The output of the D/A is 5 V because there is a 1 on the 2^2 bit input and on the 2^0 bit input, so 4 V + 1 V = 5 V.

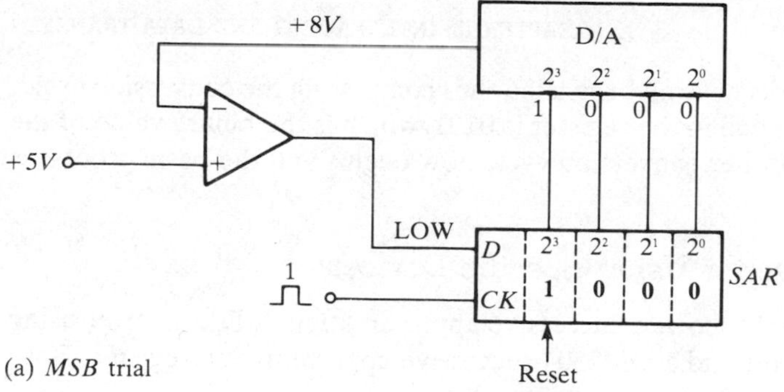

(a) *MSB* trial

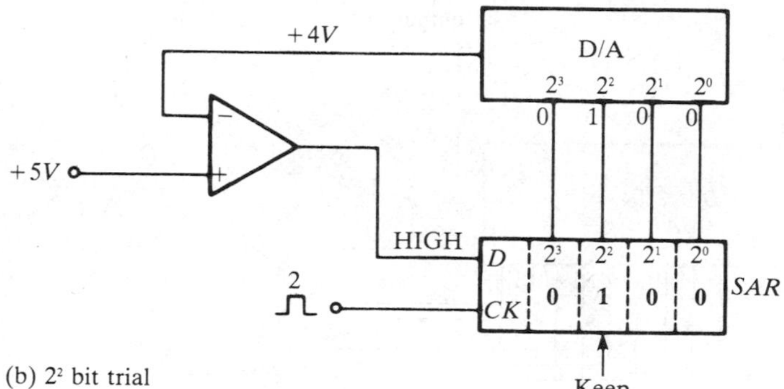

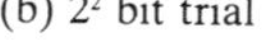
(b) 2^2 bit trial

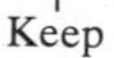

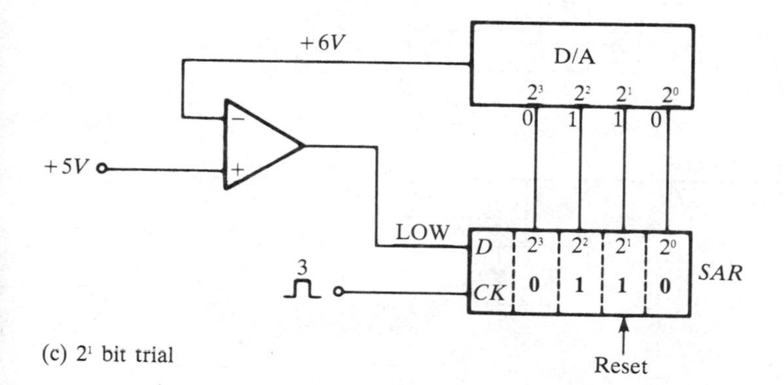

(c) 2^1 bit trial

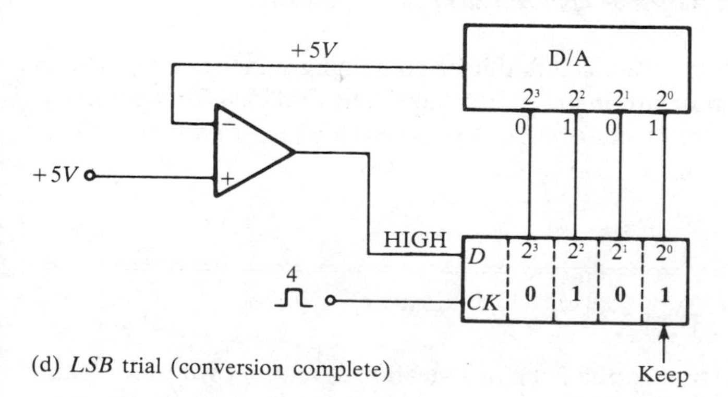

(d) *LSB* trial (conversion complete)

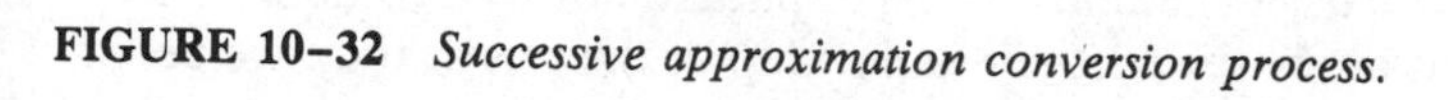

FIGURE 10–32 *Successive approximation conversion process.*

The four bits have all been tried thus completing the conversion cycle. At this point the binary code in the register is 0101_2 which is the binary value of the analog input of 5 V. Another conversion cycle now begins and the basic process is repeated.

SA Converter Using Specific Devices

Figure 10–33 shows a successive approximation A/D converter using an AD1200 D/A converter and a DM2504 successive approximation register. Both devices are available from National Semiconductor.

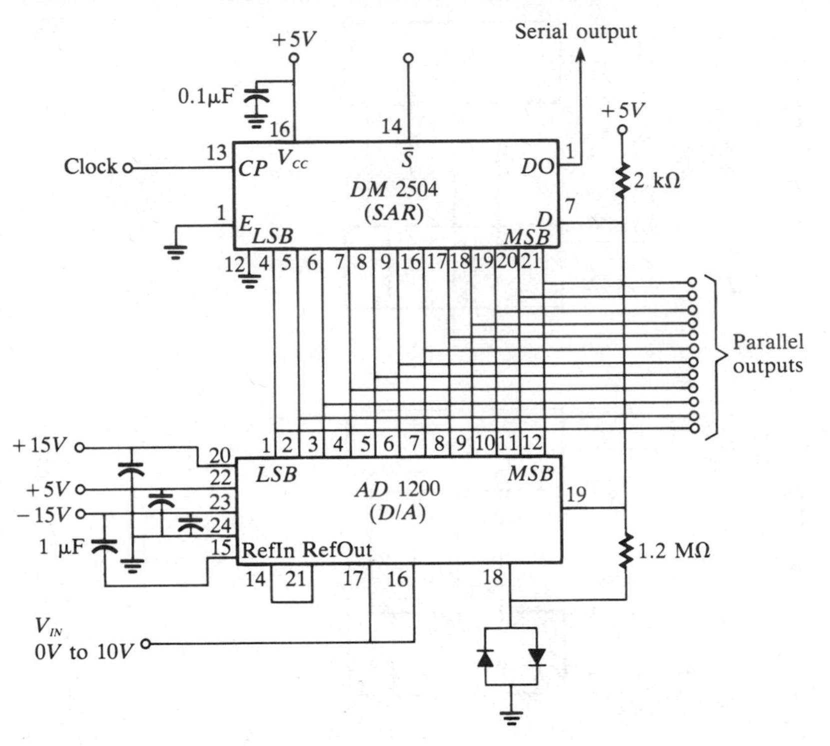

FIGURE 10–33 *A twelve-bit successive approximation A/D converter.*

The AD1200 is a 12-bit D/A device containing a ladder network, a reference circuit, and a comparator in one IC package. The DM2504 SAR contains all the control logic necessary to perform A/D conversion when used with a D/A converter as illustrated in Figure 10–33.

10–6 SERIAL AND PARALLEL DATA

Data in binary form can be transferred from one location to another within a digital system either by one of two basic methods, serial or parallel, or by a combination of both. These two methods are based on the relationship of the bits as they are being removed from place to place.

Serial Data

Serial data means that the bits follow one another so that only one bit at a time is transferred on a single line, as illustrated in Figure 10–34.

The rate at which the bits are transferred from *A* to *B* is called the *baud rate* and is expressed in bits per second. For instance if 300 bits are transferred from *A* to *B* in one second, the rate is 300 baud.

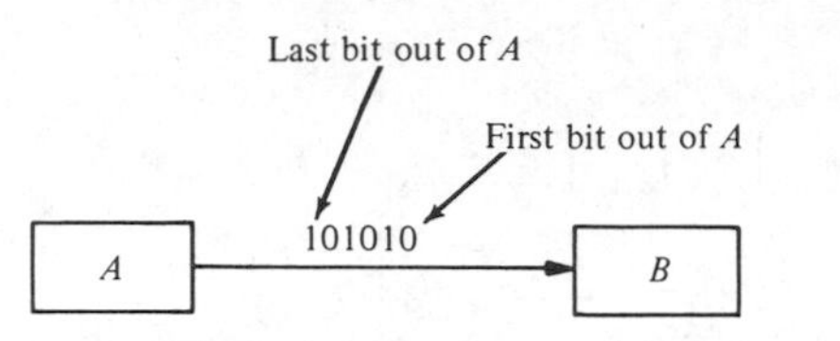

FIGURE 10–34 *Serial transfer of bits from A to B.*

Parallel Data

Parallel data means that all bits in a given group are transferred simultaneously on separate lines, as illustrated in Figure 10–35.

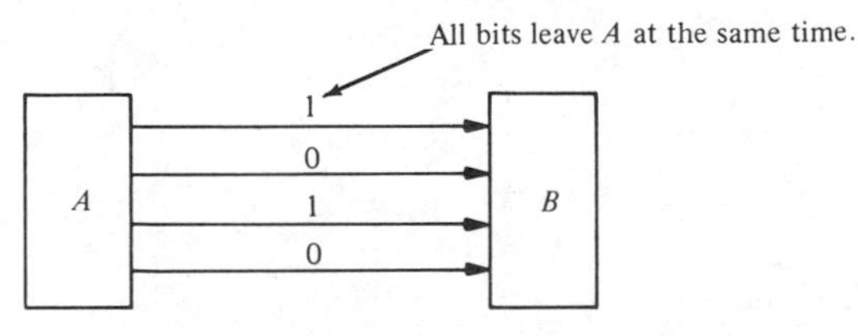

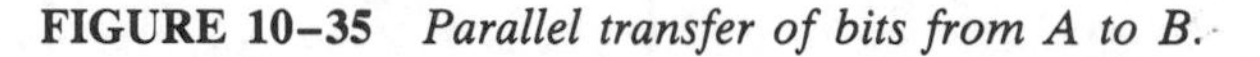

FIGURE 10–35 *Parallel transfer of bits from A to B.*

Synchronous and Asynchronous Waveforms

Two or more waveforms are *synchronous* if there is a definite time relationship between them. This fixed time relationship is usually established by a timing waveform called a *clock.*

The transitions of the synchronous waveforms occur only at the time of a clock pulse transition as shown in Figure 10–36(a). Notice that changes in waveforms *A* and *B* occur only on the positive-going edges of the clock in in this case.

Two waveforms are *asynchronous* if there is no fixed time relationship between them, as shown in Figure 10–36(b). Notice that the transitions of waveform *A* do not occur at any definite time with respect to the clock.

10–7 UNIVERSAL ASYNCHRONOUS RECEIVER TRANSMITTER (UART)

Microprocessors send and receive data as *parallel* bits on a *data bus* as discussed earlier. Often microprocessor based systems must communicate with external devices that send and/or receive *serial* data. In situations such as this, it is necessary to convert parallel data into serial data and vice versa. An interfacing device used to

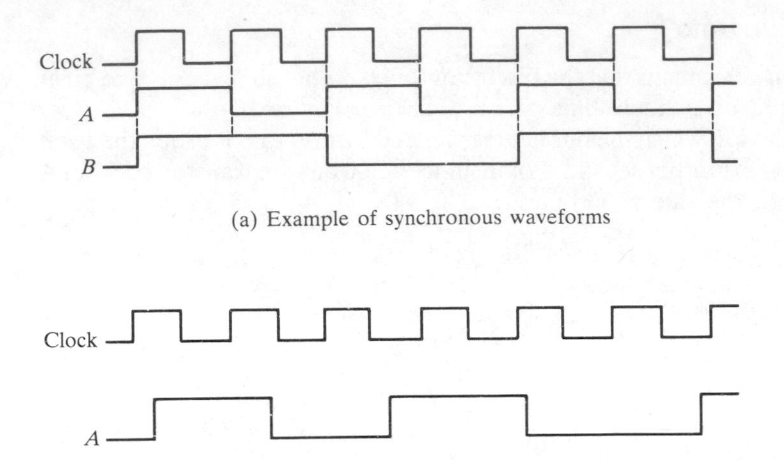

FIGURE 10–36 *Waveform relationships.*

accomplish these conversions is the UART. Figure 10–37 illustrates the UART in a microprocesser based system.

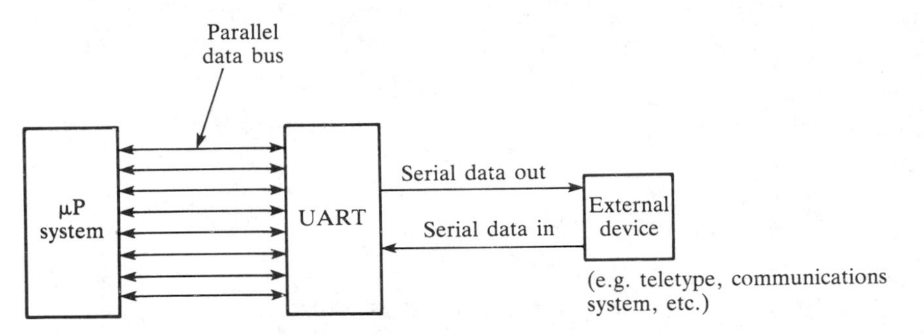

FIGURE 10–37 *UART interface.*

The UART *receives* data in serial form, converts it to parallel form, and places it on the data bus. The UART also accepts parallel data from the data bus, converts it to serial form, and transmits it to an external device. Typical examples of devices that operate with serial data are teletypes and certain communications systems. Figure 10–38 illustrates these basic conversions for an eight-bit data bus.

The basic elements of a UART are shown in Figure 10–39. The *receiver shift register* takes incoming serial data, converts them to parallel form, and stores them in the *receiver data register.* An internal bus connects the receiver data register to the data bus buffers. These are normally bidirectional three-state buffers that connect to the system data bus.

Parallel data from the system data bus go through the buffers and are stored in the *transmitter data register.* The data are then converted to serial form by the *transmitter shift register* and sent out on the line.

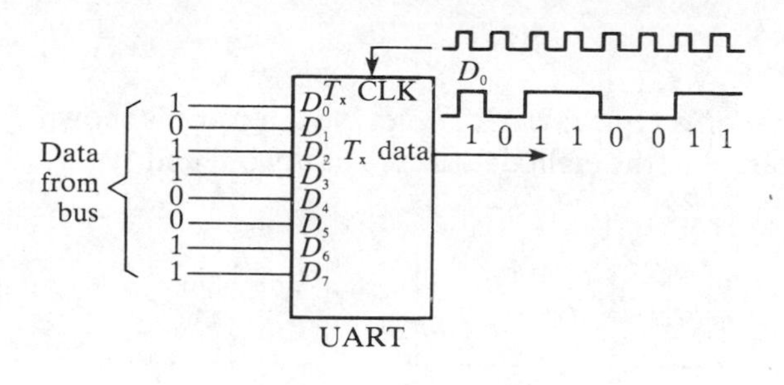

(a) Parallel-to-serial conversion (transmitting)

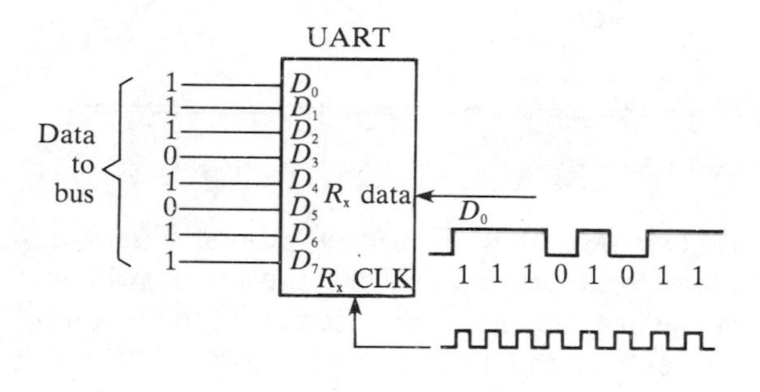

(b) Serial-to-parallel conversion (receiving)

FIGURE 10–38 *Basic operation of a UART.*

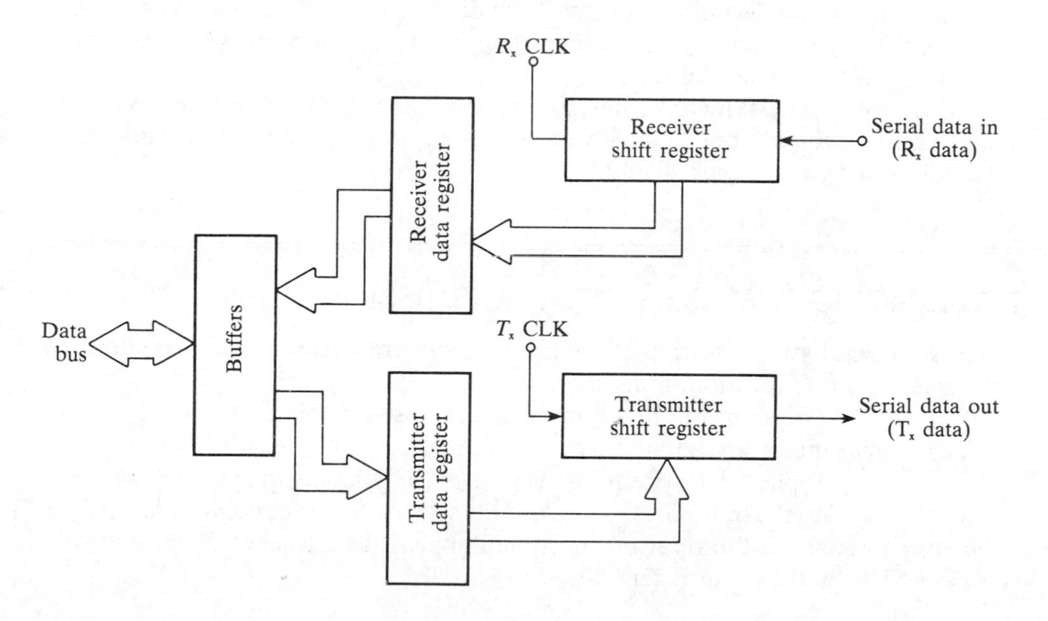

FIGURE 10–39 *Basic UART block diagram.*

Data Format

The serial data format on a typical UART/teletype interface is shown in Figure 10–40. It consists of a "start" bit, the eight data bits, a parity bit, and "stop" bits.

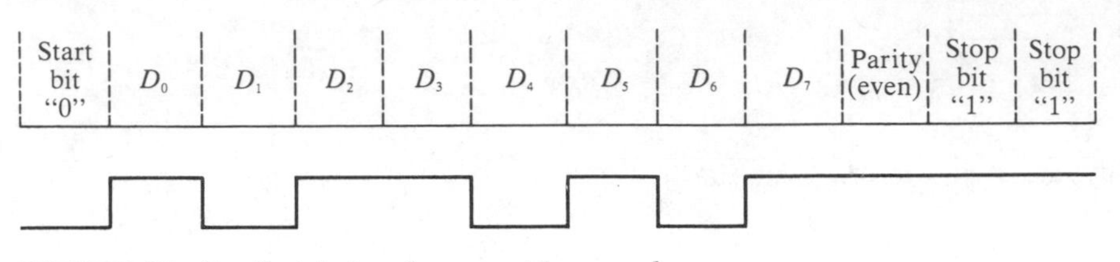

FIGURE 10–40 *Serial data format with example.*

10–8 THE SCHMITT TRIGGER AS AN INTERFACE CIRCUIT

The Schmitt trigger operates with two *threshold* or trigger points. When an increasing input signal reaches the *upper threshold point* (UTP), the circuit switches and the output goes to its HIGH level. When the input signal decreases to the *lower threshold point* (LTP), the circuit switches back and the output goes to its LOW level. This operation is illustrated in Figure 10–41(a).

As you can see, the Schmitt trigger circuit can be used to convert a sine wave into a pulse waveform. More importantly, however, in interfacing applications it is useful in reshaping pulses that have been distorted in transmission as depicted in Figure 10–41(b). Part (c) of the figure shows a simplified diagram of a data transmission system to illustrate this application.

Many standard gates are available in Schmitt trigger versions. The small *square loop* symbol on the inverter or gate symbol indicates a Schmitt trigger circuit as shown in Figure 10–42.

10–9 INTERFACING SYSTEMS WITH TRANSMISSION LINES

A data transmission line consists of two or more conductors that carry electrical signals from one location to another.

Transmission lines interconnect systems or portions of systems (subsystems) that are separated by a given physical distance.

Typically, a system that is transmitting data employs a *line driver* to feed the signal to the transmission line, and the system that is receiving data has a *line receiver* to enhance the received signal and apply it to the input of the receiving device. This is shown in Figure 10–43.

Types of Transmission Lines

Three of the most common types of transmission lines are the *coaxial cable* (coax), *shielded twisted pair,* and *ribbon cable.* These are shown in Figure 10–44.

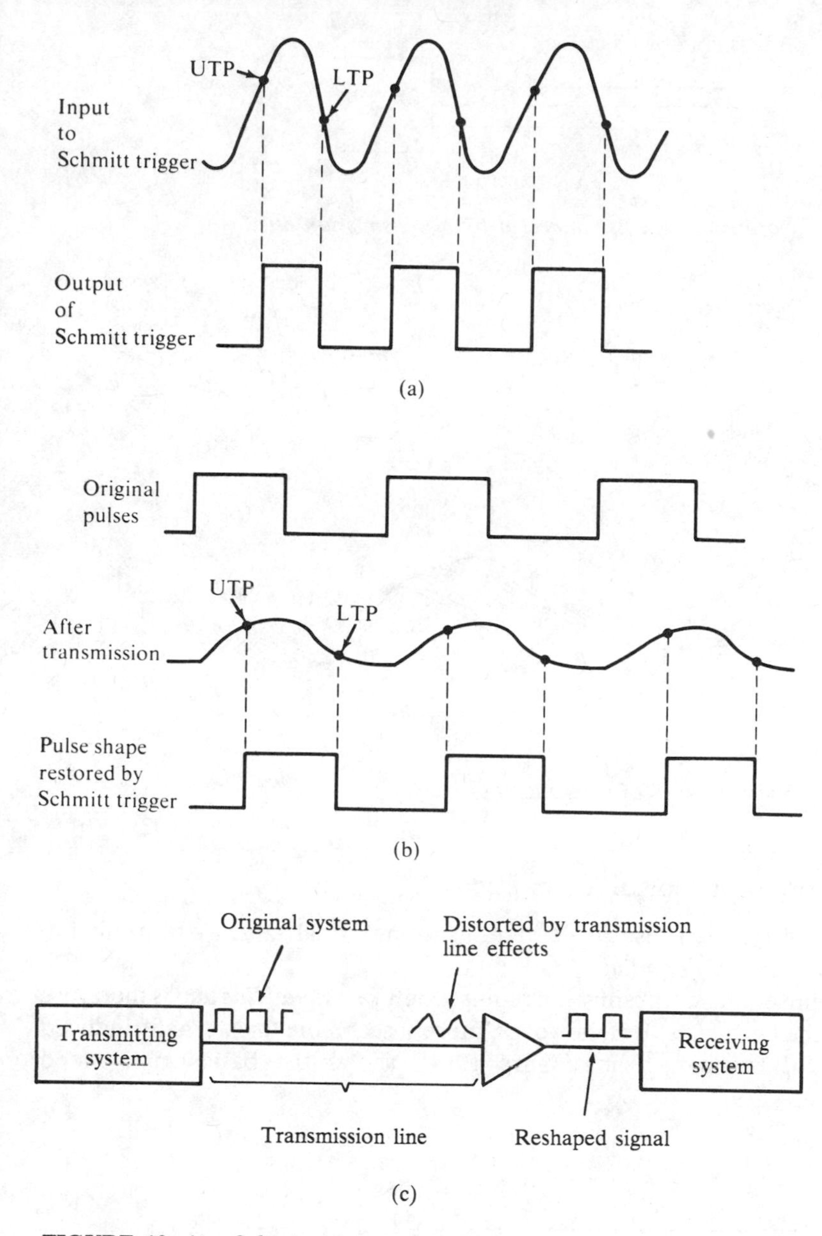

FIGURE 10–41 *Schmitt trigger operation.*

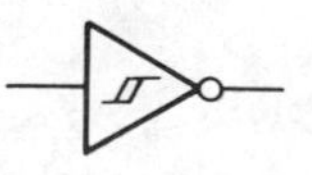

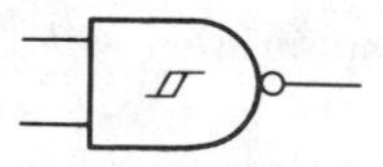

FIGURE 10–42 *Typical logic symbols for Schmitt trigger inverter and gate.*

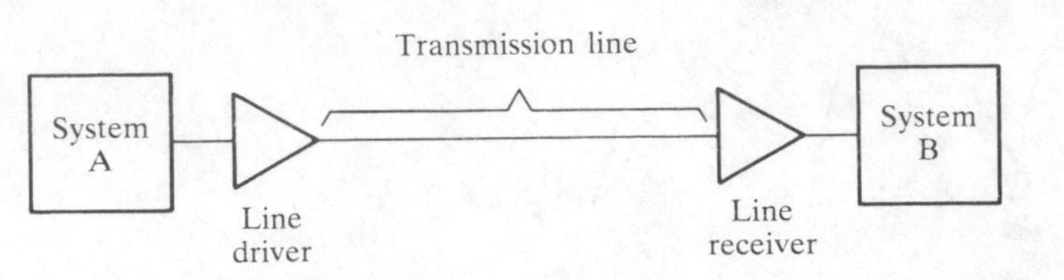

FIGURE 10–43 *Two digital systems connected by a transmission line.*

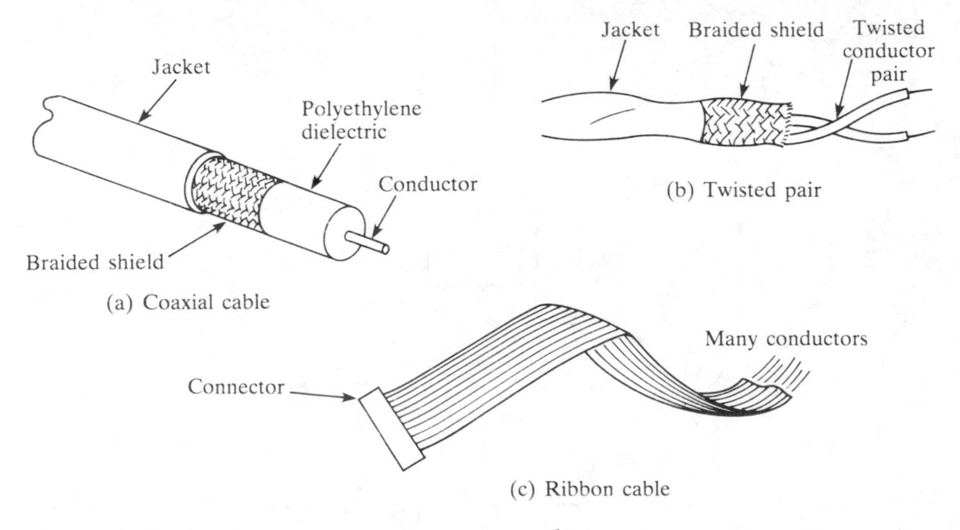

FIGURE 10–44 *Common types of transmission lines.*

Forms of Signal Transmission

There are two basic forms of transmitting digital data on a transmission line, *single-ended* and *differential.*

Single-ended transmission requires only *one signal line* and is therefore the simplest of the two forms. It is, however, more susceptable to externally induced noise than the differential form. Figure 10–45 shows the basic single-ended arrangement.

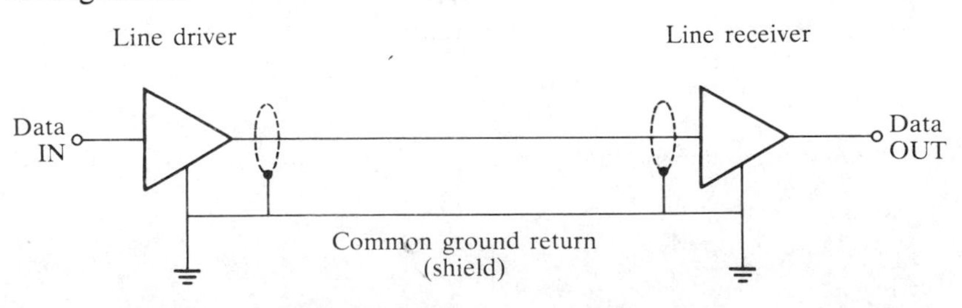

FIGURE 10–45 *Single-ended transmission with coax.*

Differential operation provides better noise immunity than does the single-ended form of operation. It uses a *differential driver* which is essentially two single-ended drivers with one producing an output that is the complement (inversion) of the other. Also, a differential line receiver and a *two-wire* transmission line is required. Figure 10–46 shows a differential arrangement.

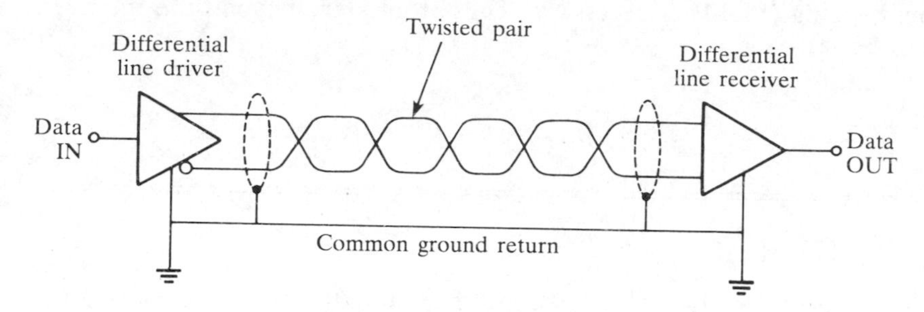

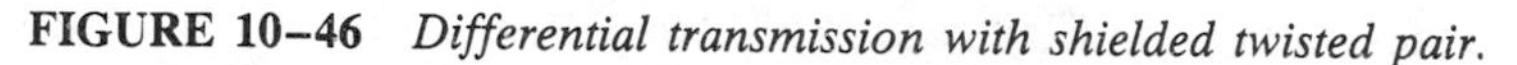
FIGURE 10–46 *Differential transmission with shielded twisted pair.*

Modes of Operation

There are three basic modes of data transmission: *simplex, half-duplex,* and *full-duplex.* Each of these modes can be employed in either single-ended or differential forms of operation.

In the *simplex* mode, data flow is one-way from line driver to line receiver, as illustrated in Figure 10–47.

The *half-duplex* mode is a nonsimultaneous two-way data flow. In this mode, there is a line driver/receiver pair (port) at each end of the transmission line, as shown in Figure 10–48. Data can flow either way from a line driver at one end to a line receiver at the other. The data flow cannot go both ways at the same time, however.

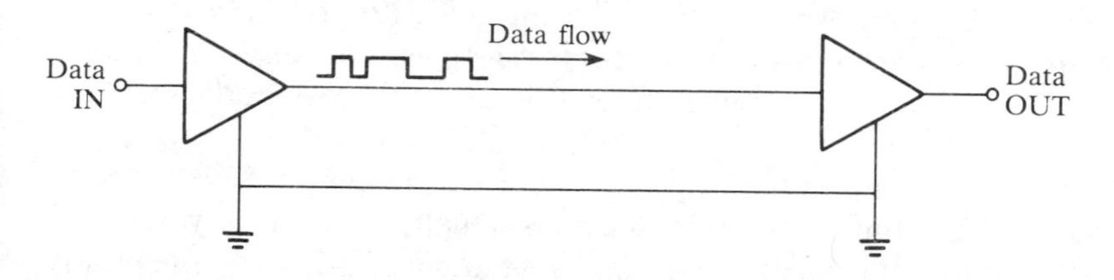

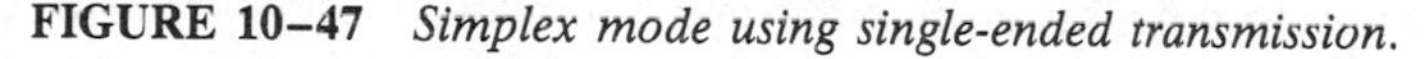
FIGURE 10–47 *Simplex mode using single-ended transmission.*

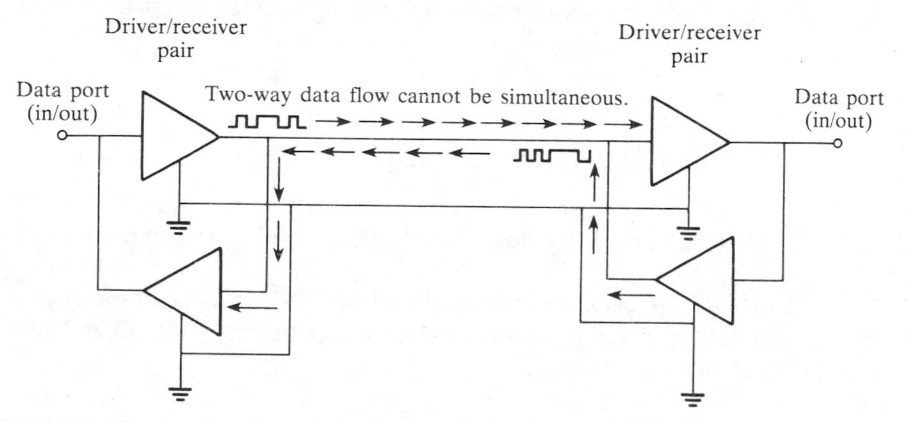

FIGURE 10–48 *Half-duplex mode using single-ended transmission.*

The *full-duplex* mode of operation is simultaneous two-way data flow. In this situation, signal separation by multiplexing techniques such as frequency division multiplexing (FDM) is necessary. These topics are beyond the intended scope of this book.

10–10 THE RS-232C INTERFACE

The most widely used interface specification for *serial* data communications is the RS–232C standard as defined by the Electronic Industries Association (EIA). This standard is used for interface between *data terminal equipment* and *data communication equipment*. Interfacing a computer and a peripheral device—printer, teletype, or CRT terminal—would be one application.

Basically, RS–232C specifies a 25-pin connector, and assigns serial signals to specific pins on the connector. Data rates up to 20 kbaud can be accomodated under this specification. Four types of data lines are defined: *data signals, control signals, timing signals,* and *grounds*.

A logic 1 data signal is defined as a voltage between $-3\ V$ and $-25\ V$. A logic 0 is a voltage between $+3V$ and $+25\ V$. Control signals are defined as *on* if they are between $+3\ V$ and $+25\ V$, and *off* if they lie between $-3\ V$ and $-25\ V$. There can be no voltages between $-3\ V$ and $+3\ V$. Special IC devices are available to translate from RS–232C levels to TTL or CMOS levels and vice versa.

Of the 25 signal lines defined by the RS–232C standard, two are grounds, four are data signals, twelve are control signals, and three are timing signals. Each signal has a particular nomenclature, abbreviation, and pin assignment.

In any given application, a piece of equipment does not have to use all signals provided for. The number of specified signals actually used will vary from one application to another. This standard, as well as the standard buses previously discussed, facilitates interfacing various equipment that conforms to the standards.

Table 10–2 provides a summary of the 25 signal lines that make up the RS–232C.

Limitations and Improvements

The maximum transmission rate permitted by the RS–232C is 20 kbaud, and the maximum permissible line length under the specification is about 15 meters.

The RS–422 is an enhancement of the RS–232C standard; it specifies a low impedance differential signal to increase the transmission rate to 10 Mbaud and the line length to about 1200 meters.

TABLE 10–2 *RS–232C signals.*

Pin Number	Signal Nomenclature	Signal Abbreviation	Signal Description	Category
1	AA	—	Protective ground	ground
2	BA	TXD	Transmitted data	data
3	BB	RXD	Received data	data
4	CA	RTS	Request to send	control
5	CB	CTS	Clear to send	control
6	CC	DSR	Data set ready	control
7	AB	—	Signal ground	ground
8	CF	DCD	Received line signal detector	control
9	—	—	—	reserved for test
10	—	—	—	reserved for test
11	—	—	—	unassigned
12	SCF	—	Secondary received line signal detector	control
13	SCB	—	Secondary clear to send	control
14	SBA	—	Secondary transmitted data	data
15	DB	—	Transmission signal element timing	timing
16	SBB	—	Secondary received data	data
17	DD	—	Receiver signal element timing	timing
18	—	—	—	unassigned
19	SCA	—	Secondary request to send	control
20	CD	DTR	Data terminal ready	control
21	CG	—	Signal quality detector	control
22	CE	—	Ring indicator	control
23	CH/CI	—	Data signal rate selector	control
24	DA	—	Transmit signal element timing	timing
25	—	—	—	unassigned

Problems

Section 10–1

10–1 Define the term *bus* and name three types of buses found in microcomputer systems.

10–2 Determine the signals on the bus line in Figure 10–49(**a**) for the input and enable waveforms in part (**b**).

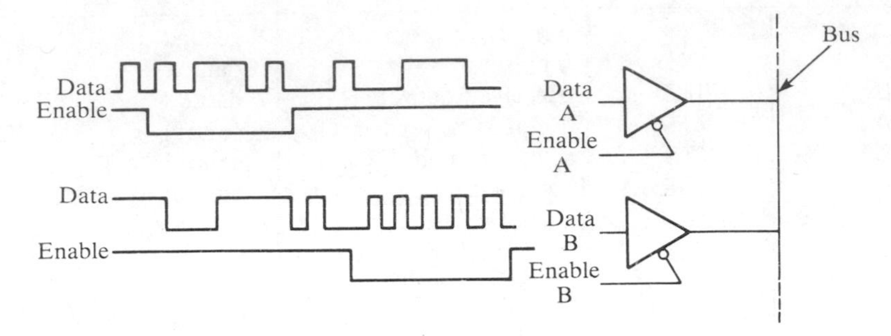

FIGURE 10–49

Section 10–2

10–3 (**a**) An S–100 bus physically consists of ______ pins.
(**b**) The S–100 bus standard is oriented around the ______ microprocessor.
(**c**) There are ______ address lines and ______ data I/O lines in the S–100.

10–4 List the three basic types of devices which can be connected to the GPIB.

10–5 Define the term *handshaking.*

Section 10–3

10–6 Distinguish between *analog* and *digital.*

Section 10–4

10–7 Determine the output of the D/A converter in Figure 10–50(**a**) if the sequence of four-bit numbers in part (**b**) is applied to the inputs.

10–8 Repeat Problem 10–7 for the inputs in Figure 10–51.

10–9 Determine the resolution for each of the following D/A converters.
(**a**) four-bit (**b**) six-bit (**c**) 18-bit

Section 10–5

10–10 Determine the binary output code of a three-bit simultaneous A/D converter for the analog input signal in Figure 10–52. The sampling rate is 100 kHz.

10–11 Repeat Problem 10–10 for the analog waveform in Figure 10–53.

10–12 For a three-bit stairstep ramp A/D converter, the reference voltage advances one step every microsecond. Determine the encoded binary sequence for the analog signal in Figure 10–54.

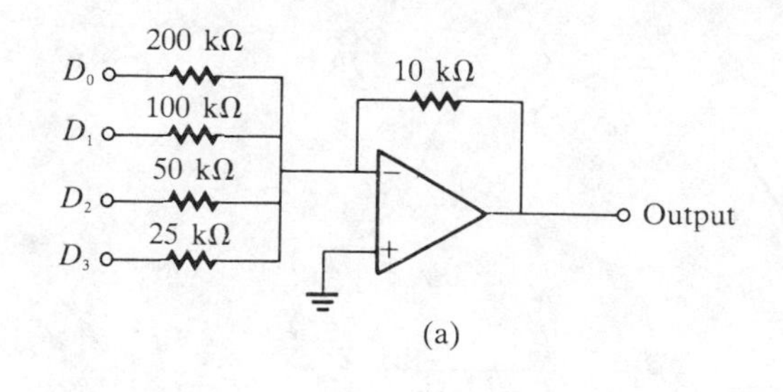

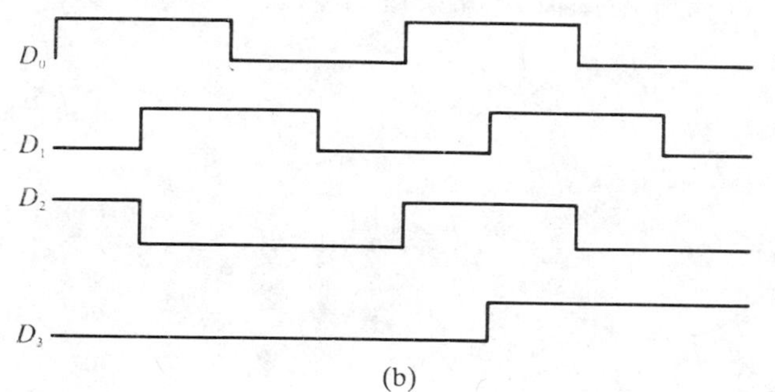

FIGURE 10–50

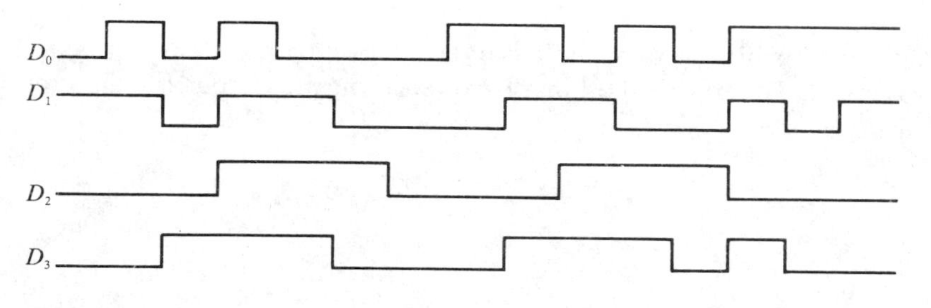

FIGURE 10–51

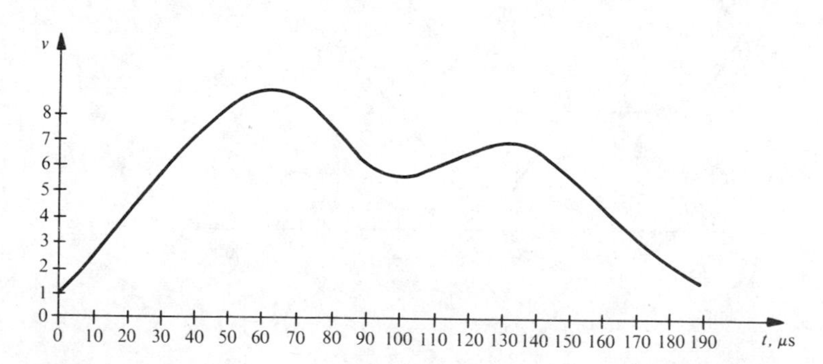

FIGURE 10–52

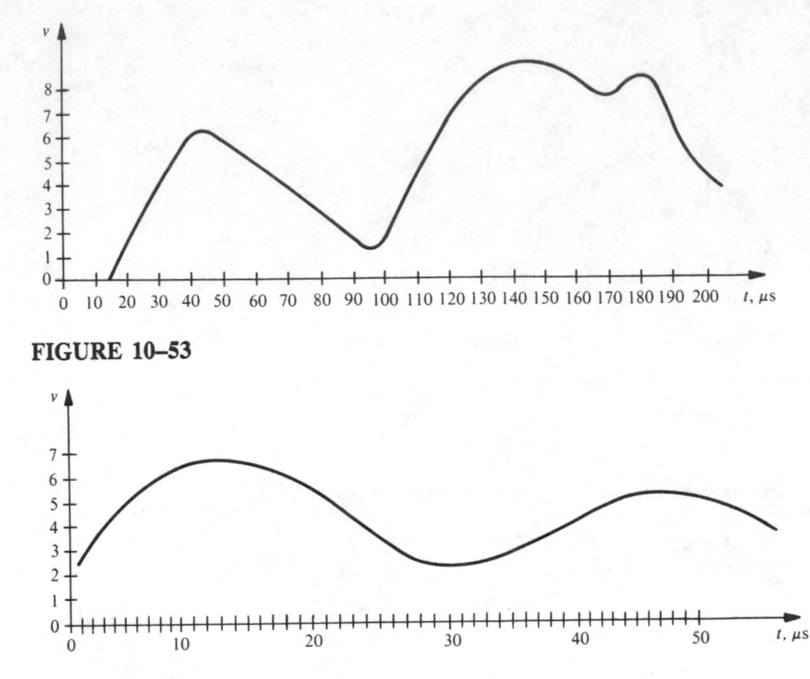

FIGURE 10–53

FIGURE 10–54

10-13 For a four-bit stairstep ramp A/D converter, assume the clock period is 1 microsecond. Determine the binary sequence on the output for the input signal in Figure 10–55.

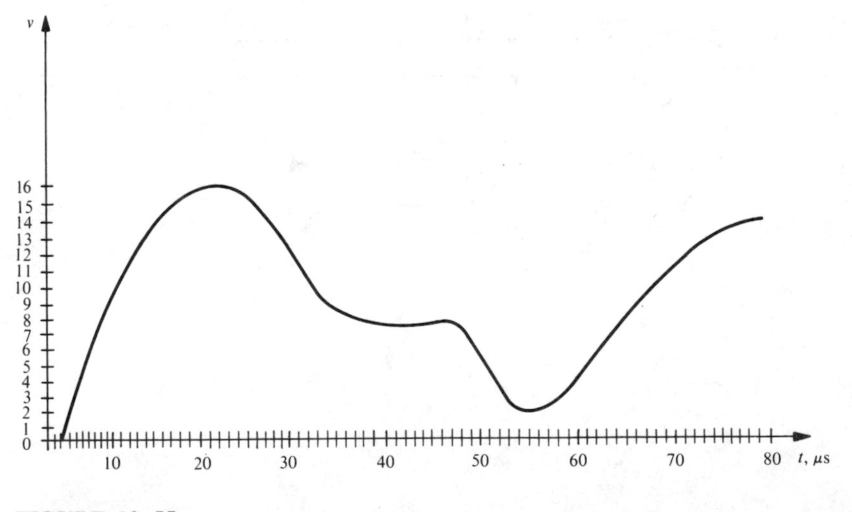

FIGURE 10–55

10–14 Repeat Problem 10–12 for a tracking A/D converter.

10–15 Repeat Problem 10–13 for a tracking A/D converter.

10–16 For a given four-bit successive approximation A/D converter, the maximum ladder output is +8 V. If a constant +6 V is applied to the analog input, determine the sequence of states for the *SA* register.

Section 10–8

10–17 For a Schmitt trigger circuit with a UTP = 1V and LTP = 0.5 V, sketch the output for the input signal in Figure 10–56.

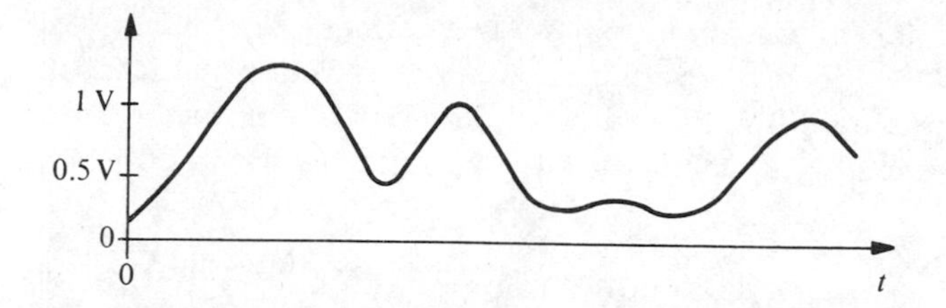

FIGURE 10–56

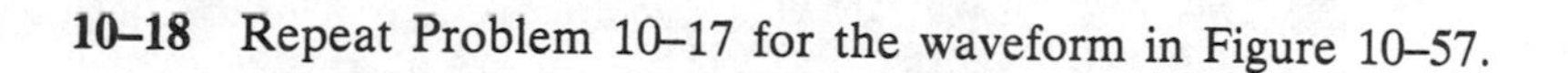

10–18 Repeat Problem 10–17 for the waveform in Figure 10–57.

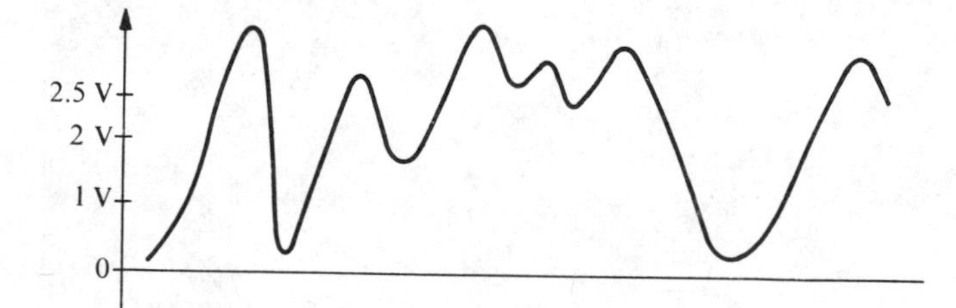

FIGURE 10–57

Section 10–9

10–19 Name the two basic forms of data transmission on a transmission line.

10–20 Describe the differences among the three modes of data transmission.

In Chapter 2 binary addition, subtraction, multiplication, and division were covered, and the 1's and 2's complements were introduced. Chapter 6 covered binary adders and explained how logic circuits are used to perform addition.

In this chapter, you will study arithmetic operations with signed binary numbers and a typical arithmetic logic unit (ALU). Other considerations of numbers in computing systems will also be introduced.

11 ARITHMETIC PROCESSES

11–1 BINARY REPRESENTATION OF SIGNED NUMBERS

The capability of handling both positive and negative numbers is a requirement of any arithmetic logic unit. A signed number consists of both *sign* and *magnitude* information. The sign indicates whether a number is positive or negative, and the magnitude is the value of the number.

In binary systems, the sign is represented by including an additional bit with the magnitude bits. Conventionally, a 0 represents a positive sign and a 1 represents a negative sign. For example, in an eight bit number the left most bit is the sign bit and the remaining seven bits are magnitude bits. This representation is as follows for decimal numbers +107 and −20:

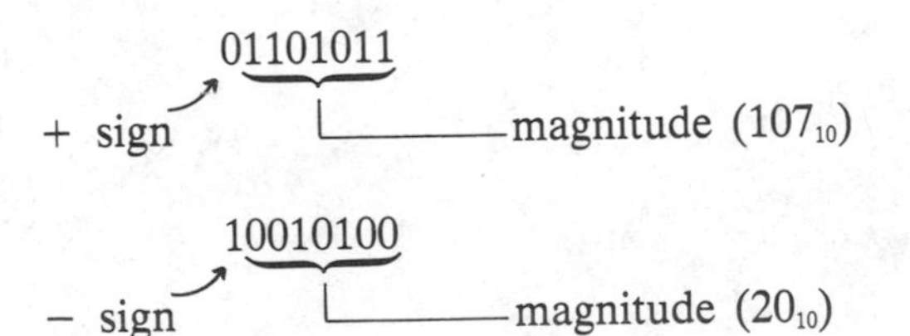

Eight bits of information (a *byte*) is very common in many microprocessor systems. With an eight-bit sign and magnitude number, a range of values from 01111111 = $+127_{10}$ to 11111111 = -127_{10} can be represented. Two or more bytes can be used to represent larger numbers. For example, 16 bits include one sign bit and 15 magnitude bits. With this format, a range of numbers from 0111111111111111 = +32,767 to 1111111111111111 = −32,767 can be represented.

1's Complement Form of Negative Numbers

Negative numbers can be represented in 1's complement form by *inverting all the magnitude bits and leaving the sign bit as is.* For example,

-53_{10} = 10110101 (true form) ⟶ 11001010 (1's complement form)

Example 11–1

Convert each negative binary number to 1's complement form.

(a) 11000100 (b) 10001001
(c) 11110110 (d) 11000011

Solution:

(a) 11000100 ⟶ 10111011
(b) 10001001 ⟶ 11110110
(c) 11110110 ⟶ 10001001
(d) 11000011 ⟶ 10111100

2's Complement Form of Negative Numbers

Most digital systems use 2's complement to represent negative numbers for arithmetic operations. One method of obtaining the 2's complement of a negative number is to add 1 to the 1's complement, as discussed in Chapter 2. Only the magnitude bits are complemented, the sign bit is left a 1.

```
11101010———→10010101 1's complement
                  +1
            10010110 2's complement
```

A second method of obtaining the 2's complement was also discussed in Chapter 2. Again, only the magnitude bits are affected. The procedure is as follows: starting with the right-most bit, the bits remain uncomplemented up to and including the first 1. The remaining magnitude bits are inverted.

```
    ┌——Complemented
    |     ┌—uncomplemented
10110110  true form
11001010  2's complement
```

Example 11–2

Convert each of the following negative binary numbers to 2's complement form using both of the methods just discussed.

(a) 10010010 (b) 10110000
(c) 11110111 (d) 10000000

Solution:

(a) Taking the 1's complement and adding 1:

```
10010010 ———→11101101   1's complement
                  + 1
             11101110   2's complement
```

Using the second method:

```
               10010010 ———→11101110
complemented—┘        └—uncomplemented
```

(b) Taking the 1's complement and adding 1:

```
10110000 ———→11001111   1's complement
                  + 1
             11010000   2's complement
```

Using the second method:

```
               10110000 ———→11010000
complemented—┘        └—uncomplemented
```

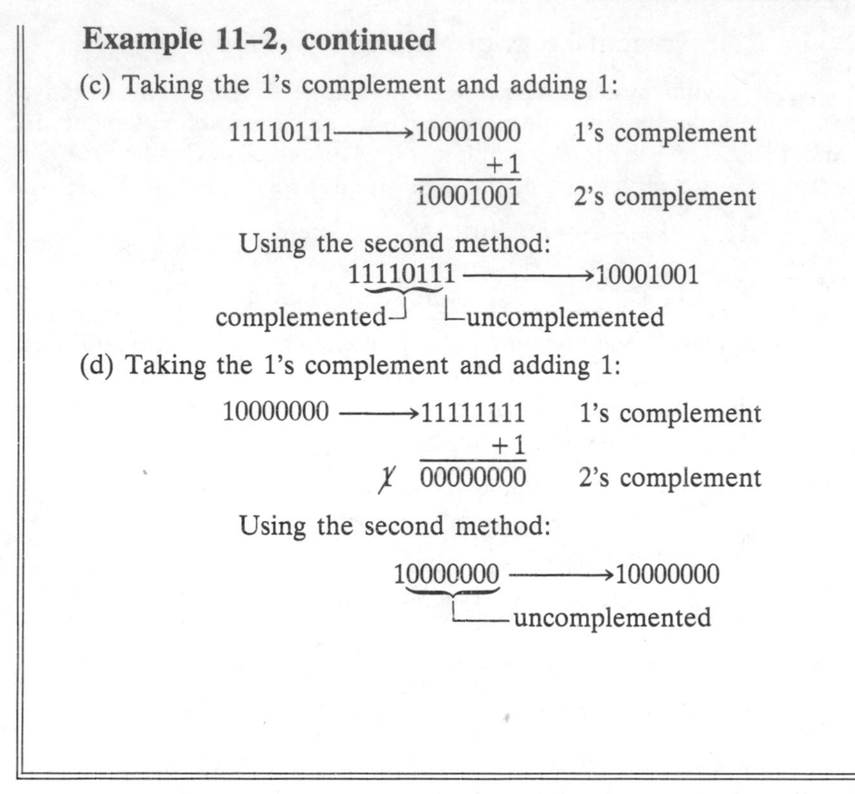

11–2 BINARY ADDITION

In this section you will see how signed binary numbers are added. The 2's complement will be used to represent negative numbers.

There are four cases that must be considered when adding two numbers: both numbers positive, positive number and smaller negative number, positive number and larger negative number, and both numbers negative. We will take one case at a time. Eight bits are used to represent each number.

Both Numbers Positive

In this case both sign bits are zero and a 2's complement is not required. To illustrate we will add +7 and +4:

7	00000111
+4	00000100
11	00001011

Positive Number and Smaller Negative Number

In this case the true binary form of the positive number is added to the 2's complement of the negative number. The sign bits are included in the addition. The result will be positive. To illustrate we will add +15 and −6:

```
     15            00001111
  +  −6            11111010
     9          1̸  00001001
                └─discard carry
```

Notice that the sign of the sum is positive (0) as it should be.

Positive Number and Larger Negative Number

Again, the true binary form of the positive number is added to the 2's complement of the negative number. The sign bits are included in the addition. The result will be negative. To illustrate we will add +16 and −24:

```
     16       00010000
  + −24       11101000
     −8       11111000     2's complement of −8
```

Notice that the result automatically comes out in 2's complement because it is a negative number.

Both Numbers Negative

In this case, the 2's complements of both numbers are added and, of course, the sum is a negative number in 2's complement form. To illustrate we will add −5 and −9:

```
     −5          11111011
  +  −9          11110111
    −14       1̸  11110010     2's complement of −14
              └─discard carry
```

Overflow

When the number of bits in the sum exceeds the number of bits in each of the numbers added, it is called *overflow* and is illustrated by the following example:

Example 11–3

```
  Decimal         Binary
     +9            01001
  +  +8          + 01000
    +17            10001
sign incorrect ─────┘└──── magnitude incorrect
```

The overflow condition can occur *only* when both numbers are positive or both numbers are negative. *It is indicated by an incorrect sign bit.*

Summary of signed addition

The following is a summary of the four cases of signed binary addition.

Both numbers positive

1. Add both numbers in *true* form, including the sign bit.
2. The sign bit of the sum will be 0 (+).
3. Overflow is possible. The sign bit and the magnitude of the sum will be incorrect.

Both numbers negative

1. Take the 1's or 2's complement of the magnitude of both numbers. Leave the sign bits as they are.
2. Add the numbers in their complement form, including sign bits.
3. Add the end-around carry in the case of the 1's complement method; drop the carry in the case of the 2's complement method. The sign bit of the sum will be 1 (−), and the magnitude of the sum will be in complement form.
4. Overflow is possible. The sign bit and magnitude of sum will be incorrect.

Larger number positive, smaller number negative

1. Take the 1's or 2's complement of the magnitude of the negative number. Leave the sign bit as is. A positive number remains in true form.
2. Add the numbers, including the sign bits.
3. Add the end-around carry for the 1's complement method; drop the carry for the 2's complement method. The sign bit of the sum will be a 0 (+), and the magnitude will be in true form.
4. No overflow is possible.

Larger number negative, smaller number positive.

1. Take the 1's or 2's complement of the magnitude of the negative number. Leave the sign bit as is. A positive number remains in true form.
2. Add the numbers, including the sign bits.
3. No carries will occur. The sum will have the proper sign bit, and the magnitude will be in complement form.
4. No overflow is possible.

Now that we have examined the basic arithmetic processes by which *two* numbers can be added, let us look at the addition of a string of numbers added two at a time. This can be accomplished by adding the first two numbers, then adding the third number to the sum of the first two, then adding the fourth number to this result, and so on. The addition of several numbers taken two at a time is illustrated in the following example, which is the basic way adders operate in microprocessor systems.

Example 11–4

Add the numbers 2, 4, 6, 5, 8, and 9.

Solution:

$$\begin{array}{rl} 2 & \\ +\ 4 & \\ \hline 6 & \leftarrow \text{first sum} \\ +\ 6 & \\ \hline 12 & \leftarrow \text{second sum} \\ +\ 5 & \\ \hline 17 & \leftarrow \text{third sum} \\ +\ 8 & \\ \hline 25 & \leftarrow \text{fourth sum} \\ +\ 9 & \\ \hline 34 & \leftarrow \text{final sum} \end{array}$$

11–3 BINARY SUBTRACTION

Subtraction is a special case of addition. For example, subtracting +6 from +9 is equivalent to *adding* −6 to +9.

Basically the subtraction operation *changes the sign of the subtrahend and adds it to the minuend.* The 2's complement method can be used in subtraction so that all operations require only addition. The four cases that were discussed in relation to the addition of signed numbers apply to the subtraction process because subtraction can be essentially reduced to an addition process. An example will illustrate.

Example 11–5

Perform each subtraction in binary for each pair of decimal numbers using the 2's complement method.

(a) 8 − 3 (b) 12 − (−9) (c) −25 − 8
(d) −58 − (−32) (e) 17 − 28

Solutions:

(a)
$$\begin{array}{r} 8 \\ -\ 3 \\ \hline +5 \end{array} \equiv \begin{array}{r} 8 \\ +\ -3 \\ \hline +5 \end{array} \qquad \begin{array}{rl} 00001000 & \\ +11111101 & \text{2's complement of } -3 \\ \hline \not{1}\,00000101 & +5 \end{array}$$

Example 11–5, continued

(b)	12	≡	12	00001100	
	$- -9$		$+ \quad 9$	$+00001001$	
	$+21$		$+21$	00010101	$+21$
(c)	-25	≡	-25	11100111	2's complement of -25
	$- \quad 8$		$+-8$	$+11111000$	2's complement of -8
	-33		-33	11011111	2's complement of -33
(d)	-58	≡	-58	11000110	2's complement of -58
	$- -32$		$+ \quad 32$	$+00100000$	
	-26		-26	11100110	2's complement of -26
(e)	17	≡	17	00010001	
	$- \quad 28$		$+-28$	$+11100100$	2's complement of -28
	-11		-11	11110101	2's complement of -11

11–4 BINARY MULTIPLICATION

In binary multiplication the two numbers are the *multiplicand* and the *multiplier* and the result is the *product*. The magnitudes of the numbers must be in true form and the sign bits are not used during the multiplication process.

The sign of the product depends on the signs of the two numbers that are being multiplied. If the two numbers have the same sign, either both positive or both negative, the product is positive. This is illustrated with decimal numbers as follows:

$$(+3)(+5) = +15 \qquad (-4)(-5) = +20$$
$$(-10)(+3) = -30 \qquad (+6)(-8) = -48$$

In the multiplication process, the sign bits of the two numbers are checked and the resulting sign of the product is stored prior to the actual multiplication. Once the product sign is determined, the multiplicand and multiplier sign bits are discarded leaving only the magnitudes to be multiplied.

The Multiplication Process

Multiplication of binary numbers in arithmetic units is often accomplished by a series of additions and shifts similar to the way you multiply in long hand. So, as in subtraction, only an adder is required. The following example illustrates binary multiplication of two unsigned seven-bit numbers.

Example 11–6

Multiply 1010011 (83_{10}) and 0111011 (59_{10}). The seven bits represent the magnitudes of the numbers and they are given in true form. The signed bits are not included here.

Solution:

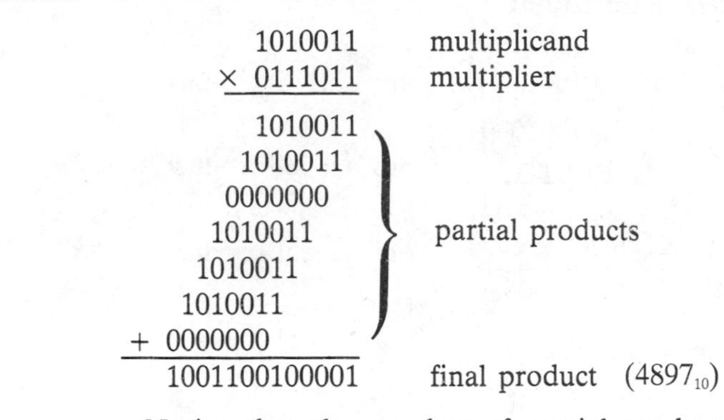

Notice that the number of partial products equals the number of multiplier bits. Also, each partial product is shifted one bit to the left with respect to the previous partial product. The final product, which is the sum of the shifted partial products, can have up to twice the number of bits of the original numbers.

Most digital arithmetic systems can add only two numbers at a time. Because of this, each partial product is added to the sum of the previous partial products and the sum of the partial products is accumulated. For example, the first partial product and the second partial product are added; then the third partial product is added to this sum, and so on.

Now we will go through the previous multiplication example as a typical arithmetic unit would handle it.

Example 11–7

Repeat the multiplication in Example 11–6 except this time keep a running total of the partial product sum so that only two numbers will have to be added at a time.

Solution:

```
     1010011    multiplicand
   × 0111011    multiplier
   ---------
     1010011    1st partial product
  + 1010011     2nd partial product
  ----------
    11111001    sum of 1st and 2nd
 + 0000000      3rd partial product
 -----------
```

Example 11–7, continued

```
      011111001     sum
   +  1010011       4th partial product
     1110010001     sum
  +  1010011        5th partial product
   100011000001     sum
 +  1010011         6th partial product
  1001100100001     sum
+  0000000          7th partial product
  1001100100001     final product
```

The multiplier and multiplicand must be in true form when multiplied. In a 2's complement system, any negative number is already in 2's complement form and must be converted to its true form before multiplying. A negative number is converted back to its true form by taking the 2's complement of the 2's complement. For example, let's convert a signed binary number from its true form to 2's complement and then back to the true form:

10101110	⟶	11010010	⟶	10101110
true		2's comp		true

To summarize the multiplication process in a typical 2's complement arithmetic unit the following steps are listed:

1. Obtain the multiplicand and multiplier.
2. If either number or both are negative convert to true form.
3. Examine both sign bits and store the sign of the product. (If both signs are the same, product is positive; if signs are different, product is negative.)
4. Examine each multiplier magnitude bit beginning with the right-most bit. When the multiplier bit is a 1, shift the partial product sum one place to right and add the multiplicand magnitude bits. This is, in effect, the same as shifting each new partial product one place to the left as in the previous example.
   If the multiplier bit is a 0, the partial product sum is shifted one place to the right and no addition is performed. This is effectively the same as adding an all zeros partial product.
5. Attach the predetermined sign bit to the final product. If the sign is negative, put product into 2's complement form.

An example will illustrate the steps in this process.

Example 11–8

Multiply the following two signed binary numbers ($+107_{10}$ and -51_{10}). The negative number is in 2's complement form.

01101011 × 11001101

Solution:

Step 1: Put the negative number in true form:

11001101 ⟶ 10110011
2's complement — true number

Step 2: Determine the sign of the product:
Because the signs are different, the product will have a negative-sign bit (1).

Step 4: Multiply the magnitudes.

	1101011	multiplicand magnitude (107_{10})
	0110011	multiplier magnitude (51_{10})
1st multiplier bit = 1	1101011	1st partial product
	1101011	shift right
2nd multiplier bit = 1	+1101011	2nd partial product
	101000001	sum
	101000001	shift right
3rd multiplier bit = 0	101000001	shift right (3rd part. prod. = 0)
4th multiplier bit = 0	101000001	shift right (4th part. prod. = 0)
5th multiplier bit = 1	+ 1101011	5th partial product
	11111110001	sum
	11111110001	shift right
6th multiplier bit = 1	+ 1101011	6th partial product
	1010101010001	final product (7th part. prod. = 0)

Step 5: Attach sign bit and take 2's complement of final product because it is negative.

$1101010101111_2 = 5457_{10}$

Multiplying by Repeated Addition

Another way to multiply two numbers is to simply add the multiplicand a number of times equal to the value of the multiplier.

For example, to multiply 4 and 8, we can add 8 four times as follows:

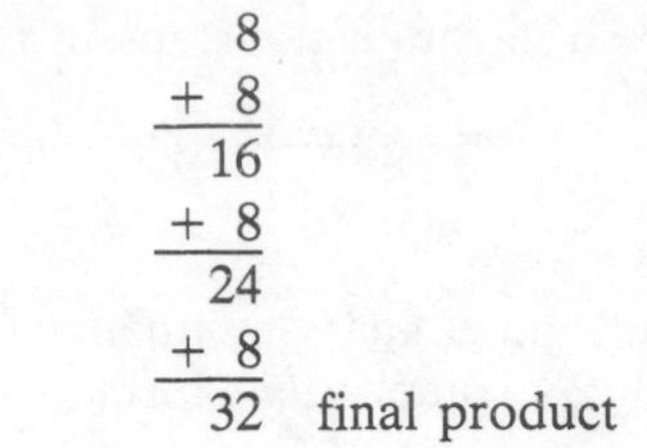

This can be done with binary numbers in digital systems as follows:

1. First, both magnitudes are put into true form and the sign of the product is stored as previously described.
2. Bring down the multiplicand and decrement the multiplier. (Subtract 1 from the multiplier by adding the 2's complement of 1.)
3. Check the multiplier for zero value. If it is zero, then the multiplication is complete. If the multiplier is not zero, add the multiplicand to the previous multiplicand sum.
4. This process continues until the multiplier equals zero. At that point, the multiplicand has been added a number of times equal to the value of the multiplicand. The final product is available and the sign is attached.

Example 11–9

Multiply 1011010 (90_{10}) and 0000011 (3_{10}). Both numbers are true magnitudes. The smaller number is the multiplier.

Solution:

Step 1: Bring down the multiplicand and decrement the multiplier:

multiplicand = 1011010

```
   0000011 = 3
 + 1111111 = 2's complement of 1
 ---------
 ̶10000010 = 2
```

Step 2: Check multiplier for zero value:

multiplier = 0000010 = 2

Step 3: Add the multiplicand to itself:

```
   1011010
 + 1011010
 ---------
  10110100
```

Step 4: Decrement the multiplier:

```
   0000010 = 2
 + 1111111 = 2's complement of 1
 ---------
 ̶10000001 = 1
```

Step 5: Check multipler for zero value:

multiplier = 0000001 = 1

Step 6: Add the multiplicand to the previous multiplicand sum:

```
  10110100
+  1011010
 100001110
```

Step 7: Decrement the multiplier:

```
  0000001 = 1
 +1111111 = 2's complement of 1
 1 0000000 = 0
```

Step 8: Check multiplier for zero value:
multiplier = 0000000 = 0

Step 9: The multiplication is complete and the final product is the last multiplicand sum:

100001110

This product is 270_{10} which is correct because the multiplicand is 90_{10} and the multiplier is 3_{10}.

BINARY DIVISION 11–5

In division, the two numbers are called the *dividend* and the *divisor*. The result is the *quotient*. The sign of the quotient is determined from the signs of the dividend and divisor. If both signs are the same, the quotient is positive. If the signs differ, the quotient is negative.

The Division Process

Division in arithmetic units can be accomplished by repeated subtractions. Since subtraction can be done with 2's complement addition, the division process requires only an adder. Keep in mind that the quotient is the number of times that the divisor goes into the dividend. The process is as follows.

1. Obtain the divisor and dividend. Initialize the quotient to zero.
2. Subtract the divisor from the dividend to get the partial remainder. If this remainder is positive, increment the quotient. (Add 1 to the quotient.) A positive partial remainder indicates that the divisor goes into the dividend (or partial remainder). If the partial remainder is negative, the divisor does not go into the dividend (or partial remainder) and therefore the division is complete.
3. If the partial remainder is positive, repeat step 2 by subtracting the divisor from the partial remainder.

An example will illustrate this process.

Example 11–10

Divide 01100100 by 00110010. The decimal equivalent of this division is $+100_{10}$ divided by $+50_{10}$.

Solution:

Step 1: Initialize the quotient to zero. Subtract the divisor from the dividend using 2's complement addition.

```
                01100100    dividend
              + 11001110    2's complement of divisor
                00110010    positive 1st partial remainder
+ sign ---------↑
```

Step 2: A positive partial remainder indicates that the divisor goes into the dividend and therefore the quotient is incremented by one. 00000001 is the new quotient.

Step 3: Subtract divisor from 1st partial remainder:

```
                00110010    1st partial remainder
              + 11001110    2's complement of divisor
                00000000    positive 2nd partial remainder
+ sign ---------↑
```

Step 4: Increment quotient to 00000010.

Step 5: Subtract divisor from 2nd partial remainder.

```
                00000000    2nd partial remainder
              + 11001110    2's complement of divisor
                11001110    negative remainder
- sign ---------↑
```

The negative remainder indicates that the divisor will not go into the 2nd partial remainder. Therefore the final quotient is $00000010 = 2_{10}$ and the division is complete.

In this particular example, the zero partial remainder indicated that the division was complete. However, a zero remainder does not always occur and therefore the check for a *negative* partial remainder is a more general test.

11–6 A SERIAL BINARY ADDER

The method of addition discussed here is termed *serial* because two numbers are added on a bit-by-bit basis, rather than simultaneously as in a parallel adder. Serial addition of two binary numbers is illustrated in Figure 11–1. The corresponding bits of each number are applied to the bit inputs of the full-adder, and the carry generated by the addition of the two previous bits is on the carry input. Every time a carry is generated, it must be delayed by one bit time so that it can be added to the next higher order bits when they come into the full-adder. During bit time 1, the A_0 and B_0 bits are added and a carry of 1 is generated. The carry bit is delayed (held) until bit time 2 and added to bits A_1 and B_1. A carry of 1 results from the addition, and is held until bit time 3, when it is added to bits A_2 and B_2; this addition results in a 0 carry. During bit time 4, A_3 and B_3 are added, completing the addition of the two four-bit numbers. This operation is illustrated in Figure 11–1.

Additional functions must be implemented in order to have a complete adder that can handle each of the four cases of addition (including subtraction). Based on the fundamental processes for sign and magnitude addition, let us itemize all of the functions that an adder circuit must perform:

1. Acquire the two numbers.
2. Check the sign bit of each number. If it is a 0, the magnitude of the number is left in true form; if it is a 1, the magnitude of the number must be complemented by the 2's complement method.
3. Determine whether addition or subtraction is to be performed. If subtraction is to be performed, the subtrahend must be complemented. If the subtrahend is negative then the complementation done in Step 2 is canceled and the negative subtrahend is actually left in true form.
4. Add the LSBs of the two numbers. Continue to add a bit at a time until all bits (including the sign bit) have been added. This is the basic serial addition as illustrated in Figure 11–1.
5. Drop the final carry in the case of 2's complement.
6. If both numbers are negative or the larger number is negative, the sum is in complement form. This must be taken into account when the next number is added to this sum—that is, the complement operation must be omitted in this case.
7. Check for the correct sign bit. If incorrect, overflow has occurred.

With these required operations in mind, let us examine the serial adder of Figure 11–2. Initially, one of the numbers to be added is shifted into *register A* and the other number into *register B*. They are held in these registers until we are ready to start the addition process. Before the numbers are added the *sign-bit comparison logic* determines if both of the numbers have the same sign. If they have the same sign, an overflow condition is possible, and the value (1 or 0) of the sign bit is stored so that it can be compared with the value of the sign bit of the sum after the numbers have been added. The logic also determines and remembers if both numbers are negative; also, if the numbers differ in sign, it activates the *magnitude comparison logic* which determines if the larger number is negative.

A negative sign bit (1) on either of the numbers activates the respective *2's complement logic*. A_S and B_S are the sign bits. The addition process starts by shifting the LSBs of each number out of the shift registers, through the 2's

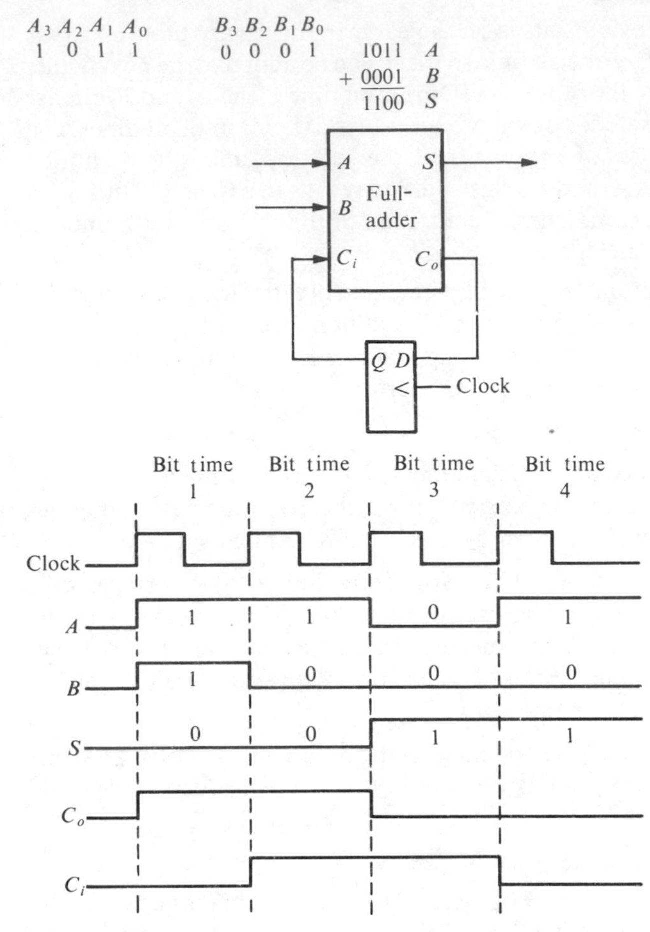

FIGURE 11–1 *Basic operation of a serial adder.*

complement logic, and onto the inputs of the full-adder. A sum bit is generated and fed back to the input of register A (sometimes called the *accumulator)*. The next clock pulse puts the first sum bit into the accumulator and shifts the next bits of the two numbers into the full-adder. The process continues until all of the bits in the two numbers have been added and the sum has been stored in the accumulator. When the addition is complete, the sign-bit comparison logic checks the sign bit of the sum against the stored sign bit of the two numbers. If they are not the same, an overflow indication is produced.

If a string of numbers is to be added, the third number is loaded into register B and will be added to the sum of the previous addition, which is now stored in the *accumulator*. Before the previous addition, the magnitude comparison logic determined if the larger of two differently signed numbers was negative, and the sign-bit comparison logic determined if both numbers were negative. In either case, the sum now stored in the accumulator is in complement form and negative. The 2's complement logic must therefore be disabled in this case, because the number is already complemented.

After the third number in the string has been loaded in register B, the entire operation is repeated and another sum is accumulated in register A. The next

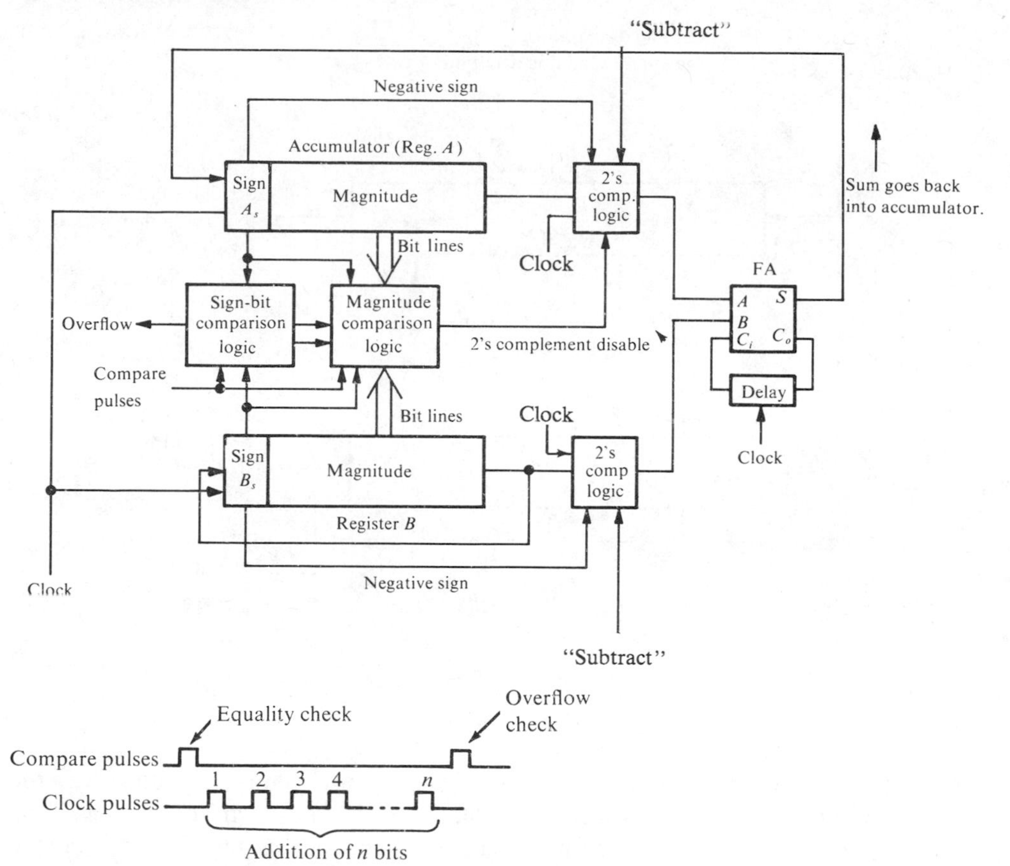

FIGURE 11–2 *Serial adder block diagram and basic timing.*

number in the string can now be loaded into register *B*, the entire addition cycle repeated, and so on.

We have discussed the general operation of a 2's complement adder at the logic block diagram level to get a basic understanding of what takes place. A 1's complement adder can be implemented, but an additional complication arises because another add cycle is required to handle the end-around carry. Now that we generally understand the operation of a serial adder, let us look at ways to implement the logic required for some of the blocks in the adder of Figure 11–2.

The Serial Adder Registers

Shift registers have already been discussed in Chapter 8, so only a brief coverage of the requirements for this particular application is necessary here. As we have seen, two shift registers are required in the serial adder. The accumulator register must have a serial input to accept the sum bits as they are fed back from the full-adder output. Each register must have a serial output so the numbers stored can be shifted out one bit at a time. New numbers can be loaded into the registers either in parallel or in series, depending on the particular system requirements. The capacity (number of stages) in each register must be equal to the number of bits in the magnitude of the

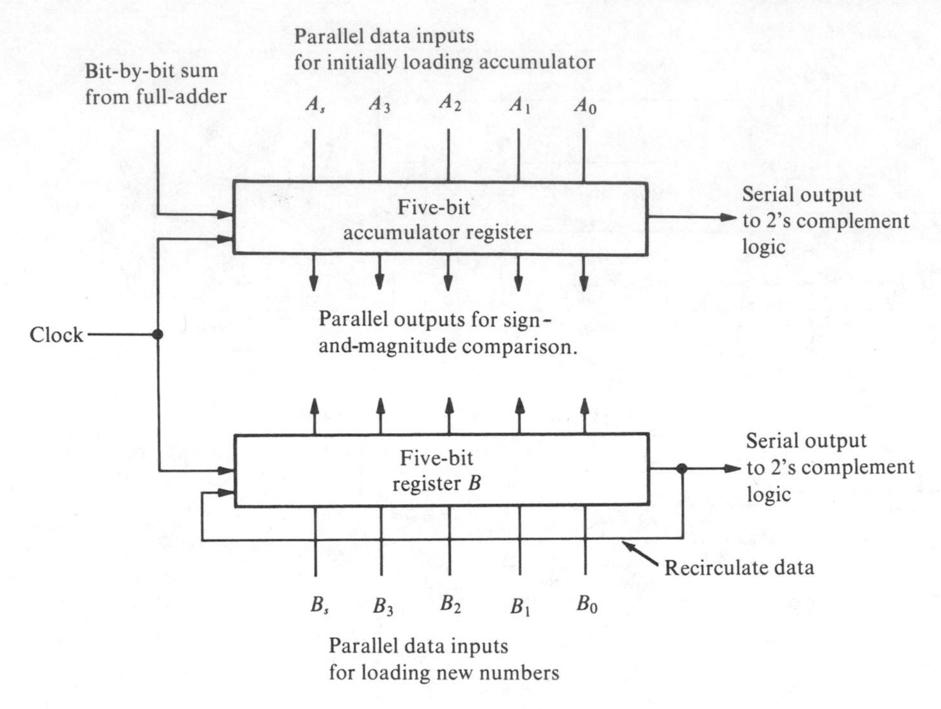

FIGURE 11–3 *Registers for a five-bit serial adder.*

numbers to be handled plus a stage to store the sign bit. Figure 11–3 illustrates the operation of the two registers with a five-bit capacity (four magnitude bits and one sign bit) and parallel loading. Register B can recirculate the output to the input so the number can be saved as it is shifted out.

Sign-Bit Comparison Logic

As we have seen, the purpose of this logic is to check the sign bits of the numbers in the registers, and if the sign bits are equal, to store the value of the sign for a possible overflow check at the end of the addition. If the sign of the sum is different from the stored sign, an overflow indication is generated. Also, if both numbers have a negative sign, the 2's complement logic for the accumulator must be disabled after the present addition has been completed and before the next addition starts. The other function of the logic is to enable the magnitude comparison logic if the two numbers differ in sign.

Figure 11–4 shows one way to implement this logic function. The operation is as follows. After the two numbers are stored in the shift registers, the sign bits are in the left-most position. The output of this stage of each register is connected to the input of an exclusive-OR gate (G_1), which performs the basic comparison of the two bits. The output of the exclusive-OR is inverted and is HIGH if the two sign bits are equal. The output of this exclusive-OR gate is stored in the flip-flop by the equality check pulse, and remains there until the end of the addition of the two numbers. When the addition is complete, the sign bit now in register A is the sign bit of the sum, and, since the content of register B was recirculated, the sign bit in this register is the sign bit of the original number. The sign of the sum should be the same as the sign of the number in register *B* if the flip-flop is SET; if they are not the same, then all inputs to the overflow check gate (G_3) are HIGH when the overflow check pulse occurs. If this condition exists, the output of the overflow check gate is HIGH, indicating an overflow condition—in which the sum is invalid. If the flip-flop was not originally SET or if the sign of the sum and the sign bit of the original numbers are the same, then no overflow indication occurs.

If the two sign bits are not equal, a HIGH appears on the exclusive-OR gate output, which is connected to the magnitude comparison logic via the sign inequality line. If the two sign bits are negative, a HIGH appears on the output of AND gate G_2, which is connected to the magnitude comparison logic via the negative signs line. The two signals are used by the magnitude comparison logic.

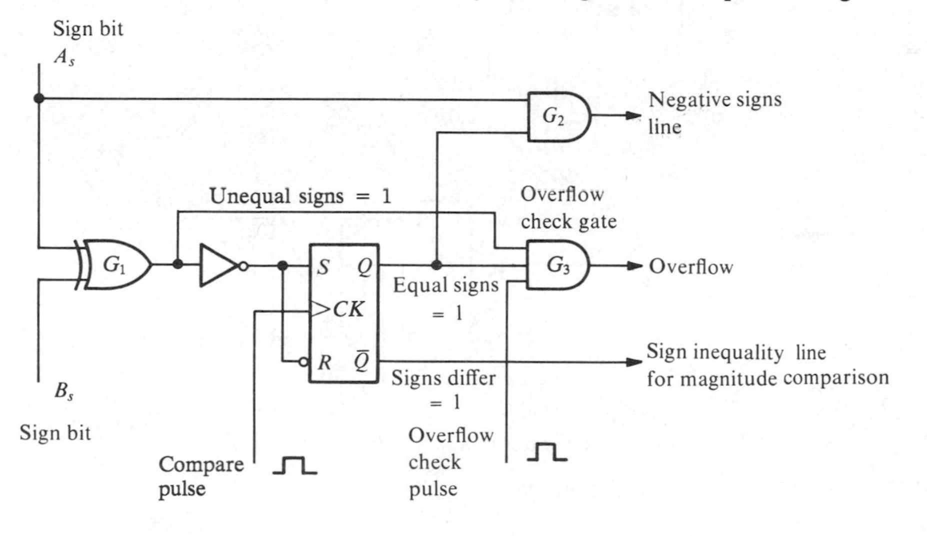

FIGURE 11–4 *Sign-bit comparison logic and overflow detection.*

Magnitude Comparison Logic

The purpose of the magnitude comparison logic is to determine if the sum resulting from an addition is in its 2's complement form as it is stored in the accumulator. As we have seen, the condition can be predicted if the two numbers to be added are both negative or if the larger number is negative. The magnitude comparison logic records the existence of either of these two conditions prior to the addition of the two numbers. After the addition of the two numbers and before the addition of the next number to the resulting sum, the magnitude comparison logic disables the 2's complement logic for register *A* because *A*'s content is negative and already in complement form. Figure 11–5 shows one way in which this function can be achieved. We are assuming for this particular case a four-bit magnitude, although the same principle can be extended to numbers of any size.

The four magnitude bits from each register are connected to the inputs of the magnitude comparator. (Comparators were discussed in Chapter 6.) The sign

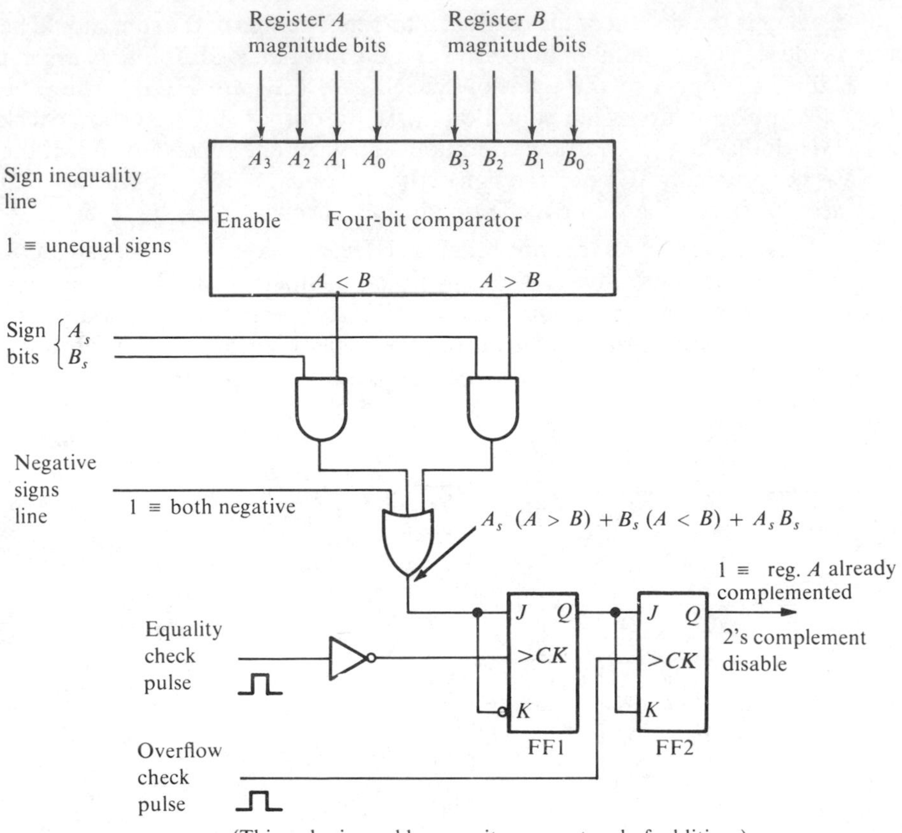

FIGURE 11–5 *Magnitude comparison logic.*

inequality line is connected to the enable input of the comparator, and if it is HIGH, a comparison of the magnitudes of the two numbers is performed. If the number in register A is larger than the number in register B, the $A > B$ output goes HIGH. If the number in register B is larger, the $A < B$ output goes HIGH. The following expression gives the conditions for the case of a larger negative number, where A_s and B_s are the sign bits:

$$A_s(A > B) + B_s(A < B)$$

This says that if $A_s = 1$ AND $(A > B) = 1$ OR if $B_s = 1$ AND $(A < B) = 1$, then the larger of the two numbers is negative. The function is implemented by the AND/OR logic as indicated in Figure 11–5. The negative signs line is also ORed with the above function, as shown. When the larger of the two numbers is negative or when both numbers are negative, the output of the OR gate is HIGH; this is stored in FF1 at the time of the equality check pulse prior to addition. The output of the flip-flop is transferred into FF2 at the end of the addition by the overflow check pulse. The output of FF2 is used to disable the 2's complement logic during the next addition.

2's Complement Logic

The 2's complement of a number can be achieved in a relatively easy fashion. Before we proceed with a description of the logic, let us review the basic process involved. It is interesting to note that if we take a binary number and (going from right to left) copy each bit up to and including the first 1 and then write the complement of each bit thereafter, we get the 2's complement of the binary number. This is illustrated by the following examples.

Example 11–11

Determine the 2's complement of each of the following binary numbers: 1011, 101100, 11011000.

Solutions:

```
1 0 1 1          true form
↓ ↓ ↓ ↓
0 1 0 1          2's complement
comp.  true

1 0 1 1 0 0      true form
↓ ↓ ↓ ↓ ↓ ↓
0 1 0 1 0 0      2's complement
comp.  true

1 1 0 1 1 0 0 0  true form
↓ ↓ ↓ ↓ ↓ ↓ ↓ ↓
0 0 1 0 1 0 0 0  2's complement
comp.     true
```

The logic required to perform this 2's complement operation is shown in Figure 11–6. Basically, the logic allows the bits to pass through to the output without being inverted until a 1 occurs; this first 1 is allowed to pass, but each bit thereafter is inverted.

The detailed operation for the particular implementation of Figure 11–6 is as follows. The data (binary numbers stored in the shift register *A or B* in Figure 11–2) are shifted serially out (LSB first) of the register by the clock pulses as indicated; these serial data are applied to AND gate G_1 in true form and to AND gate G_2 in complement form. The flip-flop is initially RESET, thus enabling gate G_1 and disabling gate G_2. The bits pass through gate G_1 and OR gate G_3 uncomplemented until after the first 1 occurs; one bit time after the first 1, the flip-flop SETS because its J input is HIGH and K is LOW. Now gate G_2 is enabled and gate G_1 disabled, allowing the complements of the remaining bits to pass through gates G_2 and G_3; this operation produces the 2's complement of the input number in serial form on the

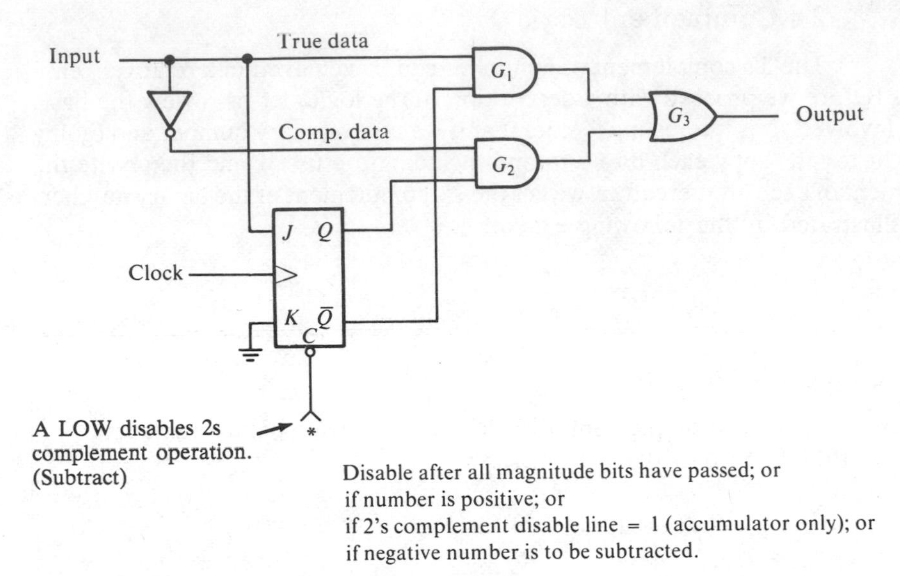

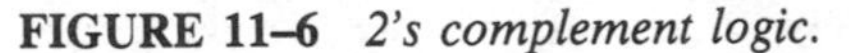
FIGURE 11–6 *2's complement logic.*

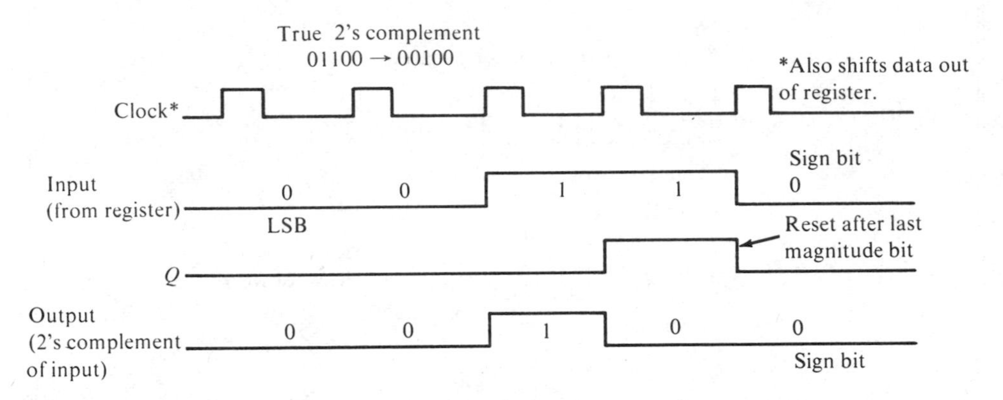

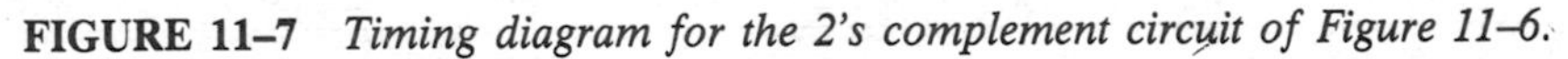
FIGURE 11–7 *Timing diagram for the 2's complement circuit of Figure 11–6.*

Example 11–12

Show the adder states on a bit-by-bit basis to illustrate the serial addition of the following two positive numbers:

$$\begin{array}{r} 01001 \\ +\ 00011 \\ \hline 01100 \end{array}$$

Solution:

For simplicity and clarity, the comparison logic is omitted. The fundamental operation in the addition process is shown in Figure 11–8.

---

output line. The timing diagram in Figure 11–7 illustrates a particular four-bit number being converted to its 2's complement. The complementation process must be discontinued after all of the magnitude bits have passed through and before the sign bit comes in, because we do not want to complement the sign bit; this can be done by providing a RESET pulse to the flip-flop immediately following the last magnitude bit, which is after the fourth bit in our example. When this is done, the sign bit will go through the 2's complement logic uncomplemented.

Serial Addition Time

Let us look at the time required to complete an addition cycle with the serial adder. One bit time is required for the sign operation prior to adding the two numbers. Bit times equal to the total number of bits in each number (including the sign bit) are required to perform the actual addition. One bit time following the addition is required to check for an overflow condition. Therefore, a total of $n + 2$ bit times is used for a complete add cycle where n is the quantity of bits in the number.

A PARALLEL BINARY ADDER 11–7

A complete five-bit parallel binary adder is shown in Figure 11–9 in block diagram form. Negative numbers are handled easily in 1's complement form by inverting each bit of the negative number before adding, and connecting the carry output (C_o) of the last adder stage to the carry input (C_i) of the first (LSB) stage to perform the end-around carry operation. This contrasts with the case of the serial adder where 2's complement operation is easier. Taking the 1's complement of a negative number can be done within the storage register, as we will see later. Sign-bit and magnitude comparison are also required, for the same purposes as in the serial adder.

The two numbers to be added are parallel-loaded into registers *A* and *B*. If either number has a negative sign (1), its magnitude is complemented within the register. A check is made for two negatively signed numbers and for a larger negative number. After a short time is allowed for these operations to take place and for the final sum to appear on the outputs, a pulse is applied to register *A*, causing the sum to be loaded back into the register so the next number can be added to it. Before the next addition, a check for an overflow condition is made. The next number is loaded into register *B*, and the process is repeated. The addition process in the parallel adder takes place very quickly compared to the serial adder—the major advantage is speed.

One way to accomplish the 1's complement of a negative number is by running the output of each stage to an exclusive-OR gate, as shown in Figure 11–10. When the complement control line is HIGH, the gates will complement each register output, and when the complement control line is LOW, the register outputs pass through the gates uncomplemented.

ADDITION OF SIGNED BCD NUMBERS 11–8

BCD is a numerical code, and many applications require that arithmetic operations be performed. Addition is the most important operation because subtraction,

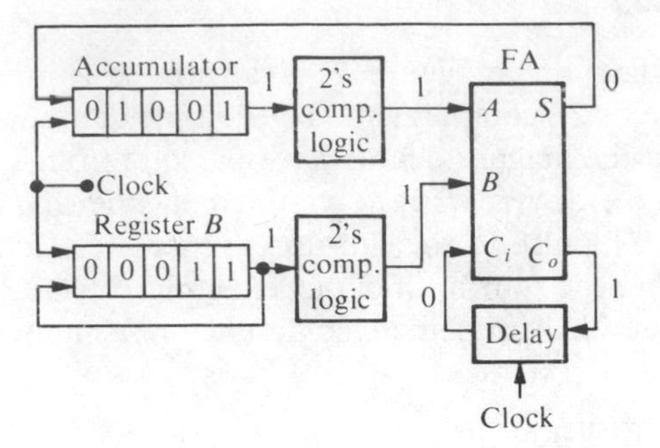

Initially

Numbers are in registers. Sign-and-magnitude comparisons are made (logic omitted here). 2s complement logic is disabled because both numbers are positive. LSBs are added first.

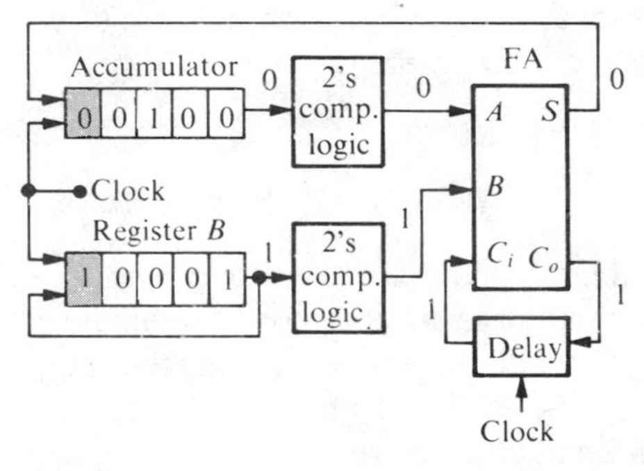

First clock pulse

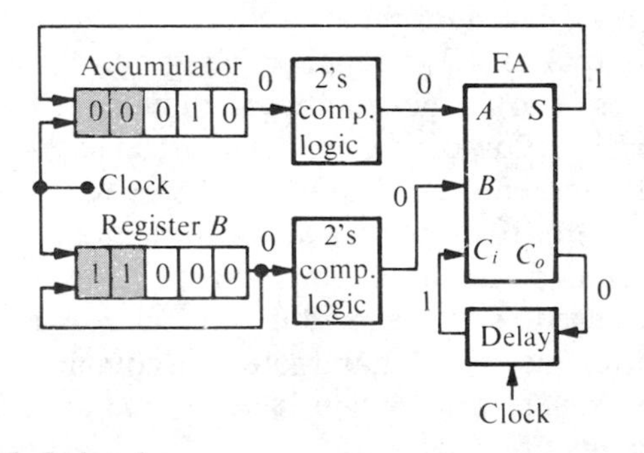

Shifts both numbers one place to the right. Least significant sum bit is now in first stage of accumulator (shaded cell). Least significant bit of number in register B is recirculated. Next LSBs are added, then second bits are added.

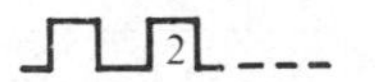

Second clock pulse

Shift both numbers one place to the right. Next least significant sum bit is entered into accumulator (sum bits in shaded cells). Next bit in register B is recirculated. Third bits are added.

FIGURE 11–8

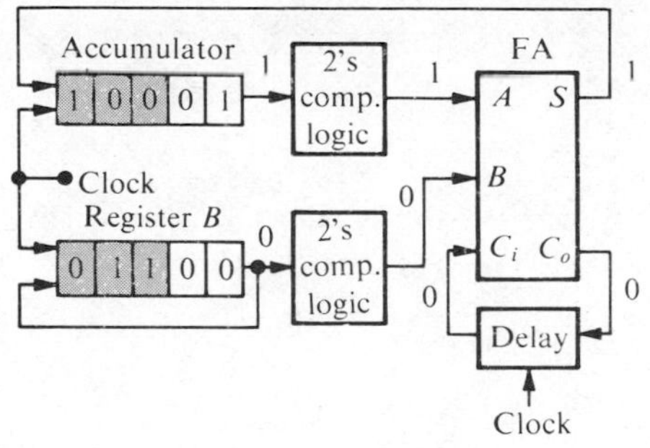

Third clock pulse

3

Shifts both numbers one place to the right. Third sum bit is entered into accumulator (sum bits in shaded cells). Third bit in register B is recirculated. MSBs are added.

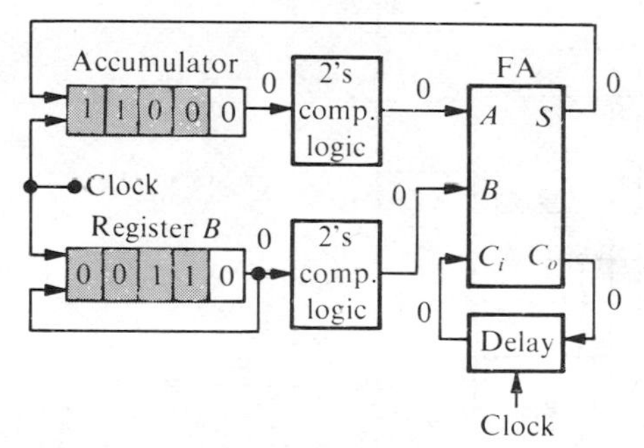

Fourth clock pulse

4

Shifts both numbers one place to the right. Most significant sum bit is entered into accumulator (sum bits is shaded cells). MSB in register B is recirculated. Sign bits are added.

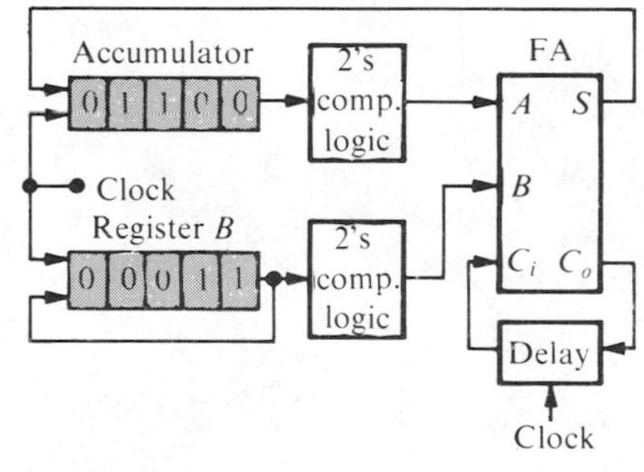

Fifth clock pulse

5

Shifts last bits out of registers. Accumulator now contains final sum. Original contents of register B have been restored. Sign check can now be made for overflow. Next number to be added can now be loaded into register B.

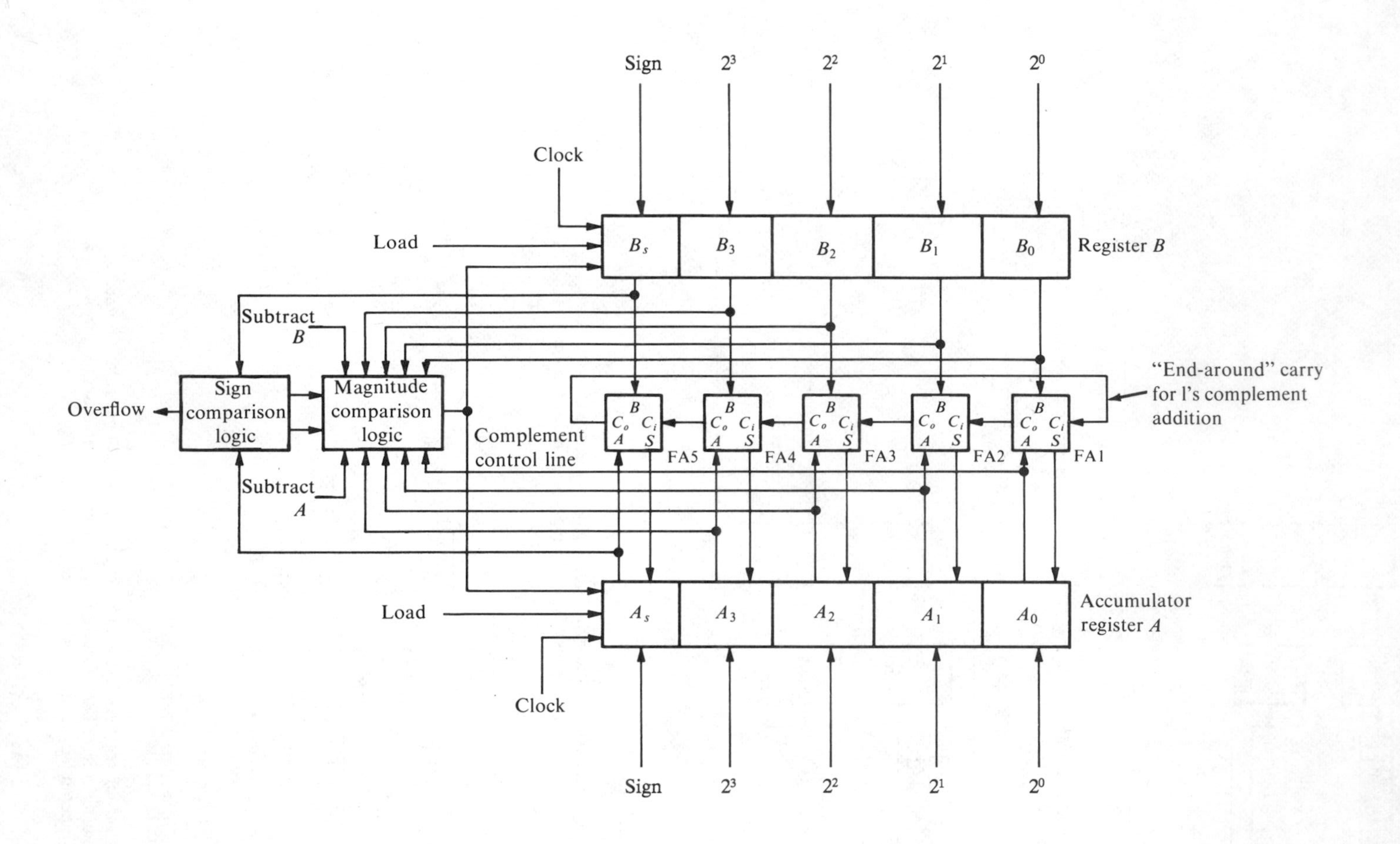

FIGURE 11–9 *A five-bit parallel binary adder.*

multiplication, and division can be accomplished using addition. Here is how two BCD (8421) numbers are added:

1. Add the two numbers using binary addition.
2. If the four-bit sum is equal to or less than 9, it is *valid.*
3. If the four-bit sum is greater than 9 or if a carry is generated from the four-bit sum, the sum is *invalid.*
4. To adjust the invalid sum, add 0110_2 to the four-bit sum. If a carry results from this addition, add it to the next higher order BCD digit.

Excess-3 is also a form of BCD code. The rules for excess-3 addition are:

1. Add the excess-3 numbers using binary addition.
2. If there is *no carry* from a four-bit sum, *subtract* 0011_2 from that sum to adjust to the excess-3 code.
3. If there is a carry from a four-bit sum, *add* 0011_2 to that sum to adjust to the excess-3 code for the digit and add 0011_2 to any new column (digit) generated by the last carry.

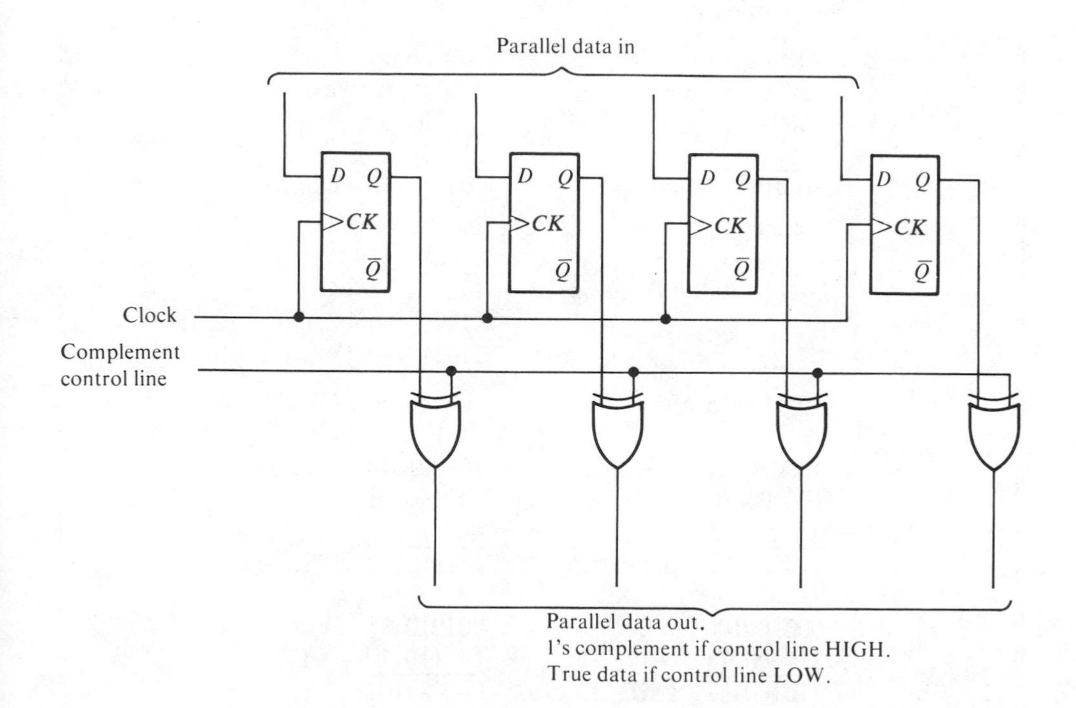

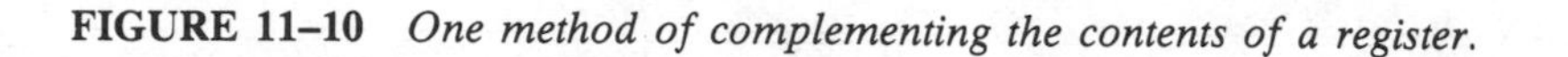

FIGURE 11–10 *One method of complementing the contents of a register.*

Addition of signed BCD numbers can be performed by using 9's or 10's complement methods. A negative BCD number can be expressed by taking the 9's or 10's complement. Recall that the 10's complement is the 9's complement plus 1.

Since it is easier, from a hardware standpoint, to convert excess-3 code than 8421 BCD to the 9's complement, excess-3 is sometimes used in BCD arithmetic processes.

Example 11–13

Decimal	*BCD (8421)*		*Excess-3*	
5	10's comp of −3:		10's comp of −3:	
+ (−3)	10011	true BCD	10110	true excess-3
2	10110	9's comp	11001	9's comp
	+ 1		+ 1	
	10111	10's comp	11010	10's comp
	00101	+5	01000	+5
	+ 10111	−3	+ 11010	−3
	11100	invalid	~~1~~ 00010	invalid excess-3
	+ 0110	add 6	+ 00011	add 3 to correct
	~~1~~ 00010	+2	00101	+2 in excess-3
	↑ drop carry			
−8	10's comp of −8:		10's comp of −8:	
+ 6	11000	true BCD	11011	true excess-3
−2	10001	9's comp	10100	9's comp
	+ 1		+ 1	
	10010	10's comp	10101	10's comp
	10010	−8	10101	−8
	+ 00110	+6	+ 01001	+6
	11000	−2(10's comp)	11110	no carry
			+ 11101	subt 3 (2's comp)
			11011	−2(10's comp)
24	10's comp of −17:		10's comp of −17:	
+(−17)				
+ 7	100010111	true BCD	101001010	true excess-3
	100000010	9's comp	110110101	9's comp
	+ 1		+ 1	
	110000011	10's comp	110110110	10's comp
	000100100	+24	001010111	+24
	+ 110000011	−17	110110110	−17
	110100111	MSD invalid	000001101	
	+ 0110	add 6	000111101	add 3/subt 3
	000000111	+7	000111010	+7

Addition of signed BCD numbers basically follows the rules listed except for the complement operation for negative numbers. The following example shows addition of both 8421 BCD and excess-3 signed numbers using the 10's complement method.

A BCD ADDER 11–9

Figure 11–11 shows a BCD adder. Functionally, the system operates as follows. Two four-bit BCD digits are entered into the adder along with any input carry. The resulting sum is compared to 9_{10} (1001_2), and if it is equal to or less than 9 and no

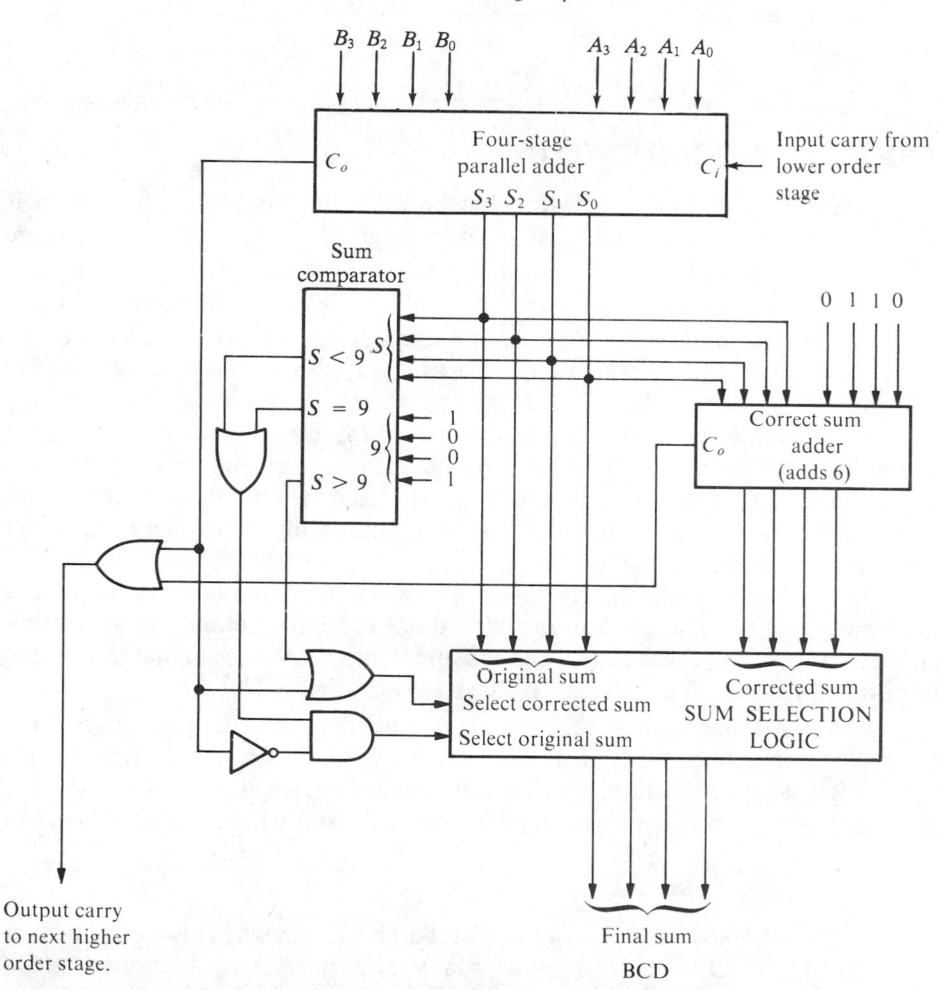

FIGURE 11–11 *Block diagram for single-digit BCD adder.*

output carry occurs, it is gated through the sum selection logic to the final outputs. If the sum is greater than 9 or if an output carry occurs, the sum is invalid. The correct sum adder is enabled, adding 6 (0110) to the invalid sum; the corrected sum is then gated through the sum selection logic to the final outputs. The carry outputs of both adders are ORed and go to the next higher stage (if any).

Example 11–14

Show the states within a BCD adder for addition of the digits 5_{10} and 3_{10}, and for 9_{10} and 4_{10}. Refer to Figure 11–11 for a complete diagram.

Solutions:

See Figure 11–12.

11–10 ARITHMETIC LOGIC UNIT (ALU)

The arithmetic logic unit is a device that performs all the arithmetic and logic operations on binary data. An ALU is an integral part of any microprocessor system, as illustrated in Figure 11–13.

The figure shows that the arithmetic or logic operations are performed on two eight-bit numbers which is common in many microprocessors. The control unit receives an externally generated instruction that tells it what operation is to be performed. It then puts the ALU in the appropriate arithmetic or logic mode.

Basically, there are two registers associated with the ALU, the *accumulator* and the *data register*. Each register can store an eight-bit number. The numbers are then applied to the inputs of the ALU and the operation selected by the control unit is performed on the numbers. The final result of the operation is then stored back in the accumulator.

For example, a given instruction might require that two positive numbers be added. The control unit puts the ALU into the *add mode,* the two numbers in the registers are added, and the sum is put into the accumulator replacing the original number. This process is illustrated in Figure 11–14.

As mentioned, the ALU is an important integral part of a microprocessor and in Chapter 12, we will study the 6800 microprocessor and examine its internal operation. ALUs are also availale as single devices in LSI form. The 74181 is a good example of this type of ALU and we will look at it next.

The 74181 ALU

The block and pin diagram for the 74181 is shown in Figure 11–15 as an example of an *IC* ALU. This device is capable of performing 16 binary arithmetic operations on two four-bit numbers. Each of these operations is selected by a four-bit code on the function select inputs (S_0, S_1, S_2, S_3). The data inputs are A_0 through A_3 and B_0 through B_3.

Subtraction in this particular device uses the 1's complement method with the 1's complement of the subtrahend internally generated. There is also a comparison operation. When two four-bit numbers are equal in magnitude, the $A = B$ output goes HIGH. The ALU must be in the subtract mode and the C_n input must be HIGH to function as a comparator.

When an arithmetic operation is performed, the result appears on the four function output ($F_0F_1F_2F_3$) and on the carry output, C_{n+4}.

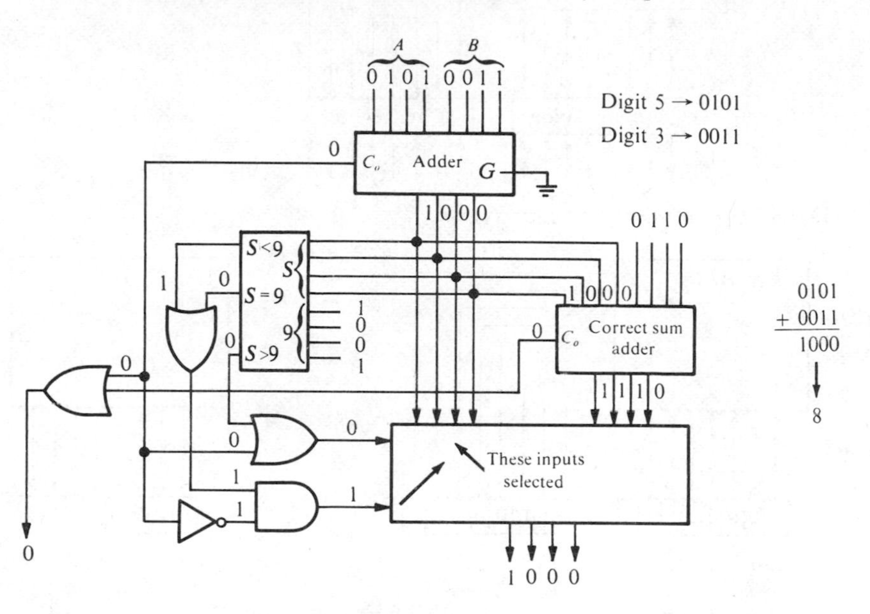

(a) Addition of 5_{10} and 3_{10}.

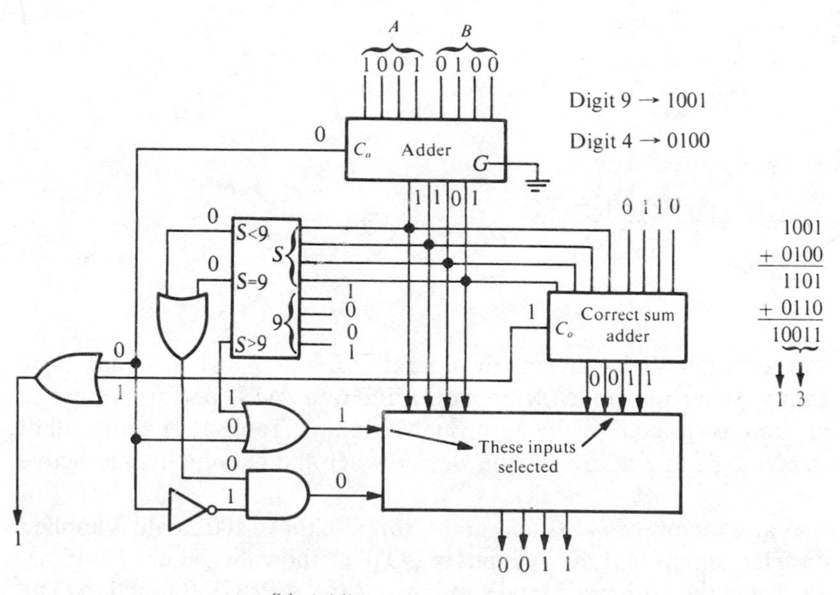

(b) Addition of 9_{10} and 4_{10}.

FIGURE 11–12

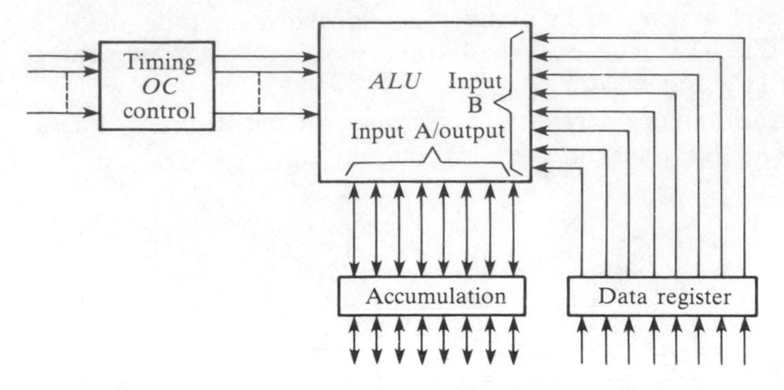

FIGURE 11–13 *ALU in a microprocessor system.*

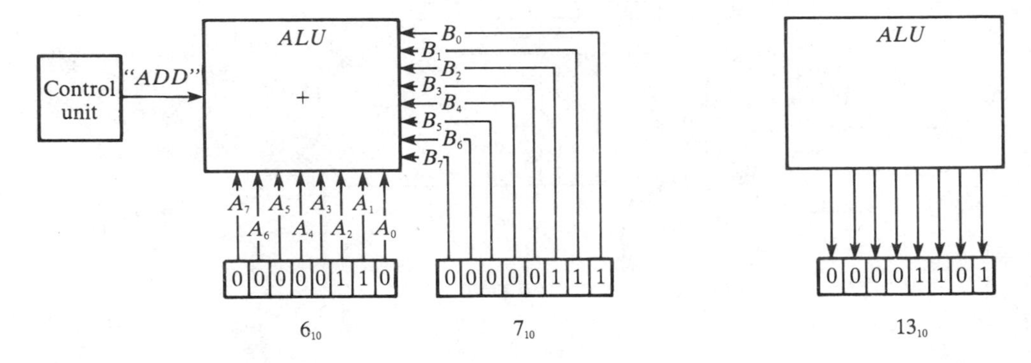

FIGURE 11–14 *Example of ALU addition operation (7 + 6 = 13).*

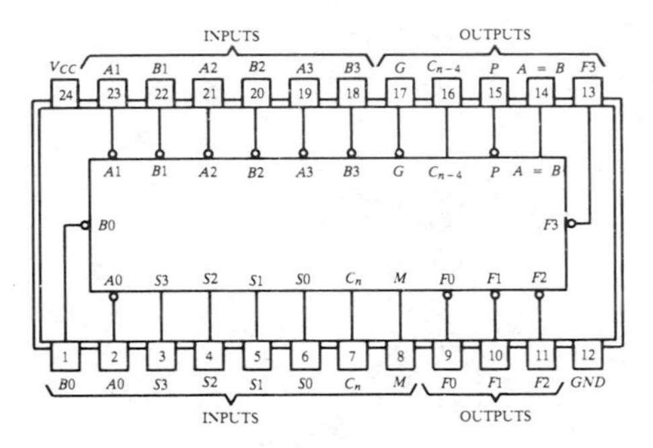

FIGURE 11–15 *The 74181 ALU.*

In addition to its arithmetic operations, the 74181 also has logic function capability. When the mode *(M)* input is HIGH, the ALU performs any one of 16 logic functions as selected by the S_0 through S_3 inputs. Table 11–1 shows all of the arithmetic and logic operations for this device when the data inputs are active HIGH.

As an example of 74181 operation, suppose that a 1001 state is applied to the function select inputs and the *M* input is LOW as shown in Figure 11–16(a). This sets the device in the arithmetic mode and selects the *A* PLUS *B* function. The two numbers on the data inputs are added and the sum appears on the function outputs as shown in the figure.

TABLE 11–1

$S_3S_2S_1S_0$	Logic $M = 1$	$M = 0$ Arithmetic Operations Arithmetic $C_n = 1$ (no carry)	Arithmetic $C_n = 0$ (carry)
0000	$F = \overline{A}$	$F = A$	$F = A$ PLUS 1
0001	$F = \overline{A + B}$	$F = A + B$	$F = (A + B)$ PLUS 1
0010	$F = AB$	$F = A + \overline{B}$	$F = (A + \overline{B})$ PLUS 1
0011	$F = 0$	$F =$ MINUS 1 (2's comp)	$F = 0$
0100	$F = \overline{AB}$	$F = A$ PLUS $A\overline{B}$	$F = A$ *PLUS* $A\overline{B}$ PLUS 1
0101	$F = \overline{B}$	$F = (A + B)$ PLUS $A\overline{B}$	$F = (A + B)$ PLUS $A\overline{B}$ PLUS 1
0110	$F = A \oplus B$	$F = A$ MINUS B MINUS 1	$F = A$ MINUS B
0111	$F = A\overline{B}$	$F = A\overline{B}$ MINUS 1	$F = A\overline{B}$
1000	$F = \overline{A} + B$	$F = A$ PLUS AB	$F = A$ PLUS AB PLUS 1
1001	$F = \overline{A \oplus B}$	$F = A$ PLUS B	$F = A$ PLUS B PLUS 1
1010	$F = B$	$F = (A + \overline{B})$ PLUS AB	$F = (A + \overline{B})$ PLUS AB PLUS 1
1011	$F = AB$	$F = AB$ MINUS 1	$F = AB$
1100	$F = 1$	$F = A$ PLUS A*	$F = A$ PLUS A PLUS 1
1101	$F = A + \overline{B}$	$F = (A + B)$ PLUS A	$F = (A + B)$ PLUS A PLUS 1
1110	$F = A + B$	$F = (A + \overline{B})$ PLUS A	$F = (A + \overline{B})$ PLUS A PLUS 1
1111	$F = A$	$F = A$ MINUS 1	$F = A$

* Each bit is shifted to the next more significant position

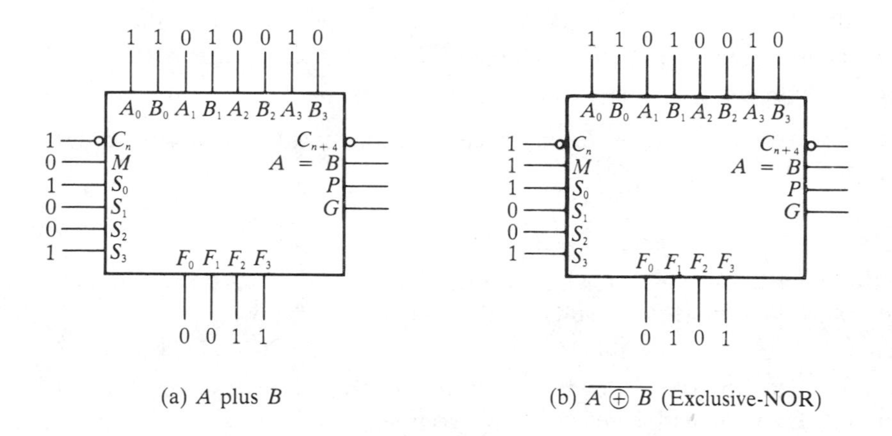

(a) A plus B

(b) $\overline{A \oplus B}$ (Exclusive-NOR)

FIGURE 11–16 *74181 ALU operation.*

If the M input is taken HIGH, the device is put into the logic mode and $\overline{A \oplus B}$ and is produced on the outputs as shown in Figure 11–16(b) when $S_3S_2S_1S_0 =$ 1001.

Problems

Section 11–1

11–1 Express each of the following decimal numbers as signed true binary numbers (8 bits).
(a) +17 **(b)** +34 **(c)** −69 **(d)** +110 **(e)** −125

11–2 Convert each sign and true magnitude binary number to 1's complement.
(a) 10010001 **(b)** 11100111 **(c)** 10000101 **(d)** 11110111

11–3 Convert each number in Problem 11–2 to 2's complement.

11–4 Convert each sign and true magnitude binary number to 2's complement.
(a) 10000110 **(b)** 11111000 **(c)** 10101010 **(d)** 10001111

Section 11–2

11–5 Perform each of the following additions of signed numbers by using the 2's complement method as needed. Leave negative results in 2's complement form.
(a) 33 + 15 **(b)** 56 + (−27) **(c)** 99 + (−75)
(d) −46 + 25 **(e)** −110 + (−84)

11–6 Perform each of the following additions of signed numbers using the 2's complement method as needed. Leave negative results in 2's complement form.
(a) 59 +21 **(b)** 76 + (−35) **(c)** −90 + 23
(d) −19 + (−41) **(e)** −121 + 99

11–7 Perform the following additions using the 2's complement method. The numbers are sign and true magnitude. Leave negative results in 2's complement form.
(a) 01001 + 10110 **(b)** 00110 + 10011 **(c)** 10100 + 10010
(d) 11101 + 10001 **(e)** 11000 + 00111 **(f)** 10101 + 01010

11–8 Repeat Problem 11–7 using the 1's complement method.

11–9 Explain overflow in an addition process and name the conditions under which it can occur.

Section 11–3

11–10 Perform each of the following subtractions using the 2's complement method. Leave negative results in 2's complement form.
(a) 43 − 27 **(b)** 20 − (−14) **(c)** 30 − (−45)

11–11 Perform each of the following subtractions using the 2's complement method. Leave negative results in 2's complement form.
(a) 39 − 12 **(b)** 55 − 72 **(c)** −18 − 12

Section 11–4

11–12 Multiply the following signed binary numbers. The negative number is already in 2's complement form.
(a) 000101100 × 100110110 **(b)** 110001000 × 010001011

11–13 Multiply by repeated addition. Left-most bit is sign.
multiplicand = 011110110 multiplier = 100000100

Section 11–5

11–14 Divide 010000100 by 001100011.

Section 11–6

11–15 Show the adder states on a bit-by-bit basis to illustrate the serial addition of the following two numbers with four magnitude bits and a sign bit. Refer to Example 11–12.

00111 + 00100

11–16 Repeat Problem 11–15 for the following two numbers:

11000 + 00110

11–17 Explain how a string of several numbers is added with a serial adder.

11–18 Using the 2's complement logic of Figure 11–6, show the bit-by-bit conversion of the following binary numbers:
(a) 10110100 **(b)** 11110000 **(c)** 10010011

11–19 Repeat Problem 11–18 for the following numbers:
(a) 11000110 **(b)** 10000101 **(c)** 10000000

11–20 Illustrate an overflow condition and its detection by showing the states within the circuit of Figure 11–4.

11–21 Explain why the serial adder takes longer to complete an addition than a parallel adder.

Section 11–8

11–22 Convert the following signed decimal numbers to BCD (8421) and add using 10's complement.
(a) 8 + (−4) **(b)** −6 + 5 **(c)** 36 + (−21)

11–23 Add the following BCD numbers:
(a) 0110 + 0010 **(b)** 1000 + 0001 **(c)** 0101 + 0111
(d) 0111 + 0101 **(e)** 1001 + 0100 **(f)** 1001 + 1001

11–24 Convert the following decimal numbers to BCD, and perform the indicated addition:
(a) 7 + 2 **(b)** 12 + 9 **(c)** 49 + 36
(d) 9 + 5 **(e)** 25 + 23 **(f)** 192 + 148

Section 11–9

11–25 Show the stages within the BCD adder of Figure 11–11 for each of the following additions:
(a) 3 + 2 **(b)** 7 + 6 **(c)** 9 + 9 **(d)** 5 + 4
(e) 9 + 8 **(f)** 5 + 5

11–26 Repeat Problem 11–25 for the following additions:
(a) 1 + 4 **(b)** 6 + 7 **(c)** 5 + 1 **(d)** 2 + 7
(e) 3 + 9 **(f)** 9 + 0

Section 11–10

11–27 The result of an arithmetic or logic operation in a typical ALU is placed in a register called the ________.

11–28 Set up the 74181 ALU with the proper input levels so that 8_{10} is subtracted from 13_{10}.

This chapter provides an introduction to the microprocessor, a digital device that has had a major impact on the electronics industry in recent years. The purpose of this coverage is to give you a fundamental understanding of what a microprocessor is and how it functions.

Although there are many microprocessors on the market, our coverage here focuses on the Motorola 6800. This device and its descendents form one of the most popular and widely used microprocessor families. From the study of this microprocessor, you can understand the common features found on most microprocessors regardless of the manufacturer, and you will have the basis for learning in detail any specific device you may be working with.

12

THE MICROPROCESSOR

12–1 DEFINITION OF A MICROPROCESSOR (μP)

A microprocessor is a large scale integrated circuit (LSI) that contains the processing portion of a *microcomputer*. Physically, it is a single chip device housed in a package, as shown in Figure 12–1.

FIGURE 12–1 *Microprocessor package. (Courtesy of Motorola)*

The Microcomputer

Since a microprocessor is used as part of a computer, let's first get a basic idea of what a microcomputer is and distinguish the difference between it and a microprocessor. Figure 12–2 shows a block diagram of a basic microcomputer.

As you can see, it consists of four functional blocks: *microprocessor, memory, input/output (I/O) interface,* and *I/O device.* The microprocessor unit within the microcomputer is interconnected to the memory and I/O interface with an *address bus* and a *data bus.*

The function of the *memory* is to store binary data that are to be used or processed by the microprocessor. The function of the *I/O interface* is to get information into and out of the memory or microprocessor from the I/O device. The I/O device can be a keyboard, video terminal, card reader, printer, magnetic disk or tape unit, or other type of equipment.

The *address bus* provides a path from microprocessor to memory and I/O interface. It allows the microprocessor to select the memory address from which to acquire data or in which data are to be stored. Also it provides for communication with an I/O device for inputting or outputting data.

The *data bus* provides a path over which data are transferred between the microprocessor, memory, and I/O interface.

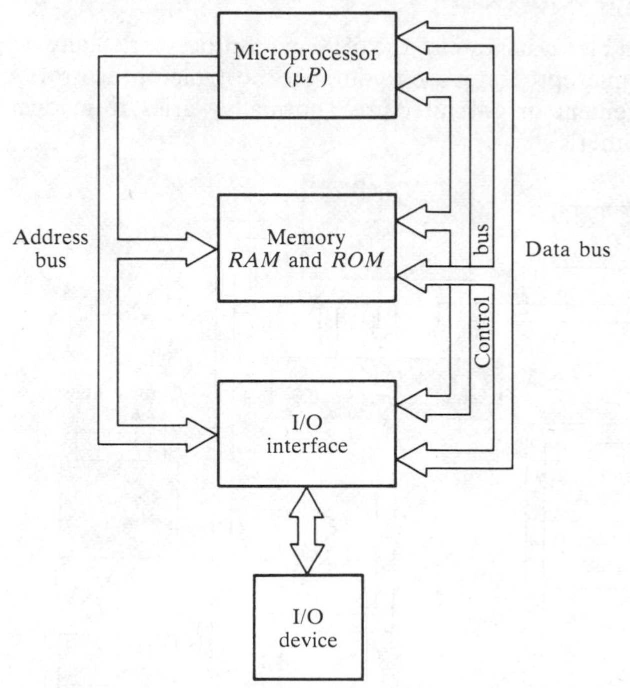

FIGURE 12–2 *Basic microcomputer.*

There are two basic trends in the physical make-up of microcomputers. One is a multichip approach where the microprocessor, memory, and I/O interface units are housed in separate packages. The other is a single-chip approach where the microprocessor, memory, and I/O are integrated on a single chip and housed in a single package.

At this point, the basic difference between a microprocessor and a microcomputer should be clear. The microprocessor is simply the processing unit of the microcomputer and has no memory or I/O of its own. To construct a microcomputer system, memory chips and I/O chips must be connected to the microprocessor chip.

Words and Bytes

A *complete* unit of binary information or data is called a *word.* For instance, the number 200_{10} can be represented by eight bits as 11001000_2. This, then, is an eight-bit word because it completely represents the decimal number 200. Now consider the decimal number 32,768. This number cannot be represented with eight bits but requires 16 bits—1000000000000000_2. So a 16-bit word is required for representing decimal numbers from 0 to 65,535.

Most microprocessors handle bits in eight-bit groups called *bytes.* A byte may be a word or only part of a word. For instance, the *eight-bit* word mentioned above consists of *one byte.* However, the *16-bit* word consists of *two bytes.*

12–2 ELEMENTS OF A MICROPROCESSOR

Figure 12–3 shows a *simplified* block diagram of the 6800 microprocessor. Many of the elements contained in this microprocessor are common to most microprocessors, although the internal arrangement or "architecture" normally varies from one manufacturer's device to another's.

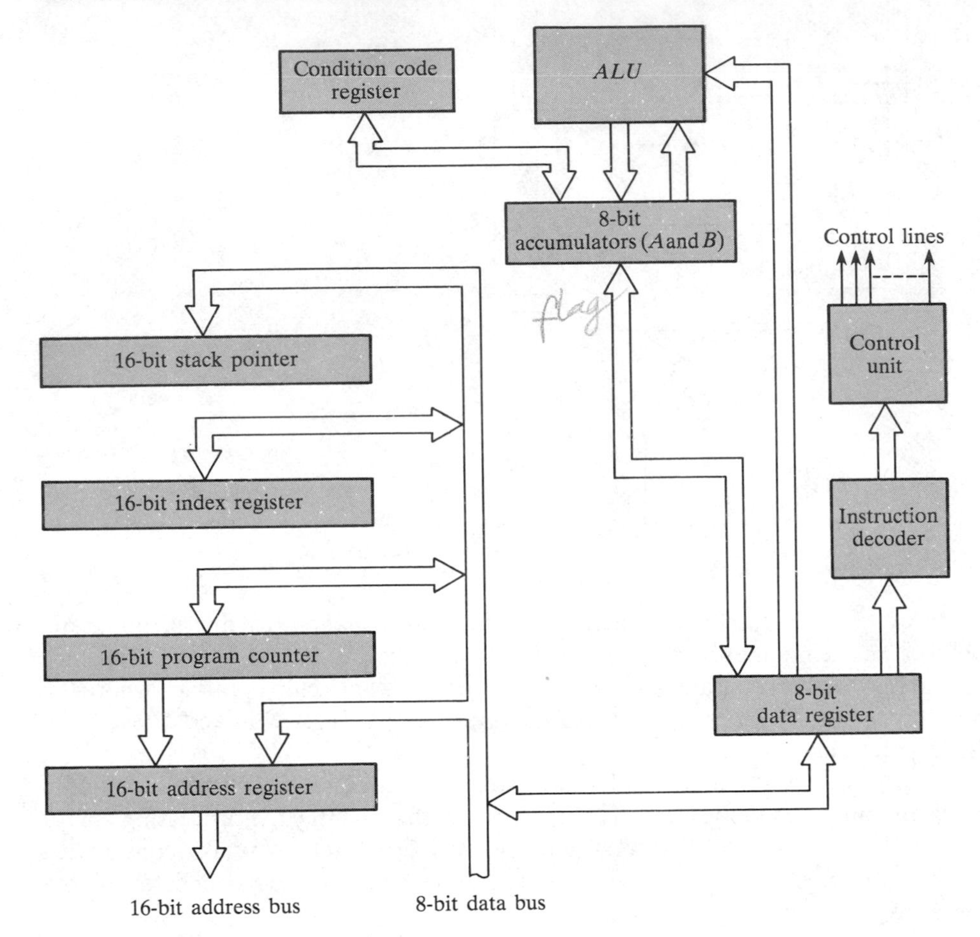

FIGURE 12–3 *6800 microprocessor simplified block diagram.*

We will now examine the elements of this microprocessor to understand the function of each.

Arithmetic Logic Unit (ALU)

This part of the microprocessor contains the logic to perform arithmetic and logic operations. Data are brought into the ALU from one or both of the registers called the *accumulators* and from the *data register*. The 6800 has two accumulators, *A* and *B*. Both accumulators and the data register are eight-bit registers that store one byte of data. Each byte brought into the ALU is called an *operand.* This is the group of bits to be operated on by the ALU.

For example, Figure 12–4 shows an eight-bit number from accumulator *A* being *added* to an eight-bit number from the data register. The result of this addition (sum) is put into an accumulator replacing the original operand that was stored there.

When the ALU performs an operation on two operands, the result of the operation always goes into an accumulator and replaces the previous operand.

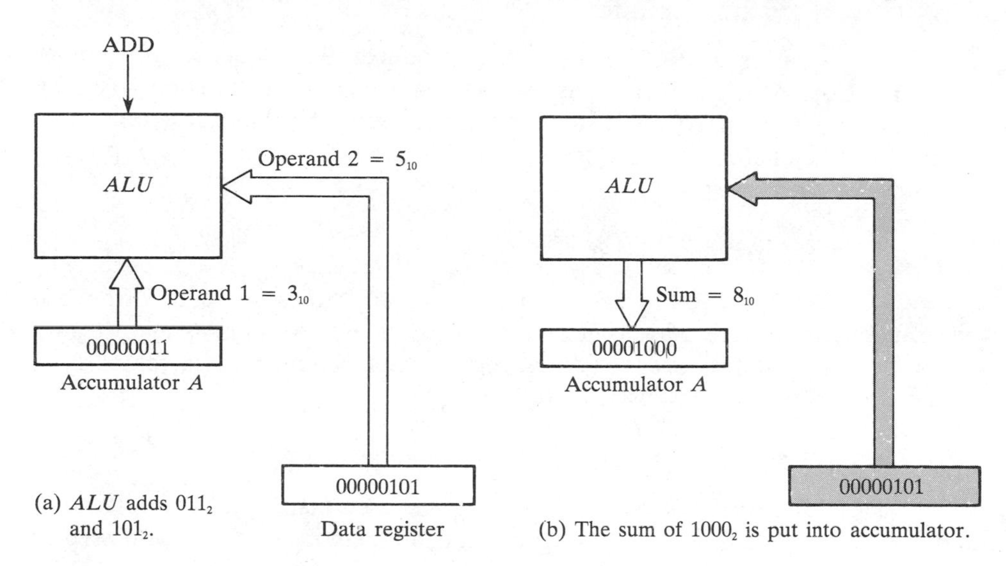

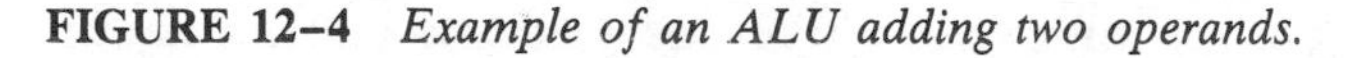
FIGURE 12–4 *Example of an ALU adding two operands.*

Accumulator and Data Register

As you have seen, one function of the accumulator is to store an operand prior to an operation by the ALU. The other function is to store the *result* of the operation after it has been performed.

The data register temporarily stores a byte of data that is to be put onto the data bus or that has been taken off of the data bus. As mentioned, both of these registers are eight-bit registers in the 6800 microprocessor.

Instruction Decoder and Control Unit

An *instruction* is a binary code that tells the microprocessor what it is to do. An orderly arrangement of many different instructions makes up a typical *program. A program is a step-by-step procedure used by the microprocessor to carry out a specified task.*

The *instruction decoder* within the microprocessor decodes an instruction code that has been transferred on the data bus from the memory. The instruction code is commonly called an *op code* for operation code. When the op code is decoded, the instruction decoder provides the control unit with this information so that it can produce the proper signals and timing sequence to *execute* the instruction.

Condition Code Register

This element is sometimes called a *flag* register or *status* register. Its basic purpose is to indicate the status of the contents of the accumulator or certain other conditions within the microprocessor. For instance, it can indicate a zero result, a negative result, the occurrence of a carry, and the occurrence of an overflow from the accumulator.

Program Counter

This is a 16-bit counter that produces the sequence of memory addresses from which the program instructions are taken. The content of the program counter is always the memory address from which the next byte is to be taken. In some microprocessors, the program counter is known as the *instruction pointer.*

Address Register

This is a 16-bit register that temporarily stores an address from the program counter in order to put it onto the address bus. As soon as the program counter loads an address into the address register, it is incremented (increased by 1) to the address of the next instruction.

Stack Pointer and Index Register

The *stack pointer* is a 16-bit register used mainly during subroutines and interrupts. It is used in conjunction with the memory stack, as discussed in Chapter 9.

The *index register* is also a 16-bit register used as one means of addressing the memory. It is used with a mode of addressing called *indexed addressing.*

Address Bus and Data Bus

The *address bus* consists of 16 parallel lines to accommodate a 16-bit address code. This allows the microprocessor to address up to $2^{16} = 65{,}536$ bytes of memory. Some microprocessors have more address bits so that larger memories can be addressed.

The *data bus* consists of eight parallel lines. One byte of data can be transferred to or from the memory or I/O on this bus at any given time. Some microprocessors have a 16-bit data bus and are referred to as *16-bit microprocessors.* The 6800 microprocessor described in this chapter is, of course, an *eight-bit* device.

12–3 THE MICROPROCESSOR AND THE MEMORY

As we mentioned, the microprocessor is connected to a memory with the address bus and the data bus. In addition, there are certain control signals that must be sent between the microprocessor and the memory such as the *read* and the *write* controls. This is illustrated in Figure 12–5.

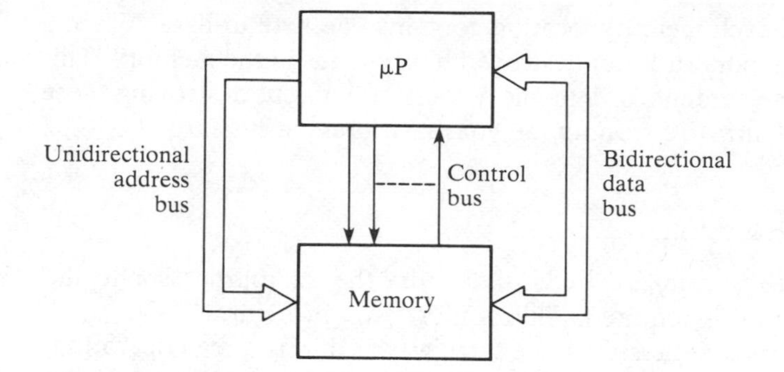

FIGURE 12–5 *A microprocessor with memory.*

As indicated, the address bus is unidirectional. This means that the address bits go only one way—from the microprocessor to the memory. The data bus is bidirectional. This provides for data bits to be transferred from the memory to the microprocessor or from the microprocessor to the memory.

Read Operation

To transfer a byte of data from the memory to the microprocessor, a *read* operation must be performed. This is done as illustrated in Figure 12–6.

To begin, the program counter contains the 16-bit address of the byte to be read from the memory. This address is loaded into the address register and put onto the address bus. The program counter is then advanced by one (incremented) to the next address and waits.

Once the address code is on the bus, the microprocessor control unit sends a *read* signal to the memory. At the memory, the address bits are decoded and the desired memory location is selected. The *read* signal causes the *contents* of the selected address to be put on the data bus. The data byte is then loaded into the data register to be used by the microprocessor. This completes the read operation.

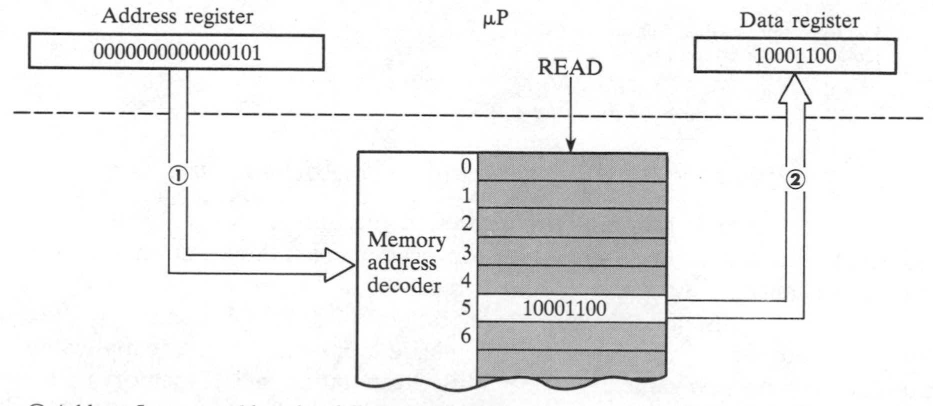

① Address 5_{10} put on address bus followed by Read signal.

② Contents of address 5_{10} in memory put on data bus and stored by data register.

FIGURE 12–6 *Illustration of the read operation.*

Note that each memory location contains one byte of data. When a byte is read from the memory, it is not destroyed but remains in the memory. This process of "copying" the contents of a memory location without destroying those contents is called *nondestructive readout,* as you have previously learned.

Write Operation

In order to transfer a byte of data from the microprocessor to the memory, a *write* operation is required. This is illustrated in Figure 12–7.

The memory is addressed in the same way as during a read operation. A data byte being held in the data register is put onto the data bus and the microprocessor sends the memory a *write* signal. This causes the byte on the data bus to be stored at the selected location in the memory as specified by the 16-bit address code. The existing contents of that particular memory location are *replaced* by the new data byte. This completes the write operation.

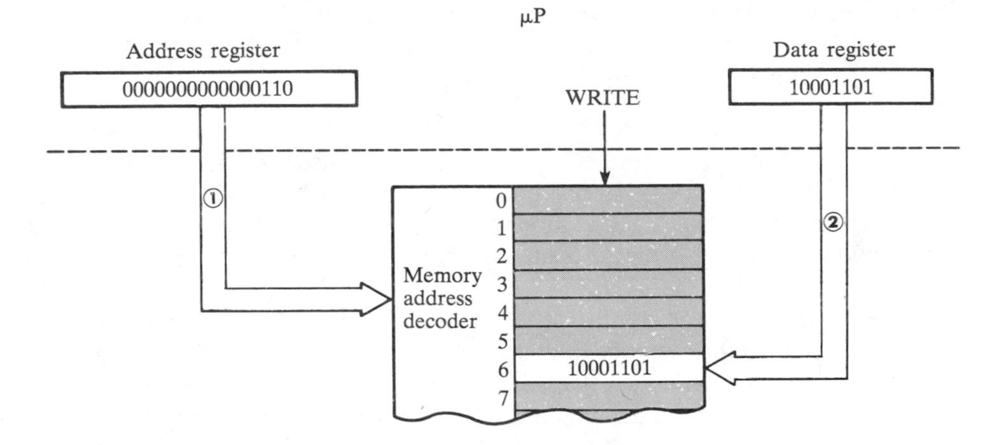

① Address put on address bus.

② Data put on data bus followed by Write signal.
Data stored in address 6_{16}.

FIGURE 12–7 *Illustration of the write operation.*

Hexadecimal Representation of Address and Data

The only things a microprocessor recognizes are combinations of 1s and 0s. However, most literature on microprocessors uses the hexadecimal number system to simplify the representation of binary quantities.

For instance, the binary address 0000000000001111 can be written as 000F in hexadecimal. A 16-bit address can have a *minimum* hexadecimal value of 0000_{16} and a *maximum* value of $FFFF_{16}$. With this notation, a 64K memory (actually 65,536) can be shown in block form as in Figure 12–8. The lowest memory address is 0000_{16} and the highest address is $FFFF_{16}$.

A data byte can also be represented in hexadecimal. A data byte can be either an *instruction, operand,* or *address.* A data byte is eight bits and can represent decimal numbers from 0_{10} to 255_{10}, or it can represent up to 256_{10} instructions. For example, a microprocessor code that is 10001100 in binary is written as 8C in hexadecimal.

Address (hexadecimal)	Contents
0000	
0001	
0002	
0003	
0004	
0005	
0006	
0007	
FFFB	
FFFC	
FFFD	
FFFE	
FFFF	

FIGURE 12–8 *Representation of a 64K memory.*

FETCH AND EXECUTE 12–4

When a program (list of instructions arranged to perform a specified task) is being run, the microprocessor goes through a repetitive sequence consisting of two fundamental phases, as shown in Figure 12–9.

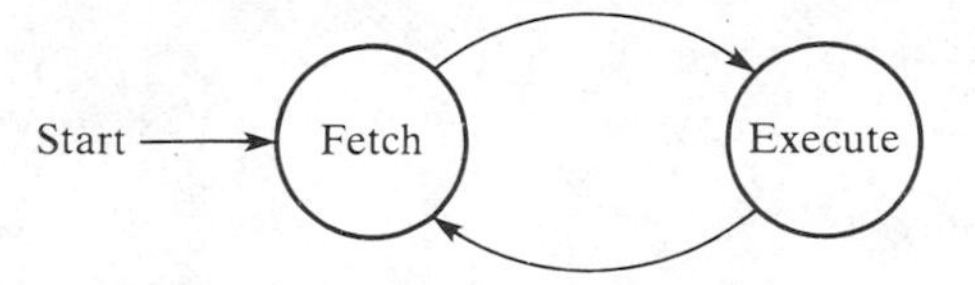

FIGURE 12–9

One phase is called *fetch* and the other *execute*. During the *fetch* phase, *an instruction is read from the memory and decoded* by the instruction decoder.

During the *execute* phase, the microprocessor *carries out the sequence of operations indicated by that particular instruction.* As soon as one instruction has been executed, the microprocessor returns to the fetch phase to get the next instruction from the memory.

You will see a fetch/execute sequence illustrated later when we talk about specific instructions.

12–5 INSTRUCTION FORMAT

As you have seen, a microprocessor must address the memory in order to obtain data or store data. Microprocessors have several ways in which to generate an address when an instruction is being executed. These are called the *addressing modes* of the microprocessor, and they provide for wide programming flexibility.

The 6800 microprocessor has several addressing modes. Each of the instructions available to the microprocessor has a certain addressing mode associated with it. These are *inherent, immediate, direct, extended, relative,* and *indexed.* These addressing modes are covered in the next section.

Instructions

Microprocessors typically have at least 50 instructions that make up what is called the *instruction set.* The 6800 has 72 instructions. Although instruction sets vary from one manufacturer to another, many instructions are common to all microprocessors.

Right now, in order to illustrate the various addressing modes, we will use four instructions from the 6800 instruction set as examples. These are the *load accumulator* instruction (LDAA), the *store accumulator* instruction (STAA), the *addition* instruction (ADDA), and the *clear accumulator* instruction (CLRA). The *A* on the end of each of these instructions indicates that they apply to accumulator *A.* There are similar instructions for operating with accumulator *B.*

The shorthand designations for instructions—LDAA, STAA, ADDA, and CLRA—are called *mnemonics.*

Format

A microprocessor instruction can consist of one, two, or three bytes depending on the type. A *one-byte instruction* format is illustrated in Figure 12–10(a). This type of instruction requires one memory location. The eight-bit

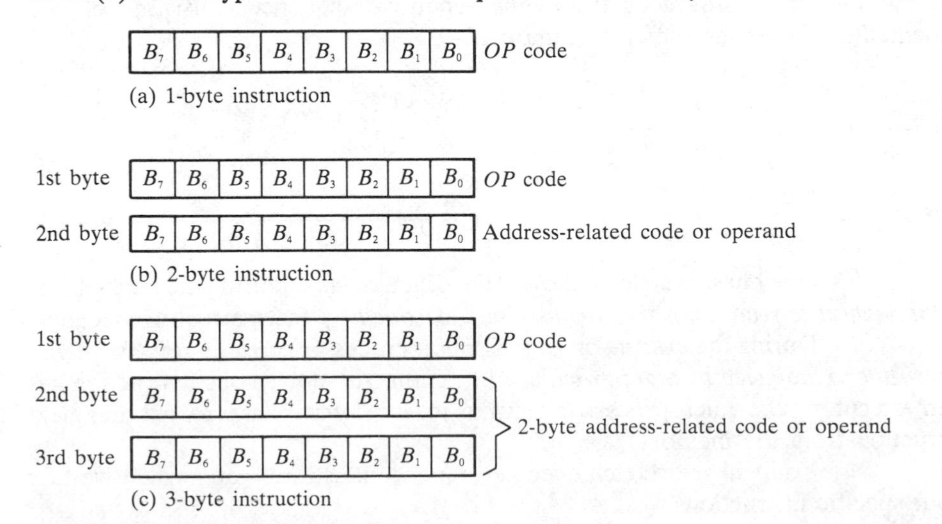

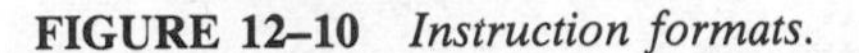
FIGURE 12–10 *Instruction formats.*

instruction code is called the *op code* (operation code) and uniquely identifies that instruction.

A *two-byte instruction* format is illustrated in Figure 12–10(b). The first byte is the op code, and the second byte can be either an operand or a code associated with a memory address. This type of instruction is stored in two *consecutive* memory locations.

A *three-byte instruction* format is shown in Figure 12–10(c). In this case, the first byte is the op code, and the second and third bytes can be an operand or an address-associated code. This type of instruction is stored in three *consecutive* memory locations.

ADDRESSING MODES 12–6

We will now look at the addressing modes mentioned in the previous section.

Inherent Addressing

This is sometimes known as *implied* addressing. One-byte instructions fall into this category. These instructions either require no operand or the operand is implied by the op code.

For example, CLRA is a one-byte instruction with the operand implied. Since this instruction *clears* the accumulator, the operand is eight 0s or 00_{16}. This implied operand ends up in the accumulator after this instruction is executed.

In the 6800 microprocessor, the op code for the CLRA is $4F_{16}$, or 01001111_2 in binary.

Another example is the *halt* or *wait* instruction used to indicate the end of a program. WAI is a one-byte instruction that requires *no* operand; it simply tells the microprocessor to stop all operations. (It actually does more, but we will not go into that now.) The op code for WAI in the 6800 is $3E_{16}$, or 00111110_2 in binary.

The sequence that the microprocessor goes through when handling a one-byte instruction (inherent addressing) is shown in Figure 12–11, using the WAI instruction as an example.

Immediate Addressing

Immediate addressing is used in conjunction with many two-byte instructions and some three-byte instructions. For a two-byte instruction using immediate addressing, the *first byte is the op code and the second byte is the operand.*

The LDAA and the ADDA instructions can both use immediate addressing. For instance, the *LDAA immediate* instruction has an op code of 86_{16}, or 10000110_2 in binary. This op code is stored in one byte in memory (a single location), and an operand is stored in the memory location immediately following the op code. This is illustrated in Figure 12–12(a) where the two bytes are at memory addresses 0000_{16} and 0001_{16}. The example operand is 0005_{16}.

When decoded, this *LDAA immediate* instruction will tell the microprocessor to get the *contents of the next memory location (operand) and load it into the accumulator.* The fetch and execute cycle for this instruction is illustrated in Figure 12–12(b), (c), and (d).

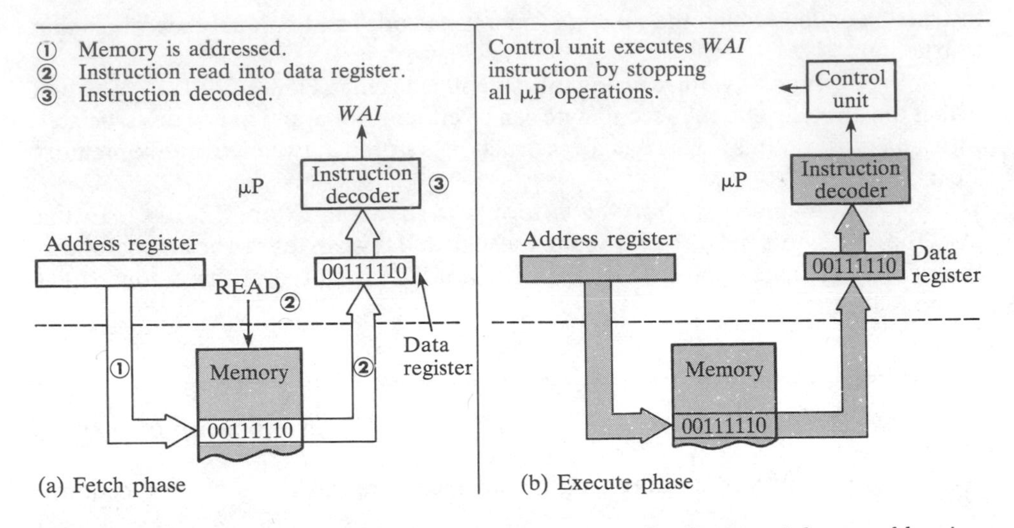

FIGURE 12–11 *Fetch/execute for the WAI instruction. This illustrates inherent addressing.*

Also, as another example, the ADDA instruction can be used with immediate addressing. In this case, the op code $8B_{16}$ and an operand occupy two consecutive memory bytes. When the *ADDA immediate* instruction is decoded, it tells the microprocessor to get the operand from the next memory location and add it to the present contents of the accumulator.

Notice that these immediate addressed instructions involve two read operations, one during the fetch phase to get the op code and the other during the execute phase to get the operand. Compare this to the single read operation required for an inherent addressed instruction such as WAI.

Direct Addressing

Instructions that use direct addressing are two-byte instructions. The first byte is the op code and the second byte is the *address of the operand,* not the operand itself.

The LDAA, STAA, and ADDA can all use direct addressing. For instance, LDAA using direct addressing has a different op code (96_{16}) than LDAA immediate (86_{16}).

To illustrate how the microprocessor handles an instruction with direct addressing, let's again use LDAA as an example. Figure 12–13(a) shows the *LDAA direct* instruction in memory addreses 0000_{16} and 0001_{16}. The first byte is the op code 96_{16} (10010110_2), and the second byte is the *operand address,* 0005_{16}.

When the *LDAA direct* instruction is decoded, it tells the microprocessor to load the operand located at the memory address specified by the second byte of the instruction. Figure 12–13(b), (c), (d), and (e) illustrate the fetch and execute phases for the *LDAA direct* instruction.

Extended Addressing

This type of addressing is the same as direct except that it uses a *two-byte operand address* rather than a one-byte address as in direct.

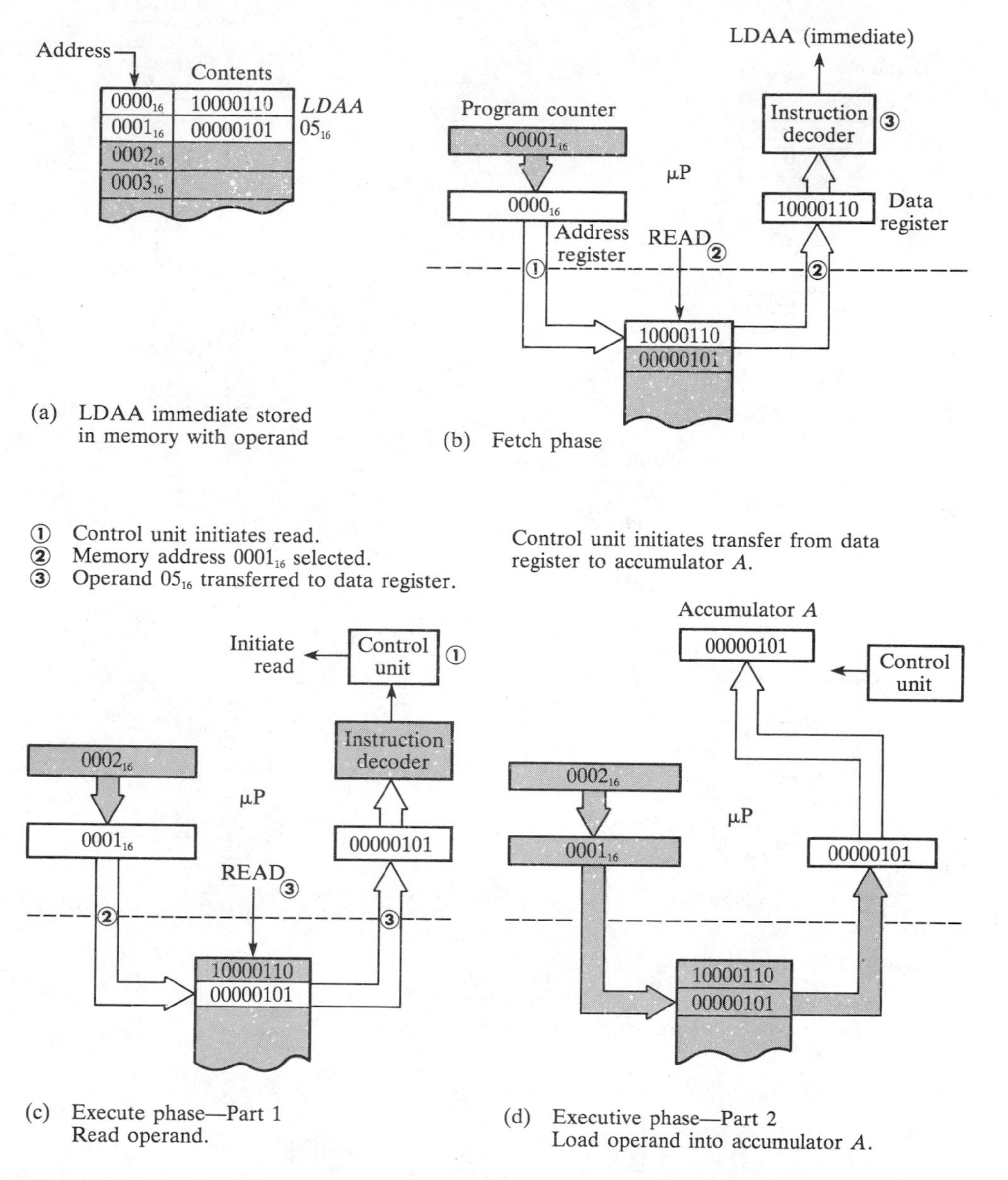

(a) LDAA immediate stored in memory with operand

(b) Fetch phase

(c) Execute phase—Part 1
Read operand.

(d) Executive phase—Part 2
Load operand into accumulator A.

FIGURE 12–12 *Fetch/execute for the LDAA immediate instruction. This illustrates immediate addressing.*

Direct addressing is limited to the first 256_{10} memory locations (00_{16} through FF_{16}) because of its eight-bit operand address. Its advantage is that it takes only two bytes of memory to store.

Extended addressing allows the microprocessor to address up to $2^{16} = 65{,}536$ bytes of memory. An instruction using extended addressing is a *three-byte*

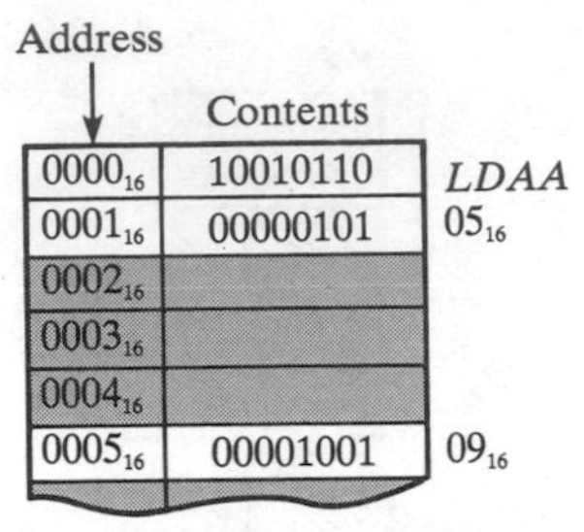

(a) *LDAA* direct stored in memory with operand address and operand

① Memory address 0000_{16} selected.
② *LDAA OP* code transferred to data register.
③ *LDAA* instruction decoded.

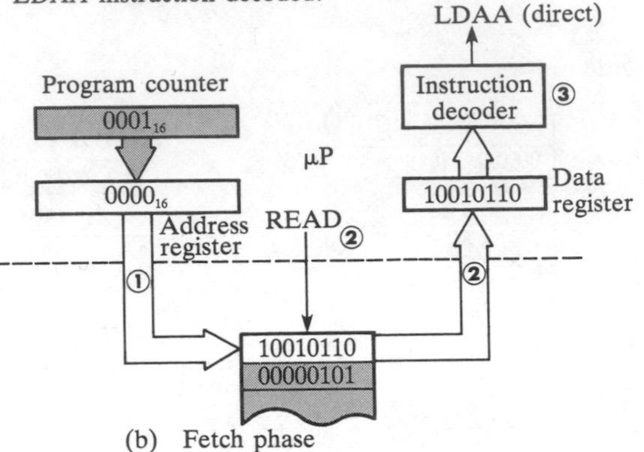

(b) Fetch phase

① Control unit initiates read.
② Address 0001_{16} selected.
③ Operand address transferred to data register.

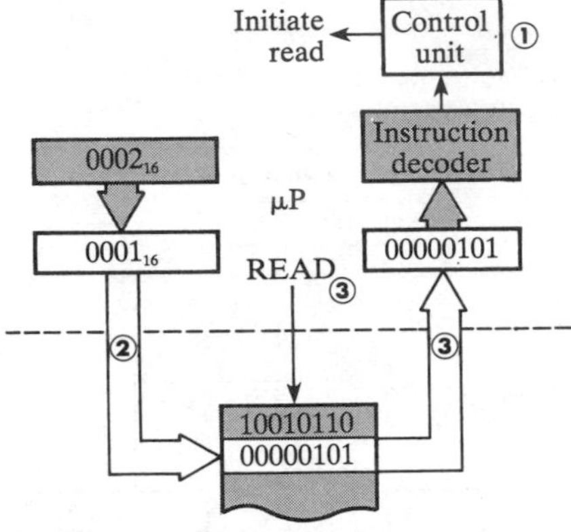

(c) Execute phase—Part 1
Read operand address.

Control unit initiates transfer of operand address to address register.

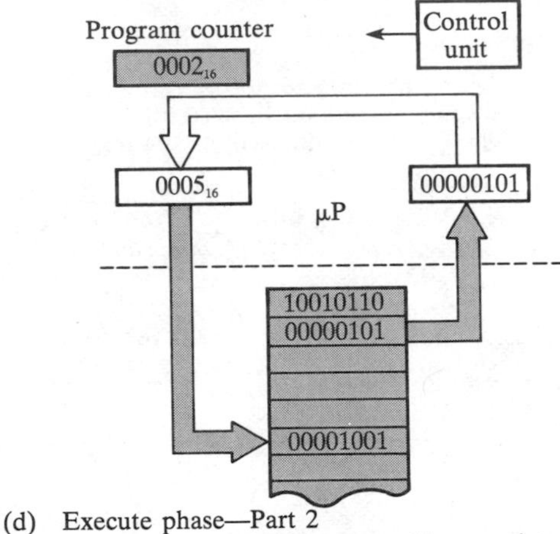

(d) Execute phase—Part 2
Transfer operand address to address register.

① Control unit initiates read.
② Address 0005_{16} selected.
③ Operand transferred to data register.
④ Operand transferred to accumulator *A*.

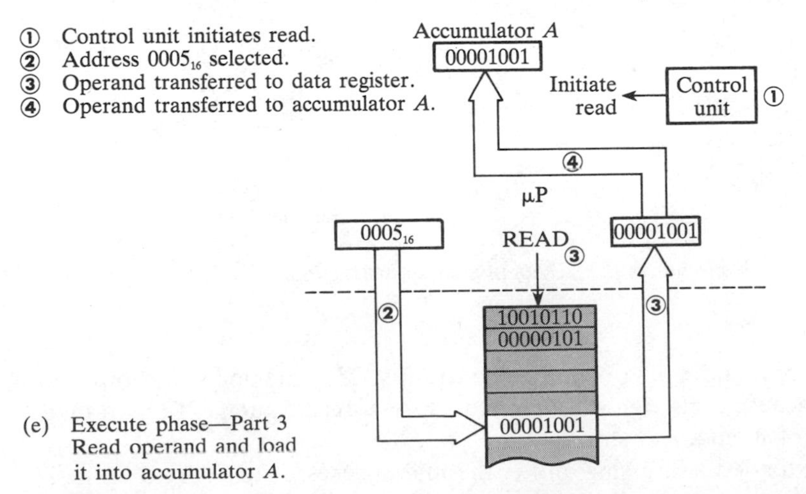

(e) Execute phase—Part 3
Read operand and load it into accumulator *A*.

FIGURE 12–13 *Fetch/execute for the LDAA direct instruction. This illustrates direct addressing.*

1st byte	1	0	1	1	0	1	1	1	*OP* code	
2nd byte	0	0	0	0	0	0	0	1	HIGH order bits	2-byte address
3rd byte	1	1	1	1	1	1	1	1	LOW order bits	

FIGURE 12–14 *Extended addressing instruction format (STAA extended).*

instruction. The first byte is the op code, the second byte is the higher eight bits of the operand address, and the third byte is the lower eight bits.

This format is shown in Figure 12–14 where B7$_{16}$ is the op code for *STAA extended,* and the two address bytes represent memory address 01FF$_{16}$ LDAA, STAA, and ADDA can all be used with extended addressing.

Indexed Addressing

This mode of addressing is used in conjunction with the *index register.* An instruction using indexed addressing consists of *two bytes.* The first byte is the op code and the second byte is called the *offset address.* This format is shown in Figure 12–15. The LDAA, STAA, and ADDA instruction can all use indexed addressing.

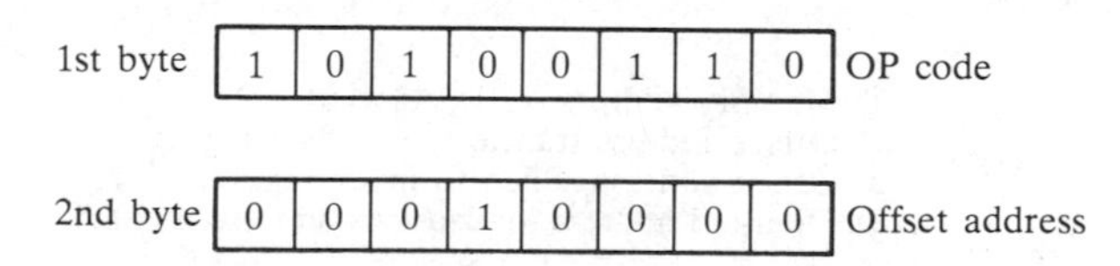

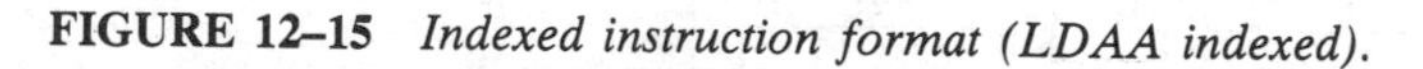

FIGURE 12–15 *Indexed instruction format (LDAA indexed).*

When an indexed instruction is executed, *the offset address is added to the contents of the index register.* This gives the *operand address.*

Figure 12–16 illustrates indexed addressing, again using the LDAA instruction. The LDAA indexed instruction has an op code of A6$_{16}$ in the 6800 microprocessor. In this example, the op code is stored at memory location 0000$_{16}$, and the offset address of 10$_{16}$ is stored at address 0001$_{16}$. The operand is stored at address 001A$_{16}$ as shown in Figure 12–16(a).

When the LDAA indexed instruction is executed, it tells the microprocessor to load accumulator *A* with the operand located at the address specified by the sum of the index register contents and the offset address. In this case the sum of 000A$_{16}$ (contents of index register) and 10$_{16}$ (offset address) is 001A$_{26}$. This is the address in memory where the operand is stored.

Figure 12–16(b), (c), and (d) illustrate the fetch and execute phases for the LDAA indexed instruction.

Relative Addressing

This form of addressing is used by a class of instructions known as *branch instructions.* Basically, a branch instruction allows the microprocessor to go back or skip ahead a specified number of addresses in a program. Branching instructions are used to form program *loops,* as you will see later.

The format of a branch instruction is shown in Figure 12–17. The first byte is the op code, and the second byte is the *relative address.* When a branch

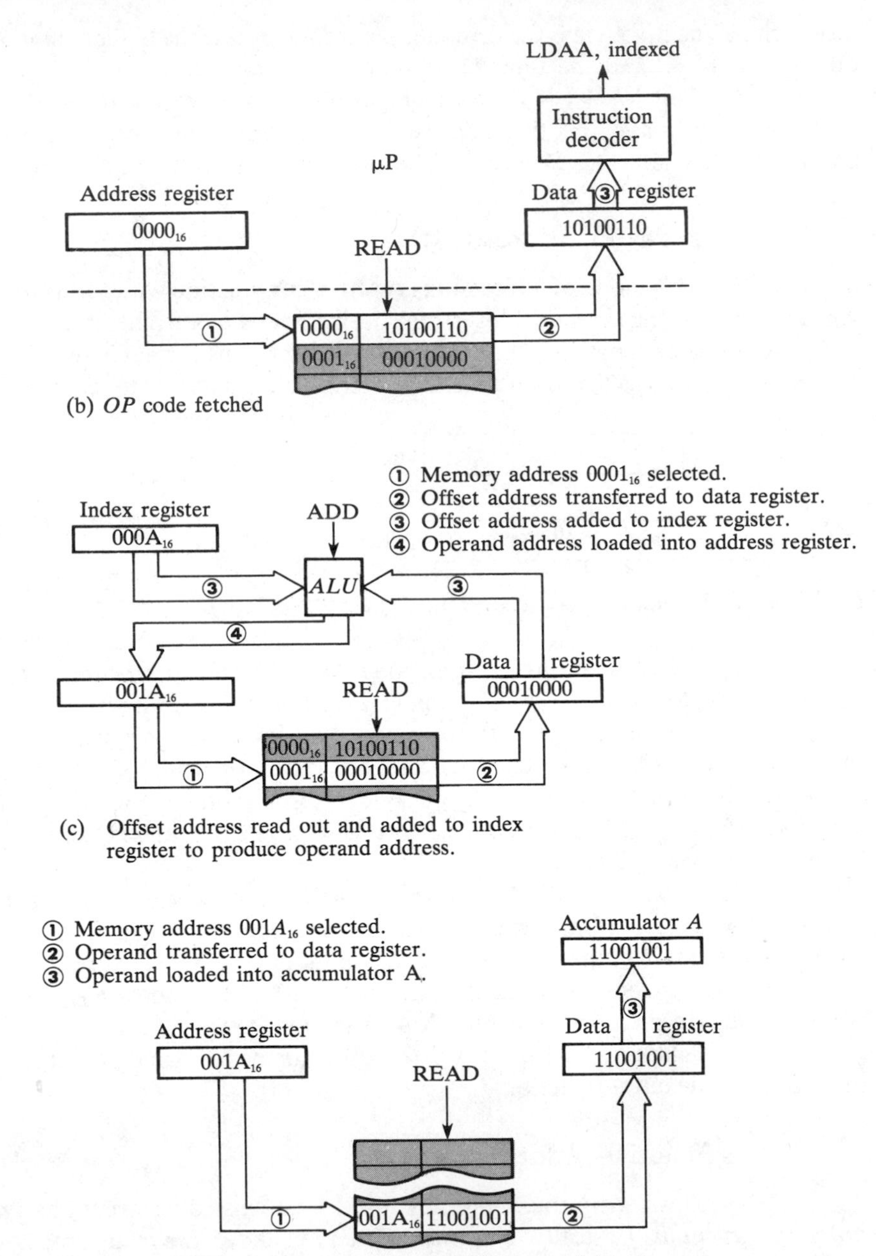

FIGURE 12–16 *Fetch/execute for the LDAA indexed instruction. This illustrates indexed addressing.*

instruction is executed, *the relative address is added to the contents of the program counter* to form the address to which the program is to branch. The particular branch instruction in Figure 12–17 is called *branch always* (BRA) with an op code of 20_{16}.

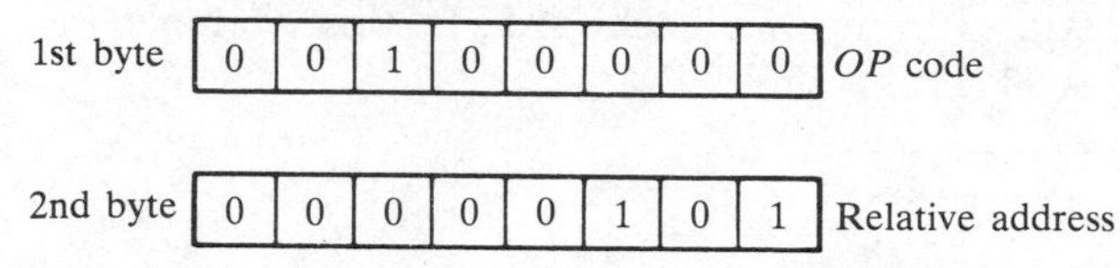

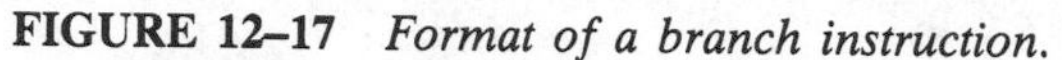

FIGURE 12–17 *Format of a branch instruction.*

To *branch forward,* the relative address is specified as a positive number. This is illustrated in Figure 12–18 where the BRA instruction op code and relative address 05_{16} are at memory addresses 0007_{16} and 0008_{16} respectively. After the op code and relative address are read from the memory, the program counter contains the next address which is 0009_{16}.

When the BRA instruction is executed, the relative address is added to the program count (05_{16} + 0009_{16} = $000E_{16}$). This result is put into the program counter to update it to the address to which the program is to branch. This address is then selected during the next fetch phase.

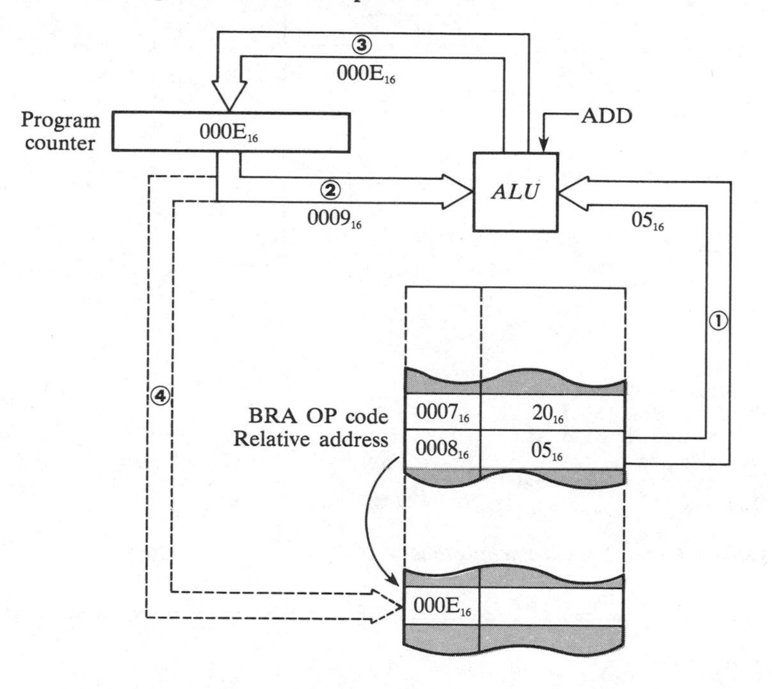

① Relative address transferred from memory address 0008_{16} to ALU.
② Program count transferred to ALU and added to relative address.
③ Result loaded into program counter.
④ Memory address $000E_{16}$ selected.

FIGURE 12–18 *Example of branching forward.*

To *branch backward,* the relative address is specified as a negative number in 2's complement form. When the relative address is added to the program count, an effective subtraction results and the new address is less than the present address. This causes the microprocessor to go back in the program to some previous address.

Figure 12–19 illustrates branching backward. In this example the relative address is -5_{10} which is 11111011_2 or FB_{16}.

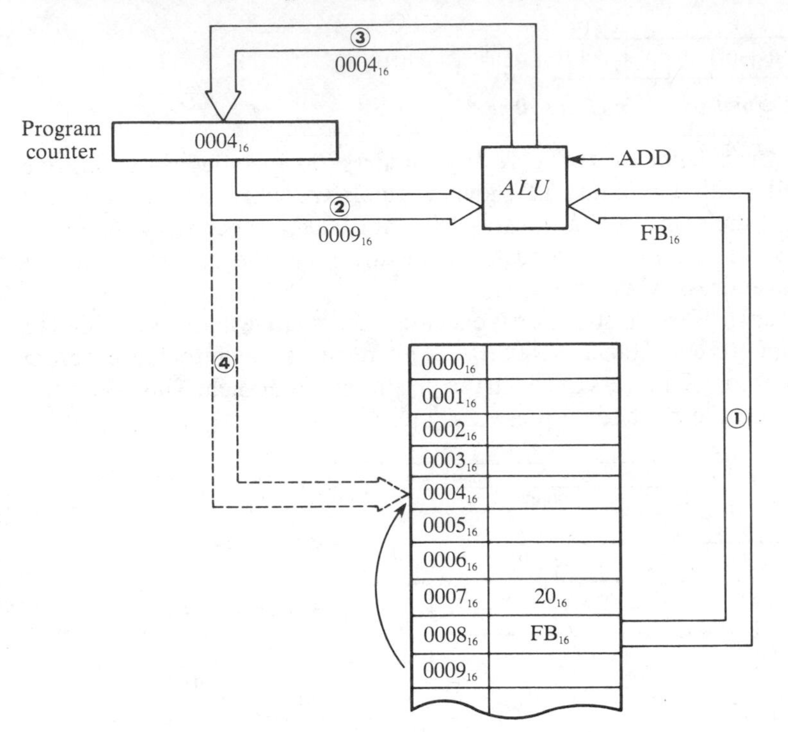

① Relative address transferred from memory address 0008_{16} to ALU.
② Program count transferred to ALU and added to relative address.
③ Result loaded into program counter.
④ Memory address 0004_{16} selected.

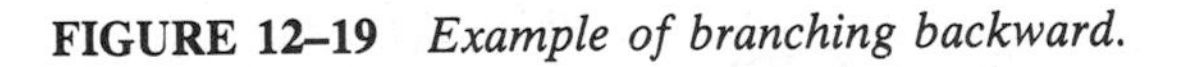

FIGURE 12–19 *Example of branching backward.*

12–7 A SIMPLE PROGRAM

In this section, a very simple program will be used to illustrate how a program is run on the microprocessor. This program adds two numbers and stores the sum in the memory. Initially, the program is stored at memory addresses 0000_{16} thrugh 0006_{16}

Figure 12–20 shows the program using immediate *load* (LDAA) and *add* (ADDA) instructions. The two numbers to be added are 8_{10} and 12_{10}. The first operand (8_{10}) follows the LDAA instruction in memory, and the second operand (12_{10}) follows the ADDA instruction. The *store* (STAA) instruction uses direct addressing to store the sum of the two operands at memory address 0007_{16}.

Memory Address	Memory Contents		Mneumonic/ Contents
	Binary	Hexadecimal	
0000_{16}	10000110	86	LDAA immediate
0001_{16}	00001000	08	8_{10}
0002_{16}	10001011	8B	ADDA immediate
0003_{16}	00001100	0C	12_{10}
0004_{16}	10010111	97	STAA direct
0005_{16}	00000111	07	7_{10}
0006_{16}	00111110	3E	WAI
0007_{16}			Reserved for sum

FIGURE 12–20 *A program to add 8_{10} and 12_{10} and store the sum in memory*

We will now follow the microprocessor through its sequence of operations in the process of running this program. Figure 12–21 shows each step. In part *(a)*, the microprocessor *fetches* the LDAA immediate instruction from memory by performing a read operation at address 0000_{16}. The program counter is then advanced to 0001_{16}.

Part *(b)*, illustrates the *execution* of the LDAA immediate instruction. The operand 8_{10} is read from address 0001_{16} and loaded into the accumulator. The program counter then advances to 0002_{16}.

In part *(c)*, the microprocessor *fetches* the ADDA immediate instruction from memory address 0002_{16}. The program counter then advances to 0003_{16}.

Part *(d)* shows the *execution* of the ADDA immediate instruction. This envolves reading the operand 12_{10} from address 0003_{16} and then adding it to the first operand (8_{10}) which is stored in the accumulator. The sum of these two numbers is then loaded into the accumulator, replacing the first operand. The program counter then advances to 0004_{16}.

Next the microprocessor *fetches* the STAA direct instruction as shown in part *(e)* and the program counter advances to 0005_{16}. Parts *(f)* and *(g)*, show the *execution* of STAA direct, in which 07_{16} is read from memory address 0005_{16} and loaded into the address register. The sum in the accumulator is then stored at address 07_{16} in memory. The program counter then advances to 0006_{16}.

Part *(h)* shows the fetch and execution of the WAI instruction which ends the program.

SUBROUTINES 12–8

A subroutine is basically a program within a main program. A subroutine is normally used when a specific operation is required repeatedly during the running of a main program.

The 6800 microprocessor has several instructions that can be used for subroutines. These include JMP (jump), JSR (jump to subroutine), and RTS (return from subroutine).

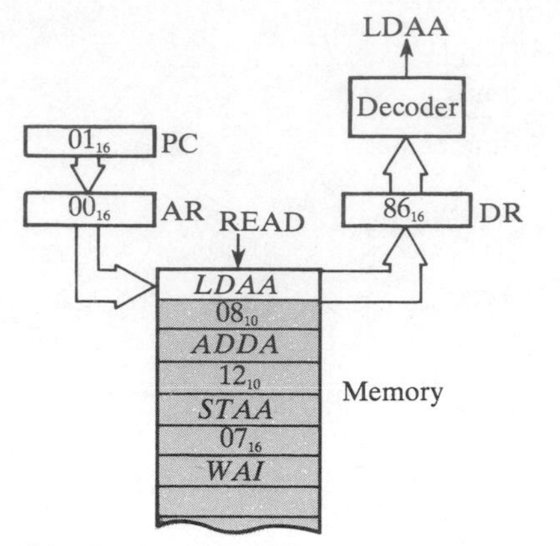

(a) Fetch *LDAA OP* code.

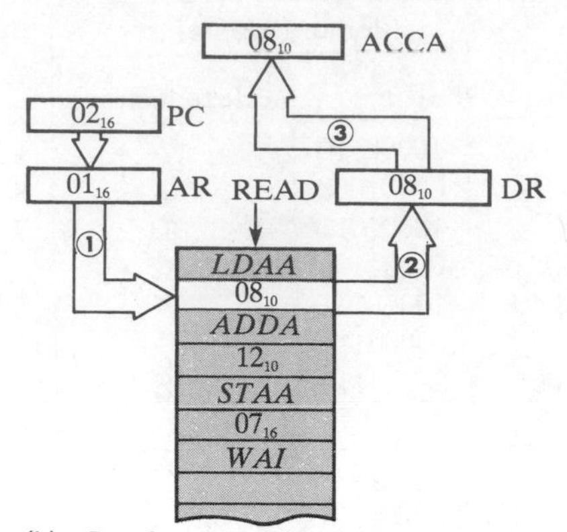

(b) Load operand 08_{10} into accumulator A.

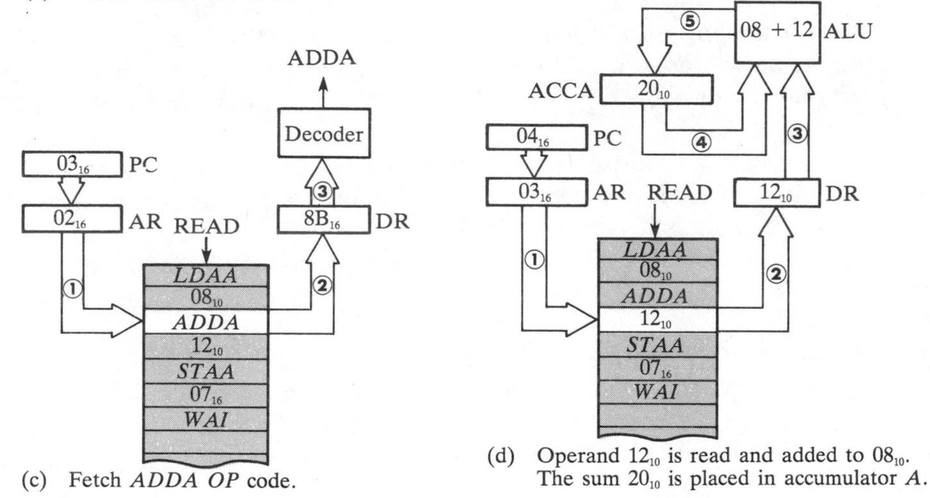

(c) Fetch *ADDA OP* code.

(d) Operand 12_{10} is read and added to 08_{10}. The sum 20_{10} is placed in accumulator *A*.

FIGURE 12–21 *Illustration of program from Figure 12–20 being run.*

JMP Instruction

This instruction allows the microprocessor to jump from one point in a program to another point. This is a three-byte instruction, as shown in Figure 12–22. The first byte is the opcode ($7E_{16}$). The second and third bytes are the MSB and LSB, respectively, of the address to which the microprocessor jumps.

As an example, Figure 12–22 shows the JMP instruction as part of a main program. This causes a jump to address $B000_{16}$, which is the beginning of the subroutine. At the end of the subroutine, another JMP causes the microprocessor to jump back to where it left off in the main program.

The JMP instruction is useful where a program is to be repeated in an endless loop. A limitation of the JMP instruction is that the microprocessor can return to only *one* address in the main program. So a jump from several points in the main program to a single subroutine and back to the appropriate address is not feasible. The JSR and RTS instructions overcome this limitation.

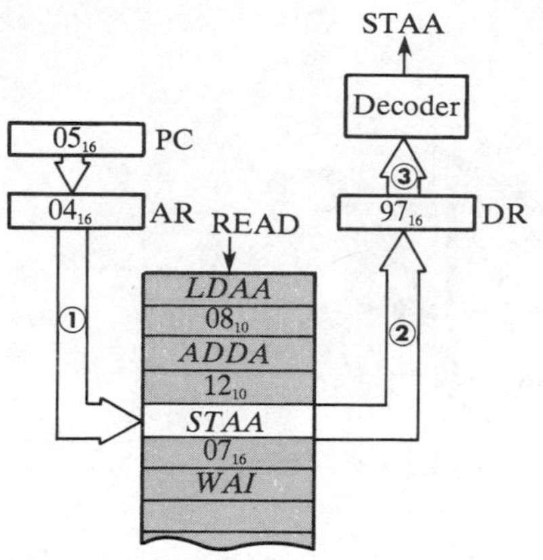

(e) Fetch *STAA OP* code.

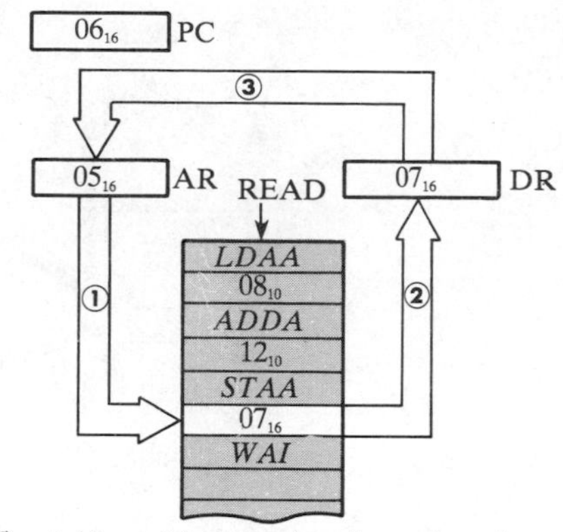

(f) Address 06_{16} is read and transferred into address register.

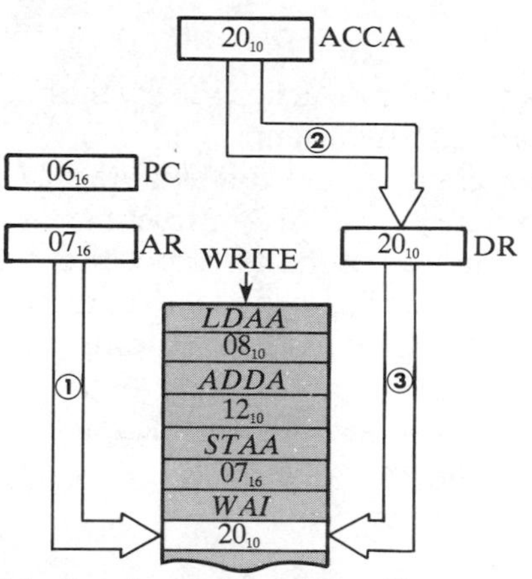

(g) Sum is stored at address 07_{16}.

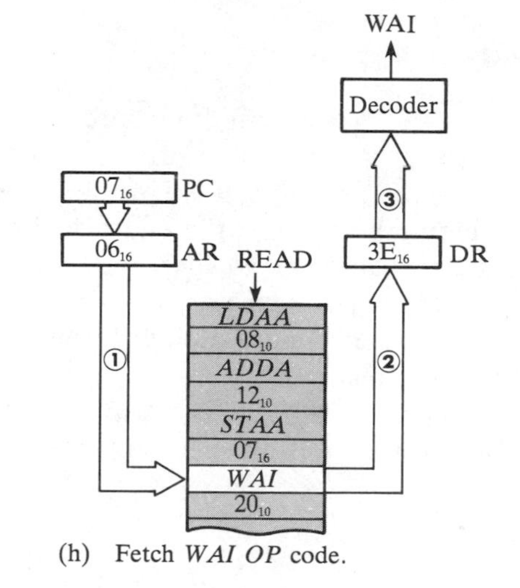

(h) Fetch *WAI OP* code.

FIGURE 12–21, continued

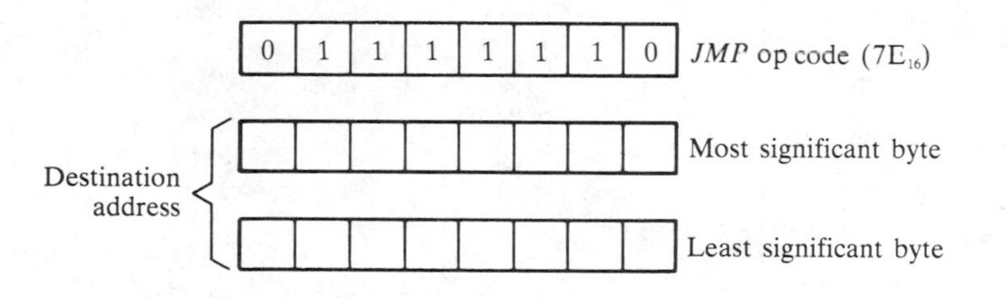

FIGURE 12–22 *JMP instruction format.*

JSR and RTS Instructions

The JSR is a three-byte instruction with a format similar to JMP. RTS is a single byte instruction.

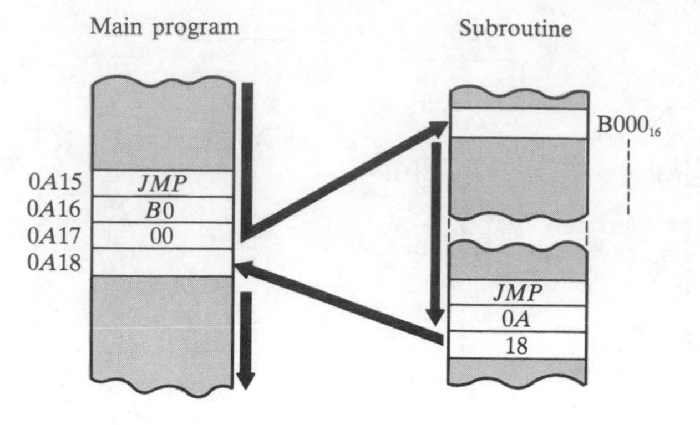

FIGURE 12–23 *A jump to a subroutine and back.*

The second and third bytes of the JSR instruction specify the beginning address of the subroutine to which the program is to jump.

When JSR is executed, the program counter contents are *pushed* into the memory stack. At the end of the subroutine, the RTS instruction causes the program count to be *pulled* from the stack, which causes the microprocessor to return to the address in the main program where it left off. This allows the subroutine to be "called" several times during a main program where each time a different return address is required.

Figure 12–24 illustrates use of the JSR and RTS instructions for calling a subroutine three times during the main program.

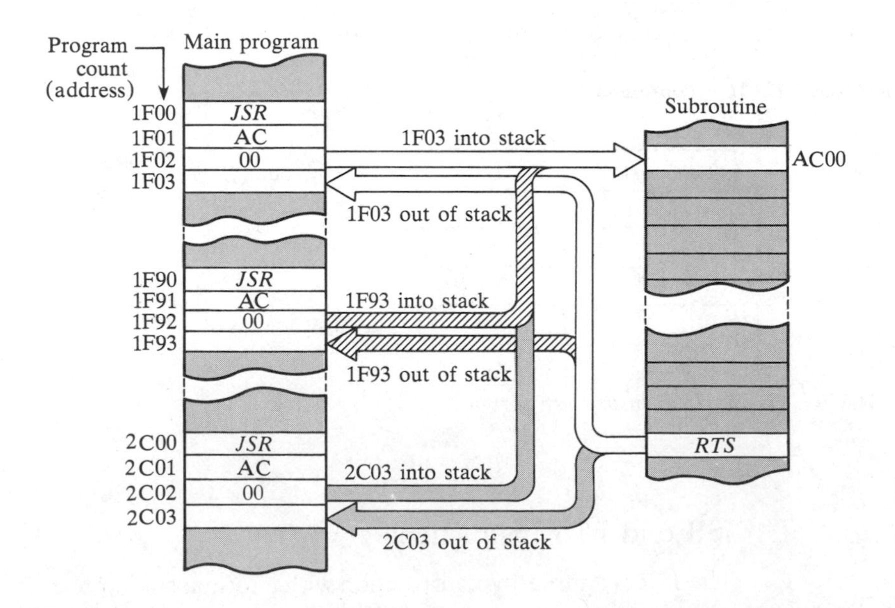

FIGURE 12–24 *Example of JSR and RTS used to call a subroutine several times.*

INPUT/OUTPUT OPERATIONS 12–9

To be useful, the microprocessor system must be able to accept data from external devices, such as keyboards, and to send results to external devices, such as video terminals or printers.

In the 6800 microprocessor, input and output (I/O) operations are handled as transfers to or from a memory address. That is, an I/O device is assigned an address and treated as a memory location.

For example, a keyboard can be assigned a two-byte address. To transfer a byte of data representing a keyboard character into the accumulator, the microprocessor executes an LDAA direct instruction. The second two bytes of this instruction make the address assigned to the keybord.

When the microprocessor puts this address on the address bus, a keyboard decoder enables the data from the keyboard onto the data bus to be transferred into the microprocessor system. Figure 12–25 illustrates this operation for a keyboard address of $C000_{16}$.

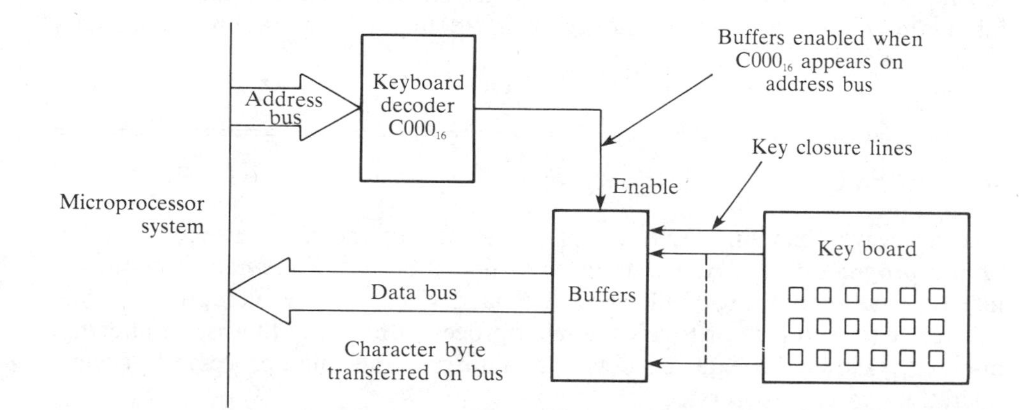

FIGURE 12–25 *Example of input of data byte from keyboard.*

An output device is also assigned a memory address. To transfer a byte of data from accumulator *A* to the external device, the microprocessor executes an STAA direct instruction. The second two bytes of this instruction make the address assigned to the device.

When STAA instruction is executed, the output device address is placed on the address bus and the data byte on the data bus. The device decoder responds to the address code by enabling the output buffers thus effecting the transfer to the device.

For example, Figure 12–26 shows an output operation in which the external device is a printer assigned an address of $D000_{16}$. When the address code is placed on the address bus, the printer's decoder enables the data transfer.

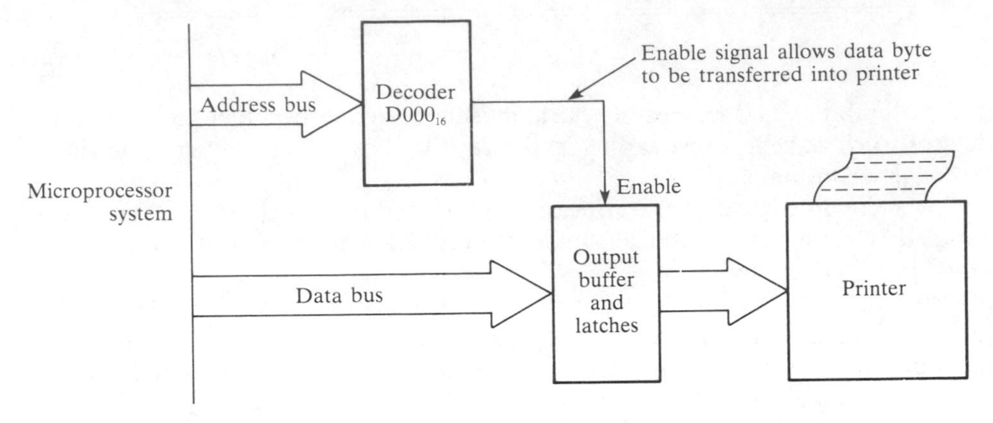

FIGURE 12–26 *Example of output of data byte to printer.*

As you have seen, the 6800 microprocessor handles input/output operations as data transfers to and from memory using the STAA direct and the LDAA direct instructions. These are programmed or *software* I/O operations, and the method is called *memory-mapped* I/O. Other microprocessors often have special input and output instructions rather than treating I/O operations as memory transfers.

12–10 INTERRUPTS

Another way of handling I/O transfers uses a concept known as *interrupts*. Rather than a *program controlled* I/O transfer as discussed in the previous section, the interrupt method in the 6800 allows an external device to tell the microprocessor that it is ready to send data or to ask the microprocessor for data: that is, an interrupt operation allows the external device to interrupt the microprocessor's normal operations to request service.

The 6800 microprocessor has available several types of interrupts. These are *Reset, NMI (nonmaskable interrupt), IRQ (interrupt request),* and *SWI (software interrupt)*. All of these are based on a principle called *vectored interrupts*.

The idea behind the concept of vectored interrupts is that, when any interrupt is initiated by an external device, the microprocessor must know what to do to *service* that certain device. To accomplish this, *interrupt vectors* are stored in the eight highest addresses of ROM. An interrupt vector is simply the beginning address of a service routine.

When the microprocessor gets a particular type of interrupt, it finishes the current instruction and goes to the address in ROM containing the appropriate interrupt vector. The interrupt vector tells the microprocessor where to go in RAM to begin the sequence of steps required to service the interrupting device. This process is illustrated in Figure 12–27.

Reset Interrupt

This is a *hardware* interrupt because a prescribed signal on the RESET line of the 6800 will initiate a RESET interrupt sequence. The purpose is to load the

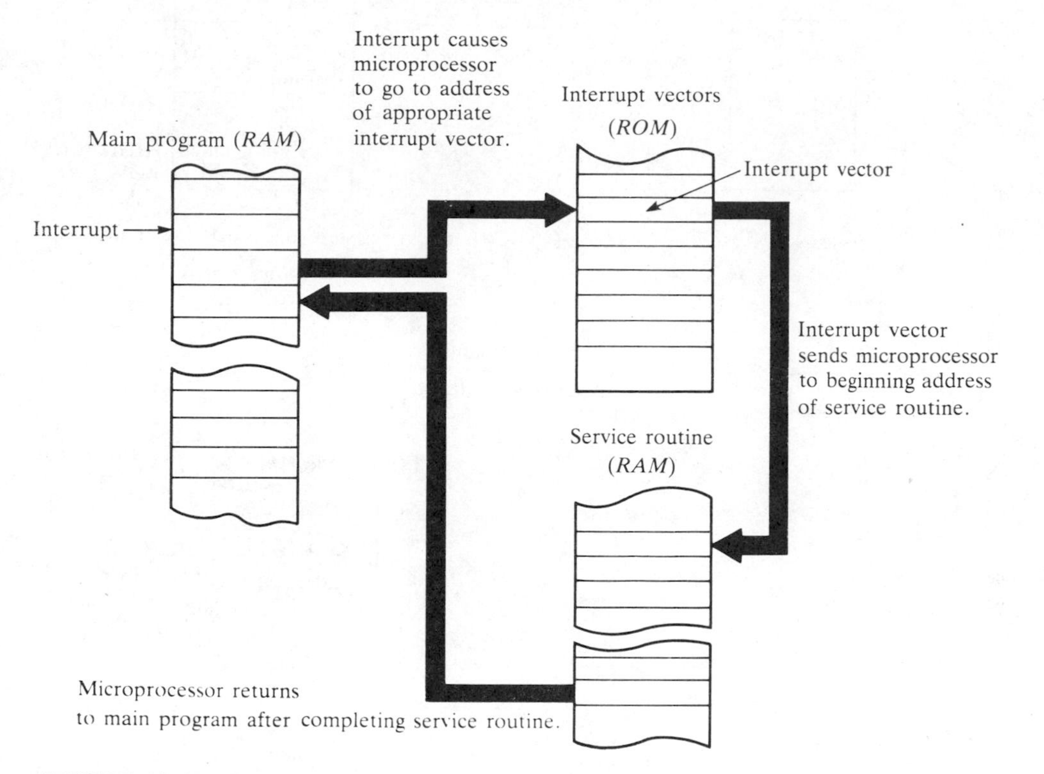

FIGURE 12–27 *Interrupt sequence.*

first address of a program to be executed into the program counter. This initializing program count is taken from the memory address specified by the interrupt vector. In most cases the microprocessor starts with a *monitor* program that initializes all registers and starts up the system.

NMI (Non-Maskable Interrupt)

This is also a hardware interrupt initiated by a signal on the $\overline{\text{NMI}}$ line. The term *nonmaskable* means that the microprocessor cannot ignore an interrupt signal on this line.

A HIGH to LOW transition on the $\overline{\text{NMI}}$ line initiates a nonmaskable interrupt sequence which first causes the microprocessor to *push* the contents of the program counter, index register, accumulators, and condition code register into the stack. This information is saved so that the microprocessor can return to where it left off in the main program at the time it was interrupted.

Next, the *interrupt mask bit* in the condition code register is SET. This allows the microprocessor to ignore any additional interrupts that might occur while it is servicing the current one. Finally, the microprocessor goes to the appropriate interrupt vector in ROM and gets the beginning address of the service routine for the interrupting device. When the interrupt service program is completed, the stack contents are *pulled* and the microprocessor resumes normal operation. Figure 12–28 demonstrates the NMI sequence.

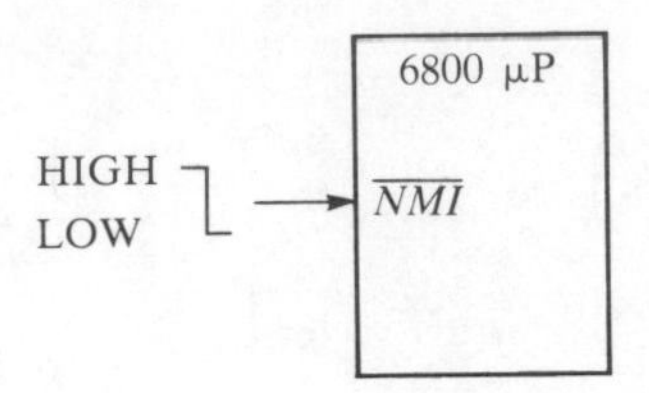

(a) Micro processor gets $\overline{NMI}$ signal.

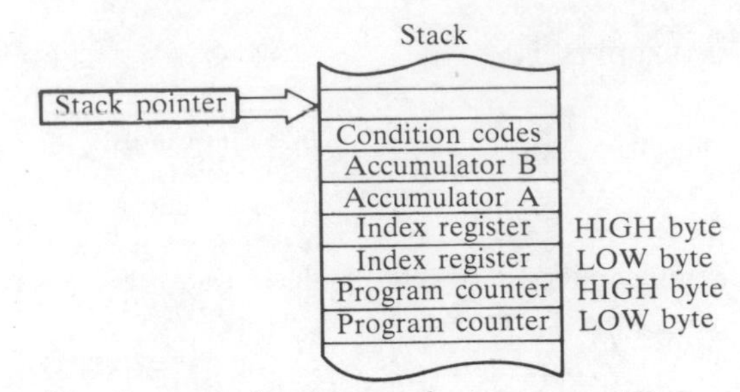

(b) Contents of registers and program counter pushed into stack.

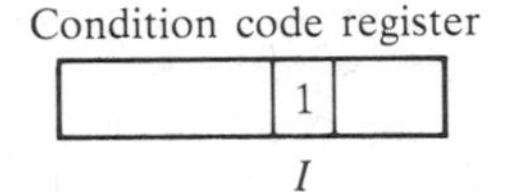

(c) Interrupt mask bit set.

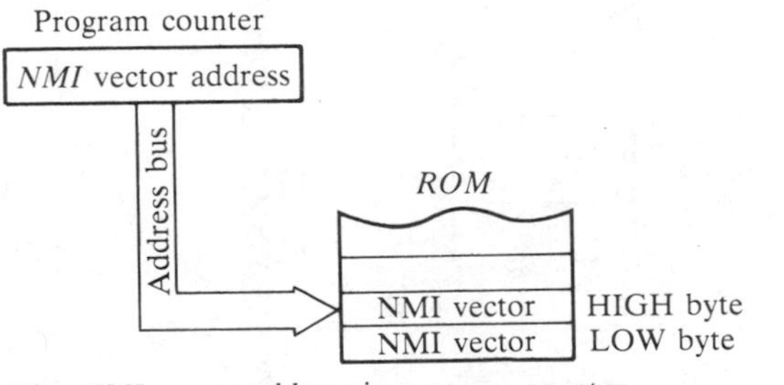

(d) *NMI* vector address in program counter.
NMI vector read out and placed in program counter.

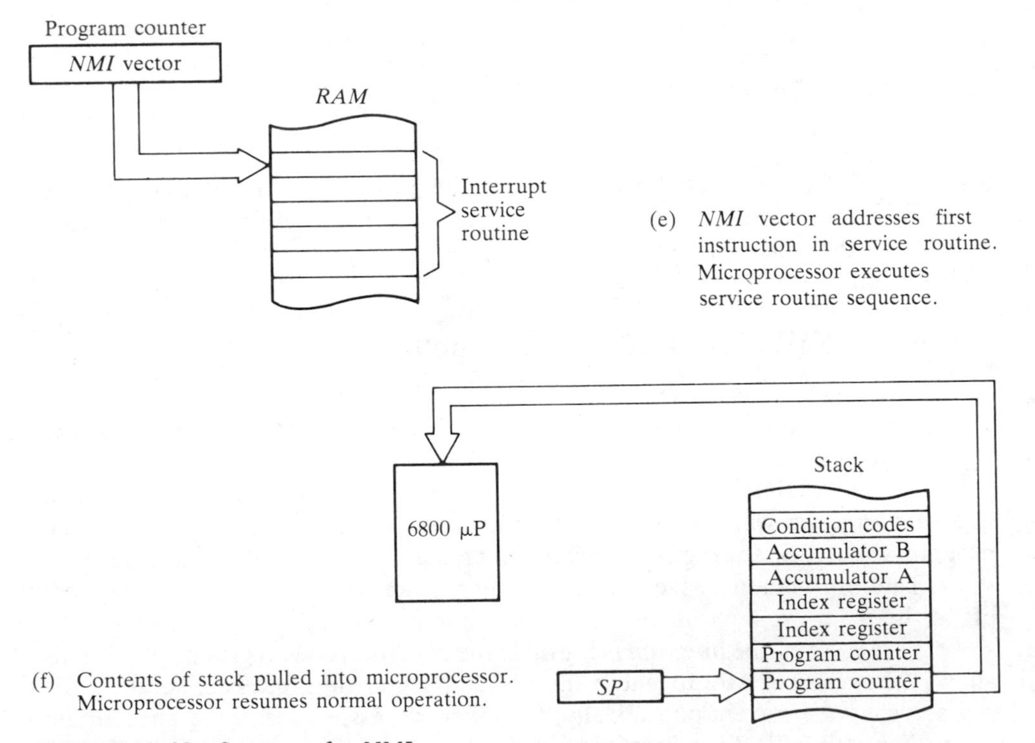

(e) *NMI* vector addresses first instruction in service routine. Microprocessor executes service routine sequence.

(f) Contents of stack pulled into microprocessor. Microprocessor resumes normal operation.

FIGURE 12–28 *Sequence for NMI.*

We can envision one application where a nonmaskable interrupt could be used. Suppose a microprocessor system is used to control a precision machine tool. Suppose, also, that a sensor picks up a hazardous condition such as someone's hand too close to the machine. The sensor generates a $\overline{NMI}$ signal to the microprocessor and a shutdown sequence is initiated in the order previously described.

IRQ (Interrupt Request)

This is also a hardware interrupt similar to the NMI. The difference is that the microprocessor may ignore the IRQ, depending on the state of the interrupt mask bit in the condition code register. The interrupt mask bit can be SET with a software instruction (SEI) and RESET with another instruction (CLI).

SWI (Software Interrupt)

SWI is an *instruction* that allows the microprocessor to be programmed to initialize an interrupt sequence similar to the hardware interrupts.

Problems

Section 12–1

12–1 Draw the basic block diagram of a microcomputer.

12–2 Specify the purpose of the following:
(a) address bus **(b)** data bus **(c)** I/O interface

12–3 How many bytes are required to represent the following decimal numbers?
(a) 125 **(b)** 253 **(c)** 478 **(d)** 28,952 **(e)** 63,175

Section 12–2

12–4 Which of the following is used to store the results of an ALU operation?
(a) data register **(b)** index register **(c)** program counter
(d) accumulator **(e)** condition code register

12–5 Bytes of data operated on in the ALU are called
(a) digits **(b)** op codes **(c)** operands **(d)** mnemonics

12–6 The hexadecimal representations of two numbers are $0A_{16}$ and $1C_{16}$. If these are added in the ALU, what is the hexadecimal representation of the accumulator's contents after the operation?

12–7 Which one of the following statements is true?
(a) The 6800 is a complete microcomputer.
(b) The 6800 has an eight-bit address bus.
(c) The 6800 has two accumulators.
(d) The 6800 has a 16-bit data bus.

12-8 The binary code for a microprocessor instruction is called
(a) operand **(b)** byte **(c)** op code **(d)** mnemonic

12–9 Name four indications that are provided by the condition code register.

12–10 If the program counter contains $00FA_{16}$, what is the address of the previously addressed byte in memory?

Section 12–3

12–11 The address register contains $00AF_{16}$. Address $00AF_{16}$ in memory contains $1C_{16}$. After a read operation, what are the contents of the data register? The program counter?

12–12 The address register contains $013E_{16}$. The data register contains 08_{16}. After a write operation, what are the contents of the selected memory location? The program counter?

Section 12–4

12–13 Name the two phases of a microprocessor cycle.

12–14 During which phase is the instruction op code read from memory?

12–15 During which phase is the op code decoded?

12–16 During which phase is the instruction carried out?

Section 12–5

12–17 The first byte of an instruction is always the
(a) address **(b)** data **(c)** op code **(d)** operand

12–18 An instruction can consist of
(a) only one byte **(b)** two bytes **(c)** three bytes
(d) all of the above

Section 12–6

12–19 Which of the following instructions use *inherent* addressing?
(a) ADDA **(b)** CLRA **(c)** STAA **(d)** LDAA **(e)** WAI

12–20 The program sequence in Figure 12–29 appears in memory. Both instructions use *immediate* addressing. What is the result of this sequence of operations and where can the result be found?

Address	Contents
00F0	*LDAA*
00F1	05_{16}
00F2	*ADDA*
00F3	02_{16}

FIGURE 12–29

12–21 The program sequence in Figure 12–30 appears in memory. Both instructions use *direct* addressing. What is the result in the hexadecimal and where can it be found?

Address	Contents
00DF	*LDAA*
00E0	E3
00E1	*ADDA*
00E2	E4
00E3	08
00E4	1B

FIGURE 12–30

12–22 Repeat Problem 12–21 for the following sequence using *extended* addressing in Figure 12–31.

Address	Contents
01B3	*LDAA*
01B4	01
01B5	BC
01B6	*ADDA*
01B7	01
01B8	BD
01B9	*STAA*
01BA	01
01BB	BE
01BC	0A
01BD	1B
01BE	

FIGURE 12–31

12–23 The index register contains 0005_{16}. When the following program sequence in Figure 12–32 is run, what will be the contents of the accumulator? Both instructions use *indexed* addressing.

Address	Contents
0000	*LDAA*
0001	01
0002	*ADDA*
0003	01
0004	03
0005	A4
0006	CD
0007	9E
0008	2D
0009	1C

FIGURE 12–32

12–24 To what address will a program branch when the BRA op code located at address $00FA_{16}$ is followed by $7C_{16}$?

Section 12–8

12–25 Explain the difference between the JMP and the JSR instructions.

12–26 Write the instruction format for a jump to address $14A2_{16}$.

12–27 The microprocessor encounters a JSR op code at address $0F12_{16}$. What goes into the stack when this instruction is executed?

A

INTEGRATED CIRCUIT TECHNOLOGIES

A–1 INTEGRATED CIRCUIT INVERTERS

TTL Inverter

Transistor-transistor logic (TTL or T²L) is one of the most widely used integrated circuit technologies. The basic element in TTL circuits is the *bipolar transistor*. TTL has replaced earlier circuit technologies such as *resistor-transistor logic* (RTL) and *diode-transistor logic* (DTL).

The *logic function* of an inverter or any type of gate is always the same regardless of the type of circuit technology that is used. TTL is only *one* way to build an inverter or any other logic function. Figure A–1 shows a TTL circuit for an inverter.

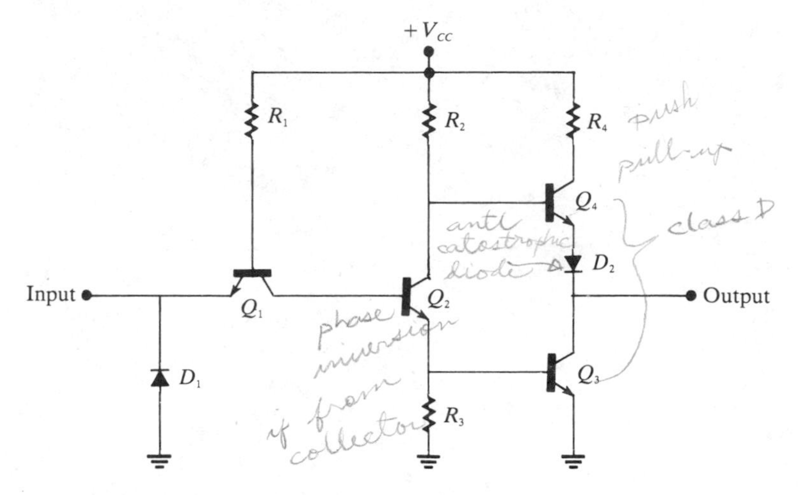

FIGURE A–1 *TTL Inverter circuit.*

Q_1 is the input coupling transistor, and D_1 is the input clamp diode. Transistor Q_2 is called a *phase splitter,* and the combination of Q_3 and Q_4 forms the output circuit often referred to as a *totem-pole* arrangement. The operation of each of these components in the circuit including diode D_2 will now be discussed.

When the input is a HIGH, the base-emitter junction of Q_1 is reverse-biased, and the base-collector junction is forward-biased. This permits current through R_1 and the base-collector junction of Q_1 into the base of Q_2, thus driving Q_2 into saturation. As a result, Q_3 is turned on by Q_2 and its collector voltage, which is the output, is near ground potential. We therefore have a LOW output for a HIGH input. At the same time, the collector of Q_2 is at a sufficiently low voltage level to keep Q_4 off.

When the input is LOW the base-emitter junction of Q_1 is forward-biased and the base-collector junction is reverse-biased. There is current through R_1 and the base-emitter junction of Q_1 out to the LOW input. A LOW provides a path to ground for the current. There is no current into the base of Q_2, so it is off. The collector of Q_2 is HIGH, thus turning Q_4 on. A saturated Q_4 provides a low impedance path from V_{CC} to the output; we therefore have a HIGH on the output for a LOW on the input. At the same time, the emitter of Q_2 is at ground potential, keeping Q_3 off.

The purpose of diode D_1 in the TTL circuit is to prevent negative spikes of voltage on the input from damaging Q_1. Diode D_2 insures that Q_4 will turn off when Q_2 is on (HIGH input). In this condition the collector voltage of Q_2 is equal to the V_{BE} of Q_3 plus the V_{CE} of Q_2. Diode D_2 provides an additional V_{BE} equivalent drop in series with the base-emitter junction of Q_4 to insure its turn-off when Q_2 is on.

The operation of the TTL inverter for the two input states is illustrated in Figure A–2. In the upper circuit, the base of Q_1 is 2.1V above ground so that Q_2 and Q_3 are on. In the lower circuit, the base of Q_1 is about 0.7V above ground—not enough to turn Q_2 and Q_3 on.

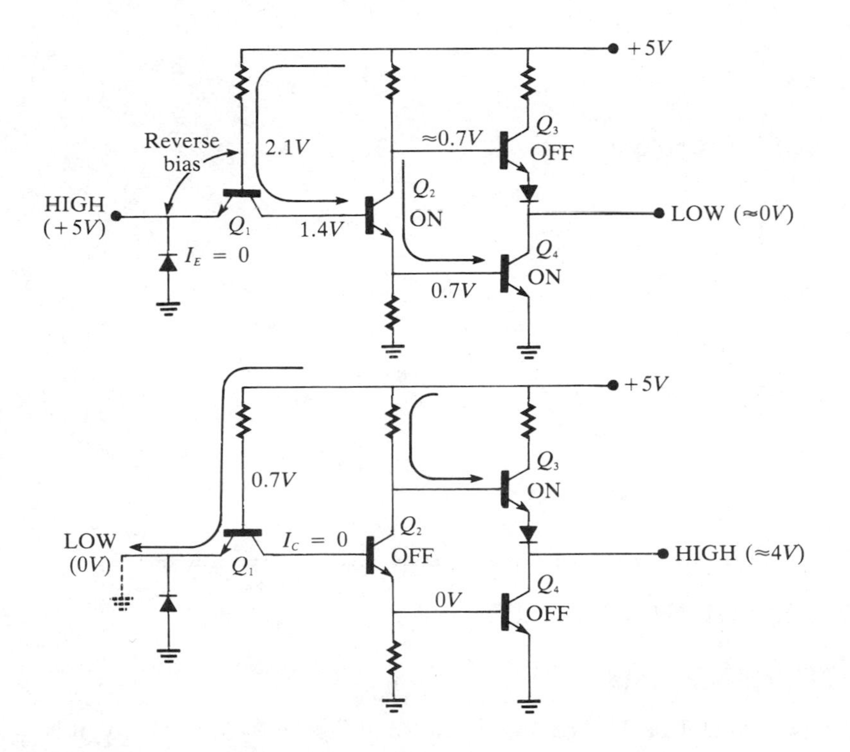

FIGURE A–2 *TTL inverter operation.*

CMOS Inverter

Another common integrated circuit technology uses the MOSFET as its basic element and is called *complementary* MOS logic. CMOS uses both p and n channel enhancement MOSFETs, as shown in the inverter circuit in Figure A–3.

When a HIGH is applied to the input, the p-channel MOSFET Q_1 is off, and the n-channel MOSFET Q_2 is on. This connects the output to ground through the *on* resistance of Q_2, resulting in a LOW output. When a LOW is applied to the input, Q_1 is on and Q_2 is off. This connects the output to $+V_{DD}$ through the *on* resistance of Q_1, resulting in a HIGH output. The operation is illustated in Figure A–4.

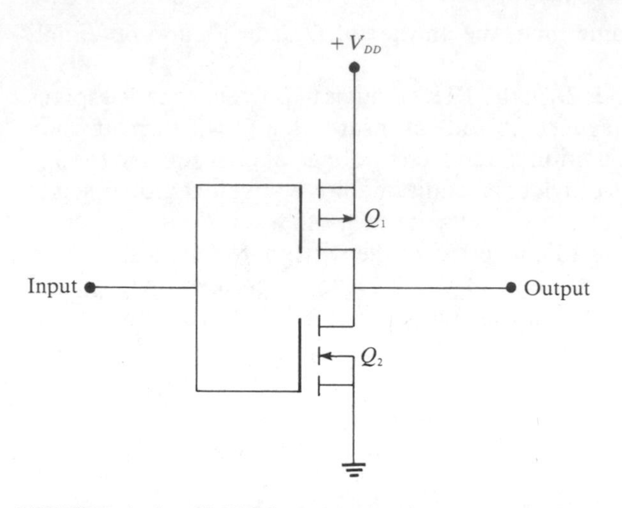

FIGURE A–3 *CMOS inverter circuit.*

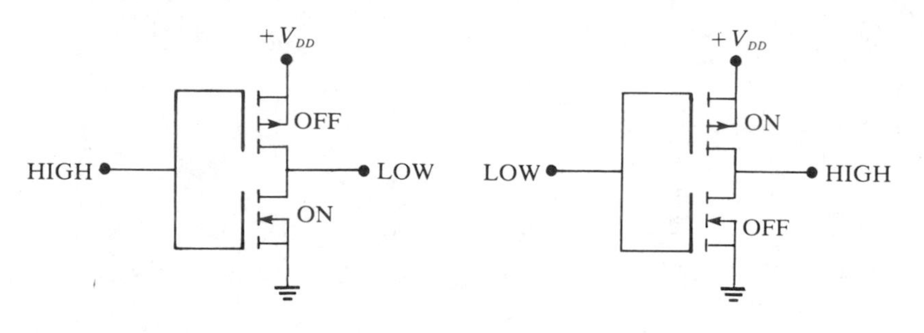

FIGURE A–4 *CMOS inverter operation.*

A–2 INTEGRATED CIRCUIT NAND GATES

TTL NAND Gate

A two-input TTL NAND gate is shown in Figure A–5. Basically, it is the same as the inverter circuit except for the additional input emitter of Q_1. TTL technology utilizes *multiple-emitter transistors* for the input devices. These can be compared to the diode arrangement, as shown in Figure A–6.

Perhaps you can understand the operation of this circuit better by visualizing Q_1 in Figure A–5 replaced by the diode arrangement. A LOW on either input A or input B forward-biases the respective diode and reverse-biases D_3 (Q_1 base-collector junction). This action keeps Q_2 off and results in a HIGH output in the same way as described for the TTL inverter. Of course, a LOW on both inputs will do the same thing.

A HIGH on *both* inputs reverse-biases both input diodes and forward-biases D_3 (Q_1 base-collector junction). This action turns Q_2 on and results in a LOW output in the same way as described for the TTL inverter.

You should recognize this operation as that of the NAND function: the output is LOW if and only if all inputs are HIGH.

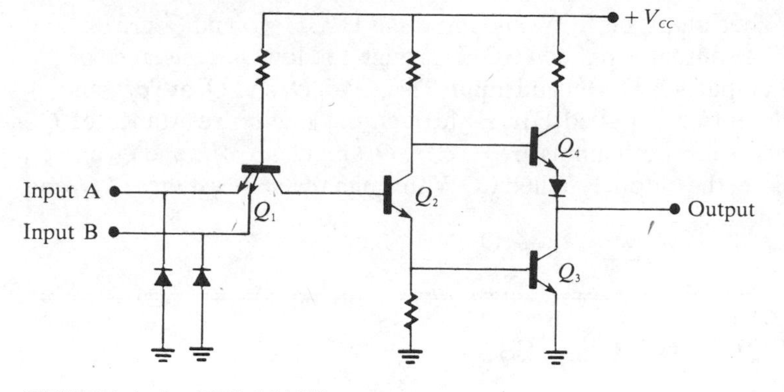

FIGURE A–5 *TTL NAND gate circuit.*

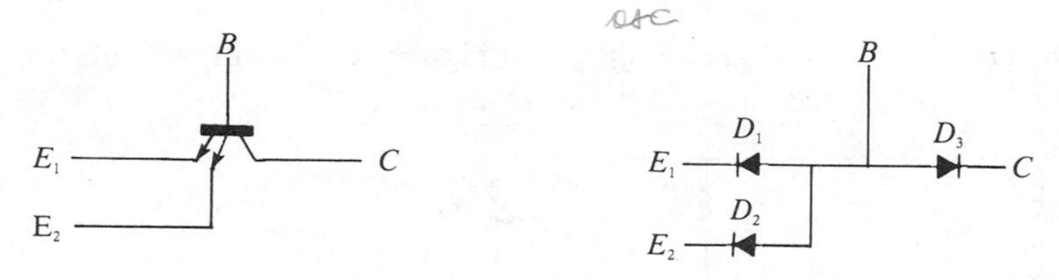

FIGURE A–6 *Diode equivalent of a TTL multiple-emitter transistor.*

CMOS NAND Gate

Figure A–7 shows a CMOS NAND Gate with two inputs. Notice the arrangement of the complementary pairs (*n*- and *p*-channel MOSFETs).

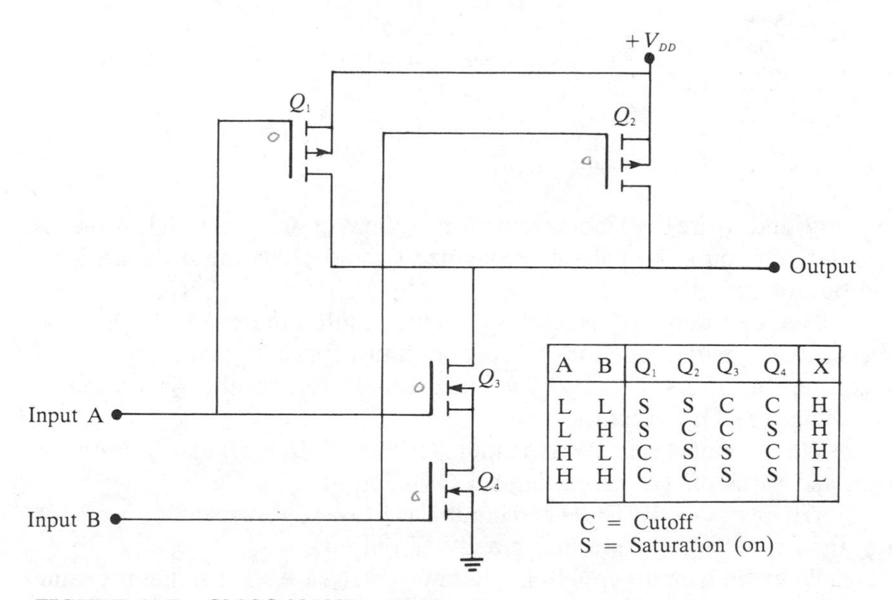

A	B	Q_1	Q_2	Q_3	Q_4	X
L	L	S	S	C	C	H
L	H	S	C	C	S	H
H	L	C	S	S	C	H
H	H	C	C	S	S	L

FIGURE A–7 *CMOS NAND gate circuit.*

The operation is as follows. When both inputs are LOW, Q_1 and Q_2 are *on*, and Q_3 and Q_4 are *off*. The output is pulled HIGH through the *on* resistance of Q_1

and Q_2 in parallel. When input A is LOW and input B is HIGH, Q_1 and Q_4 are *on,* and Q_2 and Q_3 are *off.* The output is pulled HIGH through the low *on* resistance of Q_1.

When input A is HIGH and input B is LOW, Q_1 and Q_4 are *off,* and Q_2 and Q_3 are *on.* The output is pulled HIGH through the low *on* resistance of Q_2.

Finally, when both inputs are HIGH, Q_1 and Q_2 are *off,* and Q_3 and Q_4 are low *on.* In this case, the output is pulled LOW through the *on* resistance of Q_3 and Q_4 in series to ground.

A–3 INTEGRATED CIRCUIT NOR GATES

TTL NOR Gate

A two-input TTL NOR gate is shown in Figure A–8. Compare this to the NAND gate circuit and notice the additional transistors.

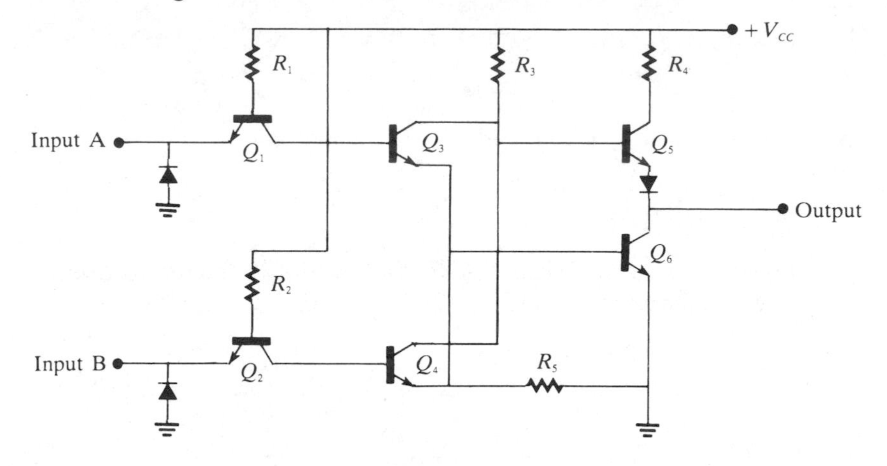

FIGURE A–8 *TTL NOR gate circuit.*

Q_1 and Q_2 are both input transistors. Q_3 and Q_4 are in parallel and act as a phase splitter. Of course you should recognize Q_5 and Q_6 as the ordinary TTL totem-pole output circuit.

The operation is as follows. When both inputs are LOW, the forward-biased base-emitter junctions of the input transistors pull current away from the phase-splitter transistors Q_3 and Q_4, keeping them *off.* As a result, Q_5 is *on* and Q_6 is *off,* producing a HIGH output.

When input A is LOW and input B is HIGH, Q_3 is *off* and Q_4 is *on.* Q_4 turns on Q_6 and turns off Q_5, producing a LOW output.

When input A is HIGH and input B is LOW, Q_3 is *on* and Q_4 is off. Q_3 turns on Q_6 and turns off Q_5, producing a LOW output.

When both inputs are HIGH, both Q_3 and Q_4 are *on.* This has the same effect as either one being *on,* turning Q_6 *on* and Q_5 *off.* The result is still a LOW output. You should recognize this operation as that of the NOR function. (The output is LOW when any of the inputs are HIGH.)

CMOS NOR Gate

Figure A–9 shows a CMOS NOR gate with two inputs. Notice the arrangement of the complementary pairs.

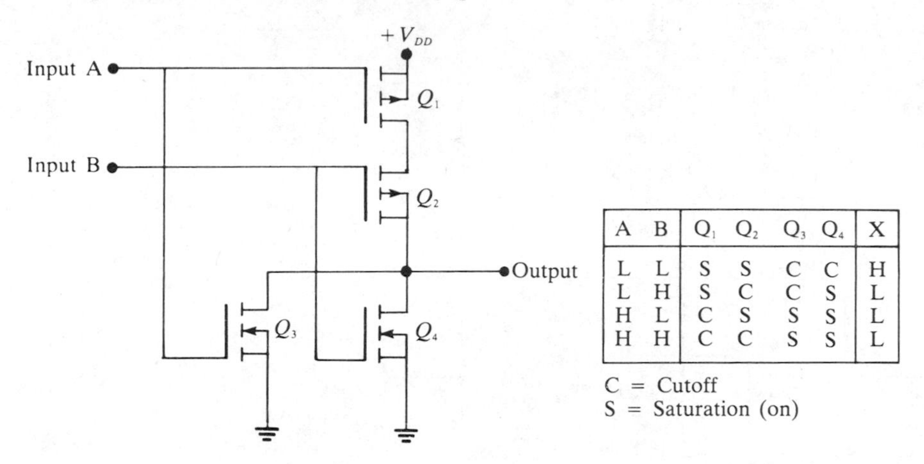

A	B	Q_1	Q_2	Q_3	Q_4	X
L	L	S	S	C	C	H
L	H	S	C	C	S	L
H	L	C	S	S	S	L
H	H	C	C	S	S	L

C = Cutoff
S = Saturation (on)

FIGURE A–9 *CMOS NOR gate circuit.*

The operation is as follows. When both inputs are LOW, Q_1 and Q_2 are *on*, and Q_3 and Q_4 are *off*. As a result the output is pulled HIGH through the low *on* resistance of Q_1 and Q_2 in series.

When input A is LOW and input B is HIGH, Q_1 and Q_3 are *on*, and Q_2 and Q_4 are *off*. The output is pulled LOW through the low *on* resistance of Q_3 to ground.

When input A is HIGH and input B is LOW, Q_1 and Q_3 are *off*, and Q_2 and Q_4 are *on*. The output is pulled LOW through the *on* resistance of Q_4 to ground.

When both inputs are HIGH, Q_1 and Q_2 are *off*, and Q_3 and Q_4 are *on*. The output is pulled LOW through the *on* resistance of Q_3 and Q_4 in parallel to ground.

TTL AND GATES AND OR GATES A–4

Figure A–10 shows a TTL two-input AND gate and a two-input OR gate. Compare these to the NAND and NOR gate circuits and notice the additional circuitry that is shaded. In each case, this transistor arrangement provides an inversion so the NAND becomes AND and the NOR becomes OR.

OPEN-COLLECTOR GATES A–5

The TTL gates described in the previous sections all had the totem-pole output circuit; another type of output available in TTL integrated circuits is the

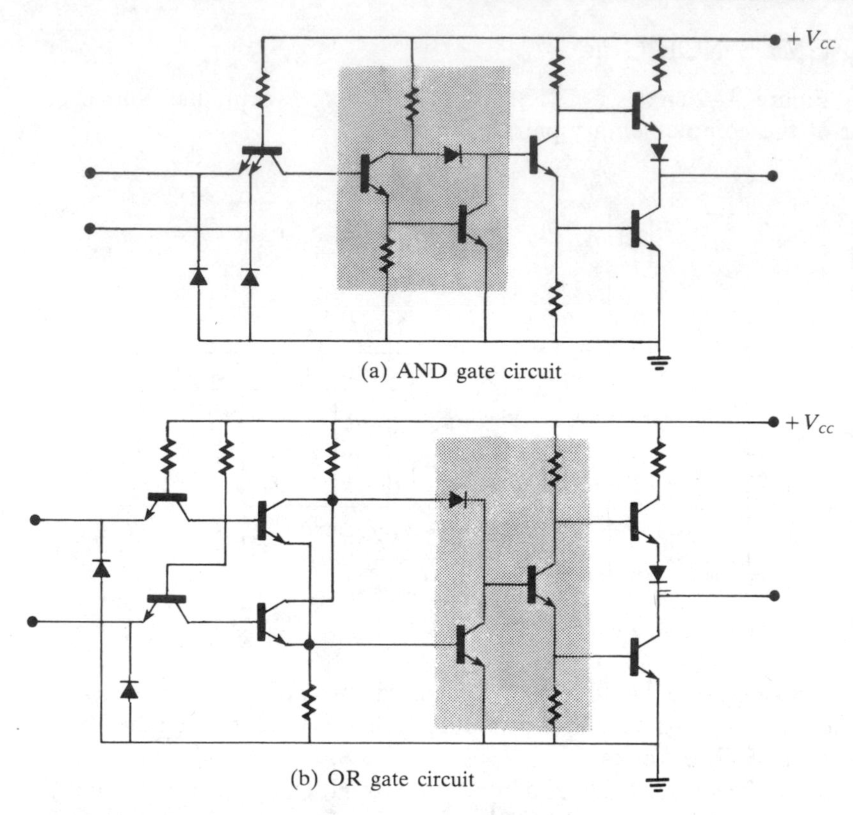

FIGURE A–10 *TTL AND gate and OR gate circuits.*

open-collector. A TTL inverter with an open collector output is shown in Figure A–11(a). The other gates are also available with this type of output.

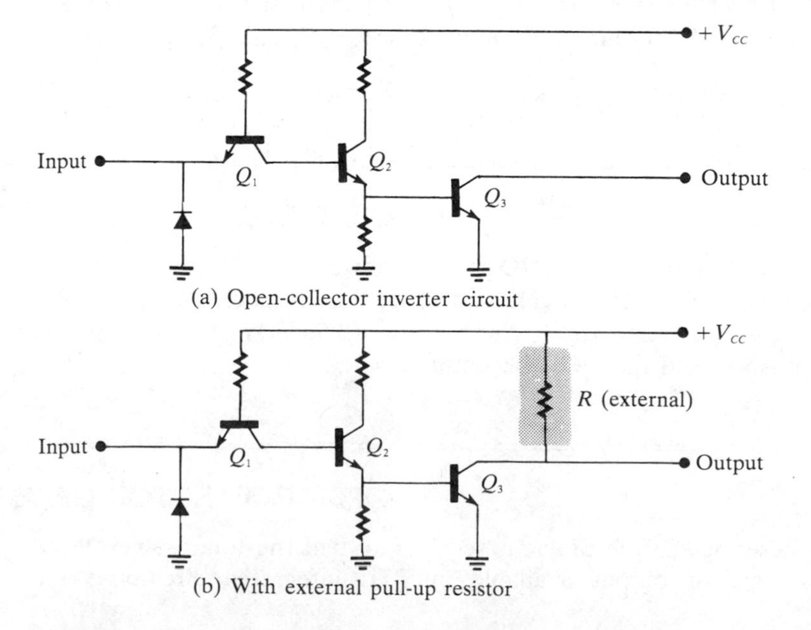

FIGURE A–11 *TTL inverter with open collector output.*

Notice that the output is the collector of transistor Q_3 with nothing connected to it; hence the name *open-collector*. In order to get the proper HIGH and LOW logic levels out of the circuit, an *external pull-up resistor* must be connected to V_{CC} from the collector of Q_3, as shown in Figure A–11 (b). When Q_3 is *off,* the output is pulled up to V_{CC} through the external resistor. When Q_3 is *on,* the output is connected to near-ground through the saturated transistor.

Wired-NOR Operation

The primary advantage of an open collector gate is that the outputs of several of these gates can be connected together to achieve a *wired logic* function. This cannot be done with the totem-pole outputs because damage to the output transistors will result.

When several open collector inverters are connected as shown in Figure A–12, a *wired-NOR* operation is created. Part (b) of the figure is the equivalent logic diagram. The dashed NOR gate symbol is not a physical gate, but represents the effective logic function created by connecting the open-collector outputs together.

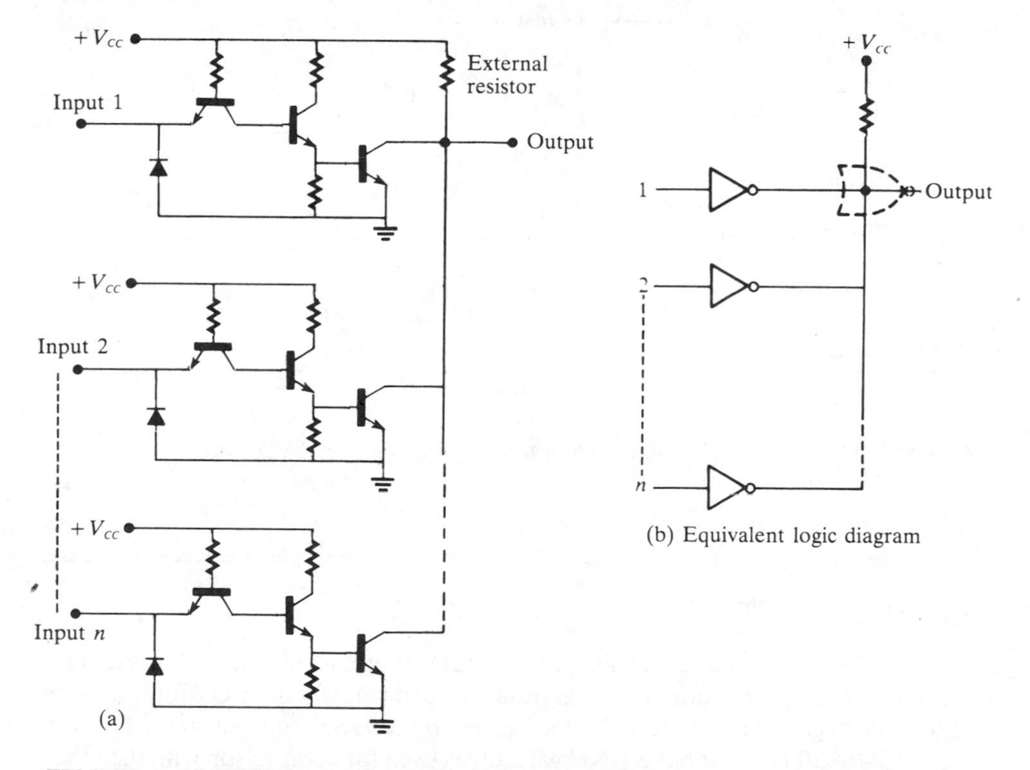

FIGURE A–12 *TTL inverters connected for wired NOR.*

The operation is as follows. When *any* of the inputs are HIGH, the corresponding output transistors saturate, thus pulling the output LOW. It makes no difference whether any of the other output transistors are *on* or *off,* as long as at least one is *on* the output is LOW. The output is HIGH only when all of the output transistors are *off* as a result of all of the inputs being LOW. Of course, you recognize this operation as that of the NOR gate.

Wired-AND Operation

By connecting the outputs of open-collector AND gates together, a wired-AND function is achieved as shown in Figure A–13. Only the logic diagram is shown because the connections to the output transistors are the same as for the inverters.

As you can see, if any AND gate input is LOW, the output of the associated AND gate is LOW, thus pulling the output of the wire-AND configuration LOW—regardless of whether the output transistors of the other gates are *on* or *off*. When all inputs are HIGH, all output transistors are *off*, and the output is pulled up to V_{CC} through the external resistors. This produces an AND function to which other inputs can be added by simply connecting additional open-collector AND gates to the circuit.

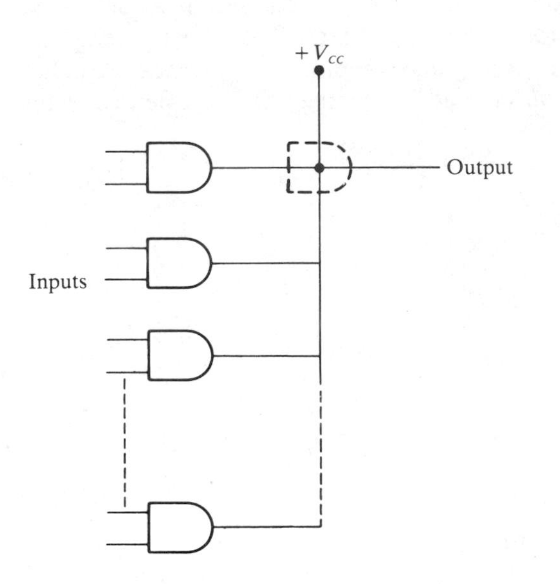

FIGURE A–13 *Wired-AND configuration using open-collector AND gates.*

A–6 MORE TTL CIRCUITS

Figure A–14 shows the basic TTL gate circuit that was discussed earlier. It is a *current sinking* type of logic that draws current from the load when in the LOW output state and sources negligible current to the load when in the HIGH output state. This is a *standard* 54/74 TTL two-input NAND gate and is used for comparison with the other types of TTL circuits.

Low Power TTL (54L/74L)

The 54L/74L series of TTL circuits is designed for low power consumption. A typical gate circuit is shown in Figure A–15. Notice that the circuit's

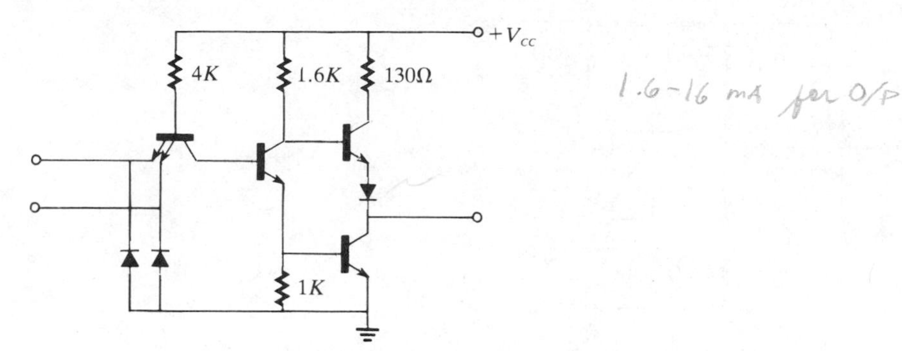

FIGURE A–14 *A 54/74 series standard TTL NAND gate.*

resistor values are considerably higher than those of the standard gate in Figure A–14. This, of course, results in less current and therefore less power but increases the switching time of the gate. The typical power dissipation of a standard 54/74 gate is 10 mW, and that of a 54L/74L gate is 1 mW. The savings in power, however, is paid for in loss of speed. A typical standard 54/74 gate has a propagation delay time of 10 ns, compared to 33 ns for a 54L/74L gate.

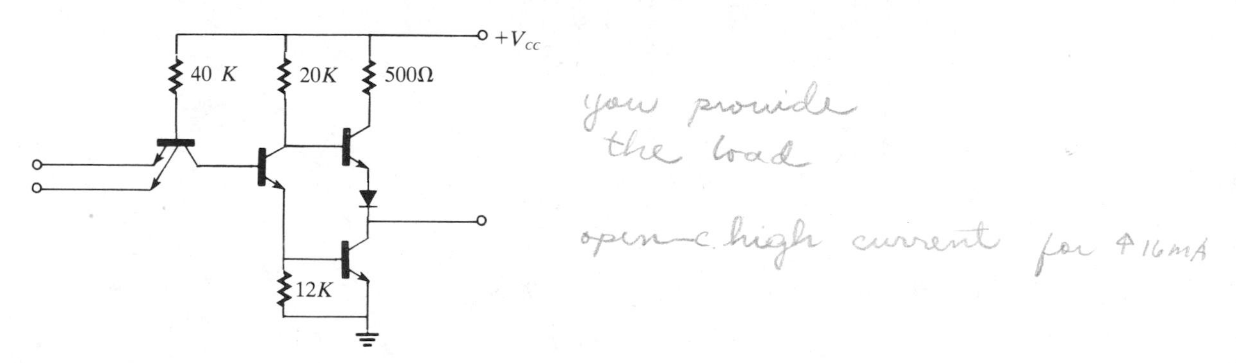

FIGURE A–15 *A 54L/74L series TTL NAND gate.*

High-Speed TTL (54H/74H)

The 54H/74H series of TTL circuits is designed to provide less propagation delay than the standard gates, but the cost is higher power consumption. Notice that the 54H/74H circuit in Figure A–16 has lower resistor values than standard TTL. Typical propagation delay for a 54H/74H gate is 6 ns, but the power consumption is 22 mW.

Schottky TTL (54S/74S)

The 54S/74S series of TTL circuits provides a faster switching time than the 54H/74H series but requires much less power. These circuits incorporate Schottky diodes to prevent the transistors from going into saturation, thereby decreasing the time for a transistor to turn on or off.

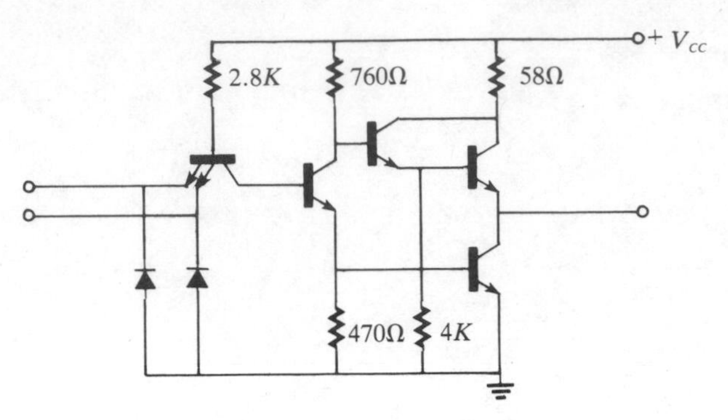

FIGURE A–16 *A 54H/74H series TTL NAND gate.*

Typical propagation delay for a 54S/74S gate is 3 ns, and the power dissipation is 19 mW. Figure A–17 shows a Schottky gate circuit.

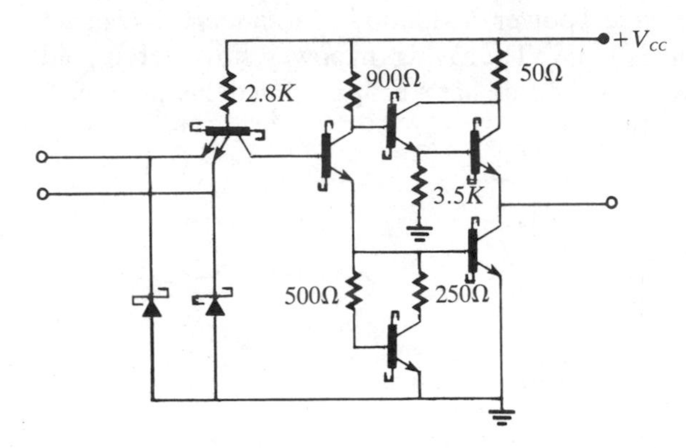

FIGURE A–17 *54S/74S series TTL NAND gate.*

Low-Power Schottky TTL (54LS/74LS)

This TTL series compromises speed to obtain a lower power dissipation. A NAND gate circuit is shown in Figure A–18. Notice that the input uses Schottky diodes rather than the conventional transistor input. The typical power dissipation for a gate is 2 mW, and the propagation delay is 9.5 ns.

TTL Comparison

A comparison of the performance of the various TTL families just discussed is given in Table A–1 for both gates and flip-flops.

More on TTL Loading

TTL loading considerations were discussed in Chapter 3. Some additional information will now be covered concerning the differences in loading among the five TTL series just discussed. Recall that a single gate input connected to

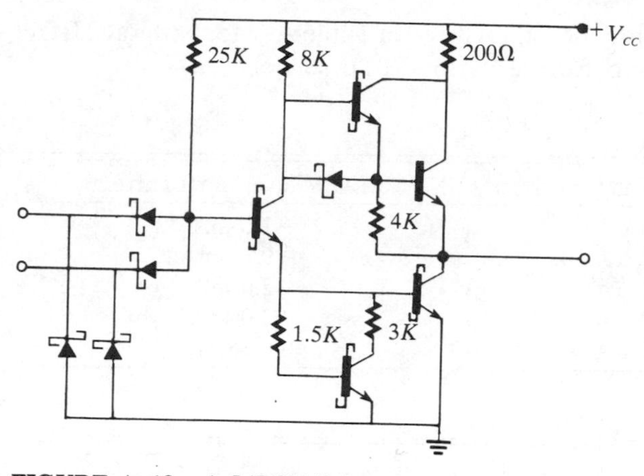

FIGURE A–18 *A 54LS/74LS series TTL NAND gate.*

TABLE A–1 *Typical TTL performance characteristics.*

Series	Gates		Flip-flops
	Propagation delay time	Power dissipation	Clock frequency range
54/74	10 ns	10 mW	dc to 35 MHz
54L/74L	33 ns	1 mW	dc to 3 MHz
54H/74H	6 ns	22 mW	dc to 50 MHz
54S/74S	3 ns	19 mW	dc to 125 MHz
54LS/74LS	9.5 ns	2 mW	dc to 45 MHz

a gate output of the same type of circuit represents a *unit load* to the driving gate. It was stated that the current into the output of a driving gate at the LOW level from one unit load was 1.6 mA. This was for the *standard* series 54/74 TTL. A single unit load connection is illustrated in Figure A–19. If unused inputs of the load gate are connected to the input being driven, it is still considered one unit load in the LOW state, because the current is essentially determined by the internal pull-up resistor of the load gate. This situation is also illustrated in Figure A–19.

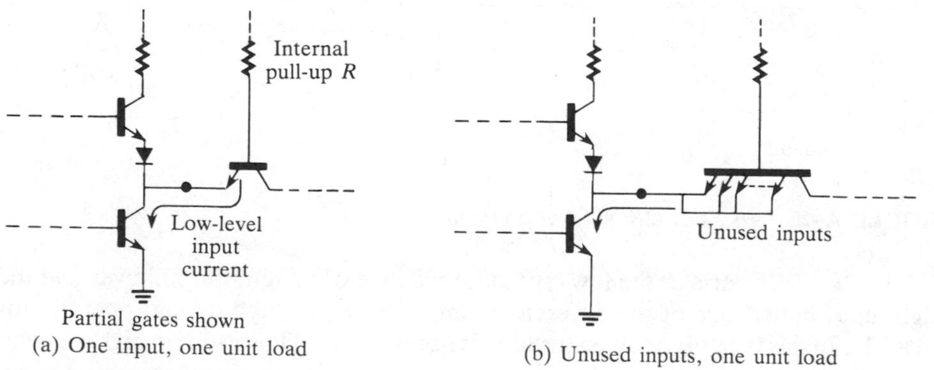

(a) One input, one unit load

(b) Unused inputs, one unit load

FIGURE A–19 *TTL loading.*

The load current for one load is listed in Table A–2 for both the HIGH state and the LOW state outputs for the various TTL series.

TABLE A–2

Series	HIGH level input current	LOW level input current
54/74	40 μA	−1.6 mA
54L/74L	10 μA	−0.18 mA
54H/74H	50 μA	−2.0 mA
54S/74S	50 μA	−2.0 mA
54LS/74LS	20 μA	−0.36 mA

A–7 ECL CIRCUITS

ECL stands for *emitter-coupled logic* which, like TTL, is a bipolar technology. The typical ECL circuit, as shown in Figure A–22(a) consists of a differential amplifier input circuit, a bias circuit, and emitter follower outputs.

The emitter follower outputs provide the OR logic function and its NOR complement, as indicated by Figure A–20(b).

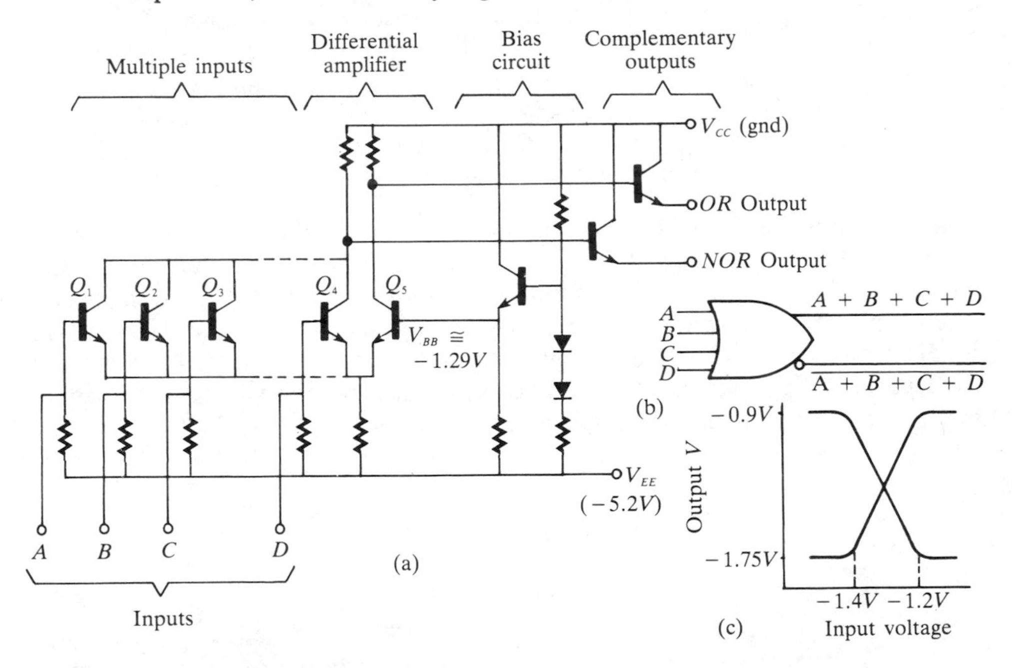

FIGURE A–20 *An ECL OR/NOR gate circuit.*

Because of the low output impedance of the emitter follower and the high input impedance of the differential amplifier input, high fan-out operation is possible. In this type of circuit, saturation is not possible. This results in higher power consumption and limited voltage swing (less than 1V). The lack of saturation in the ECL circuits permits high frequency switching.

The V_{CC} pin is normally connected to ground, and −5.2V from the power supply is connected to V_{EE} for best operation. Notice in Figure A–20(c) that the output varies from a LOW level of −1.75V to a HIGH level of −0.9 V with respect to ground. In positive logic, a 1 is the HIGH level (less negative) and a 0 is the LOW level (more negative).

ECL Circuit Operation

Beginning with LOWs on all inputs (typically −1.75 V), assume that Q_1 through Q_4 are off because the base-emitter junctions are reverse-biased, and that transistor Q_5 is conducting (but not saturated). The bias circuit holds the base of Q_5 at −1.29 V, and therefore its emitter is approximately 0.8 V below the base at −2.09 V. The voltage differential from base to emitter of the input transistors Q_1 through Q_4 is −2.09 V − (−1.75 V) = −0.34 V. This is less than the forward-bias voltage of these transistors, and they are therefore off. This condition is shown in Figure A–21(a).

When any one (or all) of the inputs are raised to the HIGH level (−0.9 V), that transistor (or transistors) will conduct. When this happens, the voltage at the emitters of Q_1 through Q_5 increases from −2.09 V to −1.7 V (one base-emitter drops below the −0.9 V base). Since the base of Q_5 is held at a constant −1.29 V by the bias circuit, Q_5 turns off. The resulting collector voltages are coupled through the emitter followers to the output terminals. Because of the differential action of Q_1 through Q_4 with Q_5, Q_5 is off when one or more of the Q_1 through Q_4 transistors conduct, thus providing simultaneous complementary outputs. This is shown in Figure A–21(b).

When all of the inputs are returned to the LOW state, Q_1 through Q_4 are again cut off and Q_5 conducts.

Noise Margin

As you know, the noise margin of a gate is the measure of its immunity to undesired voltage fluctuations (noise). Typical ECL circuits have noise margins from about 0.2 V to 0.25 V. This is less than that for TTL and makes ECL unreliable in high noise environments.

Comparison of ECL with Schottky TTL

Table A–3 shows a comparison of typical values of key parameters for Schottky TTL and ECL.

TABLE A–3

	Propagation delay time	Power dissipation per gate	Voltage swing	dc supply voltage	Flip-flop clock freq
54S/74S	3 ns	19 mW	3.0 *V*	+5 *V*	125 MHz
ECL	1 ns	60 mW	0.85 *V*	−5.2 *V*	500 MHz

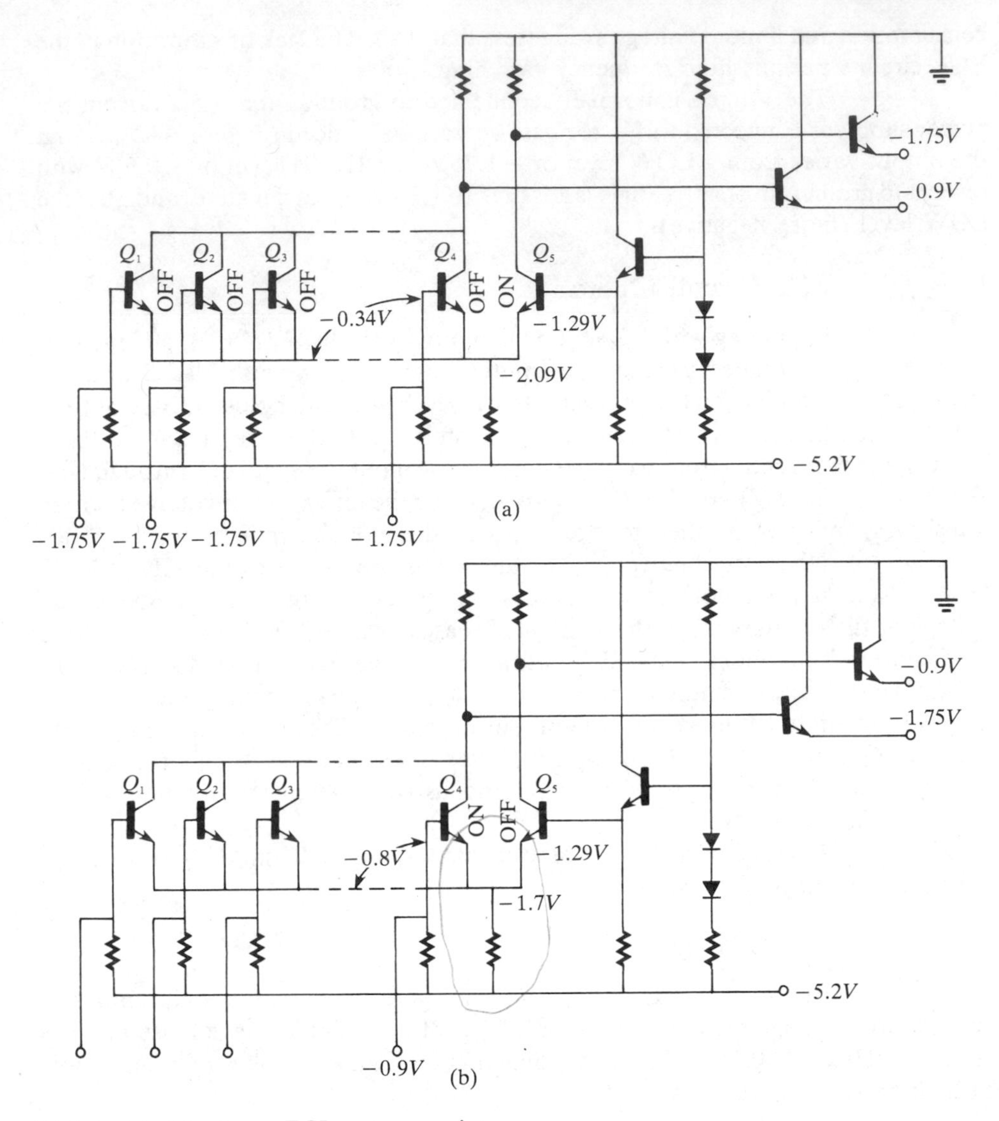

FIGURE A–21 *Basic ECL gate operation.*

A–8 I²L CIRCUITS

Integrated injection logic (I²L) is a bipolar technology that allows extremely high component densities on a chip (up to ten times that of TTL). I²L is being used for complex LSI functions such as microprocessors and is simpler to fabricate than either TTL or MOS. It also has a low power requirement and reasonably good switching speeds that are improving all the time.

The basic I²L gate is extremely simple, as indicated in Figure A–22. Transistor Q_1 acts as a current source and active pull-up, and the multiple-collector transistor Q_2 operates as an inverter.

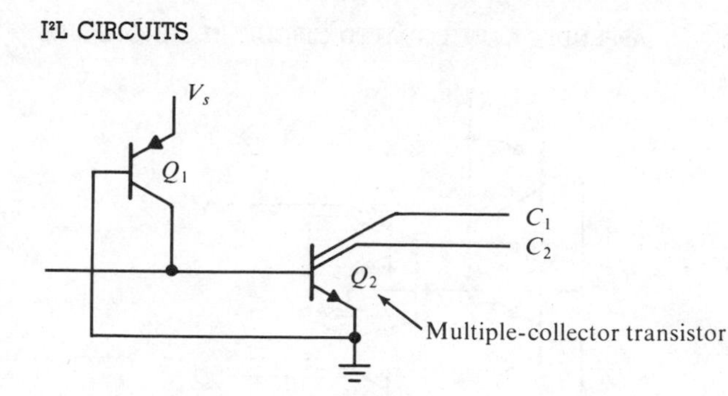

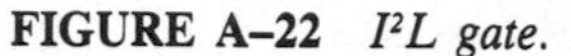
FIGURE A–22 *I²L gate.*

Because the base of Q_1 and the emitter of Q_2 are common and the collector of Q_1 and the base of Q_2 are common, the entire I²L gate (constructed on a silicon chip) takes only the space of a single TTL multiple-emitter transistor.

Transistor Q_1 is called a *current-injector transistor* because, when its emitter is connected to an external power source, it can supply current into the base of Q_2. Switching action of Q_2 is accomplished by steering the injector current as follows. A LOW on the base of Q_2 will pull the injector current away from the base of Q_2 and through the low impedance path(s) provided by the driving gate(s), thus turning Q_2 OFF. When the output transistor of the driving gate is OFF (open), it corresponds to a HIGH input; this causes the injector current to be steered into the base of Q_2, which turns it ON. This action is illustrated in Figure A–23. Figure A–24 shows an example of an I²L implementation of a logic function.

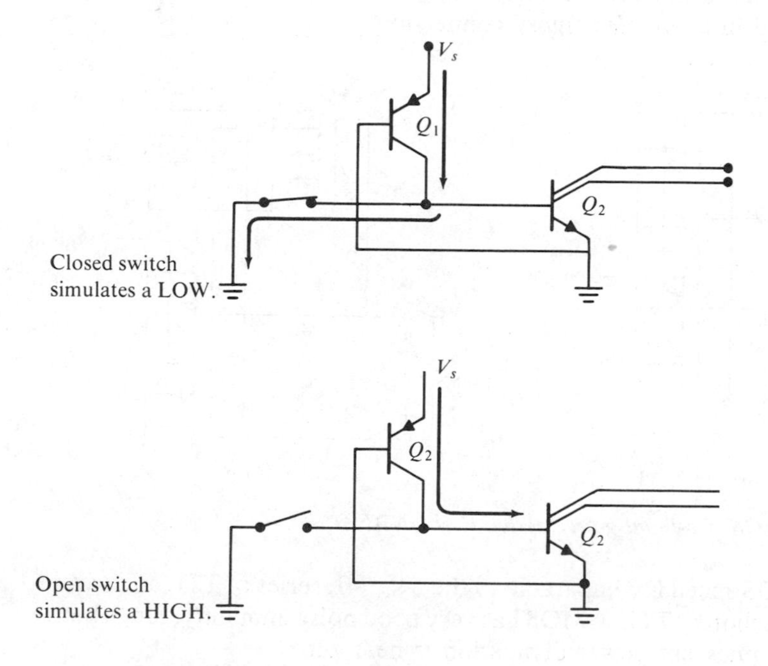

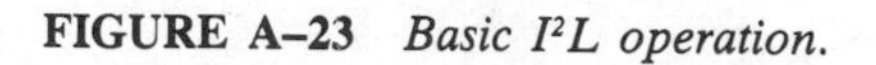
FIGURE A–23 *Basic I²L operation.*

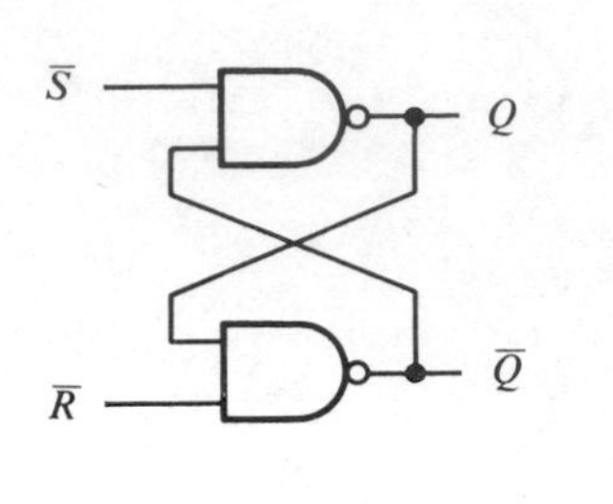

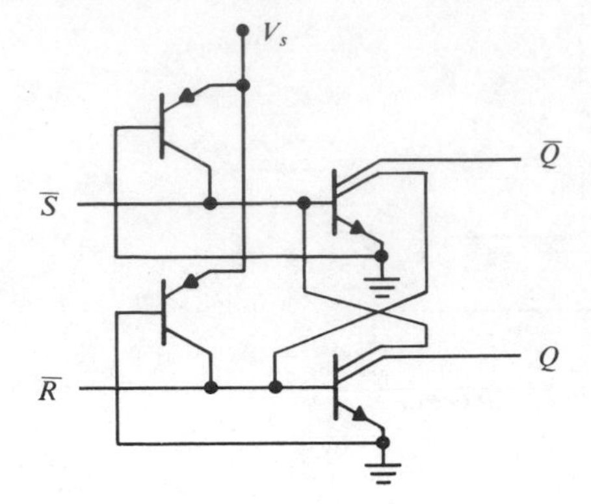

FIGURE A–24 *An I^2L latch.*

A–9 CMOS CIRCUITS

CMOS stands for *complementary metal oxide semiconductor.* In Chapter 3, CMOS gates were introduced and compared to the TTL circuits. CMOS is a popular type of MOS logic and is available in some LSI and MSI functions as well as SSI. Many of these devices are pin-compatible with equivalent TTL devices. In this section we will look further at MOS technology.

Figure A–25 shows a CMOS two-input NOR gate and a two-input NAND gate. As you recall, CMOS consists of both *n*-channel and *p*-channel MOS transistors arranged in a *complementary* connection.

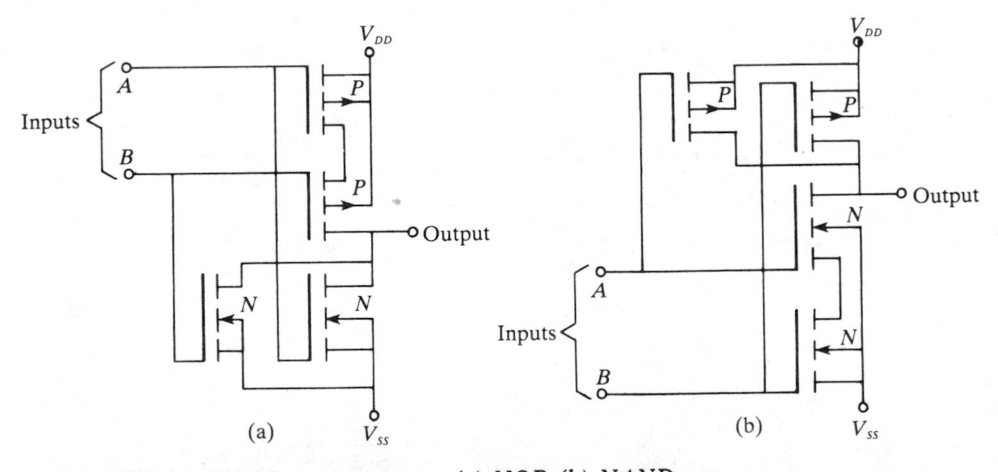

FIGURE A–25 *CMOS two-input gates (a) NOR (b) NAND.*

CMOS speed is comparable to the 54L/74L series of TTL and several times slower than Schottky TTL. CMOS has very good noise immunity, much better than TTL and its quiescent power dissipation is near zero.

Power Dissipation

With no input signals, the *p*-channel and the *n*-channel transistors are not conducting simultaneously, thus there is only leakage current from the positive (V_{DD}) to the negative power supply connection (V_{SS}). Typically this leakage current is 0.5 nA per gate, producing a power of 2.5 nW for a 5 V supply.

When signals are applied to the inputs, additional power is consumed to charge and discharge on-chip parasitic capacitances as well as load capacitance. Also, there is a short time during transitions when both the *n*- and *p*-type transistors are partially conducting. Dynamic power consumption occurs when these conditions exist and is proportional to the frequency at which the circuit is switching, the load capacitance, and to the square of the supply voltage.

It has been found that the power consumption of a CMOS gate exceeds that of a Schottky low power gate somewhere between 500 kHz and 2 MHz.

Supply Voltage

Unlike TTL, which is intolerant of large ranges in dc supply voltage, CMOS can be operated over a supply voltage range of 3 to 15 V. Low power consumption and wide supply voltage range make CMOS ideal for battery operated equipment.

Propagation Delay

CMOS devices are slow compared to TTL and very sensitive to capacitive loading. TTL with an output impedance of about 100 Ω is not affected much by an increase in capacitive loading. CMOS, however, has an output impedance of about 1 kΩ and is ten times more sensitive to capacitive loading. Also, propagation delay increases with an increase in supply voltage.

Noise Immunity

The noise immunity of a CMOS gate is approximately 45 percent of the supply voltage. For example, the noise margin is 2.25 V for a 5 V supply, 4.5 V for a 10 V supply, and so forth. In comparison, TTL exhibits a noise immunity that is typically 1 V and which can be as low as 0.4 V.

Also, the inherent capacitance of CMOS circuits produces delays and as a result acts as a noise filter. Ten-nanosecond voltage spikes tend to disappear in a chain of CMOS gates but are amplified by the faster switching TTL gates.

Because of these features, CMOS is widely used in industrial equipment where there is a great deal of electrically and electromagnetically produced noise.

Interfacing with TTL

When CMOS is operated with a 5 V power supply, interfacing with TTL is relatively easy. The input impedance of CMOS is high enough so that TTL can drive it without any loss of fan-out capability in the LOW state. However, the HIGH level voltage of TTL is typically 3.5 V which is insufficient to drive CMOS reliably. A pull-up resistor connected from the output of the TTL device to V_{CC} as shown in Figure A–26 will increase the HIGH level sufficiently to drive the CMOS device.

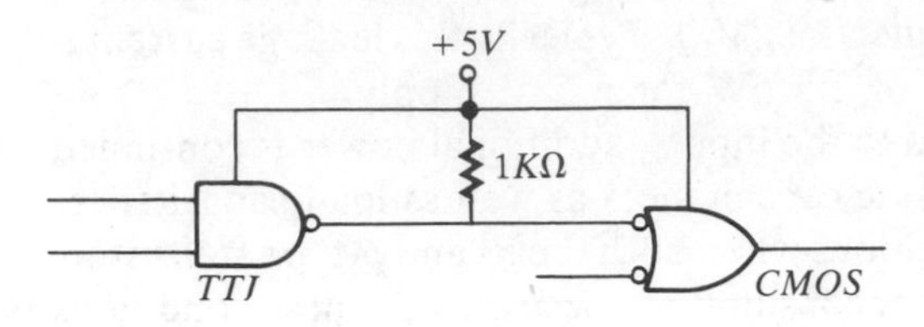

FIGURE A–26 *TTL to CMOS interface for 5 V operation.*

When CMOS is operated with supply voltages higher than 5 V you cannot interface directly with TTL. In this situation, level translator circuits which convert TTL levels to CMOS levels must be used. These circuits are available in IC form.

Precautions for Handling CMOS

All MOS devices are subject to damage from electrostatic charges. Because of this, handling must be done with special care as itemized below.

1. MOS devices are shipped in conductive foam to prevent charge buildup. When they are removed do not touch the pins.
2. The devices should be placed with pins down on a grounded surface, such as a metal plate, when removed from protective material. Do not place MOS devices in polystyrene foam or plastic trays.
3. All tools, test equipment, and metal work benches should be earth-grounded. Personnel working with MOS devices should, in certain environments, have their wrist grounded with a length of cable and a large-value resistor. The resistor prevents fatal shock should the person come in contact with a voltage source.
4. Do not insert MOS devices into sockets or pc boards with the power on.
5. All unused inputs must be connected to V_{SS} or V_{DD}, whichever provides the proper logic level. Never leave unused pins open because they can acquire static charge and "float" to unpredicted levels.
6. After assembly on pc boards, protection should be provided by storing or shipping boards with their connectors in conductive foam. MOS input and output pins may also be protected with large valued resistors connected to ground.

Comparison of CMOS with TTL

Table A–4 gives a comparison of key parameters of CMOS with TTL technologies.

TABLE A–4 *Key parameters of CMOS and some TTL circuits.*

Parameter	54/74	54L/74L	54LS/74LS	CMOS (5 *V*)	CMOS (10 *V*)
Propagation delay	10 ns	33 ns	9.5 ns	40 ns	20 ns
Quiescent power	10 mW	1 mW	2 mW	10 nW	10 nW
Noise immunity	1*V*	1*V*	0.8 *V*	2*V*	4*V*
Fan out	10	10	20	50+	50+
Flip-flop Clock frequency	35 MHz	3 MHz	45 MHz	8 MHz	16 MHz

PMOS AND NMOS A–10

MOS circuits are used largely in LSI (large-scale integration) functions such as long shift registers, large memories, and microprocessor products. This is because of the low power consumption and very small chip area required for MOS transistors. Of all the MOS technologies, CMOS is the only one that provides SSI functions at the gate and flip-flop level in a variety comparable to TTL.

PMOS

PMOS was one of the first high density MOS circuit technologies to be produced. It utilizes enhancement mode *p*-channel MOS transistors to form the basic gate building blocks. Figure A–27 shows a basic PMOS gate that produces the NOR function in positive logic.

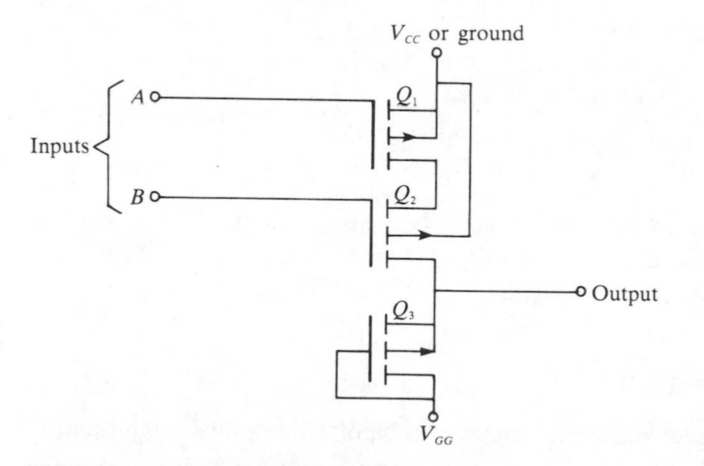

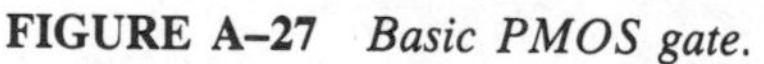
FIGURE A–27 *Basic PMOS gate.*

The operation of the PMOS gate is as follows. The supply voltage V_{GG} is a negative voltage, and V_{CC} is a positive voltage or ground (0 V). Transistor Q_3 is permanently biased to create a constant drain to source resistance. Its sole purpose is to function as a current limiting resistor. If a HIGH (V_{CC}) is applied to input A or B, then Q_1 or Q_2 is off and the output is pulled down to a voltage near V_{GG}, which represents a LOW. When a LOW voltage (V_{GG}) is applied to both input A and input B, both transistors Q_1 and Q_2 are turned on. This causes the output to go to a HIGH level (near V_{CC}). Since a LOW output occurs when either or both inputs are HIGH, and a HIGH output occurs only when all inputs are LOW, we have a NOR gate.

NMOS

NMOS devices were developed as processing technology improved, and now most memories and microprocessors use NMOS. The *n*-channel MOS transistor is used in NMOS circuits, as shown in Figure A–28 for a NAND gate and a NOR gate.

In Figure A–28 (a), Q_3 acts as a resistor to limit current. When a LOW (V_{GG} or ground) is applied to one or both inputs, then at least one of the transistors (Q_1 or Q_2) is off, and the output is pulled up to a HIGH level near V_{CC}. When HIGHs (V_{CC}) are applied to both inputs A and B, both Q_1 and Q_2 conduct and the output is LOW. This action, of course identifies this circuit as a NAND gate.

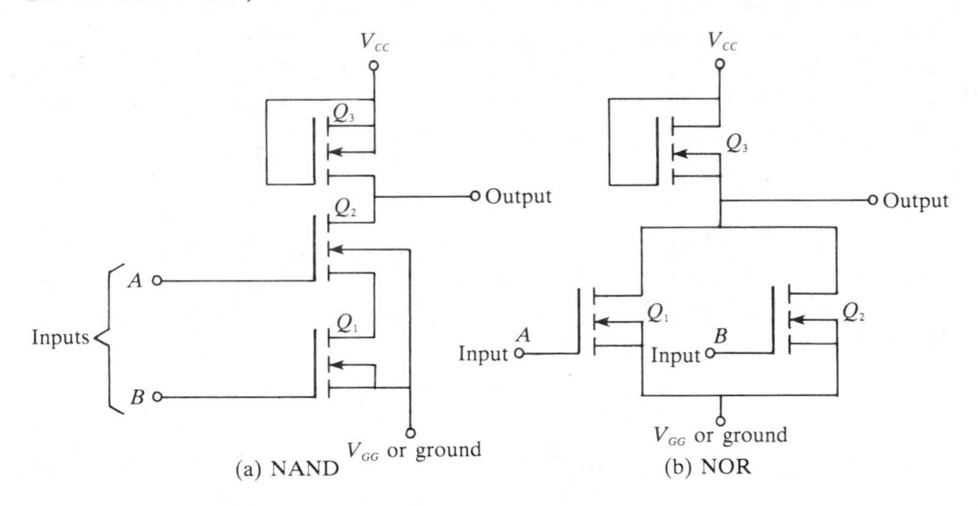

FIGURE A–28 *NMOS gates.*

In Figure A–28 (b), Q_3 again acts as a resistor. A HIGH on either input turns Q_1 or Q_2 on pulling the output LOW. When both inputs are LOW, both transistors are off and the output is pulled up to a HIGH level.

Problems

A–1 Describe one major difference between a bipolar integrated circuit and an MOS integrated circuit.

A–2 Explain why an open TTL input acts as a HIGH.

A–3 List the five series of TTL circuits.

A–4 The fan-out of a standard TTL gate is 10 unit loads; this, of course, means that it can drive 10 other *standard* TTL gates. How many 54LS/74LS gates can the standard gate drive?

A–5 If two unused inputs of a TTL gate are connected to an input being driven by another TTL gate, how many other inputs can be driven by this gate. Assume the fan-out is 10 unit loads.

A–6 How much current does a 54/74 gate sink if it is driving seven 54/74 loads?

A–7 What is the chief advantage of ECL over other IC technologies? What is a second advantage?

A–8 In what type of application should ECL not be used? Why?

A–9 If the frequency of operation of a CMOS device is increased, what happens to the dynamic power consumption?

A–10 Does CMOS or TTL perform better in a high noise environment? Why?

A–11 Why are MOS devices shipped in conductive foam?

A–12 What type of MOS technology is predominant in LSI devices such as memories and microprocessors?

B
DATA
SHEETS

recommended operating conditions

	54 FAMILY 74 FAMILY	SERIES 54 SERIES 74 '00, '04, '10, '20, '30			SERIES 54H SERIES 74H 'H00, 'H04, 'H10, 'H20, 'H30			SERIES 54L SERIES 74L 'L00, 'L04, 'L10, 'L20, 'L30			SERIES 54LS SERIES 74LS 'LS00, 'LS04, 'LS10, 'LS20, 'LS30			SERIES 54S SERIES 74S 'S00, 'S04, 'S10, 'S20, 'S30, 'S133			UNIT
		MIN	NOM	MAX	MIN	NOM	MAX	MIN	NOM	MAX	MIN	NOM	MAX	MIN	NOM	MAX	
Supply voltage, V_{CC}	54 Family	4.5	5	5.5	4.5	5	5.5	4.5	5	5.5	4.5	5	5.5	4.5	5	5.5	V
	74 Family	4.75	5	5.25	4.75	5	5.25	4.75	5	5.25	4.75	5	5.25	4.75	5	5.25	
High-level output current, I_{OH}	54 Family			−400			−500			−100			−400			−1000	μA
	74 Family			−400			−500			−200			−400			−1000	
Low-level output current, I_{OL}	54 Family			16			20			2			4			20	mA
	74 Family			16			20			3.6			8			20	
Operating free-air temperature, T_A	54 Family	−55		125	−55		125	−55		125	−55		125	−55		125	°C
	74 Family	0		70	0		70	0		70	0		70	0		70	

electrical characteristics over recommended operating free-air temperature range (unless otherwise noted)

PARAMETER	TEST FIGURE	TEST CONDITIONS†		SERIES 54 SERIES 74 '00, '04, '10, '20, '30			SERIES 54H SERIES 74H 'H00, 'H04, 'H10, 'H20, 'H30			SERIES 54L SERIES 74L 'L00, 'L04, 'L10, 'L20, 'L30			SERIES 54LS SERIES 74LS 'LS00, 'LS04, 'LS10, 'LS20, 'LS30			SERIES 54S SERIES 74S 'S00, 'S04, 'S10, 'S20, 'S30, 'S133			UNIT
				MIN	TYP‡	MAX	MIN	TYP‡	MAX	MIN	TYP‡	MAX	MIN	TYP‡	MAX	MIN	TYP‡	MAX	
V_{IH} High-level input voltage	1, 2			2			2			2			2			2			V
V_{IL} Low-level input voltage	1, 2		54 Family			0.8			0.8			0.7			0.7			0.8	V
			74 Family			0.8			0.8			0.7			0.8			0.8	
V_{IK} Input clamp voltage	3	V_{CC} = MIN, I_I = §				−1.5			−1.5						−1.5			−1.2	V
V_{OH} High-level output voltage	1	V_{CC} = MIN, V_{IL} = V_{IL} max, I_{OH} = MAX	54 Family	2.4	3.4		2.4	3.5		2.4	3.3		2.5	3.4		2.5	3.4		V
			74 Family	2.4	3.4		2.4	3.5		2.4	3.2		2.7	3.4		2.7	3.4		
V_{OL} Low-level output voltage	2	V_{CC} = MIN, V_{IH} = 2 V, I_{OL} = MAX	54 Family		0.2	0.4		0.2	0.4		0.15	0.3		0.25	0.4			0.5	V
			74 Family		0.2	0.4		0.2	0.4		0.2	0.4		0.25	0.5			0.5	
		V_{CC} = MIN, V_{IH} = 2 V, I_{OL} = 4 mA	Series 74LS												0.4				
I_I Input current at maximum input voltage	4	V_{CC} = MAX	V_I = 5.5 V			1			1			0.1						1	mA
			V_I = 7 V												0.1				
I_{IH} High-level input current	4	V_{CC} = MAX	V_{IH} = 2.4 V			40			50			10							μA
			V_{IH} = 2.7 V												20			50	
I_{IL} Low-level input current	5	V_{CC} = MAX	V_{IL} = 0.3 V									−0.18							mA
			V_{IL} = 0.4 V			−1.6			−2						−0.4				
			V_{IL} = 0.5 V															−2	
I_{OS} Short-circuit output current◆	6	V_{CC} = MAX	54 Family	−20		−55	−40		−100	−3		−15	−20		−100	−40		−100	mA
			74 Family	−18		−55	−40		−100	−3		−15	−20		−100	−40		−100	
I_{CC} Supply current	7	V_{CC} = MAX		See table on next page															mA

†For conditions shown as MIN or MAX, use the appropriate value specified under recommended operating conditions.
‡All typical values are at V_{CC} = 5 V, T_A = 25°C.
§I_I = −12 mA for SN54'/SN74', −8 mA for SN54H'/SN74H', and −18 mA for SN54LS'/SN74LS' and SN54S'/SN74S'.
◆Not more than one output should be shorted at a time, and for SN54H'/SN74H', SN54LS'/SN74LS', and SN54S'/SN74S', duration of short-circuit should not exceed 1 second.

POSITIVE-NAND GATES AND INVERTERS WITH TOTEM-POLE OUTPUTS

supply current¶

TYPE	I_{CCH} (mA) Total with outputs high		I_{CCL} (mA) Total with outputs low		I_{CC} (mA) Average per gate (50% duty cycle)
	TYP	MAX	TYP	MAX	TYP
'00	4	8	12	22	2
'04	6	12	18	33	2
'10	3	6	9	16.5	2
'20	2	4	6	11	2
'30	1	2	3	6	2
'H00	10	16.8	26	40	4.5
'H04	16	26	40	58	4.5
'H10	7.5	12.6	19.5	30	4.5
'H20	5	8.4	13	20	4.5
'H30	2.5	4.2	6.5	10	4.5
'L00	0.44	0.8	1.16	2.04	0.20
'L04	0.66	1.2	1.74	3.06	0.20
'L10	0.33	0.6	0.87	1.53	0.20
'L20	0.22	0.4	0.58	1.02	0.20
SN54L30	0.11	0.33	0.29	0.51	0.20
SN74L30	0.11	0.2	0.29	0.51	0.20
'LS00	0.8	1.6	2.4	4.4	0.4
'LS04	1.2	2.4	3.6	6.6	0.4
'LS10	0.6	1.2	1.8	3.3	0.4
'LS20	0.4	0.8	1.2	2.2	0.4
'LS30	0.35	0.5	0.6	1.1	0.48
'S00	10	16	20	36	3.75
'S04	15	24	30	54	3.75
'S10	7.5	12	15	27	3.75
'S20	5	8	10	18	3.75
'S30	3	5	5.5	10	4.25
'S133	3	5	5.5	10	4.25

¶Maximum values of I_{CC} are over the recommended operating ranges of V_{CC} and T_A; typical values are at V_{CC} = 5 V, T_A = 25°C.

switching characteristics at V_{CC} = 5 V, T_A = 25°C

TYPE	TEST CONDITIONS#	t_{PLH} (ns) Propagation delay time, low-to-high-level output			t_{PHL} (ns) Propagation delay time, high-to-low-level output		
		MIN	TYP	MAX	MIN	TYP	MAX
'00, '10	C_L = 15 pF, R_L = 400 Ω		11	22		7	15
'04, '20			12	22		8	15
'30			13	22		8	15
'H00	C_L = 25 pF, R_L = 280 Ω		5.9	10		6.2	10
'H04			6	10		6.5	10
'H10			5.9	10		6.3	10
'H20			6	10		7	10
'H30			6.8	10		8.9	12
'L00, 'L04, 'L10, L20	C_L = 50 pF, R_L = 4 kΩ		35	60		31	60
'L30			35	60		70	100
'LS00, 'LS04 'LS10, 'LS20	C_L = 15 pF, R_L = 2 kΩ		9	15		10	15
'LS30			8	15		13	20
'S00, 'S04	C_L = 15 pF, R_L = 280 Ω		3	4.5		3	5
'S10, 'S20	C_L = 50 pF, R_L = 280 Ω		4.5			5	
'S30, 'S133	C_L = 15 pF, R_L = 280 Ω		4	6		4.5	7
	C_L = 50 pF, R_L = 280 Ω		5.5			6.5	

#Load circuits and voltage waveforms are shown on pages 3-10 and 3-11.

schematics (each gate)

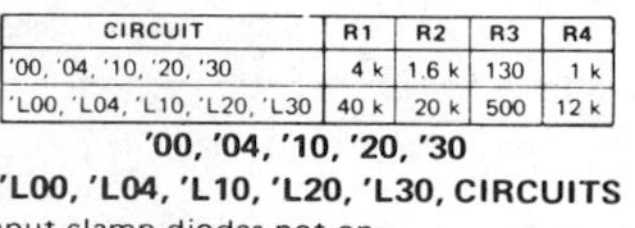

CIRCUIT	R1	R2	R3	R4
'00, '04, '10, '20, '30	4 k	1.6 k	130	1 k
'L00, 'L04, 'L10, 'L20, 'L30	40 k	20 k	500	12 k

'00, '04, '10, '20, '30
'L00, 'L04, 'L10, 'L20, 'L30, CIRCUITS
Input clamp diodes not on SN54L'/SN74L' circuits.

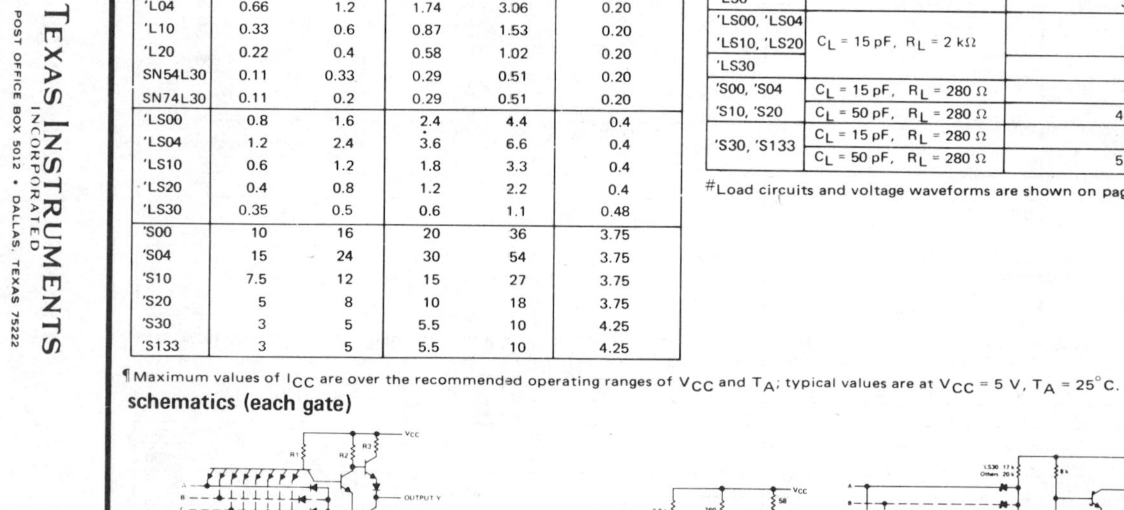

'H00, 'H04, 'H10, 'H20, 'H30 CIRCUITS

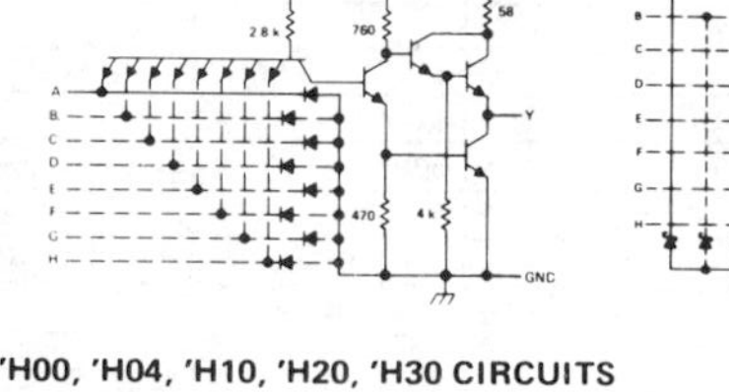

'LS00, 'LS04, 'LS10, 'LS20, 'LS30 CIRCUITS
*The 12-kΩ resistor is not on 'LS30.

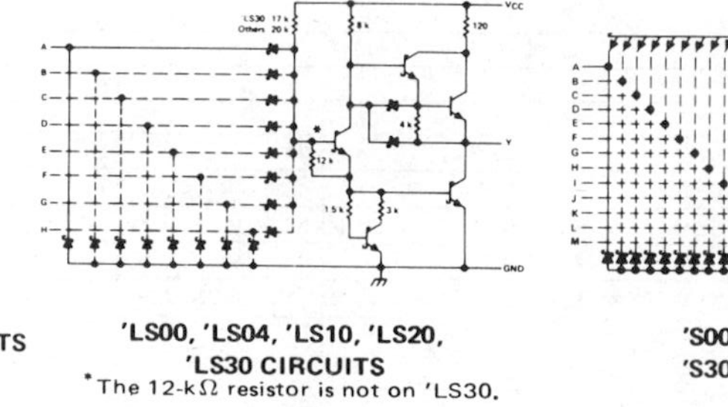

'S00, 'S04, 'S10, 'S20, 'S30, 'S133 CIRCUITS

Resistor values shown are nominal and in ohms.

recommended operating conditions

	SERIES 54/74	'70 MIN	'70 NOM	'70 MAX	'72, '73, '76, '107 MIN	NOM	MAX	'74 MIN	NOM	MAX	'109 MIN	NOM	MAX	'110 MIN	NOM	MAX	'111 MIN	NOM	MAX	UNIT
Supply voltage, V_{CC}	Series 54	4.5	5	5.5	4.5	5	5.5	4.5	5	5.5	4.5	5	5.5	4.5	5	5.5	4.5	5	5.5	V
	Series 74	4.75	5	5.25	4.75	5	5.25	4.75	5	5.25	4.75	5	5.25	4.75	5	5.25	4.75	5	5.25	V
High-level output current, I_{OH}				−400			−400			−400			−800			−800			−800	μA
Low-level output current, I_{OL}				16			16			16			16			16			16	mA
Pulse width, t_w	Clock high	20			20			30			20			25			25			ns
	Clock low	30			47			37			20			25			25			ns
	Preset or clear low	25			25			30			20			25			25			ns
Input setup time, t_{su}		20↑			0↑			20↑			10↑			20↑			0↑			ns
Input hold time, t_h		5↑			0↓			5↑			6↑			5↑			30↑			ns
Operating free-air temperature, T_A	Series 54	−55		125	−55		125	−55		125	−55		125	−55		125	−55		125	°C
	Series 74	0		70	0		70	0		70	0		70	0		70	0		70	°C

↑↓The arrow indicates the edge of the clock pulse used for reference: ↑ for the rising edge, ↓ for the falling edge.

electrical characteristics over recommended operating free-air temperature range (unless otherwise noted)

PARAMETER		TEST CONDITIONS†	'70 MIN	'70 TYP‡	'70 MAX	'72, '73, '76, '107 MIN	TYP‡	MAX	'74 MIN	TYP‡	MAX	'109 MIN	TYP‡	MAX	'110 MIN	TYP‡	MAX	'111 MIN	TYP‡	MAX	UNIT
V_{IH} High-level input voltage			2			2			2			2			2			2			V
V_{IL} Low-level input voltage					0.8			0.8			0.8			0.8			0.8			0.8	V
V_{IK} Input clamp voltage		V_{CC} = MIN, I_I = −12 mA			−1.5			−1.5			−1.5			−1.5			−1.5			−1.5	V
V_{OH} High-level output voltage		V_{CC} = MIN, V_{IH} = 2 V, V_{IL} = 0.8 V, I_{OH} = MAX	2.4	3.4		2.4	3.4		2.4	3.4		2.4	3.4		2.4	3.4		2.4	3.4		V
V_{OL} Low-level output voltage		V_{CC} = MIN, V_{IH} = 2 V, V_{IL} = 0.8 V, I_{OL} = 16 mA		0.2	0.4		0.2	0.4		0.2	0.4		0.2	0.4		0.2	0.4		0.2	0.4	V
I_I Input current at maximum input voltage		V_{CC} = MAX, V_I = 5.5 V			1			1			1			1			1			1	mA
I_{IH} High-level input current	D, J, K, or $\overline{K}$	V_{CC} = MAX, V_I = 2.4 V			40			40			40			40			40			40	μA
	Clear				80			80			120			160			160			80	μA
	Preset				80			80			80			80			160			80	μA
	Clock				40			80			80			80			40			120	μA
I_{IL} Low-level input current	D, J, K, or $\overline{K}$	V_{CC} = MAX, V_I = 0.4 V			−1.6			−1.6			−1.6			−1.6			−1.6			−1.6	mA
	Clear ★				−3.2			−3.2			−3.2			−4.8			−3.2			−3.2	mA
	Preset ★				−3.2			−3.2			−1.6			−3.2			−3.2			−3.2	mA
	Clock				−1.6			−3.2			−3.2			−3.2			−1.6			−4.8	mA
I_{OS} Short-circuit output current◆	Series 54	V_{CC} = MAX	−20		−57	−20		−57	−20		−57	−30		−85	−20		−57	−20		−57	mA
	Series 74		−18		−57	−18		−57	−18		−57	−30		−85	−18		−57	−18		−57	mA
I_{CC} Supply current (Average per flip-flop)		V_{CC} = MAX, See Note 1		13	26		10	20		8.5	15		9	15		20	34		14	20.5	mA

†For conditions shown as MIN or MAX, use the appropriate value specified under recommended operating conditions.
‡All typical values are at V_{CC} = 5 V, T_A = 25°C.
◆Not more than one output should be shorted at a time.
★Clear is tested with preset high and preset is tested with clear high.

NOTE 1: With all outputs open, I_{CC} is measured with the Q and $\overline{Q}$ outputs high in turn. At the time of measurement, the clock input is at 4.5 V for the '70, '110, and '111; and is grounded for all the others.

switching characteristics, $V_{CC} = 5\ V$, $T_A = 25°C$

PARAMETER¶	FROM (INPUT)	TO (OUTPUT)	TEST CONDITIONS	'70			'72, '73 '76, '107			'74			'109			'110			'111			UNIT
				MIN	TYP	MAX	MIN	TYP	MAX	MIN	TYP	MAX	MIN	TYP	MAX	MIN	TYP	MAX	MIN	TYP	MAX	
f_{max}				20	35		15	20		15	25		25	33		20	25		20	25		MHz
t_{PLH}	Preset	Q	$C_L = 15\ pF$,			50		16	25			25		10	15		12	20		12	18	ns
t_{PHL}	(as applicable)	$\bar{Q}$	$R_L = 400\ \Omega$,			50		25	40			40		23	35		18	25		21	30	
t_{PLH}	Clear	$\bar{Q}$	See Note 2			50		16	25			25		10	15		12	20		12	18	ns
t_{PHL}	(as applicable)	Q				50		25	40			40		17	25		18	25		21	30	
t_{PLH}	Clock	Q or $\bar{Q}$			27	50		16	25		14	25		10	16		20	30		12	17	ns
t_{PHL}					18	50		25	40		20	40		18	28		13	20		20	30	

¶ f_{max} ≡ maximum clock frequency; t_{PLH} ≡ propagation delay time, low-to-high-level output; t_{PHL} ≡ propagation delay time, high-to-low-level output.
NOTE 2: Load circuit and voltage waveforms are shown on page 3-10.

functional block diagrams

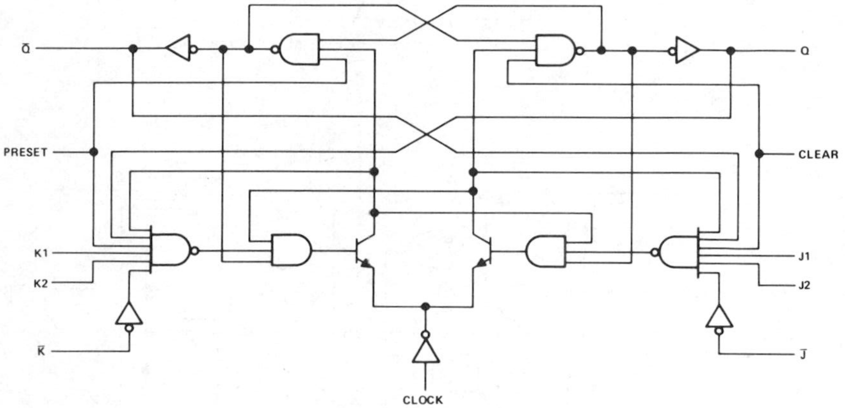

'70—GATED J-K WITH CLEAR AND PRESET

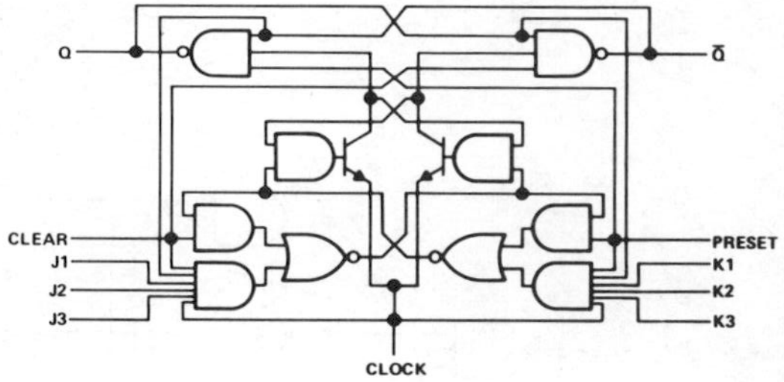

'72—GATED J-K WITH CLEAR AND PRESET

See following pages for:
'73—DUAL J-K WITH CLEAR
'74—DUAL D WITH CLEAR AND PRESET
'76—DUAL J-K WITH CLEAR AND PRESET
'107—DUAL J-K WITH CLEAR
'109—DUAL J-$\bar{K}$ WITH CLEAR AND PRESET
'110—GATED J-K WITH CLEAR AND PRESET
'111—DUAL J-K WITH CLEAR AND PRESET

functional block diagrams (continued)

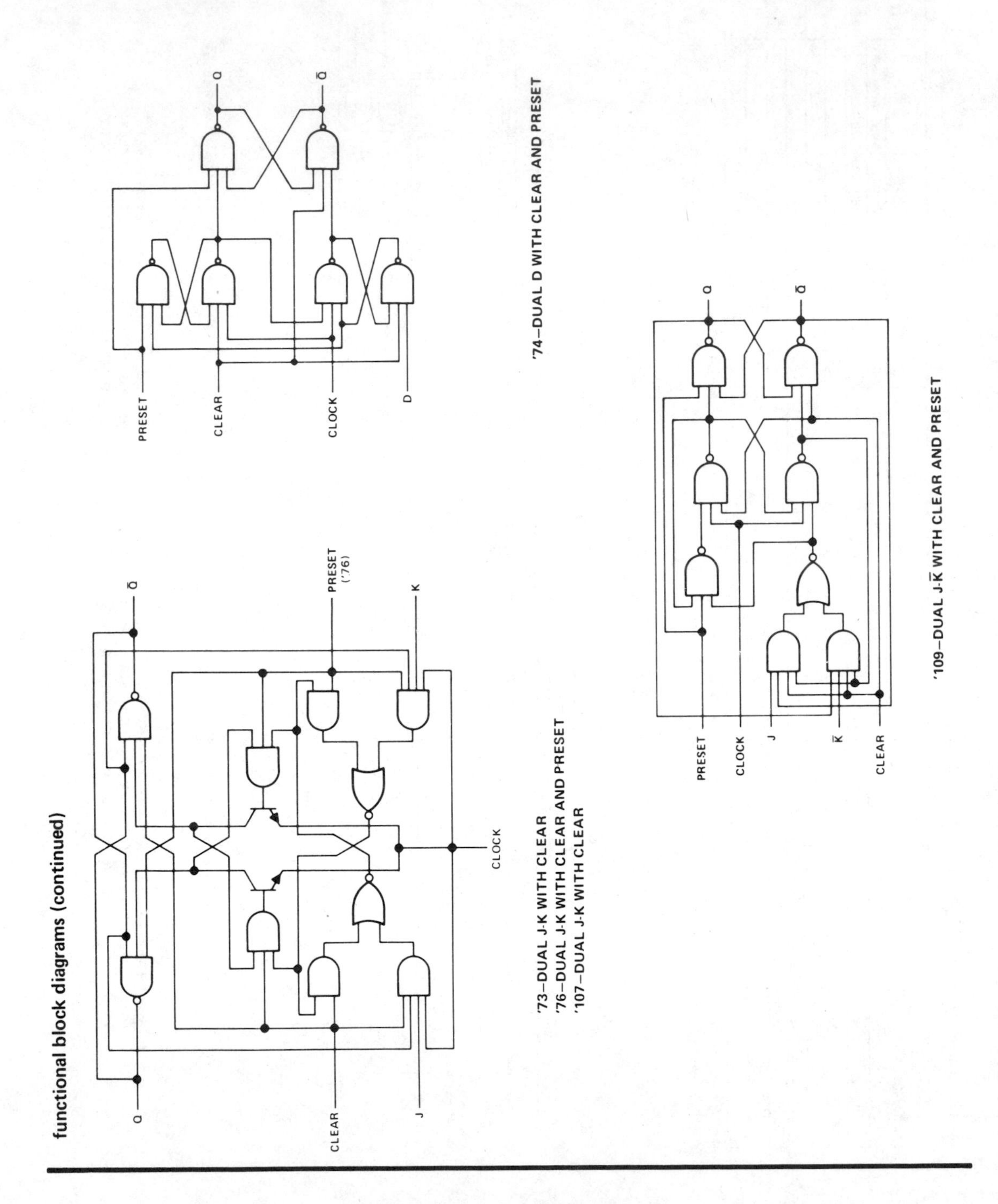

'74—DUAL D WITH CLEAR AND PRESET

'73—DUAL J-K WITH CLEAR
'76—DUAL J-K WITH CLEAR AND PRESET
'107—DUAL J-K WITH CLEAR

'109—DUAL J-$\overline{K}$ WITH CLEAR AND PRESET

functional block diagrams (continued)

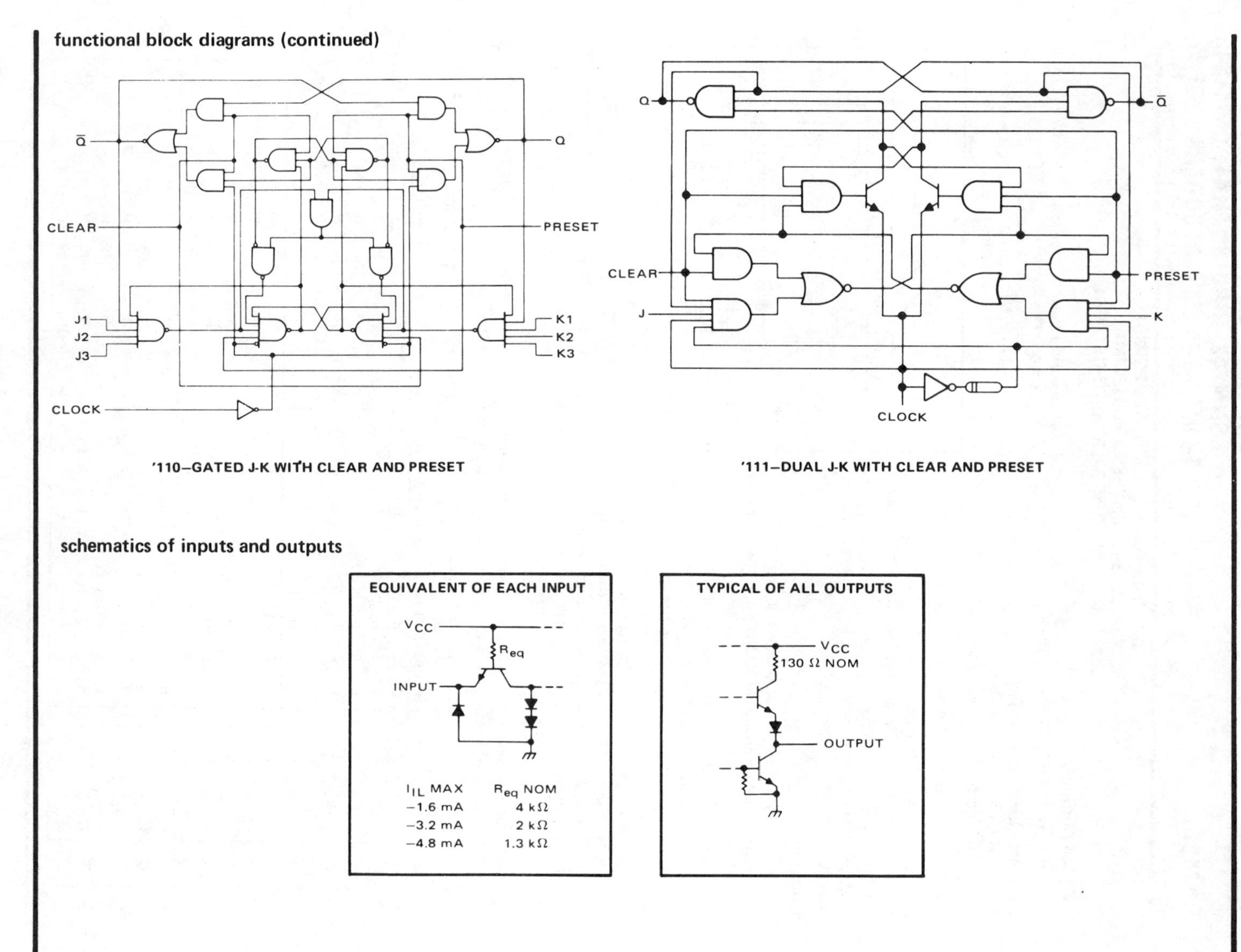

'110—GATED J-K WITH CLEAR AND PRESET

'111—DUAL J-K WITH CLEAR AND PRESET

schematics of inputs and outputs

TYPES SN5442A THRU SN5444A, SN54L42 THRU SN54L44, SN54LS42, SN7442A THRU SN7444A, SN74L42 THRU SN74L44, SN74LS42
4-LINE-TO-10-LINE DECODERS (1-OF-10)

TTL
MSI

BULLETIN NO. DL-S 7611861, MARCH 1974–REVISED OCTOBER 1976

'42A, 'L42, 'LS42 . . . BCD-TO-DECIMAL
'43A, 'L43 . . . EXCESS-3-TO-DECIMAL
'44A, 'L44 . . . EXCESS-3-GRAY-TO-DECIMAL

- All Outputs Are High for Invalid Input Conditions
- Also for Application as
  4-Line-to-16-Line Decoders
  3-Line-to-8-Line Decoders
- Diode-Clamped Inputs

TYPES	TYPICAL POWER DISSIPATION	TYPICAL PROPAGATION DELAYS
'42A, '43A, '44A	140 mW	17 ns
'L42, 'L43, 'L44	70 mW	49 ns
'LS42	35 mW	17 ns

SN5442A THRU SN5444A, SN54LS42 . . . J OR W PACKAGE
SN54L42 THRU SN54L44 . . . J PACKAGE
SN7442A THRU SN7444A,
SN74L42 THRU SN74L44, SN74LS42 . . . J OR N PACKAGE
(TOP VIEW)

positive logic: see function table

description

These monolithic decimal decoders consist of eight inverters and ten four-input NAND gates. The inverters are connected in pairs to make BCD input data available for decoding by the NAND gates. Full decoding of valid input logic ensures that all outputs remain off for all invalid input conditions.

The '42A, 'L42, and 'LS42 BCD-to-decimal decoders, the '43A and 'L43 excess-3-to-decimal decoders, and the '44A and 'L44 excess-3-gray-to-decimal decoders feature inputs and outputs that are compatible for use with most TTL and other saturated low-level logic circuits. D-c noise margins are typically one volt.

Series 54, 54L, and 54LS circuits are characterized for operation over the full military temperature range of −55°C to 125°C; Series 74, 74L, and 74LS circuits are characterized for operation from 0°C to 70°C.

FUNCTION TABLE

NO.	'42A, 'L42, 'LS42 BCD INPUT				'43A, 'L43 EXCESS-3-INPUT				'44A, 'L44 EXCESS-3-GRAY INPUT				ALL TYPES DECIMAL OUTPUT									
	D	C	B	A	D	C	B	A	D	C	B	A	0	1	2	3	4	5	6	7	8	9
0	L	L	L	L	L	L	H	H	L	L	H	L	L	H	H	H	H	H	H	H	H	H
1	L	L	L	H	L	H	L	L	L	H	H	L	H	L	H	H	H	H	H	H	H	H
2	L	L	H	L	L	H	L	H	L	H	H	H	H	H	L	H	H	H	H	H	H	H
3	L	L	H	H	L	H	H	L	L	H	L	H	H	H	H	L	H	H	H	H	H	H
4	L	H	L	L	L	H	H	H	L	H	L	L	H	H	H	H	L	H	H	H	H	H
5	L	H	L	H	H	L	L	L	H	H	L	L	H	H	H	H	H	L	H	H	H	H
6	L	H	H	L	H	L	L	H	H	H	L	H	H	H	H	H	H	H	L	H	H	H
7	L	H	H	H	H	L	H	L	H	H	H	H	H	H	H	H	H	H	H	L	H	H
8	H	L	L	L	H	L	H	H	H	H	H	L	H	H	H	H	H	H	H	H	L	H
9	H	L	L	H	H	H	L	L	H	L	H	L	H	H	H	H	H	H	H	H	H	L
INVALID	H	L	H	L	H	H	L	H	H	L	H	H	H	H	H	H	H	H	H	H	H	H
INVALID	H	L	H	H	H	H	H	L	H	L	L	H	H	H	H	H	H	H	H	H	H	H
INVALID	H	H	L	L	H	H	H	H	H	L	L	L	H	H	H	H	H	H	H	H	H	H
INVALID	H	H	L	H	L	L	L	L	L	L	L	L	H	H	H	H	H	H	H	H	H	H
INVALID	H	H	H	L	L	L	L	H	L	L	L	H	H	H	H	H	H	H	H	H	H	H
INVALID	H	H	H	H	L	L	H	L	L	L	H	H	H	H	H	H	H	H	H	H	H	H

H = high level, L = low level

TYPES SN5442A THRU SN5444A, SN54L42 THRU SN54L44, SN54LS42, SN7442A THRU SN7444A, SN74L42 THRU SN74L44, SN74LS42 4-LINE-TO-10-LINE DECODERS (1-OF-10)

REVISED OCTOBER 1976

functional block diagrams and schematics of inputs and outputs

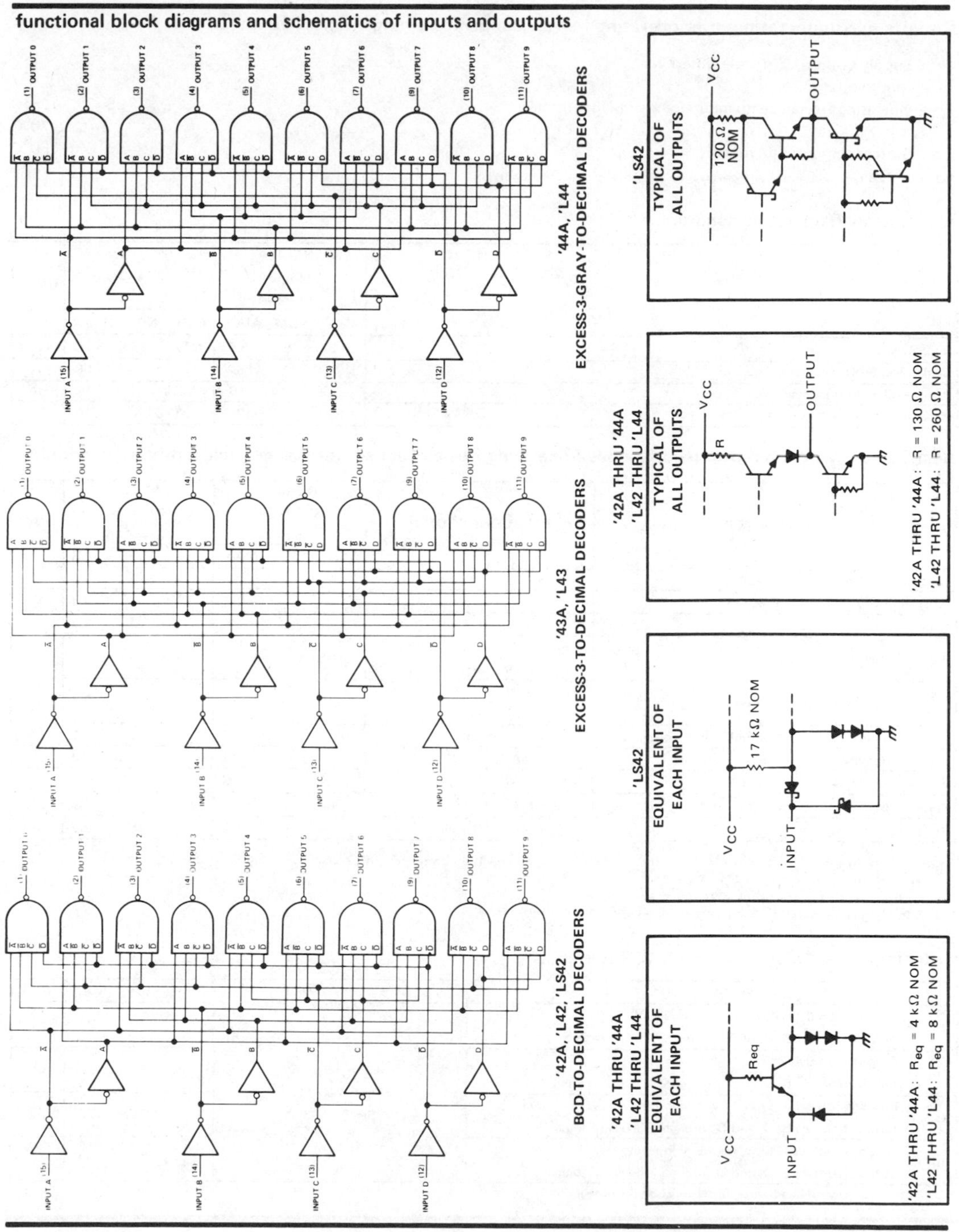

absolute maximum ratings over operating free-air temperature range (unless otherwise noted)

Supply voltage, V_{CC} (see Note 1) . . . . 7 V
Input voltage . . . . 5.5 V
Operating free-air temperature range: SN54' Circuits . . . . −55°C to 125°C
SN74' Circuits . . . . 0°C to 70°C
Storage temperature range . . . . −65°C to 150°C

NOTE 1: Voltage values are with respect to network ground terminal.

recommended operating conditions

	SN5442A SN5443A SN5444A			SN7442A SN7443A SN7444A			UNIT
	MIN	NOM	MAX	MIN	NOM	MAX	
Supply voltage, V_{CC}	4.5	5	5.5	4.75	5	5.25	V
High-level output current, I_{OH}			−800			−800	μA
Low-level output current, I_{OL}			16			16	mA
Operating free-air temperature, T_A	−55		125	0		70	°C

electrical characteristics over recommended operating free-air temperature range (unless otherwise noted)

PARAMETER		TEST CONDITIONS†	SN5442A SN5443A SN5444A			SN7442A SN7443A SN7444A			UNIT
			MIN	TYP‡	MAX	MIN	TYP‡	MAX	
V_{IH}	High-level input voltage		2			2			V
V_{IL}	Low-level input voltage				0.8			0.8	V
V_{IK}	Input clamp voltage	V_{CC} = MIN, I_I = −12 mA			−1.5			−1.5	V
V_{OH}	High-level output voltage	V_{CC} = MIN, V_{IH} = 2 V, V_{IL} = 0.8 V, I_{OH} = −800 μA	2.4	3.4		2.4	3.4		V
V_{OL}	Low-level output voltage	V_{CC} = MIN, V_{IH} = 2 V, V_{IL} = 0.8 V, I_{OL} = 16 mA		0.2	0.4		0.2	0.4	V
I_I	Input current at maximum input voltage	V_{CC} = MAX, V_I = 5.5 V			1			1	mA
I_{IH}	High-level input current	V_{CC} = MAX, V_I = 2.4 V			40			40	μA
I_{IL}	Low level input current	V_{CC} = MAX, V_I = 0.4 V			−1.6			−1.6	mA
I_{OS}	Short-circuit output current§	V_{CC} = MAX	−20		−55	−18		−55	mA
I_{CC}	Supply current	V_{CC} = MAX, See Note 2		28	41		28	56	mA

†For conditions shown as MIN or MAX, use the appropriate values specified under recommended operating conditions.
‡All typical values are at V_{CC} = 5 V, T_A = 25°C.
§Not more than one output should be shorted at a time.
NOTE 2: I_{CC} is measured with all outputs open and all inputs grounded.

switching characteristics, V_{CC} = 5 V, T_A = 25°C

PARAMETER		TEST CONDITIONS	MIN	TYP	MAX	UNIT
t_{PHL}	Propagation delay time, high-to-low-level output from A, B, C, or D through 2 levels of logic	C_L = 15 pF, R_L = 400 Ω, See Note 3		14	25	ns
t_{PHL}	Propagation delay time, high-to-low-level output from A, B, C, or D through 3 levels of logic			17	30	ns
t_{PLH}	Propagation delay time, low-to-high-level output from A, B, C, and D through 2 levels of logic			10	25	ns
t_{PLH}	Propagation delay time, low-to-high-level output from A, B, C, and D through 3 levels of logic			17	30	ns

NOTE 3: Load circuits and waveforms are shown on page 3-10.

TYPES SN5483A, SN54LS83A, SN7483A, SN74LS83A
4-BIT BINARY FULL ADDERS WITH FAST CARRY

BULLETIN NO. DL-S 7611853, MARCH 1974—REVISED OCTOBER 1976

- Full-Carry Look-Ahead across the Four Bits
- Systems Achieve Partial Look-Ahead Performance with the Economy of Ripple Carry
- SN54283/SN74283 and SN54LS283/SN74LS283 Are Recommended For New Designs as They Feature Supply Voltage and Ground on Corner Pins to Simplify Board Layout

TYPE	TYPICAL ADD TIMES: TWO 8-BIT WORDS	TYPICAL ADD TIMES: TWO 16-BIT WORDS	TYPICAL POWER DISSIPATION PER 4-BIT ADDER
'83A	23 ns	43 ns	310 mW
'LS83A	25 ns	45 ns	95 mW

SN5483A, SN54LS83A . . . J OR W PACKAGE
SN7483A, SN74LS83A . . . J OR N PACKAGE
(TOP VIEW)

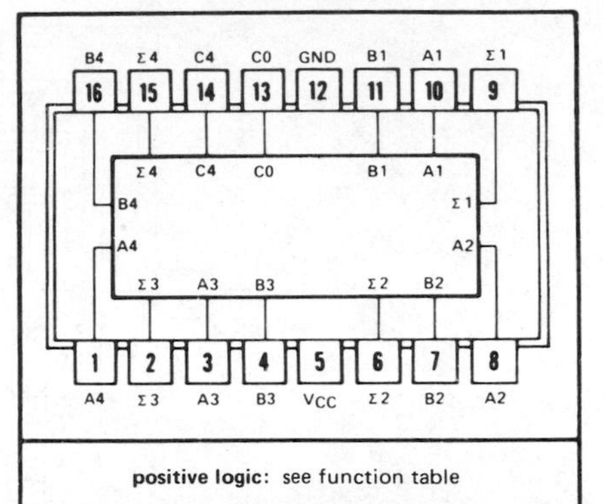

positive logic: see function table

description

These improved full adders perform the addition of two 4-bit binary numbers. The sum (Σ) outputs are provided for each bit and the resultant carry (C4) is obtained from the fourth bit. These adders feature full internal look ahead across all four bits generating the carry term in ten nanoseconds typically. This provides the system designer with partial look-ahead performance at the economy and reduced package count of a ripple-carry implementation.

The adder logic, including the carry, is implemented in its true form meaning that the end-around carry can be accomplished without the need for logic or level inversion.

Designed for medium-speed applications, the circuits utilize transistor-transistor logic that is compatible with most other TTL families and other saturated low-level logic families.

Series 54 and 54LS circuits are characterized for operation over the full military temperature range of −55°C to 125°C, and Series 74 and 74LS circuits are characterized for operation from 0°C to 70°C.

FUNCTION TABLE

INPUT				OUTPUT					
				WHEN C0 = L / WHEN C2 = L			WHEN C0 = H / WHEN C2 = H		
A1 / A3	B1 / B3	A2 / A4	B2 / B4	Σ1 / Σ3	Σ2 / Σ4	C2 / C4	Σ1 / Σ3	Σ2 / Σ4	C2 / C4
L	L	L	L	L	L	L	H	L	L
H	L	L	L	H	L	L	L	H	L
L	H	L	L	H	L	L	L	H	L
H	H	L	L	L	H	L	H	H	L
L	L	H	L	L	H	L	H	H	L
H	L	H	L	H	H	L	L	L	H
L	H	H	L	H	H	L	L	L	H
H	H	H	L	L	L	H	H	L	H
L	L	L	H	L	H	L	H	H	L
H	L	L	H	H	H	L	L	L	H
L	H	L	H	H	H	L	L	L	H
H	H	L	H	L	L	H	H	L	H
L	L	H	H	L	L	H	H	L	H
H	L	H	H	H	L	H	L	H	H
L	H	H	H	H	L	H	L	H	H
H	H	H	H	L	H	H	H	H	H

H = high level, L = low level

NOTE: Input conditions at A1, B1, A2, B2, and C0 are used to determine outputs Σ1 and Σ2 and the value of the internal carry C2. The values at C2, A3, B3, A4, and B4 are then used to determine outputs Σ3, Σ4, and C4.

absolute maximum ratings over operating free-air temperature range (unless otherwise noted)

Supply voltage, V_{CC} (see Note 1)	7 V
Input voltage: '83A	5.5 V
'LS83A	7 V
Interemitter voltage (see Note 2)	5.5 V
Operating free-air temperature range: SN5483A, SN54LS83A	−55°C to 125°C
SN7483A, SN74LS83A	0°C to 70°C
Storage temperature range	−65°C to 150°C

NOTES: 1. Voltage values, except interemitter voltage, are with respect to network ground terminal.
2. This is the voltage between two emitters of a multiple-emitter transistor. This rating applies for the '83A only between the following pairs: A1 and B1, A2 and B2, A3 and B3, A4 and B4.

REVISED OCTOBER 1976

functional block diagram

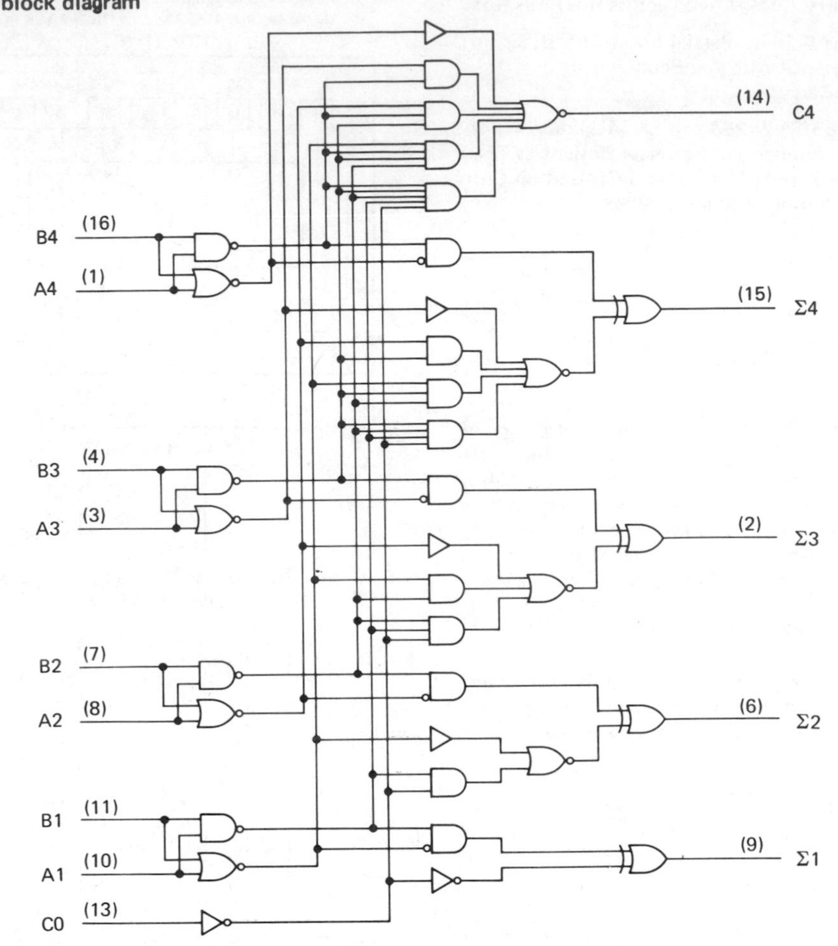

schematics of inputs and outputs

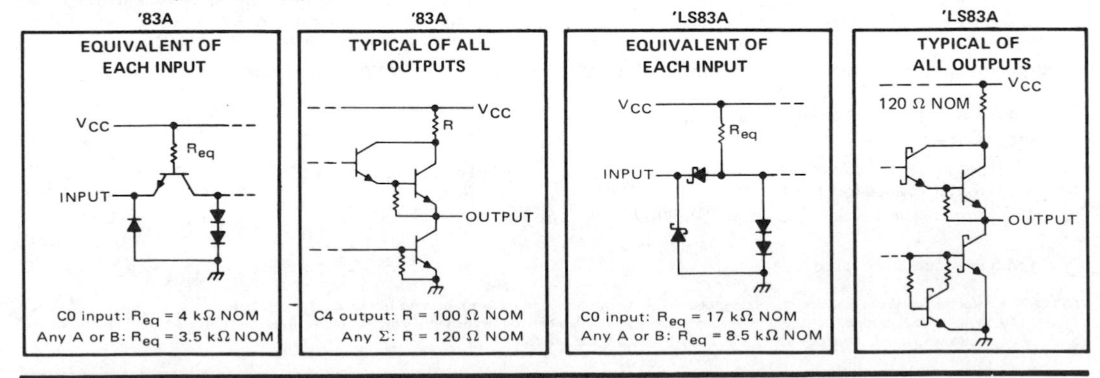

recommended operating conditions

		SN5483A MIN	SN5483A NOM	SN5483A MAX	SN7483A MIN	SN7483A NOM	SN7483A MAX	UNIT
Supply Voltage, V_{CC}		4.5	5	5.5	4.75	5	5.25	V
High-level output current, I_{OH}	Any output except C4			−800			−800	μA
	Output C4			−400			−400	
Low-level output current, I_{OL}	Any output except C4			16			16	mA
	Output C4			8			8	
Operating free-air temperature, T_A		−55		125	0		70	°C

electrical characteristics over recommended operating free-air temperature range (unless otherwise noted)

PARAMETER			TEST CONDITIONS†		SN5483A MIN	SN5483A TYP‡	SN5483A MAX	SN7483A MIN	SN7483A TYP‡	SN7483A MAX	UNIT
V_{IH}	High-level input voltage				2			2			V
V_{IL}	Low-level input voltage						0.8			0.8	V
V_{IK}	Input clamp voltage		V_{CC} = MIN, I_I = −12 mA				−1.5			−1.5	V
V_{OH}	High-level output voltage		V_{CC} = MIN, V_{IH} = 2 V, V_{IL} = 0.8 V, I_{OH} = MAX		2.4	3.4		2.4	3.4		V
V_{OL}	Low-level output voltage		V_{CC} = MIN, V_{IH} = 2 V, V_{IL} = 0.8 V, I_{OL} = MAX			0.2	0.4		0.2	0.4	V
I_I	Input current at maximum input voltage		V_{CC} = MAX, V_I = 5.5 V				1			1	mA
I_{IH}	High-level input current		V_{CC} = MAX, V_I = 2.4 V				40			40	μA
I_{IL}	Low-level input current		V_{CC} = MAX, V_I = 0.4 V				−1.6			−1.6	mA
I_{OS}	Short-circuit output current§	Any output except C4	V_{CC} = MAX		−20		−55	−18		−55	mA
		Output C4			−20		−70	−18		−70	
I_{CC}	Supply current		V_{CC} = MAX, Outputs open	All B low, other inputs at 4.5 V		56			56		mA
				All inputs at 4.5 V		66	99		66	110	

†For conditions shown as MIN or MAX, use the appropriate value specified under recommended operating conditions.
‡All typical values are at V_{CC} = 5 V, T_A = 25°C.
§Only one output should be shorted at a time.

switching characteristics, V_{CC} = 5 V, T_A = 25°C

PARAMETER¶	FROM (INPUT)	TO (OUTPUT)	TEST CONDITIONS	MIN	TYP	MAX	UNIT
t_{PLH}	C0	Any Σ	C_L = 15 pF, R_L = 400 Ω, See Note 3		14	21	ns
t_{PHL}					12	21	
t_{PLH}	A_i or B_i	Σ_i			16	24	ns
t_{PHL}					16	24	
t_{PLH}	C0	C4	C_L = 15 pF, R_L = 780 Ω, See Note 3		9	14	ns
t_{PHL}					11	16	
t_{PLH}	A_i or B_i	C4			9	14	ns
t_{PHL}					11	16	

¶t_{PLH} ≡ Propagation delay time, low-to-high-level output
t_{PHL} ≡ Propagation delay time, high-to-low-level output
NOTE 3: Load circuit and voltage waveforms are shown on page 3-10.

TTL
MSI

TYPES SN5490A, SN5492A, SN5493A, SN54L90, SN54L93, SN54LS90, SN54LS92, SN54LS93, SN7490A, SN7492A, SN7493A, SN74L90, SN74L93, SN74LS90, SN74LS92, SN74LS93 DECADE, DIVIDE-BY-TWELVE, AND BINARY COUNTERS

BULLETIN NO. DL-S 7611807, MARCH 1974—REVISED OCTOBER 1976

'90A, 'L90, 'LS90 . . . DECADE COUNTERS

'92A, 'LS92 . . . DIVIDE-BY-TWELVE COUNTERS

'93A, 'L93, 'LS93 . . . 4-BIT BINARY COUNTERS

TYPES	TYPICAL POWER DISSIPATION
'90A	145 mW
'L90	20 mW
'LS90	45 mW
'92A, '93A	130 mW
'LS92, 'LS93	45 mW
'L93	16 mW

description

Each of these monolithic counters contains four master-slave flip-flops and additional gating to provide a divide-by-two counter and a three-stage binary counter for which the count cycle length is divide-by-five for the '90A, 'L90, and 'LS90, divide-by-six for the '92A and 'LS92, and divide-by-eight for the '93A, 'L93, and 'LS93.

All of these counters have a gated zero reset and the '90A, 'L90, and 'LS90 also have gated set-to-nine inputs for use in BCD nine's complement applications.

To use their maximum count length (decade, divide-by-twelve, or four-bit binary) of these counters, the B input is connected to the Q_A, output. The input count pulses are applied to input A and the outputs are as described in the appropriate function table. A symmetrical divide-by-ten count can be obtained from the '90A, 'L90, or 'LS90 counters by connecting the Q_D output to the A input and applying the input count to the B input which gives a divide-by-ten square wave at output Q_A.

SN54', SN54LS' . . . J OR W PACKAGE
SN54L' . . . J OR T PACKAGE
SN54', SN74L', SN74LS' . . . J OR N PACKAGE

'90A, 'L90, 'LS90 (TOP VIEW)

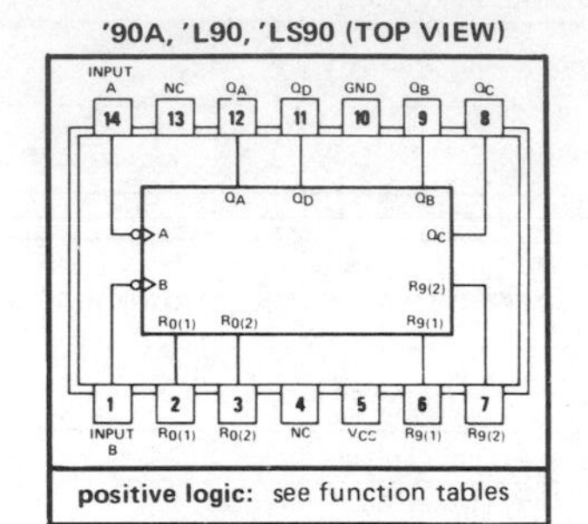

positive logic: see function tables

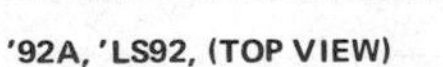

'92A, 'LS92, (TOP VIEW)

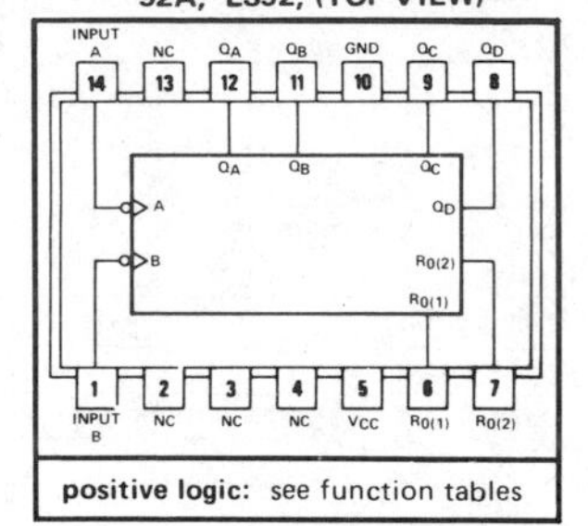

positive logic: see function tables

'93A, 'LS93 (TOP VIEW)

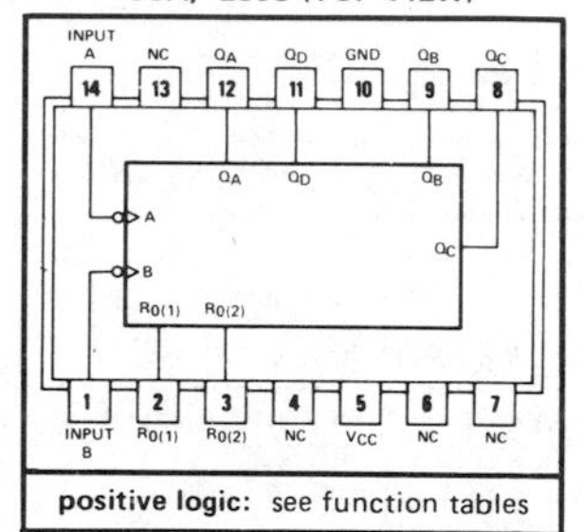

positive logic: see function tables

'L93 (TOP VIEW)

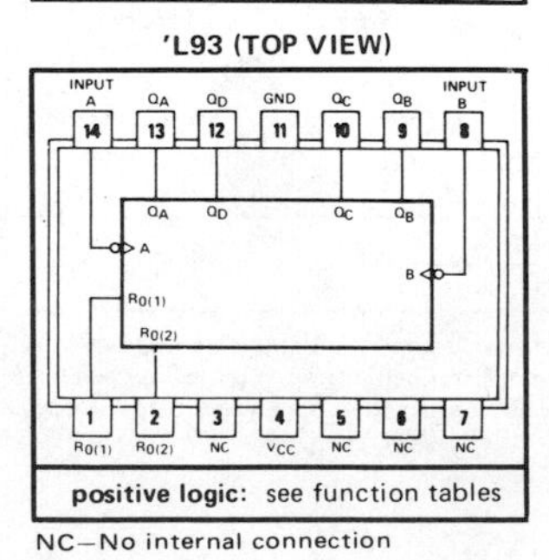

positive logic: see function tables

NC—No internal connection

'90A, 'L90, 'LS90
BCD COUNT SEQUENCE
(See Note A)

COUNT	OUTPUT			
	Q_D	Q_C	Q_B	Q_A
0	L	L	L	L
1	L	L	L	H
2	L	L	H	L
3	L	L	H	H
4	L	H	L	L
5	L	H	L	H
6	L	H	H	L
7	L	H	H	H
8	H	L	L	L
9	H	L	L	H

'90A, 'L90, 'LS90
BI-QUINARY (5-2)
(See Note B)

COUNT	OUTPUT			
	Q_A	Q_D	Q_C	Q_B
0	L	L	L	L
1	L	L	L	H
2	L	L	H	L
3	L	L	H	H
4	L	H	L	L
5	H	L	L	L
6	H	L	L	H
7	H	L	H	L
8	H	L	H	H
9	H	H	L	L

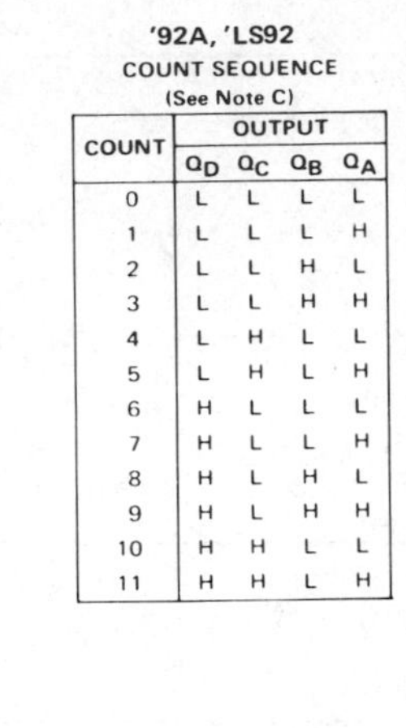

'92A, 'LS92
COUNT SEQUENCE
(See Note C)

COUNT	OUTPUT			
	Q_D	Q_C	Q_B	Q_A
0	L	L	L	L
1	L	L	L	H
2	L	L	H	L
3	L	L	H	H
4	L	H	L	L
5	L	H	L	H
6	H	L	L	L
7	H	L	L	H
8	H	L	H	L
9	H	L	H	H
10	H	H	L	L
11	H	H	L	H

'93A, 'L93, 'LS93
COUNT SEQUENCE
(See Note C)

COUNT	OUTPUT			
	Q_D	Q_C	Q_B	Q_A
0	L	L	L	L
1	L	L	L	H
2	L	L	H	L
3	L	L	H	H
4	L	H	L	L
5	L	H	L	H
6	L	H	H	L
7	L	H	H	H
8	H	L	L	L
9	H	L	L	H
10	H	L	H	L
11	H	L	H	H
12	H	H	L	L
13	H	H	L	H
14	H	H	H	L
15	H	H	H	H

'90A, 'L90, 'LS90
RESET/COUNT FUNCTION TABLE

RESET INPUTS				OUTPUT			
$R_{0(1)}$	$R_{0(2)}$	$R_{9(1)}$	$R_{9(2)}$	Q_D	Q_C	Q_B	Q_A
H	H	L	X	L	L	L	L
H	H	X	L	L	L	L	L
X	X	H	H	H	L	L	H
X	L	X	L	COUNT			
L	X	L	X	COUNT			
L	X	X	L	COUNT			
X	L	L	X	COUNT			

'92A, 'LS92, '93A, 'L93, 'LS93
RESET/COUNT FUNCTION TABLE

RESET INPUTS		OUTPUT			
$R_{0(1)}$	$R_{0(2)}$	Q_D	Q_C	Q_B	Q_A
H	H	L	L	L	L
L	X	COUNT			
X	L	COUNT			

NOTES: A. Output Q_A is connected to input B for BCD count.
B. Output Q_D is connected to input A for bi-quinary count.
C. Output Q_A is connected to input B.
D. H = high level, L = low level, X = irrelevant

functional block diagrams

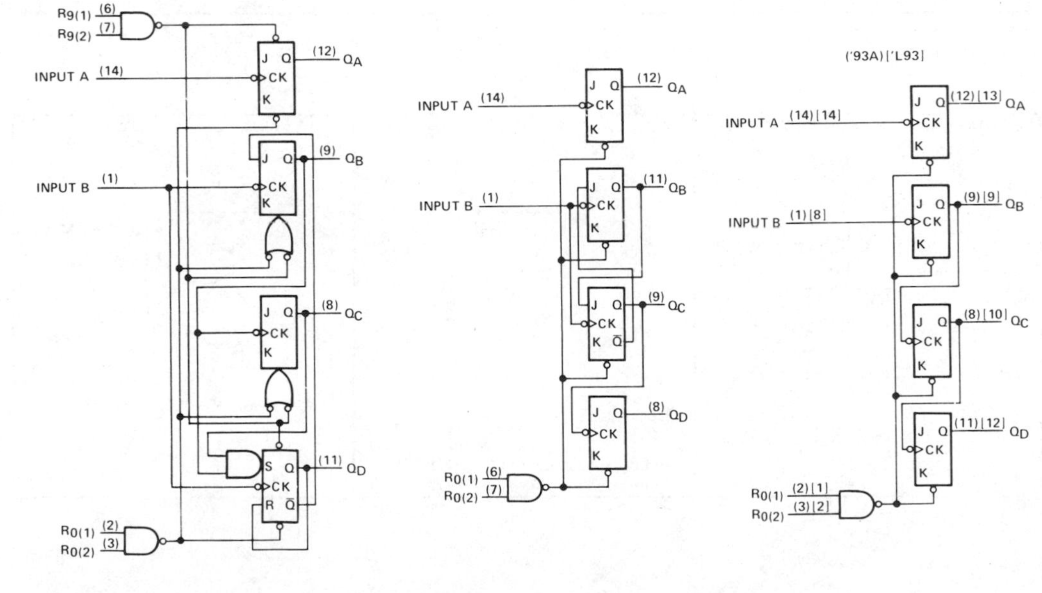

The J and K inputs shown without connection are for reference only and are functionally at a high level.

schematics of inputs and outputs

'90A, '92A, '93A

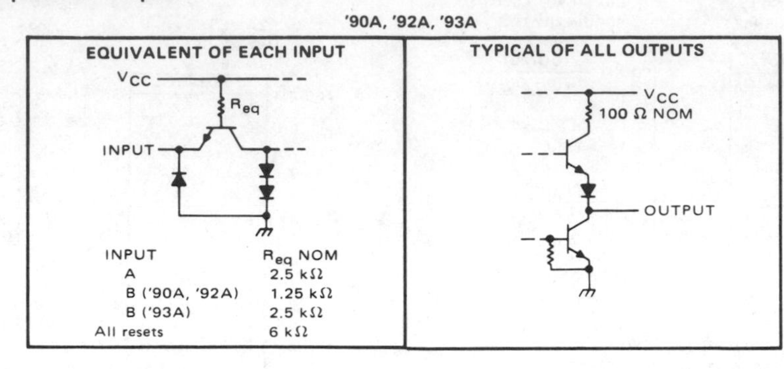

INPUT	Req NOM
A	2.5 kΩ
B ('90A, '92A)	1.25 kΩ
B ('93A)	2.5 kΩ
All resets	6 kΩ

'L90, 'L93

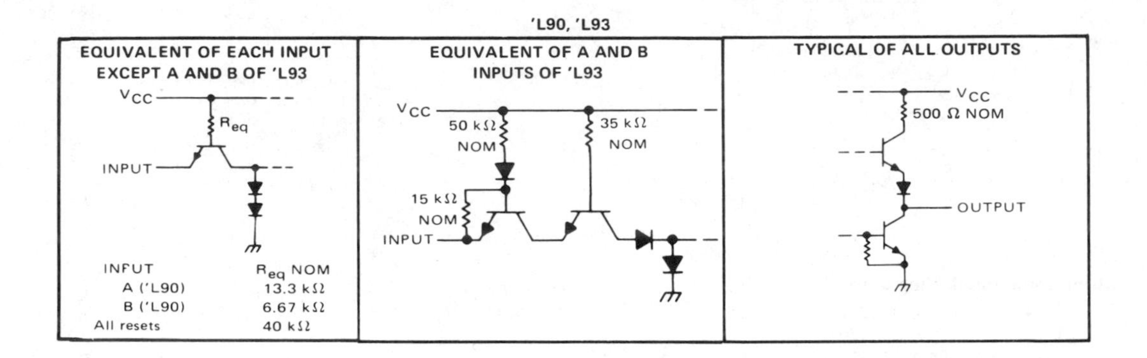

INPUT	Req NOM
A ('L90)	13.3 kΩ
B ('L90)	6.67 kΩ
All resets	40 kΩ

'LS90, 'LS92, 'LS93

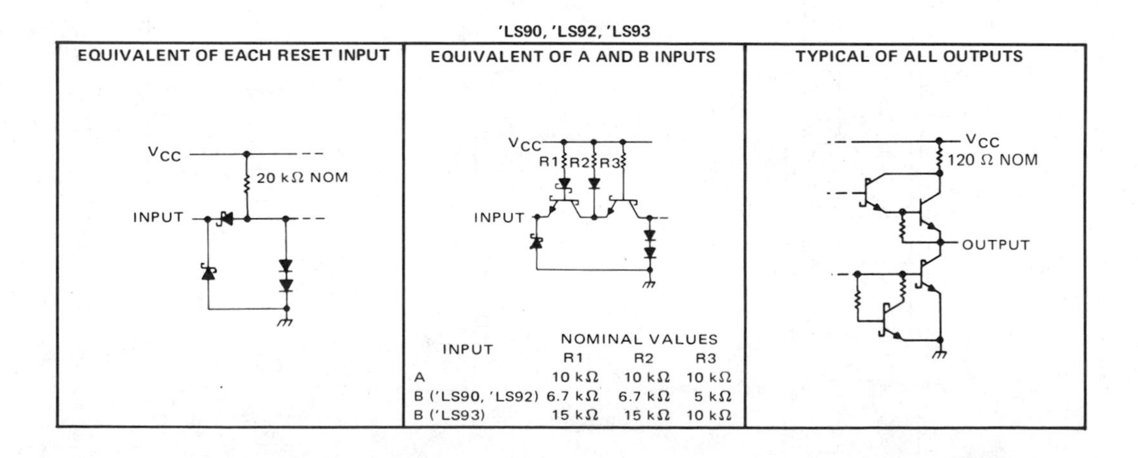

INPUT	NOMINAL VALUES R1	R2	R3
A	10 kΩ	10 kΩ	10 kΩ
B ('LS90, 'LS92)	6.7 kΩ	6.7 kΩ	5 kΩ
B ('LS93)	15 kΩ	15 kΩ	10 kΩ

absolute maximum ratings over operating free-air temperature range (unless otherwise noted)

Supply voltage, V_{CC} (see Note 1) . . . 7 V
Input voltage . . . 5.5 V
Interemitter voltage (see Note 2) . . . 5.5 V
Operating free-air temperature range: SN5490A, SN5492A, SN5493A . . . −55°C to 125°C
SN7490A, SN7492A, SN7493A . . . 0°C to 70°C
Storage temperature range . . . −65°C to 150°C

NOTES: 1. Voltage values, except interemitter voltage, are with respect to network ground terminal.
2. This is the voltage between two emitters of a multiple-emitter transistor. For these circuits, this rating applies between the two R_0 inputs, and for the '90A circuit, it also applies between the two R_9 inputs.

recommended operating conditions

		SN5490A, SN5492A SN5493A			SN7490A, SN7492A SN7493A			UNIT
		MIN	NOM	MAX	MIN	NOM	MAX	
Supply voltage, V_{CC}		4.5	5	5.5	4.75	5	5.25	V
High-level output current, I_{OH}				−800			−800	μA
Low-level output current, I_{OL}				16			16	mA
Count frequency, f_{count} (see Figure 1)	A input	0		32	0		32	MHz
	B input	0		16	0		16	
Pulse width, t_w	A input	15			15			ns
	B input	30			30			
	Reset inputs	15			15			
Reset inactive-state setup time, t_{su}		25			25			ns
Operating free-air temperature, T_A		−55		125	0		70	°C

electrical characteristics over recommended operating free-air temperature range (unless otherwise noted)

PARAMETER		TEST CONDITIONS†		'90A MIN	'90A TYP‡	'90A MAX	'92A MIN	'92A TYP‡	'92A MAX	'93A MIN	'93A TYP‡	'93A MAX	UNIT
V_{IH} High-level input voltage				2			2			2			V
V_{IL} Low-level input voltage						0.8			0.8			0.8	V
V_{IK} Input clamp voltage		V_{CC} = MIN, I_I = −12 mA				−1.5			−1.5			−1.5	V
V_{OH} High-level output voltage		V_{CC} = MIN, V_{IH} = 2 V, V_{IL} = 0.8 V, I_{OH} = −800 μA		2.4	3.4		2.4	3.4		2.4	3.4		V
V_{OL} Low-level output voltage		V_{CC} = MIN, V_{IH} = 2 V, V_{IL} = 0.8 V, I_{OL} = 16 mA¶			0.2	0.4		0.2	0.4		0.2	0.4	V
I_I Input current at maximum input voltage		V_{CC} = MAX, V_I = 5.5 V				1			1			1	mA
I_{IH} High-level input current	Any reset	V_{CC} = MAX, V_I = 2.4 V				40			40			40	μA
	A input					80			80			80	
	B input					120			120			80	
I_{IL} Low-level input current	Any reset	V_{CC} = MAX, V_I = 0.4 V				−1.6			−1.6			−1.6	mA
	A input					−3.2			−3.2			−3.2	
	B input					−4.8			−4.8			−3.2	
I_{OS} Short-circuit output current§		V_{CC} = MAX	SN54'	−20		−57	−20		−57	−20		−57	mA
			SN74'	−18		−57	−18		−57	−18		−57	
I_{CC} Supply current		V_{CC} = MAX, See Note 3			29	42		26	39		26	39	mA

†For conditions shown as MIN or MAX, use the appropriate value specified under recommended operating conditions.
‡All typical values are at V_{CC} = 5 V, T_A = 25°C.
§Not more than one output should be shorted at a time.
¶Q_A outputs are tested at I_{OL} = 16 mA plus the limit value for I_{IL} for the B input. This permits driving the B input while maintaining full fan-out capability.
NOTE 3: I_{CC} is measured with all outputs open, both R_0 inputs grounded following momentary connection to 4.5 V, and all other inputs grounded.

switching characteristics, V_{CC} = 5 V, T_A = 25°C

PARAMETER¶	FROM (INPUT)	TO (OUTPUT)	TEST CONDITIONS	'90A			'92A			'93A			UNIT
				MIN	TYP	MAX	MIN	TYP	MAX	MIN	TYP	MAX	
f_{max}	A	Q_A	C_L = 15 pF, R_L = 400 Ω, See Figure 1	32	42		32	42		32	42		MHz
	B	Q_B		16			16			16			
t_{PLH}	A	Q_A			10	16		10	16		10	16	ns
t_{PHL}					12	18		12	18		12	18	
t_{PLH}	A	Q_D			32	48		32	48		46	70	ns
t_{PHL}					34	50		34	50		46	70	
t_{PLH}	B	Q_B			10	16		10	16		10	16	ns
t_{PHL}					14	21		14	21		14	21	
t_{PLH}	B	Q_C			21	32		10	16		21	32	ns
t_{PHL}					23	35		14	21		23	35	
t_{PLH}	B	Q_D			21	32		21	32		34	51	ns
t_{PHL}					23	35		23	35		34	51	
t_{PHL}	Set-to-0	Any			26	40		26	40		26	40	ns
t_{PLH}	Set-to-9	Q_A, Q_D			20	30							ns
t_{PHL}		Q_B, Q_C			26	40							

¶ f_{max} ≡ maximum count frequency
t_{PLH} ≡ propagation delay time, low-to-high-level output
t_{PHL} ≡ propagation delay time, high-to-low-level output

PARAMETER MEASUREMENT INFORMATION

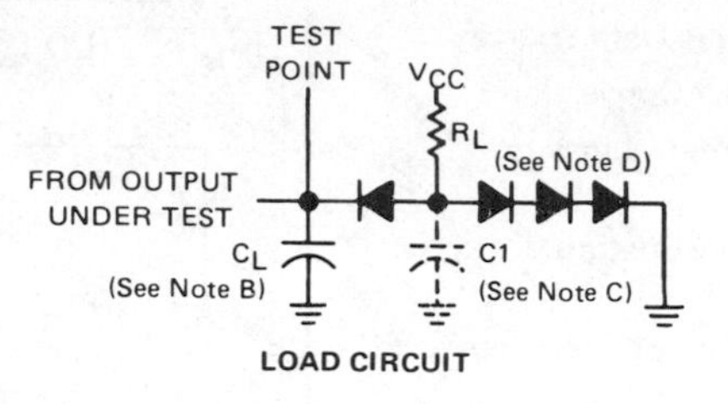

LOAD CIRCUIT

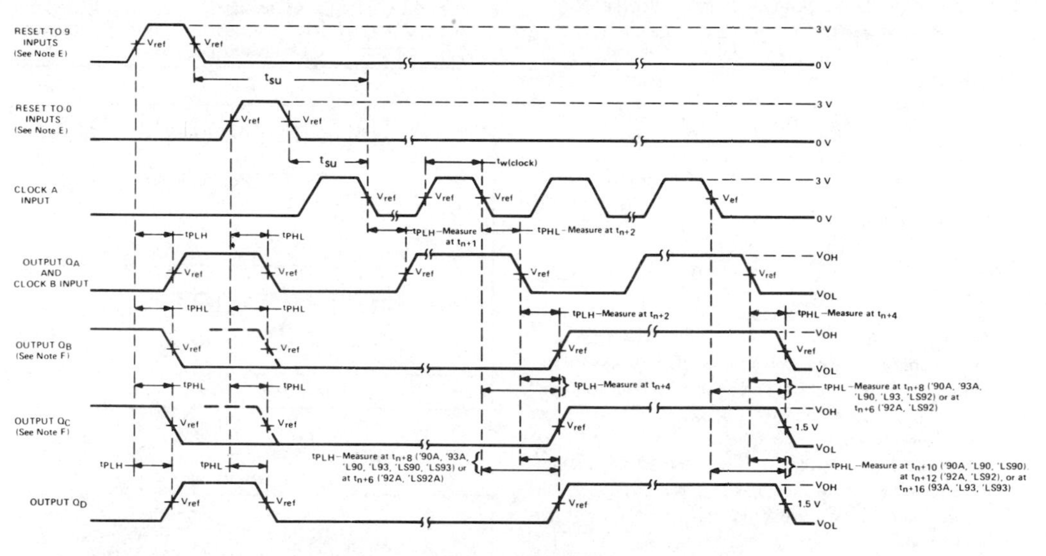

VOLTAGE WAVEFORMS

NOTES: A. Input pulses are supplied by a generator having the following characteristics:
for '90A, '92A, '93A, $t_r \leq 5$ ns, $t_f \leq 5$ ns, PRR = 1 MHz, duty cycle = 50%, $Z_{out} \approx 50$ ohms;
for 'L90, 'L93, $t_r \leq 15$ ns, $t_f \leq 15$ ns, PRR = 500 kHz, duty cycle = 50%, $Z_{out} \approx 50$ ohms;
for 'LS90, 'LS92, 'LS93, $t_r \leq 15$ ns, $t_f \leq 5$ ns, PRR = 1 MHz, duty cycle = 50%, $Z_{out} \approx 50$ ohms.

B. C_L includes probe and jig capacitance.

C. C1 (30 pF) is applicable for testing 'L90 and 'L93.

D. All diodes are 1N916 or 1N3064.

E. Each reset input is tested separately with the other reset at 4.5 V.

F. Reference waveforms are shown with dashed lines.

G. For '90A, '92A, and '93A; V_{ref} = 1.5 V. For 'L90, 'L93, 'LS90, 'LS92, and 'LS93; V_{ref} = 1.3 V.

FIGURE 1

TYPES SN54150, SN54151A, SN54152A, SN54LS151, SN54LS152, SN54S151, SN74150, SN74151A, SN74LS151, SN74S151 DATA SELECTORS/MULTIPLEXERS

BULLETIN NO. DL-S 7611819, DECEMBER 1972–REVISED OCTOBER 1976

- '150 Selects One-of-Sixteen Data Sources
- Others Select One-of-Eight Data Sources
- Performs Parallel-to-Serial Conversion
- Permits Multiplexing from N Lines to One Line
- Also For Use as Boolean Function Generator
- Input-Clamping Diodes Simplify System Design
- Fully Compatible with Most TTL and DTL Circuits

TYPE	TYPICAL AVERAGE PROPAGATION DELAY TIME DATA INPUT TO W OUTPUT	TYPICAL POWER DISSIPATION
'150	11 ns	200 mW
'151A	8 ns	145 mW
'152A	8 ns	130 mW
'LS151	11 ns†	30 mW
'LS152	11 ns†	28 mW
'S151	4.5 ns	225 mW

†Tentative data

description

These monolithic data selectors/multiplexers contain full on-chip binary decoding to select the desired data source. The '150 selects one-of-sixteen data sources; the '151A, '152A, 'LS151, 'LS152, and 'S151 select one-of-eight data sources. The '150, '151A, 'LS151, and 'S151 have a strobe input which must be at a low logic level to enable these devices. A high level at the strobe forces the W output high, and the Y output (as applicable) low.

The '151A, 'LS151, and 'S151 feature complementary W and Y outputs whereas the '150, '152A, and 'LS152 have an inverted (W) output only.

The '151A and '152A incorporate address buffers which have symmetrical propagation delay times through the complementary paths. This reduces the possibility of transients occurring at the output(s) due to changes made at the select inputs, even when the '151A outputs are enabled (i.e., strobe low).

SN54150 . . . J OR W PACKAGE
SN74150 . . . J OR N PACKAGE
(TOP VIEW)

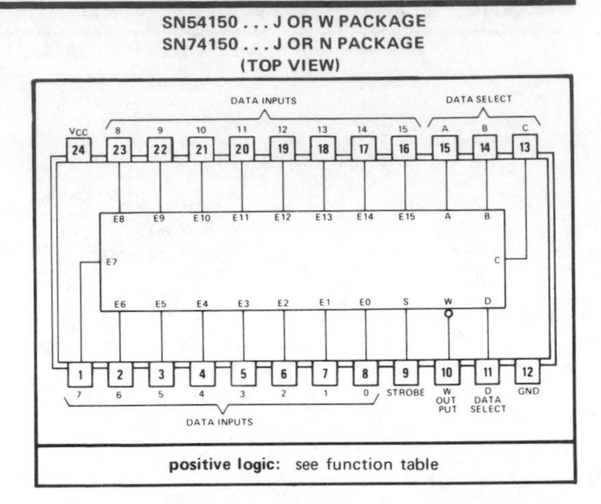

positive logic: see function table

SN54151A, SN54LS151, SN54S151 . . . J OR W PACKAGE
SN74151A, SN74LS151, SN74S151 . . . J OR N PACKAGE
(TOP VIEW)

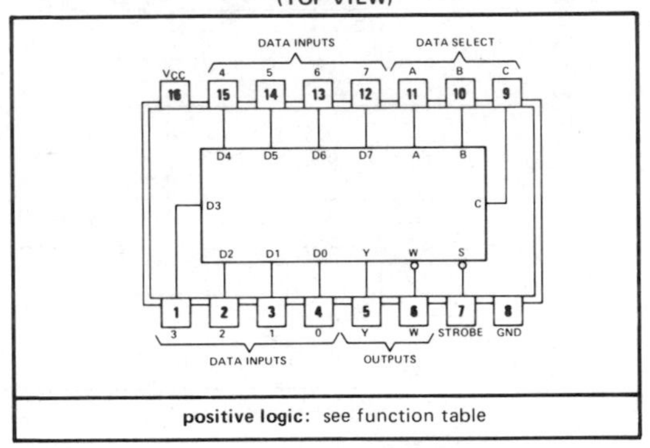

positive logic: see function table

SN54152A, SN54LS152 . . . W PACKAGE
(TOP VIEW)

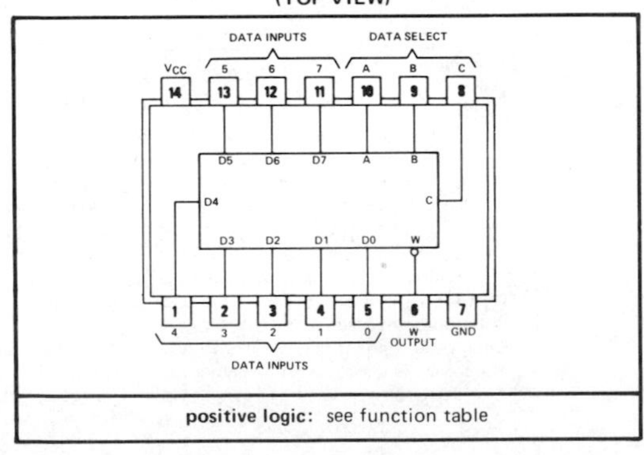

positive logic: see function table

logic

'150

FUNCTION TABLE

INPUTS					OUTPUT
SELECT				STROBE	W
D	C	B	A	S	
X	X	X	X	H	H
L	L	L	L	L	$\overline{E0}$
L	L	L	H	L	$\overline{E1}$
L	L	H	L	L	$\overline{E2}$
L	L	H	H	L	$\overline{E3}$
L	H	L	L	L	$\overline{E4}$
L	H	L	H	L	$\overline{E5}$
L	H	H	L	L	$\overline{E6}$
L	H	H	H	L	$\overline{E7}$
H	L	L	L	L	$\overline{E8}$
H	L	L	H	L	$\overline{E9}$
H	L	H	L	L	$\overline{E10}$
H	L	H	H	L	$\overline{E11}$
H	H	L	L	L	$\overline{E12}$
H	H	L	H	L	$\overline{E13}$
H	H	H	L	L	$\overline{E14}$
H	H	H	H	L	$\overline{E15}$

'151A, 'LS151, 'S151

FUNCTION TABLE

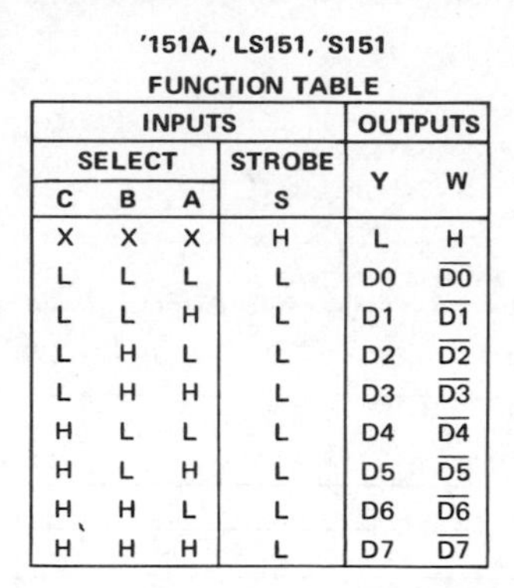

INPUTS				OUTPUTS	
SELECT			STROBE	Y	W
C	B	A	S		
X	X	X	H	L	H
L	L	L	L	D0	$\overline{D0}$
L	L	H	L	D1	$\overline{D1}$
L	H	L	L	D2	$\overline{D2}$
L	H	H	L	D3	$\overline{D3}$
H	L	L	L	D4	$\overline{D4}$
H	L	H	L	D5	$\overline{D5}$
H	H	L	L	D6	$\overline{D6}$
H	H	H	L	D7	$\overline{D7}$

'152A, 'LS152

FUNCTION TABLE

SELECT INPUTS			OUTPUT
C	B	A	W
L	L	L	$\overline{D0}$
L	L	H	$\overline{D1}$
L	H	L	$\overline{D2}$
L	H	H	$\overline{D3}$
H	L	L	$\overline{D4}$
H	L	H	$\overline{D5}$
H	H	L	$\overline{D6}$
H	H	H	$\overline{D7}$

H = high level, L = low level, X = irrelevant

$\overline{E0}$, $\overline{E1}$ . . . $\overline{E15}$ = the complement of the level of the respective E input

D0, D1 . . . D7 = the level of the D respective input

functional block diagrams

'151A, 'LS151, 'S151

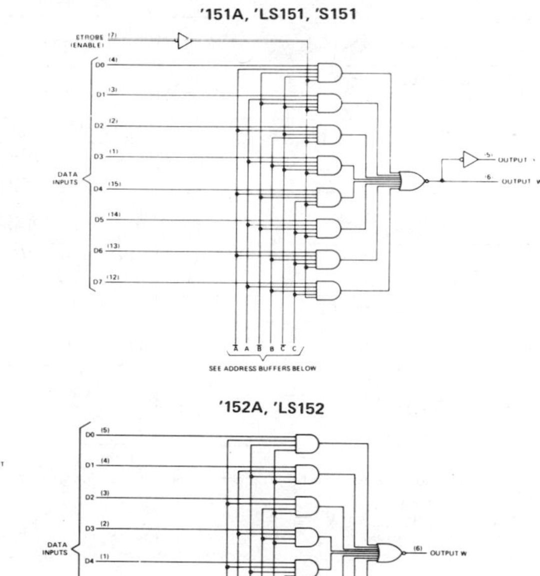

'150

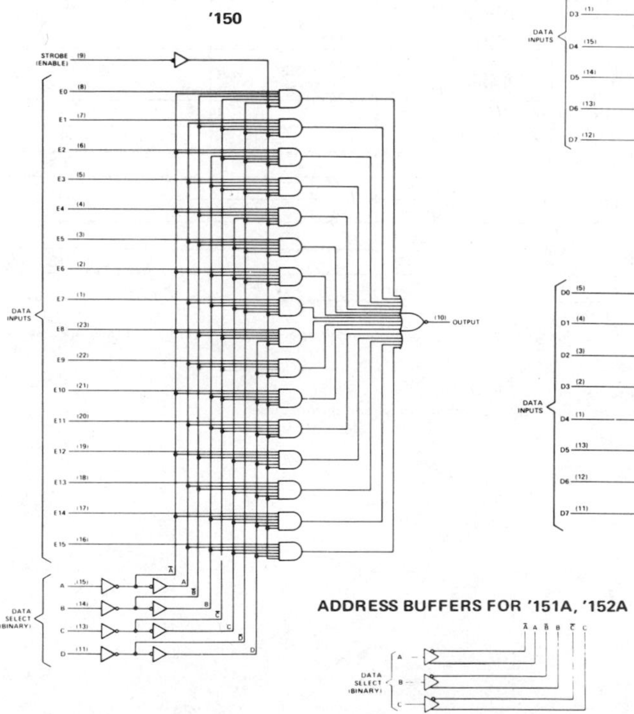

ADDRESS BUFFERS FOR '151A, '152A

ADDRESS BUFFERS FOR 'LS151, 'S151, 'LS152

absolute maximum ratings over operating free-air temperature range (unless otherwise noted)

Supply voltage, V_{CC} (see Note 1) . . . . . 7 V
Input voltage (see Note 2) . . . . . 5.5 V
Operating free-air temperature range: SN54' Circuits . . . . . −55°C to 125°C
SN74' Circuits . . . . . 0°C to 70°C
Storage temperature range: . . . . . −65°C to 150°C

NOTES: 1. Voltage values are with respect to network ground terminal.
2. For the '150, input voltages must be zero or positive with respect to network ground terminal.

recommended operating conditions

	SN54'			SN74'			UNIT
	MIN	NOM	MAX	MIN	NOM	MAX	
Supply voltage, V_{CC}	4.5	5	5.5	4.75	5	5.25	V
High-level output current, I_{OH}			−800			−800	μA
Low-level output current, I_{OL}			16			16	mA
Operating free-air temperature, T_A	−55		125	0		70	°C

electrical characteristics over recommended operating free-air temperature range (unless otherwise noted)

PARAMETER		TEST CONDITIONS†		'150			'151A, '152A			UNIT
				MIN	TYP‡	MAX	MIN	TYP‡	MAX	
V_{IH}	High-level input voltage			2			2			V
V_{IL}	Low-level input voltage					0.8			0.8	V
V_{IK}	Input clamp voltage	V_{CC} = MIN, I_I = −8 mA							−1.5	V
V_{OH}	High-level output voltage	V_{CC} = MIN, V_{IH} = 2 V, V_{IL} = 0.8 V, I_{OH} = −800 μA		2.4	3.4		2.4	3.4		V
V_{OL}	Low-level output voltage	V_{CC} = MIN, V_{IH} = 2 V, V_{IL} = 0.8 V, I_{OL} = 16 mA			0.2	0.4		0.2	0.4	V
I_I	Input current at maximum input voltage	V_{CC} = MAX, V_I = 5.5 V				1			1	mA
I_{IH}	High-level input current	V_{CC} = MAX, V_I = 2.4 V				40			40	μA
I_{IL}	Low-level input current	V_{CC} = MAX, V_I = 0.4 V				−1.6			−1.6	mA
I_{OS}	Short-circuit output current§	V_{CC} = MAX	SN54'	−20		−55	−20		−55	mA
			SN74'	−18		−55	−18		−55	
I_{CC}	Supply current	V_{CC} = MAX, See Note 3	'150		40	68				mA
			'151A					29	48	
			'152A					26	43	

†For conditions shown as MIN or MAX, use the appropriate value specified under recommended operating conditions for the applicable device type.
‡All typical values at V_{CC} = 5 V, T_A = 25°C.
§Not more than one output of the '151A should be shorted at a time.
NOTE 3: I_{CC} is measured with the strobe and data select inputs at 4.5 V, all other inputs and outputs open.

switching characteristics, V_{CC} = 5 V, T_A = 25°C

PARAMETER¶	FROM (INPUT)	TO (OUTPUT)	TEST CONDITIONS	'150			'151A, '152A			UNIT
				MIN	TYP	MAX	MIN	TYP	MAX	
t_{PLH}	A, B, or C (4 levels)	Y	C_L = 15 pF, R_L = 400 Ω, See Note 4					25	38	ns
t_{PHL}								25	38	
t_{PLH}	A, B, C, or D (3 levels)	W			23	35		17	26	ns
t_{PHL}					22	33		19	30	
t_{PLH}	Strobe	Y						21	33	ns
t_{PHL}								22	33	
t_{PLH}	Strobe	W			15.5	24		14	21	ns
t_{PHL}					21	30		15	23	
t_{PLH}	D0 thru D7	Y						13	20	ns
t_{PHL}								18	27	
t_{PLH}	E0 thru E15, or D0 thru D7	W			8.5	14		8	14	ns
t_{PHL}					13	20		8	14	

¶ t_{PLH} ≡ propagation delay time, low-to-high-level output
t_{PHL} ≡ propagation delay time, high-to-low-level output

NOTE 4: Load circuit and voltage waveforms are shown on page 3-10.

schematics of inputs and outputs

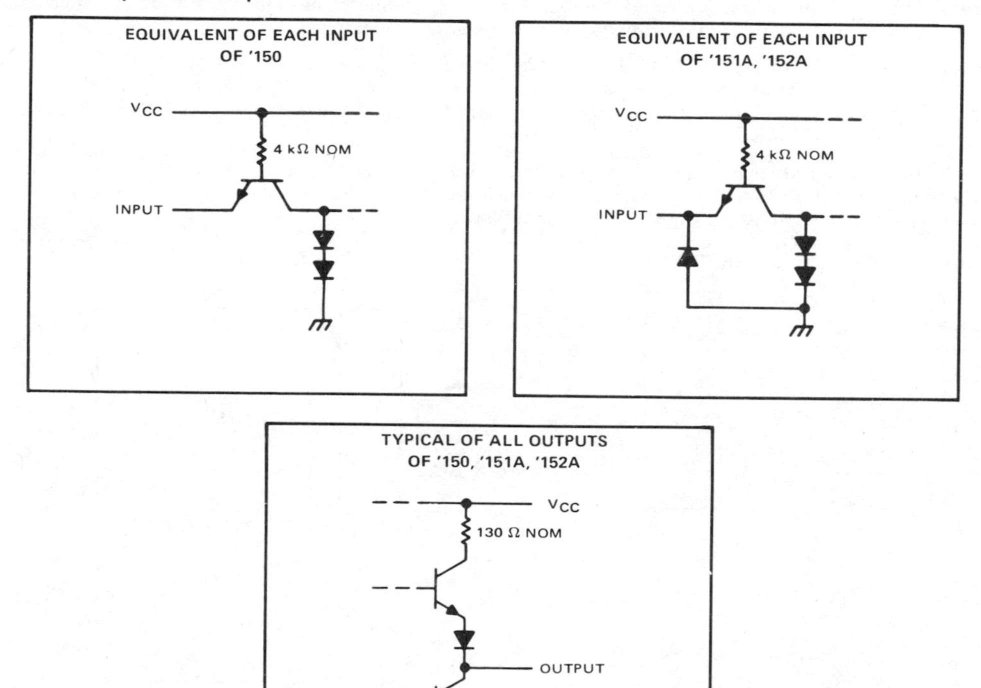

intel®

2147
4096 X 1 BIT STATIC RAM

	2147-3	2147	2147L	2147-6
Max. Access Time (ns)	55	70	70	85
Max. Active Current (mA)	180	160	140	160
Max. Standby Current (mA)	30	20	10	20

- **HMOS Technology**
- **Completely Static Memory — No Clock or Timing Strobe Required**
- **Equal Access and Cycle Times**
- **Single +5V Supply**
- **Automatic Power-Down**
- **High Density 18-Pin Package**
- **Directly TTL Compatible — All Inputs and Output**
- **Separate Data Input and Output**
- **Three-State Output**

The Intel® 2147 is a 4096-bit static Random Access Memory organized as 4096 words by 1-bit using HMOS, a high-performance MOS technology. It uses a uniquely innovative design approach which provides the ease-of-use features associated with non-clocked static memories and the reduced standby power dissipation associated with clocked static memories. To the user this means low standby power dissipation without the need for clocks, address setup and hold times, nor reduced data rates due to cycle times that are longer than access times.

$\overline{CS}$ controls the power-down feature. In less than a cycle time after $\overline{CS}$ goes high — deselecting the 2147 — the part automatically reduces its power requirements and remains in this low power standby mode as long as $\overline{CS}$ remains high. This device feature results in system power savings as great as 85% in larger systems, where the majority of devices are deselected.

The 2147 is placed in an 18-pin package configured with the industry standard pinout. It is directly TTL compatible in all respects: inputs, output, and a single +5V supply. The data is read out nondestructively and has the same polarity as the input data. A data input and a separate three-state output are used.

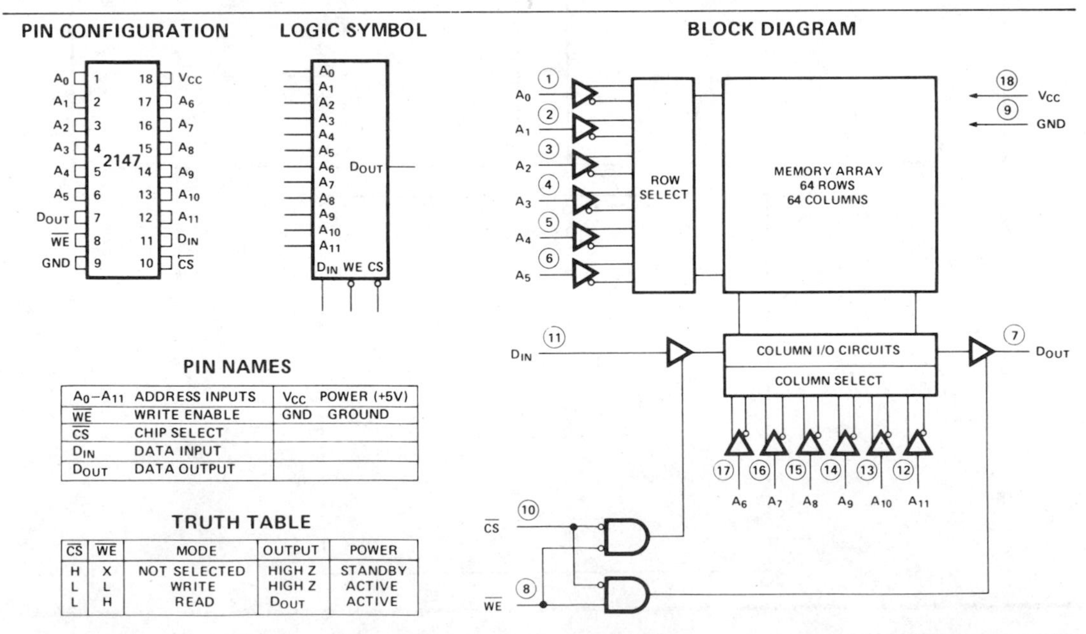

PIN NAMES

A_0–A_{11}	ADDRESS INPUTS	V_{CC}	POWER (+5V)
$\overline{WE}$	WRITE ENABLE	GND	GROUND
$\overline{CS}$	CHIP SELECT		
D_{IN}	DATA INPUT		
D_{OUT}	DATA OUTPUT		

TRUTH TABLE

$\overline{CS}$	$\overline{WE}$	MODE	OUTPUT	POWER
H	X	NOT SELECTED	HIGH Z	STANDBY
L	L	WRITE	HIGH Z	ACTIVE
L	H	READ	D_{OUT}	ACTIVE

JULY 1979

ABSOLUTE MAXIMUM RATINGS*

Temperature Under Bias .... -10°C to 85°C
Storage Temperature .... -65°C to +150°C
Voltage on Any Pin With
Respect to Ground .... -1.5V to +7V
Power Dissipation 1.2W
D.C. Output Current 20mA

**COMMENT: Stresses above those listed under "Absolute Maximum Ratings" may cause permanent damage to the device. This is a stress rating only and functional operation of the device at these or any other conditions above those indicated in the operational sections of this specification is not implied. Exposure to absolute maximum rating conditions for extended periods may affect device reliability.*

D.C. AND OPERATING CHARACTERISTICS [1]

T_A = 0°C to 70°C, V_{CC} = +5V ±10%, unless otherwise noted.

Symbol	Parameter	2147-3 Min.	2147-3 Typ. [2]	2147-3 Max.	2147, 2147-6 Min.	2147, 2147-6 Typ. [2]	2147, 2147-6 Max.	2147L Min.	2147L Typ. [2]	2147L Max.	Unit	Test Conditions
I_{LI}	Input Load Current (All Input Pins)		0.01	10		0.01	10		0.01	10	μA	V_{CC}=MAX, V_{IN}=GND to V_{CC}
$\lvert I_{LO} \rvert$	Output Leakage Current		0.1	50		0.1	50		0.1	50	μA	$\overline{CS}$=V_{IH}, V_{CC}=Max., V_{OUT}=GND to 4.5V
I_{CC}	Operating Current		120	170		100	150		100	135	mA	T_A=25°C; V_{CC}=Max., $\overline{CS}$=V_{IL}, Outputs Open
				180			160			140	mA	T_A=0°C; V_{CC}=Max., $\overline{CS}$=V_{IL}, Outputs Open
I_{SB}	Standby Current		18	30		12	20		7	10	mA	V_{CC}=Min. to Max. $\overline{CS}$=V_{IH}
I_{PO} [3]	Peak Power-On Current		35	70		25	50		15	30	mA	V_{CC}=GND to V_{CC} Min., $\overline{CS}$=Lower of V_{CC} or V_{IH} Min.
V_{IL}	Input Low Voltage	-1.0		0.8	-1.0		0.8	-1.0		0.8	V	
V_{IH}	Input High Voltage	2.0		6.0	2.0		6.0	2.0		6.0	V	
V_{OL}	Output Low Voltage			0.4			0.4			0.4	V	I_{OL} = 8mA
V_{OH}	Output High Voltage	2.4			2.4			2.4			V	I_{OH} = -4.0mA
I_{OS} [4]	Output Short Circuit Current	-120		120	-120		120	-120		120	mA	V_{OUT}=GND to V_{CC}

Notes:
1. The operating ambient temperature range is guaranteed with transverse air flow exceeding 400 linear feet per minute.
2. Typical limits are at V_{CC} = 5V, T_A = +25°C, and specified loading.
3. I_{CC} exceeds I_{SB} maximum during power on, as shown in Graph 7. A pull-up resistor to V_{CC} on the $\overline{CS}$ input is required to keep the device deselected; otherwise, power-on current approaches I_{CC} active.
4. Duration not to exceed one minute.

A.C. TEST CONDITIONS

Input Pulse Levels	GND to 3.5 Volts
Input Rise and Fall Times	10 nsec
Input and Output Timing Reference Levels	1.5 Volts
Output Load	See Figure 1

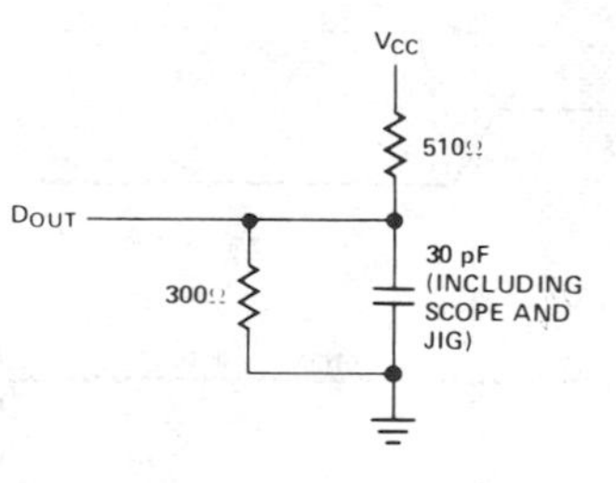

Figure 1. Output Load

CAPACITANCE [5]

T_A = 25°C, f = 1.0MHz

Symbol	Parameter	Max.	Unit	Conditions
C_{IN}	Input Capacitance	5	pF	V_{IN} = 0V
C_{OUT}	Output Capacitance	6	pF	V_{OUT} = 0V

Note 5. This parameter is sampled and not 100% tested.

A.C. CHARACTERISTICS

T_A = 0°C to 70°C, V_{CC} = +5V±10%, unless otherwise noted.

READ CYCLE

Symbol	Parameter	2147-3 Min.	2147-3 Max.	2147, 2147L Min.	2147, 2147L Max.	2147-6 Min.	2147-6 Max.	Unit
t_{RC}	Read Cycle Time	55		70		85		ns
t_{AA}	Address Access Time		55		70		85	ns
t_{ACS1}[1]	Chip Select Access Time		55		70		85	ns
t_{ACS2}[2]	Chip Select Access Time		65		80		85	ns
t_{OH}	Output Hold from Address Change	5		5		5		ns
t_{LZ}[3]	Chip Selection to Output in Low Z	10		10		10		ns
t_{HZ}[4]	Chip Deselection to Output in High Z	0	40	0	40	0	40	ns
t_{PU}	Chip Selection to Power Up Time	0		0		0		ns
t_{PD}	Chip Deselection to Power Down Time		30		30		30	ns

WAVEFORMS

READ CYCLE NO. 1[4,5]

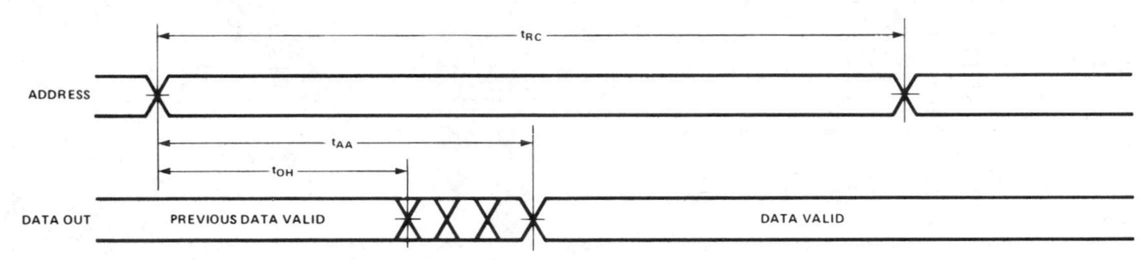

READ CYCLE NO. 2[4,6]

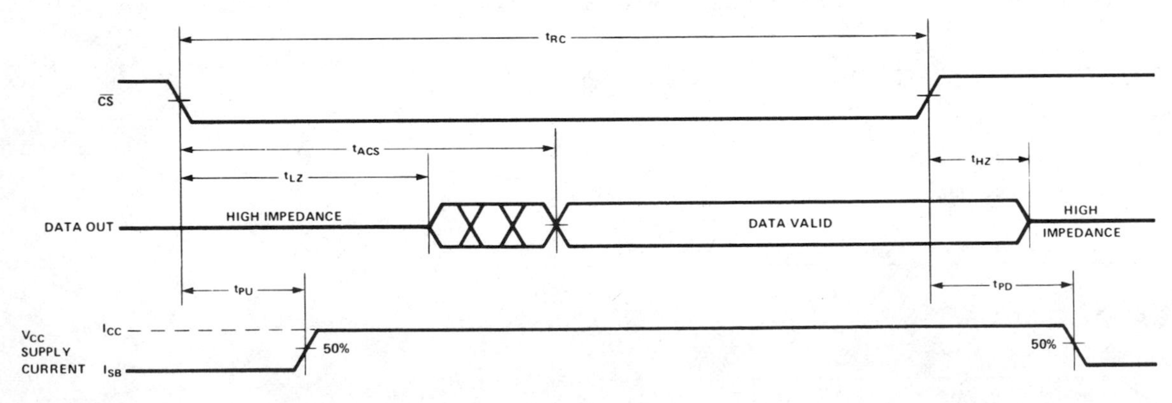

Notes:
1. Chip deselected for greater than 55ns prior to selection.
2. Chip deselected for a finite time that is less than 55ns prior to selection. (If the deselect time is 0ns, the chip is by definition selected and access occurs according to Read Cycle No. 1.)
3. At any given temperature and voltage condition, t_{HZ} max is less than t_{LZ} min both for a given device and from device to device.
4. $\overline{WE}$ is high for Read Cycles.
5. Device is continuously selected, $\overline{CS} = V_{IL}$.
6. Addresses valid prior to or coincident with $\overline{CS}$ transition low.

A.C. CHARACTERISTICS WRITE CYCLE (Continued)

Symbol	Parameter	2147-3 Min.	2147-3 Max.	2147, 2147L Min.	2147, 2147L Max.	2147-6 Min.	2147-6 Max.	Unit
t_{WC}	Write Cycle Time	55		70		85		ns
t_{CW}	Chip Selection to End of Write	45		55		65		ns
t_{AW}	Address Valid to End of Write	45		55		65		ns
t_{AS}	Address Setup Time	0		0		0		ns
t_{WP}	Write Pulse Width	35		40		45		ns
t_{WR}	Write Recovery Time	10		15		20		ns
t_{DW}	Data Valid to End of Write	25		30		30		ns
t_{DH}	Data Hold Time	10		10		10		ns
t_{WZ}	Write Enabled to Output in High Z	0	30	0	35	0	40	ns
t_{OW}	Output Active from End of Write	0		0		0		ns

WAVEFORMS

WRITE CYCLE #1 ($\overline{WE}$ CONTROLLED)

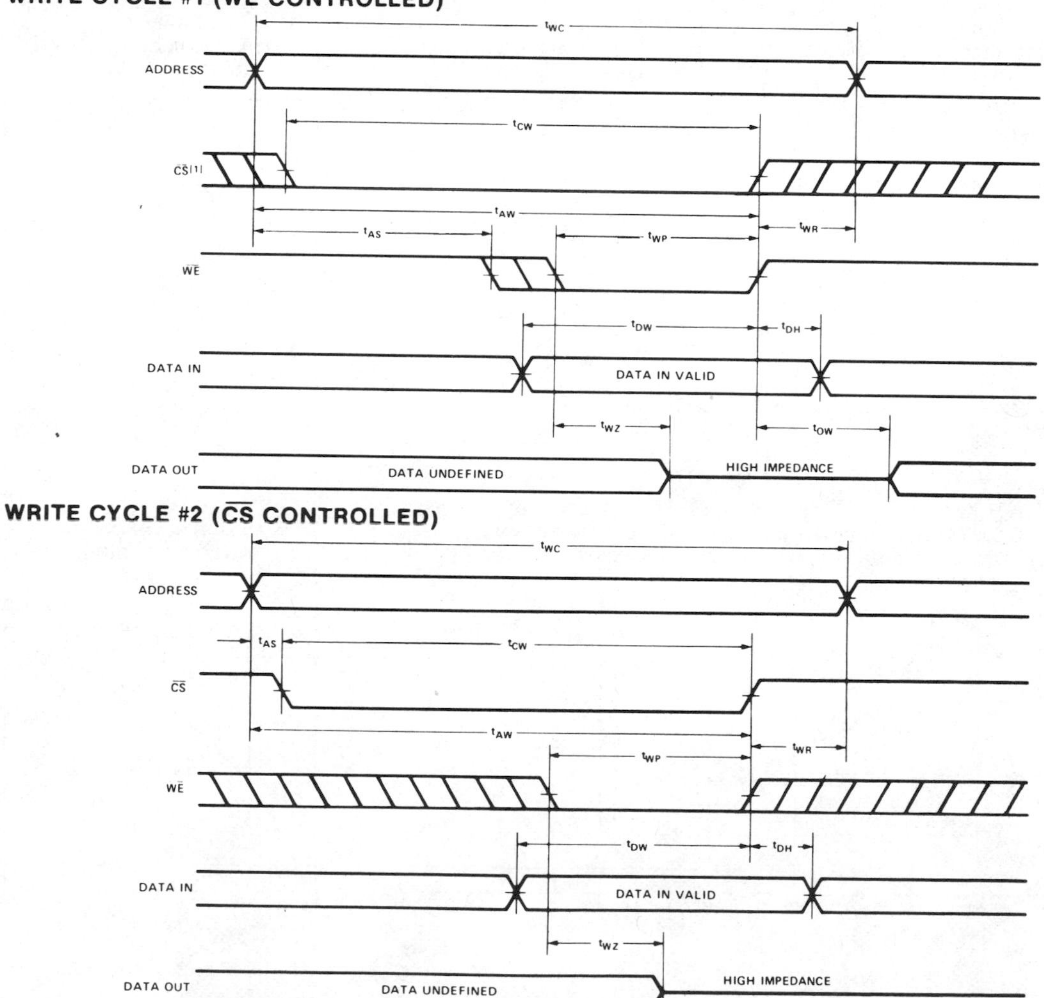

WRITE CYCLE #2 ($\overline{CS}$ CONTROLLED)

Note: 1. If $\overline{CS}$ goes high simultaneously with $\overline{WE}$ high, the output remains in a high impedance state.

TYPICAL D.C. AND A.C. CHARACTERISTICS

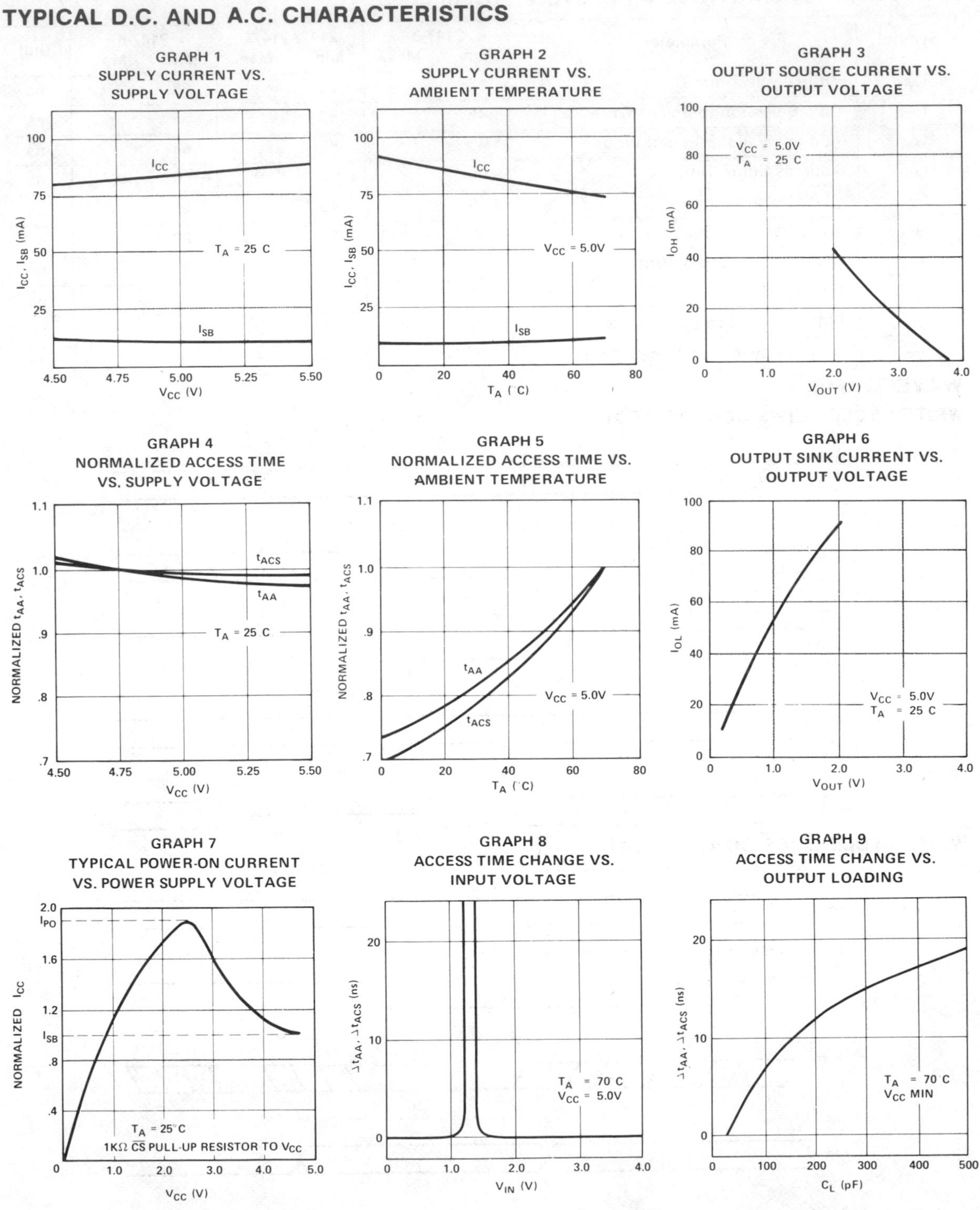

Note 1. The supply current curves shown in Graphs 1 and 2 are for the 2147.
The supply current curves for the 2147L and 2147-3 can be calculated by scaling proportionately.

DEVICE DESCRIPTION

The 2147 is produced with HMOS, a new high-performance MOS technology which incorporates on-chip substrate bias generation combined with device scaling to achieve high-performance. The speed-power product of this process has been measured at 1pj, approximately four times better than previous MOS processes.

This process, combined with new design ideas, gives the 2147 its unique features. High speed, low power and ease-of-use have been obtained in a single part. The low-power feature is controlled with the Chip Select input, which is not a clock and does not have to be cycled. Multiple read or write operations are possible during a single select period. Access times are equal to cycle times, resulting in data rates of 14.3 MHz and 18 MHz for the 2147 and 2147-3, respectively. This is considerably higher performance than for clocked static designs.

Whenever the 2147 is deselected, it automatically reduces its power requirements to a fraction of the active power, as shown in Figure 1. This is achieved by switching off the power to unnecessary portions of the internal peripheral circuitry. This feature adds up to significant system power savings. The average power per device declines as system size grows because a continually higher portion of the memory is deselected. Device power dissipation asymptotically approaches the standby power level, as shown in Figure 2.

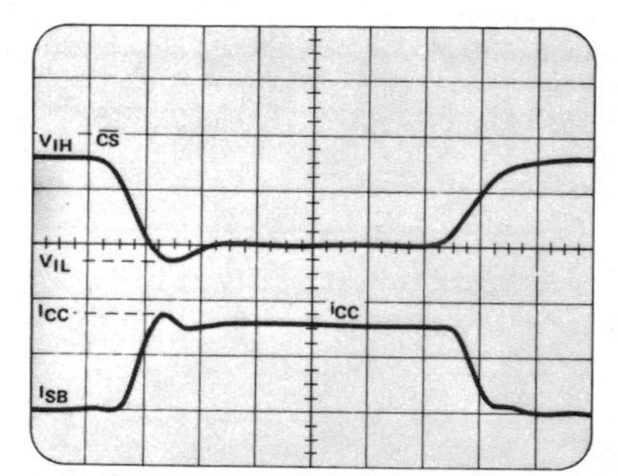

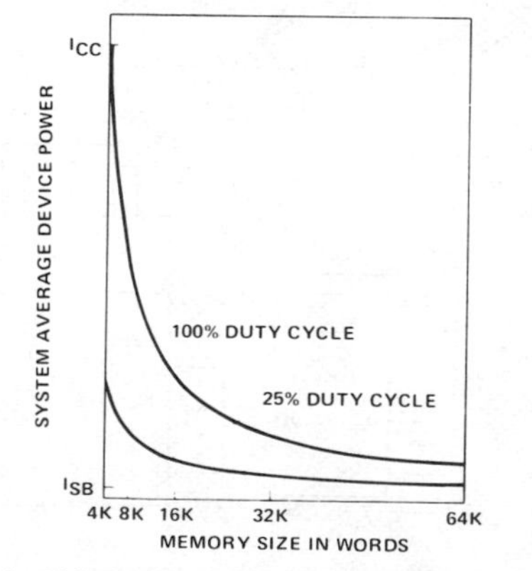

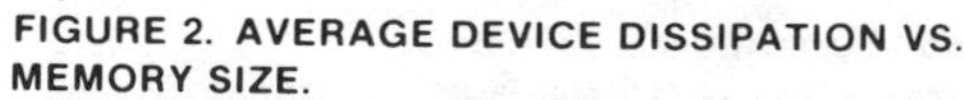
FIGURE 2. AVERAGE DEVICE DISSIPATION VS. MEMORY SIZE.

There is no functional constraint on the amount of time the 2147 is deselected. However, there is a relationship between deselect time and Chip Select access time. With no compensation, the automatic power switch would cause an increase in Chip Select access time, since some time is lost in repowering the device upon selection. A feature of the 2147 design is its ability to compensate for this loss. The amount of compensation is a function of deselect time, as shown in Figure 3. For short deselect times, Chip Select access time becomes slower than address access time, since full compensation typically requires 40ns. For longer deselect times, Chip Select access time actually becomes faster than address access time because the compensation more than offsets the time lost in powering up. The spec accounts for this characteristic by specifying two Chip Select access times, t_{ACS1} and t_{ACS2}.

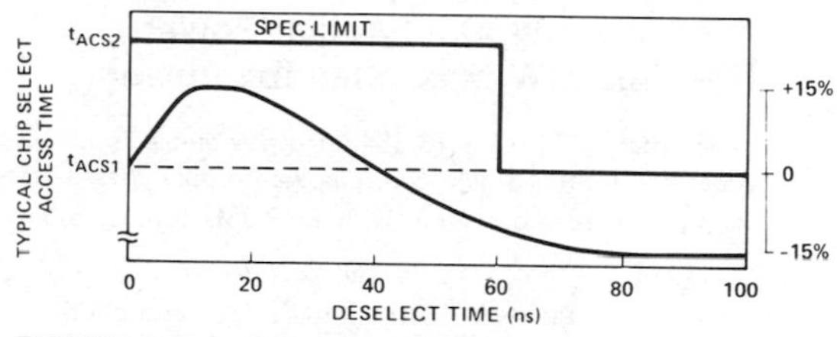

FIGURE 3. t_{ACS} VS. DESELECT TIME.

The power switching characteristic of the 2147 requires more careful decoupling than would be required of a constant power device. It is recommended that a 0.1μF to 0.3μF ceramic capacitor be used on every other device, with a 22μF to 47μF bulk electrolytic decoupler every 16 devices. The actual values to be used will depend on board layout, trace widths and duty cycle. Power supply gridding is recommended for PC board layout. A very satisfactory grid can be developed on a two-layer board with vertical traces on one side and horizontal traces on the other, as shown in Figure 4.

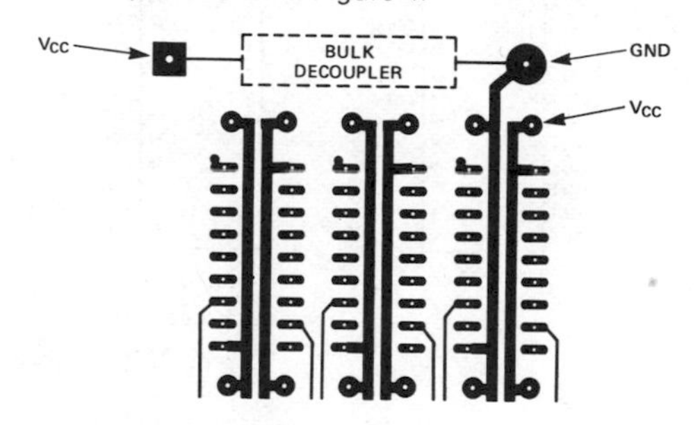

FIGURE 4. PC LAYOUT.

Terminations are recommended on input signal lines to the 2147 devices. In high speed systems, fast drivers can cause significant reflections when driving the high impedance inputs of the 2147. Terminations may be required to match the impedance of the line to the driver. The type of termination used depends on designer preference and may be parallel resistive or resistive-capacitive. The latter reduces terminator power dissipation.

INTEL CORPORATION, 3065 Bowers Avenue, Santa Clara, California 95051 (408) 246-7501

Printed in U.S.A./E-206/0779/5K/BA-BL

intel®

2716*
16K (2K × 8) UV ERASABLE PROM

- **Fast Access Time**
  - **— 350 ns Max. 2716-1**
  - **— 390 ns Max. 2716-2**
  - **— 450 ns Max. 2716**
  - **— 490 ns Max. 2716-5**
  - **— 650 ns Max. 2716-6**
- **Single +5V Power Supply**
- **Low Power Dissipation**
  - **— 525 mW Max. Active Power**
  - **— 132 mW Max. Standby Power**
- **Pin Compatible to Intel® 2732 EPROM**
- **Simple Programming Requirements**
  - **— Single Location Programming**
  - **— Programs with One 50 ms Pulse**
- **Inputs and Outputs TTL Compatible during Read and Program**
- **Completely Static**

The Intel® 2716 is a 16,384-bit ultraviolet erasable and electrically programmable read-only memory (EPROM). The 2716 operates from a single 5-volt power supply, has a static standby mode, and features fast single address location programming. It makes designing with EPROMs faster, easier and more economical.

The 2716, with its single 5-volt supply and with an access time up to 350 ns, is ideal for use with the newer high performance +5V microprocessors such as Intel's 8085 and 8086. A selected 2716-5 and 2716-6 is available for slower speed applications. The 2716 is also the first EPROM with a static standby mode which reduces the power dissipation without increasing access time. The maximum active power dissipation is 525 mW while the maximum standby power dissipation is only 132 mW, a 75% savings.

The 2716 has the simplest and fastest method yet devised for programming EPROMs – single pulse TTL level programming. No need for high voltage pulsing because all programming controls are handled by TTL signals. Program any location at any time—either individually, sequentially or at random, with the 2716's single address location programming. Total programming time for all 16,384 bits is only 100 seconds.

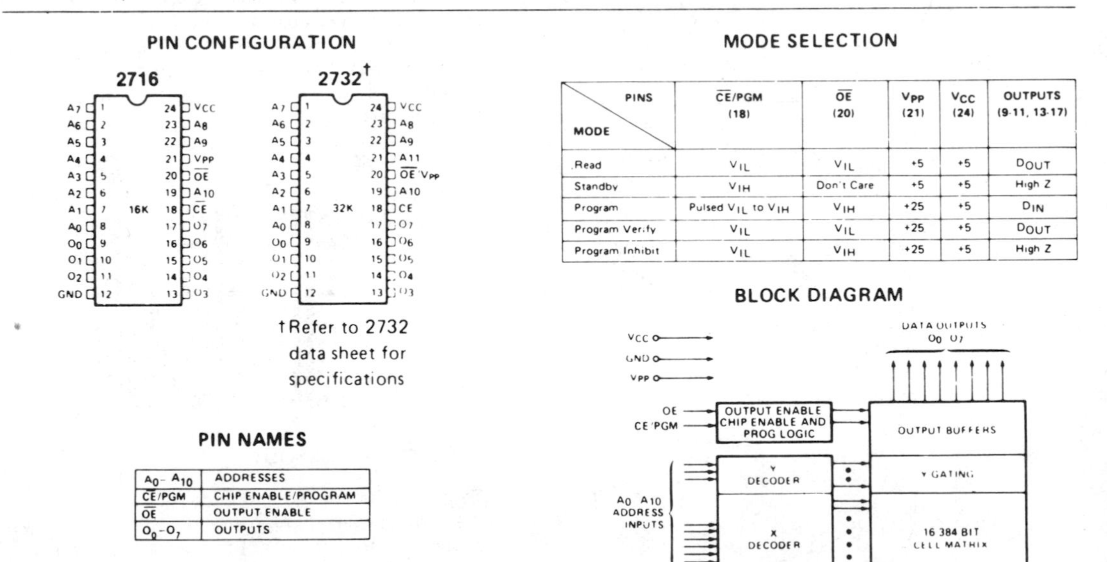

MODE SELECTION

MODE \ PINS	$\overline{CE}$/PGM (18)	$\overline{OE}$ (20)	V_{PP} (21)	V_{CC} (24)	OUTPUTS (9-11, 13-17)
Read	V_{IL}	V_{IL}	+5	+5	D_{OUT}
Standby	V_{IH}	Don't Care	+5	+5	High Z
Program	Pulsed V_{IL} to V_{IH}	V_{IH}	+25	+5	D_{IN}
Program Verify	V_{IL}	V_{IL}	+25	+5	D_{OUT}
Program Inhibit	V_{IL}	V_{IH}	+25	+5	High Z

PIN NAMES

A_0–A_{10}	ADDRESSES
$\overline{CE}$/PGM	CHIP ENABLE/PROGRAM
$\overline{OE}$	OUTPUT ENABLE
O_0–O_7	OUTPUTS

*PART(S) ALSO AVAILABLE IN EXTENDED TEMPERATURE RANGE FOR MILITARY AND INDUSTRIAL GRADE APPLICATONS.

Intel Corporation, 3065 Bowers Avenue, Santa Clara, California 95051

OCTOBER 1980

PROGRAMMING

The programming specifications are described in the Data Catalog PROM/ROM Programming Instructions Section.

Absolute Maximum Ratings*

Temperature Under Bias . . . . -10°C to +80°C
Storage Temperature . . . . -65°C to +125°C
All Input or Output Voltages with
Respect to Ground . . . . +6V to -0.3V
V_{PP} Supply Voltage with Respect
to Ground During Program . . . . +26.5V to -0.3V

**COMMENT:* Stresses above those listed under "Absolute Maximum Ratings" may cause permanent damage to the device. This is a stress rating only and functional operation of the device at these or any other conditions above those indicated in the operational sections of this specification is not implied. Exposure to absolute maximum rating conditions for extended periods may affect device reliability.

DC and AC Operating Conditions During Read

	2716	2716-1	2716-2	2716-5	2716-6
Temperature Range	0°C – 70°C	0°C – 70°C	0°C – 70°C	0°C – 70°C	0°C – 70°C
V_{CC} Power Supply [1,2]	5V ±5%	5V ±10%	5V ±5%	5V ±5%	5V ±5%
V_{PP} Power Supply [2]	V_{CC}	V_{CC}	V_{CC}	V_{CC}	V_{CC}

READ OPERATION

D.C. and Operating Characteristics

Symbol	Parameter	Limits Min.	Limits Typ. [3]	Limits Max.	Unit	Conditions
I_{LI}	Input Load Current			10	μA	V_{IN} = 5.25V
I_{LO}	Output Leakage Current			10	μA	V_{OUT} = 5.25V
I_{PP1} [2]	V_{PP} Current			5	mA	V_{PP} = 5.25V
I_{CC1} [2]	V_{CC} Current (Standby)		10	25	mA	$\overline{CE} = V_{IH}$, $\overline{OE} = V_{IL}$
I_{CC2} [2]	V_{CC} Current (Active)		57	100	mA	$\overline{OE} = \overline{CE} = V_{IL}$
V_{IL}	Input Low Voltage	-0.1		0.8	V	
V_{IH}	Input High Voltage	2.0		V_{CC}+1	V	
V_{OL}	Output Low Voltage			0.45	V	I_{OL} = 2.1 mA
V_{OH}	Output High Voltage	2.4			V	I_{OH} = -400 μA

NOTES:
1. V_{CC} must be applied simultaneously or before V_{PP} and removed simultaneously or after V_{PP}.
2. V_{PP} may be connected directly to V_{CC} except during programming. The supply current would then be the sum of I_{CC} and I_{PP1}.
3. Typical values are for T_A = 25°C and nominal supply voltages.
4. This parameter is only sampled and is not 100% tested.

Typical Characteristics

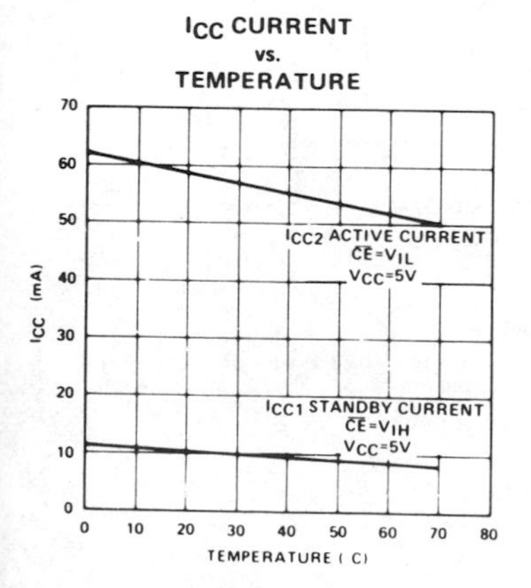

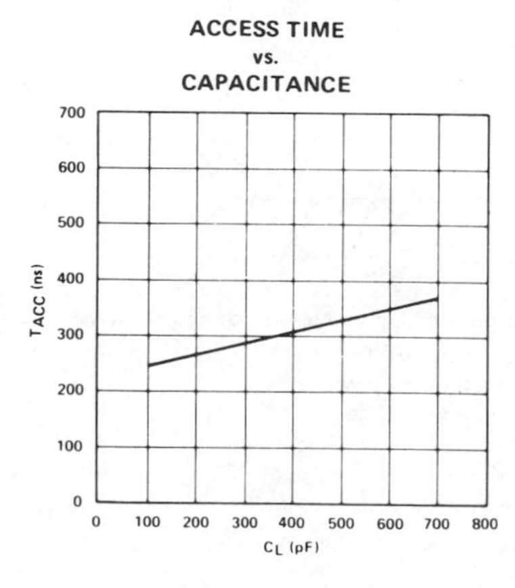

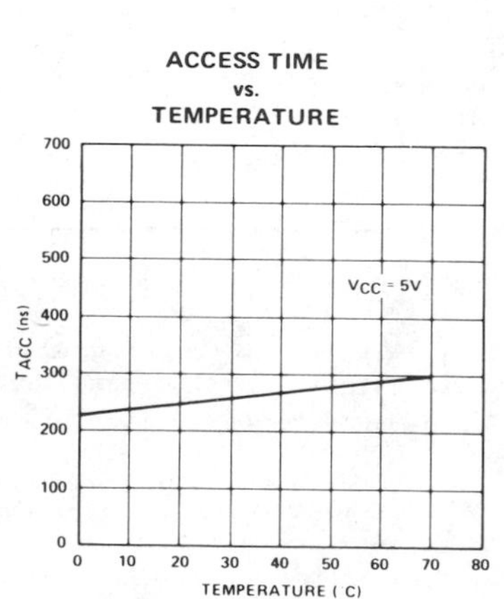

A.C. Characteristics

Symbol	Parameter	Limits (ns)										Test Conditions
		2716		2716-1		2716-2		2716-5		2716-6		
		Min.	Max.	Min.	Max.	Min.	Max.	Min.	Max.	Min.	Max.	
t_{ACC}	Address to Output Delay		450		350		390		450		450	$\overline{CE} = \overline{OE} = V_{IL}$
t_{CE}	$\overline{CE}$ to Output Delay		450		350		390		490		650	$\overline{OE} = V_{IL}$
t_{OE}	Output Enable to Output Delay		120		120		120		160		200	$\overline{CE} = V_{IL}$
t_{DF}	Output Enable High to Output Float	0	100	0	100	0	100	0	100	0	100	$\overline{CE} = V_{IL}$
t_{OH}	Output Hold from Addresses, $\overline{CE}$ or $\overline{OE}$ Whichever Occurred First	0		0		0		0		0		$\overline{CE} = \overline{OE} = V_{IL}$

Capacitance [4] $T_A = 25°C$, f = 1 MHz

Symbol	Parameter	Typ.	Max.	Unit	Conditions
C_{IN}	Input Capacitance	4	6	pF	$V_{IN} = 0V$
C_{OUT}	Output Capacitance	8	12	pF	$V_{OUT} = 0V$

A.C. Test Conditions:

Output Load: 1 TTL gate and $C_L = 100$ pF
Input Rise and Fall Times: ≤20 ns
Input Pulse Levels: 0.8V to 2.2V
Timing Measurement Reference Level:
Inputs 1V and 2V
Outputs 0.8V and 2V

A. C. Waveforms [1]

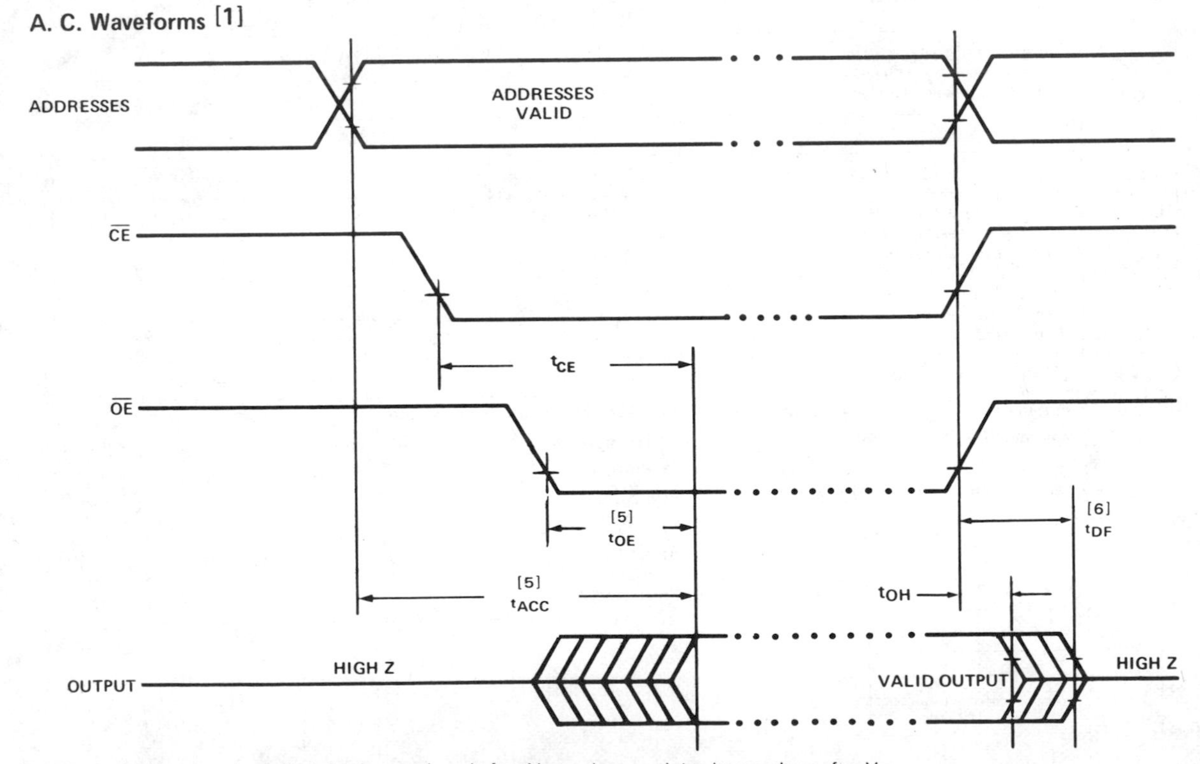

NOTE:

1. V_{CC} must be applied simultaneously or before V_{PP} and removed simultaneously or after V_{PP}.
2. V_{PP} may be connected directly to V_{CC} except during programming. The supply current would then be the sum of I_{CC} and I_{PP1}.
3. Typical values are for $T_A = 25°C$ and nominal supply voltages.
4. This parameter is only sampled and is not 100% tested.
5. This parameter is only sampled and is not 100% tested.
6. OE may be delayed up to $t_{ACC} - t_{OE}$ after the falling edge of CE without impact on t_{ACC}.
7. t_{DF} is specified from OE or CE, whichever occurs first.

ERASURE CHARACTERISTICS

The erasure characteristics of the 2716 are such that erasure begins to occur when exposed to light with wavelengths shorter than approximately 4000 Angstroms (Å). It should be noted that sunlight and certain types of fluorescent lamps have wavelengths in the 3000–4000Å range. Data show that constant exposure to room level fluorescent lighting could erase the typical 2716 in approximately 3 years, while it would take approximatley 1 week to cause erasure when exposed to direct sunlight. If the 2716 is to be exposed to these types of lighting conditions for extended periods of time, opaque labels are available from Intel which should be placed over the 2716 window to prevent unintentional erasure.

The recommended erasure procedure (see Data Catalog PROM/ROM Programming Instruction Section) for the 2716 is exposure to shortwave ultraviolet light which has a wavelength of 2537 Angstroms (Å). The integrated dose (i.e., UV intensity X exposure time) for erasure should be a minimum of 15 W-sec/cm^2. The erasure time with this dosage is approximately 15 to 20 minutes using an ultra violet lamp with a 12000 $\mu W/cm^2$ power rating. The 2716 should be placed within 1 inch of the lamp tubes during erasure. Some lamps have a filter on their tubes which should be removed before erasure.

DEVICE OPERATION

The five modes of operation of the 2716 are listed in Table I. It should be noted that all inputs for the five modes are at TTL levels. The power supplies required are a +5V V_{CC} and a V_{PP}. The V_{PP} power supply must be at 25V during the three programming modes and must be at 5V in the other two modes.

TABLE I. MODE SELECTION

MODE \ PINS	$\overline{CE}$/PGM (18)	$\overline{OE}$ (20)	V_{PP} (21)	V_{CC} (24)	OUTPUTS (9-11, 13-17)
Read	V_{IL}	V_{IL}	+5	+5	D_{OUT}
Standby	V_{IH}	Don't Care	+5	+5	High Z
Program	Pulsed V_{IL} to V_{IH}	V_{IH}	+25	+5	D_{IN}
Program Verify	V_{IL}	V_{IL}	+25	+5	D_{OUT}
Program Inhibit	V_{IL}	V_{IH}	+25	+5	High Z

READ MODE

The 2716 has two control functions, both of which must be logically satisfied in order to obtain data at the outputs. Chip Enable ($\overline{CE}$) is the power control and should be used for device selection. Output Enable ($\overline{OE}$) is the output control and should be used to gate data to the output pins, independent of device selection. Assuming that addresses are stable, address access time (t_{ACC}) is equal to the delay from $\overline{CE}$ to output (t_{CE}). Data is available at the outputs 120 ns (t_{OE}) after the falling edge of $\overline{OE}$ assuming that $\overline{CE}$ has been low and addresses have been stable for at least $t_{ACC} - t_{OE}$.

STANDBY MODE

The 2716 has a standby mode which reduces the active power dissipation by 75%, from 525 mW to 132 mW. The 2716 is placed in the standby mode by applying a TTL high signal to the $\overline{CE}$ input. When in standby mode the outputs are in a high impedence state, independent of the $\overline{OE}$ input.

OUTPUT OR-TIEING

Because 2716's are usually used in larger memory arrays, Intel has provided a 2 line control function that accomodates this use of multiple memory connections. The two line control function allows for:

a) the lowest possible memory power dissipation, and
b) complete assurance that output bus contention will not occur.

To most efficiently use these two control lines, it is recommended that $\overline{CE}$ (pin 18) be decoded and used as the primary device selecting function, while $\overline{OE}$ (pin 20) be made a common connection to all devices in the array and connected to the READ line from the system control bus. This assures that all deselected memory devices are in their low power standby mode and that the output pins are only active when data is desired from a particular memory device.

PROGRAMMING

Initially, and after each erasure, all bits of the 2716 are in the "1" state. Data is introduced by selectively programming "0's" into the desired bit locations. Although only "0's" will be programmed, both "1's" and "0's" can be presented in the data word. The only way to change a "0" to a "1" is by ultraviolet light erasure.

The 2716 is in the programming mode when the V_{PP} power supply is at 25V and OE is at V_{IH}. The data to be programmed is applied 8 bits in parallel to the data output pins. The levels required for the address and data inputs are TTL.

When the address and data are stable, a 50 msec, active high, TTL program pulse is applied to the $\overline{CE}$/PGM input. A program pulse must be applied at each address location to be programmed. You can program any location at any time – either individually, sequentially, or at random. The program pulse has a maximum width of 55 msec. The 2716 must not be programmed with a DC signal applied to the $\overline{CE}$/PGM input.

Programming of multiple 2716s in parallel with the same data can be easily accomplished due to the simplicity of the programming requirements. Like inputs of the paralleled 2716s may be connected together when they are programmed with the same data. A high level TTL pulse applied to the $\overline{CE}$/PGM input programs the paralleled 2716s.

PROGRAM INHIBIT

Programming of multiple 2716s in parallel with different data is also easily accomplished. Except for $\overline{CE}$/PGM, all like inputs (including $\overline{OE}$) of the parallel 2716s may be common. A TTL level program pulse applied to a 2716's $\overline{CE}$/PGM input with V_{PP} at 25V will program that 2716. A low level $\overline{CE}$/PGM input inhibits the other 2716 from being programmed.

PROGRAM VERIFY

A verify should be performed on the programmed bits to determine that they were correctly programmed. The verify may be performed wth V_{PP} at 25V. Except during programming and program verify V_{PP} must be at 5V.

C

ERROR DETECTION AND CORRECTION

C–1 ERROR DETECTION CODES

The *2-out-of-5* code is sometimes used in communications work. It utilizes five bits to represent the ten decimal digits, so it is a form of BCD code. Each code word has exactly two 1s, which facilitates decoding and provides for better error detection than the single parity bit method. If other than two 1s appear, an error is indicated.

The *63210 BCD* code is also characterized by having exactly two 1s in each of the five-bit groups. Like the 2-out-of-5 code, it provides reliable error detection and is used in some applications.

The *biquinary* code (two-five) is used in certain counters and is composed of a two-bit group and a five-bit group, each with a single 1. Its weights are 50 43210. The two-bit group, having weights 50, indicates whether the number represented is less than, equal to, or greater than 5. The five bit-group indicates the count above or below 5.

The *ring counter* code has ten bits, one for each decimal digit, and a single 1 makes error detection possible. It is easy to decode, but wastes bits and requires more circuitry to implement than the four- or five-bit codes. The name is derived from the fact that the code is generated by a certain type of shift register, or "ring counter." Its weights are 9876543210.

Each of these codes is listed in Table C–1. You should realize that this is not an exhaustive coverage of all codes, but simply an introduction to some of them.

TABLE C–1 *Some codes with error detection properties.*

Decimal	2-out-of-5	63210	5043210	9876543210
0	00011	00110	01 00001	0000000001
1	00101	00011	01 00010	0000000010
2	00110	00101	01 00100	0000000100
3	01001	01001	01 01000	0000001000
4	01010	01010	01 10000	0000010000
5	01100	01100	10 00001	0000100000
6	10001	10001	10 00010	0001000000
7	10010	10010	10 00100	0010000000
8	10100	10100	10 01000	0100000000
9	11000	11000	10 10000	1000000000

C–2 HAMMING ERROR CORRECTION CODE

This section discusses a method, generally known as the *Hamming code,* which not only provides for the detection of a bit error, but also identifies which bit is in error so it can be corrected. The code uses a number of parity bits (dependent on the number of information bits) located at certain positions in the code group.

The Hamming code is constructed as follows for *single-error* correction.

Number of Parity Bits

If the number of information bits is designated *m*, then the number of parity bits, *p*, is determined by the following relationship:

$$2^p \geq m + p + 1 \qquad \text{(C–1)}$$

For example, if we have four information bits, then *p* is found by trial and error using Equation (C–1). Let $p = 2$. Then,

$$2^p = 2^2 = 4$$

and

$$m + p + 1 = 4 + 2 + 1 = 7$$

Since 2^p has to be equal to or greater than $m + p + 1$, the relationship in Equation (C–1) is *not* satisfied. We have to try again. Let $p = 3$. Then,

$$2^p = 2^3 = 8$$

and

$$m + p + 1 = 4 + 3 + 1 = 8$$

This value of *p* satisfies the relationship of Equation (C–1), and therefore three parity bits are required to provide single-error correction for four information bits. It should be noted here that error detection and correction are provided for *all* bits, both parity and information, in a code group.

Placement of the Parity Bits in the Code

Now that we have found the number of parity bits required in our particular example, we must arrange the bits properly in the code. At this point you should realize that, in this example, the code is composed of the four information bits and the three parity bits. The left-most bit is designated *bit* 1, the next bit is *bit* 2, and so on, as shown below:

bit 1, bit 2, bit 3, bit 4, bit 5, bit 6, bit 7

The parity bits are located in the positions that are numbered corresponding to ascending powers of two: 1, 2, 4, 8, etc., as indicated:

$$P_1, P_2, M_1, P_3, M_2, M_3, M_4$$

The symbol P_n designates a particular parity bit, and M_n designates a particular information bit where *n* is the position number.

Assignment of Parity Bit Values

Finally, we must properly assign a 1 or 0 value to each parity bit. Since each parity bit provides a check on certain other bits in the total code, we have to know the value of these others in order to assign the parity-bit value. To do this, first number each bit position in *binary;* that is, write the binary number for each decimal position number (as shown in the second two rows of Table C–2). Next, indicate the parity and information bit locations, as shown in the first row of Table C–2. Notice that the binary position number of parity bit P_1 has a 1 for its right-most digit. This parity bit checks all bit positions, including itself, that have *1s in the same location in the binary position numbers*. Therefore, parity bit P_1 checks bit positions 1, 3, 5, and 7.

TABLE C–2 *Bit position table for a seven-bit error correcting code.*

Bit Designation Bit Position Binary Position Number	P_1 1 001	P_2 2 010	M_1 3 011	P_3 4 100	M_2 5 101	M_3 6 110	M_4 7 111
Information Bits (M_n)							
Parity Bits (P_n)							

The binary position number for parity bit P_2 has a 1 for its middle bit. It checks all bit positions, including itself, that have 1s in this same position. Therefore, parity bit P_2 checks bit positions 2, 3, 6, and 7.

The binary position number for parity bit P_3 has a 1 for its left-most bit. It checks all bit positions, including itself, that have 1s in this same position. Therefore, parity bit P_3 checks bit positions 4, 5, 6, and 7.

In each case *the parity bit is assigned a value to make the quantity of 1s in the set of bits that it checks odd or even,* depending on which is specified. The following examples should make this procedure clear.

Example C–1

Determine the single-error correcting code for the BCD number 1001 (information bits) using even parity.

Solution:

Step 1. Find the number of parity bits required. Let $p = 3$. Then,

$$2^p = 2^3 = 8$$

$$m + p + 1 = 4 + 3 + 1 = 8$$

Three parity bits are sufficient.
Total code bits = 4 + 3 = 7.

Step 2. Construct a bit position table.

Bit Designation Bit Position Binary Position Number	P_1 1 001	P_2 2 010	M_1 3 011	P_3 4 100	M_2 5 101	M_3 6 110	M_4 7 111
Information Bits			1		0	0	1
Parity Bits	0	0		1			

Parity bits are determined in the following steps:

Step 3. Determine the parity bits as follows:
P_1 checks bit positions 1, 3, 5, and 7 and must be a 0 in order to have an even number of 1s (2) in this group.

P_2 checks bit positions 2, 3, 6, and 7 and must be a 0 in order to have an even number of 1s (2) in this group.
P_3 checks bit positions 4, 5, 6, and 7 and must be a 1 in order to have an even number of 1s (2) in this group.

Step 4. These parity bits are entered into the table, and the resulting combined code is 0011001.

Example C–2

Determine the single-error correcting code for the information code 10110 for odd parity.

Solution:

Step 1. Determine the number of parity bits required. In this case the number of information bits, *m*, is five.
From the previous example we know that $p = 3$ will not work.
Try $p = 4$.

$$2^p = 2^4 = 16$$

$$m + p + 1 = 5 + 4 + 1 = 10$$

Four parity bits are sufficient.
Total code bits = 5 + 4 = 9.

Step 2. Construct a bit position table.

Bit Designation	P_1	P_2	M_1	P_3	M_2	M_3	M_4	P_4	M_5
Bit Position	1	2	3	4	5	6	7	8	9
Binary Position Number	0001	0010	0011	0100	0101	0110	0111	1000	1001
Information Bits			1		0	1	1		0
Parity Bits	1	0		1				1	

Parity bits are determined in the following steps:

Step 3. Determine the parity bits as follows:
P_1 checks bit positions 1, 3, 5, 7, and 9 and must be a 1 to have an odd number of 1s (3) in this group.
P_2 checks bit positions 2, 3, 6, and 7 and must be a 0 to have an odd number of 1s (3) in this group.
P_3 checks bit positions 4, 5, 6, and 7 and must be a 1 to have an odd number of 1s (3) in this group.
P_4 checks bit positions 8 and 9 and must be a 1 to have an odd number of 1s (1) in this group.

Step 4. These parity bits are entered into the table, and the resulting combined code is 101101110.

C–3 DETECTING AND CORRECTING AN ERROR

Now that a method for constructing an error correcting code has been covered, how do we use it to locate and correct an error? Each parity bit, along with its corresponding group of bits, must be checked for the proper parity. If there are three parity bits in a code word, then three parity checks are made. If there are four parity bits, four checks have to be made, and so on. Each parity check will yield a good or a bad result. The total result of all the parity checks indicates the bit, if any, that is in error, as follows:

Step 1. Start with the group checked by P_1.
Step 2. Check the group for proper parity. A 0 represents a good parity check and 1 represents a bad check.
Step 3. Repeat Step 2 for each parity group.
Step 4. The binary number formed by the results of each parity check designates the position of the code bit that is in error. This is the *error position code.* The first parity check generates the least significant bit (LSB). If all checks are good, there is no error.

Example C–3

Assume that the code word in Example C–1 (0011001) is transmitted, and that 0010001 is received. The receiver does not "know" what was transmitted and must look for proper parities to determine if the code is correct. Designate any error that has occurred in transmission if *even parity* is used.

Solution:

First, make up a bit position table.

Bit Designation	P_1	P_2	M_1	P_3	M_2	M_3	M_4
Bit Position	1	2	3	4	5	6	7
Binary Position Number	001	010	011	100	101	110	111
Received Code	0	0	1	0	0	0	1

First parity check:
P_1 checks positions 1, 3, 5, and 7.
There are two 1s in this group.
Parity check is good. ⟶ 0 (LSB)

Second parity check:
P_2 checks positions 2, 3, 6, and 7.
There are two 1s in this group.
Parity check is good. ⟶ 0

Third parity check:

P_3 checks positions 4, 5, 6, and 7.
There is one 1 in this group.
Parity check is bad. ⟶ 1 (MSB)

Result:

The error position code is 100 (binary 4). This says that the bit in the number 4 position is in error. It is a 0 and should be a 1. The corrected code is 0011001, which agrees with the transmitted code.

Example C–4

The code 101101010 is received. Correct any errors. There are four parity bits and odd parity is used.

Solution:

First, make a bit position table.

Bit Designation	P_1	P_2	M_1	P_3	M_2	M_3	M_4	P_4	M_5
Bit Position	1	2	3	4	5	6	7	8	9
Binary Position Number	0001	0010	0011	0100	0101	0110	0111	1000	1001
Received Code	1	0	1	1	0	1	0	1	0

First parity check:
P_1 checks positions 1, 3, 5, 7, and 9.
There are two 1s in this group.
Parity check is bad. ⟶ 1 (LSB)

Second parity check:
P_2 checks positions 2, 3, 6, and 7.
There are two 1s in this group.
Parity check is bad. ⟶ 1

Third parity check:
P_3 checks positions 4, 5, 6, and 7.
There are two 1s in this group.
Parity check is bad. ⟶ 1

Fourth parity check:
P_4 checks positions 8 and 9.
There is one 1 in this group.
Parity check is good. ⟶ 0 (MSB)

Result:

The error position code is 0111 (binary 7). This says that the bit in the number 7 position is in error. The corrected code is therefore 101101110.

D CONVERSIONS

Decimal	BCD(8421)	Octal	Binary	Decimal	BCD(8421)	Octal	Binary
0	0000	0	0	25	00100101	31	11001
1	0001	1	1	26	00100110	32	11010
2	0010	2	10	27	00100111	33	11011
3	0011	3	11	28	00101000	34	11100
4	0100	4	100	29	00101001	35	11101
5	0101	5	101	30	00110000	36	11110
6	0110	6	110	31	00110001	37	11111
7	0111	7	111	32	00110010	40	100000
8	1000	10	1000	33	00110011	41	100001
9	1001	11	1001	34	00110100	42	100010
10	00010000	12	1010	35	00110101	43	100011
11	00010001	13	1011	36	00110110	44	100100
12	00010010	14	1100	37	00110111	45	100101
13	00010011	15	1101	38	00111000	46	100110
14	00010100	16	1110	39	00111001	47	100111
15	00010101	17	1111	40	01000000	50	101000
16	00010110	20	10000	41	01000001	51	101001
17	00010111	21	10001	42	01000010	52	101010
18	00011000	22	10010	43	01000011	53	101011
19	00011001	23	10011	44	01000100	54	101100
20	00100000	24	10100	45	01000101	55	101101
21	00100001	25	10101	46	01000110	56	101110
22	00100010	26	10110	47	01000111	57	101111
23	00100011	27	10111	48	01001000	60	110000
24	00100100	30	11000	49	01001001	61	110001

Decimal	BCD(8421)	Octal	Binary	Decimal	BCD(8421)	Octal	Binary
50	01010000	62	110010	75	01110101	113	1001011
51	01010001	63	110011	76	01110110	114	1001100
52	01010010	64	110100	77	01110111	115	1001101
53	01010011	65	110101	78	01111000	116	1001110
54	01010100	66	110110	79	01111001	117	1001111
55	01010101	67	110111	80	10000000	120	1010000
56	01010110	70	111000	81	10000001	121	1010001
57	01010111	71	111001	82	10000010	122	1010010
58	01011000	72	111010	83	10000011	123	1010011
59	01011001	73	111011	84	10000100	124	1010100
60	01100000	74	111100	85	10000101	125	1010101
61	01100001	75	111101	86	10000110	126	1010110
62	01100010	76	111110	87	10000111	127	1010111
63	01100011	77	111111	88	10001000	130	1011000
64	01100100	100	1000000	89	10001001	131	1011001
65	01100101	101	1000001	90	10010000	132	1011010
66	01100110	102	1000010	91	10010001	133	1011011
67	01100111	103	1000011	92	10010010	134	1011100
68	01101000	104	1000100	93	10010011	135	1011101
69	01101001	105	1000101	94	10010100	136	1011110
70	01110000	106	1000110	95	10010101	137	1011111
71	01110001	107	1000111	96	10010110	140	1100000
72	01110010	110	1001000	97	1000111	141	1100001
73	01110011	111	1001001	98	10011000	142	1100010
74	01110100	112	1001010	99	10011001	143	1100011

ANSWERS TO SELECTED ODD-NUMBERED PROBLEMS

CHAPTER 1

1–1 One (1), zero (0)

1–3 11010001

1–5 4 ms

1–7 Periodic

1–9 A flip-flop uses feedback to retain a binary state. A gate does not have storage capability.

1–11 AND gate

1–13 Comparison: The process of comparing the magnitudes of two numbers.
Arithmetic: Addition, subtraction, multiplication, and division.
Decoding: The process of converting coded information into decimal or other common forms.
Encoding: The process of converting common information into coded form.
Counting: The process of producing a sequence of binary states in response to input pulses.
Registers: Temporary storage devices for binary information.
Multiplexing: The process of switching digital data from several sources onto one common line.

1–15 Demultiplexing is essentially the reverse of the multiplexing process described in Answer 1–13.

1–17 Dual in-line package (DIP)

1–19 A set of parallel conductors over which data are transmitted from one point to another under specified conditions.

1–21 Logic probe, logic pulser, current tracer, logic clip, logic analyzer.

CHAPTER 2

2–1 (a) 6 (b) 4 (c) 1
(d) 87 (e) 82 (f) 74
(g) 50 (h) 13 (i) 872
(j) 618 (k) 309 (l) 8645

2–3 (a) 23 (b) 34 (c) 98
(d) 118 (e) 137 (f) 187

2–5 (a) 3_{10} (b) 4_{10} (c) 7_{10}
(d) 8_{10} (e) 9_{10} (f) 12_{10}
(g) 11_{10} (h) 15_{10}

2–7 (a) 51.75_{10} (b) 42.25_{10} (c) 65.875_{10}
(d) 120.625_{10} (e) 92.65625_{10} (f) 113.0625_{10}
(g) 90.625_{10} (h) 127.96875_{10}

2–9 (a) 5 bits (b) 6 bits (c) 6 bits
(d) 7 bits (e) 7 bits (f) 7 bits
(g) 8 bits (h) 8 bits (i) 9 bits

2–11 (a) 100_2 (b) 100_2 (c) 1000_2
(d) 1101_2 (e) 1110_2 (f) 11000_2

2–13 (a) 1001_2 (b) 1000_2 (c) 100011_2
(d) 110110_2 (e) 10101001_2
(f) 10110110_2

2–15 (a) 010 (b) 001 (c) 0101
(d) 00101000 (e) 0001010
(f) 11110

2–17 (a) 10 (b) 001 (c) 0111
(d) 0011 (e) 00100 (f) 01101

2–19 (a) 1010_2 (b) 10001_2 (c) 11000_2
(d) 110000_2 (e) 111101_2
(f) 1011101_2 (g) 1111101_2
(h) 10111010_2 (i) 100101010_2

2–21 (a) 10_{10} (b) 23_{10} (c) 46_{10}
(d) 52_{10} (e) 67_{10} (f) 367_{10}
(g) 115_{10} (h) 532_{10} (i) 4085_{10}

2–23 (a) 001011_2 (b) 101111_2
(c) 001000001_2 (d) 011010001_2
(e) 101100000_2 (f) 100110101011_2
(g) 001011010111001_2
(h) 100101110000000_2
(i) 0010000000010001011_2
(j) 001000011.100101_2

2–25 (a) 00111000_2 (b) 01011001_2
(c) 101000010100_2 (d) 010111001000_2
(e) 0100000100000000_2
(f) 1111101100010111_2
(g) 10001010.1001_2

2–27 (a) 35_{10} (b) 146_{10} (c) 26_{10}
(d) 141_{10} (e) 243_{10} (f) 235_{10}
(g) 1474_{10} (h) 1792_{10}

2–29 (a) 60_{16} (b) $10B_{16}$ (c) $1BA_{16}$

2–31 (a) 00010000 (b) 00010011
(c) 00011000 (d) 00100001

	(e) 00100101	**(f)** 00110110
	(g) 01000100	**(h)** 01010111
	(i) 01101001	**(j)** 10011000
	(k) 000100100101	**(l)** 000101010110

2–33 **(a)** 000100000100 **(b)** 000100101000
(c) 000100110010 **(d)** 000101010000
(e) 000110000110 **(f)** 001000010000
(g) 001101011001 **(h)** 010101000111
(i) 0001000001010001
(j) 0010010101100011

2–35 **(a)** 80_{10} **(b)** 237_{10} **(c)** 346_{10}
(d) 421_{10} **(e)** 754_{10} **(f)** 800_{10}
(g) 978_{10} **(h)** 1683_{10} **(i)** 9018_{10}
(j) 6667_{10}

2–37 **(a)** 00010100 **(b)** 00010010
(c) 00010111 **(d)** 00010110
(e) 01010010 **(f)** 000100001001
(g) 000110010101 **(h)** 0001001001101001

2–39 **(a)** **(b)** 0110 **(c)** 1001
(d) 01000011 **(e)** 01001011
(f) 01011100 **(g)** 10001001
(h) 10101000 **(i)** 010001101001
(j) 010001111100 **(k)** 010101100100
(l) 100000110011
(m) 0100010110000100
(n) 0101011010101100
(o) 1001101101110100

2–41 **(a)** 0011 **(b)** 0110 **(c)** 1001
(d) 1100 **(e)** 01000110
(f) 01011000 **(g)** 01111010
(h) 10100101 **(i)** 010001101001
(j) 010010101011

CHAPTER 3

3–1

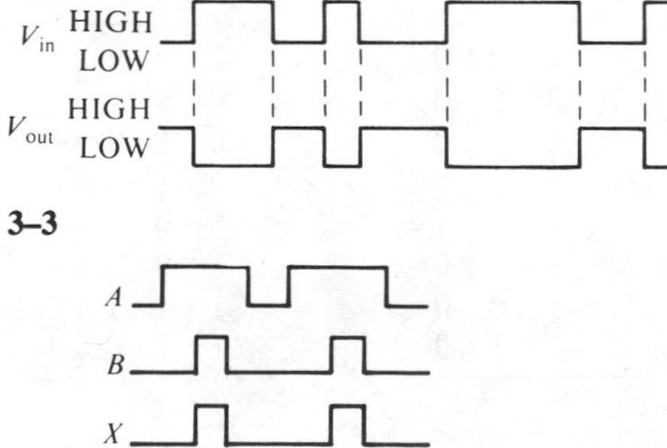

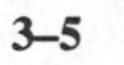
3–5

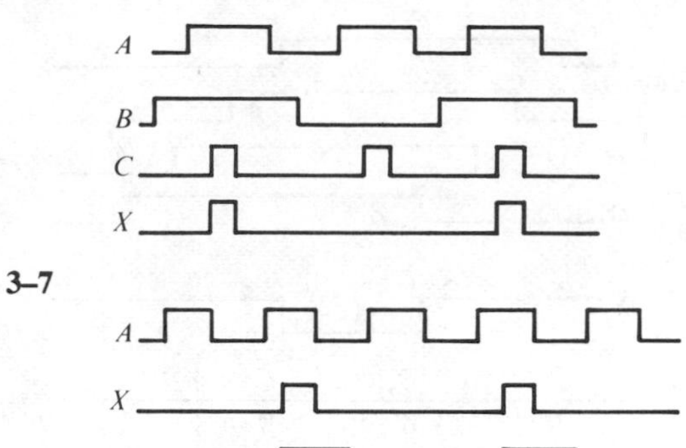

3–7

A

X

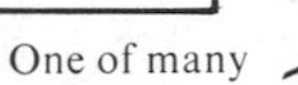
One of many possibilities.

3–9

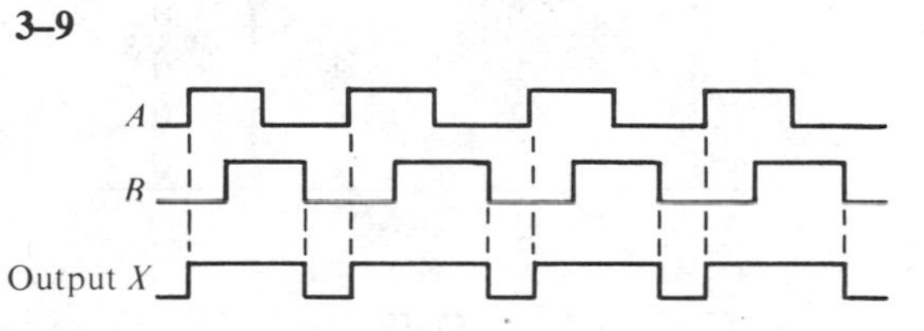

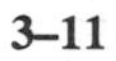
3–11

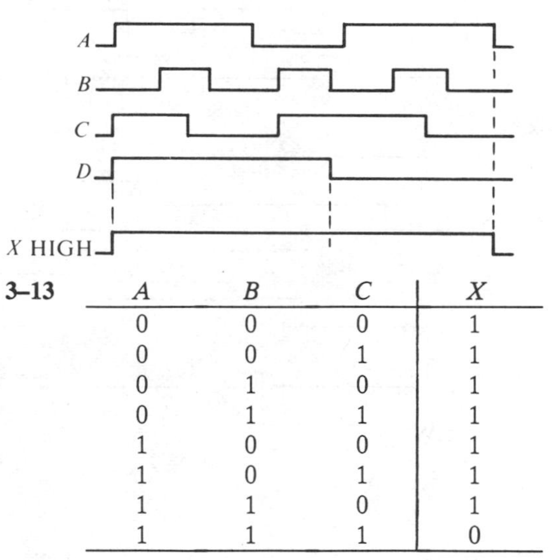

3–13

A	B	C	X
0	0	0	1
0	0	1	1
0	1	0	1
0	1	1	1
1	0	0	1
1	0	1	1
1	1	0	1
1	1	1	0

3–15

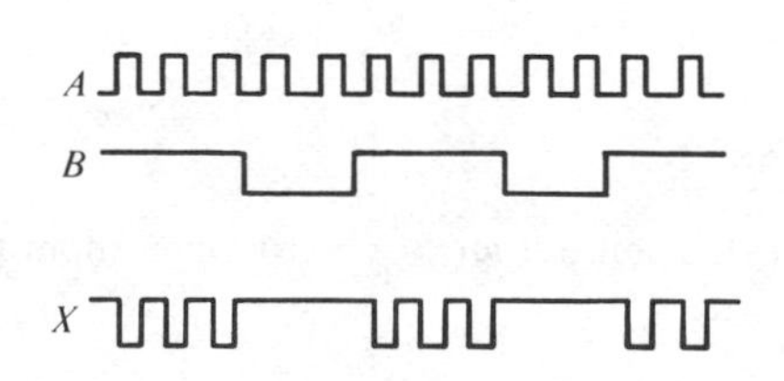

3–17

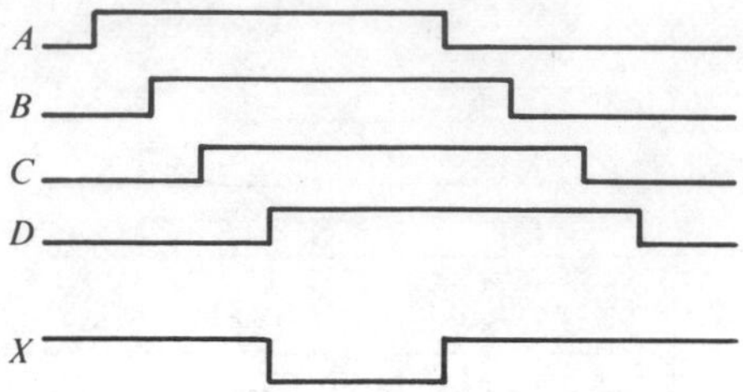

A
B
C
D
X

3–19

A	B	C	X
0	0	0	1
0	0	1	0
0	1	0	0
0	1	1	0
1	0	0	0
1	0	1	0
1	1	0	0
1	1	1	0

3–21

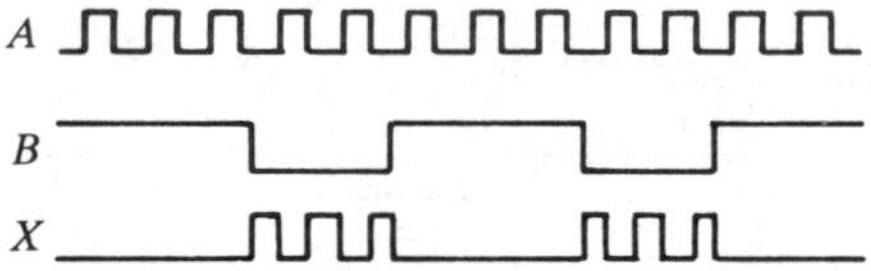

A
B
X

3–23

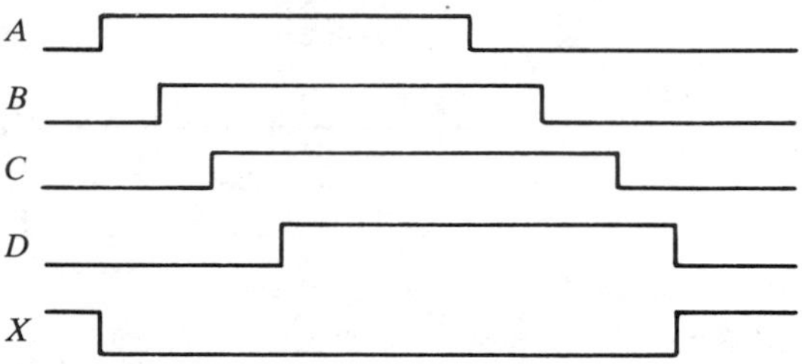

A
B
C
D
X

3–25 20 mW

3–27 40 mW

3–29 0.3 V

3–31 400 μA

3–33 11 ns

3–35 (**a**) Open output (**b**) Open input

CHAPTER 4

4–1 (**a**)

A	B	X
0	0	0
0	1	0
1	0	0
1	1	1

(**b**)

A	B	C	X
0	0	0	0
0	0	1	0
0	1	0	0
0	1	1	0
1	0	0	0
1	0	1	0
1	1	0	0
1	1	1	1

(**c**)

A	B	X
0	0	0
0	1	1
1	0	1
1	1	1

(**d**)

A	B	C	X
0	0	0	0
0	0	1	1
0	1	0	1
0	1	1	1
1	0	0	1
1	0	1	1
1	1	0	1
1	1	1	1

(**e**)

A	B	C	X
0	0	0	0
0	0	1	1
0	1	0	0
0	1	1	1
1	0	0	0
1	0	1	1
1	1	0	1
1	1	1	1

(**f**)

A	B	X
0	0	1
0	1	1
1	0	0
1	1	1

(**g**)

A	B	C	X
0	0	0	0
0	0	1	0
0	1	0	0
0	1	1	0
1	0	0	1
1	0	1	0
1	1	0	0
1	1	1	0

(**h**)

A	B	C	X
0	0	0	0
0	0	1	1
0	1	0	0
0	1	1	1
1	0	0	0
1	0	1	0
1	1	0	1
1	1	1	1

(i)

A	B	C	X
0	0	0	0
0	0	1	0
0	1	0	0
0	1	1	0
1	0	0	0
1	0	1	1
1	1	0	1
1	1	1	1

(j)

A	B	C	X
0	0	0	1
0	0	1	1
0	1	0	1
0	1	1	0
1	0	0	0
1	0	1	0
1	1	0	0
1	1	1	0

4–3 **(a)** $X = AB$ **(b)** $X = \overline{A}$
(c) $X = A + B$ **(d)** $X = A + B + C$

4–5 **(a)**

A	B	X
0	0	0
0	1	1
1	0	1
1	1	1

(b)

A	B	X
0	0	0
0	1	0
1	0	0
1	1	1

(c)

A	B	C	X
0	0	0	0
0	0	1	0
0	1	0	0
0	1	1	1
1	0	0	0
1	0	1	0
1	1	0	1
1	1	1	1

(d)

A	B	C	X
0	0	0	0
0	0	1	0
0	1	0	0
0	1	1	1
1	0	0	0
1	0	1	1
1	1	0	0
1	1	1	1

(e)

A	B	C	X
0	0	0	0
0	0	1	0
0	1	0	0
0	1	1	1
1	0	0	1
1	0	1	1
1	1	0	0
1	1	1	1

4–7 **(a)** $\overline{A} + B + \overline{C}D$
(b) $\overline{A} + \overline{B} + (\overline{C} + \overline{D})(\overline{E} + \overline{F})$
(c) $\overline{A}B\overline{C}D + \overline{A} + \overline{B} + \overline{C} + D$
(d) $\overline{A} + B + C + D + A\overline{B}\overline{C}D$
(e) $AB + (\overline{C} + \overline{D})(E + \overline{F}) + ABCD$

4–9 **(a)** $AC + AD + BC + BD$
(b) $AD + \overline{B}CD$
(c) $ABC + ACD$

4–11 **(a)** Sum-of-products
(b) Sum-of-products
(c) Product-of-sums
(d) Product-of-sums
(e) Sum-of-products
(f) Sum-of-products
(g) Sum-of-products
(h) Product-of-sums

4–13 $X = AB + \overline{A}\,\overline{B}$

4–15 $X = \overline{A}\overline{B}C + \overline{A}B\overline{C} + A\overline{B}\overline{C} + \overline{A}\,\overline{B}\,\overline{C}$

4–17

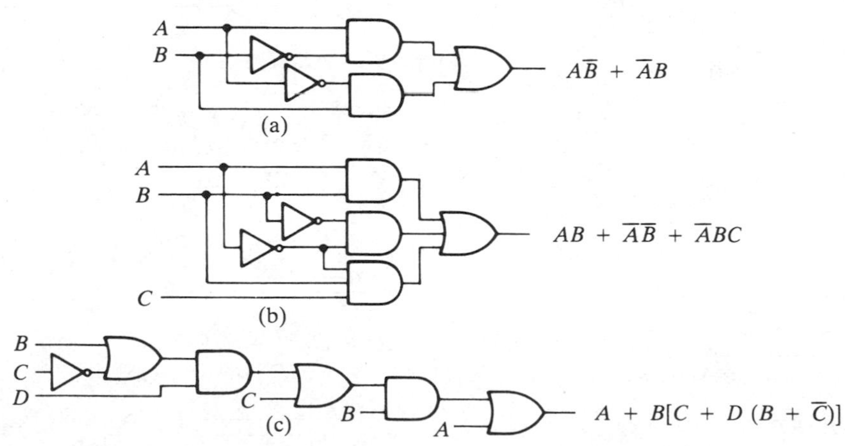

4–19 **(a)** $A + \overline{B}C$ **(b)** $\overline{A}B$ **(c)** $A + C$
(d) AB **(e)** $AB + \overline{A}C + \overline{B}C$

4–21 **(a)** 4 **(b)** 8 **(c)** 16 **(d)** 32

4–23 **(a)** $X = \overline{A}\,\overline{B} + \overline{B}C$ **(b)** $X = AC$
(c) $X = B$ **(d)** $X = \overline{C}$
(e) $X = A + \overline{B}C$. No simplification possible.

4–25 **(a)** No simplification
(b) $X = \overline{A}\,\overline{B}\,\overline{C} + ABC$
(c) $X = B\overline{C} + A\overline{C}D$
(d) $X = \overline{B}C$
(e) $X = \overline{B} + \overline{D}$

CHAPTER 5

5–1 **(a)**

A	B	X
0	0	0
0	1	0
1	0	0
1	1	1

(b)

A	B	X
0	0	0
0	1	1
1	0	0
1	1	1

(c)

A	B	X
0	0	1
0	1	1
1	0	0
1	1	1

(d)

A	B	X
0	0	0
0	1	1
1	0	1
1	1	1

5–3

A	B	X
0	0	0
0	1	1
1	0	1
1	1	0

5–5 **(a)** $X = (AB + C)D + E$
(b) $X = \overline{\overline{(\overline{A} + B)}\,\overline{BC}} + D$
(c) $X = (AB + \overline{C})D + \overline{E}$
(d) $X = \overline{(AB + CD)\,(EF + GH)}$
(e) $X = \overline{[\overline{(\overline{\overline{AB}C})D}]\,[\overline{\overline{EF}G}]}$

5–7

A	B	C	D	X
0	0	0	0	0
0	0	0	1	0
0	0	1	0	0
0	0	1	1	0
0	1	0	0	0
0	1	0	1	0
0	1	1	0	1
0	1	1	1	1
1	0	0	0	0
1	0	0	1	0
1	0	1	0	1
1	0	1	1	0
1	1	0	0	0
1	1	0	1	0
1	1	1	0	1
1	1	1	1	0

5–9 **(a)** $X = AB$

(b) $X = A + B$

(c) $X = AB + C$

(d) $X = ABC + D$

(e) $X = A + B + C$

(f) $X = ABCD$

(g) $X = A(CD + B)$

(h) $X = AB(C + DEF) + CE(A + B + F)$

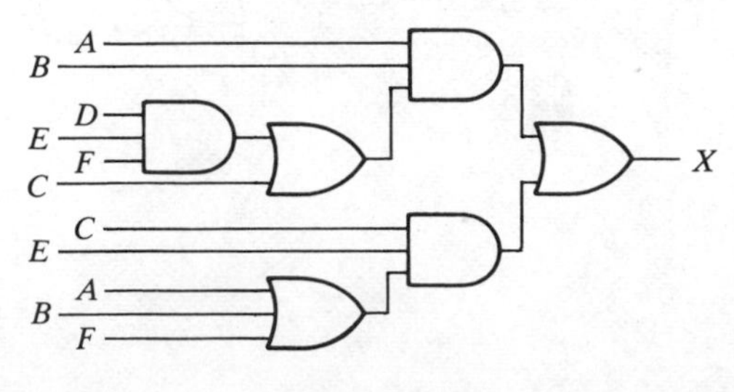

5–11

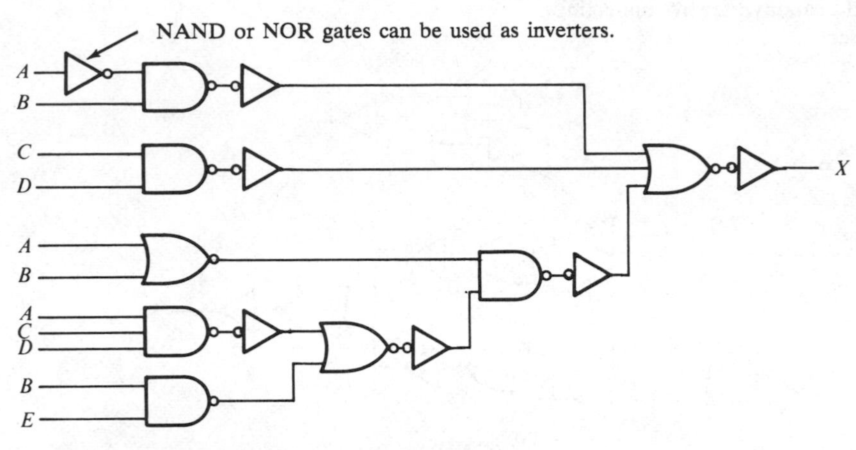

(a) This is an implementation of the expression *as is* without modification or simplification. It can be done in a simpler way.

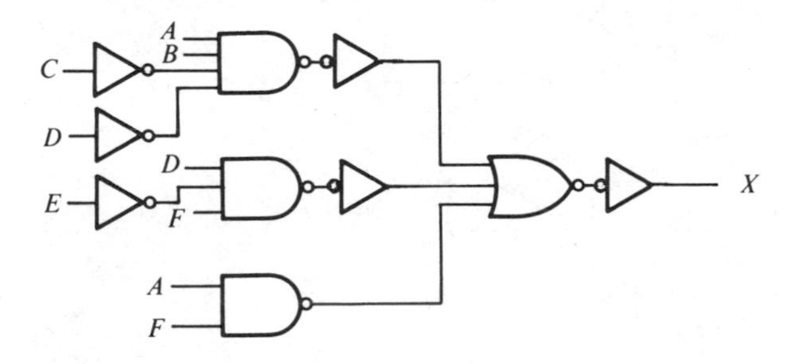

(b) Again, this is a straightforward implementation of the expression as it "reads."

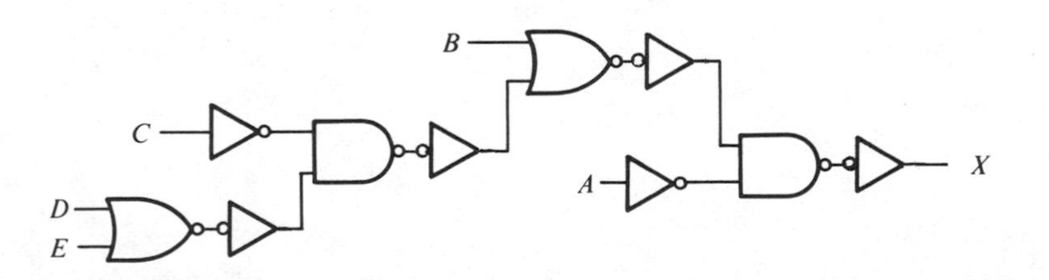

(c) Again, not the simplest— but the most straightforward.

5–13 In each circuit an inverter symbol is shown for a NOR gate used as an inverter by connecting all inputs together.

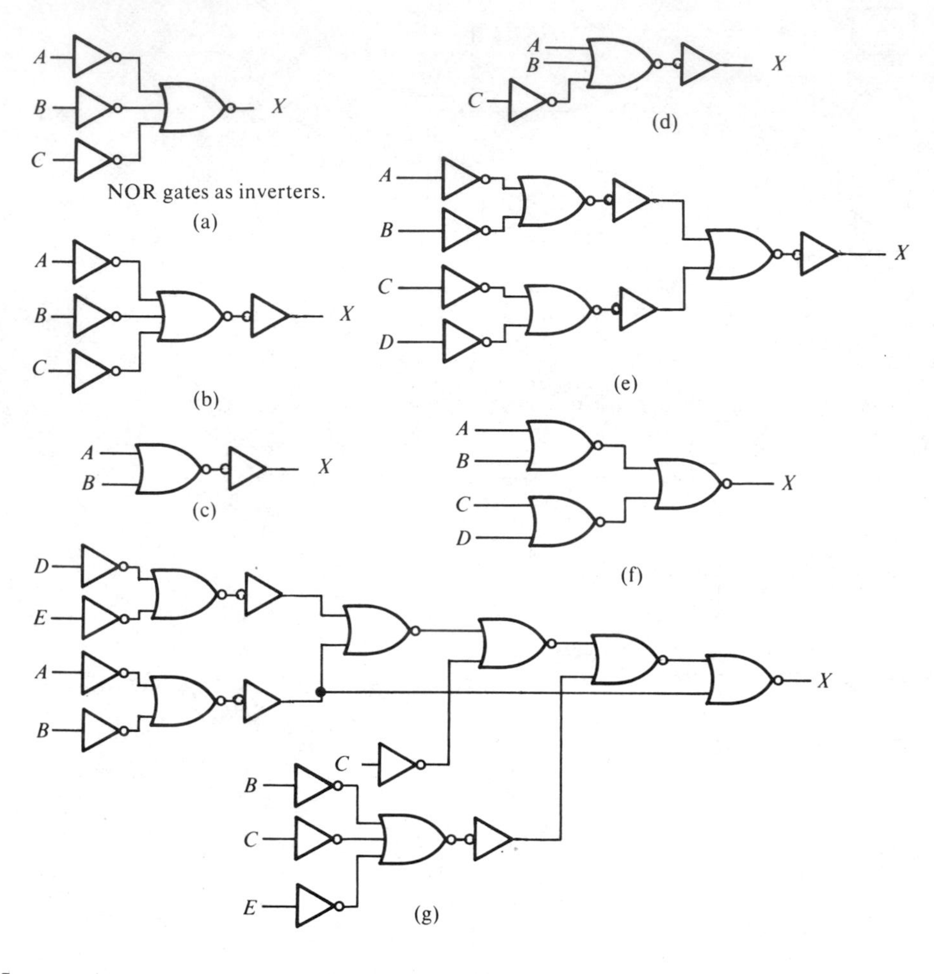

5–15

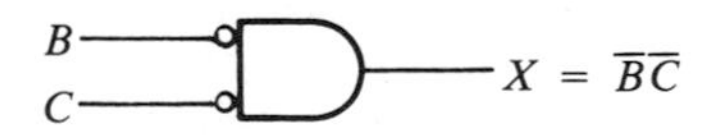

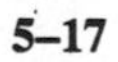

5–17

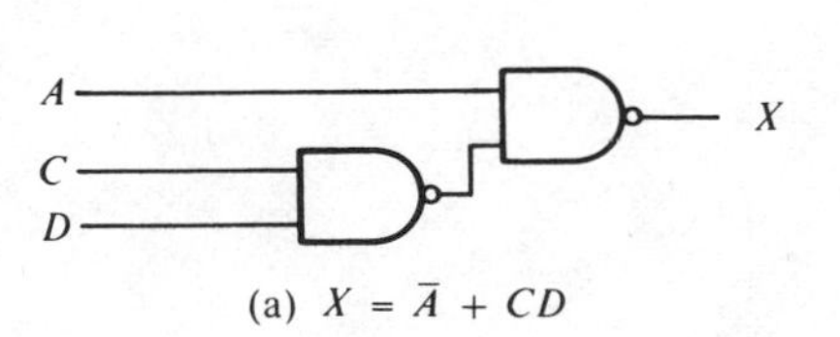

(a) $X = \overline{A} + CD$

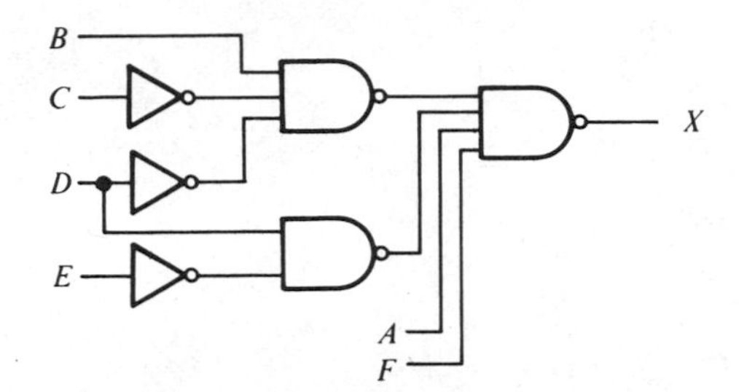

(b) $X = \overline{A} + \overline{F} + D\overline{E} + B\overline{C}\overline{D}$

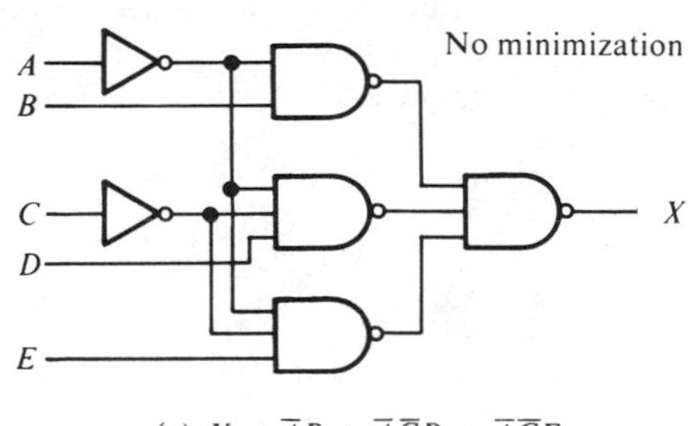

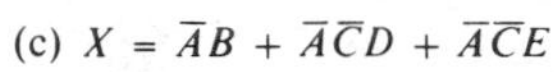

(c) $X = \overline{A}B + \overline{A}\overline{C}D + \overline{A}\overline{C}E$

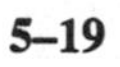

5–19

$X = \overline{A}C + A\overline{C}\overline{D} + BCD$

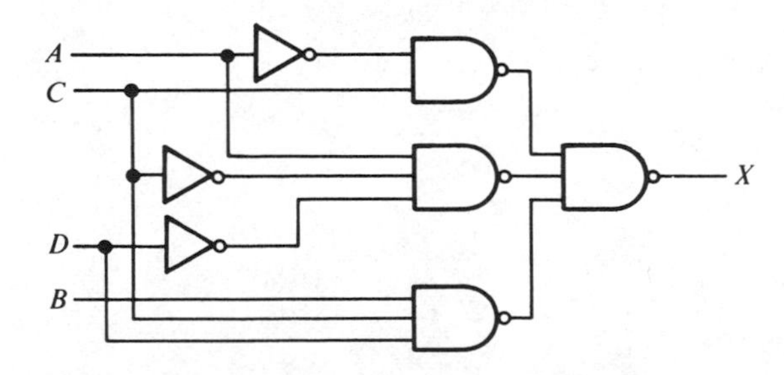

Inverter symbols represent NAND gates connected as inverters (all inputs connected).

5–21

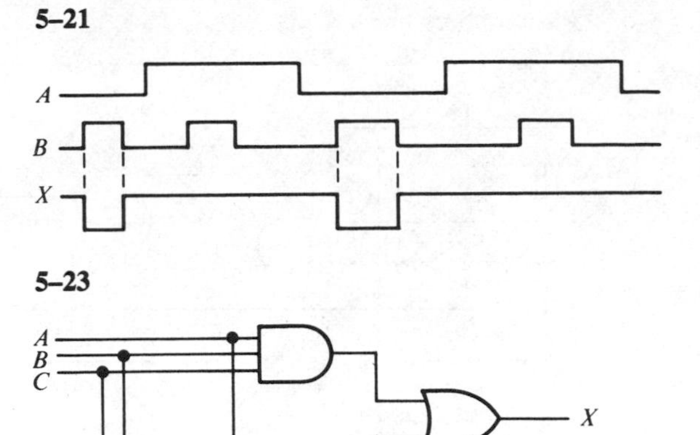

5–23

5–25

$X \equiv$ lamp ON.

$A \equiv$ front door switch ON.

$B \equiv$ back door switch ON.

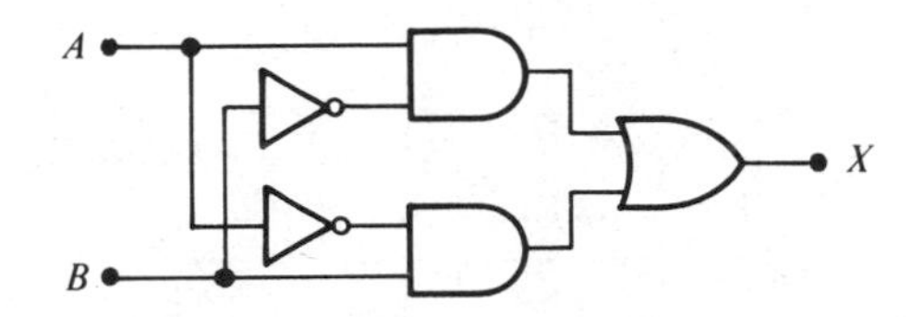

5–27 The output waveform is incorrect.

5–29 Gate G_2 failed with output high.

5–31 Pin numbers are circled.

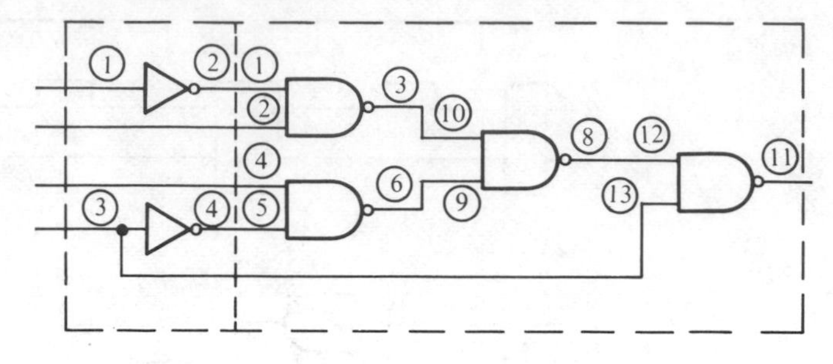

5–33 G_2 output failed high.

CHAPTER 6

6–1 A full adder handles two bits and an input carry. A half adder handles two bits but no input carry.

6–3

(a)	(b)	(c)
G_1 output = 0	G_1 output = 1	G_1 output = 1
G_2 output = 0	G_2 output = 1	G_2 output = 0
G_3 output = 0	G_3 output = 0	G_3 output = 1
G_4 output = 0	G_4 output = 0	G_4 output = 0
G_5 output = 0	G_5 output = 0	G_5 output = 1

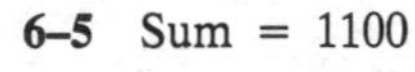

6–5 Sum = 1100

6–7

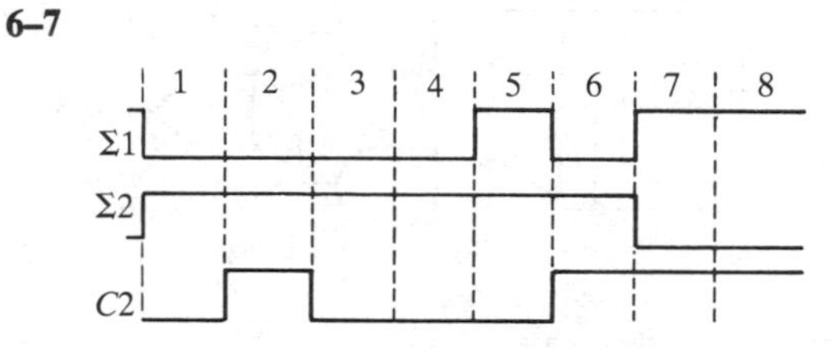

6–9 Not functioning properly. Pin 2 should be LOW.

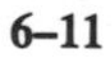

6–11

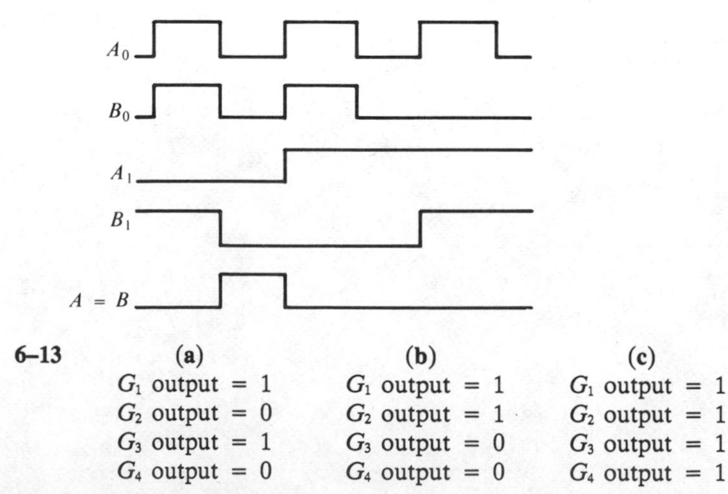

6–13

(a)	(b)	(c)
G_1 output = 1	G_1 output = 1	G_1 output = 1
G_2 output = 0	G_2 output = 1	G_2 output = 1
G_3 output = 1	G_3 output = 0	G_3 output = 1
G_4 output = 0	G_4 output = 0	G_4 output = 1

(a)	(b)	(c)
G_5 output = 0	G_5 output = 0	G_5 output = 1
G_6 output = 0	G_6 output = 0	G_6 output = 0
G_7 output = 1	G_7 output = 0	G_7 output = 0
G_8 output = 0	G_8 output = 0	G_8 output = 0
G_9 output = 0	G_9 output = 0	G_9 output = 0
G_{10} output = 1	G_{10} output = 0	G_{10} output = 0
G_{11} output = 0	G_{11} output = 0	G_{11} output = 0
G_{12} output = 0	G_{12} output = 0	G_{12} output = 0
G_{13} output = 0	G_{13} output = 1	G_{13} output = 0
G_{14} output = 0	G_{14} output = 0	G_{14} output = 0
G_{15} output = 0	G_{15} output = 1	G_{15} output = 0
$A > B$	$A < B$	$A = B$

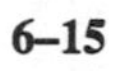

6–15

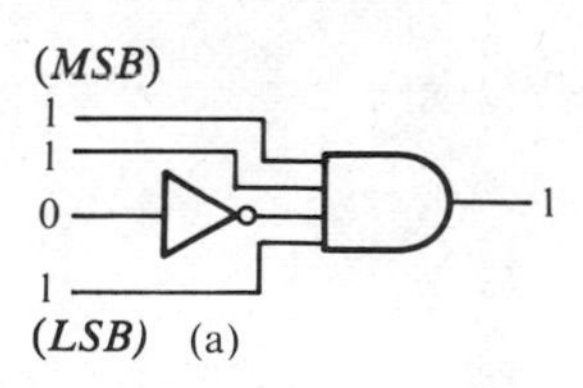

(a)

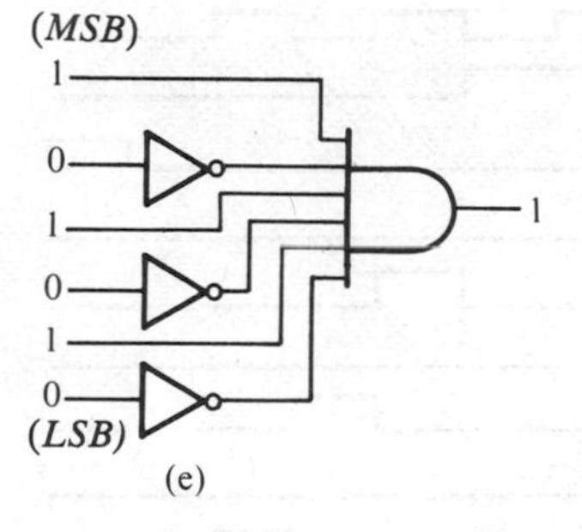

(e)

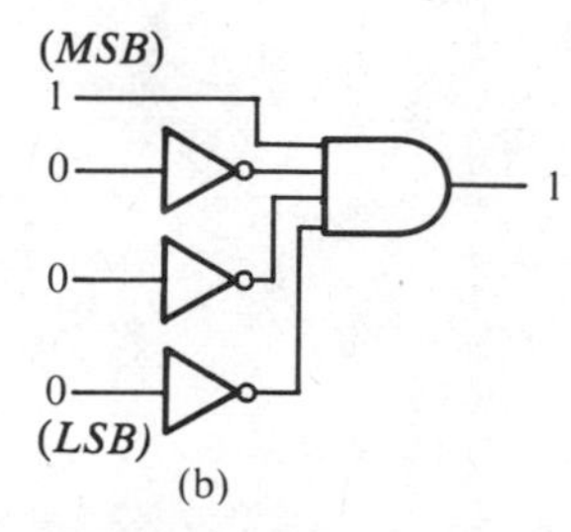

(b)

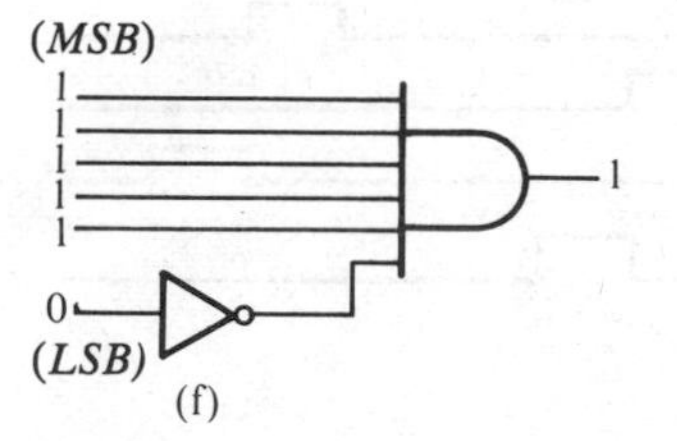

(f)

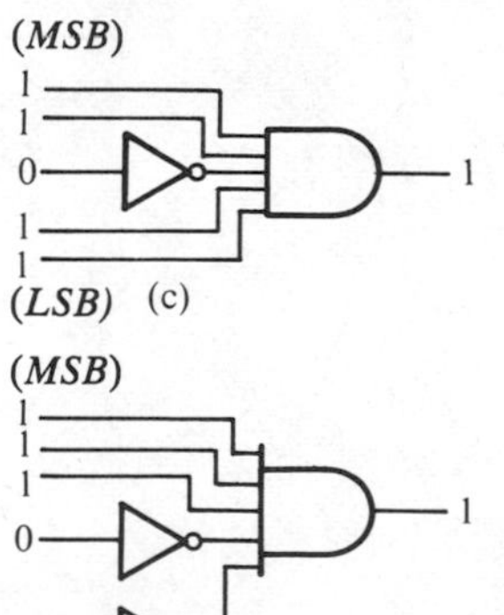

(c)

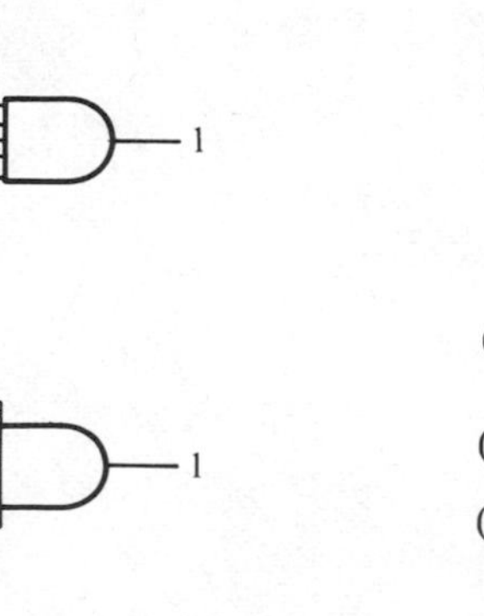

(d)

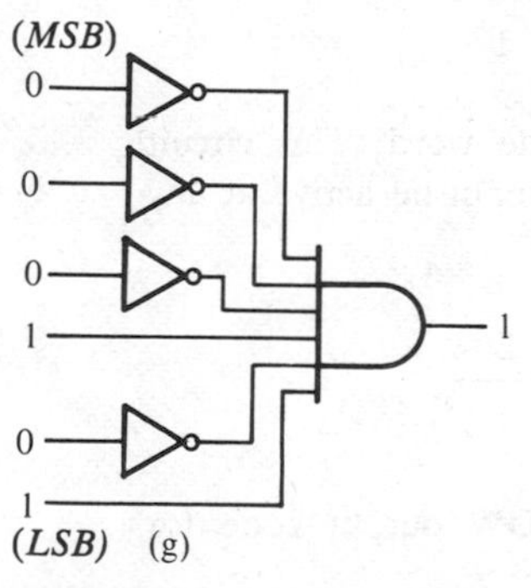

(g)

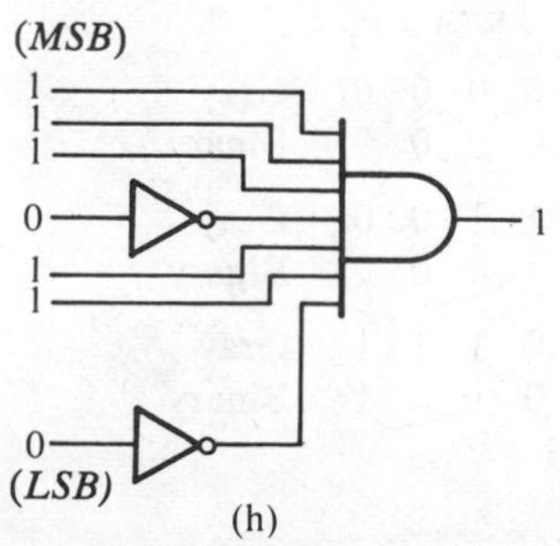

(h)

6–17

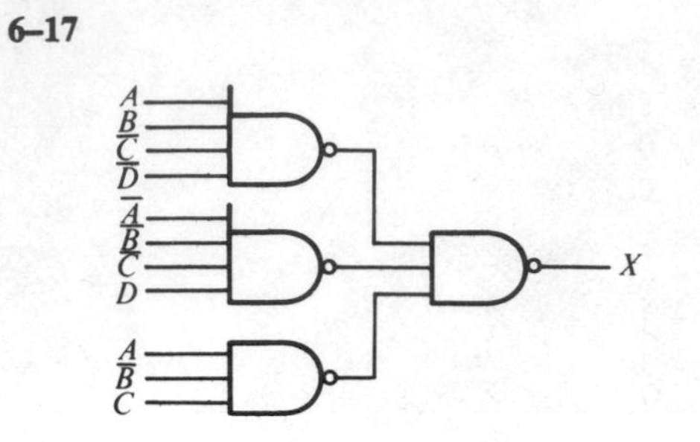

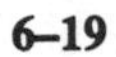

6–19

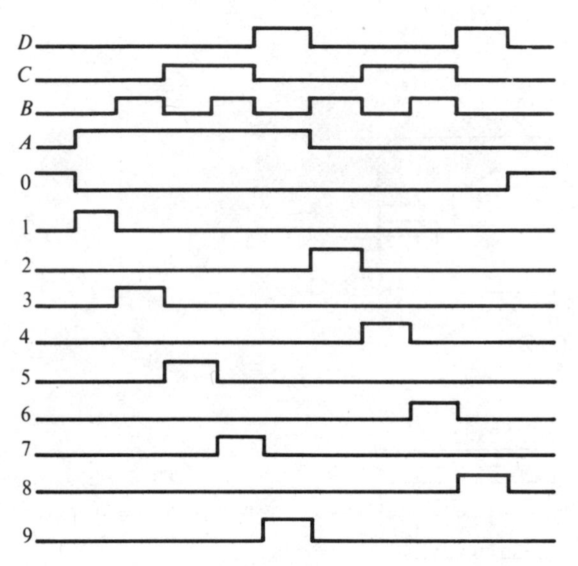

6–21 Output code:

A	*B*	*C*	*D*
1	0	1	1

Not a valid BCD code word. This circuit requires that a *single* input be active at any given time.

6–23

A	*B*	*C*	*D*
1	0	0	1

6–25 0111 is the active LOW output code for decimal 8 (1000).

6–27

(a)	1	0	1	0	0	0	0	0	0	0	Gray
	1	1	0	0	0	0	0	0	0	0	Binary
(b)	0	0	1	1	0	0	1	1	0	0	Gray
	0	0	1	0	0	0	1	0	0	0	Binary
(c)	1	1	1	1	0	0	0	1	1	1	Gray
	1	0	1	0	0	0	0	1	0	1	Binary

(d) 0 0 0 0 0 0 0 0 0 1 Gray
0 0 0 0 0 0 0 0 0 1 Binary

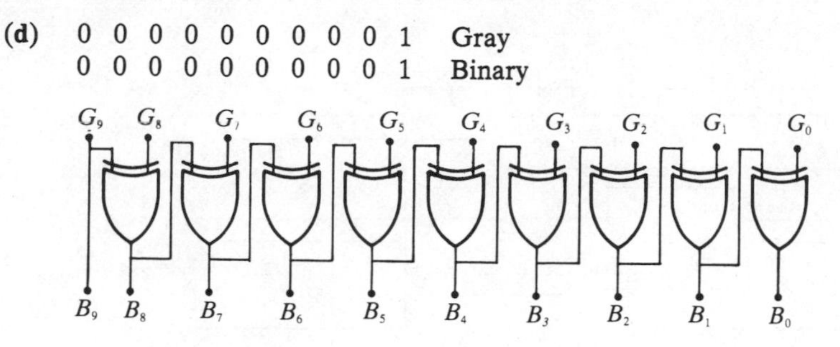

6–29 The 10_2 on the S_1, S_0 inputs selects the D_2 input and therefore the output is HIGH.

6–31

6–33

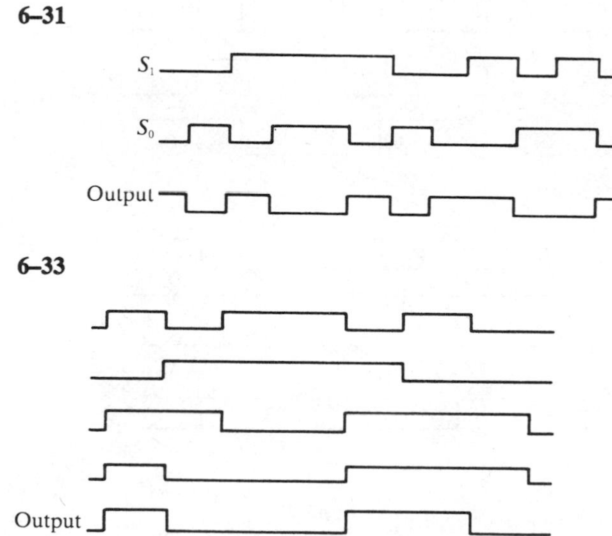

CHAPTER 7

7–1 Assuming Q starts LOW, the waveform is as shown.

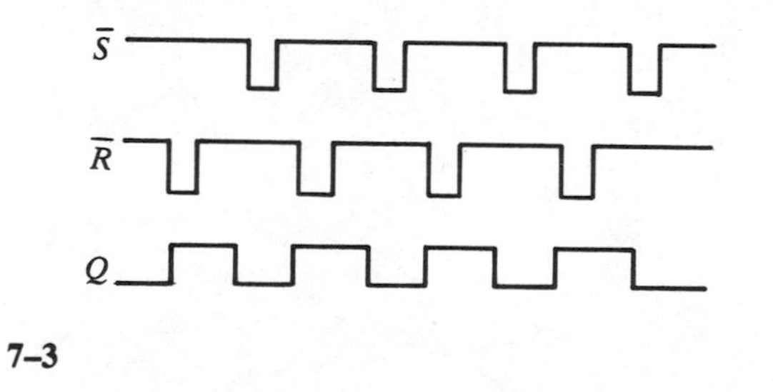

7–3

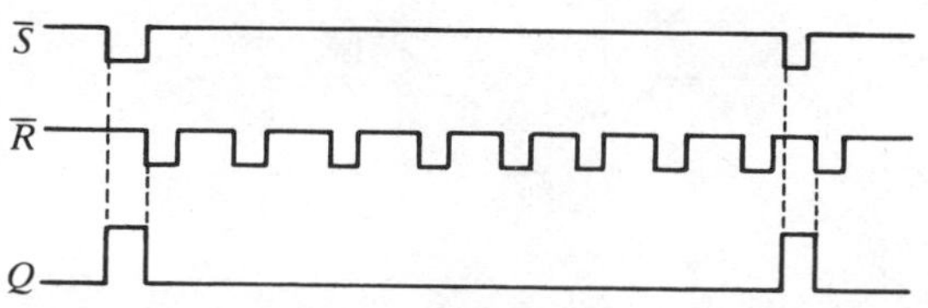

7–5

7–7

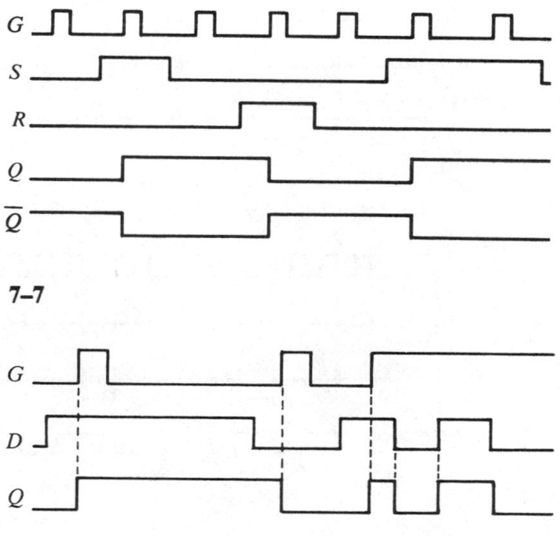

7–9 In an S–R latch, the state of the output is determined by both S and R. In a D latch, the state of the D input controls the output.

7–11 A gated S–R latch responds to the states of the S and R inputs as long as the gate input is in its active state. The edge-triggered S–R flip-flop responds only on the edge of a clock pulse.

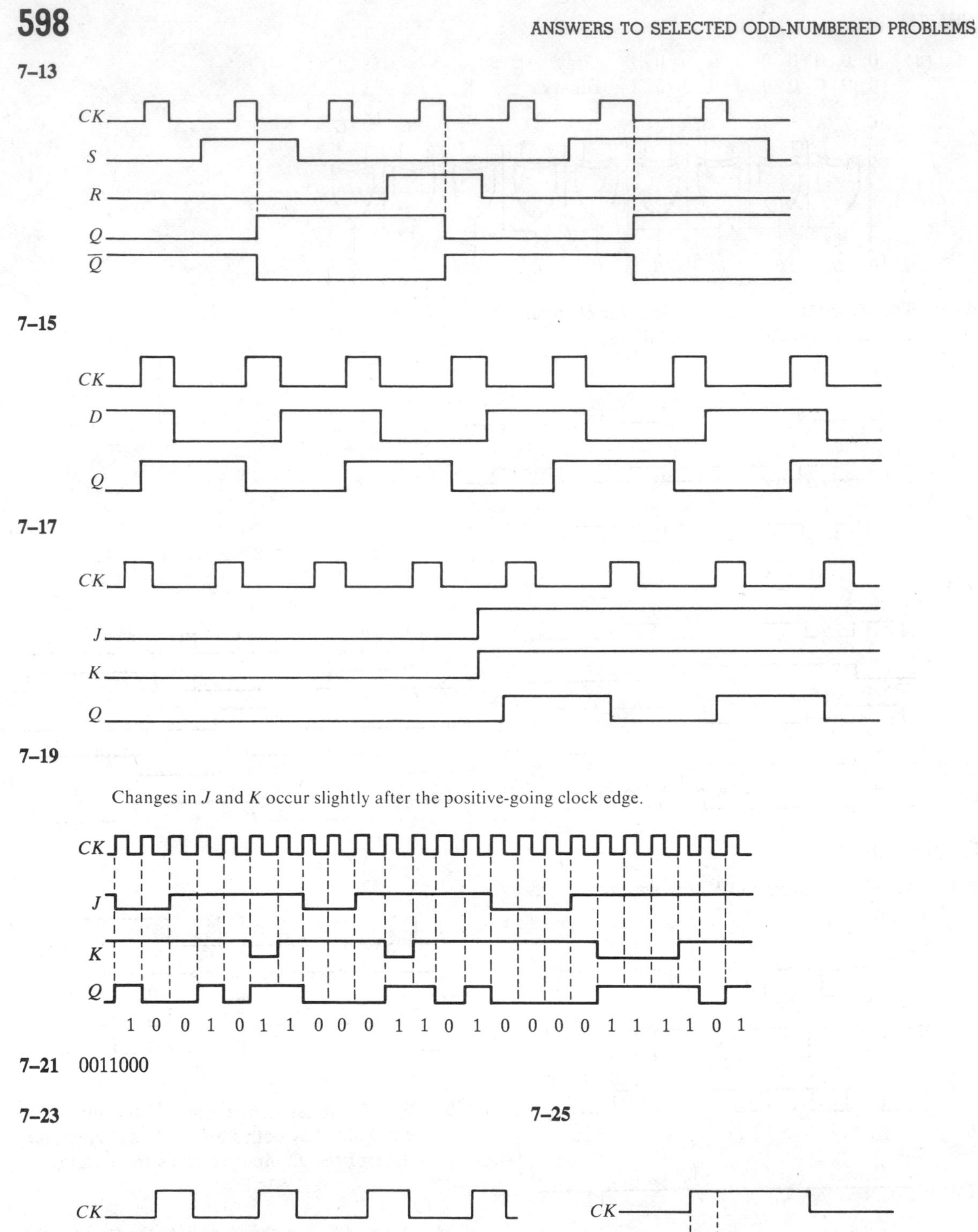

7–13

7–15

7–17

7–19

Changes in J and K occur slightly after the positive-going clock edge.

7–21 0011000

7–23

7–25

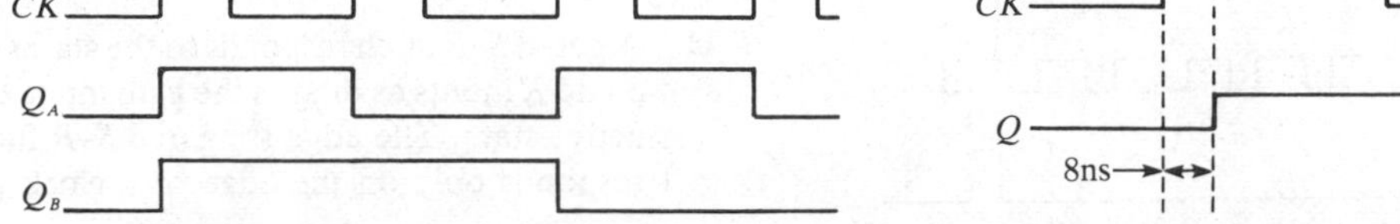

7–27 14.29 MHz

7–29 1.848 μs

7–31 A monostable has one stable state. An astable has no stable states.

7–33 $R_1 = 0$, $R_2 = 36\ k\Omega$.

CHAPTER 8

8–1 (a) Reset (b) Set (c) Reset (d) Set

8–3 (a) 2 (b) 2 (c) 3 (d) 4 (e) 4 (f) 5 (g) 6 (h) 7

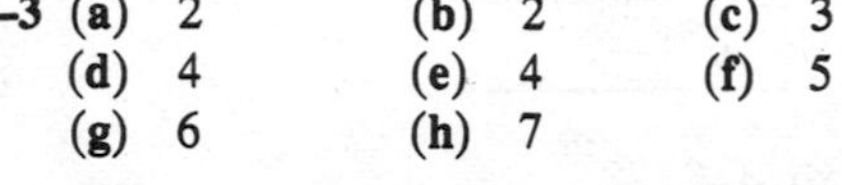

8–5

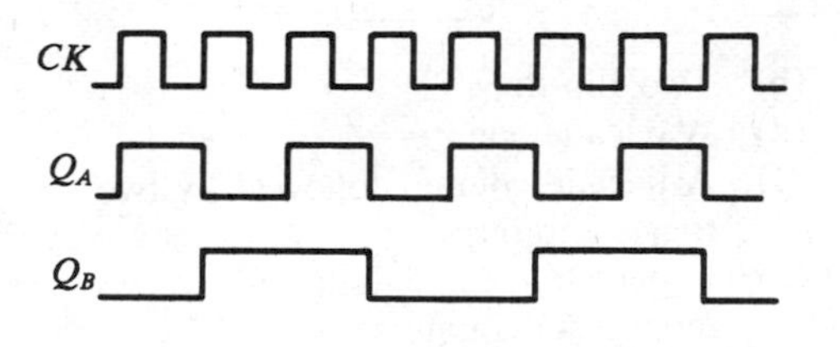

8–7

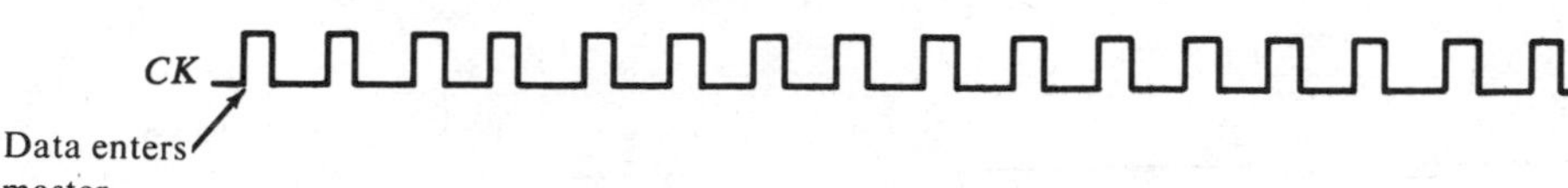

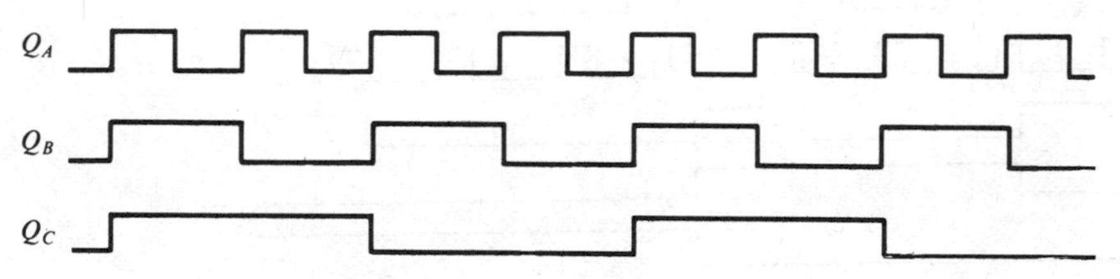

8–9

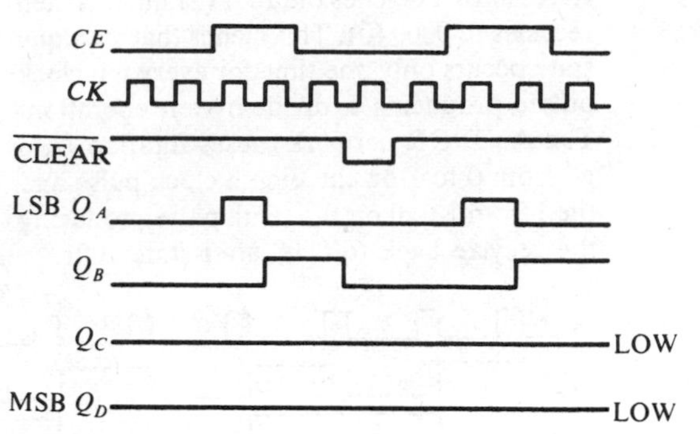

8–11

CK

Q_A: 1 0 1 0 1 0 1 0 1 0 1 0 1 0 1 0 1 0 1 0 1 0 1 0 1 0 1 0 1 0 1 0

Q_B: 0 1 1 0 0 1 1 0 0 1 1 0 0 1 1 0 0 1 1 0 0 1 1 0 0 1 1 0 0 1 1 0

Q_C: 0 0 0 1 1 1 1 0 0 0 0 1 1 1 1 0 0 0 0 1 1 1 1 0 0 0 0 1 1 1 1 0

Q_D: 0 0 0 0 0 0 0 1 1 1 1 1 1 1 1 0 0 0 0 0 0 0 0 1 1 1 1 1 1 1 1 0

Q_E: 0 0 0 0 0 0 0 0 0 0 0 0 0 0 0 1 1 1 1 1 1 1 1 1 1 1 1 1 1 1 1 0

0 2 4 6 8 10 12 14 16 18 20 22 24 26 28 30 0

8–13 The counter sequences through the binary states until it reaches the 1001 (9) state. It then recycles to 0000 (0). This means that a unique state occurs only one time for every ten clock pulses producing a divide-by-ten operation. The AND/OR network causes flip-flop D to go from 0 to 1 on the eighth clock pulse and then from 1 to 0 on the tenth pulse producing the recycle back to 0000 after state 1001.

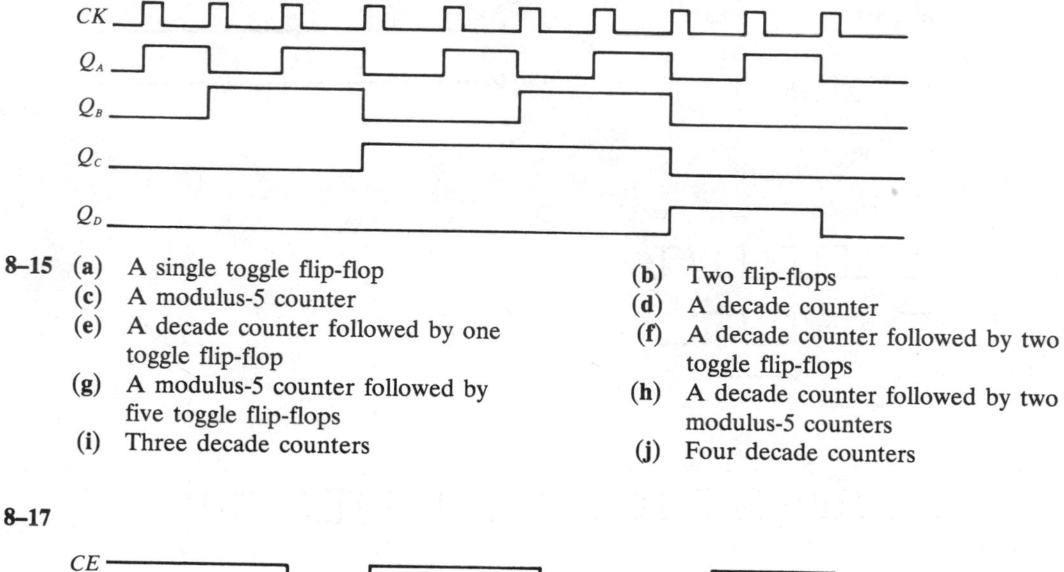

8–15 (a) A single toggle flip-flop
(b) Two flip-flops
(c) A modulus-5 counter
(d) A decade counter
(e) A decade counter followed by one toggle flip-flop
(f) A decade counter followed by two toggle flip-flops
(g) A modulus-5 counter followed by five toggle flip-flops
(h) A decade counter followed by two modulus-5 counters
(i) Three decade counters
(j) Four decade counters

8–17

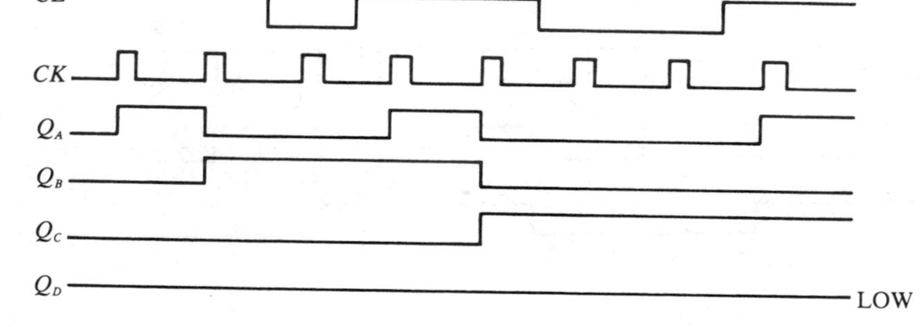

8–19

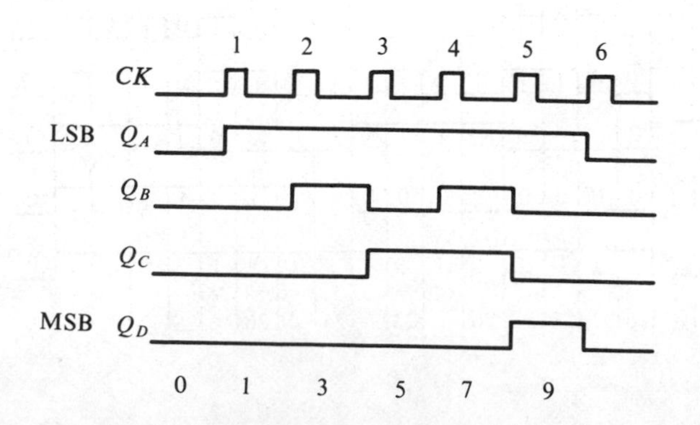

8–21

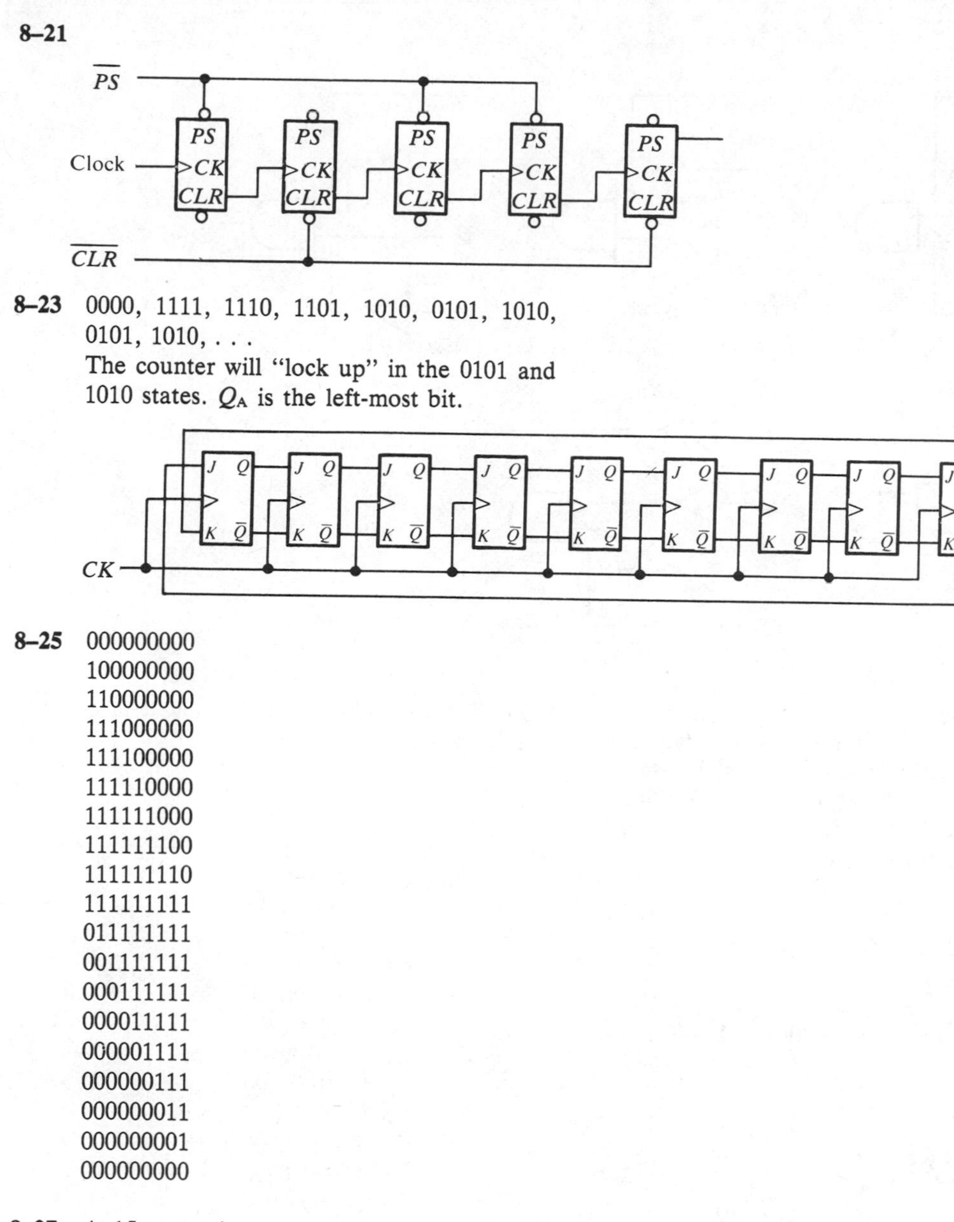

8–23 0000, 1111, 1110, 1101, 1010, 0101, 1010, 0101, 1010, . . .
The counter will "lock up" in the 0101 and 1010 states. Q_A is the left-most bit.

8–25 000000000
100000000
110000000
111000000
111100000
111110000
111111000
111111100
111111110
111111111
011111111
001111111
000111111
000011111
000001111
000000111
000000011
000000001
000000000

8–27 A 15-stage ring counter with the following stages SET: 4, 8, and 13.

8–29

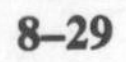

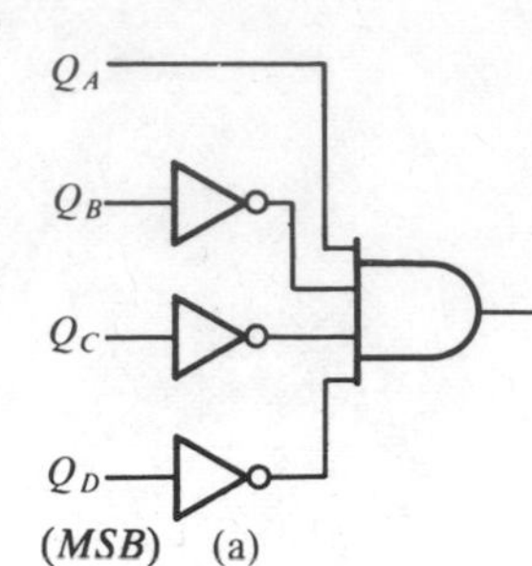

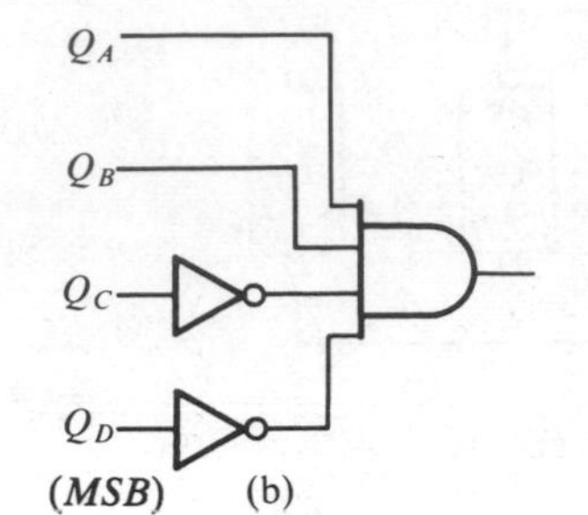

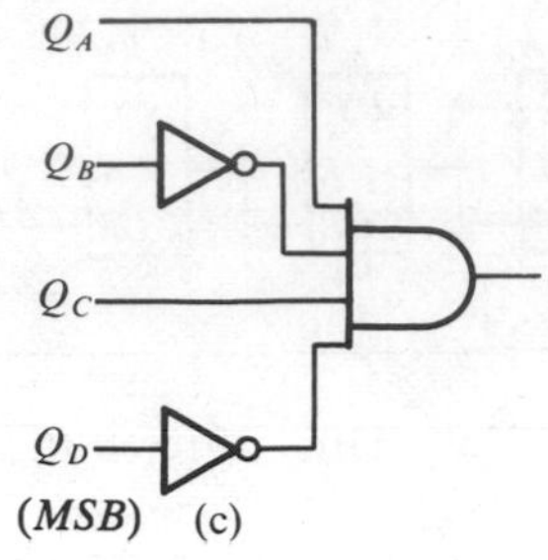

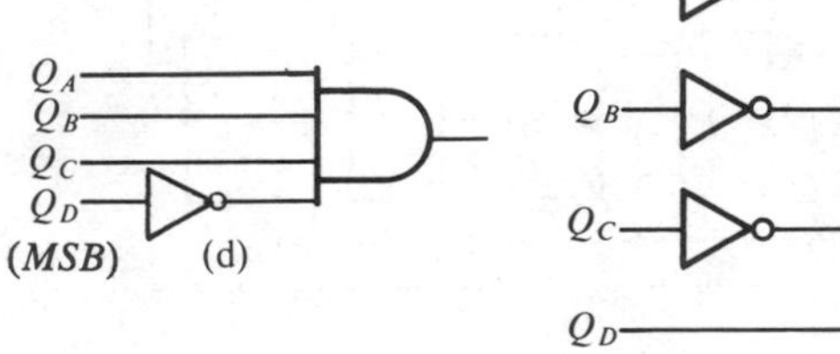

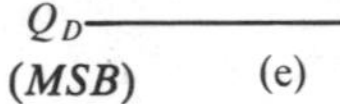

8–31 When the flip-flop is SET, it is in one of its stable states which represents a binary 1. When it is RESET, it is in its other stable state which represents a binary 0. Since the flip-flop can reman indefiitely in either state, it acts as a memory or storage element.

8–33

After clock pulse	Q_A	Q_B	Q_C	Q_D
1	0	1	1	1
2	1	0	1	1
3	1	1	0	1
4	0	1	1	0
5	1	0	1	1
6	0	1	0	1
7	1	0	1	0
8	0	1	0	1
9	0	0	1	0

8–35 The output looks like the input delayed by 10 clock pulses.

8–37 11011010

8–39

Initially	01001100
After *CK1*	00100110
CK2	00010011
CK3	00001001
CK4	00010010
CK5	00100100
CK6	01001000
CK7	00100100
CK8	01001000
CK9	10010000
*CK*10	00100000
*CK*11	00010000
*CK*12	00001000

CHAPTER 9

9–1 (a)

Address		0	1	2	3
(D_0) Register	1	0	0	0	0
(D_1) Register	2	0	0	1	0
(D_2) Register	3	0	0	1	0
(D_3) Register	4	0	0	0	0

(b)

Address		0	1	2	3
(D_0) Register	1	0	1	0	0
(D_1) Register	2	0	0	1	0
(D_2) Register	3	0	1	1	0
(D_3) Register	4	0	1	0	0

(c)

Address		0	1	2	3
(D_0) Register	1	0	1	0	0
(D_1) Register	2	0	0	1	0
(D_2) Register	3	0	1	1	1
(D_3) Register	4	0	1	0	1

9–3

Address		0	1	2	3
(D_0) Register	1	0	0	1	1
(D_1) Register	2	1	0	1	0
(D_2) Register	3	0	1	0	1
(D_3) Register	4	1	1	1	0

9–5 Row 2, column 9

9–7 HGFED = 11001 selects row 25.
Y = 10011101

9–9 A static RAM utilizes bistable storage cells which can retain data indefinitely. A dynamic RAM utilizes capacitive storage which must be refreshed (recharged) periodically to retain data.

9–11

A_1	A_0	O_3	O_2	O_1	O_0
0	0	0	0	0	0
0	1	1	0	1	1
1	0	1	1	0	0
1	1	0	1	0	1

9–13

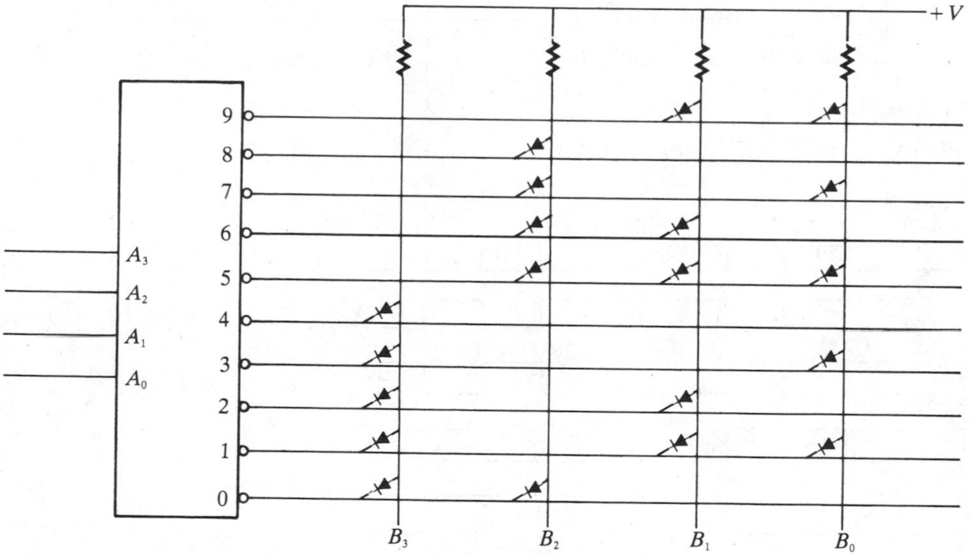

9–15

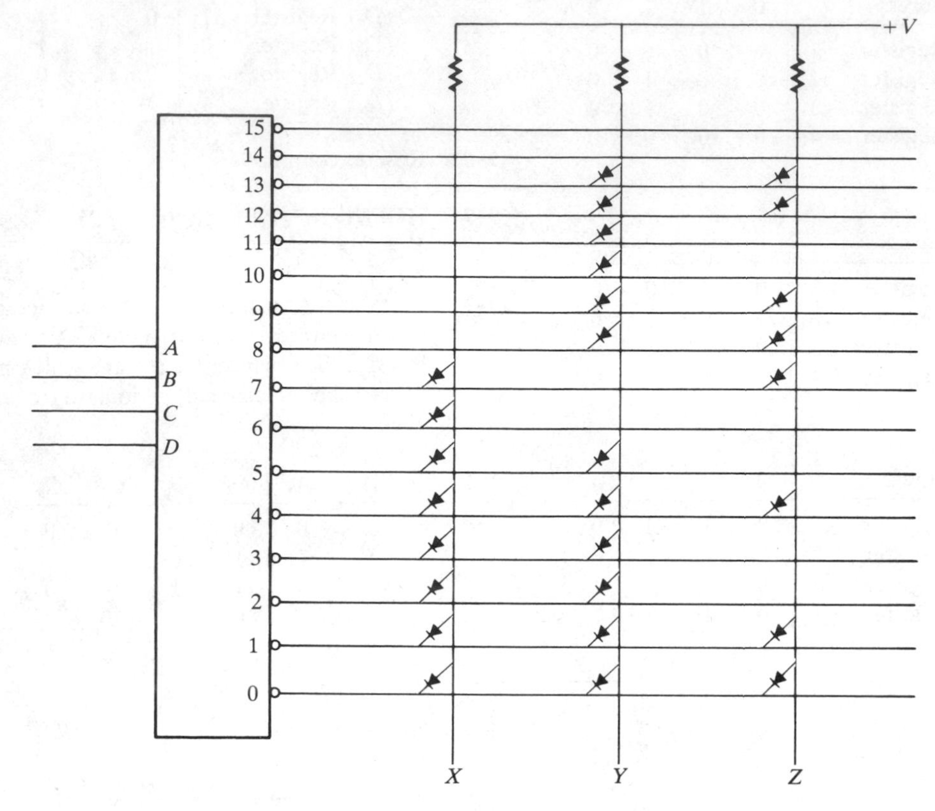

9–17 (**a**) FPGA (**b**) PAL (**c**) FPLA

9–19

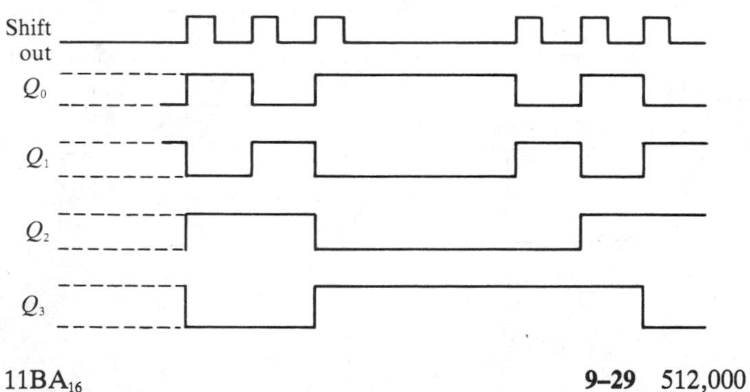

9–21 $11BA_{16}$

9–23 $00DD_{16}$

9–25 Garnet

9–27 (**a**) 10111001 (left to right)
(**b**) 01101001
(**c**) 11011000

9–29 512,000

9–31 The sum of the data bytes in a record is the checksum. It is used to detect an error.

CHAPTER 10

10–1 A bus is the interconnections and set of specified signals and levels that permit two systems or subsystems to properly operate together.
S–100, GPIB (IEEE 488), and Multibus.

10–3 (a) 100 (b) 8080A (c) 16, 8

10–5 The procedure by which two systems or subsystems communicate in order to transfer data or perform other tasks.

10–9 (a) 6.25% (b) 1.56% (c) 0.00038%

10–11 000, 001, 100, 101, 101, 100, 100, 011, 001, 001, 100, 110, 111, 111, 111, 111, 111, 111, 110, 100

10–13 0000, 0000, 0000, 1110, 1100, 0111, 0110, 0011, 0010, 1100

10–19 Single-ended, Differential.

CHAPTER 11

11–1 (a) 00010001 (b) 00100010
(c) 11000101 (d) 01101110
(e) 11111101

11–3 (a) 11101111 (b) 10011001
(c) 11111011 (d) 10001001

11–5 (a) 0110000 (b) 0011101
(c) 00011000 (d) 1101011
(e) 100111110

11–7 (a) 00011 (b) 00011
(c) 11010 (d) 10010
(e) 10001 (f) 00101

11–9 Overflow occurs when the number of bits in the sum exceeds the number of bits in each number. Overflow can occur when both numbers are positive or when both are negative.

11–11 (a) 0011011 (b) 11101111
(c) 11100010

11–13 1111011000

11–15

	Accumulator	Reg B	FA Sum	FA C_0
Initially	00111	00100	1	0
After $CK1$	10011	00010	1	0
$CK2$	11001	00001	0	1
$CK3$	01100	10000	1	0
$CK4$	10110	01000	0	0
$CK5$	01011	00100		
	True results			

11–21 In a serial adder, the bits of the numbers are added two at a time (one bit from each number) and, therefore, the add cycle time depends on the quantity of bits in the numbers. In a parallel adder, all bits are added simultaneously and a faster add time results.

11–23 (a) 1000 (b) 1001
(c) 00010010 (d) 00010010
(e) 00010011 (f) 00011000

11–25

(a)	Original sum = 0101	Corrected sum = 1011	Final sum = 0101	Carry out = 0
(b)	Original sum = 1101	Corrected sum = 0011	Final sum = 0011	Carry out = 1
(c)	Original sum = 0010	Corrected sum = 1000	Final sum = 1000	Carry out = 1
(d)	Original sum = 1001	Corrected sum = 1111	Final sum = 1001	Carry out = 0
(e)	Original sum = 0001	Corrected sum = 0111	Final sum = 0111	Carry out = 1
(f)	Original sum = 1010	Corrected sum = 0000	Final sum = 0000	Carry out = 1

Other points within the adder are left to the reader.

11–27 Accumulator

CHAPTER 12

12–3 (a) 1 (b) 1 (c) 2 (d) 2
(e) 2

12–5 (c) operands

12–7 (c) The 6800 has two accumulators.

12–9 Zero result, overflow, negative result, and carry.

12–11 $1C_{16}$, $00B0_{16}$

12–13 Fetch and execute.

12–15 Fetch.

12–17 (c) Op code.

12–19 (b) CLRA and (e) WAI

12–21 23_{16}, accumulator *A*

12–23 $16B_{16}$

12–25 JMP returns to only one address regardless of the number of times a subroutine is executed. JSR/RTS provides for return to the proper point in the main program each time a subroutine is executed.

12–27 $0F15_{16}$

APPENDIX A

A–1 MOS normally consumes less power and is slower than bipolar.

A–3 Standard, low power (L), Schottky (S), low-power Schottky (LS), and high speed (H).

A–5 8

A–7 ECL is faster; higher fan-out.

A–9 Increases.

A–11 Static charge protection.

GLOSSARY

A

Accumulator A register in a microprocessor in which the result of a given operation is stored temporarily.

A/D conversion The process of converting an analog signal into digital form.

Addend In addition, the number added to a second number called the *augend.*

Adder A digital circuit that performs the addition of numbers.

Address The location of a given storage cell in a memory.

Alphanumeric A system of symbols consisting of both numerals and alphabetic characters.

ALU Arithmetic logic unit, generally a part of the central processing unit in computers and microprocessors.

Amplitude In terms of pulse waveforms, the height or maximum value of the pulse.

Analog Being continuous or having a continuous range of values, as opposed to a discrete set of values.

AND gate A digital logic circuit in which a HIGH output occurs if and only if all the inputs are HIGH.

ANSI American National Standards Institute.

Arithmetic Related to the four operations of add, subtract, multiply, and divide.

Astable Having no stable state. A type of multivibrator that oscillates between two quasistable states.

Asynchronous Having no fixed time relationship.

Augend In addition, the number to which the addend is added.

B

Base One of the three regions in a bipolar transistor. Also, the number of symbols in a number system. The decimal system has a base of ten because there are ten digits.

BCD Binary Coded Decimal, a digital code.

Binary Having two values or states. The binary number sytem has two digits.

Binary fractional notation A method of binary notation where the magnitude of a number is represented by the bits to the right of the binary point.

Bipolar Referring to a junction type of semiconductor device. A pnp or npn transistor.

Bistable Having two stable states. A type of multivibrator commonly known as a flip-flop.

Bit Binary digit. A 1 or a 0.

Boolean Algebra A mathematics of logic.

Bubble memory A type of memory that uses tiny magnetic bubbles to store 1s and 0s.

Byte A group of eight bits.

C

Carry save A method of binary addition where the carries are "saved" until the total sum is formed and then added to the sum.

Cascade A configuration in which one device drives another.

CCD Charge coupled device. A type of semiconductor technology.

Cell A single storage element in a memory.

Character A symbol, letter, or numeral.

Circuit A combination of electrical and/or electronic components connected together to perform a specified function.

Clear To reset, as in the case of a flip-flop, counter, or register.

Clock The basic timing signal in a digital system.

CMOS Complementary metal oxide semiconductor.

Code A combination of binary digits that represents information such as numbers, letters, and other symbols.

Code converter An electronic digital circuit that converts one type of coded information into another coded form.

Collector One of the three regions in a bipolar transistor.

Combinational logic A combination of gate networks, having no storage capability, used to generate a specified function. Sometimes called *combinatorial logic.*

Comparator A digital device that compares the magnitudes of two digital quantities and produces an output indicating the relationship of the quantities.

Complement In Boolean algebra, the inverse function. The complement of a 1 is a 0, and vice versa.

Computer A digital electronic system that can be programmed to perform various tasks, such as mathematical computations, at extremely high speed, and that can store large amounts of data.

Core A magnetic memory element.

Counter A digital circuit capable of counting electronic events, such as pulses, by progressing through a sequence of binary states.

CPU Central processing unit, a main component in all computers.

D

D/A conversion A process whereby information in digital form is converted into analog form.

Data Information in numeric, alphabetic, or other form.

D flip-flop A type of bistable multivibrator in which the output follows the state of the *D* input.

Decade counter A digital counter having ten states.

Decode To determine the meaning of coded information.

Decoder A digital circuit that converts coded information into a familiar form.

Decrement To decrease the contents of a register or counter by one.

Delay The time interval between the occurrence of an event at one point in a circuit and the corresponding occurrence of a related event at another point.

DeMorgan's theorems (1) The complement of a product of terms is equal to the sum of the complements of each term. (2) The complement of a sum of terms is equal to the product of the complements of each term.

Difference The result of a subtraction.

Digit A symbol representing a given quantity in a number system.

Digital Related to digits or discrete quantities.

DIP Dual in-line package. A type of integrated circuit package.

Dividend In a division operation, the quantity that is being divided.

Divisor In a division operation, the quantity that is divided into the dividend.

DMA Direct memory access.

Don't care A condition in a logic network in which the output is independent of the state of a given input.

Down count A counter sequence in which each successive state is less than the previous state in binary value.

Drain One of the three terminals of an FET.

DTL Diode-transistor logic.

Duplex Bidirectional transmission of data along a transmission line.

Dynamic memory A memory having cells that tend to lose stored information over a period of time and therefore must be "refreshed." Typically, the storage elements are capacitors.

E

EAPROM Electrically alterable programmable read-only memory.

ECL Emitter-coupled logic.

Edge-triggered flip-flop A type of flip-flop in which input data are entered and appear on the output on the same clock edge.

EEPROM Electrically erasable programmable read-only memory.

Emitter One of the three regions of a bipolar transistor.

Enable To activate or put into an operational mode.

Encode To convert information into coded form.

Encoder A digital circuit that converts information into coded form.

End-around carry The final carry that is added to the result in a 1's or 9's complement addition.

EPROM Erasable programable read-only memory.

Error correction The process of correcting bit errors occurring in a digital code.

Error detection The process of detecting bit errors occurring in a digital code.

Even parity A characteristic of a group of bits having an even number of 1s.

Excess-3 A digital code where each of the decimal digits is represented by a four-bit code derived by adding 3 to each of the digits.

Exclusive-OR A logic function that is true if one but not both of the variables is true.

Execute The cycle of a CPU in which an instruction is carried out.

F

Fall time The time interval between the 10 percent point to the 90 percent point on the negative-going edge of a pulse.

Fan out The number of equivalent gate inputs that a logic gate can drive.

FET Field-effect transistor.

Fetch The cycle of a CPU in which an instruction or data byte is retrieved from the memory.

FIFO First-in–first-out memory.

Fixed point A binary point having a fixed location in a binary number.

Flat pack A type of integrated circuit package.

Flip-flop A bistable device used for storing a bit of information.

Floating point A binary point having a variable location in a binary number.

Floppy disk A magnetic storage device. Typically an eight-inch flexible mylar disk.

FPLA Field programmable logic array.

Frequency The number of pulses in one second for a periodic waveform. Expressed in Hertz (Hz) or pulses per second (p/s).

Full adder A digital circuit that adds two binary digits and an input carry to produce a sum and an output carry.

Full-duplex Simultaneous bidirectional transmission of data on a transmission line.

G

Gate A logic circuit that performs a specified logical operation such as AND, OR, NAND, NOR, and Exclusive-OR.

Generator An energy source for producing electrical or magnetic signals.

Glitch A voltage or current spike of short duration and usually unwanted.

Gray code A type of digital code characterized by a single bit change from one code word to the next.

H

Half adder A digital circuit that adds two bits and produces a sum and an output carry. It cannot handle input carries.

Hamming code A type of error detection and correction code.

Hexadecimal A number system consisting of 16 characters. A number system with a base of 16.

Hold time The time interval required for the control levels to remain on the inputs to a flip-flop after the triggering edge of the clock in order to order to reliably activate the device.

Hysteresis A characteristic of a magnetic core or a threshold triggered circuit.

I

IC Integrated circuit. A type of circuit where all the components are integrated on a single silicon chip of very small size.

I²L Integrated injection logic.

Increment To increase the contents of a register or counter by 1.

Indexing The modification of the address of the operand contained in the instruction of a microprocessor.

Index register A register used for indexed addressing in a microprocessor.

Indirect address The address of the address of an operand.

Information In a digital system, the data as represented in binary form.

Inherent address An address that is implied or inherent in the instruction itself.

Initialize To put a logic circuit in a beginning state, such as to clear a register.

Input The signal or line going into a circuit. A signal that controls the operation of a circuit.

Instruction In a microprocessor or computer system, the information that tells the machine what to do. One step in a computer program.

Interrupt The process of stopping the normal execution of a program in a computer in order to handle a higher priority task.

Inversion Conversion of a HIGH level to a LOW level or vice versa.

Inverter The digital circuit that performs inversion.

J

J–K flip-flop A type of flip-flop that can operate in the set, reset, no change, and toggle modes.

Johnson counter A type of digital counter characterized by a unique sequence of states.

Jump A computer operation in which a computer goes from one point in a program to another point or to another program by skipping the normal sequence of instructions. A type of microprocessor instruction.

Junction The boundary between an *n* region and *p* region in a semiconductor device.

K

Kansas City Standard A format for recording digital data on magnetic surface.

Karnaugh map An arrangement of cells representing the combinations of variables in a Boolean expression and used for a systematic simplification of the expression.

L

LCD Liquid crystal display.

Leading edge The first edge to occur on a pulse.

LED Light-emitting diode.

LIFO Last-in–first-out memory.

Logic In digital electronics, the decision-making capability of gate circuits in terms of yes/no or on/off type of operation.

Look-ahead carry A method of binary addition whereby carries from preceding stages are anticipated, thus avoiding carry propagation delays.

Loop A part of a computer program in which the machine repeats a segment of the program over and over. Also, in magnetic bubble memories, the tracks on which the bubbles move.

LSB Least significant bit.

LSD Least significant digit.

LSI Large-scale integration.

M

Magnetic bubble ·A tiny magnetic region in magnetic material created by an external magnetic field.

Magnetic core A memory element made of magnetic materials and capable of existing in either of two states.

Magnitude The size or value of a quantity.

Manchester A format for recording digital data on a magnetic surface.

Master reset Normally, the input to a counter, register, or other digital storage device that completely resets or clears the device.

Master-slave flip-flop A type of flip-flop in which the input data are entered into the device on the leading edge of the clock and appear on the output on the trailing edge.

Memory address The location of a storage cell in a memory array.

Memory array An arrangement of memory cells.

Memory cell An individual storage element in a memory.

Microprocessor A large-scale integrated circuit that can be programmed to perform arithmetic and logic functions and to manipulate data.

Minuend The number being subtracted from in a subtraction operation.

Modified modulus counter A counter that does not sequence through all of its "natural" states.

Mod-2 addition Exclusive-OR addition. A sum of two bits with the carry dropped.

Modulus The maximum number of states in a counter sequence.

Monostable Having only one stable state. A multivibrator characterized by one stable state and commonly called a *one-shot.*

MOS Metal oxide semiconductor.

MSI Medium-scale integration.

Multiplex To put information from several sources onto a single line or transmission path.

Multiplexer A digital circuit capable of multiplexing digital data.

Multiplicand The number being multiplied.

Multiplier The number used to multiply the multiplicand.

N

NAND gate A logic gate that performs an inverted AND operation (NOT-AND).

Natural count The maximum possible modulus of a counter.

Negative logic The system of logic where a LOW represents a 1 and a HIGH represents a 0.

Nesting Referring to the arrangement of subroutines where the program exits one subroutine and goes to another, eventually returning to the first. Several levels of nesting can occur.

Nixie tube A vacuum tube used for digital displays.

NMOS *n*-channel metal oxide semiconductor.

Noise immunity The ability of a circuit to reject unwanted signals.

Noise margin The difference between the maximum low output of a gate and the maximum acceptable low level input of an equivalent gate. Also, the difference between the minimum high output of a gate and the minimum acceptable high level input of an equivalent gate.

NOR gate A logic gate that performs an inverted OR operation (NOT-OR).

NOT circuit An inverter.

npn Referring to a junction structure of a bipolar transistor.

Numeric Related to numbers.

O

Octal A number system having a base of 8 consisting of eight digits.

Odd parity Referring to a group of binary digits having an odd number of 1s.

One-shot A monostable multivibrator.

Op code Operation code. The part of a microprocessor instruction that designates the task to be performed.

Operand A quantity being operated on in a microprocessor.

OR gate A logic gate that produces a HIGH output when any one or more of its inputs are HIGH.

Oscillator An electronic circuit that switches back and forth between two states. The astable multivibrator is an example.

P

Parallel Characteristic of two lines having an equal distance between them at all points.

Parity Referring to the oddness or evenness of the number of 1s in a specified group of bits.

Period The time required for a periodic waveform to repeat itself.

Periodic Repeating at fixed intervals.

PMOS *p*-channel metal oxide semiconductor.

pnp Referring to a junction structure of a bipolar transistor.

Positive logic The system of logic where a HIGH represents a 1 and a LOW represents a 0.

p/s Pulses per second. A measure of frequency of a pulse waveform.

Preset To initialize a digital circuit to a predetermined state.

PRF Pulse repetition frequency.

Priority encoder A digital logic circuit that produces a coded output corresponding to the highest valued input.

Product The result of a multiplication.

Product-of-sums A form of Boolean expression that is the ANDing of ORed terms.

Program A list of instructions that are arranged in a specified order to control the operation of a microprocessor system or computer. The program tells the machine what to do on a step by step basis.

Program counter A counter in a microprocessor that keeps up with the place in the program. It acts as a "bookmark" that tells the computer the next instruction to be executed.

Propagation delay The time interval between the occurrence of an input transition and the corresponding output transition.

Pulse A sudden change from one level to another followed by a sudden change back to the original level.

Pulse duration The time interval that a pulse remains at its high level (positive-going pulse) or at its low level (negative-going pulse). Typically measured between the 50 percent-points on the leading and trailing edges of the pulse.

Pulse width Pulse duration.

Q

Quotient The result of a division.

R

Race A condition in a logic network where the differences in propagation times through two or more signal paths in the network can produce an erroneous output.

Radix The base of a number system. The number of digits in a given number system.

RAM Random access memory.

Read The process of retrieving information from a memory.

Recirculate The process of retaining information in a register as it is shifted out.

Refresh The process of renewing the contents of a dynamic memory.

Regenerative Having feedback so that an initiated change is automatically continued, such as when a multivibrator switches from one state to the other.

Register A digital circuit capable of storing and moving (shifting) binary information. Typically used as a temporary storage device.

Reset The state of a flip-flop, register, or counter when 0s are stored. Equivalent to the clear function.

Ring counter A digital circuit made up of a series of flip-flops in which the contents are continuously recirculated.

Ringing A damped sinusoidal oscillation.

Ripple counter A digital counter in which each flip-flop is clocked with the output of the previous stage.

Rise time The time required for the positive-going edge of a pulse to go from 10 percent of its full value to 90 percent of its full value.

ROM Read only memory.

S

Semiconductor A material used to construct electronic devices such as integrated circuits, transistors, and diodes. Silicon is the most common semiconductor material.

Sequential logic A broad category of digital circuits whose logic states depend on a specified time sequence.

Serial An in-line arrangement where one element follows another, such as in a serial shift register. Also, the occurrence of events, such as pulses, in a time sequence rather than simultaneously.

Set The state of a flip-flop when it is in the binary 1 state.

Set-up time The time interval required for the control levels to be on the inputs to a digital circuit, such as a flip-flop, prior to the triggering edge of the clock pulse.

Shift To move binary data within a shift register or other storage device.

Shift register A digital circuit capable of storing and shifting binary data.

Silicon A semiconductor material.

Simplex A mode of data transmission whereby the data can be sent in only one direction.

SR flip-flop A set-reset flip-flop.

SSI Small scale integration.

Stack A LIFO memory consisting of registers or memory locations.

Stage One storage element in a register or counter.

Static memory A memory composed of storage elements such as flip-flops or magnetic cores that are capable of retaining information indefinitely.

Storage The memory capability of a digital device. The process of retaining digital data for later use.

Strobe A pulse used to sample the occurrence of an event at a specified point in time in relation to the event.

Subroutine A program that is normally used to perform specialized or repetitive operations during the course of a main program. A subprogram.

Subtractor One of the operands in a subtraction.

Subtrahend The other operand in a subtraction.

Sum The result of an addition.

Sum-of-products A form of Boolean expression that is the ORing of ANDed terms.

Synchronous Having a fixed time relationship.

T

Terminal count The final state of a counter sequence.

T flip-flop A type of flip-flop that toggles or changes state on each clock pulse.

Three-state logic A type of logic circuit having the normal two-state (HIGH, LOW) output and, in addition, an open state in which it is disconnected from its load.

Toggle The action of a flip-flop when it changes back and forth between its two states on each clock pulse.

Trailing edge The second transition of a pulse.

Transistor A semiconductor device exhibiting current gain or voltage gain. When used as a switching device, it can approximate an open or a closed switch.

Transition A change from one level to another.

Transmission line A cable or other physical medium over which data is sent from one point to another.

Trigger A pulse used to initiate a change in the state of a logic circuit.

TTL Transistor-transistor logic.

U

Unit load One gate input represents a unit load to a gate output within the same logic family.

Up-count A counter sequence in which each binary state has a successively higher value.

UV EPROM Ultra-violet erasable programmable read only memory.

V

Variable modulus counter A counter in which the maximum number of states can be changed.

Volatile The characteristic of a memory whereby it loses stored information if power is removed.

W

Weight The value of a digit in a number based on its position in the number.

Weighted code A digital code that utilizes weighted numbers as the individual code words.

Wire-AND An arrangement of logic circuits in which the gate outputs are physically connected to form an "implied" AND function.

Word A group of bits representing a complete piece of digital information.

Write The process of storing information in a memory.

INDEX

A

B

C

D

E

F

G

H

I

J

K

L

M

N

O

P

Q

R

S

T

U